Campbell Essential Biology

4th edition

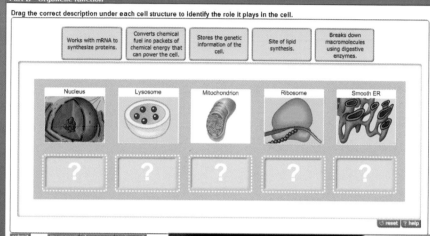
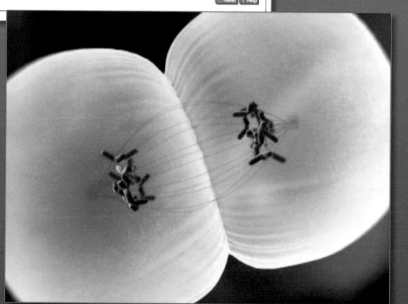

Campbell Essential Biology

4th edition

Eric J. Simon
New England College

Jane B. Reece
Berkeley, California

Jean L. Dickey
Clemson University

Benjamin Cummings

Boston   Columbus   Indianapolis   New York   San Francisco   Upper Saddle River
Amsterdam   Cape Town   Dubai   London   Madrid   Milan   Munich   Paris   Montréal   Toronto
Delhi   Mexico City   São Paulo   Sydney   Hong Kong   Seoul   Singapore   Taipei   Tokyo

Vice President and Editorial Director: *Frank Ruggirello*
Editor-in-Chief: *Beth Wilbur*
Executive Editor: *Chalon Bridges*
Executive Director of Development: *Deborah Gale*
Senior Editorial Manager: *Ginnie Simione Jutson*
Senior Development Editors: *Evelyn Dahlgren, John Burner, and Suzanne Olivier*
Senior Art Development Editor: *Hilair Chism*
Art Developmental Editor: *Carla Simmons*
Senior Supplements Project Editor: *Susan Berge*
Editorial Assistant: *Joshua Taylor*
Director of Media Development: *Lauren Fogel*
Director of Media Strategy: *Stacy Treco*
Senior Media Producers: *Deb Greco and Jonathan Ballard*
Director of Marketing: *Christy Lawrence*
Executive Marketing Manager: *Lauren Harp*
Executive Managing Editor: *Erin Gregg*
Managing Editor: *Michael Early*
Senior Production Supervisor: *Shannon Tozier*

Design Director: *Stuart Jackman*
Design Manager: *Anthony Limerick*
Senior Designer: *Ian Midson*
Publisher, DK Education: *Sophie Mitchell*
Text Design and Page Layouts: *DK Education*
Art Concepts and Direction: *DK Education*
Production Service and Composition: *S4Carlisle Publishing Services*
Illustrations: *Precision Graphics*
Cover Design: *Stuart Jackman*
Cover Production: *Seventeenth Street Studios*
Manufacturing Buyer: *Michael Penne*
Manager, Rights and Permissions: *Zina Arabia*
Manager, Visual Research: *Elaine Soares*
Image Permission Coordinator: *Cynthia Vincenti*
Photo Research: *DK Education and Kristin Piljay*
Senior Photo Editor: *Donna Kalal*
Text and Cover Printer: *Courier Kendallville*
Cover Image: *Frank Greenaway/Dorling Kindersley Media Library*

Credits and acknowledgments borrowed from other sources and reproduced, with permission, in this textbook appear on page A-5.

Library of Congress Cataloging-in-Publication Data
Simon, Eric J. (Eric Jeffrey), 1967-
  Campbell essential biology / Eric J. Simon, Jane B. Reece, Jean L.
Dickey. — 4th ed.
    p. cm.
  Rev. ed. of: Essential biology / Neil A. Campbell, Jane B. Reece, Eric
J. Simon. 3rd ed. c2007.
  Includes bibliographical references and index.
  ISBN 978-0-321-65289-8
  1. Biology—Textbooks. I. Reece, Jane B. II. Dickey, Jean. III. Campbell, Neil A., 1946-2004. Essential biology. IV. Title.
V. Title: Essential biology.
  QH308.2.C343 2010
  570—dc22

2009027251

ISBN 10: 0-321-65289-4; ISBN 13: 978-0-321-65289-8 (Student edition)
ISBN 10: 0-321-64259-7; ISBN 13: 978-0-321-64259-2 (Professional copy)
ISBN 10: 0-321-65290-8; ISBN 13: 978-0-321-65290-4 (Books a la Carte edition)

1 2 3 4 5 6 7 8 9 10—CRK—13 12 11 10 09
Manufactured in the United States of America.

Benjamin Cummings
is an imprint of

www.pearsonhighered.com

Dorling Kindersley Limited,
80 Strand,
London WC2R ORL.

About the Authors

NEIL A. CAMPBELL

(1946–2004) combined the inquiring nature of a research scientist with the soul of a caring teacher. Over his 30 years of teaching introductory biology to both science majors and nonscience majors, many thousands of students had the opportunity to learn from him and be stimulated by his enthusiasm for the study of life. While he is greatly missed by his many friends throughout the biology community, his coauthors remain inspired by his visionary dedication to education and are committed to searching for ever better ways to engage students in the wonders of biology.

ERIC J. SIMON

is an associate professor of biology at New England College, in Henniker, New Hampshire. He teaches introductory biology to science majors and nonscience majors, as well as upper-level courses in genetics, microbiology, and molecular biology. Dr. Simon received a B.A. in biology and computer science and an M.A. in biology from Wesleyan University and a Ph.D. in biochemistry from Harvard University. His research focuses on innovative ways to use technology to improve teaching and learning in the science classroom, particularly for nonscience majors. Dr. Simon is also a coauthor of *Biology: Concepts & Connections*, 6th Edition.

To the teachers and mentors who sparked and promoted my love of biology, especially Jerome Losty (North Stratfield Elementary), James Donovan (Andrew Warde High School), David Beveridge (Wesleyan University), Stephen Harrison (Harvard University), and Neil Campbell; and to my greatest fan and champion: my mother

JANE B. REECE

has worked in biology publishing since 1978, when she joined the editorial staff of Benjamin Cummings. Her education includes an A.B. in biology from Harvard University (where she was initially a philosophy major), an M.S. in microbiology from Rutgers University, and a Ph.D. in bacteriology from the University of California, Berkeley. At UC Berkeley and later as a postdoctoral fellow in genetics at Stanford University, her research focused on genetic recombination in bacteria. Dr. Reece taught biology at Middlesex County College (New Jersey) and Queensborough Community College (New York). She is the lead author of *Biology*, 8th Edition, and *Biology: Concepts & Connections*, 6th Edition.

To Chalon Bridges, Ginnie Simione Jutson, Evelyn Dahlgren, John Burner, Suzanne Olivier, Hilair Chism, Carla Simmons, and all the other heroes of our editorial team, for their essential contributions to Essential Biology

JEAN L. DICKEY

is a professor of biology at Clemson University, in South Carolina. She had no idea that science was interesting until her senior year in high school, when a scheduling problem landed her in an advanced biology course. Abandoning plans to study English or foreign languages, she enrolled in Kent State University as a biology major. After receiving her B.S. in biology, she went on to earn a Ph.D. in ecology and evolution from Purdue University. Since joining the faculty at Clemson in 1984, Dr. Dickey has specialized in teaching nonscience majors, including a course designed for pre-service elementary teachers and workshops for in-service teachers. She also developed an investigative laboratory curriculum for general biology. Dr. Dickey is the author of *Laboratory Investigations for Biology*, 2nd Edition, and is a coauthor of *Biology: Concepts & Connections*, 6th Edition.

To my mother, who taught me to love learning, and to my daughters Katherine and Jessie, the twin delights of my life

Detailed Contents

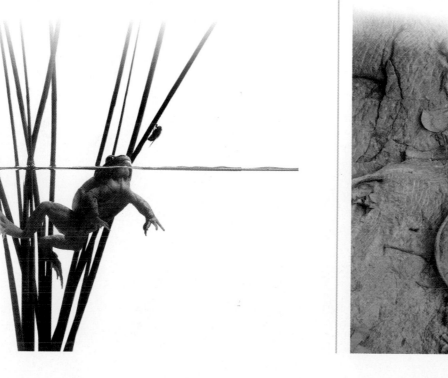

4 A Tour of the Cell 54
■■□□ Chapter Thread: BACTERIAL DISEASES AND CELLULAR STRUCTURES

5 The Working Cell 74
■■■□ Chapter Thread: MOLECULAR TECHNOLOGIES

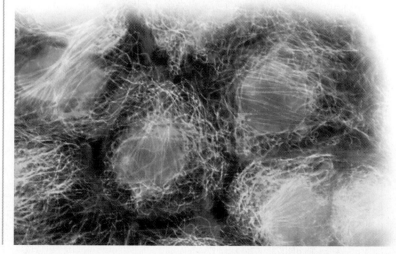

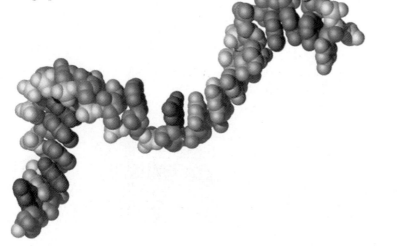
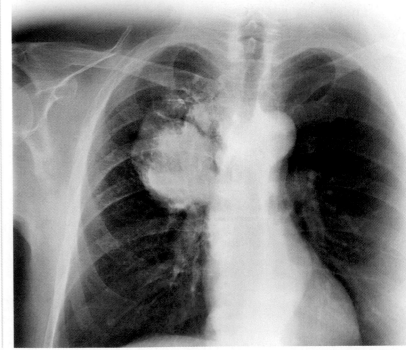

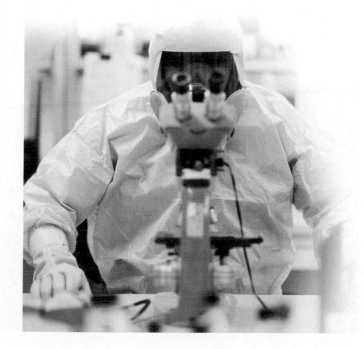

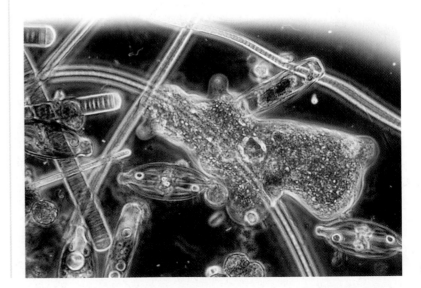

Each chapter is built around three features connected by a unifying thread

■■■ CHAPTER THREAD

The chapter thread weaves a single compelling topic throughout the chapter. For example, here the thread is **Life with and without Sex,** and you'll consistently find this thread: at the beginning, middle, and end of each chapter.

The **CHAPTER CONTENTS** gives you an overview of the chapter's features and topics.

Reproduction in the rain forest.
Botanists attempted to save the Hawaiian bellflower (visible in foreground) by means of sexual and asexual reproduction.

120

8 Cellular Reproduction: Cells from Cells

CHAPTER CONTENTS
■■■ Chapter Thread: LIFE WITH AND WITHOUT SEX

 BIOLOGY AND SOCIETY: Life with and without Sex

Rain Forest Rescue

Deep in a Hawaiian rain forest, the thrumming of a helicopter cut through the dense tropical air. After touching down in a clearing, a pair of rescuers jumped out, one of them carrying a precious bundle. A 20-minute hike through dense foliage brought them to their "patient": the wild *Cyanea kuhihewa*, a plant in the bellflower family. The rescuers were in the middle of a life-or-death battle, armed with the pollen of a species on the very brink of extinction. The stakes couldn't have been higher: The small bellflower before them was the last surviving member of its species in the wild.

The rescuers, scientists from the National Tropical Botanical Garden, were trying to promote sexual reproduction, the reproductive process that involves fertilization, the union of a sperm and an egg. Using a fine brush, the botanists transferred sperm-carrying pollen from a garden-grown plant onto the egg-containing wild bloom. Their hope was to initiate the development of a new seed, and from it, a new plant.

Unfortunately, the bellflower fertilization attempt failed. Even worse, the last remaining wild specimen died in 2003. But hope remains for *C. kuhihewa*. The garden's botanists have also propagated the plant by asexual reproduction, the production of offspring by a single parent. Many plants that normally reproduce sexually can be induced to reproduce asexually in the laboratory. Indeed, work at the botanical garden has produced several new bellflower plants in this way, using stem tissue snipped from wild plants in the 1990s.

Cell division is at the heart of organismal reproduction, whether by sexual or asexual means. The ability of organisms to reproduce their own kind is the one characteristic that best distinguishes living things from nonliving matter. And the perpetuation of life depends on the production of new cells. Plants created from cuttings result from repeated cell divisions, as does the development of a multicellular organism from a fertilized egg. And eggs and sperm themselves result from cell division of a special kind. In this chapter, we'll look at how individual cells reproduce and then see how cell reproduction underlies the process of sexual reproduction. During our discussion, we'll consider examples of asexual and sexual reproduction among both plants and animals.

121

BIOLOGY AND SOCIETY: Rain Forest Rescue

The **BIOLOGY AND SOCIETY** relates biology to your life and interests. For example, in this chapter you'll discover what sex has to do with saving endangered species.

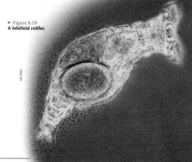

BIOLOGY AND SOCIETY: Life with and without Sex

Rain Forest Rescue

Deep in a Hawaiian rain forest, the thrumming of a helicopter cut through the dense tropical air. After touching down in a clearing, a pair of rescuers jumped out, one of them carrying a precious bundle. A 20-minute hike through dense foliage brought them to their "patient": the wild *Cyanea kuhihewa*, a plant in the bellflower family. The rescuers were in the middle of a life-or-death battle, armed with the pollen of a species on the very brink of extinction. The stakes couldn't have been higher: The small bellflower before them was the last surviving member of its species in the wild.

The rescuers, scientists from the National Tropical Botanical Garden, were trying to promote sexual reproduction, the reproductive process that involves fertilization, the union of a sperm and an egg. Using a fine brush, the botanists transferred sperm-carrying pollen from a garden-grown plant onto the egg-containing wild bloom. Their hope was to initiate the development of a new seed, and from it, a new plant.

Unfortunately, the bellflower fertilization attempt failed. Even worse, the last remaining wild specimen died in 2003. But hope remains for *C. kuhihewa*. The garden's botanists have also propagated the plant by asexual reproduction, the production of offspring by a single parent. Many plants that normally reproduce sexually can be induced to reproduce asexually in the laboratory. Indeed, work at the botanical garden has produced several new bellflower plants in this way, using stem tissue snipped from wild plants in the 1990s.

Cell division is at the heart of organismal reproduction, whether by sexual or asexual means. The ability of organisms to reproduce their own kind is the one characteristic that best distinguishes living things from nonliving matter. And the perpetuation of life depends on the production of new cells. Plants created from cuttings result from repeated cell divisions, as does the development of a multicellular organism from a fertilized egg. And eggs and sperm themselves result from cell division of a special kind. In this chapter, we'll look at how individual cells reproduce and then see how cell reproduction underlies the process of sexual reproduction. During our discussion, we'll consider examples of asexual and sexual reproduction among both plants and animals.

PROCESS OF SCIENCE: Do All Animals Have Sex?

The **PROCESS OF SCIENCE** gives you real-world examples of how the scientific method is applied. For example, in this chapter you'll learn how researchers study an animal that reproduces without sex.

THE PROCESS OF SCIENCE: Life with and without Sex

Do All Animals Have Sex?

As mentioned in the Biology and Society section, many plants can reproduce via both sexual and asexual routes. But while some animal species can also reproduce asexually, very few animals reproduce *only* asexually. In fact, evolutionary biologists have traditionally considered asexual reproduction an evolutionary dead end (for reasons we'll discuss in the Evolution Connection section).

To investigate a case where asexual reproduction seemed to be the norm, researchers from Harvard University studied a group of animals called the bdelloid rotifers (Figure 8.19). This class of nearly microscopic freshwater invertebrates includes over 300 known species. Despite hundreds of years of **observations**, no one had ever found bdelloid rotifer males or evidence of sexual reproduction. But the possibility remained that bdelloids had sex very infrequently or that the males were impossible to recognize by appearance. Thus, the Harvard research team posed the **question**, Does this entire class of animals reproduce solely by asexual means?

The researchers formed the **hypothesis** that bdelloid rotifers have indeed thrived for millions of years despite a lack of sexual reproduction. But how to prove it? In most species, the two versions of a gene in a pair of homologous chromosomes are very similar due to the constant trading of genes during sexual reproduction. If a species has survived without sex for millions of years, the researchers reasoned, then changes in the DNA sequences of homologous genes should accumulate independently, and the two versions of the genes should have significantly diverged from each other over time. This led to the **prediction** that bdelloid rotifers would display much more variation in their pairs of homologous genes than most organisms.

In a simple but elegant **experiment**, the researchers compared the sequences of a particular gene in bdelloid and non-bdelloid rotifers. Their **results** were striking. Among non-bdelloid rotifers that reproduce sexually, the two homologous versions of the gene were nearly identical, differing by only 0.5% on average. In contrast, the two versions of the same gene in bdelloid rotifers differed by 3.5–54%. These data provided strong evidence that bdelloid rotifers have evolved for millions of years without any sexual reproduction.

► Figure 8.19
A bdelloid rotifer.

EVOLUTION CONNECTION: The Advantages of Sex

The **EVOLUTION CONNECTION** concludes the chapter by demonstrating how the theme of evolution runs throughout all of biology. For example, in this chapter you'll learn about the evolutionary advantages of sex.

EVOLUTION CONNECTION: Life with and without Sex

The Advantages of Sex

Throughout this chapter, we've examined cell division within the context of reproduction. Like the Hawaiian bellflower in the Biology and Society section, many plants can reproduce both sexually and asexually (Figure 8.24). An important advantage of asexual reproduction is that there is no need for a partner. Asexual reproduction may thus confer an evolutionary advantage when plants are sparsely distributed and unlikely to be able to exchange pollen. Furthermore, if a plant is superbly suited to a stable environment, asexual reproduction has the advantage of passing on its entire genetic legacy intact.

In contrast to plants, the vast majority of animals reproduce by sexual means. There are exceptions, such as organisms that can regenerate after fragmentation (see Figure 8.1) and the bdelloid rotifers discussed in the Process of Science section. But many animals reproduce only through sex. Therefore, sex must enhance evolutionary fitness. But how? The answer remains elusive. Most hypotheses focus on the unique combinations of genes formed during meiosis and fertilization. By producing offspring of varied genetic makeup, sexual reproduction may enhance survival by speeding adaptation to a changing environment, such as one that contains evolving pathogens. Another idea is that shuffling genes during sexual reproduction might allow a population to rid itself of harmful genes more rapidly. But for now, one of biology's most basic questions—Why have sex?—remains a hotly debated topic that is the focus of much ongoing research and discussion.

◄ Figure 8.24 **Sexual and asexual reproduction.** Many plants, such as this strawberry, have the ability to reproduce both sexually (via flowers that produce fruit) and asexually (via runners).

Quickly orient yourself and find content

FIND THE CHAPTER YOU WANT
using the tabs on the left, and quickly locate units by the color of the tabs.

LOCATE STUDY QUESTIONS
in the margins.

- *CHECKPOINT QUESTIONS help you assess your understanding of the content you just read.*

- *For instant feedback, the answer appears directly under the question.*

- *The CHECKPOINT ICON within the text alerts you to stop and test yourself before you continue reading.*

CHAPTER 8
CELLULAR REPRODUCTION:
CELLS FROM CELLS

☑CHECKPOINT

Ordinary cell division produces two daughter cells that are genetically identical. Name three functions of this type of cell division.

Answer: cell replacement, growth of an organism, asexual reproduction of an organism

What Cell Reproduction Accomplishes

When you hear the word *reproduction*, you probably think of the birth of new organisms. But reproduction actually occurs much more often at the cellular level. Consider the skin on your arm. Skin cells are constantly reproducing themselves and moving outward toward the surface, replacing dead cells that have rubbed off. This renewal of your skin goes on throughout your life. And when your skin is injured, additional cell reproduction helps heal the wound.

When a cell undergoes reproduction, or **cell division**, the two "daughter" cells that result are genetically identical to each other and to the original "parent" cell. (Biologists traditionally use the word *daughter* in this context; it does not imply gender.) Before the parent cell splits into two, it duplicates its **chromosomes**, the structures that contain most of the organism's DNA. Then, during the division process, one set of chromosomes is distributed to each daughter cell. As a rule, the daughter cells receive identical sets of chromosomes.

As summarized in **Figure 8.1**, cell division plays several important roles in the lives of organisms. For example, within your body, millions of cells must divide every second to replace damaged or lost cells. Another function of cell division is growth. All of the trillions of cells in your body are the result of repeated cell divisions that began in your mother's body with a single fertilized egg cell.

Another vital function of cell division is reproduction of the organism. Single-celled organisms, such as amoebas, reproduce by dividing in half, and the offspring are genetic replicas of the parent. Because it does not involve fertilization of an egg by a sperm, this type of reproduction is called **asexual reproduction**. Offspring produced by asexual reproduction inherit all of their chromosomes from a single parent. Many multicellular organisms can reproduce asexually as well. For example, some sea star species have the ability to grow new individuals from fragmented pieces. And if you've ever grown a houseplant from a clipping, you've observed asexual reproduction in plants. In asexual reproduction, there is one simple principle of inheritance: The lone parent and each of its offspring have identical genes. The type of cell division responsible for asexual reproduction and for the growth and maintenance of multicellular organisms is called mitosis.

Sexual reproduction is different; it requires fertilization of an egg by a sperm. The production of egg and sperm cells involves a special type of cell division called meiosis, which occurs only in reproductive organs (such as testes and ovaries in humans). As we'll discuss later, a sperm or egg cell has only half as many chromosomes as the parent cell that gave rise to it.

Note that two kinds of cell division are involved in the lives of sexually reproducing organisms: meiosis for reproduction and mitosis for growth and maintenance. The remainder of the chapter is divided into two main sections, one for each type of cell division. ☑

▶ **Figure 8.1 Three functions of cell division.**

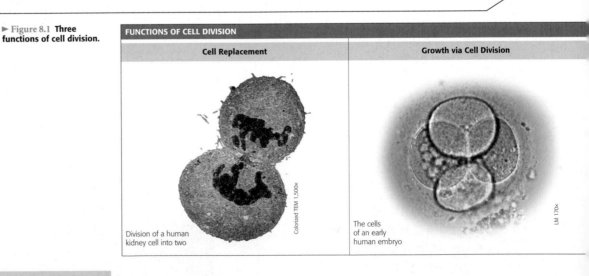

FUNCTIONS OF CELL DIVISION

| Cell Replacement | Growth via Cell Division |

Division of a human kidney cell into two

Colorized TEM 1,500x

The cells of an early human embryo

LM 170x

MAIN HEADING
at top of a page

The Cell Cycle and Mitosis

REMEMBER THE MAIN TOPIC
you are studying, using the colored tab at the top corner of each right-hand page.

BioFlix™
Mitosis

MP3 tutor sessions
Mitosis

Almost all of the genes of a eukaryotic cell—around 21,000 in humans—are located on chromosomes in the cell nucleus. (The main exceptions are genes on small DNA molecules found in mitochondria and chloroplasts.) Because chromosomes are the lead players in cell division, let's focus on them before turning our attention to the cell as a whole.

Eukaryotic Chromosomes

Each eukaryotic chromosome contains one very long DNA molecule, typically bearing thousands of genes. The number of chromosomes in a eukaryotic cell depends on the species (Figure 8.2). Chromosomes are made up of a material called **chromatin**, a combination of DNA and protein molecules. The protein molecules help organize the chromatin and help control the activity of its genes.

Most of the time, the chromosomes exist as a diffuse mass of fibers that are much longer than the nucleus they are stored in. In fact, if stretched out, the DNA in just one of your cells would be taller than you! As a cell prepares to divide, its chromatin

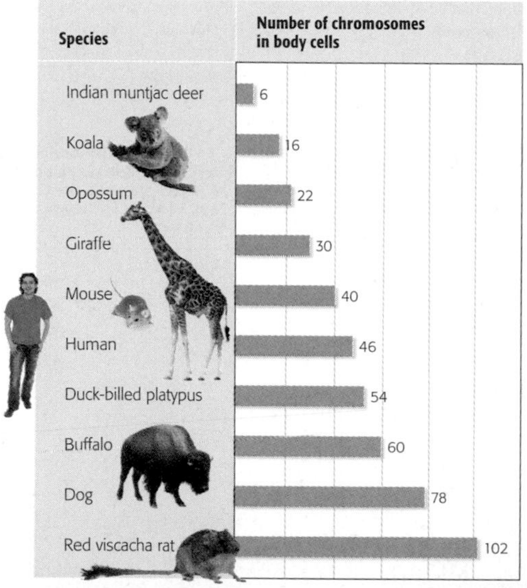
Species
Number of chromosomes in body cells
Indian muntjac deer
6
Koala
16
Opossum
22
Giraffe
30
Mouse
40
Human
46
Duck-billed platypus
54
Buffalo
60
Dog
78
Red viscacha rat
102

► Figure 8.2 **The number of chromosomes in the cells of selected mammals.** Notice that humans have 46 chromosomes and that the number of chromosomes does not correspond to the size or complexity of an organism.

Asexual Reproduction

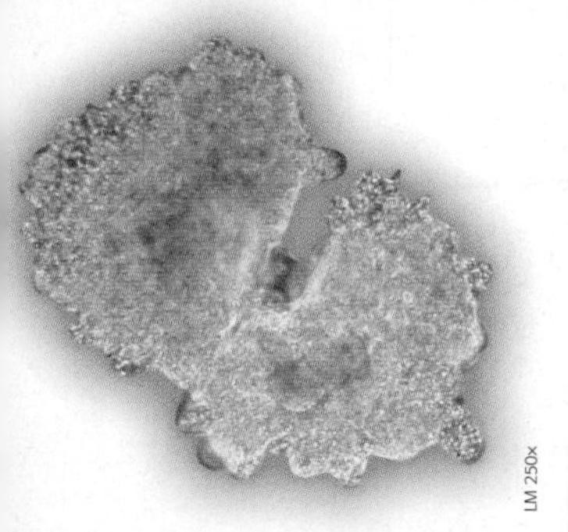
LM 250×

Reproduction of an amoeba

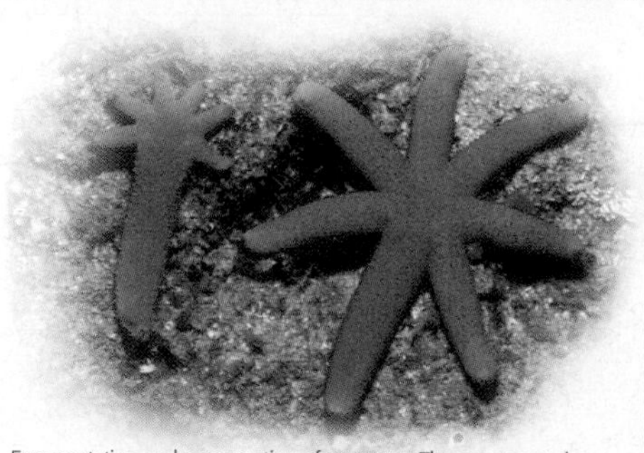

Fragmentation and regeneration of a sea star. The sea star on the right lost and replaced an arm. The severed arm grew into the new sea star on the left.

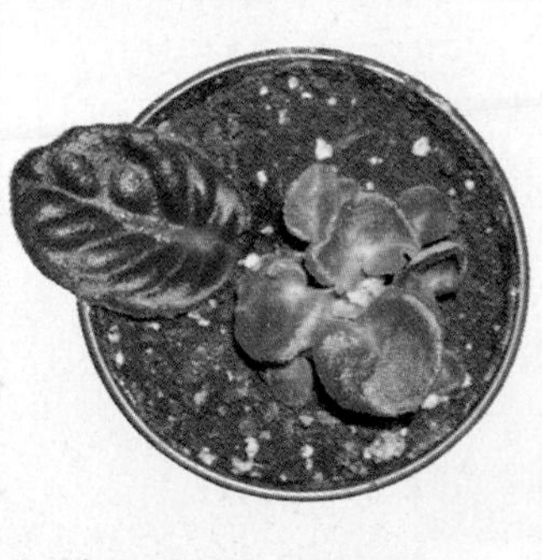

Reproduction of an African violet from a clipping (large leaf)

123

Effortlessly connect images and words

Tools for both visual and verbal learners

FIGURES
and their text descriptions appear on the same page.

GREEN FIGURE NUMBERS
help your eyes move between the text and the figures.

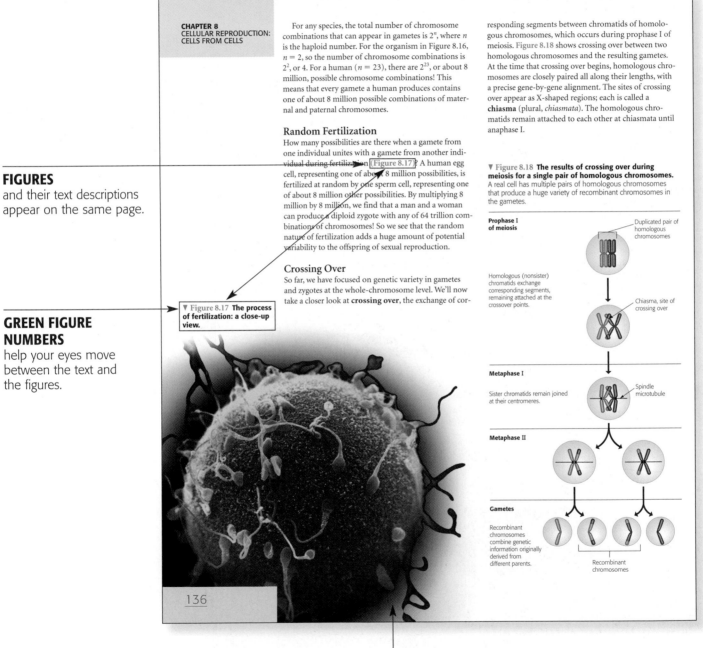

For any species, the total number of chromosome combinations that can appear in gametes is 2^n, where n is the haploid number. For the organism in Figure 8.16, $n = 2$, so the number of chromosome combinations is 2^2, or 4. For a human ($n = 23$), there are 2^{23}, or about 8 million, possible chromosome combinations! This means that every gamete a human produces contains one of about 8 million possible combinations of maternal and paternal chromosomes.

Random Fertilization

How many possibilities are there when a gamete from one individual unites with a gamete from another individual during fertilization (Figure 8.17)? A human egg cell, representing one of about 8 million possibilities, is fertilized at random by one sperm cell, representing one of about 8 million other possibilities. By multiplying 8 million by 8 million, we find that a man and a woman can produce a diploid zygote with any of 64 trillion combinations of chromosomes! So we see that the random nature of fertilization adds a huge amount of potential variability to the offspring of sexual reproduction.

Crossing Over

So far, we have focused on genetic variety in gametes and zygotes at the whole-chromosome level. We'll now take a closer look at **crossing over**, the exchange of cor-

responding segments between chromatids of homologous chromosomes, which occurs during prophase I of meiosis. Figure 8.18 shows crossing over between two homologous chromosomes and the resulting gametes. At the time that crossing over begins, homologous chromosomes are closely paired all along their lengths, with a precise gene-by-gene alignment. The sites of crossing over appear as X-shaped regions; each is called a **chiasma** (plural, *chiasmata*). The homologous chromatids remain attached to each other at chiasmata until anaphase I.

▼ Figure 8.17 **The process of fertilization: a close-up view.**

▼ Figure 8.18 **The results of crossing over during meiosis for a single pair of homologous chromosomes.** A real cell has multiple pairs of homologous chromosomes that produce a huge variety of recombinant chromosomes in the gametes.

Prophase I of meiosis

Duplicated pair of homologous chromosomes

Homologous (nonsister) chromatids exchange corresponding segments, remaining attached at the crossover points.

Chiasma, site of crossing over

Metaphase I

Sister chromatids remain joined at their centromeres.

Spindle microtubule

Metaphase II

Gametes

Recombinant chromosomes combine genetic information originally derived from different parents.

Recombinant chromosomes

136

PHOTOS
complement the illustrations.

UNIQUE VISUAL ORGANIZER FIGURES

help you to see important categories at a glance.

IDENTIFY THE SUBJECT OF THE FIGURE
by glancing at the heading.

SPOT THE CATEGORIES
at once by looking at the subheadings.

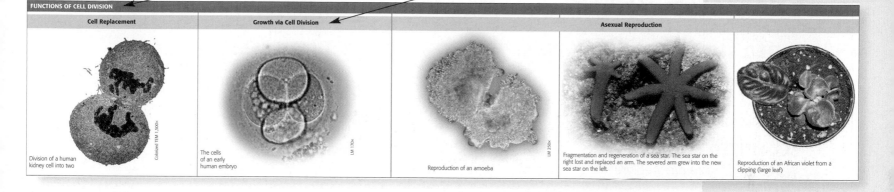

FUNCTIONS OF CELL DIVISION

Cell Replacement	Growth via Cell Division	Asexual Reproduction

Division of a human kidney cell into two
Colorized TEM 1,500x

The cells of an early human embryo
LM 170x

Reproduction of an amoeba
LM 250x

Fragmentation and regeneration of a sea star. The sea star on the right lost and replaced an arm. The severed arm grew into the new sea star on the left.

Reproduction of an African violet from a clipping (large leaf)

CHAPTER REVIEW
words and images work together to help you review the key concepts of the chapter.

Chapter Review

SUMMARY OF KEY CONCEPTS

Go to the Study Area at **www.masteringbiology.com** for practice quizzes, myEBook, BioFlix™ 3-D animations, MP3 Tutor Sessions, videos, current events, and more.

What Cell Reproduction Accomplishes

Cell reproduction, also called cell division, produces genetically identical daughter cells:

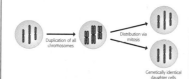

Duplication of all chromosomes

Distribution via mitosis

Genetically identical daughter cells

Some organisms use mitosis (ordinary cell division) to reproduce. This is called asexual reproduction, and it results in offspring that are genetically identical to the lone parent and to each other. Mitosis also enables multicellular organisms to grow and develop and to replace damaged or lost cells. Organisms that reproduce sexually, by the union of a sperm with an egg cell, carry out meiosis, a type of cell division that yields gametes with only half as many chromosomes as body (somatic) cells.

The Cell Cycle and Mitosis

Eukaryotic Chromosomes

The genes of a eukaryotic genome are grouped into multiple chromosomes in the nucleus. Each chromosome contains one very long DNA molecule, with many genes, that is wrapped around histone proteins. Individual chromosomes are coiled up and therefore visible with a light microscope only when the cell is in the process of dividing; otherwise, they are in the form of thin, loosely packed chromatin fibers. Before a cell starts dividing, the chromosomes duplicate, producing sister chromatids (containing identical DNA) joined together at the centromere.

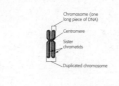

Chromosome (one long piece of DNA)

Centromere

Sister chromatids

Duplicated chromosome

The Cell Cycle

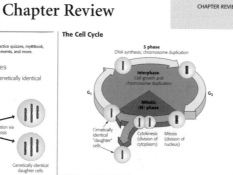

S phase
DNA synthesis; chromosome duplication

Interphase
Cell growth and chromosome duplication

G₁

G₂

Mitotic (M) phase

Genetically identical "daughter" cells

Cytokinesis (division of cytoplasm)

Mitosis (division of nucleus)

Mitosis and Cytokinesis

Mitosis is divided into four phases: prophase, metaphase, anaphase, and telophase. At the start of mitosis, the chromosomes coil up and the nuclear envelope breaks down (prophase). Then a mitotic spindle made of microtubules moves the chromosomes to the middle of the cell (metaphase). The sister chromatids then separate and are moved to opposite poles of the cell (anaphase), where two new nuclei form (telophase). Cytokinesis overlaps the end of mitosis. In animals, cytokinesis occurs by cleavage, which pinches the cell in two. In plants, a membranous cell plate divides the cell in two. Mitosis and cytokinesis produce genetically identical cells.

Cancer Cells: Growing Out of Control

When the cell cycle control system malfunctions, a cell may divide excessively and form a tumor. Cancer cells may grow to form malignant tumors, invade other tissues (metastasize), and even kill the organism. Surgery can remove tumors, and radiation and chemotherapy are effective as treatments because they interfere with cell division. You can increase the likelihood of surviving some forms of cancer through lifestyle changes and regular screenings.

Meiosis, the Basis of Sexual Reproduction

Homologous Chromosomes

The somatic cells (body cells) of each species contain a specific number of chromosomes; human cells have 46, made up of 23 pairs of homologous chromosomes. The chromosomes of a homologous pair carry genes for the same characteristics at the same places. Mammalian males have X and Y sex chromosomes (only partly homologous), while females have two X chromosomes.

Choose the study tools that fit your learning style

VIDEO TUTOR SESSION TOPICS

MITOSIS AND MEIOSIS
Review the similarities and differences between these two processes.

SEX-LINKED PEDIGREES
Walk through a family pedigree.

DNA STRUCTURE
Review the levels of DNA structure and how they relate to each other.

DNA PROFILING TECHNIQUES
Learn important concepts of DNA technology as you watch a DNA profile be generated.

BIODIVERSITY
Organize information about biodiversity into a coherent framework, and remember different groups by relating them to foods that you eat.

PHYLOGENETIC TREES
Learn how to read these important figures by seeing how one is constructed.

MP3 TUTOR SESSIONS

ERIC SIMON
Author Eric Simon hosts both the **Video Tutor Sessions** and **MP3 Tutor Sessions** that provide on-the-go tutorials focused on key concepts and vocabulary.

Chromosome duplication

Sister chromatids

™

BIOFLIX 3-D ANIMATIONS AND ACTIVITIES

These dynamic 3-D animations and activities help you to visualize and learn the toughest topics in biology. BioFlix topics include Mechanisms of Evolution, DNA Replication, Carbon Cycle, and more.

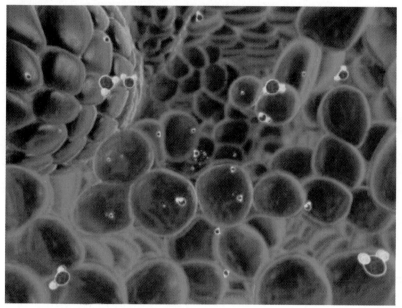

Water Transport In Plants

Membrane Transport

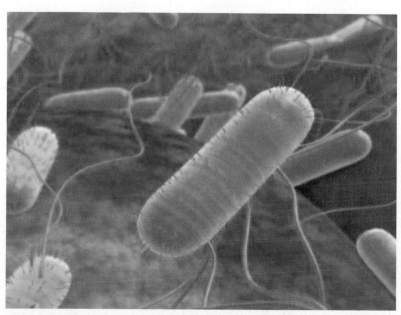

DNA Replication

Mastering**BIOLOGY**™

www.masteringbiology.com

 **myeBook** gives you access to the text whenever and wherever you can access the Internet. The eBook pages look exactly like the printed text, plus you get powerful interactive and customization functions.

HIGHLIGHT FUNCTION
Lets you highlight what you want to remember.

ANNOTATION FUNCTION
Allows you to take notes.

GOOGLE-BASED
search function.

ZOOM
Lets you zoom in and out for better viewing.

HYPERLINKS
Link to quizzes, tests, activities, and animations.

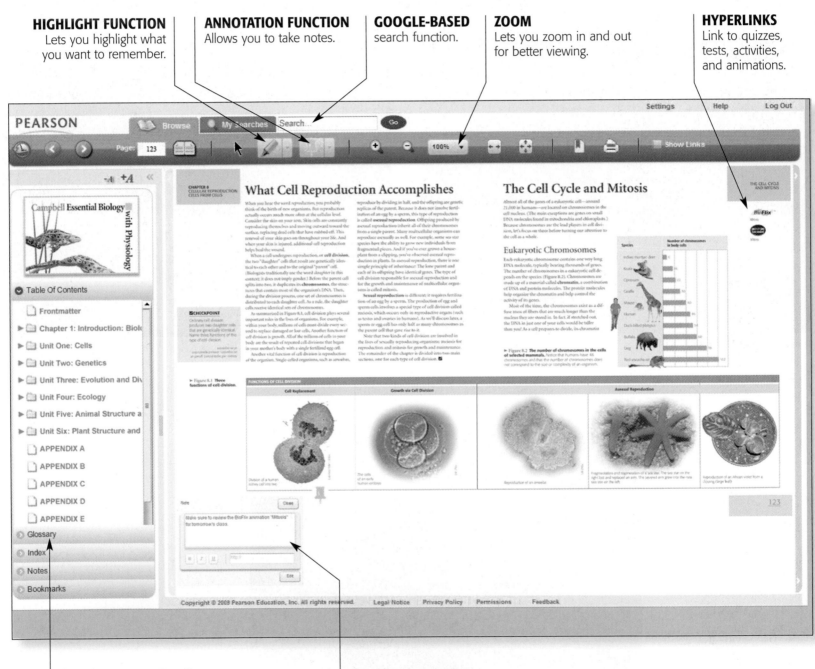

INTERACTIVE GLOSSARY
Provides pop-up definitions and terms.

INSTRUCTOR NOTES
Your instructor might also share his or her notes and highlights with the class.

MasteringBIOLOGY™
Make Learning Part of the Grade®

www.masteringbiology.com

The Mastering system is an online assessment and tutorial system designed to help instructors teach more efficiently, and is pedagogically proven to help students learn. With MasteringBiology you can…

• Break free from only offering multiple-choice assessment but still get automatically graded assignments.
• Personalize your course by editing pre-loaded assignments or uploading your own.
• Take the guesswork out of lectures, quizzes, and tests by utilizing current data on your students' performance.
• Assign auto-graded activities in MasteringBiology that appeal to diverse learning styles like visual and auditory learners: BioFlix™ Activities, Video Tutor Sessions, MP3 Tutor Sessions, *New York Times* Articles, *Discovery Channel* Videos, Animation Activities, GraphIt! Activities, Pre-Lecture Quiz, Post-Lecture Quiz, and Test Bank Questions.

COLOR BAR DATA
Green indicates correct answers. Red shows the percentage of students who requested an answer. Orange indicates the average number of wrong answers per student.

PRE-LECTURE QUIZZES
Help students arrive at lecture prepared and keep on track all term.

CUSTOMIZE CONTENT
Questions and answers can be easily edited.

AT-A-GLANCE STATISTICS
Reveal data for your class as well as national results.

WRONG ANSWER SUMMARY
Gives unique insight into your students' misunderstandings and allows for just-in-time teaching adjustments.

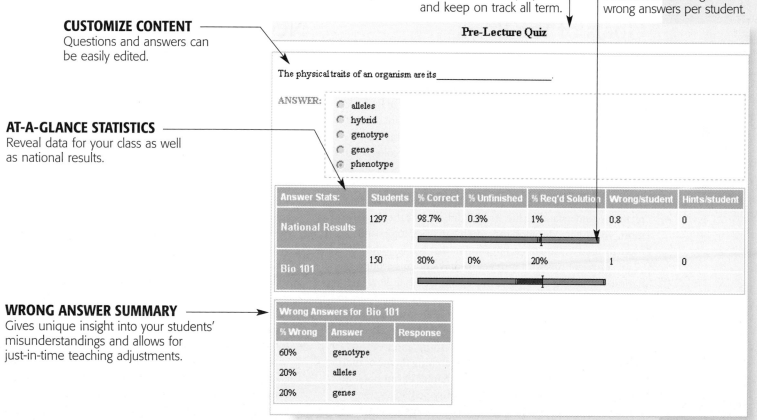

Pre-Lecture Quiz

The physical traits of an organism are its _____.

ANSWER:
○ alleles
○ hybrid
○ genotype
○ genes
● phenotype

Answer Stats:	Students	% Correct	% Unfinished	% Req'd Solution	Wrong/student	Hints/student
National Results	1297	98.7%	0.3%	1%	0.8	0
Bio 101	150	80%	0%	20%	1	0

Wrong Answers for Bio 101		
% Wrong	Answer	Response
60%	genotype	
20%	alleles	
20%	genes	

ADDITIONAL TOOLS TO HELP YOU SHINE IN LECTURE
Instructor Resource DVD, Instructor Guide, Electronic Test Bank, and course management system cartridges including Blackboard, WebCT, and CourseCompass™. Go to www.pearsonhighered.com for more information on any of these resources and a complete list of all supplements.

Preface

It is not an exaggeration to say that there has never been a better time to study biology. Nearly every one of us is curious about the natural world; that is why our popular culture is filled with books, movies, TV shows, comic strips, and video games that display biological wonders and challenge us to think about important biological concepts. While some people *say* that they don't like biology (or, more often, science in general), nearly all will admit that the subject *does* have a significant impact on their life through its connections to medicine, agriculture, environmental issues, psychology, forensics, and myriad other areas. And yet, despite this inherent interest, it can be a struggle for nonscientists to engage in the subject. One reason is that today's students have a wide variety of learning styles. Some enjoy reading, while others prefer to study pictures; still others learn best from listening to explanations. In writing *Campbell Essential Biology*, the authors hope to help instructors motivate and educate the next generation of citizens by tapping into the inherent love of nature that we all share and the various learning styles that we all display.

Goals of the Book

While our world is rich with "teachable moments" and learning opportunities, the ongoing growth of knowledge threatens to suffocate a curious person under an avalanche of information. Neil Campbell conceived of *Essential Biology* as a tool for helping instructors and students focus on the most important areas of biology within a single semester by organizing the material within four core areas: cells, genes, evolution, and ecology. Neil's vision, which we carry on and extend in this edition, has enabled us to keep *Campbell Essential Biology* manageable in size without being superficial in developing the concepts that are most fundamental to understanding life. As Neil did, we take the "less is more" mantra in education today to mean fewer topics and more focused explanations, not content that is more diluted.

In conversations with instructors and students around the nation, we have noticed some important trends in how biology is taught. In particular, many instructors voice a desire to achieve three goals in their course: to engage students by relating the core content to their lives, to clarify the process of science (that is, the scientific method in action), and to demonstrate how evolution is the overarching theme of biology. To help achieve these goals, each chapter of our book includes three important features. First, an opening section called Biology and Society highlights a connection between the chapter subject and students' lives. Second, a section called The Process of Science describes the key steps in the scientific process using a classic or modern experiment (this new edition has one such section in every chapter). And third, a concluding Evolution Connection section relates the chapter to biology's unifying theme.

New to This Edition

In this edition of *Campbell Essential Biology*, we go even further to help students grapple with the pedagogical goals of this edition:

- **Chapter Threads** Every chapter in *Campbell Essential Biology* has its own unifying chapter thread woven throughout. A chapter thread is a high-interest topic that helps to demonstrate the relevance of the chapter subject. For example, Chapter 3 ("The Molecules of Life") opens with a Biology and Society section on lactose intolerance. This topic is expanded upon at several points in the chapter. At an appropriate point, a Process of Science section explains how biologists tracked an important mutation related to lactose tolerance to a chromosomal spot outside the gene itself. Finally, the chapter ends with an Evolution Connection section that describes the recent evolution of lactose tolerance and intolerance. By relating this compelling topic to our three pedagogical goals for this edition, the chapter thread reinforces the relevance of the chapter's content.

- **Appealing New Look** We all know that learning begins with engagement. Students only learn when their attention is focused on the material—in other words, to learn the content, you must want to read the book! Accordingly, we have significantly updated the look and layout of *Campbell Essential Biology* through a new partnership with the acclaimed publishing house Dorling Kindersley Ltd. ("DK"). DK is famous for books that grab and keep readers' attention, and we believe that our partnership with DK has significantly improved the look of *Campbell Essential Biology*, providing a more inviting reading experience.

- **Easier Navigation of the Book** We all know that biology is a big topic with a lot of details. Another goal of our new DK design is to help readers find the information they want quickly so that they spend less time searching, more time learning. The layout of the book has been revised in several ways that increase the efficiency of studying. For example, colored tabs help readers quickly locate chapters and the main topics within them; major sections nearly always begin at the top of a left-hand page (the most logical place); and frequent Checkpoint boxes—linked to a check icon within the text—are located in the corners of pages and the end of sections. The Checkpoints help readers assess their learning as they read.

- **Ecology Revamped** Ecology is gaining importance within the biology curriculum and is one of the best ways to interest students in the world around them. In this edition of *Campbell Essential Biology*, the ecology unit has been largely rewritten by new coauthor Jean Dickey, of Clemson University. Jean used her expertise as an experienced teacher and writer to breathe fresh life into this topic. By making a direct connection between ecological principles and environmental issues, the ecology chapters show students how *they* can have an impact on the environment and our Earth's future.

- **Improved Illustration Program** Through conversations with students and instructors, we know that many students describe themselves as visual learners. To try to help those of us who learn best from pictures, we have created a new style of figure. Called *visual organizer figures*, these new figures group information into categories using strong images and short descriptive phrases, allowing students to see at a glance how information can be organized. Additionally, references to figures within the text are located near the figure itself and are highlighted in a bold green color, enabling students to navigate quickly between the text and figures. And 75% of the photographs in this edition are new, helping to create an engaging visual experience for the reader.

- **Better eBook** For students who prefer to study online, an improved eBook provides full access to the text wherever there is an Internet connection. The online textbook is identical to the printed version in appearance, and it allows students to add annotations, consult an interactive glossary, follow links to new information, and use a Google-based search function.

- **More BioFlix™ 3-D Animations** With this edition, we continue to expand and update the resources and supplements available to instructors and students. Accordingly, we have expanded our suite of BioFlix 3-D Animations, which use dynamic, cutting-edge, three-dimensional graphics to help students visualize complex processes. Our newest BioFlix animations focus on the mechanisms of evolution, the carbon cycle, population ecology, and homeostasis.

- **Video Tutor Sessions** For the last edition, we introduced a set of MP3 Tutor Sessions intended to help students learn vocabulary by hearing it used in context by lead author Eric Simon. New to this edition is a suite of Video Tutor Sessions hosted by Eric Simon. Featuring graphics clearly explained by Eric's narration, these new video podcasts walk students through difficult topics—for example, demonstrating how a DNA profile is generated, working through genetics problems, and manipulating models of DNA. Because both the MP3 Tutor Sessions and Video Tutor Sessions are hosted by the textbook's lead author, students can learn from a single, consistent voice.

- **MasteringBiology™** Many instructors wish to assess student learning outside the classroom. To help address this need, we have created a new version of MasteringBiology, a powerful study and assessment tool that is proving highly successful with the Campbell majors textbook. The *Campbell Essential Biology* version of MasteringBiology is loaded with content from the book and can provide instructors with previously unavailable insights into student effort and comprehension.

The simple fact that nonscientists greatly outnumber scientists reminds us of the importance of students who are not science majors. And attitudes about science and scientists are often shaped by a single required science course—*this* course. We hope we can tap into the innate love of nature that nearly all of us feel and nurture it into a genuine love of biology. In this spirit, we hope that this textbook and its supplements will help every reader make biological perspectives a part of his or her personal worldview. Please let us know how we are doing and how we can improve the next edition of *Campbell Essential Biology*.

ERIC SIMON
Department of Biology
New England College
Henniker, NH 03242
esimon@nec.edu

JANE REECE
c/o Pearson Benjamin Cummings
1301 Sansome Street
San Francisco, CA 94111
JaneReece@cal.berkeley.edu

JEAN DICKEY
Department of Biology
Clemson University
Clemson, SC 29634
dickeyj@clemson.edu

Acknowledgments

Throughout our time writing *Campbell Essential Biology*, the author team has had the great fortune of collaborating with an extremely talented group of publishing professionals and educators. While the responsibility for any shortcomings lies solely with us, the merits of the book and its supplements reflect the contributions of a great many dedicated colleagues.

First and foremost, we must acknowledge our huge debt to Neil Campbell, the original author of this book and a source of ongoing inspiration for each of us. Although this edition of the book has been carefully and thoroughly revised—to update its science, its connections to students' lives, its pedagogy, and its currency—it remains infused with Neil's vision and his commitment to share biology with introductory students. We have therefore changed the book's title to honor and acknowledge Neil Campbell's galvanizing vision and impact.

This book could not have been completed without the efforts of the *Campbell Essential Biology* team at Pearson Benjamin Cummings. The leading visionary of that team remains executive editor Chalon Bridges. All of us are continuously humbled and inspired by Chalon's limitless dedication to the nonmajors course, its students, and its professors. Chalon's positive energy suffuses every aspect of our publishing effort. Of course, publishing excellence flows from the top and we are grateful to Linda Davis, president of Pearson Math and Science, and Paul Corey, president of Pearson Benjamin Cummings, for the leadership they provide to the entire company. Further guidance was provided by vice president and editorial director Frank Ruggirello, editor-in-chief Beth Wilbur, executive director of development Deborah Gale, and vice president and director of media development Lauren Fogel. All Pearson Benjamin Cummings authors benefit from such supportive leadership.

It is no exaggeration to say that the fingerprints of the best editorial team in the industry appear on every page of this book. In dealing with the complexities of creating this edition, the authors were masterfully and kindly shepherded, scheduled, persuaded, and overseen by senior editorial manager Ginnie Simione Jutson. We owe Ginnie a deep debt of gratitude for her talents, patience, and hard work. Our three senior developmental editors—Evelyn Dahlgren, John Burner, and Suzanne Olivier—worked tirelessly to hone the words that convey our love of biology. Senior art developmental editor Hilair Chism and art developmental editor Carla Simmons skillfully and creatively put the "art" in "teaching through art." Editorial assistant Joshua Taylor worked hard to keep it all flowing. The dedication shown by the entire editorial team inspired the authors to do our very best.

Once we formulated our words and images, the production and manufacturing teams transformed them into the book you hold in your hands. We were lucky to forge an alliance with Dorling Kindersley Ltd. ("DK"), whose innovative staff carefully considered how to make every image you see better and improved the text design and layout of literally every page of this edition. For this we wish to thank our colleagues at DK Education in London: publisher Sophie Mitchell, design director Stuart Jackman, design manager Anthony Limerick, and designer Ian Midson. At Pearson Benjamin Cummings, senior production supervisor Shannon Tozier oversaw the production process and kept everyone and everything on track. We also wish to thank executive managing editor Erin Gregg and managing editor Michael Early.

For the production and composition of the book, we thank project editor Norine Strang and composition supervisor Holly Paige, of S4Carlisle Publishing Services, whose professionalism and commitment to the quality of the finished product eased our authorial burdens tremendously. The authors owe much to copyeditor Janet Greenblatt and proofreader Joanna Dinsmore for polishing our words and making us seem like better writers. If you notice one new thing about this edition, it very well may be the many beautiful photographs. For that we thank senior photo editor Donna Kalal, photo researcher Kristin Piljay, and the designers at DK Education, Anthony Limerick and Ian Midson. We thank production manager Kristina Seymour and the artists at Precision Graphics for creating many beautiful illustrations. We also thank permissions editor Sue Ewing for keeping us on the straight and narrow. In the final stages of production, the talents of manufacturing buyer Michael Penne shone.

Most instructors view the textbook as just one piece of the learning puzzle. Filling in the gaps are this book's supplements and media. We are lucky to have a *Campbell Essential Biology* supplements team that is fully committed to the core goals of accuracy and readability. Senior supplements project editor Susan Berge, with assistance from assistant editor Brady Golden, expertly coordinated the supplements, a difficult task given their number and variety, and production supervisor Jane Brundage handled the production of these materials (no small feat!). We owe particular gratitude to the supplements authors, especially the indefatigable and eagle-eyed Ed Zalisko, of Blackburn College, who wrote the Study Guide and Instructor Guide and revised the PowerPoint® Lectures; the highly skilled and knowledgeable Jim Newcomb, of New England College, who revised the Study Card, Quiz Show, and the Study Area quizzes; Richard Myers, who also revised the Study Area quizzes; and Mike Tveten, of Pima Community College, who revised the active learning Clicker Questions. We thank copyeditor John Hammett for his work on the PowerPoint Lectures.

We are grateful to Tom Owens, of Cornell University, for coauthoring the amazing BioFlix™ 3-D Animations, which so beautifully illustrate complex biological topics, as well as developmental manager Pat Burner and storyboard artist Russell Chun, whose commitment to perfection shine through every animation. We also thank the multitalented Jennifer Yeh, who coauthored BioFlix 3-D Animations, wrote the BioFlix Student Activities, revised the Test Bank, and contributed to the MasteringBiology™ Study Area. Developmental editor Karen Gulliver ensured that the BioFlix Activities would be useful to students and professors alike.

We wish to thank the talented group of publishing professionals who worked on the comprehensive media program that accompanies *Campbell Essential Biology*, most notably vice president and director of media strategy Stacy Treco and the team dedicated to MasteringBiology: our inspirational senior media producer Deb Greco, director of editorial content Tania Mlawer, and developmental editor Sarah Jensen. We are enormously grateful to the entire Mastering Team (in alphabetical order): Chad Arthur, Victoria Bell, Ruth Berry, Andrew Billeb, Florin Bocaneala, Lewis Costas, Daphne Dor-Ner, Katie Foley, Keith Hedger, Julia Henderson, Joe Ignazi, Jessica Kadar, Ram Kelath, Jeff King, David Kokorowski, Helio Leal, Mary Lee, Claire Masson, Nissi Mathew, Adam Morton, Fred Mueller, Ian Nordby, Maria Panos, Caroline Power, Jamie Reckelhoff, John Rofior, Wendy Romaniecki, Karen Sheh, Sarah Smith, Doug Stevenson, Kristen Sutton, Margaret Trombley, and Rasil Warnakulasooriya. We are also grateful to the MasteringBiology Faculty Advisory Board of Norris Armstrong, Janet Gaston, Eileen Gregory, James Murphy, Julie Olsen, and Sukanya V. Subramanian for their valuable contributions. If you find MasteringBiology to be as useful a tool as we do, you have these folks to thank. We thank senior media producer Jonathan Ballard for tirelessly working to bring our collective vision to fruition. Vital contributions were also made by media project editor Brienn Buchanan, associate web developer Josh Gentry, and senior supplements production supervisor Liz Winer.

As educators and writers, we are very lucky to have a crack marketing team; after all, "market" is just another way of saying "you, the students and instructors we are trying to serve." Christy Lawrence, director of marketing, and Lauren Harp, executive marketing manager, helped us achieve our authorial goals by keeping us constantly focused on the needs of students and instructors. For their amazing efforts in marketing, we also thank market development manager Brooke Suchomel, market development coordinator Cassandra Cummings, creative director Lillian Carr, copywriter supervisor Jane Campbell, and marketing communication specialist Jessica Perry. Jay Jenkins was our marketing manager before departing for a new position supporting our essential field effort. We also send a hearty thank you to the Pearson Benjamin Cummings sales reps for representing *Campbell Essential Biology* on campuses. These representatives are our lifeline to the greater educational community, telling us what you like (and don't like) about this book and the accompanying supplements and media. Their enthusiasm for helping students makes them not only ideal ambassadors, but our partners in education.

Contributing substantial experience and wisdom to the planning of this edition were an invaluable "brain trust" of colleagues: Jay Withgott; Jim Newcomb, of New England College; Lisa Weasel, of Portland State University; and Beverly Brown, of Nazareth College. Eric Simon would like to thank the following colleagues for specific content suggestions and other support: Alexander J. Travis (Cornell University College of Veterinary Medicine), David Mark Welch (Marine Biological Laboratory), William Nash, Mark J. Daly (Center for Human Genetic Research, Harvard Medical School), Kevin McMahon (New Hampshire State Police Forensics Laboratory), Marshall Simon, Nikos Kyrpides (Genomes OnLine Database), Michael C. Ain (Johns Hopkins Children's Center), and Jamey Barone. Eric Simon would also like to thank his colleagues at New England College for their support and for providing a model of excellence in education. Finally, Eric Simon thanks Amanda Marsh for her expert eye, keen attention to detail, tireless commitment, constant support, compassion, and wisdom. Furthermore, at the end of these acknowledgments you'll find a list of the many instructors who provided valuable information about their courses, reviewed chapters, and/or conducted class tests of *Campbell Essential Biology* with their students. All of our best ideas spring from the classroom, so we thank them for their efforts and support.

Most of all, we thank our families, friends, and colleagues who continue to tolerate our obsession with doing our best for science education.

ERIC SIMON, JANE REECE, JEAN DICKEY

REVIEWERS OF THIS EDITION

Shireen Alemadi
Minnesota State University, Moorhead

Heather Ashworth
Utah Valley University

Neil Baker
Ohio State University

Andrew Baldwin
Mesa Community College

Verona Barr
Heartland Community College

Judy Bluemer
Morton College

David Boose
Gonzaga University

Carol A. Britson
University of Mississippi

Steven Brumbaugh
Green River Community College

Wilbert Butler
Tallahassee Community College

Thomas F. Chubb
Villanova University

Erica Corbett
Southeastern Oklahoma State University

Laurie-Ann Crawford
Hawkeye Community College

Marirose T. Ethington
Genesee Community College

Brandon Foster
Wake Technical Community College

J. L. Henriksen
Bellevue University

W. Wyatt Hoback
University of Nebraska at Kearney

Elizabeth Hodgson
York College of Pennsylvania

Dianne Jennings
Virginia Commonwealth University

Arnold J. Karpoff
University of Louisville

Cindy Klevickis
James Madison University

Holly Kupfer
Central Piedmont Community College

Brenda Leady
University of Toledo

Eric Lovely
Arkansas Tech University

Lisa Maranto
Prince George's Community College

James Newcomb
New England College

Michael Nosek
Fitchburg State College

Martha Powell
University of Alabama

Kerri Skinner
University of Nebraska at Kearney

Joyce Stamm
University of Evansville

Marshall D. Sundberg
Emporia State University

Nathan Trueblood
California State University, Sacramento

Helen Walter
Diablo Valley College

Kristen Walton
Missouri Western State University

Daniel Williams
Winston-Salem University

Judy A. Williams
Southeastern Oklahoma State University

REVIEWERS OF PREVIOUS EDITIONS

Marilyn Abbott
Lindenwood College

Tammy Adair
Baylor University

Felix O. Akojie
Paducah Community College

William Sylvester Allred, Jr.
Northern Arizona University

Estrella Z. Ang
University of Pittsburgh

David Arieti
Oakton Community College

C. Warren Arnold
Allan Hancock Community College

Mohammad Ashraf
Olive-Harvey College

Bert Atsma
Union County College

Yael Avissar
Rhode Island College

Barbara J. Backley
Elgin Community College

Gail F. Baker
LaGuardia Community College

Kristel K. Bakker
Dakota State University

Linda Barham
Meridian Community College

Charlotte Barker
Angelina College

S. Rose Bast
Mount Mary College

Sam Beattie
California State University, Chico

Rudi Berkelhamer
University of California, Irvine

Penny Bernstein
Kent State University, Stark Campus

Suchi Bhardwaj
Winthrop University

Donna H. Bivans
East Carolina University

Andrea Bixler
Clarke College

Brian Black
Bay de Noc Community College

Allan Blake
Seton Hall University

Karyn Bledsoe
Western Oregon University

Judy Bluemer
Morton College

Sonal Blumenthal
University of Texas at Austin

Lisa Boggs
Southwestern Oklahoma State University

Dennis Bogyo
Valdosta State University

Virginia M. Borden
University of Minnesota, Duluth

James Botsford
New Mexico State University

Cynthia Bottrell
Scott Community College

Richard Bounds
Mount Olive College

Cynthia Boyd
Hawkeye Community College

Robert Boyd
Auburn University

B. J. Boyer
Suffolk County Community College

Mimi Bres
Prince George's Community College

Patricia Brewer
University of Texas at San Antonio

Jerald S. Bricker
Cameron University

Carol A. Britson
University of Mississippi

George M. Brooks
Ohio University, Zanesville

Janie Sue Brooks
Brevard College

Steve Browder
Franklin College

Evert Brown
Casper College

Mary H. Brown
Lansing Community College

Richard D. Brown
Brunswick Community College

Steve Brumbaugh
Green River Community College

Joseph C. Bundy
University of North Carolina at Greensboro

Carol T. Burton
Bellevue Community College

Rebecca Burton
Alverno College

Warren R. Buss
University of Northern Colorado

Miguel Cervantes-Cervantes
*Lehman College, City University
of New York*

Bane Cheek
Polk Community College

Thomas F. Chubb
Villanova University

Reggie Cobb
Nash Community College

Pamela Cole
Shelton State Community College

William H. Coleman
University of Hartford

Jay L. Comeaux
McNeese State University

James Conkey
Truckee Meadows Community College

Karen A. Conzelman
Glendale Community College

Ann Coopersmith
Maui Community College

James T. Costa
Western Carolina University

Pat Cox
University of Tennessee, Knoxville

Pradeep M. Dass
Appalachian State University

Paul Decelles
Johnson County Community College

Galen DeHay
Tri County Technical College

Cynthia L. Delaney
University of South Alabama

Jean DeSaix
University of North Carolina at Chapel Hill

Elizabeth Desy
Southwest State University

Edward Devine
Moraine Valley Community College

Dwight Dimaculangan
Winthrop University

Deborah Dodson
Vincennes Community College

Diane Doidge
Grand View College

Don Dorfman
Monmouth University

Richard Driskill
Delaware State University

Lianne Drysdale
Ozarks Technical Community College

Terese Dudek
Kishawaukee College

Shannon Dullea
North Dakota State College of Science

David A. Eakin
Eastern Kentucky University

Brian Earle
Cedar Valley College

Ade Ejire
Johnston Community College

Dennis G. Emery
Iowa State University

Virginia Erickson
Highline Community College

Carl Estrella
Merced College

Marirose T. Ethington
Genesee Community College

Paul R. Evans
Brigham Young University

Zenephia E. Evans
Purdue University

Jean Everett
College of Charleston

Dianne M. Fair
*Florida Community College
at Jacksonville*

Joseph Faryniarz
Naugatuck Valley Community College

Phillip Fawley
Westminster College

Lynn Fireston
Ricks College

Jennifer Floyd
Leeward Community College

Dennis M. Forsythe
The Citadel

Carl F. Friese
University of Dayton

Suzanne S. Frucht
Northwest Missouri State University

Edward G. Gabriel
Lycoming College

Anne M. Galbraith
University of Wisconsin, La Crosse

Kathleen Gallucci
Elon University

Gregory R. Garman
Centralia College

Gail Gasparich
Towson University

Kathy Gifford
Butler County Community College

Sharon L. Gilman
Coastal Carolina University

Mac Given
Neumann College

Patricia Glas
The Citadel

Ralph C. Goff
Mansfield University

Marian R. Goldsmith
University of Rhode Island

Andrew Goliszek
*North Carolina Agricultural and Technical
State University*

Tamar Liberman Goulet
University of Mississippi

Curt Gravis
Western State College of Colorado

Larry Gray
Utah Valley State College

Tom Green
West Valley College

Robert S. Greene
Niagara University

Ken Griffin
Tarrant County Junior College

Denise Guerin
Santa Fe Community College

Paul Gurn
Naugatuck Valley Community College

Peggy J. Guthrie
University of Central Oklahoma

Henry H. Hagedorn
University of Arizona

Blanche C. Haning
Vance-Granville Community College

Laszlo Hanzely
Northern Illinois University

Reba Harrell
Hinds Community College

Sherry Harrel
Eastern Kentucky University

Frankie Harris
Independence Community College

Lysa Marie Hartley
Methodist College

Janet Haynes
Long Island University

Michael Held
St. Peter's College

Consetta Helmick
University of Idaho

Michael Henry
Contra Costa College

Linda Hensel
Mercer University

Jana Henson
Georgetown College

James Hewlett
Finger Lakes Community College

Richard Hilton
Towson University

Juliana Hinton
McNeese State University

Phyllis C. Hirsch
East Los Angeles College

A. Scott Holaday
Texas Tech University

R. Dwain Horrocks
Brigham Young University

Howard L. Hosick
Washington State University

Carl Huether
University of Cincinnati

Celene Jackson
Western Michigan University

John Jahoda
Bridgewater State College

Richard J. Jensen
Saint Mary's College

Tari Johnson
Normandale Community College

Tia Johnson
Mitchell Community College

Greg Jones
Santa Fe Community College

John Jorstad
Kirkwood Community College

Tracy L. Kahn
University of California, Riverside

Robert Kalbach
Finger Lakes Community College

Mary K. Kananen
Pennsylvania State University, Altoona

Thomas C. Kane
University of Cincinnati

Arnold J. Karpoff
University of Louisville

John M. Kasmer
Northeastern Illinois University

Valentine Kefeli
Slippery Rock University

Dawn Keller
Hawkeye College

John Kelly
Northeastern University

Cheryl Kerfeld
University of California, Los Angeles

Henrik Kibak
California State University, Monterey Bay

Kerry Kilburn
Old Dominion University

Joyce Kille-Marino
College of Charleston

Peter King
Francis Marion University

Peter Kish
Oklahoma School of Science and Mathematics

Robert Kitchin
University of Wyoming

Richard Koblin
Oakland Community College

H. Roberta Koepfer
Queens College

Michael E. Kovach
Baldwin-Wallace College

Jocelyn E. Krebs
University of Alaska, Anchorage

Ruhul H. Kuddus
Utah Valley State College

Nuran Kumbaraci
Stevens Institute of Technology

Gary Kwiecinski
The University of Scranton

Roya Lahijani
Palomar College

James V. Landrum
Washburn University

Lynn Larsen
Portland Community College

Siu-Lam Lee
University of Massachusetts, Lowell

Thomas P. Lehman
Morgan Community College

William Leonard
Central Alabama Community College

Shawn Lester
Montgomery College

Leslie Lichtenstein
Massasoit Community College

Barbara Liedl
Central College

Harvey Liftin
Broward Community College

David Loring
Johnson County Community College

Lewis M. Lutton
Mercyhurst College

Maria P. MacWilliams
Seton Hall University

Mark Manteuffel
St. Louis Community College

Michael Howard Marcovitz
Midland Lutheran College

Angela M. Mason
Beaufort County Community College

Roy B. Mason
Mt. San Jacinto College

John Mathwig
College of Lake County

Lance D. McBrayer
Georgia Southern University

Bonnie McCormick
University of the Incarnate Word

Katrina McCrae
Abraham Baldwin Agricultural College

Tonya McKinley
Concord College

Mary Anne McMurray
Henderson Community College

Ed Mercurio
Hartnell College

Timothy D. Metz
Campbell University

David Mirman
Mt. San Antonio College

Nancy Garnett Morris
Volunteer State Community College

Angela C. Morrow
University of Northern Colorado

Patricia S. Muir
Oregon State University

James Newcomb
New England College

Jon R. Nickles
University of Alaska, Anchorage

Jane Noble-Harvey
University of Delaware

Jeanette C. Oliver
Flathead Valley Community College

David O'Neill
Community College of Baltimore County

Sandra M. Pace
Rappahannock Community College

Lois H. Peck
University of the Sciences, Philadelphia

Kathleen E. Pelkki
Saginaw Valley State University

Jennifer Penrod
Lincoln University

Rhoda E. Perozzi
Virginia Commonwealth University

John S. Peters
College of Charleston

Pamela Petrequin
Mount Mary College

Paula A. Piehl
Potomac State College of West Virginia University

Bill Pietraface
State University of New York Oneonta

Gregory Podgorski
Utah State University

Rosamond V. Potter
University of Chicago

Karen Powell
Western Kentucky University

Elena Pravosudova
Sierra College

Hallie Ray
Rappahannock Community College

Jill Raymond
Rock Valley College

Dorothy Read
University of Massachusetts, Dartmouth

Nathan S. Reyna
Howard Payne University

Philip Ricker
South Plains College

Todd Rimkus
Marymount University

Lynn Rivers
Henry Ford Community College

Jennifer Roberts
Lewis University

Laurel Roberts
University of Pittsburgh

April Rottman
Rock Valley College

Maxine Losoff Rusche
Northern Arizona University

Michael L. Rutledge
Middle Tennessee State University

Mike Runyan
Lander University

Travis Ryan
Furman University

Tyson Sacco
Cornell University

Sarmad Saman
Quinsigamond Community College

Leba Sarkis
Aims Community College

Walter Saviuk
Daytona Beach Community College

Neil Schanker
College of the Siskiyous

Robert Schoch
Boston University

John Richard Schrock
Emporia State University

Julie Schroer
Bismarck State College

Karen Schuster
Florida Community College at Jacksonville

Brian W. Schwartz
Columbus State University

Michael Scott
Lincoln University

Eric Scully
Towson State University

Lois Sealy
Valencia Community College

Sandra S. Seidel
Elon University

Wayne Seifert
Brookhaven College

Patty Shields
George Mason University

Cara Shillington
Eastern Michigan University

Brian Shmaefsky
Kingwood College

Rainy Inman Shorey
Ferris State University

Cahleen Shrier
Azusa Pacific University

Jed Shumsky
Drexel University

Greg Sievert
Emporia State University

Jeffrey Simmons
West Virginia Wesleyan College

Frederick D. Singer
Radford University

Anu Singh-Cundy
Western Washington University

Sandra Slivka
Miramar College

Margaret W. Smith
Butler University

Thomas Smith
Armstrong Atlantic State University

Deena K. Spielman
Rock Valley College

Minou D. Spradley
San Diego City College

Robert Stamatis
Daytona Beach Community College

Eric Stavney
Highline Community College

Bethany Stone
University of Missouri, Columbia

Mark T. Sugalski
New England College

Marshall D. Sundberg
Emporia State University

Adelaide Svoboda
Nazareth College

Sharon Thoma
Edgewood College

Kenneth Thomas
Hillsborough Community College

Sumesh Thomas
Baltimore City Community College

Betty Thompson
Baptist University

Paula Thompson
Florida Community College

Linda Tichenor
University of Arkansas, Fort Smith

John Tjepkema
University of Maine at Orono

Bruce L. Tomlinson
State University of New York, Fredonia

Leslie R. Towill
Arizona State University

Bert Tribbey
California State University, Fresno

Robert Turner
Western Oregon University

Michael Twaddle
University of Toledo

Virginia Vandergon
California State University, Northridge

William A. Velhagen, Jr.
Longwood College

Jonathan Visick
North Central College

Michael Vitale
Daytona Beach Community College

Lisa Volk
Fayetteville Technical Community College

Stephen M. Wagener
Western Connecticut State University

James A. Wallis
St. Petersburg Community College

Jennifer Warner
University of North Carolina at Charlotte

Dave Webb
St. Clair County Community College

Harold Webster
Pennsylvania State University, DuBois

Ted Weinheimer
California State University, Bakersfield

Lisa A. Werner
Pima Community College

Joanne Westin
Case Western Reserve University

Wayne Whaley
Utah Valley State College

Joseph D. White
Baylor University

Quinton White
Jacksonville University

Leslie Y. Whiteman
Virginia Union University

Rick Wiedenmann
New Mexico State University at Carlsbad

Peter J. Wilkin
Purdue University North Central

Judy A. Williams
Southeastern Oklahoma State University

Dwina Willis
Freed Hardeman University

David Wilson
University of Miami

Mala S. Wingerd
San Diego State University

E. William Wischusen
Louisiana State University

Darla J. Wise
Concord College

Michael Womack
Macon State College

Bonnie Wood
University of Maine at Presque Isle

Mark L. Wygoda
McNeese State University

Shirley Zajdel
Housatonic Community College

Samuel J. Zeakes
Radford University

Uko Zylstra
Calvin College

1 Introduction: Biology Today

Biology is all around us. Like this hiker in the Monteverde Cloud Forest Reserve of Costa Rica, we are all surrounded by biology.

CHAPTER CONTENTS

BIOLOGY AND SOCIETY: Biology in Our Everyday Lives

Biology All Around Us

We are living in a golden age of biology. The largest and best-equipped community of scientists in history is beginning to solve biological puzzles that once seemed unsolvable. We are moving ever closer to understanding how a single cell becomes a plant or animal; how plants trap solar energy and store that energy in food; how living creatures form networks in biological communities such as forests and coral reefs; and how the great diversity of life on Earth evolved from the first microbes. Exploring life has never been more exhilarating. Welcome to the big adventure of the 21st century!

Modern biology is as important as it is inspiring, with exciting breakthroughs changing our very culture. Genetics and cell biology are revolutionizing medicine and agriculture. Molecular biology is providing new tools for investigating ancestry and solving crimes. Ecology is helping us evaluate environmental issues, such as the causes and consequences of global climate change. Neuroscience and evolutionary biology are reshaping psychology and sociology. These are just a few examples of how biology is woven into the fabric of society as never before. If you think about it, you can easily find a dozen ways each day that biology affects your life.

We wrote this book to help students who are not biology majors develop an appreciation for the science of life and apply that understanding as they evaluate social issues. We believe that such a biological perspective is essential for any educated person, which is why we named our book *Essential Biology*. So, whatever your reasons for taking this course, even if only to meet your school's science requirement, you'll soon discover that this is the best time ever to study biology.

To help you get started, this first chapter of *Essential Biology* defines biology and then expands on important concepts within this definition. First, we'll survey the properties of life and the scope of life. Next, we'll introduce evolution as the theme that unifies all of biology. Finally, we'll set the study of life in the broader context of science as a process of inquiry. Throughout the chapter, we'll provide examples of how biology intersects everyday life, highlighting the relevance of this topic to society and everyone in it.

The Scope of Life

Biology is the scientific study of life. It's a subject of enormous scope that gets bigger every year. To start our investigation of life, let's look at the properties that are shared by all living things.

the appearance of this katydid has evolved in a way that camouflages the animal in its environment. Evolutionary change has been a central, unifying feature of life since life arose nearly 4 billion years ago.

The Properties of Life

Our definition of biology as the scientific study of life raises some obvious questions: What is life? What distinguishes living things from nonliving things? The phenomenon of **life** seems to defy a simple, one-sentence definition. Yet almost any child perceives that a dog or a bug or a plant is alive, while a rock is not. We recognize life largely by what living things do.

Figure 1.1 highlights some of the properties and processes associated with life: (a) *Order*. All living things exhibit complex but ordered organization, as seen in the structure of a pinecone. (b) *Regulation*. The environment outside an organism may change drastically, but the organism can adjust its internal environment, keeping it within appropriate limits. When a lizard basks on a rock, its body absorbs solar energy and is warmed to an appropriate internal temperature. (c) *Growth and development*. Information carried by genes controls the pattern of growth and development in all organisms, including the green mamba snake. (d) *Energy utilization*. Organisms take in energy and use it to perform all of life's activities. A puffin obtains energy by eating fish and then uses this energy to power swimming and other work. (e) *Response to the environment*. All organisms respond to environmental stimuli. A Venus flytrap closes its trap rapidly in response to the environmental stimulus of an insect touching sensory hairs. (f) *Reproduction*. Organisms reproduce their own kind. Thus, hippos reproduce only hippos—never snakes or puffins. (g) *Evolution*. Reproduction underlies the capacity of populations to change (evolve) over time. For example,

Life at Its Many Levels

In *Essential Biology*, we will probe life all the way down to the microscopic scale of molecules such as DNA, the chemical responsible for inheritance. At the other extreme of biological size and complexity, our exploration will take us up to the global scale of the entire **biosphere**, which consists of all the environments on Earth that support life—including soil; oceans, lakes, and other bodies of water; and the lower atmosphere. **Figure 1.2** takes you on a tour of these levels of biological organization, starting with the biosphere and working down through smaller and smaller levels.

If you follow Figure 1.2 in the opposite direction, you can see life's hierarchy from molecules to the biosphere. It takes many molecules to build a cell, many cells to make a tissue, multiple tissues to make an organ, and so on. At each new level, novel properties emerge, properties that were not part of the components of the preceding level. For example, life emerges at the level of the cell, but a test tube full of molecules is not alive. Such properties illustrate an important theme of biology, called emergent properties. The familiar saying that "the whole is greater than the sum of its parts" captures this idea. The emergent properties of the whole result from the specific arrangement and interactions of the component parts.

From the interactions within the biosphere to the molecular machinery within cells, biologists are investigating life at its many levels. In the next section, we'll take a closer look at two biological levels near opposite ends of the size scale: ecosystems and cells. ✔

✔CHECKPOINT

1. Define biology.
2. Which properties of life apply to a car? Which do not?
3. What is the smallest level of biological organization that can display all the characteristics of life?

Answers: 1. Biology is the scientific study of life. 2. A car demonstrates order, regulation, energy utilization, and response to the environment. But a car does not grow, reproduce, or evolve. 3. a cell

---

▶ Figure 1.1 **Some properties of life.**

(a) Order

(b) Regulation

(c) Growth and development

(d) Energy utilization

▼ Figure 1.2 **Zooming in on life.**

1 Biosphere
Earth's biosphere includes all the environments on Earth that support life.

2 Ecosystems
A tide pool in the Galápagos Islands is an example of an ecosystem. An ecosystem consists of all organisms living in a particular area, as well as the nonliving, physical components of the environment that affect the organisms, such as water, air, soil, and sunlight.

3 Communities
All organisms in the tide pool (iguanas, crabs, seaweed, bacteria, and others) are collectively called a community.

4 Populations
Within communities are various populations, groups of interacting individuals of one species, such as a group of iguanas.

5 Organisms
An organism is an individual living thing, like this iguana.

6 Organ Systems and Organs
An organism's body consists of several organ systems, each of which contains two or more organs. For example, the iguana's circulatory system includes its heart and blood vessels.

10 Molecules and Atoms
Finally, we reach molecules, the chemical level in the hierarchy. Molecules are clusters of even smaller chemical units called atoms. Each cell consists of an enormous number of chemicals that function together to give the cell the properties we recognize as life. DNA, the molecule of inheritance and the substance of genes, is shown here as a computer graphic. Each sphere in the DNA model represents a single atom.

7 Tissues
Each organ is made up of several different tissues, such as the heart muscle tissue shown here. A tissue consists of a group of similar cells performing a specific function.

9 Organelles
Organelles are functional components of cells, such as the nucleus that houses the DNA.

8 Cells
The cell is the smallest unit that can display all the characteristics of life.

Nucleus

← Atom

(e) Response to the environment

(f) Reproduction

(g) Evolution

Ecosystems

Life does not exist in a vacuum. Each organism interacts continuously with its environment, which includes other organisms as well as nonliving factors. The roots of a tree, for example, absorb water and minerals from the soil. Leaves take in carbon dioxide gas from the air. Chlorophyll, the green pigment of the leaves, absorbs sunlight, which drives the plant's production of sugar from carbon dioxide and water. This food production is called photosynthesis. The tree releases oxygen to the air, and its roots help form soil by breaking up rocks. Both organism and environment are affected by the interactions between them. The tree also interacts with other living things, including microscopic organisms in the soil that are associated with the plant's roots and animals that eat its leaves and fruit. The interactions between organisms and their environment take place within an **ecosystem**, such as the tide pool shown in Figure 1.2.

The dynamics of any ecosystem depend on two main processes (**Figure 1.3**). The first major process is the cycling of nutrients. For example, minerals that plants take up from the soil can eventually be recycled to the soil by microorganisms that decompose leaf litter and other organic refuse. Producers are photosynthetic organisms, such as plants; consumers are the organisms, such as animals, that feed on plants, either directly (by eating plants) or indirectly (by eating animals that eat plants); decomposers, such as fungi and many bacteria, decompose waste products and the remains of deceased organisms, changing complex dead material into simple nutrients that are recycled. The action of decomposers ensures that nutrients are recycled within an ecosystem.

The second major process in an ecosystem is the flow of energy. In contrast to recycling nutrients, an ecosystem gains and loses energy constantly. Most ecosystems are solar powered. The energy that enters an ecosystem as sunlight is captured by producers when plants and other photosynthesizers absorb the sun's energy and convert it to the chemical energy of sugars and other complex molecules. Chemical energy is then passed through a series of consumers and, eventually, decomposers, powering each organism in turn. In the process of these energy conversions between and within organisms, some energy is converted to heat, which is then lost from the system. Thus, energy flows through an ecosystem, entering as light and exiting as heat.

The biosphere is enriched by a great variety of ecosystems. A tropical rain forest in South America is an ecosystem. Very different from tropical forests are the ecosystems of deserts, such as those of the southwestern United States. A coral reef, such as the Great Barrier Reef off the eastern coast of Australia, is an ecosystem, and so is any small pond that may exist on your campus or in your city. Even a woodland patch in New York's Central Park qualifies as an ecosystem, small and artificial as it is by forest standards and disrupted as it is by human visitors. The fact is, humans are organisms that now have some presence, often disruptive, in all ecosystems. And the collective clout of 6 billion humans and their machines affects the entire biosphere. For example, our fuel burning and forest chopping are changing the atmosphere and the planet's climate in ways that we do not yet fully understand, though we already know that our actions jeopardize the diversity of life on Earth. ☑

☑**CHECKPOINT**

Within an ecosystem,
_____ are recycled,
whereas _____
flows through.

Answer: nutrients; energy

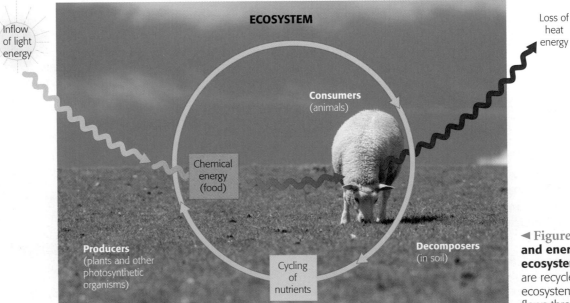

ECOSYSTEM

Inflow of light energy

Loss of heat energy

Consumers
(animals)

Chemical energy (food)

Producers
(plants and other photosynthetic organisms)

Cycling of nutrients

Decomposers
(in soil)

◀ Figure 1.3 **Nutrient and energy flow in an ecosystem.** Nutrients are recycled within an ecosystem, whereas energy flows through an ecosystem.

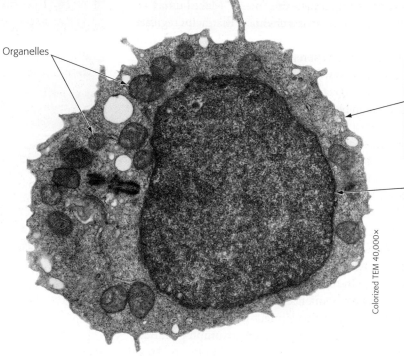

Prokaryotic cell (bacterium)
- Smaller
- Simpler structure
- DNA concentrated in nucleoid region, which is not enclosed by membrane
- Lacks most organelles

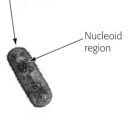

Nucleoid region

Organelles

Eukaryotic cell
- Larger
- More complex structure
- Nucleus enclosed by membrane
- Contains many types of organelles

Nucleus

Colorized TEM 40,000×

◄ **Figure 1.4 Two main kinds of cells: prokaryotic and eukaryotic.** The number near the bottom right corner represents the magnification of the image. In this case, the images shown are approximately 40,000 times bigger than the actual cells.

Cells and Their DNA

Let's downsize now from ecosystems to cells. The cell has a special place in the hierarchy of biological organization: It is the level at which the properties of life emerge—the lowest level of structure that can perform all activities required for life.

All organisms are composed of cells. They occur singly as a great variety of unicellular (single-celled) organisms, mostly microscopic. Cells are also the subunits that make up the tissues and organs of plants, animals, and other multicellular organisms. In either case, the cell is the organism's basic unit of structure and function. The ability of cells to divide to form new cells is the basis for all reproduction and for the growth and repair of multicellular organisms, including humans.

We can distinguish two major kinds of cells: prokaryotic and eukaryotic (**Figure 1.4**). The prokaryotic cell is much simpler and usually much smaller than the eukaryotic cell. The cells of the ubiquitous microorganisms called bacteria are prokaryotic. Most other forms of life, including plants and animals, are composed of eukaryotic cells. In contrast to prokaryotic cells, a eukaryotic cell is subdivided by internal membranes into many different functional compartments, or organelles. For example, the nucleus, the largest organelle in most eukaryotic cells, houses DNA, the heritable material that directs the cell's many activities. Prokaryotic cells also have DNA, but it is not packaged within a nucleus.

Though very different in structural complexity, prokaryotic and eukaryotic cells have much in common at the molecular level. Most importantly, all cells use DNA as the chemical material of genes, the discrete units of hereditary information. Of course, bacteria and humans inherit different genes, but that information is encoded in a chemical language common to all organisms. In fact, the language of life has an alphabet of just four letters. The chemical names of DNA's four molecular building blocks are abbreviated as A, G, C, and T (**Figure 1.5**).

An average-sized gene may be hundreds or thousands of chemical "letters" long. A gene's meaning to a cell is encoded in its specific sequence of these letters, just as the message of a sentence is encoded in its arrangement of letters selected from the 26 letters of the English alphabet.

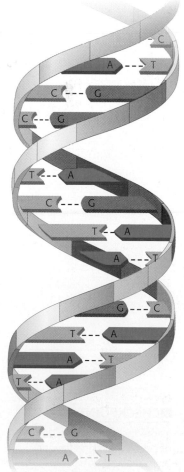

► **Figure 1.5 The language of DNA.** Every molecule of DNA is constructed from four kinds of chemical building blocks that are chained together, shown here as simple shapes and letters.

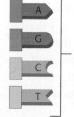

A
G
C
T

— The four chemical building blocks of DNA

A DNA molecule

One gene may be translated as "Build a blue pigment in a bacterial cell." Another gene may mean "Make human insulin in this cell." Insulin is a chemical that helps regulate your body's use of sugar as a fuel. Some people who have the disease diabetes must regulate their sugar levels by injecting themselves with insulin, which can now be produced by genetically engineered bacteria. These bacteria can make insulin because a gene for human insulin production has been transplanted into their DNA. This example of genetic engineering was one of the earliest successes of biotechnology, a field that has transformed the pharmaceutical industry and extended millions of lives (Figure 1.6). And it is only possible because biological information is written in the universal chemical language of DNA.

The entire "book" of genetic instructions that an organism inherits is called its genome. The nucleus of each human cell packs a genome that is about 3 billion chemical letters long. In recent years, scientists have tabulated virtually the entire sequence of these letters, and the press and world leaders have acclaimed this international achievement as the greatest scientific triumph ever. But unlike past cultural zeniths, such as the landing of Apollo astronauts on the moon, the sequencing of the human genome is more a commencement than a climax. As the quest continues, biologists will learn the functions of thousands of genes and how their activities are coordinated in the development and functioning of an organism. Additionally, the genomes of other organisms (such as *E. coli* bacteria, dogs, and monkeys) have been sequenced, allowing scientists to compare the genomes of different species. The emerging field of genomics—a branch of biology that studies whole genomes—is a striking example of human curiosity about life at its many levels. ☑

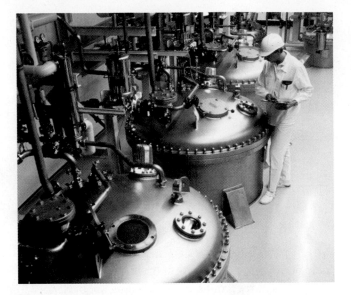

▲ Figure 1.6 **DNA technology in the drug industry.** In this biotechnology facility in Germany, genetically modified bacteria produce large quantities of human proteins.

☑CHECKPOINT

Name the two major kinds of cells. Which kind is your body made up of?

Answer: *prokaryotic and eukaryotic; eukaryotic*

Life in Its Diverse Forms

Diversity is a hallmark of life. The iguana shown in Figure 1.2 is just one of about 1.8 million species that biologists have identified and named (the scientific name for this iguana is *Amblyrhynchus cristatus*). The diversity of known life includes at least 290,000 plants, 52,000 vertebrates (animals with backbones), and 1 million insects (more than half of all known forms of life). Biologists add thousands of newly identified species to the list each year. Estimates of the total number of species range from 10 million to over 100 million. Whatever the actual number, the vast diversity of life gives biology a very wide scope.

Grouping Species: The Basic Concept

Biological diversity can be something to relish and preserve (Figure 1.7), but it can also be a bit overwhelming.

▼ Figure 1.7 **A small sample of biological diversity.** Fish called yellow-ribbon sweetlips swim around a coral reef in Indonesia.

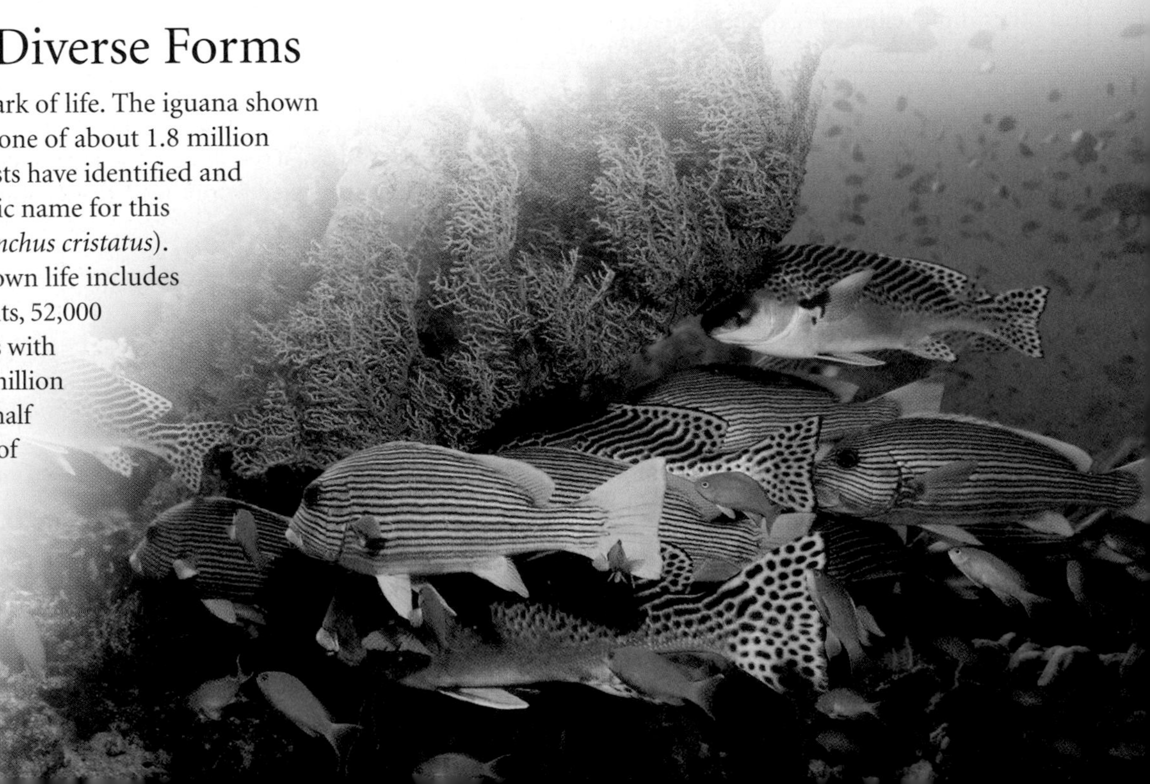

Confronted with complexity, people are inclined to categorize diverse items into a smaller number of groups. Grouping species that are similar is natural for us. We may speak of "squirrels" and "butterflies," even though we recognize that each group actually includes many different species. We may even sort groups into broader categories, such as rodents (which include squirrels) and insects (which include butterflies). Taxonomy, the branch of biology that names and classifies species, formalizes this hierarchical ordering according to a scheme you will learn about in Chapter 14. Here we consider only the broadest units of classification.

The Three Domains of Life

Biologists divide the diversity of life into three main groups. The three groups, called domains, are Bacteria, Archaea, and Eukarya (**Figure 1.8**). The first two domains, Bacteria and Archaea, identify two very different groups of organisms that have prokaryotic cells. All the eukaryotes (organisms with eukaryotic cells) are placed within the domain Eukarya, which is further divided into smaller categories called kingdoms. Most members of three of the kingdoms—Plantae, Fungi, and Animalia—are multicellular. These three kingdoms are distinguished partly by how the organisms obtain food. Plants produce their own sugars and other foods by photosynthesis. Fungi are mostly decomposers, obtaining food by digesting dead organisms. Animals obtain food by ingesting (eating) and digesting other organisms. (This is, of course, the kingdom to which we belong.) Those eukaryotes that do not fit into the other three kingdoms are referred to as the protists. Protists are generally single-celled; they include microscopic protozoans, such as amoebas. But protists also include certain multicellular forms, such as seaweeds. Scientists are in the process of organizing protists into multiple kingdoms, although they do not yet agree on exactly how to do this.

Unity in the Diversity of Life

If life is so diverse, how can biology have any unifying themes? What, for instance, can a tree, a mushroom, and a human possibly have in common? As it turns out, a great deal! Underlying the diversity of life is a striking unity, especially at the lower levels of biological organization. One of biology's major goals is to explain how such diversity arises while also accounting for characteristics common to different species. We have already seen one example: the universal genetic language of DNA. That fundamental language connects all kingdoms of life, even uniting prokaryotes such as bacteria with eukaryotes such as humans. What can account for this combination of unity and diversity in life? The scientific explanation is the biological process called evolution. ☑

▼ Figure 1.8 **The three domains of life.**

DOMAIN BACTERIA

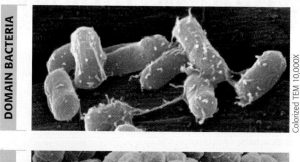

Colorized TEM 10,000X

DOMAIN ARCHAEA

TEM 18,500X

DOMAIN EUKARYA

Kingdom Plantae

Kingdom Fungi

Kingdom Animalia

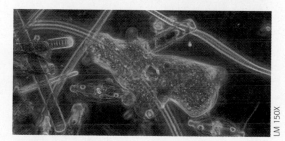

LM 150X

Protists (multiple kingdoms)

☑ **CHECKPOINT**

1. Name the three domains of life. To which do you belong?
2. Name three kingdoms found within the domain to which humans belong. Name a fourth group within this domain.

Answers: 1. Bacteria, Archaea, Eukarya; Eukarya 2. Plantae, Fungi, Animalia; the protists

Evolution: Biology's Unifying Theme

The history of life, as documented by fossils and other evidence, is a saga of a restless Earth billions of years old, inhabited by a changing cast of living forms (Figure 1.9). Life evolves. Just as each individual has a family history, each species is one twig of a branching tree of life extending back in time through ancestral species more and more remote. Species that are very similar, such as the brown bear and the polar bear, share a common ancestor that represents a relatively recent branch point on the tree of life (Figure 1.10). But through an ancestor that lived much farther back in time, all bears are also related to squirrels, humans, and all other mammals. Hair and milk-producing mammary glands are two uniquely mammalian traits. Such similarities are what we would expect if all mammals descended from a common ancestor, a prototypical mammal. And mammals, reptiles, and all other vertebrates share a common ancestor even more ancient. Evidence of a still broader relationship can be found in similarities that are seen within all eukaryotic cells. Trace life back far enough, and there are only fossils of the primeval prokaryotes that inhabited Earth over 3 billion years ago. All of life is connected. And the basis for this kinship is evolution, the process that has transformed life on Earth from its earliest beginnings to the extensive diversity we see today. Evolution is the theme that unifies all of biology.

▲ Figure 1.10 **An evolutionary tree of bears.** This tree is based on both the fossil record and a comparison of DNA sequences among modern bears.

Ancestral bear

Common ancestor of polar bear and brown bear

Giant panda

Spectacled bear

Sloth bear

Sun bear

American black bear

Asiatic black bear

Polar bear

Brown bear

30   25   20   15   10   5
Millions of years ago

▼ Figure 1.9 **Digging into the past.** A paleontologist carefully excavates a dinosaur bone at Dinosaur National Monument, Utah.

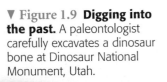

The Darwinian View of Life

The evolutionary view of life came into focus in 1859 when British biologist Charles Darwin published *The Origin of Species*. His book developed two main points. First, Darwin marshaled the available evidence in support of the evolutionary view that species living today descended from ancestral species. Darwin called this process "descent with modification." It is an insightful phrase, as it captures the duality of life's unity (descent) and diversity (modification). In the Darwinian view, for example, the diversity of bears is based on different modifications of a common ancestor from which all bears descended. As the second main point in *The Origin of Species*, Darwin proposed a mechanism for descent with modification. He called this process natural selection. ✔

Natural Selection

As you'll learn in Chapter 13, Charles Darwin gathered important evidence for his theories during an around-the-world voyage. He was particularly struck by the diversity of animals on the Galápagos Islands, off the coast of Ecuador (Figure 1.11). Darwin regarded adaptation to the environment and the origin of new species as closely related processes. If some geographic barrier—an ocean separating islands, for instance—isolated two populations of a single species from each other, the populations could diverge more and more in appearance as each adapted to local environmental conditions. Over many generations, the two populations could become dissimilar enough to be designated separate species. The labeled branch point near the lower right of Figure 1.10 represents the point in time when two populations of one bear species began to diverge from each other, adapting to different climates and resulting in the evolution of the modern brown bear and the polar bear as two distinct species. Darwin realized that explaining such adaptation was the key to understanding descent with modification, or evolution. This focus on adaptation helped Darwin envision his concept of natural selection as the mechanism of evolution.

Darwin's Inescapable Conclusion

Darwin synthesized the theory of natural selection from two observations that by themselves were neither profound nor original. Others had the pieces of the puzzle, but Darwin saw how they fit together. As the evolutionary biologist Stephen Jay Gould put it, Darwin based his mechanism of natural selection on "two undeniable facts and an inescapable conclusion." Let's look at his logic:

Observation 1: **Overproduction and competition.**
Any population of a species has the potential to produce far more offspring than the environment can possibly support with available resources such as food and shelter. This overproduction leads to competition among the varying individuals of a population for these limited resources.

Observation 2: **Individual variation.**
Individuals in a population of any species vary in many inherited traits. No two individuals in a population are exactly alike. You know this variation to be true of human populations; careful observers find variation in populations of all species.

Conclusion: **Unequal reproductive success.**
In the struggle for existence, those individuals with traits best suited to the local environment will, on average, have the greatest reproductive success: They will leave the greatest number of surviving, fertile offspring. Therefore, the very traits that enhance survival and reproductive success will be disproportionately represented in succeeding generations of a population.

▶ Figure 1.11 **Charles Darwin (1809–1882),** *The Origin of Species,* **and blue-footed boobies on the Galápagos Islands.**

It is this unequal reproductive success that Darwin called **natural selection**. And the product of natural selection is adaptation, the accumulation of favorable variations in a population over time. Returning to the example of bear evolution in Figure 1.10, one familiar adaptation is fur color. Polar bears and brown (grizzly)

▼ Figure 1.12 **Natural selection.**

Population with varied inherited traits. Initially, the population varies extensively in the coloration of individual beetles, from very light gray to charcoal.

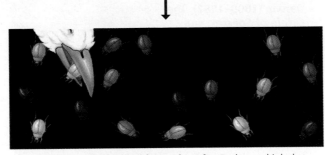

Elimination of individuals with certain traits. For hungry birds that prey on the beetles, it is easiest to spot the beetles that are lightest in color.

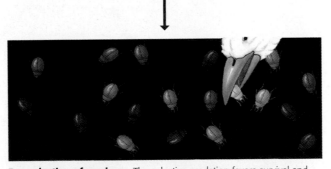

Reproduction of survivors. The selective predation favors survival and reproductive success of the darker beetles. Thus, genes for dark color are passed along to the next generation in greater frequency than genes for light color.

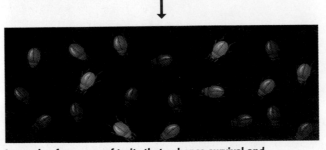

Increasing frequency of traits that enhance survival and reproductive success. Generation after generation, the beetle population adapts to its environment through natural selection.

bears, as closely related as they are, each display an evolutionary adaptation (white and brown fur, respectively) that resulted from natural selection operating in their respective environments. Presumably, natural selection tended to favor the fur color that provided each bear lineage with the best camouflage in their home territories.

As the process leading to adaptation, natural selection is also the mechanism of evolution. **Figure 1.12** presents a hypothetical example of a beetle population that colonizes a location where the soil has been blackened by a recent brush fire. Notice that natural selection works on preexisting variation (in shell color, in this case), enhancing the reproductive success of individuals with beneficial traits.

Observing Artificial Selection

Darwin found convincing evidence for the power of unequal reproduction in examples of artificial selection, the selective breeding of domesticated plants and animals by humans. We humans have been modifying other species for millennia by selecting breeding stock with certain traits. The plants we grow for food bear little resemblance to their wild ancestors; this is because humans have customized crop plants through many generations of artificial selection by selecting different parts of the plant to accentuate as food. All the vegetables shown in **Figure 1.13a** (and more) have a common ancestor in one species of wild mustard (shown in the center of the figure). The power of selective breeding is especially apparent in our pets, which have been bred for fancy and for utility. For example, people in different cultures have customized hundreds of dog breeds as different as basset hounds and Saint Bernards, all descended from wolves (**Figure 1.13b**). The tremendous variety of modern dogs reflects thousands of years of artificial selection by humans. Darwin could see that in artificial selection, humans were substituting for the environment in screening the heritable traits of populations.

Observing Natural Selection

If artificial selection could achieve so much change so rapidly, Darwin reasoned, then natural selection should be capable of considerable adaptation of species over hundreds or thousands of generations. We now recognize many examples of natural selection in action. A classic example involves the finches (a kind of bird) of the Galápagos Islands. Over a span of two decades, researchers measured changes in beak size in a population of a species of ground finches that eats mostly small seeds. In dry years, when all seeds are in short supply, the birds must eat more large seeds. Birds with larger, stronger beaks have a feeding advantage and greater

▼ Figure 1.13 **Examples of artificial selection.**

(a) Vegetables descended from wild mustard

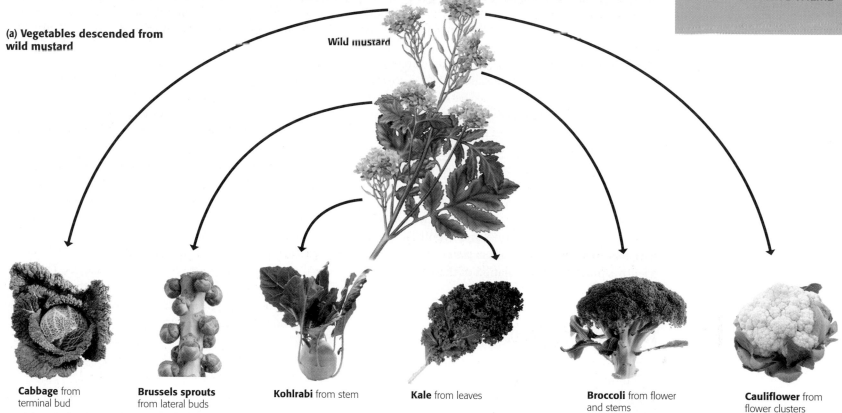

Wild mustard

Cabbage from terminal bud

Brussels sprouts from lateral buds

Kohlrabi from stem

Kale from leaves

Broccoli from flower and stems

Cauliflower from flower clusters

(b) Domesticated dogs descended from wolves

Domesticated dogs

Gray wolves

reproductive success, and the average beak depth for the population increases. During wet years, smaller beaks are more efficient for eating the now abundant small seeds, and the average beak depth decreases. Such changes are measurable evidence of natural selection in action. At the end of this chapter, you'll read about another example: how the evolution of antibiotic resistance among bacteria is changing the way doctors prescribe drugs.

Darwin's publication of *The Origin of Species* fueled an explosion in biological research and knowledge that continues today. Over the past century and a half, a tremendous amount of evidence has accumulated in support of Darwin's theory of evolution by natural selection, making it one of biology's best-demonstrated, most comprehensive, and longest-lasting theories. In every chapter of *Essential Biology*, we will highlight connections to evolution—the unifying theme of biology. ✓

✓CHECKPOINT

What mechanism did Darwin propose for evolution? What three-word phrase summarizes this mechanism?

Answer: natural selection; unequal reproductive success

The Process of Science

Recall the first definition from the start of this chapter: Biology is the scientific study of life. Now that we have explored the question *What is life?* we can turn our attention to the next obvious question: *What is science?* And how do we tell the difference between science and other ways of trying to make sense of nature?

The word *science* is derived from a Latin verb meaning "to know." **Science** is a way of knowing, one that is based on inquiry. It developed from people's curiosity about themselves and the world around them. This basic human drive to understand is manifest in two main scientific approaches: discovery science, which is mostly about *describing* nature, and hypothesis-driven science, which is mostly about *explaining* nature. Most scientists practice a combination of these two forms of inquiry.

Discovery Science

Scientists seek natural causes for natural phenomena. This limits the scope of science to the study of structures and processes that we can observe and measure, either directly or indirectly with the help of tools, such as microscopes. This dependence on observations that other people can confirm demystifies nature and distinguishes science from belief in the supernatural. Science can neither prove nor disprove that angels, ghosts, deities, or spirits, whether benevolent or evil, cause storms, rainbows, illnesses, or cures, for such explanations are outside the bounds of science.

Verifiable observations and measurements are the data of **discovery science** (Figure 1.14). In our quest to describe nature accurately, we discover its structure. In biology, discovery science enables us to describe life at its many levels, from ecosystems down to cells and molecules. Darwin's careful description of the diverse plants and animals he collected in South America is an example of discovery science. A more recent example is the sequencing of the human genome, a detailed dissection and description of our genetic material.

Discovery science can lead to important conclusions based on a type of logic called inductive reasoning. An inductive conclusion is a generalization that summarizes a large number of observations. "All organisms are made of cells" is an example. That induction was based on two centuries of biologists discovering cells in every biological specimen they observed with microscopes. The careful observations of discovery science and the inductive conclusions they sometimes produce are fundamental to our understanding of nature.

Hypothesis-Driven Science

The observations of discovery science stimulate us to ask questions and seek explanations. Ideally, such investigation makes use of what is called the scientific method. As a formal process of inquiry, the **scientific method** consists of a series of steps (Figure 1.15). These steps guide scientific investigations, but working scientists typically do not follow them rigidly; different scientists proceed through the scientific method in different ways.

Most modern scientific investigations can be described as **hypothesis-driven science**. A **hypothesis** is a proposed explanation for a set of observations—an idea on trial. We all use hypotheses in solving everyday problems. Let's say, for example, that your flashlight fails during a campout. That's an observation. The question is

▼ Figure 1.14 **Careful observation and measurement: the raw data for discovery science.** Dr. Jane Goodall spent decades recording her observations of chimpanzee behavior during field research in the jungles of Gambia.

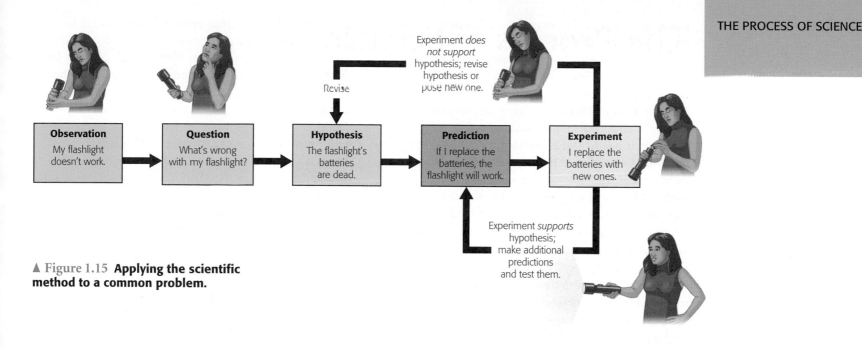

Experiment *does not support* hypothesis; revise hypothesis or pose new one.

Revise

Observation	Question	Hypothesis	Prediction	Experiment
My flashlight doesn't work.	What's wrong with my flashlight?	The flashlight's batteries are dead.	If I replace the batteries, the flashlight will work.	I replace the batteries with new ones.

Experiment *supports* hypothesis; make additional predictions and test them.

▲ Figure 1.15 **Applying the scientific method to a common problem.**

obvious: Why doesn't the flashlight work? A reasonable hypothesis based on past experience is that the batteries in the flashlight are dead.

Once a hypothesis is formed, an investigator can use deductive logic to test it. Deduction contrasts with induction, which, remember, is reasoning from a set of specific observations to reach a general conclusion. In deduction, the reasoning flows in the opposite direction, from the general to the specific. From general premises, we extrapolate to the specific results we should expect if the premises are true. For example, if all organisms are made of cells (premise 1), and humans are organisms (premise 2), then humans are composed of cells (deductive prediction about a specific case).

In the process of science, the deduction usually takes the form of predictions about what experimental results or observations we should expect if a particular hypothesis (premise) is correct. We then test the hypothesis by performing an experiment to see whether or not the results are as predicted. This deductive testing takes the form of "If . . . then" logic:

Observation: My flashlight doesn't work.

Question: What's wrong with my flashlight?

Hypothesis: The flashlight's batteries are dead.

Prediction: If I replace the batteries, the flashlight will work.

Experiment: I replace the batteries with new ones.

Predicted result: The flashlight should work.

Let's say the flashlight still doesn't work. We can test an alternative hypothesis if new flashlight bulbs are available. We could also blame the dead flashlight on campground ghosts playing tricks, but that hypothesis is untestable and therefore outside the realm of science. ✓

☑**CHECKPOINT**

If you spend a summer observing the squirrels on your campus and collecting data on their dietary habits, what kind of science are you performing? If you come up with a tentative explanation for their dietary behavior and then test your idea, what kind of science are you performing?

Answer: discovery science; hypothesis-driven science

THE PROCESS OF SCIENCE: Biology in Our Everyday Lives

Is Trans Fat Bad for You?

One way to better understand how the process of science can be applied to real-world problems is to examine a case study, an in-depth examination of an actual investigation. The rest of this section is a case study about the effects of one kind of dietary fat. Later chapters will include other case studies in the process of science, illustrating how discovery and hypothesis-driven science have been used in experiments both modern and classic. To emphasize how the scientific method has been employed in each case, the key steps will appear in blue text.

Dietary fat, a major component of the food we eat, comes in several different forms. Trans fat is a nonnatural form of fat produced through manufacturing processes. Because trans fat adds texture, increases shelf life, and is inexpensive to prepare, it was added to foods (in particular, margarine, shortening, and foods cooked in partially hydrogenated oils) in increasing amounts during the 20th century.

Dietary fat was one component examined in the Nurses' Health Study, a landmark study of over 120,000 female U.S. nurses begun in 1976. Every two years, the study participants completed questionnaires about their health status and dietary habits. In 1994, researchers examining the trove of collected data discovered that study participants who ate high levels of trans fat had nearly double the risk of heart disease compared to participants who ate little trans fat. This is a good example of discovery science: examination of a set of data gathered without a preconceived notion of what it might reveal.

While the data from the Nurses' Health Study were suggestive, some researchers remained skeptical about the link between trans fat and heart health. In a hypothesis-driven study published in 2004, an Australian research team started with the **observation** that human body fat (adipose tissue) retains traces of consumed dietary fats. This observation raised a **question**: Would the adipose tissue of heart attack patients be measurably different from adipose tissue in a similar group of healthy patients? The researchers' **hypothesis** was that it would be, leading to the **prediction** that healthy patients' body fat would contain less trans fat than the body fat in heart attack victims.

The researchers then set up an **experiment** to determine the amounts of different kinds of fat in the adipose tissue of 79 patients who had experienced a heart attack. These amounts were compared with data for 167 patients who had not experienced a heart attack but were similar to the first group in most other respects. Since trans fat is not produced by the body, the

researchers assumed that any trans fat found in the adipose tissue had been consumed in the diet. Their **results** showed significantly higher levels of trans fat in the bodies of the heart attack patients than in the healthier group **(Figure 1.16)**. These results added to the growing body of evidence that trans fats are unhealthy and should not be added to foods. Indeed, several countries (such as Australia and Denmark) have banned trans fats altogether, and some American municipalities (such as New York City) have followed suit. You would do well to read nutrition labels and avoid trans fat as much as possible in your own diet.

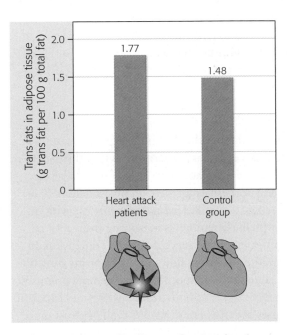

▲ Figure 1.16 **Levels of trans fat.** Results of an experiment that measured the levels of trans fat in the adipose tissue of 79 patients who had experienced a heart attack and 167 patients who had not (control group).

The trans fat study described in the Process of Science section is an example of a **controlled experiment**. An experiment of this type is designed to compare an experimental group (the patients who had a heart attack, in this case) with a control group (patients who did not have a heart attack). Ideally, a control group and an experimental group differ only in one variable—in our example, the occurrence of a heart attack. Other factors—such as age, weight, and sex of the patients—were matched between the two sets of patients. Thus, the control group canceled out the effects of all variables other than the one being tested. The use of a controlled experiment enabled researchers to draw conclusions about the effect of just one variable—trans fat—on heart attack risk. ☑

Theories in Science

Many people associate facts with science, but accumulating facts is not the primary goal of science. A telephone book is an impressive catalog of factual information, but it has little to do with science. It is true that facts, in the form of verifiable observations and repeatable experimental results, are the prerequisites of science. What really advances science, however, are new theories that tie together a number of observations that previously seemed unrelated. The cornerstones of science are the explanations that apply to the greatest variety of phenomena. People like Newton, Darwin, and Einstein stand out in the history of science not because they discovered a great many facts, but because their theories had such broad explanatory power.

What is a scientific theory, and how is it different from a hypothesis? A **theory** is much broader in scope than a hypothesis. This is a hypothesis: "White fur is an evolutionary adaptation that helps polar bears survive in an arctic habitat." But this is a theory: "Adaptations evolve by natural selection."

Because theories are so comprehensive, they only become widely accepted in science if they are supported by an accumulation of extensive and varied evidence. The use of the term *theory* in science for a comprehensive explanation supported by abundant evidence contrasts with our everyday usage, which equates theories more with speculations or hypotheses. Natural selection qualifies as a scientific theory because of its broad application and because it has been validated by a huge number of observations and experiments.

Scientific theories are not the only way of "knowing nature," of course. A comparative religion course would be a good place to learn about the diverse legends that tell of a supernatural creation of Earth and its life. Science and religion are two very different ways of trying to make sense of nature. Art is still another way. A broad education should include exposure to these different ways of viewing the world. Each of us synthesizes our worldview by integrating our life experiences and multidisciplinary education. As a science textbook and part of that broad education, *Essential Biology* showcases life in the scientific context of evolution, the one theme that continues to hold all of biology together, no matter how big and complex the subject becomes. ☑

The Culture of Science

It is not unusual for several scientists to ask the same questions. Such convergence contributes to the progressive and self-correcting qualities of science. Scientists build on what has been learned from earlier research, and they pay close attention to contemporary scientists working on the same problem. Scientists share information through publications, seminars, meetings, and personal communication. The Internet has added a new medium for this exchange of ideas and data.

Both cooperation and competition characterize the scientific culture (Figure 1.17). Scientists working in the same research field subject one another's work to careful scrutiny. It is common for scientists to check the conclusions of others by attempting to repeat experiments. This obsession with evidence and confirmation helps characterize the scientific style of inquiry. Scientists are generally skeptics.

We have seen that science has two key features that distinguish it from other styles of inquiry: (1) a dependence on observations and measurements that others can verify and (2) the requirement that ideas (hypotheses) are testable by experiments that others can repeat.

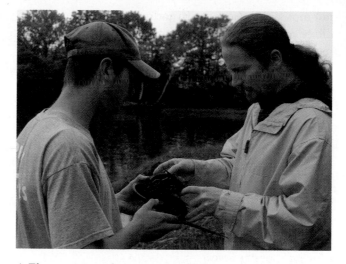

▲ Figure 1.17 **Science as a social process.** New England College neurobiologist Jim Newcomb (right) mentors a student in the methods of marine biology.

Science, Technology, and Society

Science and technology are interdependent. New technologies, such as more powerful microscopes and computers, advance science. And scientific discoveries can lead to new technologies. In most cases, technology applies scientific discoveries to the development of new goods and services. For example, just over 50 years ago, two scientists, James Watson and Francis Crick, discovered the structure of DNA through the process of science. Their discovery eventually led to a variety of DNA technologies, including the genetic engineering of microorganisms to mass-produce human insulin (see Figure 1.6) and the use of DNA profiling for investigating crimes (Figure 1.18). Perhaps Watson and Crick envisioned that their discovery would someday inform new technologies, but that probably did

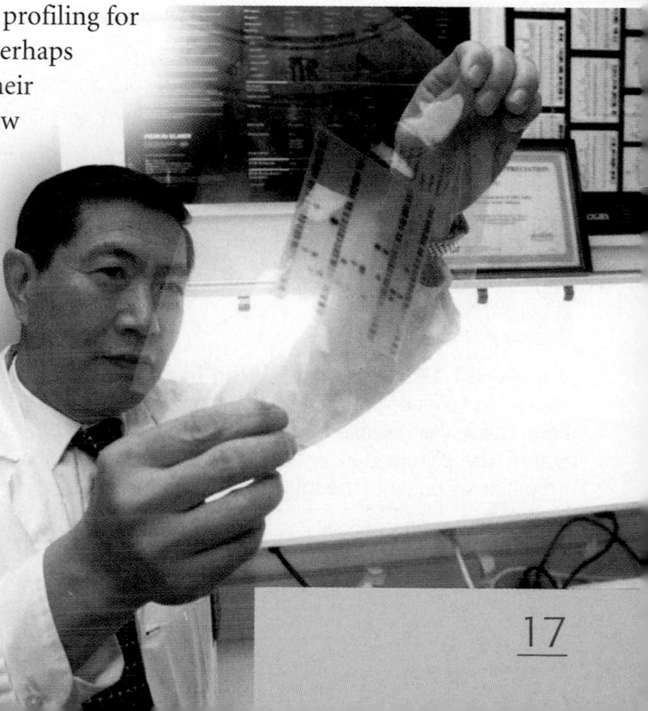

► Figure 1.18 **DNA technology and the law.** The stained bands visible in this photograph represent fragments of DNA extracted from body tissue collected at a crime scene. The pattern of bands varies from person to person.

☑CHECKPOINT

You arrange to meet a friend for dinner at 6 P.M., but she is not there. You wonder why. Another friend says, "My theory is that she forgot." If your friend were speaking like a scientist, what should she have said?

Answer: *"My hypothesis is that she forgot."*

not motivate their research, nor could they have predicted exactly what the applications would be. The direction technology takes depends less on the curiosity that drives basic science than it does on the current needs of humans and the changing cultural climate.

Technology has improved our standard of living in many ways, but it is a double-edged sword. Technology that keeps people healthier has enabled the population to grow more than tenfold in the past three centuries and to more than double to 6.6 billion in just the past 40 years. The environmental consequences are sometimes devastating. Acid rain, deforestation, global climate change, nuclear accidents, toxic wastes, and extinction of species are just a few of the repercussions of more and more people wielding more and more technology. Science can help us identify such problems and provide insight about what course of action may prevent further damage. But solutions to these problems have as much to do with politics, economics, culture, and the values of societies as with science and technology. Now that science and technology have become such powerful functions of society, each of us has the responsibility to become a "citizen scientist" by developing a reasonable amount of scientific and technological literacy. ☑

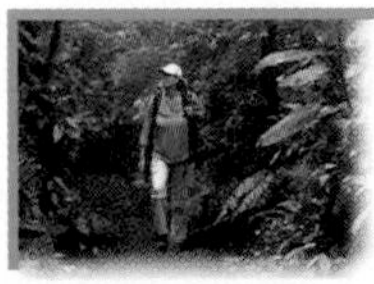

EVOLUTION CONNECTION: Biology in Our Everyday Lives

Evolution in Our Everyday Lives

Evolution is the core theme of biology. To emphasize the centrality of evolution to every area of biology, we end each chapter of *Essential Biology* with an Evolution Connection section. But is evolution connected to our everyday lives? And if so, in what ways? Evolution teaches us that the environment is a powerful selective force. Our world is rich with examples of such natural selection in action. One example that may directly affect your life is the development of resistance to antibiotics in disease-causing bacteria.

Antibiotics are drugs that help cure certain infections by impairing the bacteria that cause them. Antibiotics have saved millions of human lives, but there's a dark side to the widespread use of antibiotics: It has driven the evolution of antibiotic-resistant populations of the very bacteria the drugs are meant to kill.

When an antibiotic is taken by a patient, it will usually kill most, but not all, of the infecting bacteria. The environment, which now contains the antibiotic, selects among the varying bacteria of a population for those individuals that can survive the drug. In some cases, the mechanism of resistance in the bacteria is an ability to destroy the antibiotic—or even use the drug as food! Even though the drug kills most of the bacteria, those few bacteria that are resistant may soon multiply and become the norm in the population rather than the exception.

The evolution of antibiotic-resistant bacteria is a huge problem in public health. For example, there are now some strains of tuberculosis-causing bacteria that are resistant to all three of the antibiotics currently used to treat the disease (Figure 1.19). Unfortunate people infected with one of these resistant strains have no better chance of surviving than did tuberculosis patients a century ago.

The problem of antibiotic-resistant bacteria is influencing the way many physicians prescribe antibiotics. Doctors who understand that the abuse of these drugs is speeding the evolution of resistant bacteria are less likely to prescribe antibiotics needlessly—for instance, for a patient complaining of a common cold or flu, diseases caused by viruses (not bacteria), against which antibacterial drugs are powerless. Additionally, many farmers are reducing the use of antibiotics in animal feed, and many consumers are choosing to buy meats from animals raised without antibiotics.

It is important to note that adaptation of bacteria to an environment containing an antibiotic does not mean that the drug *created* the antibiotic resistance trait. Instead, the environment screened the heritable variations that already existed among individuals of a population and favored the individuals best suited to the present conditions. Throughout *Essential Biology*, you will learn more about how natural selection works and see other examples of how natural selection affects your life.

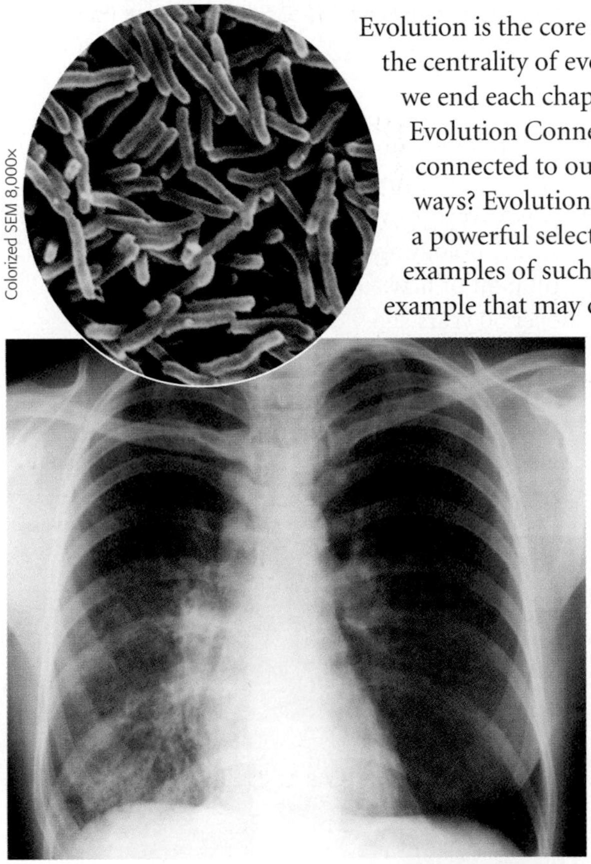

Colorized SEM 8,000x

▲ **Figure 1.19 Natural selection in action.** Antibiotic-resistant strains of tuberculosis-causing bacteria (top) have made the disease a threat again in the United States. The colorized X-ray above shows the lungs of a tuberculosis patient. The infection is shown in red.

Chapter Review

SUMMARY OF KEY CONCEPTS

(MB) Go to the Study Area at **www.masteringbiology.com** for practice quizzes, my**e**Book, BioFlix™ 3-D animations, MP3 Tutor Sessions, videos, current events, and more.

The Scope of Life

Biology is the scientific study of life. The study of biology encompasses a wide scale of size and a huge variety of life, both past and present.

The Properties of Life

All life displays a common set of characteristics:

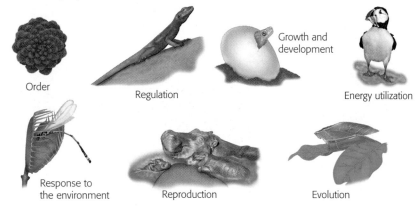

Order

Regulation

Growth and development

Energy utilization

Response to the environment

Reproduction

Evolution

Life at Its Many Levels

Within ecosystems, nutrients are recycled, but energy flows through. Cells are either prokaryotic (simple, small, and lacking membrane-bounded organelles, such as the nucleus) or eukaryotic (more complex, larger, and containing a nucleus and other membrane-bounded organelles).

Life in Its Diverse Forms

Biologists organize living organisms into three domains. The domain Eukarya is further divided into three kingdoms (distinguished partly by their means of obtaining food) and one catch-all group:

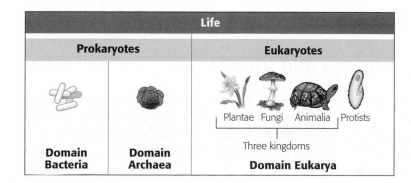

Life		
Prokaryotes	**Eukaryotes**	
	Plantae Fungi Animalia Protists	
	Three kingdoms	
Domain Bacteria	**Domain Archaea**	**Domain Eukarya**

Evolution: Biology's Unifying Theme

The Darwinian View of Life

Charles Darwin established the ideas of evolution ("descent with modification") via natural selection (unequal reproductive success) in his 1859 publication *The Origin of Species*.

Natural Selection

Darwin formulated the theory of natural selection through a logical inference: Natural selection leads to adaptations to the environment, which—when passed from generation to generation—is the mechanism of evolution.

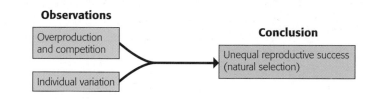

Observations

Overproduction and competition

Individual variation

Conclusion

Unequal reproductive success (natural selection)

The Process of Science

Discovery Science

Describing the natural world with verifiable data is the hallmark of discovery science.

Hypothesis-Driven Science

A scientist formulates a hypothesis (tentative explanation) to explain the natural world. The hypothesis may then be tested via the steps of the scientific method:

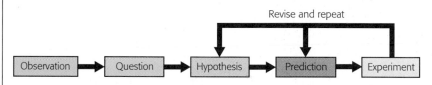

Revise and repeat

Observation → Question → Hypothesis → Prediction → Experiment

Theories in Science

A theory is a broad and comprehensive statement about the world that is supported by the accumulation of a great deal of verifiable evidence.

The Culture of Science

Scientists build on each other's work by always remaining skeptical and by seeking reproducible evidence to confirm ideas.

Science, Technology, and Society

Scientific advances promote new technologies, which in turn enable further scientific advances and useful applications to our lives.

SELF-QUIZ

1. Which of the following is *not* a characteristic of all living organisms?
   a. capable of self-reproduction
   b. composed of multiple cells
   c. complex yet organized
   d. energy utilization

2. Place the following levels of biological organization in order from smallest to largest: atom, biosphere, cell, ecosystem, molecule, organ, organism, population, tissue. Which is the smallest level capable of demonstrating all of the characteristics of life?

3. Plants use the process of photosynthesis to convert the energy in sunlight to chemical energy in the form of sugar. While doing so, they consume carbon dioxide and water and release oxygen. Explain how this process functions in both the cycling of chemical nutrients and the flow of energy through an ecosystem.

4. For each of the following organisms, match its description to its most likely domain and/or kingdom:

   a. A foot-tall organism capable of producing its own food from sunlight

   b. A microscopic, simple, nucleus-free organism found growing in a riverbed

   c. An inch-tall organism growing on the forest floor that consumes material from dead leaves

   d. A thimble-sized organism that feeds on algae growing in a pond

   1. Bacteria

   2. Eukarya/Animalia

   3. Eukarya/Fungi

   4. Eukarya/Plantae

5. How does natural selection cause a population to become adapted to its environment over time?

6. Why is it difficult to draw a conclusion from an experiment that does not include a control group?

7. Which of the following best describes the logic of the scientific method?
   a. If I generate a testable hypothesis, tests and observations will support it.
   b. If my prediction is correct, it will lead to a testable hypothesis.
   c. If my observations are accurate, they will support my hypothesis.
   d. If my hypothesis is correct, I can expect certain test results.

8. Which of the following statements best distinguishes hypotheses from theories in science?
   a. Theories are hypotheses that have been proved.
   b. Hypotheses are tentative guesses; theories are correct answers to questions about nature.
   c. Hypotheses usually are narrow in scope; theories have broad explanatory power.
   d. Hypotheses and theories mean essentially the same thing in science.

9. _____ is the core idea that unifies all areas of biology.

10. Match each of the following terms to the phrase that best describes it.

    a. Natural selection
    b. Evolution
    c. Hypothesis
    d. Biosphere

    1. A testable idea
    2. Descent with modification
    3. Unequal reproductive success
    4. All life-supporting environments on Earth

Answers to the Self-Quiz questions can be found in Appendix D.

THE PROCESS OF SCIENCE

11. The graph below shows the results of an experiment in which mice learned to run through a maze.

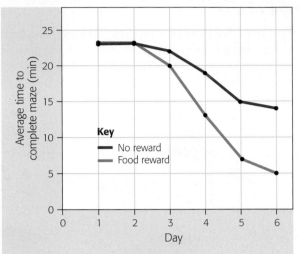

   a. State a hypothesis and prediction that you think this experiment may have been designed to test.
   b. Which was the control group and which the experimental group? Why was a control group needed?
   c. List some variables that must have been controlled so as not to affect the results.
   d. Do the data support the hypothesis you chose? Explain.

12. The fruits of wild species of tomato are tiny compared to the giant beefsteak tomatoes available today. This difference in fruit size is almost entirely due to the larger number of cells in the domesticated fruits. Plant biologists have recently discovered genes that are responsible for controlling cell division in tomatoes. Why would such a discovery be important to producers of other kinds of fruits and vegetables? To the study of human development and disease? To our basic understanding of biology?

BIOLOGY AND SOCIETY

13. The news media and popular magazines frequently report stories that are connected to biology. In the next 24 hours, record all such stories you hear or read about from three different sources and briefly describe the biological connections you perceive in each story.

14. If you pay attention, you will find yourself conducting many hypothesis-driven experiments each day. Over the next day, try to think of a good example where you performed a simple experiment to test a hypothesis about some observation. Write the experience out in a standard narrative paragraph, and then rewrite it using the steps of the scientific method (observation, question, hypothesis, prediction, and so on).

Unit 1
Cells

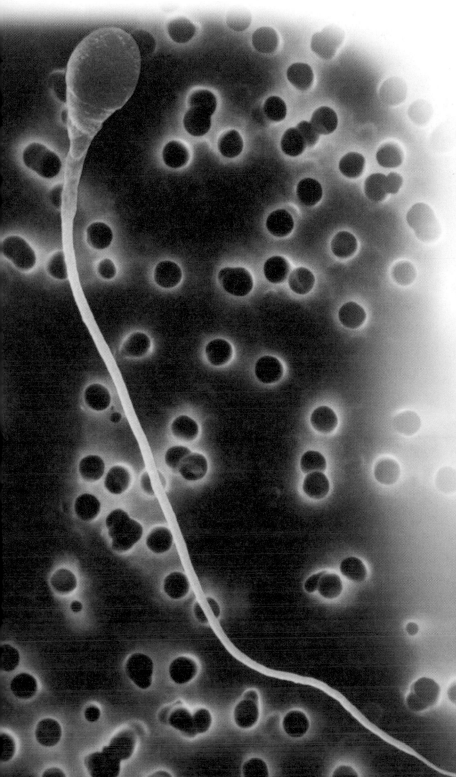

Chapter 2: **Essential Chemistry for Biology**

Chapter Thread: **Water as the Chemical of Life**

Chapter 3: **The Molecules of Life**

Chapter Thread: **Lactose Intolerance**

Chapter 4: **A Tour of the Cell**

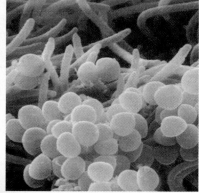

Chapter Thread: **Bacterial Diseases and Cellular Structures**

Chapter 5: **The Working Cell**

Chapter Thread: **Molecular Technologies**

Chapter 6: **Cellular Respiration: Obtaining Energy from Food**

Chapter Thread: **Life with and without Oxygen**

Chapter 7: **Photosynthesis: Using Light to Make Food**

Chapter Thread: **Harnessing Solar Energy**

2 Essential Chemistry for Biology

Ice, water.
Water, shown here in its liquid and solid forms, is essential to all life.

CHAPTER CONTENTS

BIOLOGY AND SOCIETY: Water as the Chemical of Life

Fluoride in the Water

One of the first things that people notice about you is your smile. But you might not realize that healthy teeth are much more common today than they were several decades ago. One reason for the decrease in tooth decay is the addition of fluoride-containing chemicals to drinking water and dental products.

Although the surface enamel of your teeth is lifeless, it is part of a dynamic, living system that includes the bacteria in your mouth. These oral bacteria reside in plaque, a sticky combination of food, saliva, and dead cells that coats your teeth. As these bacteria grow, they release acids that can dissolve the dental surface, causing pain and, eventually, tooth loss. Fluoride helps to prevent cavities in two ways: It affects the metabolism of oral bacteria, and it promotes the replacement of lost minerals on the surface of the teeth.

Frequent exposure to small amounts of fluoride is the best way to prevent tooth decay. How do we get fluoride in our diet? Fluoride is a common ingredient in Earth's crust. As subsurface water trickles through the crust, it dissolves fluoride (as well as other minerals). In some areas, this natural process delivers enough fluoride in drinking water to help maintain healthy teeth. But in many areas of the United States and other industrialized nations, the water supply is fluoridated—meaning that it is supplemented with small amounts of fluoride during municipal water treatment. In addition, fluoride is frequently added to dental products, such as toothpaste and mouthwash.

The ability to dissolve fluoride and many other substances is just one of several properties that make water vital for life on Earth. In fact, molecules of water participate in many chemical reactions necessary to sustain life. Living organisms are, at their most basic level, chemical systems. Many questions about life reduce to questions about chemicals and their interactions, making knowledge of chemistry essential to understanding biology. In this chapter, you will learn some essential chemistry that you can apply throughout your study of life. In the last section, we'll return to the crucial role that water plays in sustaining life on Earth.

Some Basic Chemistry

Why would a biology textbook include a chapter on chemistry? Well, take any biological system apart, and you eventually end up at the chemical level. In fact, you could think of your body as one big container of chemicals undergoing a continuous series of chemical reactions. Beginning at this basic biological level, let's explore the chemistry of life.

Matter: Elements and Compounds

All organisms and everything around them are made of matter, the physical "stuff" of the universe. Defined more formally, **matter** is anything that occupies space and has mass. Matter is found on Earth in three physical states: solid, liquid, and gas.

Matter is composed of chemical **elements**, substances that cannot be broken down into other substances. There are 92 naturally occurring elements on Earth; examples are carbon, oxygen, and gold. Each element has a symbol, the first letter or two of its English, Latin, or German name. For instance, the symbol for gold, Au, is from the Latin word *aurum*, while the symbol O stands for the English word *oxygen*. All the elements—the 92 that occur naturally and several dozen that are human-made—are listed in the periodic table of the elements, a familiar fixture in any chemistry or biology lab (**Figure 2.1**; see Appendix B for a full version).

Of the 92 naturally occurring elements, 25 are essential to life. Four of these elements—oxygen (O), carbon

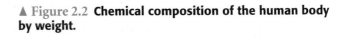

Carbon (C), hydrogen (H), and nitrogen (N)—make up about 96% of the weight of the human body, as well as most other living matter (**Figure 2.2**). Much of the remaining 4% is accounted for by 7 elements, most of which are probably familiar to you, such as calcium (Ca) and phosphorus (P). Calcium, important for building strong bones and teeth, is found abundantly in milk and dairy products as well as sardines and green, leafy vegetables (collards, kale, and broccoli, for example). Phosphorus, a component of DNA and other important biological molecules, can be obtained by eating eggs, beans, and nuts.

Less than 0.01% of your weight is made up of 14 trace elements (see Figure 2.2). **Trace elements** are required in only very small amounts, but you cannot live without them. The average human, for example, needs only a tiny speck of iodine, about 0.15 milligram (mg) each day. Iodine is an essential ingredient of a hormone produced by the thyroid gland, located in the neck. An iodine deficiency in the diet causes the thyroid gland to enlarge, a condition called goiter. Foods naturally rich in iodine include kelp and dairy products. The addition of iodine to table salt (iodized salt) has nearly eliminated goiter in many nations. Unfortunately, goiter still affects many thousands of people in developing countries (**Figure 2.3**). Fluorine, whose "fluoride" form was discussed in the Biology and Society section, is another example of a trace element; it is a needed component of healthy bones and teeth.

Carbon (C): 18.5%

Oxygen (O): 65.0%

Calcium (Ca): 1.5%
Phosphorus (P): 1.0%
Potassium (K): 0.4%
Sulfur (S): 0.3%
Sodium (Na): 0.2%
Chlorine (Cl): 0.2%
Magnesium (Mg): 0.1%

Hydrogen (H): 9.5%

Nitrogen (N): 3.3%

Trace elements: less than 0.01%
Boron (B)   Manganese (Mn)
Chromium (Cr)   Molybdenum (Mo)
Cobalt (Co)   Selenium (Se)
Copper (Cu)   Silicon (Si)
Fluorine (F)   Tin (Sn)
Iodine (I)   Vanadium (V)
Iron (Fe)   Zinc (Zn)

▲ **Figure 2.2 Chemical composition of the human body by weight.**

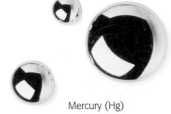

Mercury (Hg)

Copper (Cu)

Lead (Pb)

▶ **Figure 2.1 Abbreviated periodic table of the elements.** In the full periodic table (see Appendix B), each entry contains the element symbol in the center, with the atomic number above and the mass number below (both are discussed on the facing page). The element highlighted here is carbon (C).

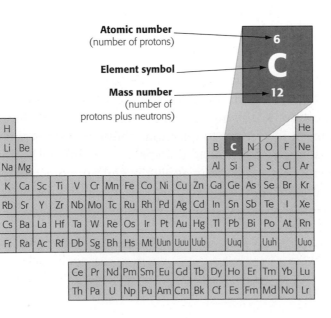

Atomic number (number of protons) — **6**

Element symbol — **C**

Mass number (number of protons plus neutrons) — **12**

H																	He
Li	Be											B	C	N	O	F	Ne
Na	Mg											Al	Si	P	S	Cl	Ar
K	Ca	Sc	Ti	V	Cr	Mn	Fe	Co	Ni	Cu	Zn	Ga	Ge	As	Se	Br	Kr
Rb	Sr	Y	Zr	Nb	Mo	Tc	Ru	Rh	Pd	Ag	Cd	In	Sn	Sb	Te	I	Xe
Cs	Ba	La	Hf	Ta	W	Re	Os	Ir	Pt	Au	Hg	Tl	Pb	Bi	Po	At	Rn
Fr	Ra	Ac	Rf	Db	Sg	Bh	Hs	Mt	Uun	Uuu	Uub		Uuq		Uuh		Uuo

Ce	Pr	Nd	Pm	Sm	Eu	Gd	Tb	Dy	Ho	Er	Tm	Yb	Lu
Th	Pa	U	Np	Pu	Am	Cm	Bk	Cf	Es	Fm	Md	No	Lr

Elements can combine to form **compounds**, substances that contain two or more elements in a fixed ratio. In everyday life, compounds are much more common than pure elements. Familiar examples are relatively simple compounds such as table salt and water. Table salt is sodium chloride, NaCl, consisting of equal parts of the elements sodium (Na) and chlorine (Cl). A molecule of water, H_2O, has two atoms of hydrogen and one atom of oxygen. Most of the compounds in living organisms contain several different elements. DNA, for example, contains carbon, nitrogen, oxygen, hydrogen, and phosphorus. ☑

Atoms

Each element consists of one kind of atom, which is different from the atoms of other elements. An **atom**, named from a Greek word meaning "indivisible," is the smallest unit of matter that still retains the properties of an element. In other words, the smallest amount of the element carbon is one carbon atom. Just how small is this "piece" of carbon? It would take about a million carbon atoms to stretch across the period at the end of this sentence.

The Structure of Atoms

Atoms are composed of subatomic particles, of which the three most important are protons, electrons, and neutrons. A **proton** is a subatomic particle with a single unit of positive electrical charge ($+$). An **electron** is a subatomic particle with a single unit of negative electrical charge ($-$). A **neutron** is electrically neutral (has no electrical charge).

Let's look at the structure of an atom of the element helium (He), the "lighter-than-air" gas used in party balloons **(Figure 2.4)**. Each atom of helium has 2 neutrons (⬤) and 2 protons (⊕) tightly packed into the **nucleus**, the atom's central core. Two electrons (⊖) move around

▼ Figure 2.3 **Why is salt "iodized"?**

Goiter, an enlargement of the thyroid gland, shown here in a Malaysian woman, can occur when a person's diet does not include enough iodine, a trace element.

Eating iodine-rich foods can prevent goiter.

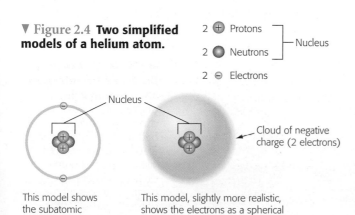

▼ Figure 2.4 **Two simplified models of a helium atom.**

2 ⊕ Protons ⎤
2 ⬤ Neutrons ⎦ Nucleus
2 ⊖ Electrons

Nucleus

Cloud of negative charge (2 electrons)

This model shows the subatomic particles in an atom of helium.

This model, slightly more realistic, shows the electrons as a spherical cloud of negative charge surrounding the nucleus.

the nucleus in a spherical cloud at nearly the speed of light. The electron cloud is much bigger than the nucleus; in fact, if the electron cloud were the size of a football stadium, the nucleus would be the size of a fly on the field. The attraction between the negatively charged electrons and the positively charged protons keeps the electrons moving. When an atom has an equal number of protons and electrons (as helium does), its net electrical charge is zero and so the atom is neutral.

Elements differ in the number of subatomic particles in their atoms. The number of protons in an atom, called the **atomic number**, determines which element it is. For example, helium always has 2 protons and thus an atomic number of 2. **Mass** is a measure of the amount of material in an object. A proton and a neutron have nearly identical mass. An electron has so little mass—only about 1/2,000 the mass of a proton—that it is usually approximated as zero. Therefore, an atom's **mass number** is just the sum of the number of protons and neutrons in its nucleus. For helium, the mass number is 4 (2 protons + 2 neutrons). Both the atomic number and the mass number can be read from the periodic table (see Figure 2.1 and Appendix B).

Isotopes

Some elements can exist in different forms called isotopes. The different **isotopes** of an element have the same numbers of protons and electrons but different numbers of neutrons; in other words, isotopes are

SOME BASIC CHEMISTRY

☑**CHECKPOINT**

1. How many of the 92 naturally occurring elements are used by living organisms? Which four are the most abundant in living cells?

2. Which of the following are compounds: water (H_2O), oxygen gas (O_2), methane (CH_4)? Why or why not?

Answers: 1. 25; oxygen, carbon, hydrogen, and nitrogen 2. H_2O and CH_4 are compounds, but O_2 is not because a compound must contain at least two different elements.

25

Table 2.1	Isotopes of Carbon		
	Carbon-12	**Carbon-13**	**Carbon-14**
Protons	6 ⎤ mass ⎟ number	6 ⎤ mass ⎟ number	6 ⎤ mass ⎟ number
Neutrons	6 ⎦ 12	7 ⎦ 13	8 ⎦ 14
Electrons	6	6	6

forms of an element that differ in mass. Table 2.1 shows the numbers of subatomic particles in the three isotopes of carbon. Carbon-12 (named for its mass number 12), with 6 neutrons and 6 protons, makes up about 99% of all naturally occurring carbon. Most of the other 1% consists of carbon-13, with 7 neutrons and 6 protons. A third isotope, carbon-14, with 8 neutrons and 6 protons, occurs in minute quantities. Notice that all three isotopes have 6 protons—otherwise, they would not be carbon. Both carbon-12 and carbon-13 are stable isotopes, meaning that their nuclei remain intact more or less forever. The isotope carbon-14, on the other hand, is unstable, or radioactive. A **radioactive isotope** is one in which the nucleus decays, giving off particles and energy.

Radioactive isotopes have many uses in biological research and medicine. Cells use radioactive isotopes the same way they use nonradioactive isotopes of the same element. Once the cell takes up the radioactive isotopes, the location and concentration of the isotopes can be detected because of the radiation they emit. This makes radioactive isotopes useful as tracers—biological spies, in effect—for monitoring the fate of atoms in living organisms. For example, a medical diagnostic tool called a PET scan works by detecting small amounts of radioactive materials introduced into the body. PET scans can diagnose heart disorders and some cancers.

Although radioactive isotopes have many beneficial uses, uncontrolled exposure to them can harm living organisms by damaging cellular molecules, especially DNA. In 1986, the explosion of a nuclear reactor at Chernobyl, Ukraine, released large amounts of radioactive isotopes, killing 30 people within a few weeks and exposing thousands to an increased risk of developing

cancer. In fact, the incidence of thyroid cancer among Ukrainian children increased tenfold in the decade following the accident.

Natural sources of radiation can also pose a threat. Radon, a radioactive gas, may cause lung cancer. Radon can contaminate buildings in regions where underlying rocks naturally contain the radioactive element uranium. Homeowners can buy a radon detector or hire a company to test their home to ensure that radon levels are safe.

Electron Arrangement and the Chemical Properties of Atoms

Of the three subatomic particles we've discussed—protons, neutrons, and electrons—electrons are the ones that primarily determine how an atom behaves when it encounters other atoms. Electrons vary in the amount of energy they possess. The farther an electron is from the nucleus, the greater its energy. Electrons do not move around an atom at just any energy level, but only at specific levels called electron shells. Depending on the number of electrons, atoms may have one, two, or more electron shells, with electrons in the outermost shell having the highest energy. Each shell can accommodate up to a specific number of electrons. The innermost shell is full with only 2 electrons, while the second and third shells can each hold up to 8 electrons.

The number of electrons in the outermost shell determines the chemical properties of an atom. Atoms whose outer shells are not full tend to interact with other atoms—that is, to participate in chemical reactions. Figure 2.5 shows the electron shells of four biologically important elements. Because the outer shells of all four atoms are not filled, these atoms react readily with other atoms. The hydrogen atom is highly reactive because it has only 1 electron in its single electron shell, which can accommodate 2 electrons. Atoms of carbon, nitrogen, and oxygen are also highly reactive because their outer shells, which can hold 8 electrons, are not filled. In contrast, the helium atom in Figure 2.4 has a single, first-level shell that is full with 2 electrons. As a result, helium is chemically unreactive. ✓

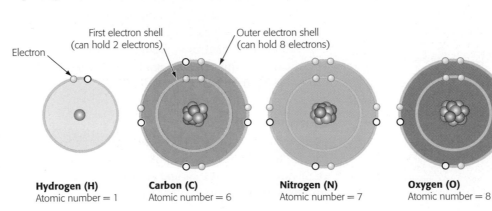

First electron shell (can hold 2 electrons)

Outer electron shell (can hold 8 electrons)

Electron

Hydrogen (H)
Atomic number = 1

Carbon (C)
Atomic number = 6

Nitrogen (N)
Atomic number = 7

Oxygen (O)
Atomic number = 8

◄ Figure 2.5 **Atoms of the four elements most abundant in living matter.** All four atoms are chemically reactive because their outermost electron shells are not filled. The small empty circles (o) in these diagrams represent unfilled "spaces" in the outer electron shells.

Chemical Bonding and Molecules

Chemical reactions enable atoms to give up or acquire electrons, thereby completing their outer shells. Atoms do this by either transferring or sharing outer electrons. These interactions usually result in atoms staying close together, held by attractions called **chemical bonds**. In this section, we will discuss three types of chemical bonds: ionic, covalent, and hydrogen bonds.

Ionic Bonds

Table salt is an example of how the transfer of electrons can bond atoms together. The two ingredients of table salt are the elements sodium (Na) and chlorine (Cl). When a chlorine atom strips an electron from a sodium atom, the electron transfer results in both atoms having full outer shells of electrons (**Figure 2.6**). Before the electron transfer, each of these atoms is electrically neutral. Because electrons are negatively charged particles, the electron transfer moves one unit of negative charge from sodium to chlorine. Both atoms are now **ions**, the term for atoms that are electrically charged as a result of gaining or losing electrons. The loss of an electron gives the sodium ion a charge of $+1$, while chlorine's gain of an electron gives it a charge of -1. The sodium ion (Na^+) and chloride ion (Cl^-) are held together by an **ionic bond**, the attraction between oppositely charged ions. Compounds, such as table salt, that are held together by ionic bonds are called ionic compounds. Fluorine in Earth's crust is often found in the form of ionic compounds such as calcium fluoride (CaF_2), the result of bonds between calcium ions (Ca^{2+}) and fluoride ions (F^-). (Note that negatively charged ions often have names ending in "-ide," like "chloride" or "fluoride.") ✔

Covalent Bonds

In contrast to the complete *transfer* of electrons that leads to ionic bonds, a **covalent bond** forms when two atoms *share* one or more pairs of outer-shell electrons. Atoms that are held together by covalent bonds form a **molecule**. For example, a covalent bond connects each hydrogen atom to the carbon in the molecule CH_4, a common gas called methane. In **Figure 2.7**, you can see that each of the four hydrogen atoms in a molecule of

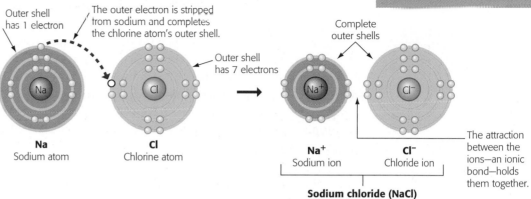

▼ Figure 2.6 **Electron transfer and ionic bonding.** When a sodium atom and a chlorine atom meet, the electron transfer between the two atoms results in two ions with opposite charges.

Outer shell has 1 electron

The outer electron is stripped from sodium and completes the chlorine atom's outer shell.

Outer shell has 7 electrons

Complete outer shells

Na Sodium atom
Cl Chlorine atom
Na⁺ Sodium ion
Cl⁻ Chloride ion

The attraction between the ions—an ionic bond—holds them together.

Sodium chloride (NaCl)

◄ Figure 2.7 **Alternative ways to represent molecules.** A molecular formula, such as CH_4 or H_2O, tells you the number of each kind of atom in a molecule but not how they are attached together. This figure shows four common ways of representing the arrangement of atoms in molecules.

Name (molecular formula)	Electron configuration Shows how each atom completes its outer shell by sharing electrons	Structural formula Represents each covalent bond (a pair of shared electrons) with a line	Space-filling model Shows the shape of a molecule by symbolizing atoms with color-coded balls	Ball-and-stick model Represents atoms with "balls" and bonds with "sticks"
Hydrogen gas (H₂)		H—H Single bond (a pair of shared electrons)		
Oxygen gas (O₂)		O=O Double bond (two pairs of shared electrons)		
Methane (CH₄)		H—C—H (with H above and H below)		

methane shares one pair of electrons with the single carbon atom.

The number of covalent bonds an atom can form is equal to the number of additional electrons needed to fill its outer shell. Note in Figure 2.7 that hydrogen (H) can form one covalent bond, oxygen (O) can form two, and carbon (C) can form four. The single covalent bond in H_2 completes the outer shells of both hydrogen atoms. In contrast, an oxygen atom needs 2 electrons to complete its outer shell. In an O_2 molecule, the two oxygen atoms share two pairs of electrons, forming a double covalent bond.

Hydrogen Bonds

Water (H_2O) is also a compound. Its structure consists of two hydrogen atoms joined to one oxygen atom by single covalent bonds (the "sticks" represent bonds between the atoms, which are shown as "balls"):

However, the electrons of the covalent bonds are not shared equally between the oxygen and hydrogen. The two yellow arrows in the following diagram indicate the stronger pull on the shared electrons that oxygen has compared with its hydrogen partners:

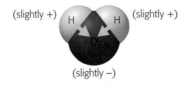

The unequal sharing of negatively charged electrons, combined with its V shape, makes a water molecule polar. A **polar molecule** has opposite charges on opposite ends. In the case of water, the oxygen end of the molecule has a slight negative charge, while the region around the two hydrogen atoms is slightly positive.

The polarity of water results in weak electrical attractions between neighboring water molecules. The molecules tend to orient such that the hydrogen atom of one molecule is near the oxygen atom of an adjacent water molecule. These weak attractions

are called **hydrogen bonds** (Figure 2.8). As you will see later in this chapter, the ability of water to form hydrogen bonds has many implications for life.

Chemical Reactions

The chemistry of life is dynamic. Your cells are constantly rearranging molecules by breaking existing chemical bonds and forming new ones. Such changes in the chemical composition of matter are called **chemical reactions**. A simple example is the reaction between oxygen gas and hydrogen gas that forms water (this is an explosive reaction, which, fortunately, does not occur in your cells):

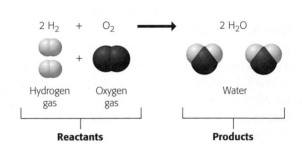

Let's translate the chemical shorthand: Two molecules of hydrogen gas ($2 H_2$) react with one molecule of oxygen gas (O_2) to form two molecules of water ($2 H_2O$). The arrows indicate the conversion of the starting materials, the **reactants** ($2 H_2$ and O_2), to the **products** ($2 H_2O$).

Notice that the same numbers of hydrogen and oxygen atoms are present in reactants and products, although they are grouped differently. Chemical reactions cannot create or destroy matter, but only rearrange it.

These rearrangements usually involve the breaking of chemical bonds in reactants and the forming of new bonds in products.

The water molecules we have built here are a good conclusion to this section on basic chemistry. Water is a substance so important in biology that we'll take a closer look at its life-supporting properties in the next section. ✔

◀ **Figure 2.8 Hydrogen bonding in water.** The charged regions of the polar water molecules are attracted to oppositely charged areas of neighboring molecules. Each molecule can hydrogen-bond to a maximum of four partners.

✔CHECKPOINT

Predict the formula for the compound that results when a molecule of sulfur trioxide (SO_3) combines with a molecule of water to produce a single molecule of product. (*Hint*: In chemical reactions, no atoms are gained or lost.)

Answer: H_2SO_4 (sulfuric acid)

Water and Life

The Properties of Water

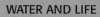

Life on Earth began in water and evolved there for 3 billion years before spreading onto land. Modern life, even land-dwelling life, is still tied to water. You've had personal experience with this dependence on water every time you seek liquids to quench your thirst and replenish your body's water. Inside your body, your cells are surrounded by a fluid that's composed mostly of water, and your cells themselves range from 70% to 95% in water content.

The abundance of water is a major reason that Earth is habitable. Water is so common in our environment that it's easy to overlook its extraordinary behavior (**Figure 2.9**). We can trace water's unique life-supporting properties to the structure and interactions of its molecules.

Water's Life-Supporting Properties

The polarity of water molecules and the hydrogen bonding that results (see Figure 2.8) explain most of water's life-supporting properties. We'll explore four of those properties here: the cohesive nature of water, the ability of water to moderate temperature, the biological significance of ice floating, and the versatility of water as a solvent.

The Cohesion of Water

Water molecules stick together as a result of hydrogen bonding. Hydrogen bonds between molecules of liquid water last for only a few trillionths of a second, yet at any instant, many of the molecules are hydrogen-bonded to others. This tendency of molecules of the same kind to stick together, called **cohesion**, is much stronger for water than for most other liquids. The cohesion of water is important in the living world. Trees, for example, depend on cohesion to help transport water from their roots to their leaves (**Figure 2.10**).

Related to cohesion is surface tension, a measure of how difficult it is to stretch or break the surface of a liquid. Hydrogen bonds give water unusually high surface tension, making it behave as though it were coated with an invisible film (**Figure 2.11**).

Evaporation from the leaves

Flow of water

▲ Figure 2.10 **Cohesion and water transport in plants.** The evaporation of water from leaves pulls water upward from the roots through microscopic tubes in the trunk of the tree. Because of cohesion, the pulling force is relayed through the tubes all the way down to the roots. As a result, water rises against the force of gravity.

SEM 130x

Microscopic tubes

Cohesion due to hydrogen bonds between water molecules

▲ Figure 2.9 **A watery world.** You can see each of the three physical states of water in this photograph: liquid water (which covers three-quarters of Earth's surface), ice, and water vapor.

▼ Figure 2.11 **A raft spider walking on water.** The cumulative strength of hydrogen bonds between water molecules allows this spider to walk on pond water without breaking the surface.

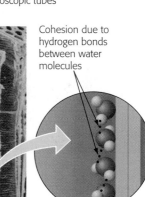

How Water Moderates Temperature

If you've ever burned your finger on a metal pot while waiting for the water in it to boil, you know that water heats up much more slowly than metal. In fact, because of hydrogen bonding, water has a stronger resistance to temperature change than most other substances.

Temperature and heat are related, but different. A swimmer crossing San Francisco Bay has a higher temperature than the water, but the bay contains far more heat because of its immense volume. **Heat** is the amount of energy associated with the movement of the atoms and molecules in a body of matter. **Temperature** measures the intensity of heat—that is, the average speed of molecules rather than the total amount of heat energy in a body of matter.

When water is heated, the heat energy first disrupts hydrogen bonds and then makes water molecules jostle around faster. The temperature of the water doesn't go up until the water molecules start to speed up. Because heat is first used to break hydrogen bonds rather than raise the temperature, water absorbs and stores a large amount of heat while warming up only a few degrees. Conversely, when water cools, hydrogen bonds form, a process that releases heat. Thus, water can release a relatively large amount of heat to the surroundings while the water temperature drops only slightly.

Earth's giant water supply—the oceans, seas, lakes, and rivers—enables temperatures to stay within limits that permit life by storing a huge amount of heat from the sun during warm periods and giving off heat to warm the air during cold conditions. That's why coastal areas generally have milder climates than inland regions. Water's resistance to temperature change also stabilizes ocean temperatures, creating a favorable environment for marine life.

Another way that water moderates temperature is by **evaporative cooling**. When a substance evaporates (changes physical state from a liquid to a gas), the surface of the liquid remaining behind cools down. This occurs because the molecules with the greatest energy (the "hottest" ones) tend to vaporize first. It's as if the five fastest runners on your track team quit school, lowering the average speed of the remaining team. On a global scale, surface evaporation cools tropical oceans. On the scale of individual organisms, evaporative cooling prevents some land-dwelling creatures from overheating. It's why sweating helps you maintain a constant body temperature, even when exercising on a hot day (**Figure 2.12**). And the old expression "It's not the heat, it's the humidity" has its basis in the difficulty of sweating water into air that is already saturated with water vapor.

The Biological Significance of Ice Floating

When most liquids get cold, their molecules move closer together. If the temperature is cold enough, the liquid freezes and becomes a solid. Water, however, behaves differently. When water molecules get cold enough, they move apart, forming ice. A chunk of ice has fewer molecules than an equal volume of liquid water; it floats because it is less dense than the liquid water around it. Like water's other life-supporting properties, floating ice is a consequence of hydrogen bonding. In contrast to the short-lived hydrogen bonds in liquid water, those in solid ice last longer, with each molecule bonded to four neighbors. As a result, ice is a spacious crystal (**Figure 2.13**).

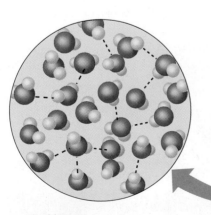

▲ Figure 2.12 **Sweating as a mechanism of evaporative cooling.**

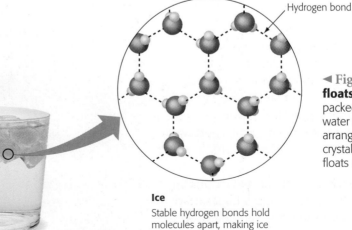

Hydrogen bond

► Figure 2.13 **Why ice floats.** Compare the tightly packed molecules in liquid water with the spaciously arranged molecules in the ice crystal. The less dense ice floats atop the denser water.

Liquid water
Hydrogen bonds constantly break and re-form.

Ice
Stable hydrogen bonds hold molecules apart, making ice less dense than water.

How does the fact that ice floats help support life on Earth? Imagine what would happen if ice sank. All ponds, lakes, and even the oceans would eventually freeze solid. During summer, only the upper few inches of the oceans would thaw. Instead, when a deep body of water cools, the floating ice insulates the liquid water below, allowing life to persist under the frozen surface.

Water as the Solvent of Life

If you've ever enjoyed a glass of sweetened ice tea or added salt to soup, you know that you can dissolve sugar or salt in water. This results in a mixture known as a **solution**, a liquid consisting of a homogeneous mixture of two or more substances. The dissolving agent is called the **solvent**, and a substance that is dissolved is called a **solute**. When water is the solvent, the resulting solution is called an **aqueous solution**.

The fluids of organisms are aqueous solutions. Water can dissolve an enormous variety of solutes necessary for life. Water is the solvent inside all cells, in blood, and in plant sap. As a solvent, it is a medium for chemical reactions. Water can dissolve salt ions, as shown in **Figure 2.14**. Each ion becomes surrounded by oppositely charged regions of water molecules. Solutes that are polar molecules, such as sugars, dissolve by orienting locally charged regions of their molecules toward water molecules in a similar way.

We have discussed four special properties of water. Next, we'll consider one way that water's unique structure can reveal important clues about the human brain. ✓

1. Explain why, if you pour very carefully, you can actually "stack" water slightly above the rim of a cup.

2. Explain why ice floats.

Answers: 1. Surface tension due to water's cohesion will keep the water from spilling over. 2. Ice is less dense than liquid water because the more stable hydrogen bonds lock the molecules into a spacious crystal.

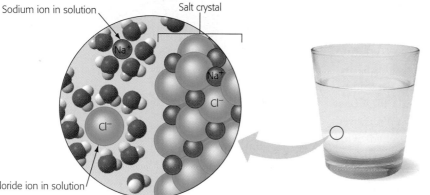

Sodium ion in solution

Salt crystal

Na⁺

Na⁺

Cl⁻

Cl⁻

Chloride ion in solution

◄ Figure 2.14 **A crystal of table salt (NaCl) dissolving in water.** As a result of electrical charge attractions, H₂O molecules surround the sodium and chloride ions, dissolving the crystal in the process.

THE PROCESS OF SCIENCE: Water as the Chemical of Life

Can Exercise Boost Your Brain Power?

Biologists use a number of imaging techniques to reveal the body's inner structure. One such technique, magnetic resonance imaging (MRI), depends on the behavior of the hydrogen atoms in water molecules. MRI first aligns the hydrogen nuclei with powerful magnets, then knocks the nuclei out of alignment with a brief pulse of radio waves. In response, the hydrogen atoms give out faint radio signals of their own, which are picked up by the MRI scanner and translated by computer into an image. Since water is a major component of all of our soft tissues, MRI allows doctors to see regions of the body that are invisible to X-rays.

MRI has been particularly useful in studying the brain. One recent study began with the **observation** that the human brain shrinks as we age: About 20% of our brain matter is lost between ages 30 and 90, and the shrinkage is most prominent in brain areas responsible for memory and learning. Because exercise had been previously associated with improved memory, researchers from the University of Illinois asked the **question**, Can aerobic exercise slow or even reverse brain loss? Their **hypothesis** was that MRI scans would reveal differences between the brains of people who regularly exercised and those who did not. They made the **prediction** that the brains of active people would shrink less than the brains of less active people.

In their **experiment**, 29 subjects in their 60s and 70s participated for six months in an exercise program of three 1-hour aerobic training sessions per week. A control group of 29 subjects (matched for age) engaged in a program of non-aerobic whole-body stretching lasting the same duration. At the start and end of the program, MRI was used to scan the brains of all participants. The **results** of the experiment showed significant increases in brain volume for the aerobic group compared to the non-aerobic controls (**Figure 2.15**). The researchers concluded that there may be a link between cardiovascular fitness and the retention of memory in aging adults. In other words, regular aerobic exercise may benefit both your body and your mind.

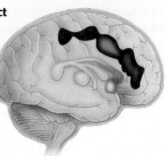

► Figure 2.15 **The effect of aerobic exercise on brain size.** The colors highlight brain regions that grew in aerobic exercise participants compared to a control group of non-aerobic exercisers.

Acids, Bases, and pH

In the aqueous solutions within organisms, most of the water molecules are intact. However, some of the water molecules actually break apart into hydrogen ions (H^+) and hydroxide ions (OH^-). A balance of these two ions is critical for the proper functioning of chemical processes within organisms.

A chemical compound that releases H^+ to a solution is called an **acid**. One example of a strong acid is hydrochloric acid (HCl), the acid in your stomach. In solution, HCl breaks apart into the ions H^+ and Cl^-. A **base** (or alkali) is a compound that accepts H^+ and removes it from solution. Some bases, such as sodium hydroxide (NaOH), do this by releasing OH^-, which combines with H^+ to form H_2O.

To describe the acidity of a solution, chemists use the **pH scale**, a measure of the hydrogen ion (H^+) concentration in a solution. The scale ranges from 0 (most acidic) to 14 (most basic). Each pH unit represents a tenfold change in the concentration of H^+ (**Figure 2.16**). For example, lemon juice at pH 2 has 100 times more H^+ than an equal amount of tomato juice at pH 4. Pure water and aqueous solutions that are neither acidic nor basic are said to be neutral; they have a pH of 7. They do contain H^+ and OH^-, but the concentrations of the two ions are equal. The pH of the solution inside most living cells is close to 7.

Even a slight change in pH can be harmful to an organism because the molecules in cells are extremely sensitive to H^+ and OH^- concentrations. Biological fluids contain **buffers**, substances that prevent harmful changes in pH by accepting H^+ when that ion is in excess and donating H^+ when it is depleted. This buffering process, however, is not

► Figure 2.16 **The pH scale.** A solution having a pH of 7 is neutral, meaning that its H^+ and OH^- concentrations are equal. The lower the pH below 7, the more acidic the solution, or the greater its excess of H^+ compared with OH^-. The higher the pH above 7, the more basic the solution, or the greater the deficiency of H^+ relative to OH^-.

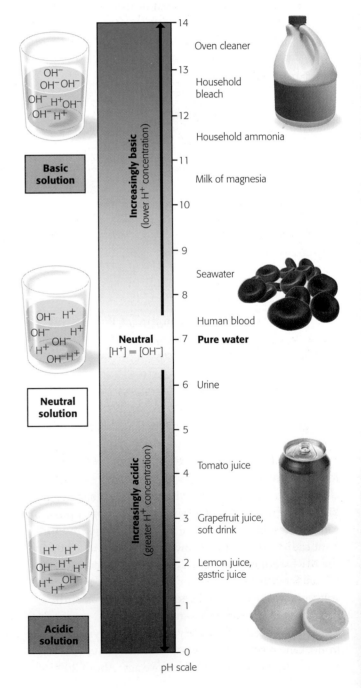

Basic solution

Neutral solution

Acidic solution

Increasingly basic (lower H^+ concentration)

Neutral $[H^+] = [OH^-]$

Increasingly acidic (greater H^+ concentration)

pH scale

- 14 Oven cleaner
- 13 Household bleach
- 12 Household ammonia
- 11 Milk of magnesia
- 10
- 9
- 8 Seawater
- Human blood
- 7 **Pure water**
- 6 Urine
- 5
- 4 Tomato juice
- 3 Grapefruit juice, soft drink
- 2 Lemon juice, gastric juice
- 1
- 0

foolproof, and changes in environmental pH can profoundly affect ecosystems. For example, much of the
human-made CO_2 released into the atmosphere is absorbed by the oceans. When it combines with seawater, a
chemical reaction produces an acid. The resulting ocean
acidification can greatly change marine environments.
This principle was illustrated by a 2008 study of the
Mediterranean Sea near Naples, Italy. There, undersea
volcanoes release high levels of CO_2 **(Figure 2.17)**. Researchers found that the resulting acidification caused
coral reef collapse and widespread death of marine organisms. This natural system suggests that high levels of dissolved human-made CO_2 could cause similar ecological
disaster.

The effects of ocean acidification are daunting re-
minders that the chemistry of life is linked to the chemistry of the environment. It reminds us, too, that
chemistry happens on a global scale, since industrial
processes in one region of the world often cause acid
precipitation to fall in another part of the world. ☑

◄ **Figure 2.17 The
release of volcanic CO_2,
causing undersea
bubbling off the coast
of Italy.**

☑**CHECKPOINT**

Compared with a solution
of pH 8, the same volume
of a solution at pH 5 has
_____ times more
hydrogen ions (H^+). This
second solution is considered a(n)_____.

Answer: 1,000; acid

EVOLUTION CONNECTION: Water as the Chemical of Life

The Search for Extraterrestrial Life

Throughout this chapter, we have highlighted the importance of water to living
systems. In fact, water is the substance that makes life as we know it possible.

One of the great questions facing humankind is, Has life evolved elsewhere
in the universe? If it has, and if this life is at all similar to ours, then it, too,
would depend on water. This explains why the search for water on other
planets (particularly Mars) is such a priority for the U.S. space program.

Researchers at the National Aeronautics and Space Administration (NASA)
have accumulated data that suggest that water was once abundant on Mars.
(Although Mars has clearly visible polar ice caps, no liquid water is known to
flow there at present.) In January 2004, NASA landed two golf-cart-sized
rovers on Mars. These robotic geologists determined the composition of
rocks and sent back images of rock formations. One of the rovers detected a
mineral that is formed only in the presence of water. Other chemical evidence
indicated that water once permeated Martian rocks. And pictures revealed
physical evidence of the effects of liquid water flowing over the Martian
surface in the past.

Images taken by NASA's *Mars Global Surveyor* spacecraft indicate that not only
did the red planet have a water-filled ancient past, but liquid water may have
flowed recently on its surface. In 2008, an analysis of Martian soil by the *Phoenix
Mars Lander* **(Figure 2.18)** provided evidence that components of the soil had at
one time been dissolved in water. As this chapter has shown, where there is water,
there may indeed be life. The evidence for water on Mars raises the tantalizing
possibility that microbial life may exist below the Martian surface.

▲ **Figure 2.18 The *Phoenix Mars Lander* probing for evidence
of water on the Martian surface.**

Chapter Review

SUMMARY OF KEY CONCEPTS

(MB) Go to the Study Area at **www.masteringbiology.com** for practice quizzes, my**e**Book, BioFlix™ 3-D animations, MP3 Tutor Sessions, videos, current events, and more.

Some Basic Chemistry

Matter: Elements and Compounds

Matter consists of elements and compounds, which are combinations of two or more elements. Of the 25 elements essential for life, oxygen, carbon, hydrogen, and nitrogen are the most abundant in living matter.

Atoms

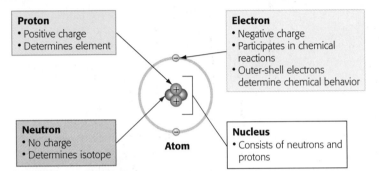

Proton
• Positive charge
• Determines element

Electron
• Negative charge
• Participates in chemical reactions
• Outer-shell electrons determine chemical behavior

Neutron
• No charge
• Determines isotope

Atom

Nucleus
• Consists of neutrons and protons

Chemical Bonding and Molecules

Transfer of one or more electrons produces attractions between oppositely charged ions:

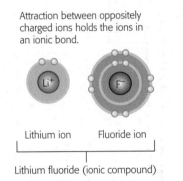

Attraction between oppositely charged ions holds the ions in an ionic bond.

Lithium ion   Fluoride ion

Lithium fluoride (ionic compound)

A molecule consists of two or more atoms connected by covalent bonds, which are formed by electron sharing:

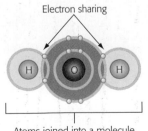

Electron sharing

Atoms joined into a molecule via covalent bonds

Water is a polar molecule; the slightly positively charged H atoms in one water molecule may be attracted to the partial negative charge of O atoms in neighboring water molecules, forming weak but important hydrogen bonds:

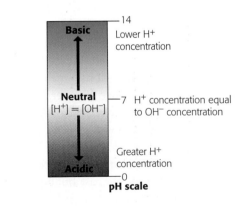

Hydrogen bond

Chemical Reactions

By breaking bonds in reactants and forming new bonds in products, chemical reactions rearrange matter.

Water and Life

Water's Life-Supporting Properties

The ability of leaves to pull water up microscopic tubes within trunks and stems is an example of how water's cohesion supports life. Water moderates temperature by absorbing heat in warm environments and releasing heat in cold environments. Evaporative cooling also helps stabilize the temperatures of oceans and organisms. The fact that ice floats because it is less dense than liquid water prevents the oceans from freezing solid. Blood and other biological fluids are aqueous solutions with a diversity of solutes dissolved in water, a versatile solvent.

Acids, Bases, and pH

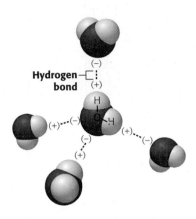

Basic — 14 — Lower H^+ concentration

Neutral $[H^+] = [OH^-]$ — 7 — H^+ concentration equal to OH^- concentration

Acidic — 0 — Greater H^+ concentration

pH scale

SELF-QUIZ

1. An atom can be changed into an ion by adding or removing
_____. An atom can be changed into a different isotope
by adding or removing _____. But if you change the
number of _____, the atom becomes a different element.

2. A nitrogen atom has 7 protons, and the most common isotope of nitrogen has 7 neutrons. A radioactive isotope of nitrogen has 9 neutrons. What are the atomic numbers and mass numbers of the stable and radioactive forms of nitrogen?

3. Why are radioactive isotopes useful as tracers in research on the chemistry of life?

4. A sulfur atom has 6 electrons in its third (outermost) shell, which can hold 8 electrons. As a result, it forms _____ covalent bonds with other atoms. (Provide a number.)

5. What is chemically nonsensical about this structure?

   H—C≡C—H

6. Why is it unlikely that two neighboring water molecules would be arranged like this?

7. Which of the following is not a chemical reaction?
   a. Sugar ($C_6H_{12}O_6$) and oxygen gas (O_2) combine to form carbon dioxide (CO_2) and water (H_2O).
   b. Sodium metal and chlorine gas unite to form sodium chloride.
   c. Hydrogen gas combines with oxygen gas to form water.
   d. Ice melts to form liquid water.

8. Some people in your study group say they don't understand what a polar molecule is. You explain that a polar molecule
   a. is slightly negative at one end and slightly positive at the other end.
   b. has an extra electron, giving it a positive charge.
   c. has an extra electron, giving it a negative charge.
   d. has covalent bonds.

9. Explain how the unique properties of water result from the fact that water is a polar molecule.

10. A can of cola consists mostly of sugar dissolved in water, with some carbon dioxide gas that makes it fizzy and makes the pH less than 7. Describe the cola using the following terms: solute, solvent, acidic, aqueous solution.

Answers to the Self-Quiz questions can be found in Appendix D.

THE PROCESS OF SCIENCE

11. Animals obtain energy through a series of chemical reactions in which sugar ($C_6H_{12}O_6$) and oxygen gas (O_2) are reactants. This process produces water (H_2O) and carbon dioxide (CO_2) as waste products. How might you use a radioactive isotope to find out whether the oxygen in CO_2 comes from sugar or oxygen gas?

12. The following diagram shows the arrangement of electrons around the nucleus of a fluorine atom (left) and a potassium atom (right). Predict what would happen if a fluorine atom and a potassium atom came into contact. What kind of bond do you think they would form?

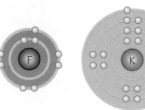

Fluorine atom   Potassium atom

BIOLOGY AND SOCIETY

13. Critically evaluate this statement: "It's paranoid and ignorant to worry about industry or agriculture contaminating the environment with chemical wastes; this stuff is just made of the same atoms that were already present in our environment."

14. A major source of the CO_2 that causes ocean acidification is emissions from coal-burning power plants. One way to reduce these emissions is to use nuclear power to produce electricity. The proponents of nuclear power contend that it is the only way that the United States can increase its energy production while reducing air pollution, because nuclear power plants emit little or no acid-precipitation-causing pollutants. What are some of the benefits of nuclear power? What are the possible costs and dangers? Do you think we ought to increase our use of nuclear power to generate electricity? Why or why not? If a new power plant were to be built near your home, would you prefer it to be a coal-burning plant or a nuclear plant? Why?

3 The Molecules of Life

Ice cream and lactose.
Lactose digestion results
from interactions between
several classes of the body's
molecules.

BIOLOGY AND SOCIETY: Lactose Intolerance

Got Lactose?

A milk mustache is meant to represent good health. Indeed, milk is a very healthy food because it's rich in protein, minerals, and vitamins. But for the majority of adults in the world, a glass of milk or a serving of milk-containing foods (such as ice cream or cheese) can cause bloating, gas, and other discomforts. Such people are exhibiting symptoms of lactose intolerance, the inability to properly digest lactose, the main sugar found in milk.

For people with lactose intolerance, the problem starts when lactose enters the small intestine. To absorb this sugar, digestive cells there must produce a molecule called lactase. Lactase is an enzyme, a protein that helps drive chemical reactions—in this case, the breakdown of lactose into smaller sugars. People with lactose intolerance produce insufficient amounts of the enzyme. Lactose that is not broken down in the small intestine passes into the large intestine, where bacteria feed on it and belch out gaseous by-products, producing uncomfortable symptoms. Sufficient lactase can thus mean the difference between delight and discomfort when someone feasts on an ice cream sundae.

What options are there for people with lactose intolerance? There is no treatment for the underlying cause: the underproduction of lactase. But symptoms can be controlled through diet. The first option is avoiding lactose-containing foods. Many substitutes are available, such as milk made from soy or milk that has been pretreated with lactase. Also, lactase in pill form can be taken along with food to ease digestion by artificially providing the enzyme that the body naturally lacks.

Lactose intolerance illustrates one way the interplay of biological molecules can affect human health. Such molecular interactions, repeated in countless variations, drive all biological processes. In this chapter, we'll explore the structure and function of large molecules that are essential to life. We'll start with an overview and then examine four classes of molecules: carbohydrates, lipids, proteins, and nucleic acids. Along the way, we'll emphasize where these molecules occur in your diet and the important roles they play in your body.

Organic Compounds

A cell is mostly water, but the rest of it consists mainly of carbon-based molecules. Carbon is unparalleled in its ability to form the skeletons of large, complex, diverse molecules that are necessary for life's functions. The study of carbon-based molecules, which are called **organic compounds**, lies at the heart of any study of life.

Carbon Chemistry

Why are carbon atoms so versatile as molecular ingredients? Recall that an atom's bonding ability is related to the number of electrons it must share to complete its outer shell. A carbon atom has 4 electrons in an outer shell that holds 8 (see Figure 2.5). Carbon completes its outer shell by sharing electrons with other atoms in four covalent bonds. Each carbon thus acts as an intersection from which an organic compound can branch off in up to four directions. And because carbon can use one or more of its bonds to attach to other carbon atoms, it is possible to construct an endless diversity of carbon skeletons varying in size and branching pattern (Figure 3.1). The carbon atoms of organic compounds can also use one or more of their bonds to partner with other elements, most commonly hydrogen, oxygen, and nitrogen.

In terms of chemical composition, the simplest organic compounds are **hydrocarbons**, which contain only carbon and hydrogen atoms. And the simplest hydrocarbon is methane, a single carbon atom bonded to four hydrogen atoms (Figure 3.2). Methane is one of the most abundant hydrocarbons in natural gas and is also produced by prokaryotes that live in swamps and in the digestive tracts of grazing animals, such as cows. Larger hydrocarbons (such as octane, with eight carbons) are the main molecules in the gasoline we burn in cars and other machines (Figure 3.3). Hydrocarbons are also important fuels in your body; the energy-rich parts of fat molecules have a hydrocarbon structure.

Each type of organic compound has a unique three-dimensional shape. Notice in Figure 3.2 that carbon's four bonds point to the corners of an imaginary tetrahedron (an object with four triangular sides). This geometric pattern occurs at each carbon "intersection" where there are four covalent bonds; thus, multicarbon organic molecules can have very elaborate shapes.

▼ Figure 3.2 **Methane, the simplest hydrocarbon.**

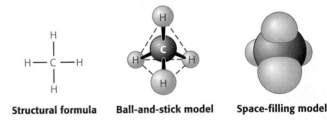

Structural formula   **Ball-and-stick model**   **Space-filling model**

▼ Figure 3.1 **Variations in carbon skeletons.** All of these examples are hydrocarbons, organic compounds consisting only of carbon and hydrogen. Notice that each carbon atom forms four bonds, and each hydrogen atom forms one bond. Remember that one line represents a single bond (sharing of one pair of electrons) and two lines represent a double bond (sharing of two pairs of electrons).

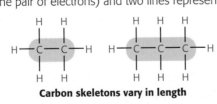

Carbon skeletons vary in length

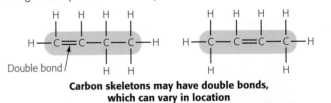

Double bond

Carbon skeletons may have double bonds, which can vary in location

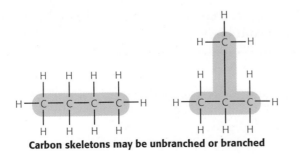

Carbon skeletons may be unbranched or branched

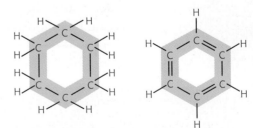

Carbon skeletons may be arranged in rings

▼ Figure 3.3 **Hydrocarbons as fuel.** Energy-rich hydrocarbons provide fuel for machines and, as components of fats, the body's cells.

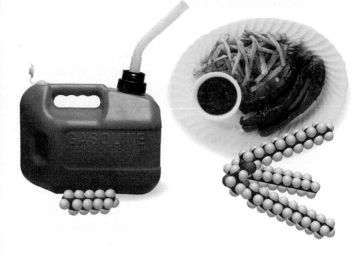

As you will see in subsequent chapters, a recurring theme in biology is the importance of the shape of molecules. Many vital processes within living organisms rely on the ability of molecules to recognize one another based on their shape.

The unique properties of an organic compound depend not only on its carbon skeleton but also on the atoms attached to the skeleton. In an organic compound, the groups of atoms that usually participate in chemical reactions are called **functional groups**. Two examples of functional groups are the hydroxyl group (—OH) and the carboxyl group (—COOH). Many biological molecules have two or more functional groups. Keeping in mind this basic scheme—carbon skeletons with functional groups—we are now ready to see how our cells make large molecules out of smaller ones.

Giant Molecules from Smaller Building Blocks

On a molecular scale, members of three categories of large biological molecules—carbohydrates (such as those found in starchy foods), proteins (such as enzymes and the molecules of your hair), and nucleic acids (such as DNA)—are gigantic; in fact, biologists call them **macromolecules** (*macro* means "big"). Despite the size of macromolecules, their structures can be easily understood because they are **polymers**, large molecules made by stringing together many smaller molecules called **monomers**. A polymer is like a pearl necklace made by joining together many pearl monomers.

Cells link monomers together through a **dehydration reaction**, a chemical reaction that removes a molecule of water (Figure 3.4a). For each monomer added to a chain, a water molecule (H_2O) is formed by the release of two hydrogen atoms and one oxygen atom. This same dehydration reaction occurs regardless of the specific monomers and the type of polymer the cell is producing.

Organisms not only make macromolecules but also break them down. For example, many molecules in your food are macromolecules. You must digest these giant molecules to make their monomers available to your cells, which can then rebuild the monomers into your own macromolecules. This digestion occurs by a process called **hydrolysis** (Figure 3.4b). Hydrolysis means to break (*lyse*) with water (*hydro*). Cells break bonds between monomers by adding water to them, a process essentially the reverse of a dehydration reaction. An example of a hydrolysis reaction is the breakdown of lactose by the enzyme lactase. Lactase promotes the hydrolysis of lactose into its monomers. ✓

▼ Figure 3.4 **Synthesis and digestion of polymers.** The only atoms shown in these diagrams are hydrogens and hydroxyl groups (—OH) in strategic locations.

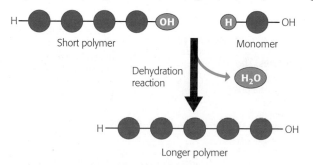

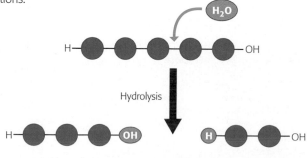

(a) Building a polymer chain. A polymer grows in length when an incoming monomer and the monomer at the end of the polymer each contribute atoms to form a water molecule. The monomers replace the lost covalent bonds with a bond to each other.

(b) Breaking a polymer chain. Hydrolysis reverses the process by adding a water molecule, which breaks the bond between two monomers.

Large Biological Molecules

The remainder of the chapter will introduce you to the four categories of large biological molecules: carbohydrates, lipids, proteins, and nucleic acids. For each category, you'll explore the structure and function of these large molecules by first learning about the smaller molecules used to build them.

Carbohydrates

Carbohydrates, commonly known as "carbs," are sugars or sugar polymers. Some examples are the small sugar molecules dissolved in soft drinks and the long starch molecules in pasta and potatoes. In animals, carbohydrates are a primary source of dietary energy; in plants, they serve as a building material for much of the plant body.

Monosaccharides

Simple sugars, or **monosaccharides** (from the Greek *mono*, single, and *sacchar*, sugar), cannot be broken down by hydrolysis into smaller sugars. Common examples are glucose, found in sports drinks, and fructose, found in fruit. Both of these simple sugars are also in honey (**Figure 3.5**). The molecular formula for glucose is $C_6H_{12}O_6$. Fructose has the same formula, but its atoms are arranged differently. Glucose and fructose are examples of **isomers**, molecules that have the same molecular formula but different structures. (Isomers are like anagrams—words that contain the same letters in a different order, such as *heart* and *earth*.) Because shape is so important, seemingly minor differences in the arrangement of atoms give isomers different properties. In this case, the rearrangement of functional groups makes fructose taste considerably sweeter than glucose.

It is convenient to draw sugars as if their carbon skeletons were linear. In water, however, many monosaccharides form rings when one part of the molecule forms a bond with another part of the molecule, as shown for glucose in **Figure 3.6**.

Monosaccharides, particularly glucose, are the main fuel molecules for cellular work. Analogous to an automobile engine consuming gasoline, your cells break down glucose molecules and extract their stored energy, giving off carbon dioxide as "exhaust." The rapid conversion of glucose to cellular energy is why an aqueous solution of glucose (often called dextrose) is injected into the bloodstream of sick or injured patients; the glucose provides an immediate energy source to tissues in need of repair. In addition to their use as an energy source, monosaccharides provide cells with carbon skeletons that can be used as a raw material for manufacturing other kinds of organic compounds. ☑

► **Figure 3.5 Monosaccharides (simple sugars).** These molecules have the two trademarks of sugars: several hydroxyl groups (—OH) and a carbonyl group $\text{C}=\text{O}$ (C=O). Glucose and fructose, which make honey sweet, are isomers.

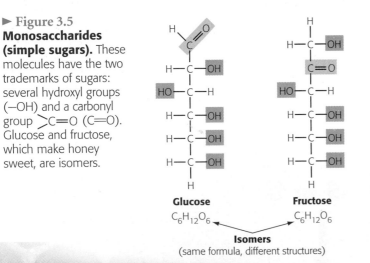

Glucose $C_6H_{12}O_6$ **Fructose** $C_6H_{12}O_6$

Isomers
(same formula, different structures)

▼ **Figure 3.6 The ring structure of glucose.**

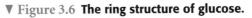

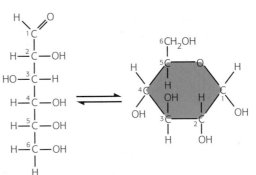

(a) Linear and ring structures. The carbon atoms are numbered so you can relate the linear and ring versions of the molecule. As the double arrows indicate, ring formation is a reversible process, but at any instant in an aqueous solution, most glucose molecules are rings.

(b) Abbreviated ring structure. From now on, we'll use the abbreviated ring symbol for glucose. Each unmarked corner represents a carbon and its attached atoms.

Disaccharides

A **disaccharide**, or double sugar, is constructed from two monosaccharides through a dehydration reaction. In the Biology and Society section, you learned about the disaccharide lactose, sometimes called "milk sugar." Lactose is made from the monosaccharides glucose and galactose (Figure 3.7). Another common disaccharide is maltose, naturally found in germinating seeds. It is used in making beer, malt whiskey and liquor, malted milk shakes, and malted milk ball candy. A molecule of maltose consists of two glucose monomers joined together.

The most common disaccharide is sucrose (table sugar), which consists of a glucose linked to a fructose. Sucrose is the main carbohydrate in plant sap, and it nourishes all the parts of the plant. Sucrose is extracted from the stems of sugarcane or the roots of sugar beets. However, sucrose is rarely used as a sweetener in processed foods in the U.S. Much more common is high-fructose corn syrup, made through a commercial process that converts natural glucose in corn syrup to the much sweeter fructose. If you read the label on a soft drink can or bottle, you're likely to find that high-fructose corn syrup is one of the first ingredients listed (Figure 3.8).

The United States is one of the world's leading markets for sweeteners, with the average American consuming about 45 kilograms (kg)—a whopping 100 pounds—per year, mainly as sucrose and high-fructose corn syrup. This national "sweet tooth" persists in spite of our growing awareness about how sugar can negatively affect our health. Sugar is a major cause of tooth decay. Moreover, high sugar consumption tends to replace eating more varied and nutritious foods. The description of sugars as "empty calories" is accurate in the sense that most sweeteners contain only negligible amounts of nutrients other than carbohydrates. For good health, we also require proteins, fats, vitamins, and minerals. And we need to include substantial amounts of complex carbohydrates—that is, polysaccharides—in our diet. Let's examine these macromolecules next.

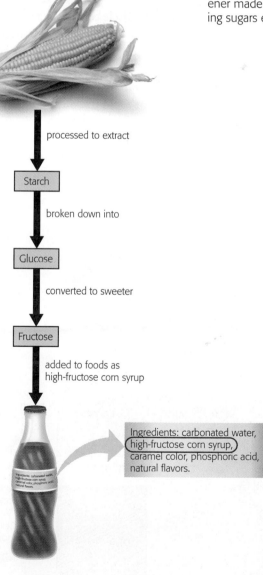

◄ Figure 3.8 **High-fructose corn syrup.** Many processed foods include high-fructose corn syrup, an artificial sweetener made by chemically treating sugars extracted from corn.

processed to extract

Starch

broken down into

Glucose

converted to sweeter

Fructose

added to foods as
high-fructose corn syrup

Ingredients: carbonated water, high-fructose corn syrup, caramel color, phosphoric acid, natural flavors.

▼ Figure 3.7 **Disaccharide (double sugar) formation.** To form a disaccharide, two simple sugars are joined by a dehydration reaction, in this case forming a bond between monomers of glucose and galactose to make the double sugar lactose. (The slight structural difference between glucose and galactose is not shown here.)

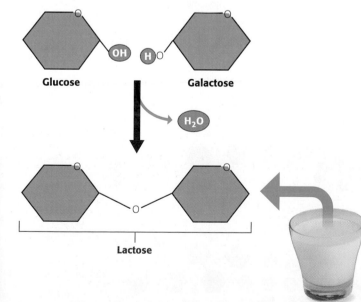

Glucose   Galactose

OH   H

H_2O

Lactose

Polysaccharides

Complex carbohydrates, or **polysaccharides**, are long chains of sugar units—polymers of monosaccharides. One familiar example of a polysaccharide is starch, found in plants. **Starch** consists of many glucose monomers strung together (**Figure 3.9a**). Plant cells store starch in granules, where it is available as a sugar stockpile that can be broken down as needed to provide energy and raw material for building other molecules. Potatoes and grains, such as wheat, corn, and rice, are the major sources of starch in the human diet. Humans and most other animals can use starch as food because their digestive systems break the bonds between glucose monomers through hydrolysis.

Animals store excess sugar in the form of a polysaccharide called **glycogen**. Glycogen and starch are similar in structure: They are both polymers of glucose monomers. But glycogen is more extensively branched (**Figure 3.9b**). Most of our glycogen is stored as granules in our liver and muscle cells, which break down the glycogen to release glucose when it is needed for energy. This is the basis for "carbo loading," the consumption of large amounts of starchy foods the night before an athletic event. The starch is converted to glycogen, which is then available for rapid use during physical activity the next day.

In addition to playing an important role in nutrition, certain polysaccharides serve as structural components. **Cellulose**, the most abundant organic compound on Earth, forms cable-like fibrils in the tough walls that enclose plant cells and is a major component of wood (**Figure 3.9c**). We take advantage of that structural strength when we use lumber as a building material.

Cellulose resembles starch and glycogen in being a polymer of glucose, but its glucose monomers are linked together in a different orientation. Unlike the glucose linkages in starch and glycogen, those in cellulose cannot be broken by most animals. The cellulose in plant foods, which passes unchanged through our digestive tract, is commonly known as dietary "fiber" or "roughage." Because it remains undigested, fiber does not serve as a nutrient, but it does appear to help keep our digestive system healthy. Most Americans do not get the recommended levels of fiber in their diet. Foods rich in fiber include fruits and vegetables, whole grains, bran, and beans. Grazing animals and wood-eating insects such as termites, which do derive nutrition from cellulose, have prokaryotes inhabiting their digestive tracts that break down the cellulose.

Monosaccharides (such as glucose or fructose) and disaccharides (such as sucrose or lactose) dissolve readily in water, forming sugary solutions, as in soft drinks. In contrast, cellulose does not dissolve in water. In spite of this difference, almost all carbohydrates are **hydrophilic** ("water-loving") molecules that adhere water to their surface. It is the hydrophilic quality of cellulose that makes a fluffy bath towel so water absorbent. ☑

☑CHECKPOINT
1. How can glucose and fructose have the same formula ($C_6H_{12}O_6$) but different properties?
2. How and why do manufacturers produce high-fructose corn syrup?

Answers: **1.** Different arrangements of atoms affect molecular shapes and properties. **2.** They convert glucose to the sweeter fructose so less syrup is needed.

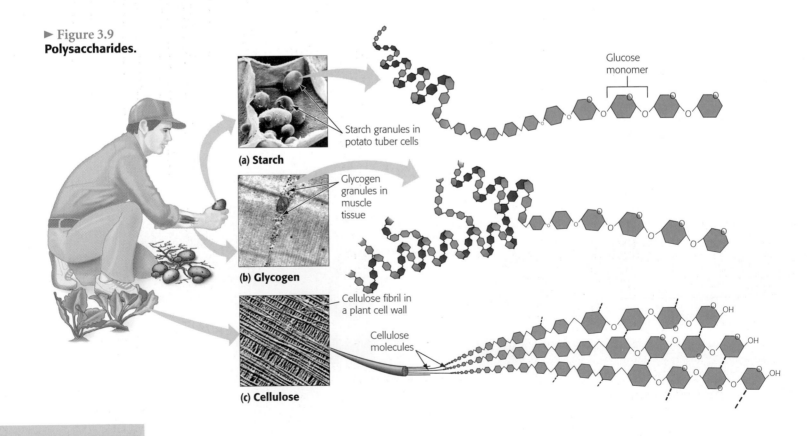

► **Figure 3.9**
Polysaccharides.

Starch granules in potato tuber cells

(a) Starch

Glycogen granules in muscle tissue

(b) Glycogen

Cellulose fibril in a plant cell wall

Cellulose molecules

(c) Cellulose

Glucose monomer

Lipids

In contrast to carbohydrates and most other biological molecules, **lipids** are organic compounds that are **hydrophobic** ("water-fearing"); they do not mix with water. You have probably observed this chemical behavior in an unshaken bottle of salad dressing: The oil, which is a type of lipid, separates from the vinegar, which is mostly water (Figure 3.10). If you shake the bottle, you can force a temporary mixture long enough to douse your salad with dressing, but what remains in the bottle will quickly separate again when you stop shaking it.

Unlike carbohydrates, proteins, and nucleic acids, lipids are neither macromolecules nor polymers. Whereas all carbohydrates, for example, are made of sugar monomers, lipids vary in structure. That is, they aren't all made of similar "building blocks." We'll look at two main types of lipids: fats and steroids.

Fats

A typical **fat** consists of a glycerol molecule joined with three fatty acid molecules via dehydration reactions (Figure 3.11). The resulting fat is called a **triglyceride**, a term you may see in the results of medical tests for fat in the blood. The major portion of a fatty acid is a long hydrocarbon that stores a lot of energy, like the hydrocarbons of gasoline. In fact, a pound of fat packs more than twice as much energy as a pound of carbohydrate. The downside to this energy efficiency is that it is very difficult for a person trying to lose weight to "burn off" excess fat. It is important to understand that a reasonable amount of body fat is both normal and healthy as a fuel reserve. We stock these long-term food stores in specialized reservoirs called adipose cells, which swell and shrink when we deposit and withdraw fat from them. This adipose tissue, or "body fat," not only stores energy but also cushions vital organs and insulates us, helping maintain a warm body temperature even when the outside air is cold.

Notice in Figure 3.11b that the bottom fatty acid bends where there is a double bond in the carbon skeleton. That fatty acid is said to be **unsaturated** because it has fewer than the maximum number of hydrogens at the location of the double bond. The other two fatty acids in the fat molecule lack double bonds in their hydrocarbon portions. Those fatty acids are **saturated**, meaning that they contain the maximum number of hydrogen atoms. A saturated fat is one with all three of its fatty acid tails saturated. If one or more of the fatty acids is unsaturated, then it's an unsaturated fat, like the one in Figure 3.11b. A polyunsaturated fat has several double bonds within its fatty acids.

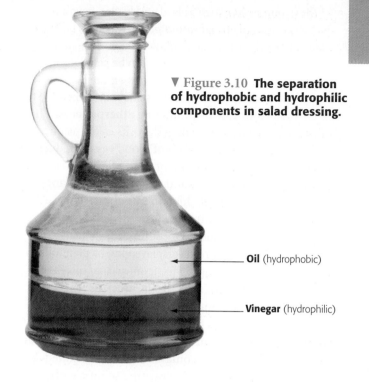

▼ Figure 3.10 **The separation of hydrophobic and hydrophilic components in salad dressing.**

Oil (hydrophobic)

Vinegar (hydrophilic)

▼ Figure 3.11 **The synthesis and structure of a fat, or triglyceride.**

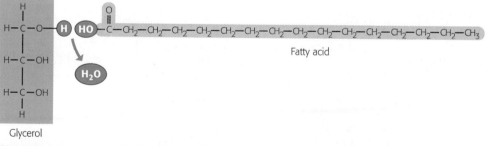

Glycerol

Fatty acid

(a) A dehydration reaction linking a fatty acid to glycerol

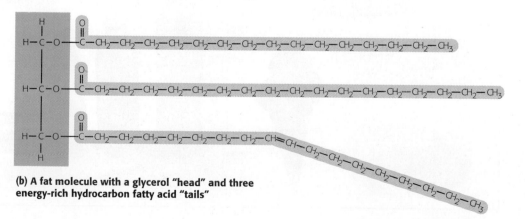

(b) A fat molecule with a glycerol "head" and three energy-rich hydrocarbon fatty acid "tails"

Most animal fats, such as lard and butter, have a relatively high proportion of saturated fatty acids. The linear shape of saturated fatty acids allows these molecules to stack easily, so saturated fats tend to be solid at room temperature, as displayed by the foods on the left side of **Figure 3.12**. Diets rich in saturated fats may contribute to cardiovascular disease by promoting **atherosclerosis**. In this condition, lipid-containing deposits called plaque build up along the inside walls of blood vessels, reducing blood flow and increasing risk of heart attacks and strokes.

In contrast to red meat and dairy products, plant and fish fats are relatively high in unsaturated fatty acids. The bent shape of unsaturated fatty acids makes them less likely to form solids, so most unsaturated fats are liquid at room temperature, as displayed by the plant oils in Figure 3.12. Oils are liquid fats. Examples of oils that are primarily unsaturated include vegetable oils (such as corn and canola oil) and fish oils (such as cod liver oil).

Although plant oils tend to be low in saturated fat, tropical plant fats are an exception. Cocoa butter, a main ingredient in chocolate, contains a mix of saturated and unsaturated fat that gives it a melting point near body temperature. Thus, chocolate stays solid at room temperature but melts in the mouth. This pleasing "mouth feel" is one of the reasons chocolate is so appealing.

Sometimes a food manufacturer wants to use a vegetable oil but needs the food product to be solid, as in the case of margarine or some peanut butters. To achieve this, the manufacturer can convert unsaturated fats to saturated fats by adding hydrogen, a process called **hydrogenation**. Unfortunately, hydrogenation also creates **trans fat**, a type of unsaturated fat that is even less healthy than saturated fat. Since 2006, the FDA has required trans fats to be listed in nutrition labels. But even if a label on a hydrogenated product says 0 grams (g) of trans fat per serving, the product might still have up to 0.5 g per serving. Any product with partially hydrogenated vegetable oil or vegetable shortening contains trans fat, as displayed by the processed foods in the center of Figure 3.12. (See the Process of Science section in Chapter 1—"Is Trans Fat Bad for You?"—for an investigation of the health effects of trans fats.) Due to their unhealthy nature, trans fats are becoming less common as food manufacturers substitute other forms of fat.

Although saturated and trans fats should generally be avoided, it is not true that *all* fats are unhealthy. In fact, some fats perform important functions within the body and are beneficial and even essential to a healthy diet. For example, fats containing omega-3 fatty acids have been shown to reduce the risk of coronary heart disease and relieve the symptoms of arthritis and inflammatory bowel disease. Some sources are nuts and oily fish such as salmon, as displayed on the right side of Figure 3.12. ✓

☑CHECKPOINT

What are "unsaturated fats"? What two kinds of fat are least healthy? What fats are especially healthy?

Answer: fats without the maximum number of hydrogens because of double bonds between some carbons; trans fats and saturated fats; fats built from omega-3 fatty acids

▼ Figure 3.12 **Types of fats.**

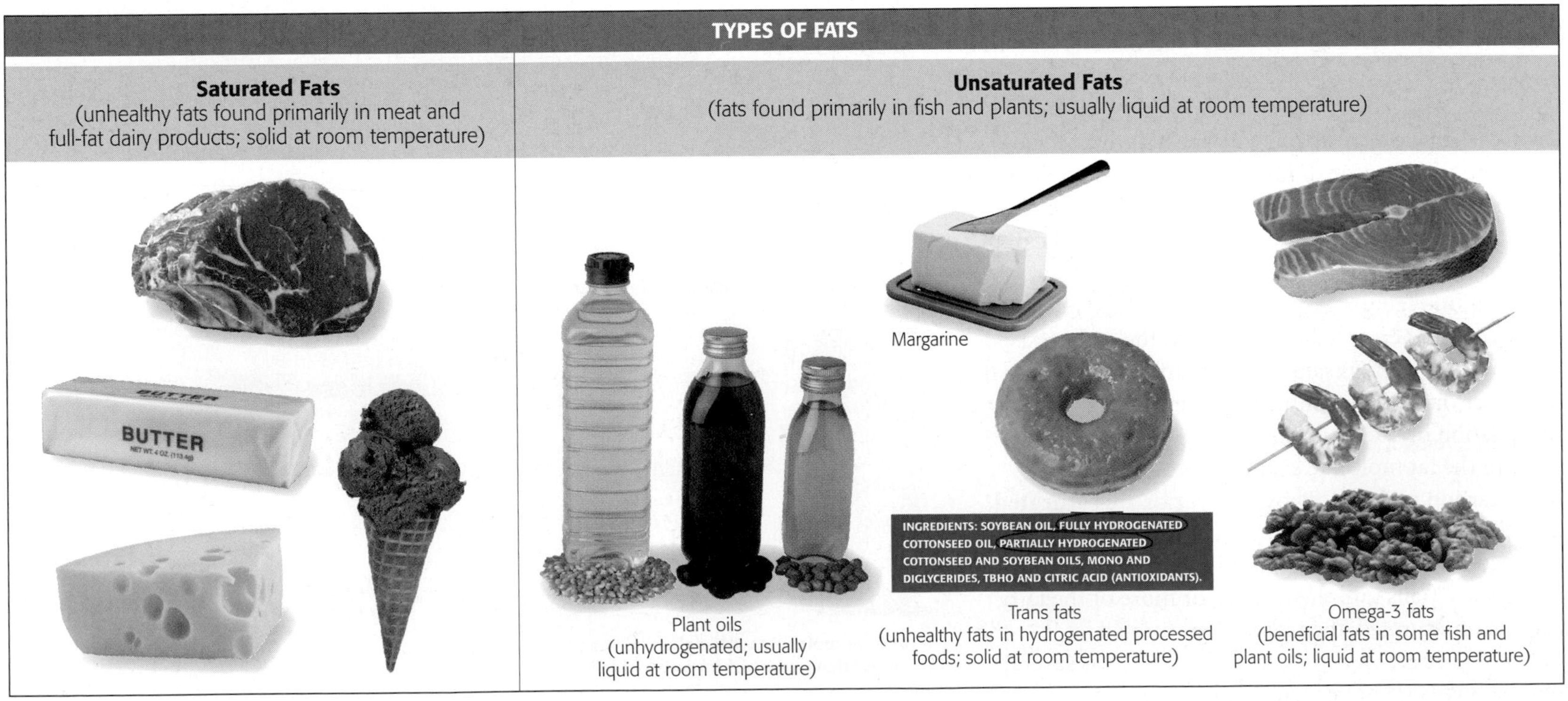

TYPES OF FATS

Saturated Fats
(unhealthy fats found primarily in meat and full-fat dairy products; solid at room temperature)

Unsaturated Fats
(fats found primarily in fish and plants; usually liquid at room temperature)

BUTTER
NET WT. 4 OZ (113 kg)

Margarine

INGREDIENTS: SOYBEAN OIL, FULLY HYDROGENATED COTTONSEED OIL, PARTIALLY HYDROGENATED COTTONSEED AND SOYBEAN OILS, MONO AND DIGLYCERIDES, TBHO AND CITRIC ACID (ANTIOXIDANTS).

Plant oils
(unhydrogenated; usually liquid at room temperature)

Trans fats
(unhealthy fats in hydrogenated processed foods; solid at room temperature)

Omega-3 fats
(beneficial fats in some fish and plant oils; liquid at room temperature)

Steroids

Classified as lipids because they are hydrophobic, **steroids** are very different from fats in structure and function. All steroids have a carbon skeleton that is bent to form four fused rings. Different steroids vary in the functional groups attached to this core set of rings, and these variations affect their function. One common steroid is cholesterol, which has a bad reputation because of its association with cardiovascular disease. However, cholesterol is an essential molecule in your body. It is a key component of the membranes that surround your cells. It is also the "base steroid" from which your body produces other steroids, such as the sex hormones estrogen and testosterone (Figure 3.13).

The controversial drugs called anabolic steroids are synthetic variants of testosterone, the male sex hormone. Testosterone causes a general buildup in muscle and bone mass during puberty in males and maintains masculine traits throughout life. Because anabolic steroids structurally resemble testosterone, they also mimic some of its effects. Some athletes use anabolic steroids to build up their muscles quickly and enhance their performance.

In 2003, the discovery that some athletes were using a new anabolic steroid called THG rocked the sports world. THG is a chemically modified ("designer") steroid intended to avoid detection by drug tests (hence its nickname, "the clear"). New tests revealed widespread use of THG and other steroids among athletes in many sports (Figure 3.14).

Using anabolic steroids is indeed a fast way to increase body size beyond what hard work alone can produce. But at what cost? Steroid abuse can cause serious physical and mental problems, including violent mood swings ("roid rage"), depression, liver damage, high cholesterol, shrunken testicles, reduced sex drive, and infertility. These last symptoms occur because anabolic steroids often cause the body to reduce its normal output of sex hormones. Most athletic organizations now ban the use of anabolic steroids because of their many potential health hazards coupled with the unfairness of an artificial advantage. ☑

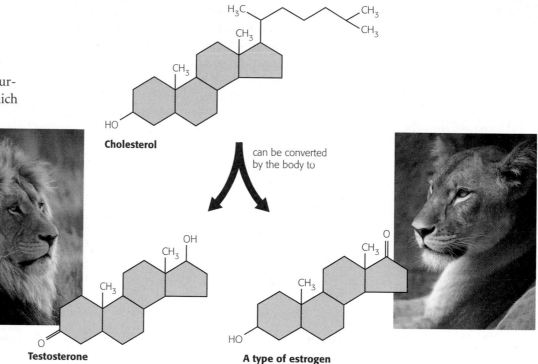

▼ Figure 3.13 **Examples of steroids.** The molecular structures of the steroids shown here are abbreviated by omitting all the atoms that make up the rings. The subtle difference between testosterone and estrogen influences the development of the anatomical and physiological differences between male and female mammals, including lions and humans.

Cholesterol

can be converted by the body to

Testosterone

A type of estrogen

▼ Figure 3.14 **Steroids and the modern athlete.** Baseball player Jose Canseco (left), an admitted user of performance-enhancing steroids, testifies before Congress during an inquiry into the abuse of drugs by professional athletes. Track-and-field star Marion Jones (right) admitted in 2007 that she had taken performance-enhancing drugs. As a result, she was stripped of her five Olympic medals. Note the similarity in chemical structure of testosterone (shown above) and THG (shown below).

THG

☑CHECKPOINT

How are dietary fats similar to human sex hormones?

Answer: Both are lipids, which are hydrophobic.

Protein Structure and Function

Proteins

Proteins are the most elaborate of life's molecules. A **protein** is a polymer constructed from amino acid monomers. Your body has tens of thousands of different kinds of proteins, each with a unique three-dimensional shape corresponding to a specific function. **Figure 3.15** shows some examples of major protein functions.

Proteins perform most of the tasks required for life. Probably their most important role is as enzymes, chemicals that change the rate of a chemical reaction without being changed in the process (as you'll see in Chapter 5). Lactase, which you read about in the Biology and Society section, is just one of the thousands of enzymes produced by cells.

The Monomers of Proteins: Amino Acids

All proteins are macromolecules constructed from a common set of 20 kinds of amino acids. Each **amino acid** consists of a central carbon atom bonded to four covalent partners (carbon, remember, always forms four covalent bonds). Three of those attachments are common to all 20 amino acids: a carboxyl group (—COOH), an amino group (—NH₂), and a hydrogen atom. The variable component of amino acids, the side group, is attached to the fourth bond of the central carbon (**Figure 3.16a**). Each type of amino acid has a unique side group, giving that amino acid its special chemical properties (**Figure 3.16b**).

▼ **Figure 3.16 Amino acids.** The 20 amino acids vary only in their side groups, which give these monomers their unique properties.

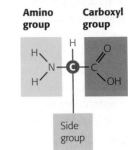

(a) The general structure of an amino acid

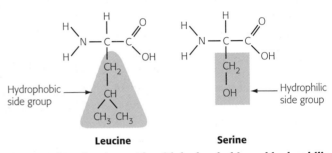

(b) Examples of amino acids with hydrophobic and hydrophilic side groups. The side group of the amino acid leucine is pure hydrocarbon. That region of leucine is hydrophobic because hydrocarbons don't mix with water. In contrast, the side group of the amino acid serine has a hydroxyl group (—OH), which is hydrophilic.

▼ **Figure 3.15 Some types of proteins.**

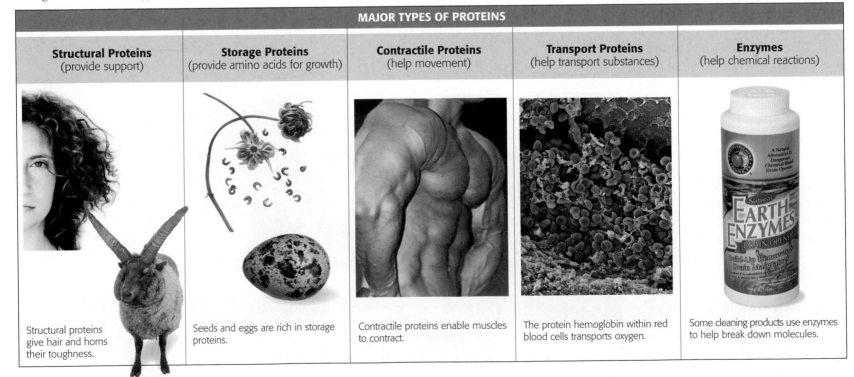

MAJOR TYPES OF PROTEINS				
Structural Proteins (provide support)	**Storage Proteins** (provide amino acids for growth)	**Contractile Proteins** (help movement)	**Transport Proteins** (help transport substances)	**Enzymes** (help chemical reactions)
Structural proteins give hair and horns their toughness.	Seeds and eggs are rich in storage proteins.	Contractile proteins enable muscles to contract.	The protein hemoglobin within red blood cells transports oxygen.	Some cleaning products use enzymes to help break down molecules.

Proteins as Polymers

Cells link amino acid monomers together by—can you guess?—dehydration reactions. The bond between adjacent amino acids is called a **peptide bond** (Figure 3.17). The resulting long chain of amino acids is called a **polypeptide**. A protein is a polymer consisting of one or more polypeptides.

Your body has tens of thousands of different kinds of proteins. How is it possible to make such a huge variety of proteins from just 20 kinds of amino acids? The answer is arrangement. You know that you can make many different words by varying the sequence of just 26 letters. Though the protein alphabet is slightly smaller (just 20 "letters"), the "words" are much longer, with a typical polypeptide being at least 100 amino acids in length. Just as each word is constructed from a unique succession of letters, each protein has a unique linear sequence of amino acids. This specific amino acid sequence is called the protein's **primary structure** (Figure 3.18).

Changing a single letter can drastically affect the meaning of a word—"tasty" versus "nasty," for instance. Similarly, even a slight change in primary structure can affect a protein's ability to function. For example, the substitution of one amino acid for another at a particular position in hemoglobin, the blood protein that carries oxygen, causes sickle-cell disease, an inherited blood disorder (Figure 3.19). ✔

▼ **Figure 3.18 The primary structure (amino acid sequence) of a protein.** This chain, drawn in serpentine fashion so that it would fit on the page, shows the primary structure of the protein lysozyme. The names of the amino acids are given as their three letter abbreviations, with the positions of lysozyme's 129 amino acids numbered along the chain.

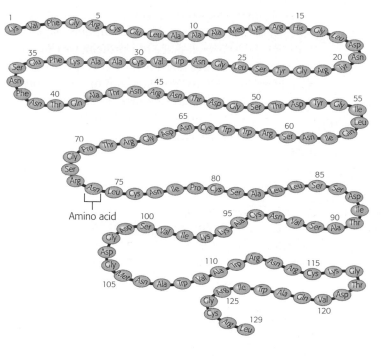

▼ **Figure 3.17 Joining amino acids.** A dehydration reaction links adjacent amino acids by a peptide bond.

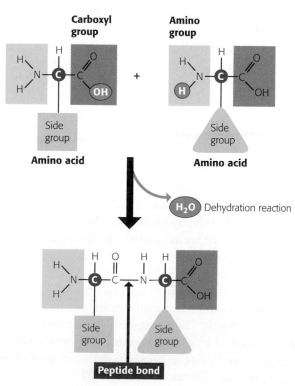

▼ **Figure 3.19 A single amino acid substitution in a protein causes sickle-cell disease.**

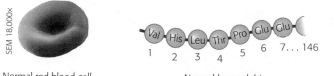

Normal red blood cell   Normal hemoglobin

(a) Normal hemoglobin. Red blood cells of humans are normally disk-shaped. Each cell contains millions of molecules of the protein hemoglobin, which transports oxygen from the lungs to other organs of the body. Next to the photograph, you can see the first 7 of the 146 amino acids in a polypeptide chain of hemoglobin.

Sickled red blood cell   Sickle-cell hemoglobin

(b) Sickle-cell hemoglobin. A slight change in the primary structure of hemoglobin causes sickle-cell disease. The inherited substitution of one amino acid—valine in place of the amino acid glutamic acid—occurs in the number 6 position of the polymer. The abnormal hemoglobin molecules tend to crystallize, deforming some of the cells into a sickle shape. The life of someone with the disease is characterized by dangerous episodes when the angular cells clog tiny blood vessels, impeding blood flow.

☑**CHECKPOINT**

What are the monomers of all proteins?

Answer: amino acids

Protein Shape

At this point, you might be thinking that a polypeptide chain is the same thing as a protein, but that's not quite true. The distinction between the two is like the relationship between a long strand of yarn and a sweater that you could knit from the yarn. A functional protein is one or more polypeptide chains precisely twisted, folded, and coiled into a molecule of unique shape. If we dissect the overall shape of a protein, we can recognize at least three levels of structure: primary, secondary, and tertiary. Proteins with more than one polypeptide chain have a fourth level: quaternary structure. Figure 3.20 shows how the levels of protein structure are related.

When a cell makes a polypeptide, the chain usually folds spontaneously to form the functional shape for that protein. It is a protein's three-dimensional shape that enables the molecule to carry out its specific function in a cell. In almost every case, a protein's function depends on its ability to recognize and bind to some other molecule. For example, the specific shape of lactase enables it to recognize and attach to lactose, its molecular target. With proteins, *function follows form*—that is, what a protein does is a consequence of its shape (Figure 3.21).

▼ Figure 3.20 **The four levels of protein structure.**

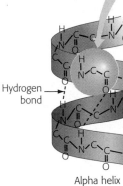

(a) Primary structure. A protein's primary structure is the unique sequence of amino acids in its polypeptide chain(s).

Amino acids

Hydrogen bond

Alpha helix

Polypeptide (single subunit)

Pleated sheet

(b) Secondary structure. Certain stretches of the polypeptide form local patterns called secondary structure. Two types are named alpha helix and pleated sheet. Secondary structure is reinforced by hydrogen bonds (dashed lines) along the polypeptide backbone. This drawing shows only the atoms of the polypeptide backbone, not the amino acid side groups.

(c) Tertiary structure. The overall three-dimensional shape of the polypeptide is called tertiary structure. It is reinforced by chemical bonds (not shown here) between the side groups of amino acids in different regions of the polypeptide chain.

A protein with four polypeptide subunits

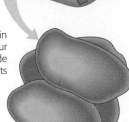

(d) Quaternary structure. Proteins with two or more polypeptide chains have a quaternary structure, which results from bonds between the chains.

► Figure 3.21 **A computer model showing a protein (purple) about to bind its target (red).**

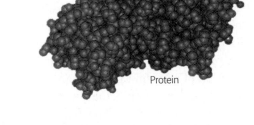

Target

Protein

What Determines Protein Shape?

A protein's shape is sensitive to the surrounding environment. An unfavorable change in temperature, pH, or some other quality of the environment can cause a protein to unravel and lose its normal shape. This is called **denaturation** of the protein. If you cook an egg, the transformation of the egg white from clear to opaque is caused by proteins in the egg white denaturing. The denatured proteins become insoluble in water and form a white solid. One of the reasons why extremely high fevers are so dangerous is that some proteins in the body become denatured above about 104°F.

The primary structure of a protein—its sequence of amino acids—causes it to fold into its functional shape. Each kind of protein has a unique primary structure and therefore a unique shape that enables it to do a certain job in a cell. What happens if it folds incorrectly? Misfolded proteins are associated with many diseases, including some severe nervous system disorders, such as Alzheimer's disease, mad cow disease, and Parkinson's disease. But what determines a protein's primary structure? As noted earlier, a protein consists of one or more polypeptide chains. The amino acid sequence of each chain is specified by a gene. And this relationship between genes and proteins brings us to this chapter's last category of large biological molecules. ✔

✔CHECKPOINT

1. Which of these is *not* made of protein? hair, muscle, cellulose, enzymes
2. Why does a denatured protein no longer function?

Answers: *1. cellulose 2. It has lost its necessary shape.*

Nucleic Acids

Nucleic acids are macromolecules that provide the directions for building proteins. The name *nucleic* comes from their location in the nuclei of eukaryotic cells. There are actually two types of nucleic acids: **DNA** (which stands for <u>d</u>eoxyribo<u>n</u>ucleic <u>a</u>cid) and **RNA** (for <u>ribon</u>ucleic <u>a</u>cid). The genetic material that humans and other organisms inherit from their parents consists of giant molecules of DNA. The DNA resides in the cell as one or more very long fibers called chromosomes. A **gene** is a specific stretch of DNA that programs the amino acid sequence of a polypeptide. Those programmed instructions, however, are written in a kind of chemical code that must be translated from "nucleic acid language" to "protein language." As you'll learn in Chapter 10, a cell's RNA molecules help make this translation (**Figure 3.22**).

Nucleic acids are polymers made from monomers called **nucleotides** (**Figure 3.23**). Each nucleotide contains three parts. At the center of each nucleotide is a five-carbon sugar, deoxyribose in DNA and ribose in RNA. Attached to the sugar is a negatively charged phosphate group containing a phosphorus atom bonded to oxygen atoms (PO_4^-). Also attached to the sugar is a nitrogenous (nitrogen-containing) base made of one or two rings. It is called a base because it accepts hydrogen ions (H^+) in aqueous solutions (see Chapter 2). The sugar and phosphate are the same in all nucleotides; only the base varies. Each DNA nucleotide has one of four nitrogenous bases: adenine (abbreviated A), guanine (G), cytosine (C), or thymine (T) (**Figure 3.24**). Thus, all genetic information is written in a four-letter alphabet.

▼ **Figure 3.22** **Building a protein.** Within the cell, a gene (a segment of DNA) provides the directions to build a molecule of RNA, which can then be translated into a protein.

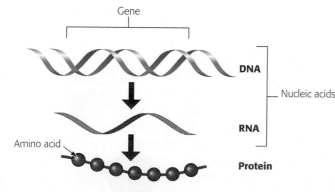

▼ **Figure 3.23** **A DNA nucleotide.** A DNA nucleotide monomer consists of three parts: a sugar (deoxyribose), a phosphate, and a nitrogenous (nitrogen-containing) base.

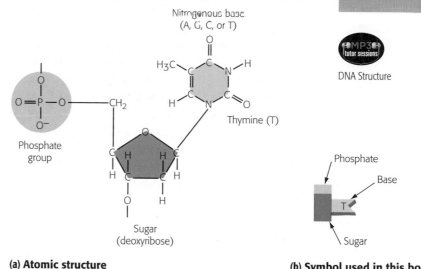

DNA Structure

(a) Atomic structure

(b) Symbol used in this book

▼ **Figure 3.24** **The nitrogenous bases of DNA.** Notice that adenine and guanine have double-ring structures. Thymine and cytosine have single-ring structures.

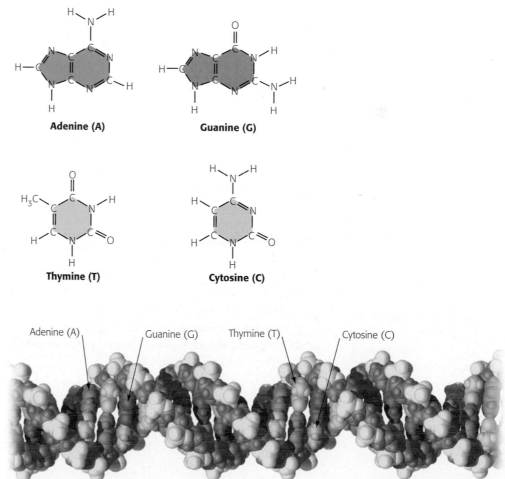

Space-filling model of DNA (showing the four bases in four different colors)

Dehydration reactions link nucleotide monomers into long chains called polynucleotides. In the case of DNA, these are called DNA strands **(Figure 3.25a)**. Nucleotides are joined by covalent bonds between the sugar of one nucleotide and the phosphate of the next. This results in a **sugar-phosphate backbone**, a repeating pattern of sugar-phosphate-sugar-phosphate, with the bases (A, T, C, or G) hanging off the backbone like appendages. With different combinations of the four bases, the number of possible polynucleotide sequences is vast. One long polynucleotide may contain many genes, each a specific series of hundreds or thousands of nucleotides. And each gene stores information in its unique sequence of nucleotide bases. This sequence is a code that provides instructions for building a specific polypeptide from amino acids.

A molecule of cellular DNA is double-stranded, with two polynucleotide strands wrapped around each other to form a **double helix (Figure 3.25b)**. In the central core of the helix, the bases along one DNA strand hydrogen-bond to bases along the other strand. The bonds are individually weak—they are hydrogen bonds, like those between water molecules—but collectively they zip the two strands together into a very stable double helix. Because of the way the functional groups hang off the bases, the base pairing in a DNA double helix is specific: The base A can pair only with T, and G can pair only with C. Thus, if you know the sequence of bases along one DNA strand, you also know the sequence along the complementary strand in the double helix. As you'll see in Chapter 10, this unique base pairing is the basis of DNA's ability to act as the molecule of inheritance.

How does RNA differ from DNA? As its name *ribonucleic acid* implies, its sugar is ribose rather than deoxyribose. Another difference between RNA and DNA is that instead of the base thymine, RNA has a similar but distinct base called uracil (U) **(Figure 3.26)**. Except for the presence of ribose and uracil, an RNA polynucleotide chain is identical to a DNA polynucleotide chain. However, RNA is usually found in single-stranded form, whereas DNA usually exists as a double helix.

Now that we've examined the structure of nucleic acids, we'll look at how a change in nucleotide sequence can affect protein production. ✔

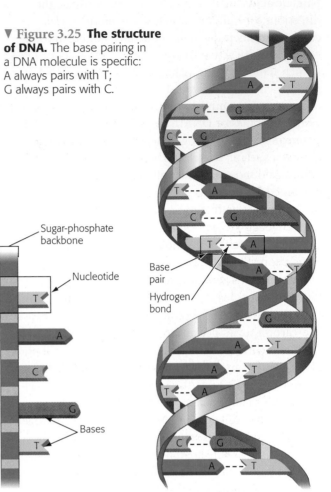

▼ **Figure 3.25 The structure of DNA.** The base pairing in a DNA molecule is specific: A always pairs with T; G always pairs with C.

Sugar-phosphate backbone

Nucleotide

Base pair

Hydrogen bond

Bases

(a) DNA strand
(polynucleotide)

(b) Double helix
(two polynucleotide strands)

▼ **Figure 3.26 An RNA nucleotide.** Notice that this RNA nucleotide differs from the DNA nucleotide in Figure 3.23 in two ways: The RNA sugar is ribose rather than deoxyribose; and the base is uracil (U) instead of thymine (T). The other three kinds of RNA nucleotides have the bases A, C, and G, as in DNA.

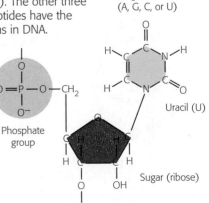

Nitrogenous base (A, G, C, or U)

Uracil (U)

Phosphate group

Sugar (ribose)

Does Lactose Intolerance Have a Genetic Basis?

The enzyme lactase, like all proteins, is encoded by a DNA gene. A reasonable hypothesis is that lactose-intolerant people have a defect in their lactase gene. However, this hypothesis is not supported by **observation**. In fact, most lactose-intolerant people have a normal version of the lactase gene. This raises the **question**, Is there a genetic basis for lactose intolerance?

In 2002, a group of Finnish and American scientists proposed the **hypothesis** that lactose intolerance can be correlated with a single nucleotide at a particular site within one chromosome. Based on their observations, they made the **prediction** that this site would be near, though not within, the lactase gene. In their **experiment**, they examined the genes of 196 lactose-intolerant people from nine Finnish families. Their **results** showed a 100% correlation between lactose intolerance and a nucleotide at a site approximately 14,000 nucleotides away (a relatively short distance in terms of the whole

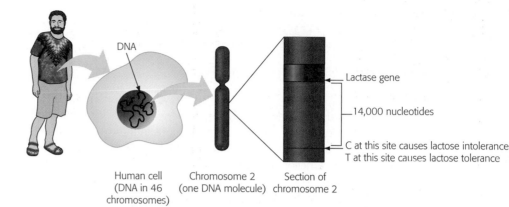

Human cell (DNA in 46 chromosomes) — Chromosome 2 (one DNA molecule) — Section of chromosome 2

DNA — Lactase gene — 14,000 nucleotides — C at this site causes lactose intolerance — T at this site causes lactose tolerance

chromosome) from the lactase gene (**Figure 3.27**). Other experiments showed that, depending on the nucleotide sequence within this region of the DNA molecule, the action of the lactase gene is ramped up or down (in a way that likely involves producing a regulatory protein that interacts with the nucleotides near the lactase gene). This study shows how a small change in a DNA nucleotide sequence can have a major effect on the production of a protein and the well-being of an organism.

▲ Figure 3.27 **A genetic cause of lactose intolerance.** A 2002 study showed a correlation between lactose intolerance and a nucleotide at a specific location on one chromosome.

Evolution and Lactose Intolerance in Humans

As you'll recall from the Biology and Society section, most of the world's population are lactose intolerant as adults and thus do not easily digest the milk sugar lactose. In fact, lactose intolerance is found in 80% of African-Americans and Native Americans and 90% of Asian-Americans, but only in about 10% of Americans of northern European descent. And as just discussed in the Process of Science section, lactose intolerance appears to have a genetic basis.

From an evolutionary perspective, it is reasonable to infer that lactose intolerance is rare among northern Europeans because the ability to tolerate lactose offered a survival advantage to their ancestors. In northern Europe's relatively cold climate, only one harvest a year is possible. Therefore, animals became a main source of food for early humans in that region. Cattle were first domesticated in northern Europe about 9,000 years ago (**Figure 3.28**). With milk and other dairy products at hand year-round, natural selection would have favored anyone with a per-

manently active lactase gene. In other cultures where dairy products were not a staple in the diet, it would be more efficient to turn off the lactase gene and not produce an unneeded enzyme.

Researchers wondered whether the genetic basis for lactose tolerance in northern Europeans might be present in other cultures who kept dairy herds. To find out, a 2006 study compared the genetic makeup and lactose tolerance of 43 ethnic groups in East Africa. The researchers identified three other genetic changes that keep the lactase gene permanently active. These genetic changes appear to have occurred beginning around 7,000 years ago, about the time that archaeological evidence shows domestication of cattle in these African regions.

Genetic changes that confer a selective advantage, such as surviving cold winters or withstanding drought by drinking milk, spread rapidly in these early peoples. Evolutionary and cultural history is thus recorded in their genes and in their continuing ability to digest milk.

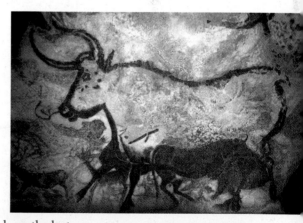

▲ Figure 3.28 **A prehistoric European cave painting of wild cattle.** This species, known as aurochs, became the first domesticated cattle in Europe. Aurochs migrated from Asia about 250,000 years ago but are now extinct.

Chapter Review

SUMMARY OF KEY CONCEPTS

(MB) Go to the Study Area at **www.masteringbiology.com** for practice quizzes, my**e**Book, BioFlix™ 3-D animations, MP3 Tutor Sessions, videos, current events, and more.

Organic Compounds

Carbon Chemistry

Carbon atoms can form large, complex, diverse molecules by bonding to four partners, including other carbon atoms. In addition to variations in the size and shape of carbon skeletons, organic compounds vary in the presence and locations of different functional groups.

Giant Molecules from Smaller Building Blocks

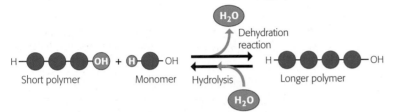

Large Biological Molecules

Large biological molecules	Functions	Components	Examples
Carbohydrates	Dietary energy; storage; plant structure	Monosaccharide	Monosaccharides: glucose, fructose Disaccharides: lactose, sucrose Polysaccharides: starch, cellulose
Lipids	Long-term energy storage (fats); hormones (steroids)	Components of a triglyceride	Fats (triglycerides); Steroids (testosterone, estrogen)
Proteins	Enzymes, structure, storage, contraction, transport, and others	Amino acid	Lactase (an enzyme), hemoglobin (a transport protein)
Nucleic acids	Information storage	Nucleotide	DNA, RNA

Carbohydrates

Simple sugars (monosaccharides) provide cells with energy and carbon skeletons for use in building other organic compounds. Double sugars (disaccharides), such as sucrose, consist of two monosaccharides joined by a dehydration reaction. Polysaccharides are macromolecules, long polymers of sugar monomers. Starch and glycogen are storage polysaccharides in plants and animals, respectively. The cellulose of plant cell walls is an example of a structural polysaccharide.

Lipids

Lipids are hydrophobic. Fats are the major form of long-term energy storage in animals. A molecule of fat, or triglyceride, consists of three fatty acids joined by a dehydration reaction to a glycerol. Most animal fats are saturated, meaning that their fatty acids have the maximum number of hydrogens. Plant oils contain mostly unsaturated fats, having fewer hydrogens in the fatty acids because of double bonding in the carbon skeletons. Steroids, including cholesterol and the sex hormones, are also lipids.

Proteins

There are 20 types of amino acids, the monomers of proteins. They are linked by dehydration reactions to form polymers called polypeptides. A protein consists of one or more polypeptides folded into a specific three-dimensional shape. Contributing to this shape are four levels of structure:

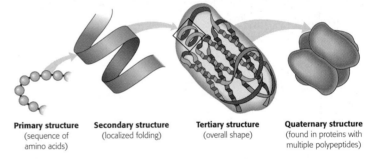

Primary structure (sequence of amino acids) **Secondary structure** (localized folding) **Tertiary structure** (overall shape) **Quaternary structure** (found in proteins with multiple polypeptides)

Shape is sensitive to environment, and if a protein is denatured (losing its shape because of an unfavorable environment), its function is also disrupted.

Nucleic Acids

Nucleic acids include RNA and DNA. DNA takes the form of a double helix, two DNA strands (polymers of nucleotides) held together by hydrogen bonds between nucleotide components called bases. There are four kinds of DNA bases: adenine (A), guanine (G), thymine (T), and cytosine (C). A always pairs with T, and G always pairs with C. These base-pairing rules enable DNA to act as the molecule of inheritance. RNA has U (uracil) instead of T.

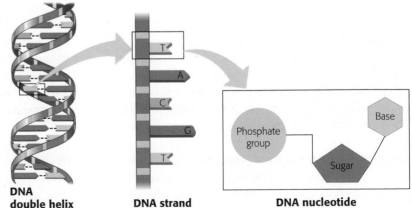

DNA double helix   **DNA strand**   **DNA nucleotide**

SELF-QUIZ

1. Draw a structural formula for C_2H_4. (*Hint*: Each carbon has four bonds; each hydrogen has one.)

2. Monomers are joined together to form larger polymers through _____ reactions. Polymers are broken down into the monomers that make them up through the chemical reaction called _____.

3. Which of the following terms includes all the others in the list?
   a. polysaccharide
   b. carbohydrate
   c. monosaccharide
   d. disaccharide

4. One molecule of dietary fat is made by joining three molecules of _____ to one molecule of _____.

5. Which of the following statements about saturated fats is true?
   a. Saturated fats contain one or more double bonds along the hydrocarbon tails.
   b. Saturated fats contain the maximum number of hydrogens along the hydrocarbon tails.
   c. Saturated fats make up the majority of most plant oils.
   d. Saturated fats are typically healthier for you than unsaturated fats.

6. Humans and other animals cannot digest wood because they
   a. cannot digest any carbohydrates.
   b. cannot chew it fine enough.
   c. lack the enzyme needed to break down cellulose.
   d. get no nutrients from it.

7. Changing one amino acid within a protein could change what about a protein?
   a. the primary structure
   b. the overall shape of the protein
   c. the function of the protein
   d. all of the above

8. Most proteins can easily dissolve in water. Knowing that, where within the overall three-dimensional shape of a protein would you most likely find hydrophobic amino acids?

9. A shortage of phosphorus in the soil would make it especially difficult for a plant to manufacture
   a. DNA.
   b. proteins.
   c. cellulose.
   d. fatty acids.

10. A glucose molecule is to _____ as a _____ is to a nucleic acid.

11. Name three similarities between DNA and RNA. Name three differences.

Answers to the Self-Quiz questions can be found in Appendix D.

THE PROCESS OF SCIENCE

12. A food manufacturer is advertising a new cake mix as fat-free. Scientists at the U.S. Food and Drug Administration (FDA) are testing the product to see if it truly lacks fat. Hydrolysis of the cake mix yields glucose, fructose, glycerol, a number of amino acids, and several kinds of molecules with long hydrocarbon chains. Further analysis shows that most of the hydrocarbon chains have a carboxyl group at one end. What would you tell the food manufacturer if you were a spokesperson for the FDA?

13. Imagine that you have produced several versions of lactase, each of which differs from normal lactase by a single amino acid. Describe a test that could indirectly determine which of the versions significantly alters the three-dimensional shape of the protein.

BIOLOGY AND SOCIETY

14. Some amateur and professional athletes take anabolic steroids to help them build strength ("bulk up"). The health risks of this practice are extensively documented. Apart from these health issues, what is your opinion about the ethics of athletes using chemicals to enhance performance? Is this a form of cheating, or is it just part of the preparation required to stay competitive in a sport where anabolic steroids are commonly used? Defend your opinion.

15. Heart disease is the leading cause of death among people in the United States and other industrialized nations. Fast food is a major source of unhealthy fats that contribute significantly to heart disease. Imagine you're a juror sitting on a trial where a fast-food manufacturer is being sued for producing a harmful product. To what extent do you think manufacturers of unhealthy foods should be held responsible for the health consequences of their products? As a jury member, how would you vote?

16. Each year, industrial chemists develop and test thousands of new organic compounds for use as insecticides, fungicides, and weed killers. In what ways are these chemicals useful and important to us? In what ways can they be harmful? Is your general opinion of such chemicals positive or negative? What influences have shaped your feelings about these chemicals?

4 A Tour of the Cell

Two kinds of cells. In this micrograph, *Staphylococcus* bacteria (yellow) cling to hairlike appendages called cilia on human skin cells.

CHAPTER CONTENTS
■■■ Chapter Thread: **BACTERIAL DISEASES AND CELLULAR STRUCTURES**

 BIOLOGY AND SOCIETY: Bacterial Diseases and Cellular Structures

Drugs That Target Bacterial Cells

Antibiotics—drugs that disable or kill infectious bacteria—are a marvel of modern medicine. Most antibiotics are naturally occurring chemicals derived from microorganisms. Penicillin, for example, was first isolated from mold in 1928 and was widely prescribed starting in the 1940s. A revolution in human health rapidly followed: Fatality rates of many diseases (such as bacterial pneumonia and surgical infections) fell drastically, saving millions of lives.

The goal of antibiotic treatment is to kill invading bacteria while minimally harming the human host. But how does the antibiotic zero in on its target among trillions of human cells? Most antibiotics achieve such precision by binding to structures found only in bacterial cells. For example, erythromycin, streptomycin, tetracycline, and chloramphenicol bind to the bacterial ribosome, a cellular structure. The ribosomes of humans and bacteria are sufficiently different that these drugs bind only to bacterial ribosomes, leaving human ribosomes unaffected. Ciprofloxacin, the antibiotic of choice against anthrax-causing bacteria, targets an enzyme that bacteria need to maintain their chromosome structure. Human cells can survive just fine without this enzyme because their chromosomes have a sufficiently different makeup. Penicillin, ampicillin, and bacitracin disrupt synthesis of the cell walls, a structural feature of most bacteria that is absent from the cells of humans and other animals.

The effect of antibiotics underscores the main point of this chapter: To understand how life works—whether in bacteria or your own body—you first need to learn about cells. Recall from Chapter 1 that cells are the smallest entities that can meet all the criteria for life. In this chapter, we'll explore the structure and function of cells. Along the way, we'll consider how cell structures are targeted by infectious bacteria and antibiotics.

Colonized TEM 15,000x

The Microscopic World of Cells

Each cell in the human body is a miniature marvel of great complexity. If a machine with millions of parts—say, a jumbo jet—were reduced to microscopic size, its complexity would not begin to rival that of a living cell.

Organisms are either single-celled, such as most prokaryotes and protists, or multicelled, such as plants, animals, and most fungi. Your own body is a cooperative society of trillions of cells of many specialized types. Three examples are the muscle cells that keep your heart beating, the nerve cells that control your muscles, and the red blood cells that carry oxygen throughout your body. Everything you do—every action and every thought—reflects processes that occur at the cellular level. For exploring this world of cells, our main tools are microscopes.

▼ Figure 4.1 **The protist** *Paramecium* **viewed with three different types of microscopes.** Photographs taken with microscopes are called micrographs. Throughout this textbook, micrographs will have size notations along the side. For example, "LM 1,000×" indicates that the micrograph was taken with a light microscope and the objects are magnified to 1,000 times their original size.

Microscopes as Windows on the World of Cells

Our understanding of nature often parallels the invention and refinement of instruments that extend human senses. For example, before microscopes were first used in the 1600s, no one knew that living organisms were composed of cells.

The type of microscope used by Renaissance scientists is the same kind of microscope you will use if your biology course includes a lab: a **light microscope (LM)**. Visible light is projected through the specimen, such as a single-celled protist (**Figure 4.1a**). Glass lenses enlarge the image and project it into a human eye or a camera.

Two important factors in microscopy are magnification and resolving power. **Magnification** is an increase in the object's apparent size compared with its actual size. The clarity of that magnified image depends on **resolving power**, the ability of an optical instrument to show two objects as separate. For example, what appears to the unaided eye as one star may be resolved as two stars with a telescope. Each optical instrument—be it an eye, a telescope, or a microscope—has a limit to its resolving power. The human eye can resolve points as close together as 0.1 millimeter (mm). (This is about the size of a very fine grain of sand.) For light microscopes, the resolving power is about 0.2 micrometer (μm), the size of a small bacterial cell. This limits the useful magnification to about 1,000×. With greater magnification, the image becomes blurry.

Cells were first described in 1665 by the British scientist Robert Hooke, who used a microscope to examine a thin slice of cork from the bark of an oak tree. For the next two centuries, scientists found cells in every organism examined with a microscope. By the mid-1800s, this accumulation of evidence led to the **cell theory**, which states that all living things are composed of cells and that all cells come from other cells.

Our knowledge of cell structure took a giant leap forward as biologists began using electron microscopes in the 1950s. Instead of using light, an **electron microscope (EM)**

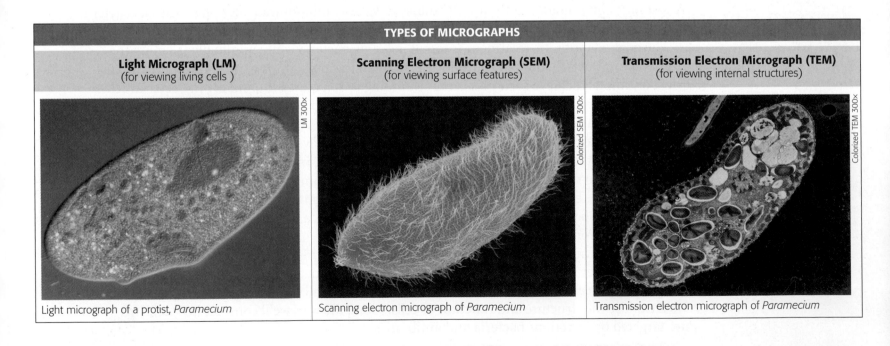

TYPES OF MICROGRAPHS

Light Micrograph (LM) (for viewing living cells)	**Scanning Electron Micrograph (SEM)** (for viewing surface features)	**Transmission Electron Micrograph (TEM)** (for viewing internal structures)

LM 300×

Colorized SEM 300×

Colorized TEM 300×

Light micrograph of a protist, *Paramecium*

Scanning electron micrograph of *Paramecium*

Transmission electron micrograph of *Paramecium*

uses a beam of electrons to resolve objects. It has much better resolving power than the light microscope. **Figures 4.1b** and **4.1c** show images made with two kinds of electron microscopes. Biologists use the **scanning electron microscope (SEM)** to study the detailed architecture of the cell surface. The **transmission electron microscope (TEM)** is especially useful for exploring the internal structure of a cell.

The most powerful modern electron microscopes can distinguish objects as small as 0.2 nanometer (nm), a thousandfold improvement over the light microscope (**Figure 4.2**). The period at the end of this sentence is about a million times bigger than an object 0.2 nm in diameter. The highest-power electron micrographs you will see in this book have magnifications of about 100,000×. Such power can reveal many details within a cell (**Figure 4.3**). However, preparing specimens for electron microscopes usually requires killing and preserving cells before they can be examined. Thus, the light microscope is still very useful as a window on living cells. ☑

▼ **Figure 4.2** **An electron microscope.** An electron micrograph is formed by a beam of electrons that is generated at the top of the microscope's column and travels through the specimen being observed.

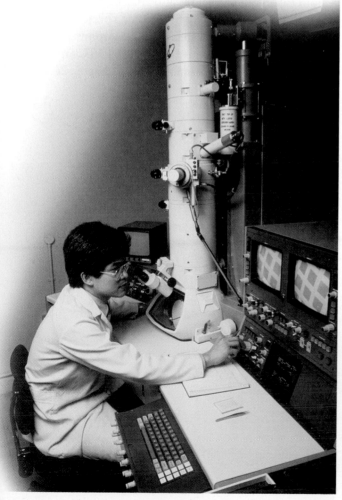

▼ **Figure 4.3** **The size range of cells.** Starting at the top of this scale with 10 m (10 meters) and going down, each reference measurement along the left side marks a tenfold decrease in size.

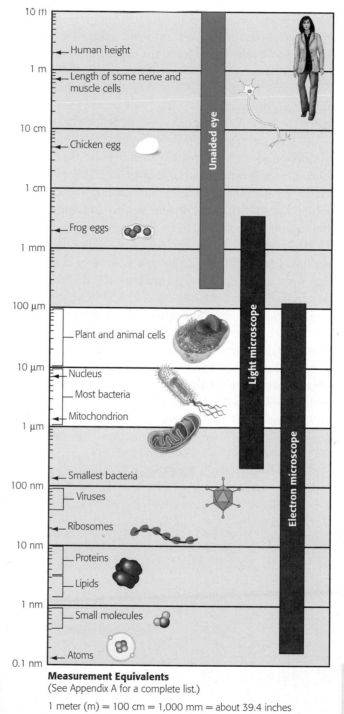

Measurement Equivalents
(See Appendix A for a complete list.)

1 meter (m) = 100 cm = 1,000 mm = about 39.4 inches

1 centimeter (cm) = 10^{-2} ($\frac{1}{100}$) m = about 0.4 inch

1 millimeter (mm) = 10^{-3} ($\frac{1}{1,000}$) m = $\frac{1}{10}$ cm

1 micrometer (μm) = 10^{-6} m = 10^{-3} mm

1 nanometer (nm) = 10^{-9} m = 10^{-3} μm

☑**CHECKPOINT**

Identify which type of microscope you would use to study (a) the internal structure of a dead human liver cell; (b) the finest details of surface texture of a human hair; and (c) the changes in shape of a living human white blood cell.

Answers: (a) transmission electron microscope; (b) scanning electron microscope; (c) light microscope

The Two Major Categories of Cells

The countless cells that exist on Earth fall into two basic categories: prokaryotic cells and eukaryotic cells. **Prokaryotic cells** characterize the organisms of the domains Bacteria and Archaea, known as prokaryotes (see Figures 1.4 and 1.8). Organisms of the domain Eukarya—protists, plants, fungi, and animals—are composed of **eukaryotic cells** and are called eukaryotes.

All cells have several basic features in common. They are all bounded by a thin outer membrane, called a **plasma membrane**, which regulates the traffic of molecules between the cell and its surroundings. All cells have DNA, and all have **ribosomes**, tiny structures that build proteins according to instructions from the DNA.

However, prokaryotic and eukaryotic cells differ in several important ways. Prokaryotes are older in an evolutionary sense: The fossil record indicates that the first prokaryotes appeared on Earth over 3.5 billion years ago, whereas the first eukaryotes did not appear until around 2.1 billion years ago. Prokaryotic cells are usually much smaller—about one-tenth the length of a typical eukaryotic cell—and are simpler in structure (see Figure 1.4). The most significant structural difference is that unlike most prokaryotic cells, eukaryotic cells have **organelles** ("little organs"), membrane-enclosed structures that perform specific functions. The most

important organelle is the **nucleus**, which houses most of a eukaryotic cell's DNA and is surrounded by a double membrane. A prokaryotic cell lacks a nucleus; its DNA is coiled into a "nucleus-like" region called the nucleoid, which is not partitioned from the rest of the cell by membranes.

Consider this analogy: A eukaryotic cell is like an office that is separated into cubicles. Within each cubicle, a specific function is performed, thus dividing the labor among many internal compartments. The cubicle boundaries within eukaryotic cells are made from membranes that help maintain a unique chemical environment inside each cubicle. In contrast, the interior of a prokaryotic cell is like an open warehouse. The spaces for specific tasks are distinct but not separated by barriers.

Figure 4.4 depicts a typical prokaryotic cell. Surrounding the plasma membrane of most prokaryotic cells is a rigid cell wall, which protects the cell and helps maintain its shape. (Recall from the Biology and Society section that bacterial ribosomes and cell walls are the targets of several antibiotics.) In some prokaryotes, another layer, a sticky outer coat called a capsule, surrounds the cell wall. Capsules provide protection and help prokaryotes stick to surfaces. Some prokaryotes have short projections called pili, which can also attach to surfaces. Many prokaryotic cells have flagella, long projections that propel them through their liquid environment.

We'll examine prokaryotes in more detail in Chapter 15. Eukaryotic cells are our main focus in this chapter. ☑

☑CHECKPOINT

How is the nucleoid region of a prokaryotic cell unlike the nucleus of a eukaryotic cell?

Answer: There is no membrane enclosing the prokaryotic nucleoid region.

▼ Figure 4.4 **An idealized prokaryotic cell.**

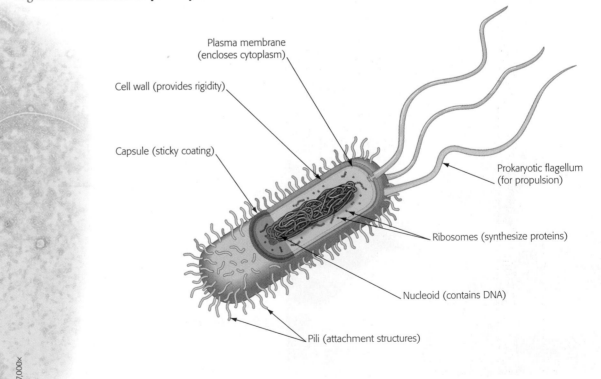

Plasma membrane (encloses cytoplasm)

Cell wall (provides rigidity)

Capsule (sticky coating)

Prokaryotic flagellum (for propulsion)

Ribosomes (synthesize proteins)

Nucleoid (contains DNA)

Pili (attachment structures)

Colorized TEM 27,000x

An Overview of Eukaryotic Cells

All eukaryotic cells—whether from animals, plants, protists, or fungi—are fundamentally similar. **Figure 4.5** provides an overview of idealized animal and plant cells. To keep from getting lost on our tour of the cell, we'll use miniature versions of these diagrams as road maps, highlighting the structure we're interested in. (Notice that the structures are color-coded; we'll use this color scheme throughout *Essential Biology*.)

The entire region of the cell between the nucleus and plasma membrane is called the **cytoplasm**. (This term is also used to refer to the interior of a prokaryotic cell.) The cytoplasm of a eukaryotic cell consists of various organelles suspended in fluid. As you can see in Figure 4.5, most organelles are found in both animal and plant cells. One important difference is the presence of chloroplasts in plant cells but not in animal cells. Chloroplasts are the organelles that convert light energy to the chemical energy of food. Another distinction is that plant cells have a protective cell wall outside the plasma membrane. We'll see other differences and similarities between plant and animal cells as we now take a closer look at the architecture of eukaryotic cells, beginning with the plasma membrane. ☑

Tour of an Animal Cell
Tour of a Plant Cell

▼ **Figure 4.5 A view of an idealized animal cell and plant cell.** For now, the labels on the drawings are just words, but these organelles will come to life as we take a closer look at how each part of the cell functions.

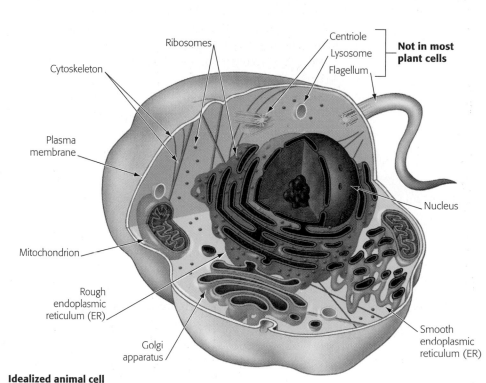

Idealized animal cell

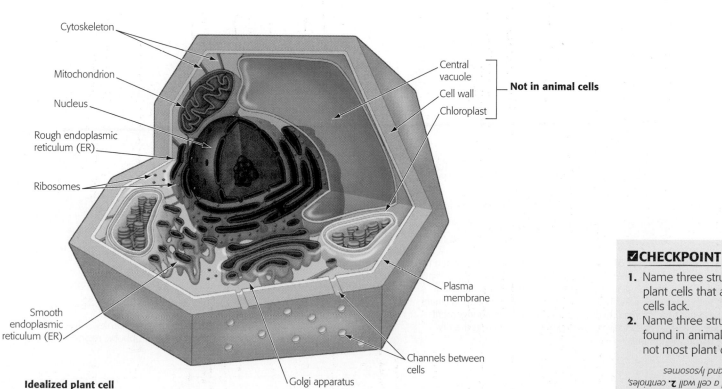

Idealized plant cell

Membrane Structure

Before we enter the cell to explore the organelles, let's make a quick stop at the surface of this microscopic world. The plasma membrane is the edge of life, the boundary that separates the living cell from its nonliving surroundings. It is a remarkable film, so thin that you would have to stack 8,000 of them to equal the thickness of the page you're reading. Yet the plasma membrane can regulate the traffic of chemicals into and out of the cell. The key to how a membrane works is its structure.

The Plasma Membrane: A Fluid Mosaic of Lipids and Proteins

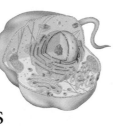

The plasma membrane and other membranes of the cell are composed mostly of lipids and proteins. The lipids belong to a special category called **phospholipids**. They are related to dietary fats but have only two fatty acid tails instead of three (see Figure 3.11b). In place of the third fatty acid, a phospholipid has a phosphate group (a combination of phosphorus and oxygen). The phosphate group is electrically charged, making it hydrophilic ("water-loving"). But the two fatty acid tails are hydrophobic ("water-fearing"). Thus, phospholipids have a chemical ambivalence in their interactions with water. The phosphate group "head" mixes with water, while the fatty acid tails avoid it. This makes phospholipids good membrane material. By forming a two-layered membrane, or **phospholipid bilayer**, the hydrophobic tails of the molecules stay away from water, while the hydrophilic heads remain surrounded by water (Figure 4.6a). Embedded in the phospholipid bilayer of most membranes are proteins that help regulate traffic across the membrane and perform other functions (Figure 4.6b). You'll learn more about membrane proteins in Chapter 5.

Membranes are not static sheets of molecules locked rigidly in place. The phospholipids and most of the proteins are free to drift about in the plane of the membrane. Thus, a membrane is a **fluid mosaic**—fluid because the molecules can move freely past one another and a mosaic because of the diversity of proteins that float like icebergs in the phospholipid sea. Next, in the Process of Science section, we'll see how some bacteria can cause illness by piercing the plasma membrane. ☑

☑**CHECKPOINT**

Why do phospholipids tend to organize into a bilayer in an aqueous solution?

Answer: The bilayer structure shields the hydrophobic tails of the phospholipids from water while exposing the hydrophilic heads to water.

▼ Figure 4.6 **Plasma membrane structure.**

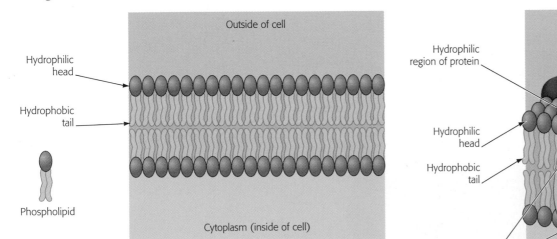

(a) **Phospholipid bilayer of membrane.** At the interface between two aqueous compartments, phospholipids arrange themselves into a bilayer. The symbol for a phospholipid that we'll use in this book looks like a lollipop with two wavy sticks. The "head" is the end with the phosphate group, and the two "tails" are hydrocarbons. The bilayer arrangement keeps the heads exposed to water while keeping the tails in the dry interior of the membrane.

(b) **Fluid mosaic model of membrane.** Membrane proteins, like the phospholipids, have both hydrophilic and hydrophobic regions.

What Makes a Superbug?

Some bacteria cause disease by rupturing the plasma membrane of human immune cells. One example is *Staphylococcus aureus*. These bacteria, commonly found on the skin and in the nose, are usually harmless but occasionally multiply and spread, resulting in a "staph infection." Staph infections typically occur in hospitals and can cause serious, even life-threatening conditions, such as surgical infections, pneumonia, or necrotizing fasciitis ("flesh-eating disease").

Most staph infections can be treated by several common antibiotics, including methicillin. But particularly dangerous strains of *S. aureus*—known as MRSA (for methicillin-resistant *S. aureus*)—are unaffected by these antibiotics. Recently, MRSA infections have become more common and have even turned up in nonhospital settings, causing alarm within the medical community.

In 2007, scientists from the National Institutes of Health (NIH) studied one deadly MRSA strain. They began with the **observation** that other bacteria use a protein called PSM to disable human immune cells by forming pores in the plasma membrane. This led them to **question** whether PSM plays a role in MRSA infection (**Figure 4.7**). Their **hypothesis** was that MRSA bacteria lacking the ability to produce PSM would be less deadly than normal MRSA strains.

In their **experiment**, they infected seven mice with a normal MRSA strain and eight mice with an MRSA strain genetically engineered to not produce PSM. The **results** over the next three days were striking: All seven mice infected with the normal MRSA strain died, while five of the eight mice infected with the strain that did not produce PSM survived. Under the electron microscope, immune cells from all of the dead mice showed pores within the plasma membrane. The researchers reached the conclusion that normal MRSA strains

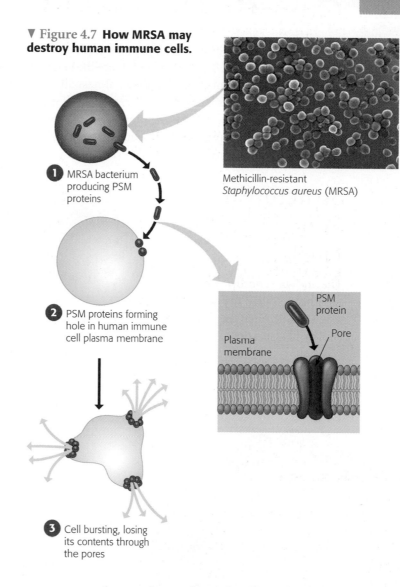

▼ Figure 4.7 **How MRSA may destroy human immune cells.**

Colorized SEM 1,300×

Methicillin-resistant *Staphylococcus aureus* (MRSA)

1 MRSA bacterium producing PSM proteins

2 PSM proteins forming hole in human immune cell plasma membrane

3 Cell bursting, losing its contents through the pores

PSM protein

Pore

Plasma membrane

appear to use the membrane-destroying PSM protein, but other factors must come into play because three mice died even in the absence of PSM. Active study of MRSA bacteria and how they interact with human cellular structures continues today.

Cell Surfaces

Plant cells have a cell wall surrounding the plasma membrane. Unlike prokaryotic cell walls, the rigid cell walls of plants are made from cellulose fibers embedded in a matrix of other molecules (see Figure 3.9c). The walls protect the cells, maintain cell shape, and keep cells from absorbing so much water that they burst. Plant cells are connected via channels that pass through the cell walls, joining the cytoplasm of each cell to that of its neighbors. These channels allow water and other small molecules to move between cells, integrating the activities of a tissue.

Although animal cells lack a cell wall, most of them secrete a sticky coat called the **extracellular matrix**. This layer holds cells together in tissues, and it can also have protective and supportive functions. In addition, the surfaces of most animal cells contain **cell junctions**, structures that connect to other cells. Cell junctions allow cells in a tissue to function in a coordinated way. ☑

☑ CHECKPOINT

What polysaccharide is the primary component of plant cell walls?

Answer: Cellulose

Cell Organelles

The Nucleus and Ribosomes: Genetic Control of the Cell

If we think of the cell as a factory, then the nucleus is its executive boardroom. The top managers are the genes, the inherited DNA molecules that direct almost all the business of the cell. As you read in Chapter 3, each gene is a stretch of DNA that stores the information necessary to produce a particular protein. Proteins then do most of the actual work of the cell.

Structure and Function of the Nucleus

The nucleus is bordered by a double membrane called the **nuclear envelope** (Figure 4.8). Each membrane of the nuclear envelope is similar in structure to the plasma membrane. Pores in the envelope allow certain materials to pass between the nucleus and the cytoplasm. Within the nucleus, long DNA molecules and associated proteins form fibers called **chromatin**. Each long chromatin fiber constitutes one **chromosome** (Figure 4.9). The number of chromosomes in a cell depends on the species; for example, each human body cell has 46 chromosomes, whereas rice cells have 24 and dog cells have 78 (see Figure 8.2 for more examples). The **nucleolus**, a prominent structure within the nucleus, is the site where the components of ribosomes are made. We'll examine ribosomes next. ✔

☑ CHECKPOINT

What is the relationship between chromosomes and chromatin?

Answer: Chromosomes are made of chromatin, which is a combination of DNA and proteins.

▼ Figure 4.8 **The nucleus.**

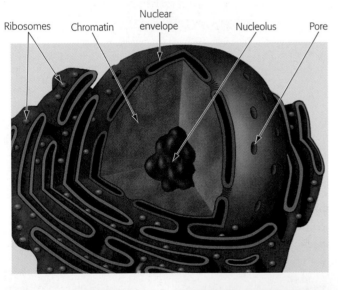

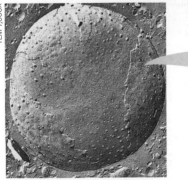

Surface of nuclear envelope

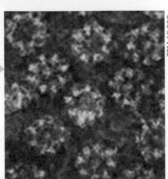

Nuclear pores

▼ Figure 4.9 **The relationship between DNA, chromatin, and a chromosome.**

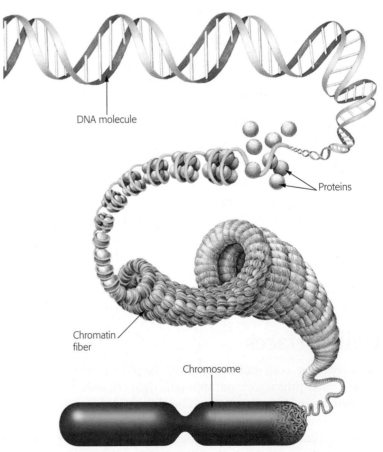

DNA molecule

Proteins

Chromatin fiber

Chromosome

Ribosomes

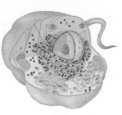

The small dots in the cells in Figure 4.5 and outside the nucleus in Figure 4.8 are the ribosomes. As you'll see in the next section, ribosomes are responsible for protein synthesis **(Figure 4.10)**. In eukaryotic cells, the components of ribosomes are made in the nucleus, then transported through the pores of the nucleus into the cytoplasm. It is in the cytoplasm that the ribosomes begin their work. Some are suspended in the fluid of the cytoplasm, making proteins that remain in the fluid **(Figure 4.11)**. Other ribosomes are attached to the outside of an organelle called the endoplasmic reticulum, making proteins that are incorporated into membranes or secreted by the cell. Recall from the Biology and Society section that there are some important structural differences between bacterial and eukaryotic ribosomes, making them a good target for antibiotic drugs.

How DNA Directs Protein Production

How do the DNA "executives" in the nucleus direct the "workers" in the cytoplasm? **Figure 4.12** shows the sequence of events in a eukaryotic cell (with the DNA and other structures being shown disproportionately large in relation to the nucleus). ❶ DNA programs protein production in the cytoplasm by transferring its coded information to a molecule called messenger RNA (mRNA). Like a middle manager, the RNA molecule then carries the order to "build this type of protein" from the nucleus to the cytoplasm. ❷ The mRNA exits through pores in the nuclear envelope and travels to the cytoplasm, where it then binds to ribosomes. ❸ As a ribosome moves along the mRNA, the genetic message is translated into a protein with a specific amino acid sequence. You'll learn how the message is translated in Chapter 10. ☑

☑**CHECKPOINT**

1. What is the function of ribosomes?
2. What is the role of mRNA in making a protein?

Answers: **1.** *protein synthesis* **2.** *A molecule of mRNA carries the genetic message from a gene (DNA) to ribosomes that translate it into protein.*

▼ Figure 4.10 **Computer model of a ribosome synthesizing a protein.**

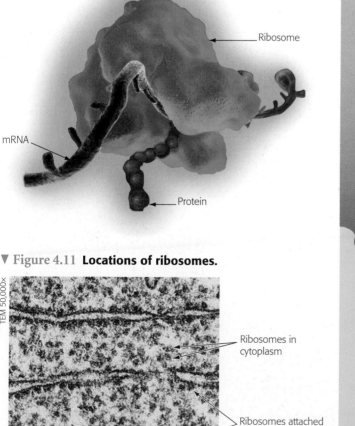

▼ Figure 4.11 **Locations of ribosomes.**

▼ Figure 4.12 **DNA → RNA → Protein.** Inherited genes in the nucleus control protein production and hence the activities of the cell.

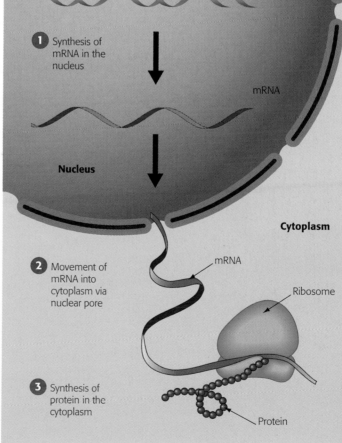

The Endomembrane System: Manufacturing and Distributing Cellular Products

The cytoplasm of a eukaryotic cell is partitioned by organelle membranes (see Figure 4.5). Some of the organelles are connected to each other, either directly by their membranes or by transfer of membrane segments between them. Together, these organelles form the **endomembrane system**. This system includes the nuclear envelope, the endoplasmic reticulum, the Golgi apparatus, lysosomes, and vacuoles.

The Endoplasmic Reticulum

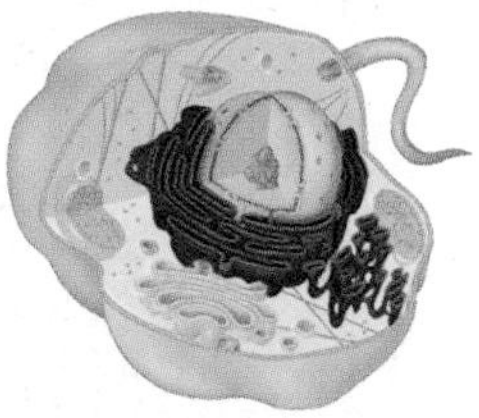

The **endoplasmic reticulum (ER)** is one of the main manufacturing facilities within a cell. It produces an enormous variety of molecules. Connected to the nuclear envelope, the ER forms a labyrinth of tubes and sacs running throughout the cytoplasm (**Figure 4.13**). The ER membrane separates its internal compartment from the surrounding fluid in the cytoplasm. There are two components that make up the ER: rough ER and smooth ER. These two types of ER are physically connected but differ in structure and function.

Rough ER

The "rough" in **rough ER** refers to the appearance of this organelle in electron micrographs (see the bottom left of Figure 4.13). The roughness is due to ribosomes that stud the outside of the ER membrane. These ribosomes produce membrane proteins and secretory proteins. Some newly manufactured membrane proteins are embedded right in the rough ER membrane. Thus, one function of rough ER is to produce new membrane. Secretory proteins are those that are exported (secreted) to the fluid outside the cell. Cells that secrete a lot of protein—such as the cells of your salivary glands, which secrete enzymes into your mouth—are especially rich in rough ER. Some products manufactured by rough ER are dispatched to other locations in the cell by means of **transport vesicles**, membranous spheres that bud from the rough ER (**Figure 4.14**).

▼ Figure 4.13 **Endoplasmic reticulum (ER).** In this drawing (top) and micrograph (bottom), the flattened sacs of rough ER and the tubes of smooth ER are connected. Notice that the ER is also connected to the nuclear envelope.

▼ Figure 4.14 **How rough ER manufactures and packages secretory proteins.**

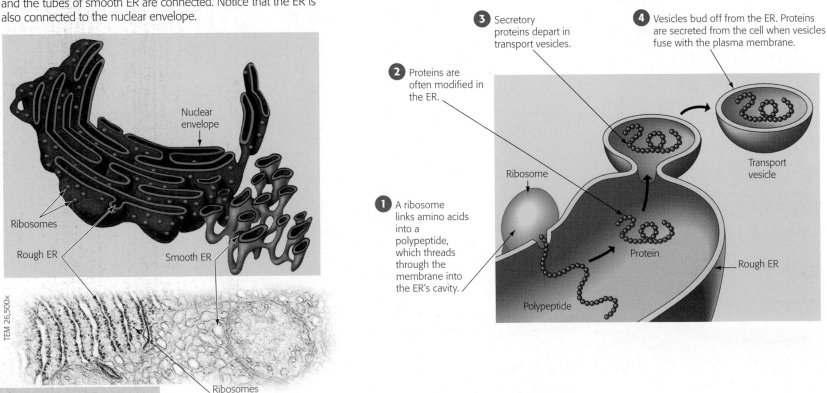

Nuclear envelope

Ribosomes

Rough ER

Smooth ER

TEM 26,500×

Ribosomes

3 Secretory proteins depart in transport vesicles.

4 Vesicles bud off from the ER. Proteins are secreted from the cell when vesicles fuse with the plasma membrane.

2 Proteins are often modified in the ER.

1 A ribosome links amino acids into a polypeptide, which threads through the membrane into the ER's cavity.

Ribosome

Protein

Transport vesicle

Rough ER

Polypeptide

Smooth ER

The "smooth" in **smooth ER** refers to the fact that this organelle lacks the ribosomes that populate the surface of rough ER (see Figure 4.13). A diversity of enzymes built into the smooth ER membrane enables this organelle to perform many functions. One is the synthesis of lipids, including steroids (see Figure 3.13). For example, the cells in ovaries or testes that produce the steroid sex hormones are enriched with smooth ER. In liver cells, enzymes of the smooth ER detoxify circulating sedatives such as barbiturates, stimulants such as amphetamines, and some antibiotics (which is why they don't persist in the bloodstream after combating an infection). As liver cells are exposed to a drug, the amounts of smooth ER and its detoxifying enzymes increase. This can strengthen the body's tolerance of the drug, meaning that higher doses will be required in the future to achieve the desired effect. The growth of smooth ER in response to one drug can also increase tolerance of other drugs. Barbiturate use, for example, may make certain antibiotics less effective by accelerating their breakdown in the liver. Furthermore, increased tolerance of drugs is one of the hallmarks of addiction—a potentially serious consequence of the continued use of certain drugs.

The Golgi Apparatus

The **Golgi apparatus**, an organelle named for its discoverer (Italian scientist Camillo Golgi), is a refinery, warehouse, and shipping center. Working in close partnership with the ER, the Golgi apparatus receives, refines, stores, and distributes chemical products of the cell (Figure 4.15). Products made in the ER reach the Golgi in transport vesicles. One side of a Golgi stack serves as a receiving dock for these vesicles. The shipping side of a Golgi stack is a depot from which finished products can be carried in transport vesicles to other organelles or to the plasma membrane. Vesicles that bind with the plasma membrane transfer proteins to it or secrete finished products to the outside of the cell. Products of the ER are usually modified by enzymes during their transit from the receiving to the shipping side of the Golgi. For example, molecular identification tags, such as phosphate groups, may be added that serve to mark and sort protein molecules into different batches for different destinations. ☑

☑CHECKPOINT

1. What makes rough ER rough?
2. What is the relationship between the Golgi apparatus and the ER in a protein-secreting cell?

Answers: 1. ribosomes attached to the membrane 2. The Golgi receives proteins from the ER via vesicles, finishes processing the proteins, and then dispatches them in vesicles.

▼ Figure 4.15 **The Golgi apparatus.** This component of the endomembrane system consists of flattened sacs arranged something like a stack of pancakes. The number of stacks in a cell (from a few to hundreds) correlates with how active the cell is in secreting proteins.

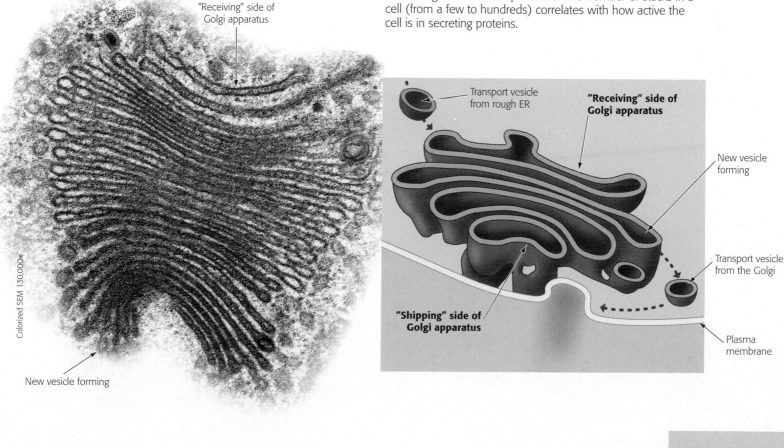

"Receiving" side of Golgi apparatus

Colorized SEM 130,000×

New vesicle forming

Transport vesicle from rough ER

"Receiving" side of Golgi apparatus

New vesicle forming

Transport vesicle from the Golgi

"Shipping" side of Golgi apparatus

Plasma membrane

Lysosomes

A **lysosome** is a sac of digestive enzymes found in animal cells. (Lysosomes are absent from most plant cells.) Lysosomes develop from vesicles that bud off from the Golgi. Enzymes within a lysosome can break down large molecules such as proteins, polysaccharides, fats, and nucleic acids. The lysosome provides a compartment where the cell can digest these molecules safely, without unleashing these digestive enzymes on the cell itself.

Lysosomes have several types of digestive functions. Many cells engulf nutrients into tiny cytoplasmic sacs called **food vacuoles**. Lysosomes fuse with the food vacuoles, exposing the food to enzymes that digest it (**Figure 4.16a**). Small molecules that result from this digestion, such as amino acids, leave the lysosome and nourish the cell. Lysosomes also help destroy harmful bacteria. Our white blood cells ingest bacteria into vacuoles, and lysosomal enzymes that are emptied into these vacuoles rupture the bacterial cell walls. Additionally, lysosomes break down the large molecules of damaged organelles. Without harming the cell, a lysosome can engulf and digest parts of another organelle, making its molecules available for the construction of new organelles (**Figure 4.16b**). Lysosomes also have sculpturing functions in embryonic development. In an early human embryo, lysosomes release enzymes that digest webbing between fingers of the developing hand.

The importance of lysosomes to cell function and human health is made strikingly clear by hereditary disorders called lysosomal storage diseases. A person with such a disease is missing one or more of the digestive enzymes normally found within lysosomes. The abnormal lysosomes become engorged with indigestible substances, and this eventually interferes with other cellular functions. Most of these diseases are fatal in early childhood. In Tay-Sachs disease, lysosomes lack a lipid-digesting enzyme. As a result, nerve cells die as they accumulate excess lipids, ravaging the nervous system. Fortunately, storage diseases are rare. ✓

✓CHECKPOINT

How can defective lysosomes result in excess accumulation of a particular chemical compound in a cell?

Answer: If the lysosomes lack an enzyme needed to break down the compound, the cell will accumulate an excess of that compound.

▼ Figure 4.16 **Two functions of lysosomes.**

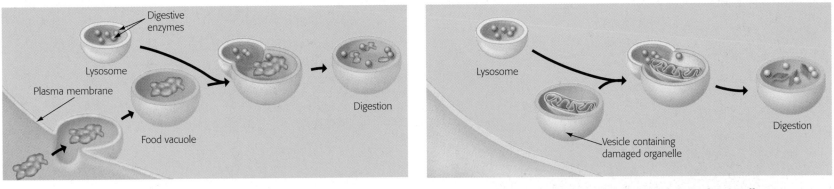

Digestive enzymes

Lysosome

Plasma membrane

Food vacuole

Digestion

(a) Lysosome digesting food

Lysosome

Vesicle containing damaged organelle

Digestion

(b) Lysosome breaking down the molecules of damaged organelles

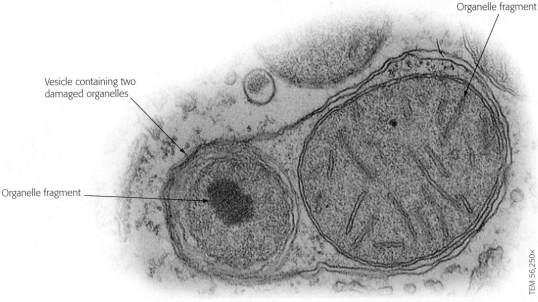

Organelle fragment

Vesicle containing two damaged organelles

Organelle fragment

TEM 56,250×

Vacuoles

Vacuoles are sacs that bud from the ER, Golgi, or plasma membrane. They come in different sizes and have a variety of functions. For example, Figure 4.16a shows a food vacuole budding from the plasma membrane. Certain freshwater protists have contractile vacuoles that pump out excess water that flows into the cell from the outside environment (**Figure 4.17a**).

Another type of vacuole is a **central vacuole**, which can account for more than half the volume of a mature plant cell (**Figure 4.17b**). The central vacuole of a plant cell is a versatile compartment. It stores organic nutrients, such as proteins stockpiled in the vacuoles of seed cells. It also contributes to plant growth by absorbing water and causing cells to expand. In the cells of flower petals, central vacuoles may contain pigments that attract pollinating insects. Central vacuoles may also contain poisons that protect against plant-eating animals.

Figure 4.18 will help you review how organelles of the endomembrane system are related. Note that a product made in one part of the endomembrane system may exit the cell or become part of another organelle without crossing a membrane. Also note that membrane made by the ER can become part of the plasma membrane through the fusion of a transport vesicle. In this way, even the plasma membrane is related to the endomembrane system. ✓

☑CHECKPOINT

Place the following cellular structures in the order they would be used in the production and secretion of a protein: Golgi apparatus, nucleus, plasma membrane, ribosome, transport vesicle.

Answer: nucleus, ribosome, transport vesicle, Golgi apparatus, plasma membrane

▼ **Figure 4.17 Two types of vacuoles.**

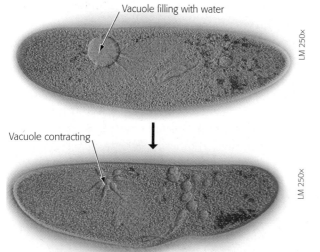

Vacuole filling with water

LM 250x

Vacuole contracting

LM 250x

(a) Contractile vacuole in *Paramecium*. A contractile vacuole fills with water and then contracts to pump the water out of the cell.

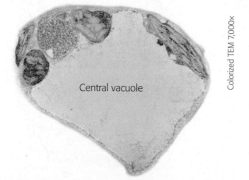

Colorized TEM 7,000x

Central vacuole

(b) Central vacuole in a plant cell. The central vacuole (the lighter-colored area in this colorized micrograph) is often the largest organelle in a mature plant cell.

▼ **Figure 4.18 Review of the endomembrane system.** The arrows show some of the pathways of cell product distribution and membrane migration via transport vesicles.

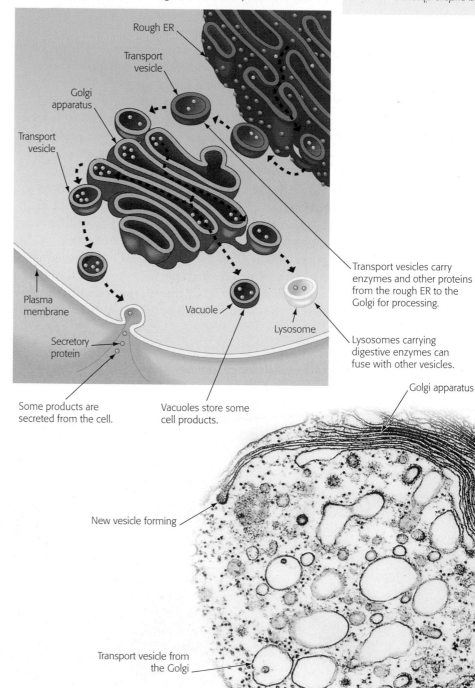

Rough ER

Transport vesicle

Golgi apparatus

Transport vesicle

Plasma membrane

Secretory protein

Some products are secreted from the cell.

Vacuole

Vacuoles store some cell products.

Lysosome

Transport vesicles carry enzymes and other proteins from the rough ER to the Golgi for processing.

Lysosomes carrying digestive enzymes can fuse with other vesicles.

Golgi apparatus

New vesicle forming

Transport vesicle from the Golgi

TEM 31,000x

Chloroplasts and Mitochondria: Energy Conversion

A cell requires a continuous energy supply to perform the work of life. Two organelles act as cellular power stations: chloroplasts and mitochondria.

Chloroplasts

Most of the living world runs on the energy provided by photosynthesis, the conversion of light energy from the sun to the chemical energy of sugar and other organic molecules. **Chloroplasts**, which are unique to the photosynthetic cells of plants and algae, are the organelles that perform photosynthesis.

A chloroplast is partitioned into three major compartments by internal membranes (**Figure 4.19**). One compartment is the space between the two membranes that envelop the chloroplast. The **stroma**, a thick fluid within the chloroplast, is the second compartment. Suspended in that fluid, the interior of a network of membrane-enclosed disks and tubes forms the third compartment. Notice in Figure 4.19 that the disks occur in interconnected stacks called **grana** (singular, *granum*). The grana are a chloroplast's solar power packs, the structures that trap light energy and convert it to chemical energy. In Chapter 7, you will learn the details of photosynthesis as it occurs within a chloroplast.

Mitochondria

Mitochondria (singular, *mitochondrion*) are the sites of cellular respiration, a process that harvests energy from sugars and other food molecules and converts it to another form of chemical energy called ATP. Cells use molecules of ATP as the direct energy source for most of their work. In contrast to chloroplasts, mitochondria are found in almost all eukaryotic cells, including your own. An envelope of two membranes encloses the mitochondrion, which contains a thick fluid called the **matrix** (**Figure 4.20**). The inner membrane of the envelope has numerous infoldings called **cristae**. Many of the enzymes and other molecules that function in cellular respiration are built into the inner membrane. By increasing the surface area of this membrane, the cristae maximize ATP output. In Chapter 6, you'll learn more about how mitochondria convert food energy to ATP energy.

Besides their ability to provide cellular energy, mitochondria and chloroplasts share another feature unique among eukaryotic organelles: They contain DNA that encodes some of their proteins. This DNA is evidence that mitochondria and chloroplasts evolved from free-living prokaryotes in the distant past (see Chapter 15 for a discussion of this hypothesis). ☑

☑ CHECKPOINT

1. What does the process of photosynthesis accomplish?
2. What is cellular respiration?

Answers: 1. the conversion of light energy to chemical energy stored in food molecules 2. a process that converts the chemical energy of sugars and other food molecules to chemical energy in the form of ATP

▼ Figure 4.19 **The chloroplast: site of photosynthesis.**

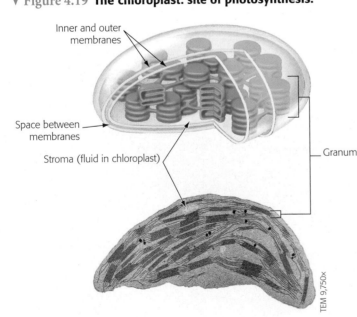

Inner and outer membranes

Space between membranes

Stroma (fluid in chloroplast)

Granum

TEM 9,750×

▼ Figure 4.20 **The mitochondrion: site of cellular respiration.**

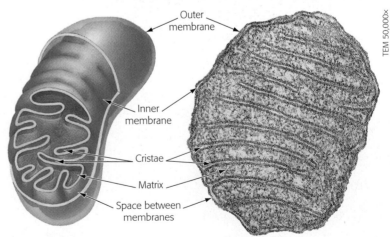

Outer membrane

Inner membrane

Cristae

Matrix

Space between membranes

TEM 50,000×

The Cytoskeleton: Cell Shape and Movement

If someone asked you to describe a house, you would most likely mention the various rooms and their locations. You probably would not think to mention the foundation and beams that support the house. Yet these structures perform an extremely important function. Similarly, cells have an infrastructure called the **cytoskeleton**, a network of fibers extending throughout the cytoplasm. The cytoskeleton serves as both skeleton and "muscles" for the cell, functioning in support and movement.

Maintaining Cell Shape

One function of the cytoskeleton is to give mechanical support to the cell and maintain its shape. This is especially important for animal cells, which lack rigid cell walls. The cytoskeleton contains several types of fibers made from different types of protein. One important type of fiber is **microtubules** (Figure 4.21a). Microtubules are straight, hollow tubes composed of proteins. The other kinds of cytoskeletal fibers, called intermediate filaments and microfilaments, are thinner and solid.

Just as the bony skeleton of your body helps fix the positions of your organs, the cytoskeleton provides anchorage and reinforcement for many organelles in a cell. For instance, the nucleus is often held in place by a cytoskeletal cage of filaments. Other organelles move along tracks made from microtubules. For example, a lysosome might reach a food vacuole by moving along a microtubule. Microtubules also guide the movement of chromosomes when cells divide.

A cell's cytoskeleton is dynamic: It can quickly dismantle in one part of the cell by removing protein subunits and re-form in a new location by reattaching the subunits. Such rearrangement can provide rigidity in a new location, change the shape of the cell, or even cause the whole cell or some of its parts to move. This process contributes to the amoeboid (crawling) movements of the protist *Amoeba* (Figure 4.21b) and some of our white blood cells. ☑

▼ Figure 4.21 **The cytoskeleton.**

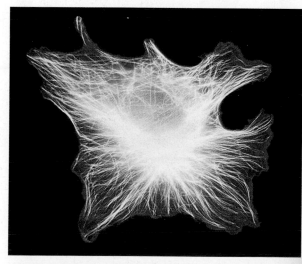

LM 700x

(a) Microtubules in the cytoskeleton. In this micrograph of an animal cell, the cytoskeleton microtubules are labeled with a fluorescent yellow dye.

(b) Microtubules and movement. The crawling movement of an *Amoeba* is due to the rapid degradation and rebuilding of microtubules.

LM 220x

Cilia and Flagella

In some eukaryotic cells, a specialized arrangement of microtubules functions in the beating of flagella and cilia. Cilia and flagella are motile appendages—extensions from a cell that aid in movement. Eukaryotic **flagella** (singular, *flagellum*) propel the cell by an undulating whiplike motion. They often occur singly, such as in the sperm cells of humans and other animals (**Figure 4.22a**). **Cilia** (singular, *cilium*) are generally shorter and more numerous than flagella and promote movement by a coordinated back-and-forth motion, like the rhythmic oars of an eight-person crew team. Both cilia and flagella propel various protists through water (**Figure 4.22b**). Though different in length, number per cell, and beating pattern, cilia and flagella have the same basic architecture, with a core of microtubules wrapped in an extension of the plasma membrane. (Although eukaryotic flagella look much like the prokaryotic flagella shown in Figure 4.4, they have a distinct internal architecture.)

Some cilia extend from nonmoving cells that are part of a tissue layer. There they move fluid over the tissue's surface. For example, the ciliated lining of your windpipe helps cleanse your respiratory system by sweeping mucus with trapped debris out of your lungs (**Figure 4.22c**). Tobacco smoke can inhibit or destroy these cilia, interfering with the normal cleansing mechanisms and allowing more toxin-laden smoke particles to reach the lungs. Frequent coughing—common in heavy smokers—then becomes the body's attempt to cleanse the respiratory system.

Because human sperm rely on flagella for movement, it's easy to understand why problems with flagella can lead to male infertility. Sperm with malfunctioning flagella cannot travel up the female reproductive tract to fertilize an egg. Interestingly, some men with a type of hereditary sterility also suffer from respiratory problems. The explanation lies in the similarities between flagella (found in sperm) and cilia (found lining the respiratory tract). Because of a defect in the structure of their flagella and cilia, their sperm do not swim (causing sterility) and their cilia do not sweep mucus out of their lungs (causing recurrent respiratory infections). ☑

☑CHECKPOINT

Compare and contrast cilia and flagella.

Answer: Cilia and flagella have the same basic structure and help move cells or move fluid over cells. Cilia are short and numerous and move back and forth. Flagella are longer, often occurring singly, and they undulate.

▼ Figure 4.22 **Flagella and cilia.**

Colorized SEM 1,300×

(a) Flagellum of a human sperm cell. A eukaryotic flagellum undulates in a snakelike motion, driving a cell such as this sperm cell through its fluid environment.

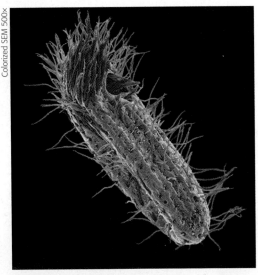

Colorized SEM 500×

(b) Cilia on a protist. Cilia are shorter and more numerous than flagella and move with a back-and-forth motion. As shown here, a dense nap of beating cilia covers *Paramecium*, a freshwater protist that can dart rapidly through its watery home.

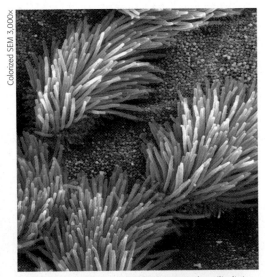

Colorized SEM 3,000×

(c) Cilia lining the respiratory tract. The cilia lining your respiratory tract sweep mucus with trapped debris out of your lungs.

The Evolution of Antibiotic Resistance

As you read in the Biology and Society section, many antibiotics disrupt the cellular structures of invading microorganisms. This mechanism is quite effective. In fact, when first introduced in the 1940s, penicillin appeared to be such a "wonder drug" that some doctors predicted the end of infections in humans (Figure 4.23).

Why hasn't that optimistic prediction come true? The answer is both simple and profound: The prediction does not take into account the force of evolution. Within a population of bacteria, random genetic changes result in slightly different proteins being produced in different bacteria. Some bacteria may produce a protein that reduces the effectiveness of an antibiotic. For example, a random change in a nucleotide sequence might lead to the production of an enzyme that deactivates or breaks down penicillin.

When an environment changes, natural selection favors individuals with beneficial genetic changes. In this case, the introduction of an antibiotic creates a pressure that favors resistant bacteria. Those individuals will survive and multiply, producing more of the same type of bacteria. Over time, drug-resistant bacteria, such as the MRSA strain discussed in the Process of Science section, become more common.

In what ways do we contribute to the problem of antibiotic resistance? Livestock producers add antibiotics to animal feed to promote growth and prevent illness. These practices may favor bacteria that resist standard antibiotics. Some doctors overprescribe antibiotics—for example, to patients with viral infections, which do not respond to such treatment. And patients misuse prescribed antibiotics by prematurely stopping their medication. This misuse allows mutant bacteria that may be killed more slowly by the drug to survive and multiply.

The evolution of antibiotic-resistant bacteria is a serious public health concern. Nearly 100,000 people die each year in the United States from infections they contract in the hospital, often from antibiotic-resistant bacteria. Penicillin, effective against many bacterial infections in the 1940s, is virtually useless today in its original form. New drugs are developed but also continue to be rendered ineffective as resistant bacteria evolve. The medical community is engaged in a race against the powerful force of bacterial evolution.

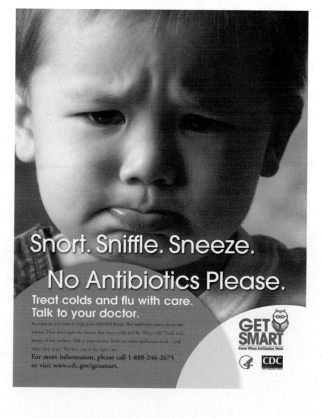

◄ Figure 4.23 **The changing role of antibiotics.** These two posters illustrate how the early promise of antibiotics (depicted in a poster from 1944) has given way to caution about their overuse (depicted in a poster from 2008).

Chapter Review

SUMMARY OF KEY CONCEPTS

Go to the Study Area at **www.masteringbiology.com** for practice quizzes, my**e**Book, BioFlix™ 3-D animations, MP3 Tutor Sessions, videos, current events, and more.

The Microscopic World of Cells

Microscopes as Windows on the World of Cells

Using early microscopes, biologists discovered that all organisms are made of cells. Resolving power limits the useful magnification of microscopes. A light microscope (LM) has useful magnifications up to about 1,000×. Electron microscopes, both scanning (SEM) and transmission (TEM), are much more powerful.

The Two Major Categories of Cells

CATEGORIES OF CELLS	
Prokaryotic Cells	**Eukaryotic Cells**
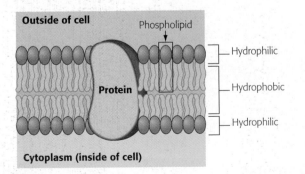	
• Smaller	• Larger
• Simpler	• More complex
• Most do not have organelles	• Have organelles
• Found in bacteria and archaea	• Found in protists, plants, fungi, animals

An Overview of Eukaryotic Cells

Many cellular functions are partitioned by membranes in the complex organization of eukaryotic cells. The largest organelle is usually the nucleus. Other organelles are located in the cytoplasm, the region between the nucleus and the plasma membrane.

Membrane Structure

The Plasma Membrane: A Fluid Mosaic of Lipids and Proteins

Outside of cell

Phospholipid

Protein

Hydrophilic

Hydrophobic

Hydrophilic

Cytoplasm (inside of cell)

Cell Surfaces

Most cells secrete an extracellular coat that helps protect and support the cell. The walls that encase plant cells support plants against the pull of grav-ity and also prevent cells from absorbing too much water. Animal cells are coated by a sticky extracellular matrix.

The Nucleus and Ribosomes: Genetic Control of the Cell

Structure and Function of the Nucleus

An envelope consisting of two membranes encloses the nucleus. Within the nucleus, DNA and proteins make up chromatin fibers; each very long fiber is a single chromosome. The nucleus also contains the nucleolus, which produces components of ribosomes.

Ribosomes

Ribosomes produce proteins in the cytoplasm.

How DNA Directs Protein Production

Genetic messages are transmitted to the ribosomes via messenger RNA, which travels from the nucleus to the cytoplasm.

The Endomembrane System: Manufacturing and Distributing Cellular Products

The Endoplasmic Reticulum

The ER consists of membrane-enclosed tubes and sacs within the cytoplasm. Rough ER, named for the ribosomes attached to its surface, makes membrane and secretory proteins. The functions of smooth ER include lipid synthesis and detoxification.

The Golgi Apparatus

The Golgi refines certain ER products and packages them in transport vesicles targeted for other organelles or export from the cell.

Lysosomes

Lysosomes, sacs containing digestive enzymes, aid digestion and recycling within the cell.

Vacuoles

These organelles include the contractile vacuoles that expel water from certain freshwater protists and the large, multifunctional central vacuoles of plant cells.

Chloroplasts and Mitochondria: Energy Conversion

Chloroplasts and Mitochondria

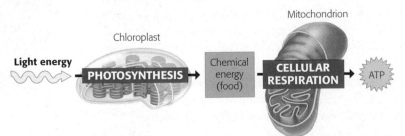

Mitochondrion

Chloroplast

Light energy

PHOTOSYNTHESIS

Chemical energy (food)

CELLULAR RESPIRATION

ATP

The Cytoskeleton: Cell Shape and Movement

Maintaining Cell Shape

Straight, hollow microtubules are an important component of the cytoskeleton, an organelle that gives support to and maintains the shape of cells.

Cilia and Flagella

Cilia and eukaryotic flagella are both motile appendages made primarily of microtubules. Cilia are short and numerous and move by coordinated beating. Flagella are long, often occur singly, and propel a cell through whiplike movements.

SELF-QUIZ

1. If you wanted to film the movement of chromosomes during cell division, the best choice for a microscope would be a
   a. light microscope, because of its magnifying power.
   b. transmission electron microscope, because of its resolving power.
   c. scanning electron microscope, because the chromosomes are on the cell surface.
   d. light microscope, because the specimen must be kept alive.

2. Using a light microscope to examine a thin section of a large spherical cell, you find that the cell is 0.3 mm in diameter. The nucleus is about one-fourth as wide. What would be the diameter of the nucleus in micrometers? (*Hint*: See Figure 4.3.)

3. You look into a light microscope and view an unknown cell. What might you see that would tell you whether the cell is prokaryotic or eukaryotic?
   a. a rigid cell wall
   b. a nucleus
   c. a plasma membrane
   d. ribosomes

4. Explain how each word in the term *fluid mosaic* describes the structure of a membrane.

5. Identify which one of the following structures includes all the others in the list: rough ER, smooth ER, endomembrane system, the Golgi apparatus.

6. The ER has two distinct regions that differ in structure and function. Lipids are synthesized within the _____, and proteins are synthesized within the _____.

7. A type of cell called a lymphocyte makes proteins that are exported from the cell. You can track the path of these proteins within the cell from production through export by labeling them with radioactive isotopes. Identify which of the following structures would be radioactively labeled in your experiment, listing them in the order in which they would be labeled: chloroplasts, Golgi apparatus, plasma membrane, smooth ER, rough ER, nucleus, mitochondria.

8. Name two similarities in the structure or function of chloroplasts and mitochondria. Name two differences.

9. Match the following organelles with their functions:
   a. ribosomes   1. movement
   b. microtubules   2. photosynthesis
   c. mitochondria   3. protein synthesis
   d. chloroplasts   4. digestion
   e. lysosomes   5. cellular respiration

10. DNA controls the cell by transmitting genetic messages that result in protein production. Place the following organelles in the order that represents the flow of genetic information from the DNA through the cell: nuclear pores, ribosomes, nucleus, rough ER, Golgi apparatus.

Answers to the Self-Quiz questions can be found in Appendix D.

THE PROCESS OF SCIENCE

11. The cells of plant seeds store oils in the form of droplets enclosed by membranes. Unlike the membranes you learned about in this chapter, the oil droplet membrane consists of a single layer of phospholipids rather than a bilayer. Draw a model for a membrane around an oil droplet. Explain why this arrangement is more stable than a bilayer.

12. Imagine that you are a pediatrician and one of your patients is a newborn who may have a lysosomal storage disease. You remove some cells from the patient and examine them under the microscope. What would you expect to see? Design a series of tests that could reveal whether the patient is indeed suffering from a lysosomal storage disease.

BIOLOGY AND SOCIETY

13. Doctors at a university medical center removed John Moore's spleen, which is a standard treatment for his type of leukemia. The disease did not recur. Researchers kept the spleen cells alive in a nutrient medium. They found that some cells produced a blood protein that showed promise as a treatment for cancer and AIDS. The researchers patented the cells. Moore sued, claiming a share in profits from any products derived from his cells. The U.S. Supreme Court ruled against Moore, stating that his lawsuit "threatens to destroy the economic incentive to conduct important medical research." Moore argued that the ruling left patients "vulnerable to exploitation at the hands of the state." Do you think Moore was treated fairly? What else would you like to know about this case that might help you decide?

5 The Working Cell

Cellular structures.
These human pancreatic cells have been stained to highlight various cellular structures.

CHAPTER CONTENTS

■■■ Chapter Thread: **MOLECULAR TECHNOLOGIES**

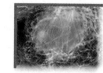

BIOLOGY AND SOCIETY: Molecular Technologies

Natural Nanotechnology

One of the most fascinating applications of modern science is nanotechnology, the manipulation of materials at the molecular scale. When designing devices of such small size, researchers often turn to living cells for inspiration. After all, you can think of a cell as a machine that continuously and efficiently performs a variety of functions. Let's consider one example of cell-based nanotechnology and see how it relates to three of the main concerns in the lives of working cells: energy, enzymes, and the plasma membrane.

All living processes depend on energy. Researchers at Cornell University are attempting to harvest the energy-producing capability of a human sperm cell. Like other cells, a sperm cell generates energy by breaking down sugars and other molecules that pass through its plasma membrane. Enzymes within the cell carry out a process called glycolysis. During glycolysis (discussed in Chapter 6), the energy released from the breakdown of glucose is used to produce molecules of ATP. Within a living sperm, the ATP produced during glycolysis and other processes provides the energy that propels the sperm through the female reproductive tract. In an attempt to harness this energy-producing system, the Cornell researchers attached three glycolysis enzymes to a computer chip. The enzymes continued to function in this artificial system. The hope is that a larger set of enzymes can eventually be used to power microscopic robots. Such nanorobots could use glucose from the bloodstream to power the delivery of drugs to body tissues.

This example of cell-based nanotechnology highlights the three main topics of this chapter. We'll explore how a cell uses energy, enzymes, and the plasma membrane to carry out the work of controlling its internal chemical environment. Along the way, we'll look at some other technologies that mimic the natural activities of living cells.

Some Basic Energy Concepts

Energy makes the world go round—both the cellular world and the larger world outside. But what exactly is energy? Our first step in understanding the working cell is to learn a few basic concepts about energy.

Conservation of Energy

Energy is defined as the capacity to perform work. Work is performed whenever an object is moved against an opposing force. In other words, work moves things in ways they would not move if left alone.

For example, imagine a diver climbing to the top of a platform and diving off (**Figure 5.1**). To get the diver to the top of the platform, work must be performed to overcome the opposing force of gravity. In the diver climbing up the steps to the diving platform, chemical energy from food is being converted to **kinetic energy**, the energy of motion. In this case, the kinetic energy takes the form of muscle movement.

What happens to the kinetic energy when the diver reaches the top of the platform? Has it disappeared?

The answer is no. You may be familiar with the principle of conservation of matter, which states that matter cannot be created or destroyed but can only be converted from one form to another. A similar principle, known as **conservation of energy**, states that it is not possible to destroy or create energy. Like matter, energy can only be converted from one form to another. A power plant, for example, does not make energy; it merely converts it from one form (such as energy stored in coal) to a more convenient form (such as electricity). That's what happens in the diver's climb up the steps. The kinetic energy of muscle movement is now stored in a form called potential energy. **Potential energy** is energy that an object has because of its location or structure, such as the energy contained by water behind a dam or by a compressed spring. In our example, the diver at the top of the platform has potential energy because of his elevated location. The act of diving off the platform into the water converts the potential energy back to kinetic energy. Life depends on the conversion of energy from one form to another. ☑

☑**CHECKPOINT**

How can an object at rest have energy?

Answer: It can have potential energy because of its location or structure.

▼ **Figure 5.1 Energy conversions during a dive.**

On the platform, the diver has more potential energy.

Climbing the steps converts kinetic energy of muscle movement to potential energy.

Diving converts potential energy to kinetic energy.

In the water, the diver has less potential energy.

Entropy

If energy cannot be destroyed, where has it gone when the diver hits the water? It has been converted to **heat**, a type of kinetic energy contained in the random motion of atoms and molecules. The friction between the body and its surroundings generated heat in the air and then in the water.

All energy conversions generate some heat. Although heat production does not destroy energy, it does make it less useful. Heat, of all energy forms, is the most difficult to "tame"—the most difficult to harness for useful work. Heat is energy in its most chaotic form, the energy of aimless molecular movement.

Entropy is a measure of the amount of disorder, or randomness, in a system. Every time energy is converted from one form to another, entropy increases. The energy conversions during the climb up the ladder and the dive from the platform increased entropy because all of the diver's potential energy was lost to the surroundings as heat. To climb up the steps again for another dive, the diver must use additional stored food energy. ✔

Chemical Energy

How can molecules derived from the food we eat provide energy for our working cells? The molecules of food, gasoline, and other fuels have a special form of potential energy called **chemical energy**, which arises from the arrangement of atoms. Carbohydrates, fats, and gasoline have structures that make them especially rich in chemical energy.

Living cells and automobile engines use the same basic process to make the chemical energy stored in their fuels available for work **(Figure 5.2)**. In both cases, this process breaks the organic fuel into smaller waste molecules that have much less chemical energy than the fuel molecules did, thereby releasing energy that can be used to perform work.

For example, the engine of an automobile mixes oxygen with gasoline in an explosive chemical reaction that breaks down the fuel molecules and pushes the pistons that eventually move the wheels. The waste products emitted from the car's exhaust pipe are mostly carbon dioxide and water. About 25% of the energy that an automobile engine extracts from its fuel is converted to

> ✔**CHECKPOINT**
>
> Which form of energy is most randomized and difficult to put to work?
>
> *Answer: heat energy*

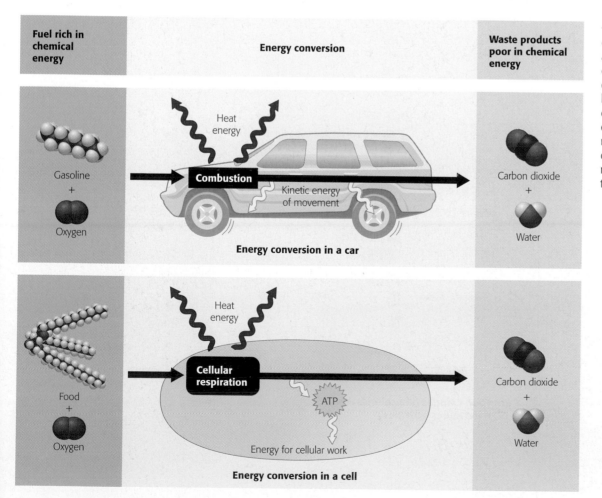

◀ **Figure 5.2 Energy conversions in a car and a cell.** In both a car and a cell, the chemical energy of organic fuel molecules is harvested using oxygen. This chemical breakdown releases energy stored in the fuel molecules and produces carbon dioxide and water. The released energy can be used to perform work.

the kinetic energy of the car's movement. Most of the rest is converted to heat—so much that the engine would melt if the car's radiator and fan did not disperse excess heat into the atmosphere.

Your cells also use oxygen to help harvest chemical energy. As in a car engine, the "exhaust" is mostly carbon dioxide and water. The "combustion" of fuel in cells is called cellular respiration, which is a more gradual and efficient "burning" of fuel compared with the explosive combustion in an automobile engine. Cellular respiration is the energy-releasing chemical breakdown of fuel molecules and the storage of that energy in a form the cell can use to perform work. We will discuss the details of cellular respiration in the next chapter. You convert about 40% of your food energy to useful work, such as the contraction of your muscles. About 60% of the energy released by the breakdown of fuel molecules generates body heat. Humans and many other animals can use this heat to keep the body at an almost constant temperature (37°C, or 98.6°F, in the case of humans), even when the surrounding air is much colder. The liberation of heat energy also explains why you feel hot after exercise. Sweating and other cooling mechanisms enable your body to lose the excess heat, much as a car's radiator keeps the engine from overheating.

Food Calories

Read any packaged food label and you'll find the number of calories in each serving of that food. Calories are units of energy. A **calorie** (cal) is the amount of energy that can raise the temperature of 1 gram (g) of water by 1°C. You could actually measure the caloric content of a peanut by burning it under a container of water to convert all of the stored chemical energy to heat and then measuring the temperature increase of the water.

Calories are tiny units of energy, so using them to describe the fuel content of foods is not practical. Instead, it's conventional to use kilocalories (kcal), units of 1,000 calories. In fact, the Calories (capital C) on a food package are actually kilocalories. That's a lot of energy. For example, one peanut has about 5 kcal. That's enough energy to increase the temperature of 1 kg (a little more than a quart) of water by 5°C in our peanut-burning experiment. And just a handful of peanuts contains enough Calories, if converted to heat, to boil 1 kg of water. In living organisms, of course, food isn't used to boil water but instead to fuel the activities of life. **Figure 5.3** shows the number of Calories in several foods and how many Calories are burned off by some typical activities. ☑

▼ Figure 5.3 **Some caloric accounting.**

Food	Food Calories
Cheeseburger	295
Spaghetti with sauce (1 cup)	241
Baked potato (plain, with skin)	220
Fried chicken (drumstick)	193
Bean burrito	189
Pizza with pepperoni (1 slice)	181
Peanuts (1 ounce)	166
Apple	81
Garden salad (2 cups)	56
Popcorn (plain, 1 cup)	31
Broccoli (1 cup)	25

(a) Food Calories (kilocalories) in various foods

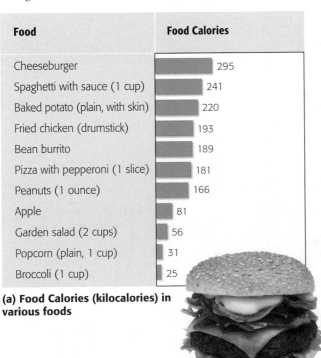

Activity	Food Calories consumed per hour by a 150-pound person*
Running (7 min/mi)	979
Dancing (fast)	510
Bicycling (10 mph)	490
Swimming (2 mph)	408
Walking (3 mph)	245
Dancing (slow)	204
Playing the piano	73
Driving a car	61
Sitting (writing)	28

*Not including energy necessary for basic functions, such as breathing and heartbeat

(b) Food Calories (kilocalories) we burn in various activities

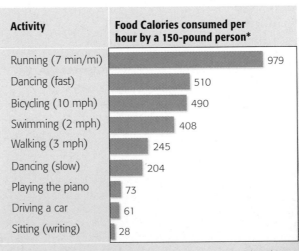

ATP and Cellular Work

The carbohydrates, fats, and other fuel molecules we obtain from food do not drive the work in our cells directly. Instead, the chemical energy released by the breakdown of organic molecules during cellular respiration is used to generate molecules of ATP. These molecules of ATP then power cellular work. ATP acts like an energy shuttle, storing energy obtained from food and then releasing it as needed at a later time.

The Structure of ATP

The abbreviation ATP stands for adenosine triphosphate. **ATP** consists of an organic molecule called adenosine plus a tail of three phosphate groups ((P)) (**Figure 5.4**). The triphosphate tail is the "business" end of ATP, the part that provides energy for cellular work. Each phosphate group is negatively charged. Negative charges repel each other. The crowding of negative charges in the triphosphate tail contributes to the potential energy of ATP. It's analogous to storing energy by compressing a spring; if you release the spring, it will "relax," and you can use that springiness to do some useful work. For ATP power, it is release of the phosphate at the tip of the triphosphate tail that makes energy available to working cells. What remains is **ADP,** adenosine diphosphate (two phosphate groups instead of three; see Figure 5.4).

Phosphate Transfer

When ATP drives work in cells, phosphate groups don't just fly off into space. ATP energizes other molecules in cells by transferring phosphate groups to those molecules. This transfer of phosphate groups helps cells perform three main kinds of work: mechanical work, transport work, and chemical work.

Imagine a bicyclist pedaling up a hill. In the muscle cells of the rider's legs, ATP transfers phosphate groups to motor proteins. The proteins then change shape, causing the muscle cells to contract and perform mechanical work (**Figure 5.5a**). ATP also enables the transport of ions and other solutes across the membranes of the rider's brain cells (**Figure 5.5b**). The transport of ions prepares the brain cells to transmit signals to muscles and other tissues. And ATP drives the chemical work of making some of a cell's large molecules (**Figure 5.5c**). Notice again in Figure 5.5 that all these types of work occur when target molecules accept a phosphate group from ATP.

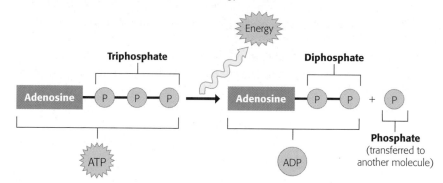

▼ **Figure 5.4 ATP power.** Each (P) in the triphosphate tail of ATP represents a phosphate group, a phosphorus atom bonded to oxygen atoms. The transfer of a phosphate from the triphosphate tail to other molecules provides energy for cellular work.

▼ **Figure 5.5 How ATP drives cellular work.** Each type of work shown here is powered when enzymes transfer phosphate from ATP to a recipient molecule.

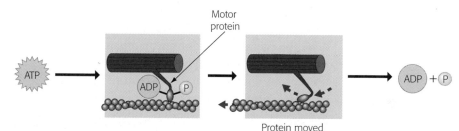

(a) Motor protein performing mechanical work

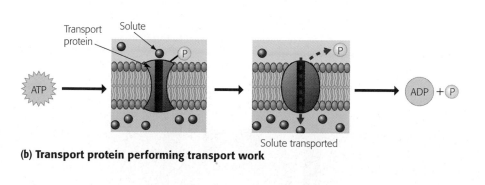

(b) Transport protein performing transport work

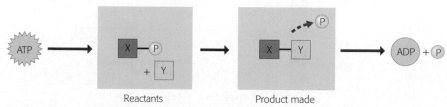

(c) Chemical reactants performing chemical work

The ATP Cycle

Your cells spend ATP continuously. Fortunately, it is a renewable resource. ATP can be restored by adding a phosphate group back to ADP. That takes energy, like recompressing a spring. And that's where food reenters the story. The chemical energy that cellular respiration harvests from sugars and other organic fuels is put to work regenerating a cell's supply of ATP. Cellular work spends ATP, which is recycled when ADP and phosphate are combined using energy released by cellular respiration (**Figure 5.6**). Thus, energy from processes that yield energy, such as the breakdown of organic fuels, is transferred to processes that consume energy, such as muscle contraction and other cellular work. The third phosphate group acts as an energy shuttle within the ATP cycle.

▼ Figure 5.6 **The ATP cycle.**

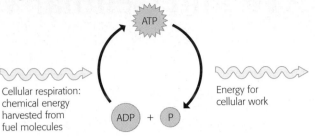

Cellular respiration: chemical energy harvested from fuel molecules

Energy for cellular work

The ATP cycle runs at an astonishing pace. A working muscle cell recycles all of its ATP about once every minute. That's about 10 million ATP molecules spent and regenerated per second per cell. ☑

Enzymes

As you've seen, a living organism contains a vast collection of chemicals, with countless chemical reactions constantly changing the organism's molecular makeup. In a sense, a living organism is a complex "chemical square dance," with the molecular "dancers" continually changing partners via chemical reactions. The total of all the chemical reactions in an organism is called **metabolism**. Interestingly, almost no metabolic reactions occur without help. Most require the assistance of **enzymes**, proteins that speed up chemical reactions. All living cells contain thousands of different enzymes, each promoting a different chemical reaction.

Activation Energy

For a chemical reaction to begin, chemical bonds in the reactant molecules must be broken. (The first step in swapping partners during a square dance is to let go of your current partner's hand.) This process requires that the molecules absorb energy from their surroundings. This energy is called **activation energy** because it activates the reactants and triggers the chemical reaction.

Enzymes enable metabolism to occur by reducing the amount of activation energy required to break the bonds of reactant molecules. If you think of the requirement for activation energy as a barrier to a chemical reaction, an enzyme's function is to lower that barrier (**Figure 5.7**). It does so by binding to reactant molecules and putting them under physical or chemical stress, making it easier to break their bonds and start a reaction. ☑

▼ Figure 5.7 **Enzymes and activation energy.**

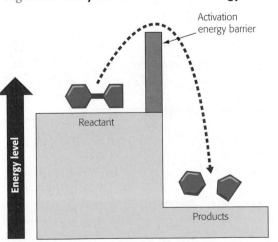

(a) Without enzyme. A reactant molecule must overcome the activation energy barrier before a chemical reaction can break the molecule into products.

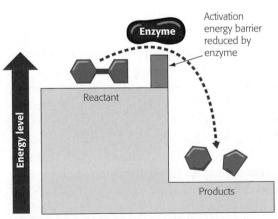

(b) With enzyme. An enzyme speeds the process by lowering the activation energy barrier.

Can Enzymes Be Engineered?

Like all other proteins, enzymes are encoded by genes. **Observations** of genetic sequences suggest that many of our genes were formed through a type of molecular evolution: One ancestral gene randomly duplicated, and the two copies diverged over time via genetic mutation, eventually becoming two distinct genes coding for two enzymes with different functions.

The natural evolution of enzymes raises a **question**: Can laboratory methods mimic this process through artificial selection? In 1997, a group of researchers at two California biotechnology companies formed the **hypothesis** that an artificial process could be used to modify the gene that codes for the enzyme lactase into a new gene coding for a new enzyme with a new function. Recall from Chapter 3 that lactase breaks down the sugar

lactose. Their **experiment** used a procedure called directed evolution. In this process, many copies of the gene for the starting lactase enzyme were mutated at random **(Figure 5.8)**. The researchers tested the enzymes resulting from these mutated genes to determine which enzymes best displayed a new activity (in this case, breaking down a different but closely related sugar). The genes for the enzymes that did show the new activity were then subjected to several more rounds of duplication, mutation, and screening.

After seven rounds, the **results** were clear: Directed evolution had produced a new enzyme with a novel function. Researchers have used similar methods to produce many artificial enzymes with desired properties. Directed evolution is another example of how scientists can mimic the natural processes of cells for useful purposes.

▼ Figure 5.8 **Directed evolution of an enzyme.** During seven rounds of directed evolution, the lactase enzyme gradually gained a new function.

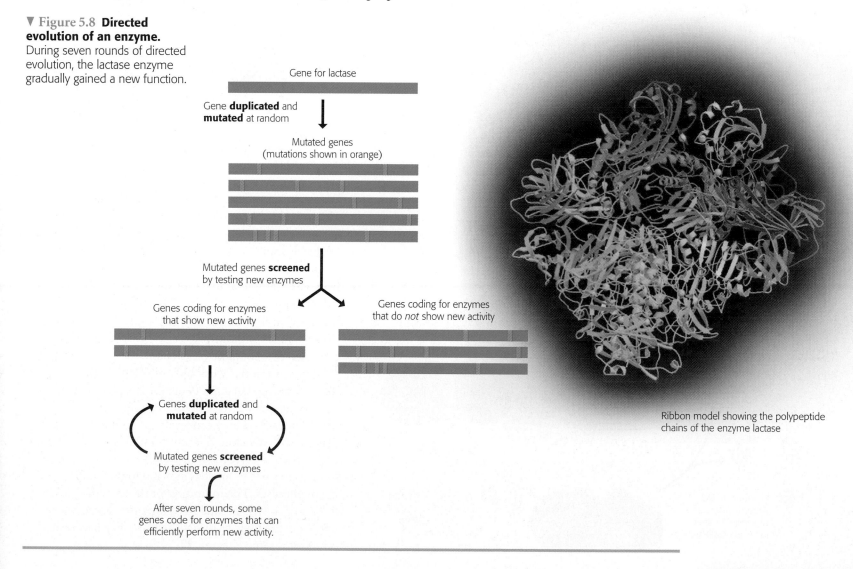

Gene for lactase

Gene **duplicated** and **mutated** at random

Mutated genes (mutations shown in orange)

Mutated genes **screened** by testing new enzymes

Genes coding for enzymes that show new activity

Genes coding for enzymes that do *not* show new activity

Genes **duplicated** and **mutated** at random

Mutated genes **screened** by testing new enzymes

After seven rounds, some genes code for enzymes that can efficiently perform new activity.

Ribbon model showing the polypeptide chains of the enzyme lactase

Induced Fit

An enzyme is very selective in the reaction it catalyzes. This specificity is based on the enzyme's ability to recognize a certain reactant molecule, which is called the enzyme's **substrate**. And the ability of the enzyme to recognize and bind to its specific substrate depends on the enzyme's shape. A region of the enzyme called the **active site** has a shape and chemistry that fit the substrate molecule. When a substrate slips into this docking station, the active site changes shape slightly to embrace the substrate and catalyze the reaction. This interaction is called **induced fit** because the entry of the substrate induces the enzyme to change shape slightly, making the fit between substrate and active site snugger. Think of a handshake: As your hand makes contact with another hand, it changes shape slightly to make a better fit.

After the products are released from the active site, the enzyme can accept another molecule of its substrate. In fact, the ability to function repeatedly is a key characteristic of enzymes. **Figure 5.9** follows the action of the enzyme sucrase, which hydrolyzes the disaccharide sucrose (the substrate). Like sucrase, many enzymes are named for their substrates, but with an *-ase* ending.

Enzyme Inhibitors

Certain molecules can inhibit a metabolic reaction by binding to an enzyme and disrupting its function (**Figure 5.10**). Some of these **enzyme inhibitors** are actually substrate imposters that plug up the active site (see Figure 5.10b). (You can't shake a person's hand if someone else puts a banana in it first!) Other inhibitors bind to the enzyme at a site remote from the active site, but the binding changes the enzyme's shape so that

▼ **Figure 5.9 How an enzyme works.** Our example is the enzyme sucrase, named for its substrate, sucrose.

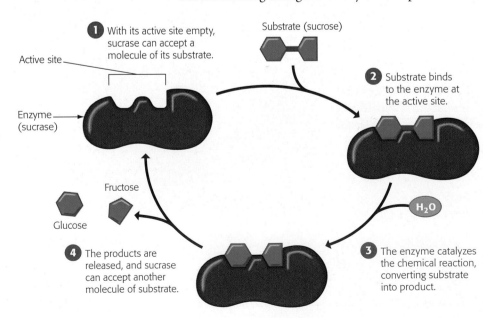

1 With its active site empty, sucrase can accept a molecule of its substrate.

Active site

Enzyme (sucrase)

Substrate (sucrose)

2 Substrate binds to the enzyme at the active site.

Fructose

Glucose

4 The products are released, and sucrase can accept another molecule of substrate.

H_2O

3 The enzyme catalyzes the chemical reaction, converting substrate into product.

▼ **Figure 5.10 Enzyme inhibitors.**

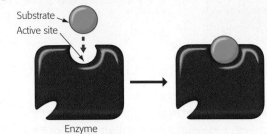

Substrate

Active site

Enzyme

(a) Enzyme and substrate binding normally

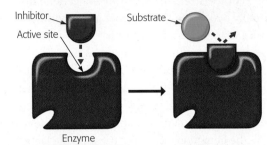

Inhibitor

Substrate

Active site

Enzyme

(b) Enzyme inhibition by a substrate imposter

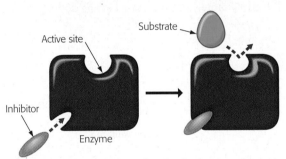

Substrate

Active site

Inhibitor

Enzyme

(c) Enzyme inhibition by a molecule that causes the active site to change shape

the active site no longer accepts the substrate (see Figure 5.10c). (Imagine being unable to shake hands because someone else is tickling your ribs, causing you to clench your hand.) In some cases, the binding is reversible, enabling certain inhibitors to regulate metabolism. For example, if metabolism is producing more of a certain product than a cell needs, that product may reversibly inhibit an enzyme required for its production. This **feedback regulation** keeps the cell from wasting resources that could be put to better use.

Many antibiotics work by inhibiting enzymes of disease-causing bacteria. Penicillin, for example, inhibits an enzyme that bacteria use in making their cell walls (see Chapter 4). Many cancer drugs inhibit enzymes that promote cell division. In addition to drugs, many toxins and poisons are irreversible inhibitors. Nerve gases bind to the active site of an enzyme vital to transmitting nerve impulses. The inhibition of this enzyme leads to rapid paralysis of vital functions and death. Many pesticides are toxic to insects because they irreversibly inhibit this same enzyme. ☑

Membrane Function

So far, we have discussed how cells control the flow of energy and the pace of chemical reactions. They must also regulate the flow of materials to and from the environment. In Chapter 4, you learned about the structure of the plasma membrane (see Figure 4.6). **Figure 5.11** describes the major functions of the proteins embedded within the phospholipid bilayer. In this section, we'll focus mainly on how the plasma membrane regulates the passage of materials into and out of the cell. Transport proteins, such as the one shown in Figure 5.11, are critical to this task. **Transport proteins** are membrane proteins that help move substances across a cell membrane. In this section, you'll learn about the most important mechanisms of transport across membranes.

Passive Transport: Diffusion across Membranes

Molecules are restless. The heat energy they contain makes them vibrate and wander randomly. One result of this motion is **diffusion**, the movement of molecules of any substance so that they spread out into the available space. Each molecule moves randomly, and yet the overall diffusion of a population of molecules is usually directional, from a region where they are more concentrated to where they are less concentrated. For example, imagine many molecules of perfume inside a bottle. If you remove the bottle top, every molecule of perfume

BioFlix ™
Membrane Transport

▼ Figure 5.11 **Primary functions of membrane proteins.** In this diagram of two adjacent membranes, all six types of membrane proteins are shown for convenience. An actual cell may have just a few of these types of proteins.

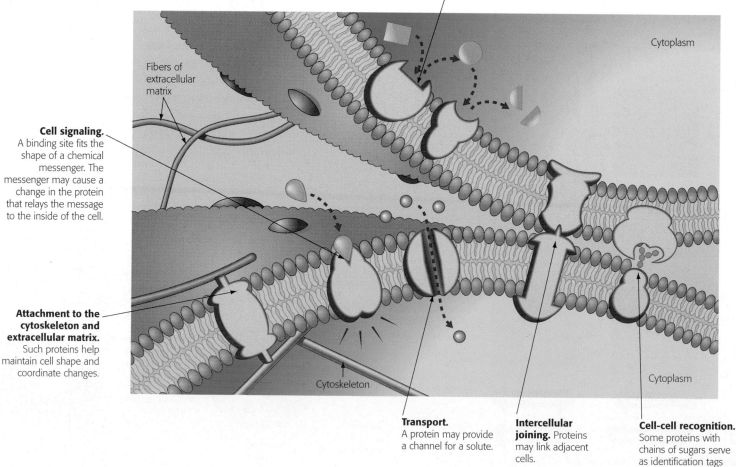

Enzymatic activity. This protein and the one next to it are enzymes, having an active site that fits a substrate. Enzymes may form a team that carries out steps of a pathway.

Cytoplasm

Fibers of extracellular matrix

Cell signaling. A binding site fits the shape of a chemical messenger. The messenger may cause a change in the protein that relays the message to the inside of the cell.

Attachment to the cytoskeleton and extracellular matrix. Such proteins help maintain cell shape and coordinate changes.

Cytoskeleton

Cytoplasm

Transport. A protein may provide a channel for a solute.

Intercellular joining. Proteins may link adjacent cells.

Cell-cell recognition. Some proteins with chains of sugars serve as identification tags recognized by other cells.

will move randomly about, but the net direction of the molecules will be out of the bottle to fill the room. You could, with great effort, return the perfume to its bottle, but the molecues would never return spontaneously.

For an example closer to a living cell, imagine a membrane separating pure water from a solution of a dye dissolved in water (**Figure 5.12**). Assume that this membrane is permeable to the dye molecules—meaning that they can pass through the membrane. Although each dye molecule moves randomly, there will be a net migration across the membrane to the side that began as pure water. The spreading of the dye across the membrane will continue until both solutions have equal concentrations of the dye. When that point is reached, there will be a dynamic equilibrium, with as many dye molecules moving per second across the membrane in one direction as the other.

Diffusion across a membrane is an example of **passive transport**—passive because the cell does not expend any energy for it to happen. However, the cell membrane does play a regulatory role by being selectively permeable. For example, small molecules such as carbon dioxide (CO_2) and oxygen (O_2) generally pass

through more readily than larger molecules such as amino acids. But the membrane is relatively impermeable to even some very small substances, such as hydrogen ions (H^+) and other inorganic ions, which are too hydrophilic to pass through the phospholipid bilayer. Passive transport is extremely important to all cells. In passive transport, a substance diffuses down its **concentration gradient**, a region in which the substance's density changes. That is, the substance moves from where it is more concentrated to where it is less concentrated. In our lungs, for example, passive transport along concentration gradients is the sole means by which O_2, essential for metabolism, enters the blood and CO_2, a metabolic waste, passes out of it.

Substances that do not cross membranes spontaneously can be transported via **facilitated diffusion** by specific transport proteins that act as selective corridors (see Figure 5.11). Without the protein, the substance does not cross the membrane or diffuses across it too slowly to be useful to the cell. Facilitated diffusion is a type of passive transport because it does not require energy. As in all passive transport, the driving force is the concentration gradient. ☑

CHECKPOINT

How is facilitated diffusion a form of passive transport?

Answer: It uses proteins to transport materials down a concentration gradient without expending energy.

▼ **Figure 5.12 Passive transport: diffusion across a membrane.** A substance will diffuse from where it is more concentrated to where it is less concentrated. Put another way, a substance tends to diffuse down its concentration gradient.

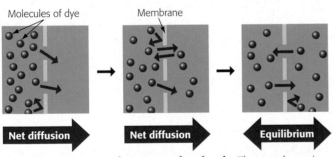

Molecules of dye   Membrane

Net diffusion   Net diffusion   Equilibrium

(a) Passive transport of one type of molecule. The membrane is permeable to these dye molecules, which diffuse down their concentration gradient. At equilibrium, the molecules are still restless, but the rate of transport is equal in both directions.

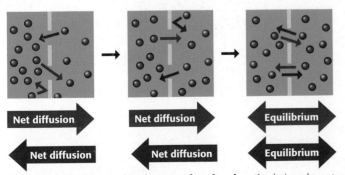

Net diffusion   Net diffusion   Equilibrium

Net diffusion   Net diffusion   Equilibrium

(b) Passive transport of two types of molecules. If solutions have two or more solutes, each will diffuse down its own concentration gradient.

Osmosis and Water Balance

The diffusion of water across a selectively permeable membrane is called **osmosis** (**Figure 5.13**). Consider the case of a membrane separating two solutions with different concentrations of a solute—say, the sugar sucrose. The membrane is permeable to water but not to the solute. The solution with a higher concentration of solute is said to be **hypertonic** to the other solution. The solution with the lower solute concentration is said

▼ **Figure 5.13 Osmosis.** A membrane separates two solutions with different sugar concentrations. Water molecules can pass through the membrane, but sugar molecules cannot.

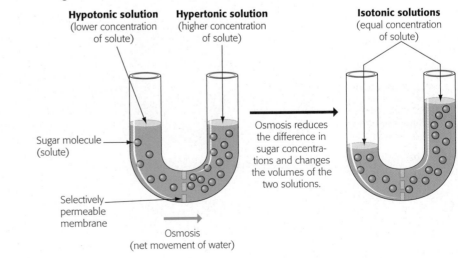

Hypotonic solution (lower concentration of solute)   **Hypertonic solution** (higher concentration of solute)   **Isotonic solutions** (equal concentration of solute)

Sugar molecule (solute)

Selectively permeable membrane

Osmosis (net movement of water)

Osmosis reduces the difference in sugar concentrations and changes the volumes of the two solutions.

to be **hypotonic** to the other. Note that the hypotonic solution, by having the lower solute concentration, has the higher water concentration. Therefore, water will diffuse across the membrane along its concentration gradient from an area of higher water concentration (hypotonic solution) to one of lower water concentration (hypertonic solution). This reduces the difference in solute concentrations and changes the volumes of the two solutions.

People can take advantage of osmosis to preserve foods. Salt is often applied to meats to cure them; the salt causes water to move out of the food toward the region of greater solute (salt) concentration. Food can also be preserved in honey because a high sugar concentration draws water out of food.

When the solute concentrations are the same on both sides of a membrane, water molecules will move at the same rate in both directions, so there will be no net change in solute concentration. Solutions of equal solute concentration are said to be **isotonic** to each other. You may have seen this term on bottles of contact lens saline solution, which is formulated to have the same solute concentration as the fluid at the surface of the human eye, making it nonirritating.

Water Balance in Animal Cells

The survival of a cell depends on its ability to balance water uptake and loss. When an animal cell, such as a red blood cell, is immersed in an isotonic solution, the cell's volume remains constant because the cell gains water at the same rate that it loses water (**Figure 5.14a**, top). In this case, the cell is isotonic to its surroundings because the two solutions have the same total concentration of solutes. Many marine animals, such as sea stars and crabs, are isotonic to seawater. But what happens if an animal cell is in contact with a hypotonic solution, which has a lower solute concentration than the cell? Due to osmosis, the cell would gain water, swell, and possibly burst (lyse) like an overfilled water balloon (**Figure 5.14b**, top). A hypertonic environment is also harsh on an animal cell; the cell shrivels and can die from water loss (**Figure 5.14c**, top).

For an animal to survive a hypotonic or hypertonic environment, the animal must have a way to balance an excessive uptake or excessive loss of water. The control of water balance is called **osmoregulation**. For example, a freshwater fish, whose environment is hypotonic to its body, has kidneys and gills that work constantly to prevent an excessive buildup of water in the body. And if you look back to Figure 4.17a, you'll see *Paramecium*'s contractile vacuole, which bails out the excess water that continuously enters the cell from the hypotonic pond water.

▼ Figure 5.14 **The behavior of animal and plant cells in different osmotic environments.**

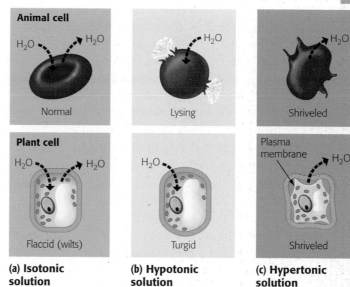

(a) **Isotonic solution**   (b) **Hypotonic solution**   (c) **Hypertonic solution**

Water Balance in Plant Cells

Problems of water balance are somewhat different for plant cells because of their rigid cell walls. A plant cell immersed in an isotonic solution is flaccid (floppy), and a plant wilts in this situation (Figure 5.14a, bottom). In contrast, a plant cell is turgid (firm) and healthiest in a hypotonic environment, with a net inflow of water (Figure 5.14b, bottom). Although the elastic cell wall expands a bit, the back pressure it exerts prevents the cell from taking in too much water and bursting, as an animal cell would in this environment. Turgor is necessary for plants to retain their upright posture and the extended state of their leaves (**Figure 5.15**). However, in a hypertonic environment, a plant cell is no better off than an animal cell. As a plant cell loses water, it shrivels, and its plasma membrane pulls away from the cell wall (Figure 5.14c, bottom). This process, called **plasmolysis**, usually kills the cell. ✔

▼ Figure 5.15 **Plant turgor.** A wilted plant regains its turgor when watered.

☑CHECKPOINT

1. An animal cell shrivels when it is _____ compared with its environment.
2. The cells of a wilted plant are _____ compared with their environment.

Answers: 1. hypertonic 2. isotonic

Active Transport: The Pumping of Molecules across Membranes

In contrast to passive transport, **active transport** requires that a cell expend energy to move molecules across a membrane. In active transport, cellular energy is used to drive a transport protein that actively pumps a solute across a membrane *against* the solute's concentration gradient—that is, away from the side where it is less concentrated and toward the side where it is more concentrated (**Figure 5.16**). Membrane proteins usually use ATP as their energy source for active transport.

Active transport enables cells to maintain internal concentrations of small solutes that differ from environmental concentrations. For example, compared with its surroundings, an animal nerve cell has a much higher concentration of potassium ions and a much lower concentration of sodium ions. The plasma membrane helps maintain these differences by pumping sodium out of the cell and potassium into the cell. This particular case of active transport (called the sodium-potassium pump) is vital in the propagation of nerve signals. ☑

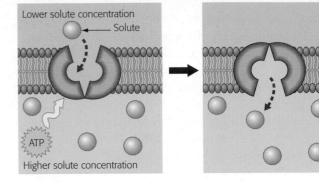

▲ **Figure 5.16 Active transport.** Like enzymes, transport proteins are specific in their recognition of atoms or molecules. This transport protein (purple) has a binding site that accepts only a certain solute. Using energy from ATP, the protein pumps the solute against its concentration gradient.

☑**CHECKPOINT**

What molecule is the usual energy source for active transport?

Answer: ATP

Exocytosis and Endocytosis: Traffic of Large Molecules

So far, we've focused on how water and small solutes enter and leave cells by moving through the plasma membrane. The story is different for macromolecules such as proteins, which are much too big to fit through the membrane itself. Their traffic into and out of the cell depends on the ability of the membrane to form sacs, thereby packaging larger molecules into vesicles. You have already seen an example in the packaging and secretion of proteins. During protein production by the cell, secretory proteins exit the cell from transport vesicles that fuse with the plasma membrane, spilling the contents outside the cell (see Figures 4.14 and 4.18). That process is called **exocytosis** (**Figure 5.17**). When you cry, for example, cells in your tear glands use exocytosis to export the salty tears. The reverse process, **endocytosis**, takes material into the cell within vesicles that bud inward from the plasma membrane (**Figure 5.18**).

There are three types of endocytosis. In **phagocytosis** ("cellular eating"), a cell engulfs a particle and packages it within a food vacuole. In **pinocytosis** ("cellular drinking"), the cell "gulps" droplets of fluid by forming vesicles. Both phagocytosis and pinocytosis are nonspecific in the substances transported. In contrast, **receptor-mediated endocytosis** is very specific: It is triggered by the binding of certain external molecules to specific receptor proteins built into the plasma membrane. This binding causes the local region of the membrane to form a vesicle that transports the specific substance into the cell. Human liver cells use receptor-mediated endocytosis to take up cholesterol particles from the blood.

▼ **Figure 5.17 Exocytosis.**

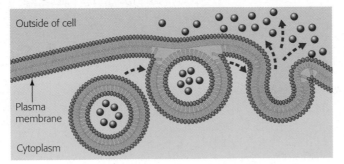

▼ **Figure 5.18 Endocytosis.**

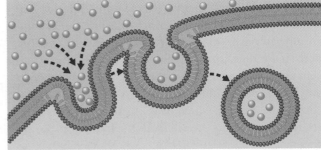

The Role of Membranes in Cell Signaling

In addition to transport, the plasma membrane plays a key role in conveying signals from the external environment into the cell and also between cells (see Figure 5.11). Communication begins when a receptor protein in the plasma membrane receives a stimulus from the outside environment or from a signaling molecule, such as a hormone. This stimulus triggers a chain reaction in one or more molecules that function in transduction (passing the signal along). The proteins and other molecules of this **signal transduction pathway** relay the signal and convert it to chemical forms that can function within the cell. This signal may lead to various responses (see also Figure 11.8).

Figure 5.19 shows an example of cell signaling: signal reception, transduction, and response. When a person gets "psyched up" for an athletic contest, certain cells in the adrenal glands secrete a hormone called epinephrine (also called adrenaline) into the bloodstream. When that hormone reaches muscle cells, it is recognized by receptor proteins in the plasma membrane. This recognition triggers responses (such as the breakdown of glycogen into glucose) in the muscle cells without the hormone even entering. This chain of events is part of the "fight-or-flight" response that enables you to attack or run when in danger—or keep alert during an intense competition. ☑

▼ Figure 5.19 **An example of cell signaling.**

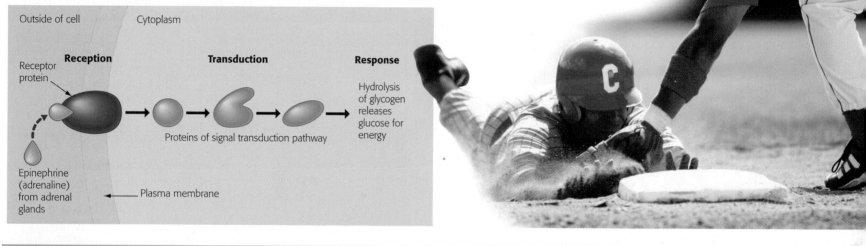

Outside of cell   Cytoplasm

Reception   **Transduction**   **Response**

Receptor protein

Proteins of signal transduction pathway

Hydrolysis of glycogen releases glucose for energy

Epinephrine (adrenaline) from adrenal glands

Plasma membrane

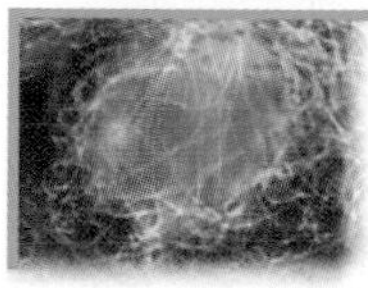

EVOLUTION CONNECTION: Molecular Technologies

The Origin of Membranes

Because all cells have a plasma membrane, it is logical to infer that membranes first formed early in the evolution of life on Earth. Phospholipids, their key ingredients, were probably among the first organic compounds that formed before life emerged. Once formed, they could self-assemble into simple membranes. When a mixture of phospholipids and water is shaken, for example, the phospholipids organize into bilayers, forming water-filled bubbles of membrane (Figure 5.20). This assembly requires neither genes nor other information beyond the properties of the phospholipids.

The tendency of lipids in water to spontaneously form membranes has led biomedical engineers to produce artificial vesicles called liposomes that can encase particular chemicals. In the future, these liposomes may deliver medications to specific sites within the body. Thus, membranes—like the other cellular components discussed in the Biology and Society and the Process of Science sections—have inspired novel technologies.

▶ Figure 5.20
The spontaneous formation of membranes: a key step in the origin of life.

Chapter Review

SUMMARY OF KEY CONCEPTS

Some Basic Energy Concepts

Conservation of Energy

Machines and organisms can transform kinetic energy (energy of motion) to potential energy (stored energy) and vice versa. In all such energy transformations, total energy is conserved. Energy cannot be created or destroyed.

Entropy

Entropy is a measure of disorder, or randomness. Every energy conversion releases some randomized energy in the form of heat.

Chemical Energy

Molecules store varying amounts of potential energy in the arrangement of their atoms. Organic compounds are relatively rich in such chemical energy.

Food Calories

Food Calories, actually kilocalories, are units used to measure the amount of energy in our foods and the amount of energy we expend in various activities.

ATP and Cellular Work

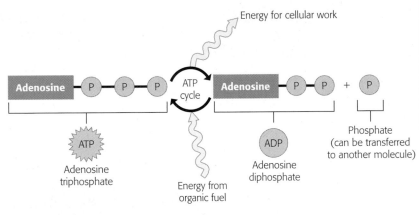

Enzymes

Activation Energy

Enzymes are biological catalysts that speed up metabolic reactions by lowering the activation energy required to break the bonds of reactant molecules.

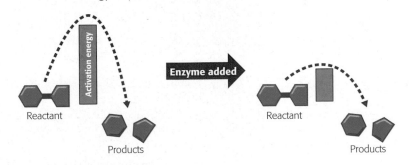

Induced Fit

The entry of a substrate into the active site of an enzyme causes the enzyme to change shape slightly, allowing for a better fit and thereby promoting the interaction of enzyme with substrate.

Enzyme Inhibitors

Enzyme inhibitors are molecules that can disrupt metabolic reactions by binding to enzymes, either at the active site or elsewhere.

Membrane Function

Proteins embedded in the plasma membrane perform a wide variety of functions, including regulating transport.

Passive Transport, Osmosis, and Active Transport

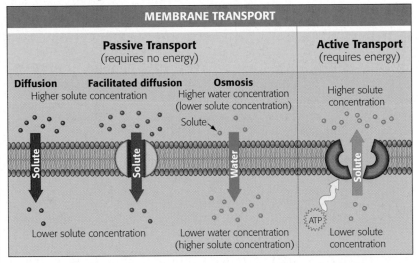

Most animal cells require an isotonic environment. Plant cells need a hypotonic environment, which keeps walled cells turgid. Osmoregulation is the control of water balance within a cell or organism.

Exocytosis and Endocytosis: Traffic of Large Molecules

Exocytosis is the secretion of large molecules within vesicles. The three kinds of endocytosis are phagocytosis ("cellular eating"), pinocytosis ("cellular drinking"), and receptor-mediated endocytosis, which enables the cell to take in specific large molecules.

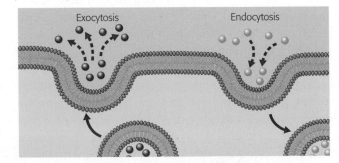

The Role of Membranes in Cell Signaling

Receptors on the cell surface trigger signal transduction pathways that control processes within the cell.

SELF-QUIZ

1. Describe the energy transformations that occur when you climb to the top of a stairway.

2. _____ is the capacity to perform work, while _____ is a measure of randomness.

3. The label on a candy bar says that it contains 150 Calories. If you could convert all of that energy to heat, you could raise the temperature of how much water by 15°C?

4. Why does removing a phosphate group from the triphosphate tail in a molecule of ATP release energy?

5. Your digestive system uses a variety of enzymes to break down large food molecules into smaller ones that your cells can assimilate. A generic name for a digestive enzyme is hydrolase. What is the chemical basis for that name? (*Hint*: Review Figure 3.4.)

6. Explain how an inhibitor can disrupt an enzyme's action without binding to the active site.

7. If someone at the other end of a restaurant smokes a cigarette, you may breathe in some smoke. The movement of smoke is similar to what type of transport?
   a. osmosis
   b. diffusion
   c. facilitated diffusion
   d. active transport

8. The total solute concentration in a red blood cell is about 2%. Sucrose cannot pass through a red blood cell's plasma membrane, but water and urea can. Osmosis will cause such a cell to shrink the most when the cell is immersed in which of the following?
   a. a hypertonic sucrose solution
   b. a hypotonic sucrose solution
   c. a hypertonic urea solution
   d. a hypotonic urea solution

9. Explain why it is not enough just to say that a solution is hypertonic.

10. What is the primary difference between passive and active transport in terms of concentration gradients?

11. Which of these types of cellular transport require(s) energy?
    a. facilitated diffusion
    b. active transport
    c. osmosis
    d. a and b

12. A _____ is a process that links the reception of a cell signal to a response within the cell.

Answers to the Self-Quiz questions can be found in Appendix D.

THE PROCESS OF SCIENCE

13. HIV, the virus that causes AIDS, depends on an enzyme called reverse transcriptase in order to multiply. Reverse transcriptase reads a molecule of RNA and creates a molecule of DNA from it. A molecule of AZT, the first drug approved to treat AIDS, has a shape very similar to that of the DNA base thymine. Propose a model for how AZT inhibits HIV.

14. Gaining and losing weight are matters of caloric accounting: Calories in the food you eat minus Calories that you spend in activity. One pound of human body fat contains approximately 3,500 Calories. Using Figure 5.3, compare ways you could burn off those Calories. How far would you have to run, swim, or walk to burn the equivalent of 1 pound of fat, and how long would it take? Which method of burning Calories appeals the most to you? The least?

BIOLOGY AND SOCIETY

15. Obesity is a serious health problem for many Americans. Several popular diet plans advocate low-carbohydrate diets. Most low-carb dieters compensate by eating more protein and fat. What are the advantages and disadvantages of such a diet? Should the government regulate the claims of diet books? How should the claims be tested? Should diet proponents be required to obtain and publish data before making claims?

16. Lead acts as an enzyme inhibitor, and it can interfere with the development of the nervous system. One manufacturer of lead-acid batteries instituted a "fetal protection policy" that banned female employees of childbearing age from working in areas where they might be exposed to high levels of lead. These women were transferred to lower-paying jobs in lower-risk areas. Some employees challenged the policy in court, claiming that it deprived women of job opportunities available to men. The U.S. Supreme Court ruled the policy illegal. But many people are uncomfortable about the "right" to work in an unsafe environment. What rights and responsibilities of employers, employees, and government agencies are in conflict? What criteria should be used to decide who can work in a particular environment?

6 Cellular Respiration: Obtaining Energy from Food

Muscles in action. Sprinters, like all athletes, depend on cellular respiration to power their muscles.

BIOLOGY AND SOCIETY: Life with and without Oxygen

Marathoners versus Sprinters

Track-and-field athletes usually have a favorite event in which they excel. For some runners, this event is a sprint, a short race of only 100 or 200 meters. For others, it may be a longer race of 1,500, 5,000, or even 10,000 m. It is unusual to find a runner who competes equally well in both 100-m and 10,000-m races; most runners just seem to feel more comfortable running races of particular lengths.

It turns out that there is a biological basis for such preferences. The muscles that move our legs contain two main types of muscle fibers: slow-twitch and fast-twitch. Slow-twitch muscle fibers can contract many times over a longer period but don't generate a lot of quick power for the body. They perform better in endurance exercises requiring slow, steady muscle activity (marathons). Fast-twitch muscle fibers can contract more quickly and powerfully than slow-twitch fibers but fatigue much more quickly. They function best in short bursts of intense activity (sprints).

All human muscles contain both slow-twitch and fast-twitch fibers, but the percentage of each fiber type in a particular muscle varies from person to person. For example, the thigh muscles of most marathon runners contain more slow-twitch fibers, whereas sprinters have more fast-twitch fibers. These differences, which are genetically determined, undoubtedly help account for varying athletic capabilities. Thus, to a certain degree, champion marathoners and sprinters are born, not trained!

What makes these two types of muscle fibers perform so differently? An important part of the answer is that they use different processes for making ATP, the molecule that supplies the energy for muscle contraction. While the cells in both types of muscle fiber break down glucose to make chemical energy for ATP production, slow-twitch fibers do it using oxygen (O_2), while fast-twitch fibers can work without oxygen.

The series of chemical reactions that provides energy to muscles is also used by other cells. In fact, all your body cells need a continuous supply of energy for you to walk, talk, and think—in short, to stay alive. In this chapter, you'll learn how cells harvest food energy and put it to work.

Energy Flow and Chemical Cycling in the Biosphere

Humans and other animals depend on plants to convert the energy of sunlight to the chemical energy of sugars and other organic molecules we consume as food. This is accomplished via photosynthesis. **Photosynthesis** uses light energy from the sun to power a chemical process that builds organic molecules (as we'll discuss in Chapter 7). But we depend on plants for more than our food. You're probably wearing clothing made of another product of photosynthesis—cotton. Most of our homes are framed with lumber, which is wood produced by photosynthetic trees. Even the text you are now reading is printed on a material (paper) that can be traced to photosynthesis in plants. But from an animal's point of view, photosynthesis is primarily about providing food.

Producers and Consumers

Plants and other **autotrophs** ("self-feeders") are organisms that make all their own organic matter—including carbohydrates, lipids, proteins, and nucleic acids—from nutrients that are entirely inorganic: carbon dioxide from the air and water and minerals from the soil. In contrast, humans and other animals are **heterotrophs** ("other-feeders"), organisms that cannot make organic molecules from inorganic ones. Therefore, we must eat organic material to get our nutrients.

Most ecosystems depend entirely on photosynthesis for food. For this reason, biologists refer to plants and other autotrophs as **producers**. Heterotrophs, in contrast, are **consumers**, because they obtain their food by eating plants or by eating animals that have eaten plants (**Figure 6.1**). We animals and other heterotrophs depend on autotrophs for organic fuel and for the raw organic materials we need to build our cells and tissues. ☑

Chemical Cycling between Photosynthesis and Cellular Respiration

The chemical ingredients for photosynthesis are carbon dioxide (CO_2), a gas that passes from the air into a plant via tiny pores, and water (H_2O), which is absorbed from the soil by the plant's roots (**Figure 6.2**). Chloroplasts in the cells of the leaves use light energy to rearrange the atoms of these ingredients to produce sugars—most importantly glucose ($C_6H_{12}O_6$)—and

▼ Figure 6.1 **Producer and consumer.** A parrot (consumer) eats a fruit produced by a photosynthetic plant (producer).

other organic molecules. A by-product of photosynthesis is oxygen gas (O_2).

Both animals and plants use the organic products of photosynthesis as sources of energy. A chemical process called cellular respiration harvests energy that is stored in sugars and other organic molecules. Cellular respiration uses O_2 to help convert energy extracted from organic fuel to another source of chemical energy, called ATP. Cells expend ATP for almost all their work. In both plants and animals, the production of ATP during cellular respiration occurs mainly in the organelles called mitochondria (see Figure 4.20).

Notice in Figure 6.2 that the waste products of cellular respiration are

CO_2 and H_2O—the very same ingredients used for photosynthesis. Plants store chemical energy via photosynthesis and then harvest this energy via cellular respiration. However, plants usually make more organic molecules than they need for fuel. This photosynthetic surplus provides the organic material for the plant to grow. It is also the source of food for humans and other consumers. (Note that plants perform *both* photosynthesis to produce fuel molecules *and* cellular respiration to burn them, while animals perform *only* cellular respiration.) Analyze nearly any food chain, and you can trace the energy and raw materials for growth back to solar-powered photosynthesis. ✔

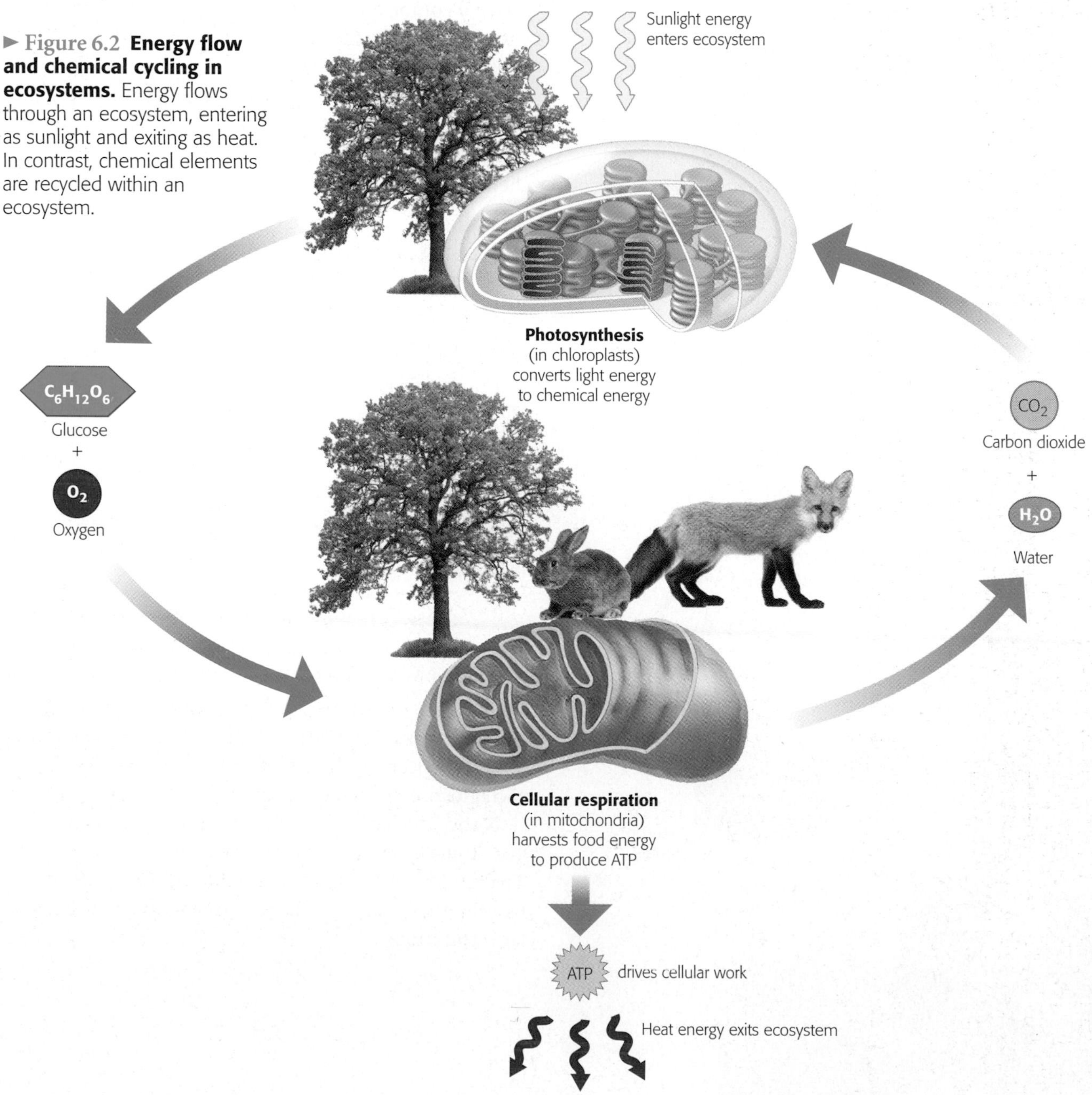

► Figure 6.2 **Energy flow and chemical cycling in ecosystems.** Energy flows through an ecosystem, entering as sunlight and exiting as heat. In contrast, chemical elements are recycled within an ecosystem.

Sunlight energy enters ecosystem

Photosynthesis
(in chloroplasts)
converts light energy
to chemical energy

$C_6H_{12}O_6$
Glucose
+
O_2
Oxygen

CO_2
Carbon dioxide
+
H_2O
Water

Cellular respiration
(in mitochondria)
harvests food energy
to produce ATP

ATP drives cellular work

Heat energy exits ecosystem

What is misleading about the following statement? "Plants perform photosynthesis, whereas animals perform cellular respiration."

Answer: It implies that cellular respiration does not also occur in plants. It does.

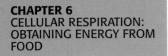

Cellular Respiration

Cellular Respiration Part 1: Glycolysis

Cellular Respiration Part 2: Citric Acid
Cycle and Electron Transport Chain

Cellular Respiration: Aerobic Harvest of Food Energy

We sometimes use the word *respiration* to mean breathing. While respiration on the organismal level should not be confused with cellular respiration, the two processes are closely related (**Figure 6.3**). Cellular respiration requires a cell to exchange two gases with its surroundings. The cell takes in oxygen in the form of the gas O_2. It gets rid of waste in the form of the gas carbon dioxide, or CO_2. Breathing results in the exchange of these same gases between your blood and the outside air. Oxygen present in the air you inhale diffuses across the lining of your lungs and into your bloodstream. And the CO_2 in your bloodstream diffuses into your lungs and exits when you exhale.

Internal combustion engines, like the ones found in cars, use O_2 (via the air intakes) to break down gasoline. A cell also requires O_2 to break down its fuel (see Figure 5.2). Cellular respiration—a living version of internal combustion—is the main way that chemical energy is harvested from food and converted to ATP energy (see Figure 5.6). Cellular respiration is an **aerobic** process, which is just another way of saying that it requires oxygen. Putting all this together, we can now define **cellular respiration** as the aerobic harvesting of chemical energy from organic fuel molecules. ☑

The Overall Equation for Cellular Respiration

A common fuel molecule for cellular respiration is glucose, a six-carbon sugar with the formula $C_6H_{12}O_6$ (see Figure 3.6). Here is the overall equation for what happens to glucose during cellular respiration:

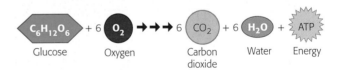

$$C_6H_{12}O_6 + 6\ O_2 \rightarrow \rightarrow \rightarrow 6\ CO_2 + 6\ H_2O + ATP$$

Glucose   Oxygen   Carbon dioxide   Water   Energy

The series of arrows indicates that cellular respiration consists of many chemical steps, not just a single chemical reaction. Remember, the main function of cellular respiration is to generate ATP for cellular work. In fact, the process can produce up to 38 ATP molecules for each glucose molecule consumed.

Notice that cellular respiration also transfers hydrogen atoms from glucose to oxygen, forming water. That hydrogen transfer turns out to be the key to why oxygen is so vital to the harvest of energy during cellular respiration.

▼ Figure 6.3 **How breathing is related to cellular respiration.** When you inhale, you breathe in O_2. The O_2 is delivered to your cells, where it is used in cellular respiration. Carbon dioxide, a waste product of cellular respiration, diffuses from your cells to your blood and travels to your lungs, where it is exhaled.

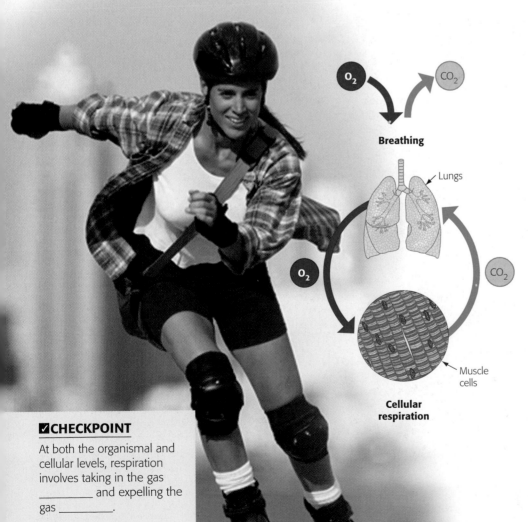

O_2

CO_2

Breathing

Lungs

O_2

CO_2

Muscle cells

Cellular respiration

☑**CHECKPOINT**

At both the organismal and cellular levels, respiration involves taking in the gas _____ and expelling the gas _____.

Answer: O_2; CO_2

The Role of Oxygen in Cellular Respiration

In tracking the transfer of hydrogen from sugar to oxygen, we are also following the transfer of electrons. The atoms of sugar and other molecules are bonded together by shared electrons (see Figure 2.7). During cellular respiration, hydrogen and its bonding electrons change partners from sugar to oxygen, forming water as a product.

Redox Reactions

Chemical reactions that transfer electrons from one substance to another substance are called oxidation-reduction reactions, or **redox reactions** for short. The loss of electrons during a redox reaction is called **oxidation**. Glucose is oxidized during cellular respiration, losing electrons to oxygen. The acceptance of electrons during a redox reaction is called **reduction**. (Note that *adding* electrons is called *reduction*; negatively charged electrons added to an atom *reduce* the amount of positive charge of that atom.) Oxygen is reduced during cellular respiration, accepting electrons (and hydrogen) lost from glucose:

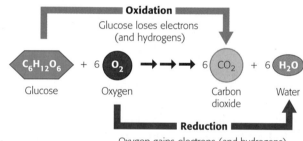

Water isn't the only thing produced when hydrogen and its bonding electrons change partners from glucose to oxygen; energy is released, too.

Why does electron transfer to oxygen release energy? In redox reactions, oxygen is an "electron grabber." An oxygen atom attracts electrons more strongly than almost any other type of atom. When electrons move (along with hydrogen) from glucose to oxygen, it is as though they are falling. They are not really falling in the sense of an apple dropping from a tree. However, in both cases, potential energy is unlocked. Instead of gravity, it is the attraction of electrons to oxygen that causes the "fall" and energy release during cellular respiration.

Figure 6.4 shows a simple example of a redox reaction: the reaction between hydrogen gas (H_2) and oxygen gas (O_2), producing water (H_2O). The reaction releases a large amount of energy as the electrons of the hydrogen "fall" into their new bonds with oxygen. This reaction is actually quite explosive, and it would be difficult for a cell to capture such a burst of energy and put it to useful work. As you'll see shortly, cellular respiration is a more controlled "fall" of electrons—more like a stepwise cascade of electrons down an energy staircase. Instead of liberating food energy in a burst, cellular respiration unlocks chemical energy in smaller amounts that cells can put to productive use. ✔

NADH and Electron Transport Chains

Let's take a closer look at the path that electrons take on their way from glucose to oxygen (Figure 6.5). The first stop is a positively charged electron acceptor called NAD^+ (nicotinamide adenine dinucleotide). The transfer of electrons from organic fuel (food) to NAD^+ reduces the NAD^+ to **NADH** (the H represents the transfer of hydrogen along with the electrons). In our staircase analogy, the electrons have now taken one baby step down in their trip from glucose to oxygen. The rest of the staircase consists of an **electron transport chain**.

▼ Figure 6.5 **The role of oxygen in harvesting food energy.** In cellular respiration, electrons (e⁻) "fall" in small steps from food to oxygen, producing water. NADH transfers electrons from food to an electron transport chain. Oxygen "pulls" the electrons down the chain.

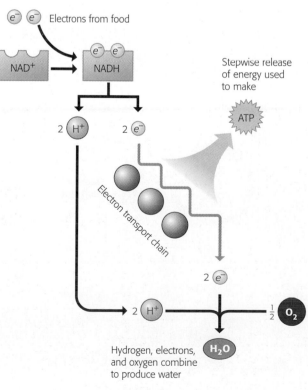

▼ Figure 6.4 **A simple redox reaction.** An all-at-once redox reaction, such as the reaction of hydrogen and oxygen to form water, releases a burst of energy.

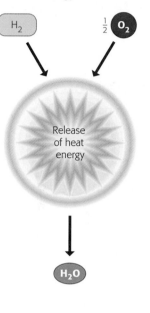

✔CHECKPOINT

During a redox reaction, the addition of electrons is called _____, while the removal of electrons is called _____.

Answer: *reduction; oxidation*

Each link in an electron transport chain is actually a molecule, usually a protein. In a series of redox reactions, each member of the chain first accepts and then donates electrons. With each transfer, the electrons give up a small amount of energy that can then be used indirectly to generate ATP. The first molecule of the chain accepts electrons from NADH. Thus, NADH carries electrons from glucose and other fuel molecules and deposits them at the top of an electron transport chain. The electrons cascade down the chain, from molecule to molecule, like an electron bucket brigade. The molecule at the bottom of the chain finally "drops" the electrons to oxygen. The oxygen also picks up hydrogen, forming water.

The overall effect of all this transfer of electrons during cellular respiration is a "downward" trip for electrons from glucose to NADH to an electron transport chain to oxygen. During the stepwise release of chemical energy during electron transport, our cells make most of their ATP. It is actually oxygen, the "electron grabber," that makes it all possible. By pulling electrons down the transport chain from fuel molecules, oxygen functions somewhat like gravity pulling objects downhill. This is how the oxygen we breathe functions in our cells and why we cannot survive more than a few minutes without it. Viewed this way, drowning is deadly because it deprives cells of the final "electron grabbers" (oxygen) needed to drive cellular respiration.

An Overview of Cellular Respiration

Cellular respiration is an example of a metabolic pathway. That means that it is not a single chemical reaction, but a series of reactions. A specific enzyme catalyzes each reaction in a metabolic pathway. More than two dozen reactions are involved in cellular respiration. We can group them into three main metabolic stages: glycolysis, the citric acid cycle, and electron transport (which you've already encountered).

Figure 6.6 is a map that will help you follow the three stages of respiration and see where each stage occurs in your cells. During **glycolysis**, a molecule of glucose is split into two molecules of a compound called pyruvic acid. The enzymes for glycolysis are located in the cytoplasm. The **citric acid cycle** (also called the Krebs cycle) completes the breakdown of glucose all the way to CO_2, one of the waste products of cellular respiration. The enzymes for the citric acid cycle are dissolved in the fluid within mitochondria. Glycolysis and the citric acid cycle generate a small amount of ATP directly. They generate much more ATP indirectly, via redox reactions that transfer electrons from fuel molecules to NAD^+, forming NADH. The third stage of cellular respiration is **electron transport**. Electrons captured from food by the NADH formed in the first two stages "fall" down electron transport chains to oxygen. The proteins and other molecules that make up electron transport chains are embedded within the inner membrane of the mitochondria. Electron transport from NADH to oxygen releases the energy your cells use to make most of their ATP. ✓

☑**CHECKPOINT**

Which stages of cellular respiration take place in the mitochondria?

Answer: *the citric acid cycle and electron transport*

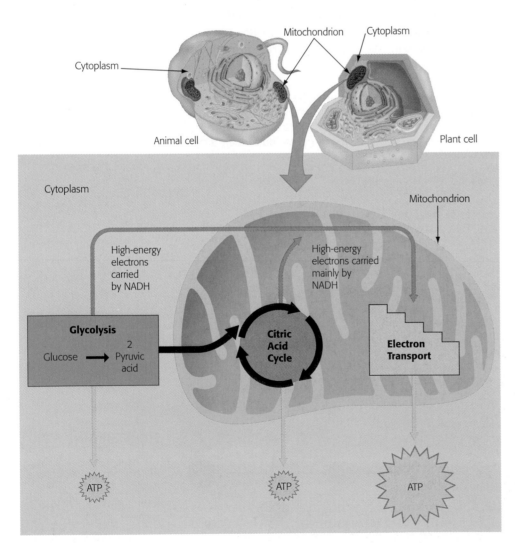

◄ Figure 6.6 **A road map for cellular respiration.**

The Three Stages of Cellular Respiration

Now that you have a big-picture view of cellular respiration, let's examine the process in more detail. A small version of Figure 6.6 will help you keep the overall process of cellular respiration in plain view as we take a closer look at its three stages.

Stage 1: Glycolysis

The word *glycolysis* means "splitting of sugar." That is exactly what happens (Figure 6.7). ❶ During glycolysis, a six-carbon glucose molecule is broken in half, forming two three-carbon molecules. Notice in Figure 6.7 that the initial split requires an energy investment of two ATP molecules per glucose. ❷ The three-carbon molecules then donate high-energy electrons to NAD⁺, the electron carrier, forming NADH. ❸ In addition to NADH, glycolysis also makes four ATP molecules directly when enzymes transfer phosphate

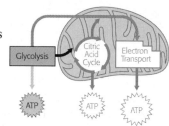

groups from fuel molecules to ADP (Figure 6.8). Glycolysis thus produces a net of two molecules of ATP per molecule of glucose. (This fact will become important during our discussion of fermentation.) What remains of the fractured glucose at the end of glycolysis are two molecules of pyruvic acid. The pyruvic acid still holds most of the energy of glucose, and that energy is harvested in the second stage of cellular respiration, the citric acid cycle.

▼ Figure 6.8 **ATP synthesis by direct phosphate transfer.** Glycolysis generates ATP when enzymes transfer phosphate groups directly from fuel molecules to ADP.

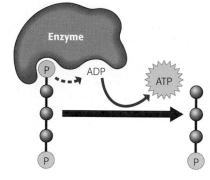

▼ Figure 6.7 **Glycolysis.** In glycolysis, a team of enzymes splits glucose, eventually forming two molecules of pyruvic acid. After investing 2 ATP at the start, glycolysis generates 4 ATP directly. More energy will be harvested later from high-energy electrons used to form NADH.

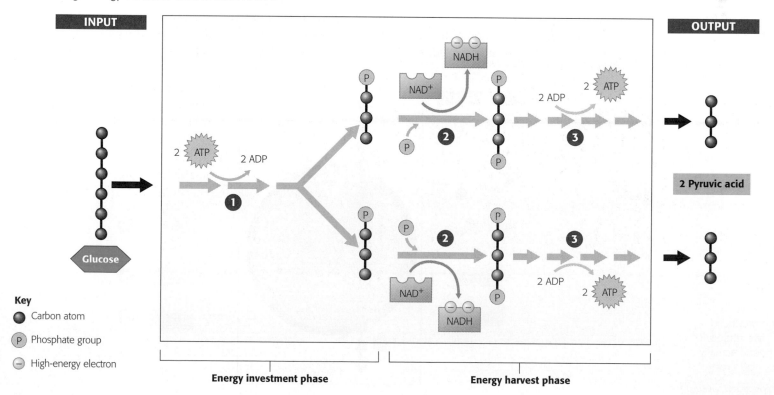

Key
- Carbon atom
- (P) Phosphate group
- (−) High-energy electron

Energy investment phase   Energy harvest phase

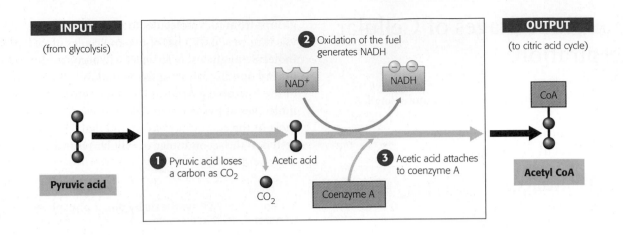

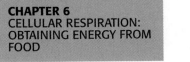

▶ Figure 6.9 **The link between glycolysis and the citric acid cycle: the conversion of pyruvic acid to acetyl CoA.** Remember that one molecule of glucose is split into two molecules of pyruvic acid. Therefore, the process shown here occurs twice.

Stage 2: The Citric Acid Cycle

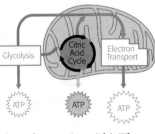

Pyruvic acid, the fuel that remains after glycolysis, is not quite ready for the citric acid cycle. (Remember that each molecule of glucose produces two molecules of pyruvic acid.) The pyruvic acid must be "prepped"—converted to a form the citric acid cycle can use (**Figure 6.9**). ❶ First, pyruvic acid loses a carbon as CO_2. This is the first of this waste product we've seen so far in the breakdown of glucose. The remaining fuel molecules, each with only two carbons left, are called acetic acid (the acid that's in vinegar). ❷ Oxidation of the fuel generates NADH. ❸ Finally, the acetic acid is attached to a molecule called coenzyme A (CoA) to form acetyl CoA. The CoA escorts the acetic acid into the first reaction of the citric acid cycle. The CoA is then stripped and recycled.

The citric acid cycle finishes extracting the energy of sugar by breaking the acetic acid molecules all the way down to CO_2 (**Figure 6.10**). ❶ Acetic acid joins a four-carbon acceptor molecule to form a six-carbon product called citric acid (for which the cycle is named). For every acetic acid molecule that enters the cycle as fuel, ❷ two CO_2 molecules eventually exit as a waste product. Along the way, the citric acid cycle harvests energy from the fuel. ❸ Some of the energy is used to produce ATP directly. However, the cycle captures much more energy in the form of ❹ NADH and ❺ a second, closely related electron carrier, $FADH_2$. ❻ All the carbon atoms that entered the cycle as fuel are accounted for as CO_2 exhaust, and the four-carbon acceptor molecule is recycled. We have tracked only one acetic acid molecule through the citric acid cycle here. But since glycolysis splits glucose in two, the citric acid cycle actually turns twice for each glucose molecule that fuels a cell. ☑

▶ Figure 6.10 **The citric acid cycle.**

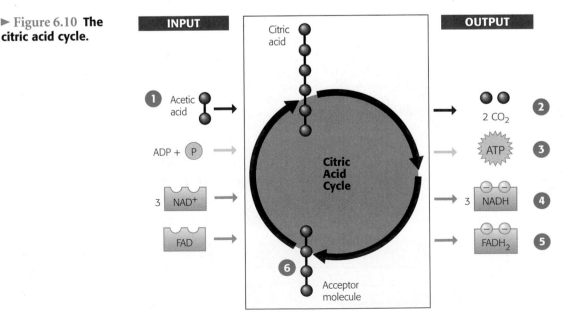

☑CHECKPOINT

Two molecules of what compound are produced by glycolysis? Two molecules of what compound actually enter the citric acid cycle?

Answer: pyruvic acid; acetic acid

Stage 3: Electron Transport

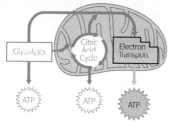

The molecules of electron transport chains are built into the inner membranes of mitochondria (see Figure 4.18). An electron transport chain functions as a chemical machine that uses the energy released by the "fall" of electrons to pump hydrogen ions (H^+) across the inner mitochondrial membrane. This pumping causes ions to become more concentrated on one side of the membrane than on the other. Such a concentration gradient stores potential energy.

The energy stored by electron transport behaves something like the elevated reservoir of water behind a dam. There is a tendency for hydrogen ions to gush back to where they are less concentrated, just as there is a tendency for water to flow downhill. The inner membrane, analogous to the dam, temporarily restrains hydrogen ions.

The energy of dammed water can be harnessed to perform work. Gates in a dam allow the water to rush downhill, turning giant turbines as it goes. The spinning turbines perform work that can be used to generate electricity. Your mitochondria have structures that act like turbines. Each of these miniature machines, called an **ATP synthase**, is constructed from protein built into

the inner mitochondrial membrane, adjacent to the proteins of the electron transport chains. **Figure 6.11** shows a simplified view of how the energy previously stored in NADH and FADH$_2$ can now be used to generate ATP. ❶ NADH and ❷ FADH$_2$ transfer electrons to an electron transport chain. ❸ The electron transport chain uses this energy supply to pump H^+ across the inner mitochondrial membrane. ❹ Oxygen pulls electrons down the transport chain. ❺ The H^+ concentrated on one side of the membrane rushes back "downhill" through an ATP synthase. This action spins a component of the ATP synthase, just as water turns the turbines in a dam. ❻ The rotation activates parts of the synthase molecule that attach phosphate groups to ADP molecules to generate ATP.

The poison cyanide produces its deadly effect by binding to one of the protein complexes in the electron transport chain (marked by ☠ in Figure 6.11). When bound there, cyanide blocks the passage of electrons to oxygen. This blockage is like turning off a faucet; electrons cease to flow through the "pipe." As a result, no H^+ gradient is generated, and no ATP is made. Cells stop working, and the organism dies. Cyanide was the lethal agent in an infamous case of product tampering: the Tylenol murders of 1982. Seven people in the Chicago area died after ingesting Tylenol capsules that had been laced with cyanide. The perpetrator of that crime was never caught. ☑

☑**CHECKPOINT**

What is the potential energy source that drives ATP production by ATP synthase?

Answer: a concentration gradient of H^+ across the inner membrane of a mitochondrion

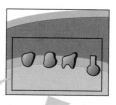

▼ Figure 6.11 **How electron transport drives ATP synthase machines.**

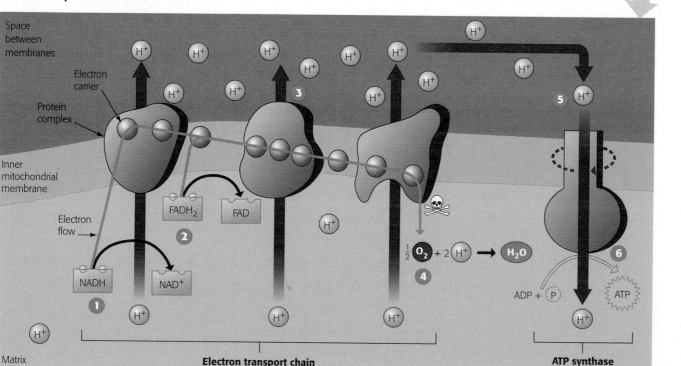

The Versatility of Cellular Respiration

We have seen that glucose can provide the energy to make the ATP our cells use for all their work. The importance of glucose is underscored by the severity of diseases in which glucose balance is disturbed. Diabetes, which affects over 20 million Americans, is caused by an inability to properly regulate glucose levels in the blood due to problems with the hormone insulin. If left untreated, a glucose imbalance can lead to cardiovascular disease, coma, and even death.

But even though we have concentrated on glucose as the fuel that is broken down during cellular respiration, respiration is a versatile metabolic furnace that can "burn" many other kinds of food molecules. Figure 6.12 diagrams some metabolic routes for the use of carbohydrates, fats, and proteins as fuel for cellular respiration.

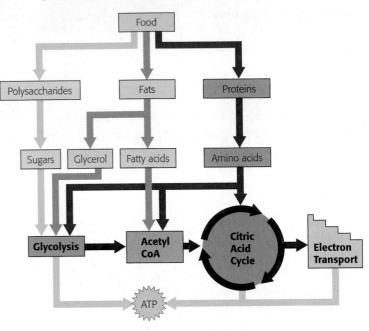

▼ Figure 6.12 **Energy from food.** The monomers from carbohydrates (polysaccharides and sugars), fats, and proteins can all serve as fuel for cellular respiration.

Adding Up the ATP from Cellular Respiration

Taking cellular respiration apart to see how all the molecular nuts and bolts of its metabolic machinery work, it's easy to lose sight of its overall function: generating up to 38 molecules of ATP per molecule of glucose (the actual number can vary by a few). Figure 6.13 will help you add up the ATP molecules. Glycolysis and the citric acid cycle each contribute 2 ATP by direct synthesis. All of the rest of the ATP molecules are produced by the ATP synthase machines, powered by the "fall" of electrons from food to oxygen. The electrons are carried from the organic fuel to electron transport chains by NADH and FADH$_2$. Each electron pair "dropped" down a transport chain from NADH can power the synthesis of up to 3 ATP. Each electron pair transferred to an electron transport chain from FADH$_2$ is worth up to 2 ATP. Next, we'll see what happens when cells harvest food energy without the help of oxygen. ☑

▼ Figure 6.13 **A summary of ATP yield during cellular respiration.**

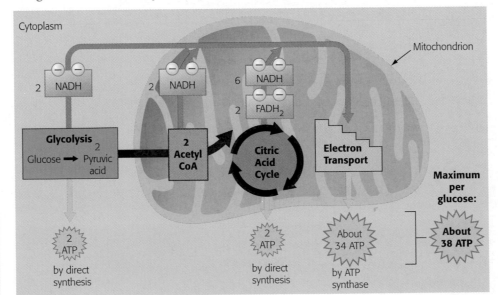

Which stage of cellular respiration produces the majority of ATP?

Answer: electron transport

Fermentation: Anaerobic Harvest of Food Energy

Although you must breathe to stay alive, some of your cells can work for short periods without oxygen. This **anaerobic** ("without oxygen") harvest of food energy is called **fermentation**.

Fermentation in Human Muscle Cells

When you walk between classes, your leg muscles require a constant supply of ATP, which is generated by cellular respiration. To keep this process going, blood provides your muscle cells with enough O_2 to keep electrons "falling" down transport chains in your mitochondria. But if you start to run because you're late for class, your muscles are forced to work under anaerobic conditions. That's because they are spending ATP at a rate that outpaces your bloodstream's delivery of O_2 from your lungs to your muscles.

After functioning anaerobically for about 15 seconds, muscle cells will begin to generate ATP by the process of fermentation. Fermentation relies on glycolysis, the same metabolic pathway that functions as the first stage of cellular respiration. Glycolysis does not require O_2 but does produce 2 ATP molecules for each glucose molecule broken down to pyruvic acid. That isn't very

efficient compared with the 38 or so ATP molecules each glucose generates during cellular respiration, but it can energize your leg muscles long enough for you to make it to class. However, your cells will have to consume more glucose fuel per second, since so much less ATP per glucose molecule is generated under anaerobic conditions.

To harvest food energy during glycolysis, NAD^+ must be present as an electron acceptor (see Figure 6.7). This is no problem under aerobic conditions, because the cell regenerates NAD^+ when NADH drops its electron cargo down electron transport chains to O_2 (see Figure 6.5). However, this recycling of NAD^+ cannot occur under anaerobic conditions because there is no O_2 to accept the electrons. Instead, NADH disposes of electrons by adding them to the pyruvic acid produced by glycolysis (**Figure 6.14**). This restores NAD^+ and keeps glycolysis working.

The addition of electrons to pyruvic acid produces a waste product called lactic acid. The lactic acid by-product is eventually transported to the liver, where liver cells convert it back to pyruvic acid. Exercise physiologists (biologists who study the effects of exercise on the body) have long speculated about the role that lactic acid plays in muscle fatigue, as you'll see next. ☑

☑**CHECKPOINT**

How many molecules of ATP can be produced from one molecule of glucose during fermentation?

Answer: 2

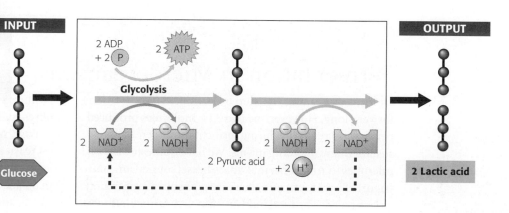

▼ **Figure 6.14 Fermentation: producing lactic acid.**
Glycolysis produces ATP even in the absence of O_2. This process requires a continuous supply of NAD^+ to accept electrons from glucose. The NAD^+ is regenerated when NADH transfers the electrons it removed from food to pyruvic acid, thereby producing lactic acid (or other waste products, depending on the species of organism).

INPUT

OUTPUT

2 ADP + 2 P   2 ATP

Glycolysis

2 NAD⁺   2 NADH   2 NADH   2 NAD⁺

2 Pyruvic acid + 2 H⁺

Glucose

2 Lactic acid

Does Lactic Acid Buildup Cause Muscle Burn?

You may have heard that the burn you feel after hard exercise ("Feel the burn! It's a *good* burn!") is due to the buildup of lactic acid in your muscles. This idea originated with the work of a British biologist named A.V. Hill. Considered one of the founders of the field of exercise physiology, Hill won a 1922 Nobel Prize for his work on understanding muscle contraction.

In 1929, Hill performed a classic experiment that began with the **observation** that muscles produce lactic acid under anaerobic conditions. Hill asked the **question**, Does the buildup of lactic acid cause muscle fatigue? To find out, Hill developed a technique for electrically stimulating dissected frog muscles in a laboratory solution. He formed the **hypothesis** that a buildup of lactic acid would cause muscle activity to stop.

Hill's **experiment** tested frog muscles under two different sets of conditions (**Figure 6.15**). First, he showed that muscle performance declined when lactic acid could not diffuse away from the muscle tissue. Next, he showed that when lactic acid was allowed to diffuse away, performance improved significantly. These **results** led Hill to the conclusion that "lactic acid accumulation is the primary cause of failure" in muscle tissue.

Given his scientific stature (he was considered the world's leading authority on muscle activity), Hill's conclusion went unchallenged for many decades. Gradually, however, evidence began to accumulate that contradicted Hill's results. For example, the effect that Hill demonstrated did not appear to occur at human body temperature. And certain individuals who are unable to accumulate lactic acid

have muscles that fatigue *more* rapidly, which is the opposite of what you would expect. Recent experiments have directly refuted Hill's conclusions. Today, the role of lactic acid in muscle fatigue remains a hotly debated topic.

The changing view of lactic acid's role in muscle fatigue illustrates an important point about the process of science: It is dynamic and subject to constant adjustment as new evidence is uncovered. This would not have surprised Hill, who himself wrote that the "built-in obsolescence" of scientific hypotheses is a necessary feature for the advancement of science.

▼ Figure 6.15 **A. V. Hill's apparatus for measuring muscle fatigue.**

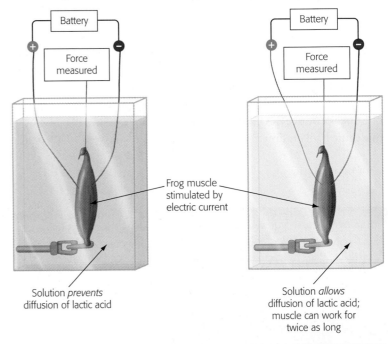

Fermentation in Microorganisms

Our muscles cannot function by lactic acid fermentation for very long. However, the two ATP molecules produced per glucose molecule during fermentation is enough to sustain many microorganisms. We have domesticated such microbes to transform milk into cheese, sour cream, and yogurt. These foods owe their sharp or sour flavor mainly to lactic acid. The food industry also uses fermentation to

produce soy sauce from soybeans, to pickle cucumbers, olives, and cabbage, and to produce meat products like sausage, pepperoni, and salami.

Yeast, a microscopic fungus, is capable of both cellular respiration and fermentation. If you keep yeast cells in an anaerobic environment, they are forced to ferment sugars and other foods to stay alive. When yeast ferment,

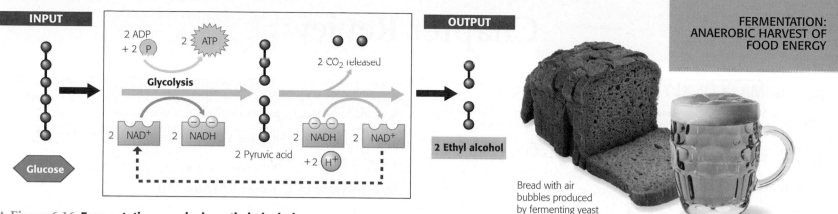

INPUT

2 ADP + 2 (P) → 2 ATP

Glycolysis

2 NAD⁺   2 NADH   2 Pyruvic acid

2 NADH   2 NAD⁺

2 CO₂ released

+2 (H⁺)

OUTPUT

2 Ethyl alcohol

▲ Figure 6.16 **Fermentation: producing ethyl alcohol.**

Bread with air bubbles produced by fermenting yeast

Beer fermentation

they produce ethyl alcohol as a waste product instead of lactic acid (**Figure 6.16**). This alcoholic fermentation also releases CO₂. For thousands of years, humans have put yeast to work producing alcoholic beverages such as beer and wine. And as every baker knows, the CO₂ bubbles from fermenting yeast also cause bread dough to rise. (The alcohol produced in fermenting bread is released during baking.) ☑

☑CHECKPOINT

What kind of acid builds up in human muscle during strenuous activity?

Answer: lactic acid

EVOLUTION CONNECTION: Life with and without Oxygen

Life before and after Oxygen

In the Biology and Society and Process of Science sections, we have seen that living cells can generate ATP either aerobically (with oxygen via cellular respiration) or anaerobically (without oxygen via fermentation). Both of these processes start with glycolysis, the splitting of glucose to form pyruvic acid. Glycolysis is thus the universal energy-harvesting process of life.

The role of glycolysis in both respiration and fermentation has an evolutionary basis. Ancient prokaryotes probably used glycolysis to make ATP long before oxygen was present in Earth's atmosphere. The oldest known fossils of bacteria date back more than 3.5 billion years, but significant levels of O₂ did not accumulate in the atmosphere until about 2.7 billion years ago (**Figure 6.17**). For almost a billion years, prokaryotes must have generated ATP exclusively from glycolysis.

The fact that glycolysis occurs in almost all organisms suggests that it evolved very early in ancestors common to all the domains of life. The location of glycolysis within the cell also implies great antiquity; the pathway does not require any of the membrane-bounded organelles of the eukaryotic cell, which evolved more than a billion years after the prokaryotic cell. Glycolysis is a metabolic heirloom from early cells that continues to function in fermentation and as the first stage in the breakdown of organic molecules by cellular respiration.

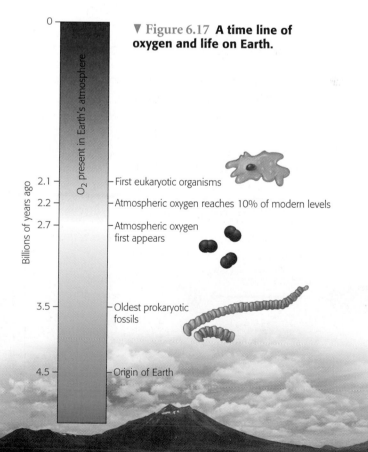

▼ Figure 6.17 **A time line of oxygen and life on Earth.**

O₂ present in Earth's atmosphere

Billions of years ago

0

2.1 — First eukaryotic organisms

2.2 — Atmospheric oxygen reaches 10% of modern levels

2.7 — Atmospheric oxygen first appears

3.5 — Oldest prokaryotic fossils

4.5 — Origin of Earth

Chapter Review

SUMMARY OF KEY CONCEPTS

Go to the Study Area at **www.masteringbiology.com** for practice quizzes, my**e**Book, BioFlix™ 3-D animations, MP3 Tutor Sessions, videos, current events, and more.

Energy Flow and Chemical Cycling in the Biosphere

Producers and Consumers

Autotrophs (producers) make organic molecules from inorganic nutrients via photosynthesis. Heterotrophs (consumers) must consume organic material and obtain energy via cellular respiration.

Chemical Cycling between Photosynthesis and Cellular Respiration

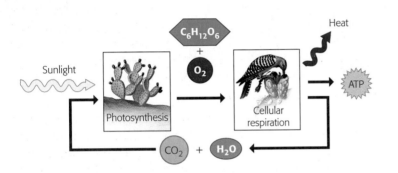

Cellular Respiration: Aerobic Harvest of Food Energy

The Overall Equation for Cellular Respiration

$$C_6H_{12}O_6 + 6\ O_2 \rightarrow \rightarrow \rightarrow 6\ CO_2 + 6\ H_2O + \text{Approx. 38 ATP}$$

The Role of Oxygen in Cellular Respiration

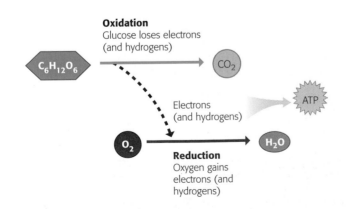

Oxidation
Glucose loses electrons (and hydrogens)

Electrons (and hydrogens)

Reduction
Oxygen gains electrons (and hydrogens)

Redox reactions transfer electrons from food molecules to an electron acceptor called NAD^+, forming NADH. The NADH then passes the high-energy electrons to an electron transport chain that eventually "drops" them to O_2. The energy released during this electron transport is used to regenerate ATP from ADP. The affinity of oxygen for electrons keeps the redox reactions of cellular respiration working.

An Overview of Cellular Respiration

You can follow the flow of molecules through the process of cellular respiration in the following diagram:

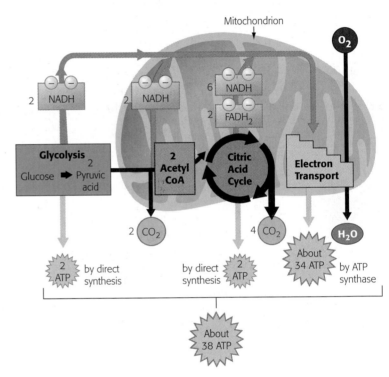

The electron transport chains pump H^+ across the inner mitochondrial membrane as electrons flow stepwise from NADH to oxygen. Backflow of H^+ across the membrane powers the ATP synthases, which attach phosphate to ADP to make ATP.

Fermentation: Anaerobic Harvest of Food Energy

Fermentation in Human Muscle Cells

When muscle cells consume ATP faster than O_2 can be supplied for cellular respiration, they regenerate ATP by fermentation. The waste product under these anaerobic conditions is lactic acid. The ATP yield per glucose is much lower during fermentation (2 ATP) than during cellular respiration (up to 38 ATP).

Fermentation in Microorganisms

Yeast and some other organisms can survive with or without O_2. Wastes from fermentation can be ethyl alcohol, lactic acid, or other compounds, depending on the species.

SELF-QUIZ

1. Which of the following statements is a correct distinction between autotrophs and heterotrophs?
   a. Only heterotrophs require chemical compounds from the environment.
   b. Cellular respiration is unique to heterotrophs.
   c. Only heterotrophs have mitochondria.
   d. Only autotrophs can live on nutrients that are entirely inorganic.

2. Why are plants called producers? Why are animals called consumers?

3. How is your breathing related to your cellular respiration?

4. Of the three stages of cellular respiration, which produces the most ATP molecules per glucose?

5. In glycolysis, _____ is oxidized and _____ is reduced.

6. The final electron acceptor of electron transport chains in mitochondria is _____.

7. The poison cyanide acts by blocking a key step in the electron transport chain. Knowing this, explain why cyanide kills so quickly.

8. Cells can harvest the most chemical energy from which of the following?
   a. an NADH molecule
   b. a glucose molecule
   c. six carbon dioxide molecules
   d. two pyruvic acid molecules

9. _____ is a metabolic pathway common to both fermentation and cellular respiration.

10. Sports physiologists at an Olympic training center wanted to monitor athletes to determine at what point their muscles were functioning anaerobically. They could do this by checking for a buildup of
    a. ADP.
    b. lactic acid.
    c. carbon dioxide.
    d. oxygen.

11. A glucose-fed yeast cell is moved from an aerobic environment to an anaerobic one. For the cell to continue to generate ATP at the same rate, approximately how much glucose must it consume in the anaerobic environment compared with the aerobic environment?

Answers to the Self-Quiz questions can be found in Appendix D.

THE PROCESS OF SCIENCE

12. Your body makes NAD$^+$ from two B vitamins, niacin and riboflavin. You need only tiny amounts of these vitamins. The U.S. Food and Drug Administration's recommended dietary allowances are 20 mg daily for niacin and 1.7 mg daily for riboflavin. These amounts are thousands of times less than the amount of glucose your body needs each day to fuel its energy requirements. How many NAD$^+$ molecules are needed for the breakdown of each glucose molecule? Why do you think your daily requirement for these substances is so small?

BIOLOGY AND SOCIETY

13. Nearly all human societies use fermentation to produce alcoholic drinks such as beer and wine. The technology dates back to the earliest civilizations. Suggest a hypothesis for how humans first discovered fermentation. In preindustrial cultures, why do you think wine was a more practical beverage than the grape juice from which it was made?

14. The consumption of alcohol by a pregnant woman can cause a series of birth defects called fetal alcohol syndrome (FAS). Symptoms of FAS include head and facial irregularities, heart defects, mental retardation, and behavioral problems. The U.S. Surgeon General's Office recommends that pregnant women abstain from drinking alcohol, and the government has mandated that a warning label be placed on liquor bottles. Imagine you are a server in a restaurant. An obviously pregnant woman orders a strawberry daiquiri. How would you respond? Is it the woman's right to make those decisions about her unborn child's health? Do you bear any responsibility in the matter? Is a restaurant responsible for monitoring the dietary habits of its customers?

7 Photosynthesis: Using Light to Make Food

Capturing solar energy. Sunlight drives the process of photosynthesis, which plants use to produce sugars.

BIOLOGY AND SOCIETY: Harnessing Solar Energy

Green Energy

What's old is sometimes new again. Throughout human history, people have burned wood to produce heat and light. Then, as societies became industrialized, wood was largely displaced as an energy source by fossil fuels such as coal, gas, and oil. But now, as fossil fuel supplies dwindle and pollution accumulates, scientists are researching better ways to use wood and other forms of *biomass* (living material) as efficient and renewable energy sources. Some researchers focus on burning plant matter directly, while others focus on using recently harvested wood or even plankton to produce *biofuels* (such as ethanol) that can be burned. When we derive energy from biomass or biofuels, we are actually tapping into the energy of the sun, which drives the process of photosynthesis in plants.

Certain types of fast-growing trees have shown promise as a source of wood fuel. Willows, for example, grow very quickly and can be cut every three years. The harvested wood is sent to power plants, where it is burned to generate electricity. The willows then resprout. Willows are a good source of biomass because each crop can produce about ten times as much wood per acre as natural forests. Other fast-growing tree species are also being tested as fuel sources.

Fuel from fast-growing trees is not only a renewable energy source; it can also be environmentally friendlier than fossil fuels. Wood has very little of the sulfur compounds (present in fossil fuels) that contribute to acid rain. Fast-growing trees also provide habitat for wildlife, reduce erosion, and help farmers diversify. Perhaps most significantly, vigorously growing young plants remove a lot of carbon dioxide from the air, potentially reducing the levels of atmospheric gases that cause global climate change. Today, biomass energy accounts for only about 4% of all energy consumed in the United States. But this renewable, relatively low-emissions resource may help satisfy our future energy needs.

Biomass fuels are just one way that people take advantage of plants' ability to capture solar energy. The topic of this chapter is photosynthesis, the process whereby plants use light to make sugars from carbon dioxide—sugars that are food for the plant and the starting point for most of our own food. Because photosynthesis can seem complex, we'll examine some basic concepts before looking at the specific mechanisms involved.

BioFlix ™

Photosynthesis

Photosynthesis

The Basics of Photosynthesis

Photosynthesis is a process whereby plants, algae (which are protists), and certain bacteria transform light energy into chemical energy, using carbon dioxide and water as starting materials. The chemical energy produced via photosynthesis is stored in the bonds of sugar molecules. As you read in Chapter 6, organisms that generate their own organic matter from inorganic ingredients are called autotrophs. Plants and other organisms that do this by photosynthesis—photosynthetic autotrophs—are the producers for most ecosystems (**Figure 7.1**). This section presents an overview of photosynthesis, focusing on plants. Later, we'll take a closer look at some details of this process.

Chloroplasts: Sites of Photosynthesis

In Chapter 4, you learned that photosynthesis in plants and algae occurs within light-absorbing organelles called **chloroplasts**. All green parts of a plant have chloroplasts and can carry out photosynthesis. In most plants, however, the leaves have the most chloroplasts (about 500,000 per square millimeter of leaf surface)

and are therefore the major locations of photosynthesis. Their green color is from **chlorophyll**, a light-absorbing pigment in the chloroplasts that plays a central role in converting solar energy to chemical energy.

Chloroplasts are concentrated in the interior cells of leaves (**Figure 7.2**). Carbon dioxide (CO_2) enters, and oxygen (O_2) exits, by way of tiny pores called **stomata** (singular, *stoma*, meaning "mouth"). In addition to carbon dioxide, photosynthesis requires water, which is absorbed by the plant's roots and transported to the leaves, where veins carry it to the photosynthetic cells.

Membranes within the chloroplast form the framework where many of the reactions of photosynthesis occur. Like a mitochondrion, a chloroplast has a double-membrane envelope (see right side of Figure 7.2). The chloroplast's inner membrane encloses a compartment filled with **stroma**, a thick fluid. Suspended in the stroma are interconnected membranous sacs called **thylakoids**. The thylakoids are concentrated in stacks called **grana** (singular, *granum*). The chlorophyll molecules that capture light energy are built into the thylakoid membranes. The structure of a chloroplast—with its stacks of disks—aids its function by providing a large surface area for the reactions of photosynthesis. ☑

▼ Figure 7.1 **Photosynthetic autotrophs: producers for most ecosystems.**

PHOTOSYNTHETIC AUTOTROPHS

Plants (mostly on land)	**Photosynthetic Protists** (aquatic)	**Photosynthetic Bacteria** (aquatic)
Forest plants	Kelp, a large alga	Micrograph of cyanobacteria

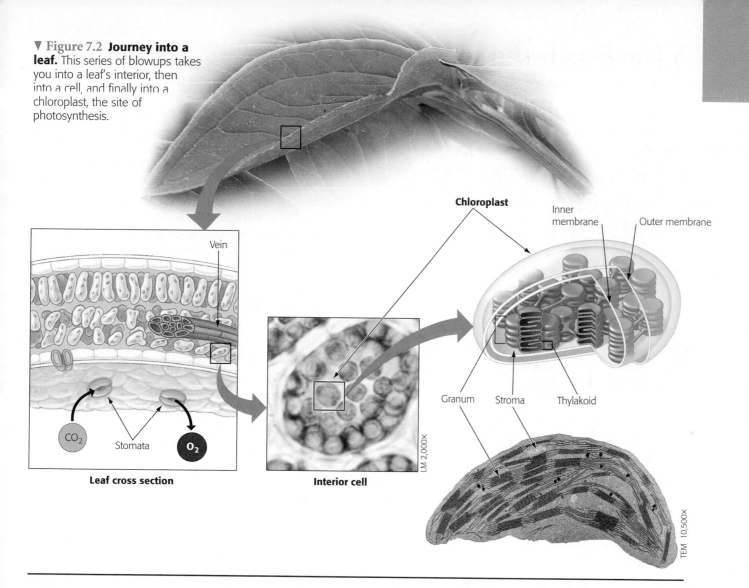

▼ Figure 7.2 **Journey into a leaf.** This series of blowups takes you into a leaf's interior, then into a cell, and finally into a chloroplast, the site of photosynthesis.

Vein

Leaf cross section

CO_2 Stomata O_2

LM 2,000×

Interior cell

Chloroplast

Inner membrane   Outer membrane

Granum   Stroma   Thylakoid

TEM 10,500×

The Overall Equation for Photosynthesis

The following chemical equation, simplified to highlight the relationship between photosynthesis and cellular respiration, provides a summary of the reactants and products of photosynthesis:

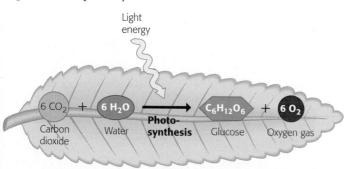

Light energy

$6 CO_2$ + $6 H_2O$ → Photo-synthesis → $C_6H_{12}O_6$ + $6 O_2$

Carbon dioxide   Water   Glucose   Oxygen gas

Notice that the reactants of photosynthesis, carbon dioxide (CO_2) and water (H_2O), are also the waste products of cellular respiration (see Figure 6.2). And photosynthesis produces what respiration uses, namely glucose ($C_6H_{12}O_6$) and oxygen (O_2). In other words,

photosynthesis takes the "exhaust" of cellular respiration and rearranges its atoms to produce food and oxygen. It's a chemical transformation that requires much energy, and sunlight absorbed by chlorophyll provides that energy.

You learned in Chapter 6 that cellular respiration is a process of electron transfer, or reduction and oxidation (redox). A "fall" of electrons from food molecules to oxygen to form water releases the energy that mitochondria can use to make ATP (see Figure 6.5). The opposite occurs in photosynthesis: Electrons are boosted "uphill" and added to carbon dioxide to produce sugar. Hydrogen is moved along with the electrons, so the redox process takes the form of hydrogen transfer from water to carbon dioxide. This transfer of hydrogen requires the chloroplast to split water molecules into hydrogen and oxygen. The hydrogen is transferred along with electrons to carbon dioxide to form sugar. The oxygen escapes through stomata into the atmosphere as O_2, a waste product of photosynthesis. ☑

A Photosynthesis Road Map

The equation for photosynthesis on the previous page is a simple summary of a complex process. Actually, photosynthesis occurs in two stages: the light reactions and the Calvin cycle (Figure 7.3).

In the **light reactions**, chlorophyll in the thylakoid membranes absorbs solar energy, which is then converted to the chemical energy of ATP and NADPH. As we have discussed previously, ATP is the molecule that drives most cellular work. **NADPH** is

an electron carrier. In photosynthesis, light drives electrons from water to NADP⁺ (the oxidized form of the carrier) to form NADPH (the reduced form of the carrier). During the light reactions, water is split, providing a source of electrons and giving off O_2 gas as a by-product. Notice that no sugar is produced during the light reactions.

The **Calvin cycle** uses the products of the light reactions to power the production of sugar from carbon dioxide. The enzymes for the Calvin cycle are dissolved in the stroma, the thick fluid within the chloroplast. ATP generated by the light reactions provides the energy for sugar synthesis. And the NADPH produced by the light reactions provides the high-energy electrons for the reduction of carbon dioxide to glucose. Thus, the Calvin cycle indirectly depends on light to produce sugar because it requires the supply of ATP and NADPH produced by the light reactions.

In the following sections, we'll take a closer look at how these two stages of photosynthesis work: the light reactions first and then the Calvin cycle. ☑

◄ Figure 7.3 **A road map for photosynthesis.** We'll use a smaller version of this road map for orientation as we take a closer look at the light reactions and the Calvin cycle.

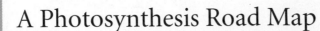

The Light Reactions: Converting Solar Energy to Chemical Energy

Chloroplasts are chemical factories powered by the sun. Let's look at how they convert sunlight into chemical energy.

The Nature of Sunlight

Sunlight is a type of energy called radiation, or electromagnetic energy. Electromagnetic energy travels through space as rhythmic waves analogous to the ripples that are made by a pebble dropped into a pond. The distance between the crests of two adjacent waves is called a **wavelength**. The full range of radiation, from the very short wavelengths of gamma rays to the very long wavelengths of radio signals, is called the **electromagnetic spectrum** (Figure 7.4). Visible light is only a small fraction of the spectrum; it consists of the wavelengths that our eyes see as different colors.

When sunlight shines on a pigmented material, certain wavelengths (colors) of the visible light are absorbed and disappear from the light that is reflected by the material. For example, we see a pair of jeans as blue because pigments in the fabric absorb the other colors, leaving only light in the blue part of the spectrum to be reflected from the fabric to our eyes.

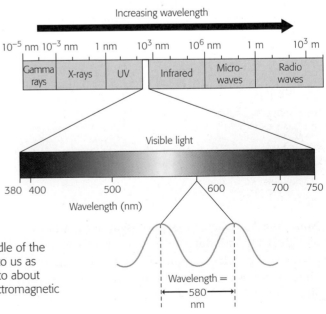

► Figure 7.4 **The electromagnetic spectrum.** The middle of the figure expands the thin slice of the spectrum that is visible to us as different colors of light, from about 380 nanometers (nm) to about 750 nm in wavelength. The bottom of the figure shows electromagnetic waves of one particular wavelength of visible light.

☑CHECKPOINT

Name the two stages of photosynthesis in their proper order.

Answer: light reactions, Calvin cycle

What Colors of Light Drive Photosynthesis?

In 1883, German biologist Theodor Engelmann made the **observation** that certain bacteria living in water tend to cluster in areas with higher oxygen concentrations. He already knew that light passed through a prism would separate into the different wavelengths (colors). Engelmann soon began to **question** whether he could use this information to determine which wavelengths of light work best for photosynthesis.

Engelmann's **hypothesis** was that oxygen-seeking bacteria would congregate near regions of algae performing the most photosynthesis (and hence producing the most oxygen). Engelmann began his **experiment** by laying a string of freshwater algal cells within a drop of water on a microscope slide. He then added oxygen-sensitive bacteria to the drop. Next, using a prism, he created a spectrum of light and shined it on the slide. His **results**, summarized in **Figure 7.5**, showed that most bacteria congregated around algae exposed to red-orange and blue-violet light, with very few bacteria moving to the area of green light. Other experiments have since verified that chloroplasts absorb light mainly in the blue-violet and red-orange part of the spectrum and that those wavelengths of light are the ones mainly responsible for photosynthesis.

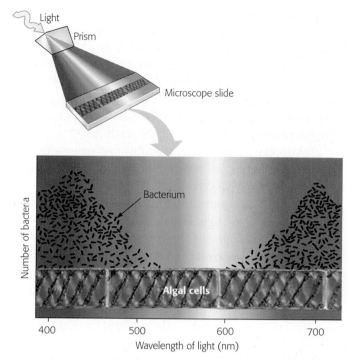

▲ Figure 7.5 **Investigating how light wavelength affects photosynthesis.** When algal cells are suspended in water on a microscope slide, oxygen-seeking bacteria migrate toward algae exposed to certain colors of light. These results suggest that blue-violet and orange-red wavelengths best drive photosynthesis, while green wavelengths hardly do so.

The selective absorption of light by leaves explains why they appear green to us; light of that color is poorly absorbed by chloroplasts and is thus reflected or transmitted toward the observer (**Figure 7.6**). Since energy cannot be destroyed, the absorbed energy must be converted to other forms. Chloroplasts contain pigments that drive the conversion of some of the solar energy they absorb to chemical energy.

▶ Figure 7.6 **Why are leaves green?** Chlorophyll and other pigments in chloroplasts reflect or transmit green light while absorbing other colors.

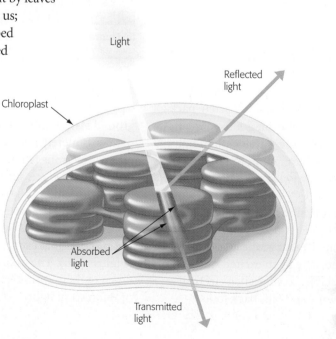

Chloroplast Pigments

Different pigments absorb light of different wavelengths, and chloroplasts contain several kinds of pigments. One, **chlorophyll *a***, absorbs mainly blue-violet and red light. Chlorophyll *a* is the pigment that participates directly in the light reactions. A very similar molecule, chlorophyll *b*, absorbs mainly blue and orange light. Chlorophyll *b* does not participate directly in the light reactions, but it broadens the range of light that a plant can use by conveying absorbed energy to chlorophyll *a*, which then puts the energy to work in the light reactions.

Chloroplasts also contain a family of yellow-orange pigments called carotenoids, which absorb mainly blue-green light. Some pass energy to chlorophyll *a*. Other carotenoids have a protective function: They absorb and dissipate excessive light energy that would otherwise damage chlorophyll. (Similar carotenoids, which we obtain from carrots and certain other plants, may help protect our eyes from bright light.) The spectacular colors of fall foliage in some parts of the world are due partly to the yellow-orange light reflected from carotenoids (**Figure 7.7**). The falling autumn temperatures cause a decrease in the levels of chlorophyll, allowing the colors of the longer-lasting carotenoids to show through.

All of these chloroplast pigments are built into the thylakoid membranes (see Figure 7.2). There the pigments are organized into light-harvesting complexes called photosystems. ☑

How Photosystems Harvest Light Energy

The theory of light as waves explains most of light's properties. However, light also behaves as discrete packets of energy called photons. A **photon** is a fixed quantity of light energy. The shorter the wavelength of light, the greater the energy of a photon. A photon of violet light, for example, packs nearly twice as much energy as a photon of red light.

When a pigment molecule absorbs a photon, one of the pigment's electrons gains energy, and we say that the electron has become "excited"; that is, the electron has been raised from a ground state to an excited state. The excited state is highly unstable, so an excited electron usually loses its excess energy and falls back to its

▼ **Figure 7.7**
Photosynthetic pigments. Falling autumn temperatures cause a decrease in the levels of green chlorophyll within the leaves of deciduous trees. This decrease allows the colors of the carotenoids to be seen.

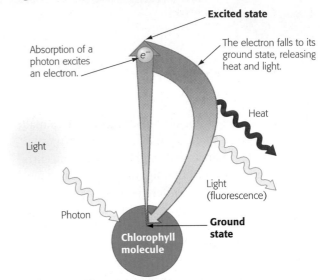

▼ Figure 7.8 **Excited electrons in pigments.**

(a) Absorption of a photon

(b) Fluorescence of a glow stick. Breaking a glass vial within a glow stick starts a chemical reaction that excites electrons within a fluorescent dye. As the electrons fall from their excited state to the ground state, the excess energy is emitted as light.

ground state almost immediately (**Figure 7.8a**). Most pigments merely release heat energy as their light-excited electrons fall back to their ground state. (That's why a dark surface, such as a black automobile hood, gets so hot on a sunny day.) But some pigments emit light as well as heat after absorbing photons. The fluorescent light emitted by a glow stick is caused by a chemical reaction that excites electrons of a fluorescent dye (**Figure 7.8b**). The excited electrons quickly fall back down to their ground state, releasing energy in the form of fluorescent light.

☑ **CHECKPOINT**

What is the specific name of the pigment that absorbs energy during the light reactions?

Answer: chlorophyll a

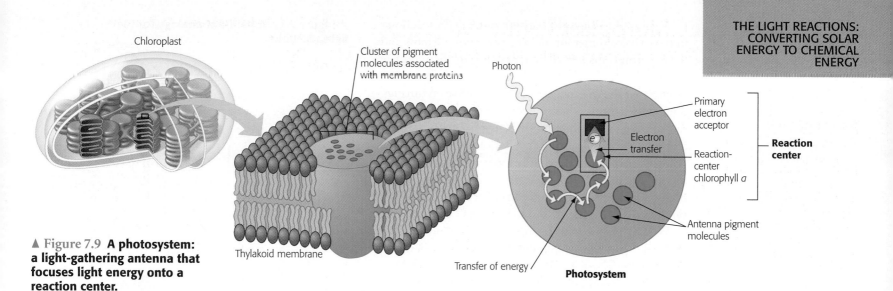

▲ Figure 7.9 **A photosystem:
a light-gathering antenna that
focuses light energy onto a
reaction center.**

In the thylakoid membrane, chlorophyll is organized with other molecules into photosystems. Each **photosystem** has a cluster of a few hundred pigment molecules, including chlorophylls *a* and *b* and some carotenoids (**Figure 7.9**). This cluster of pigment molecules functions as a light-gathering antenna. When a photon strikes one of the pigment molecules, the energy jumps from molecule to molecule until it arrives at the **reaction center** of the photosystem. The reaction center consists of a chlorophyll *a* molecule that sits next to another molecule called a **primary electron acceptor**. This primary electron acceptor traps the light-excited electron (e⁻) from the chlorophyll *a* in the reaction center. Another team of molecules built into the thylakoid membrane then uses that trapped energy to make ATP and NADPH. ☑

☑**CHECKPOINT**

What is the role of a reaction center during photosynthesis?

Answer: A reaction center transfers a light-excited photon from pigment molecules to molecules that can use this trapped energy to drive chemical reactions.

How the Light Reactions Generate ATP and NADPH

Two types of photosystems cooperate in the light reactions (**Figure 7.10**). ❶ Photons excite electrons in the chlorophyll of the water-splitting photosystem. These photons are then trapped by the primary electron acceptor. The water-splitting photosystem replaces its light-excited electrons by extracting electrons from water. This is the step that releases O_2 during photosynthesis. ❷ Energized electrons from the water-splitting photosystem pass down an electron transport chain to the NADPH-producing photosystem. The chloroplast uses the energy released by this electron "fall" to make ATP. ❸ The NADPH-producing photosystem transfers its light-excited electrons to $NADP^+$, reducing it to NADPH.

▶ Figure 7.10 **The light reactions
of photosynthesis.** The orange arrows
trace a light-driven flow of electrons
from H_2O to NADPH. These electrons
also produce ATP.

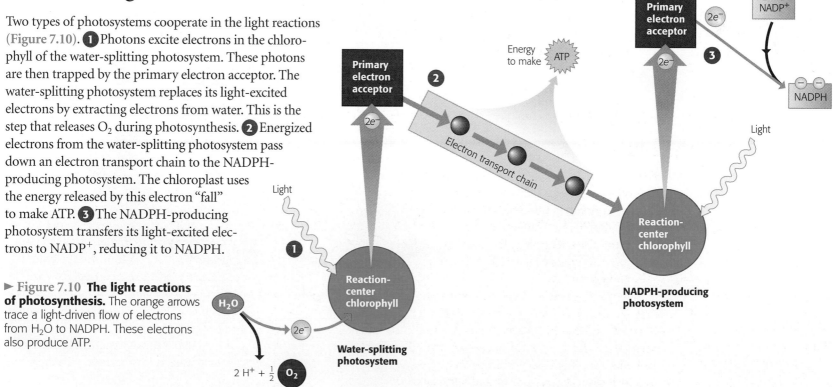

☑**CHECKPOINT**

1. Why is water required as a reactant in photosynthesis? (*Hint*: Review Figures 7.10 and 7.11.)
2. In addition to conveying electrons from the water-splitting photosystem to the NADPH-producing photosystem, the electron transport chains of chloroplasts provide the energy for the synthesis of _____.

Answers: **1.** *The splitting of water provides electrons for converting NAD⁺ to NADPH (which will be used in converting CO₂ to sugar in the Calvin cycle).* **2.** *ATP*

Figure 7.11 shows the location of the light reactions in the thylakoid membrane. The two photosystems and the electron transport chain that connects them transfer electrons from H_2O to $NADP^+$, reducing it to NADPH. Notice that the mechanism of ATP production during the light reactions is very similar to the mechanism we saw in cellular respiration (see Figure 6.11). In both cases, an electron transport chain pumps hydrogen ions (H^+) across a membrane—the inner mitochondrial membrane in the case of respiration and the thylakoid membrane in photosynthesis. And in both cases, ATP synthases use the energy stored by the H^+ gradient to make ATP. The main difference is that food provides the high-energy electrons in cellular respiration, whereas light-excited electrons flow down the transport chain during photosynthesis. The traffic of electrons shown in Figures 7.10 and 7.11 is analogous to the cartoon in **Figure 7.12**.

We have seen how the light reactions absorb solar energy and convert it to the chemical energy of ATP and NADPH. Notice again, however, that the light reactions produce no sugar. That's the job of the Calvin cycle, as we'll see next. ☑

▼ Figure 7.12 **A hard-hat analogy for the light reactions.**

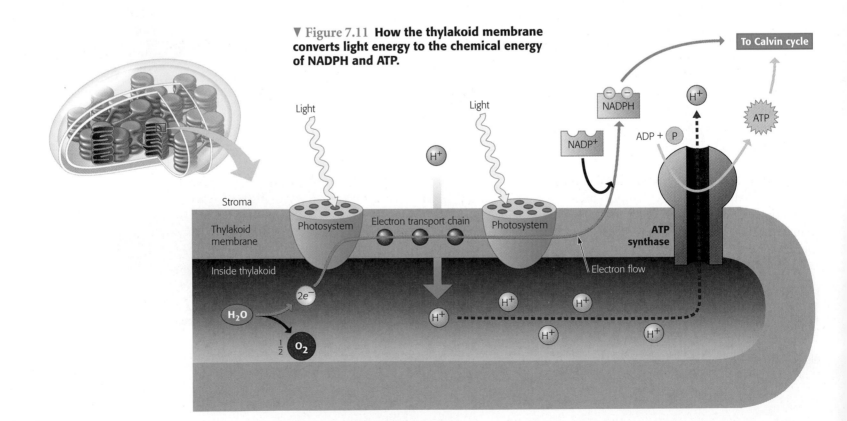

▼ Figure 7.11 **How the thylakoid membrane converts light energy to the chemical energy of NADPH and ATP.**

The Calvin Cycle: Making Sugar from Carbon Dioxide

The Calvin cycle functions like a sugar factory within the stroma of a chloroplast. It is called a cycle because, like the citric acid cycle in cellular respiration, the starting material is regenerated with each turn of the cycle. And with each turn, there are chemical inputs and outputs. The inputs are CO_2 from the air as well as ATP and NADPH produced by the light reactions. Using carbon from CO_2, energy from ATP, and high-energy electrons from NADPH, the Calvin cycle constructs an energy-rich sugar molecule called glyceraldehyde 3-phosphate (G3P). The plant cell can then use G3P as the raw material to make the glucose and other organic compounds (such as cellulose and starch) that it needs. **Figure 7.13** presents the basics of the Calvin cycle, emphasizing inputs and outputs. Each ● represents a carbon atom and each ⓟ represents a phosphate group. ☑

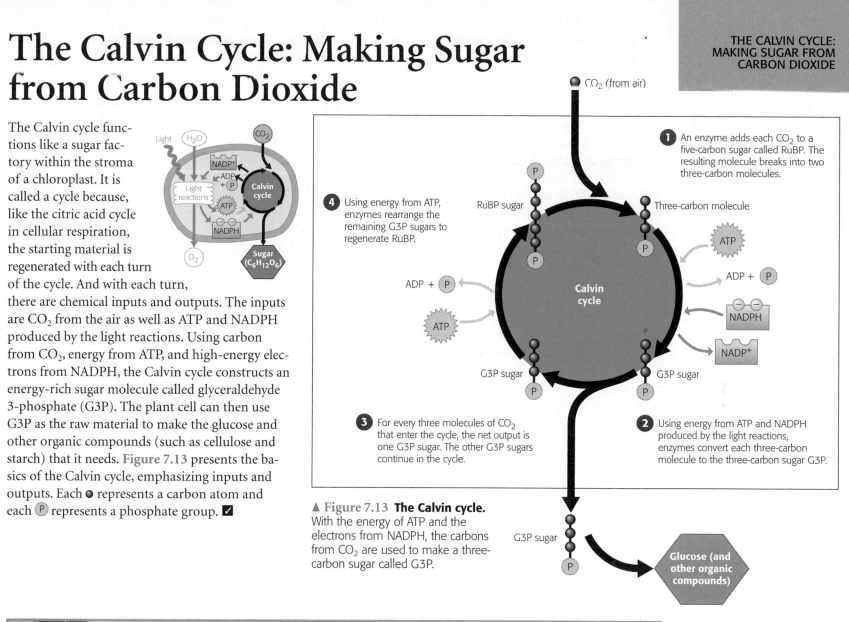

1 An enzyme adds each CO_2 to a five-carbon sugar called RuBP. The resulting molecule breaks into two three-carbon molecules.

2 Using energy from ATP and NADPH produced by the light reactions, enzymes convert each three-carbon molecule to the three-carbon sugar G3P.

3 For every three molecules of CO_2 that enter the cycle, the net output is one G3P sugar. The other G3P sugars continue in the cycle.

4 Using energy from ATP, enzymes rearrange the remaining G3P sugars to regenerate RuBP.

▲ **Figure 7.13** **The Calvin cycle.** With the energy of ATP and the electrons from NADPH, the carbons from CO_2 are used to make a three-carbon sugar called G3P.

EVOLUTION CONNECTION: Harnessing Solar Energy

Solar-Driven Evolution

Throughout this chapter, you've studied how plants convert solar energy to chemical energy via photosynthesis. In the Process of Science section, you learned that plants can best use certain wavelengths of light to drive these reactions. Sunlight strikes Earth differently in different geographic regions. Since natural selection promotes the evolution of adaptations that best suit a local environment, it's not surprising that different modes of photosynthesis have evolved in different climates.

So far, we've focused on plants that use CO_2 directly from the air to drive the Calvin cycle. Such plants are called **C_3 plants** because the first organic compound produced in the Calvin cycle is a three-carbon molecule (see Figure 7.13). C_3 plants are widely distributed and include such common crops as soybeans, oats, wheat, and rice. One problem that farmers face in growing C_3 plants, however, is that plants close their stomata on hot, dry days. This adaptation reduces water loss but also prevents CO_2 from entering the leaves. As a result, CO_2 levels get very low in the leaves, and sugar production ceases. Thus, when the weather turns hot, the productivity of C_3 crops slows significantly because of the need to save water.

In hot, dry climates, alternative modes of incorporating carbon from CO_2 have evolved in some plants, allowing them to save water without shutting down photosynthesis. These species are categorized as either C_4 plants or CAM plants.

C₄ plants are so named because they incorporate carbon from CO_2 into a four-carbon compound before proceeding to the Calvin cycle (**Figure 7.14**, left side). When the weather is hot and dry, a C₄ plant keeps its stomata mostly closed, thus conserving water. But it has an enzyme that can continue to incorporate carbon even when the leaf's CO_2 concentration is low. The resulting four-carbon compound then acts as a carbon shuttle; it donates the CO_2 to the Calvin cycle in a nearby cell, which keeps on making sugars even though the plant's stomata are mostly closed. Corn and sugarcane are examples of some agriculturally important C₄ plants.

CAM plants include pineapples, many cacti, and succulent (water-retaining) plants such as aloe and jade. Adapted to very dry climates, these species conserve water by opening their stomata and admitting CO_2 only at night (see the right side of Figure 7.14). The CO_2 is incorporated into a four-carbon compound that banks it at night and releases it to the Calvin cycle in the same cell during the day. This process keeps photosynthesis operating when the stomata are closed during the day.

The C₄ and CAM pathways are two evolutionary adaptations that maintain photosynthesis with stomata partially or completely closed on hot, dry days. The distribution of C₄ and CAM plants on Earth today reflects this evolutionary history; such plants are found in hot climates where the need to conserve water outweighs the need for continuous sugar production. During the long evolution of plants in diverse environments, natural selection has refined photosynthetic adaptations that enable these plants to continue producing food even in arid conditions.

▼ Figure 7.14 **C₄ and CAM photosynthesis.** The C₄ and CAM pathways are two evolutionary adaptations that maintain photosynthesis with stomata partially or completely closed on hot, dry days.

ALTERNATIVE PHOTOSYNTHETIC PATHWAYS

C₄ Pathway (example: sugarcane)	**CAM Pathway** (example: pineapple)

Cell type 1 · CO_2
Four-carbon compound
Cell type 2 · CO_2
Calvin cycle
Sugar
C₄ plant

CO_2 · Night
Four-carbon compound
CO_2
Calvin cycle
Sugar · Day
CAM plant

Carbon incorporation and the Calvin cycle occur in different types of cells.

Carbon incorporation and the Calvin cycle occur in the same cells at different times.

Chapter Review

SUMMARY OF KEY CONCEPTS

(MB) Go to the Study Area at **www.masteringbiology.com** for practice quizzes, myeBook, BioFlix™ 3-D animations, MP3 Tutor Sessions, videos, current events, and more.

The Basics of Photosynthesis

Photosynthesis is a process whereby light energy is transformed into chemical energy, which is stored as bonds in sugars made from carbon dioxide and water.

Chloroplasts: Sites of Photosynthesis

Chloroplasts contain a thick fluid called stroma surrounding a network of membranes called thylakoids.

The Overall Equation for Photosynthesis

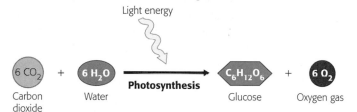

A Photosynthesis Road Map

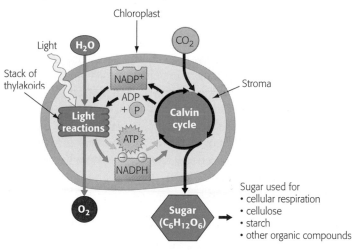

The Light Reactions: Converting Solar Energy to Chemical Energy

The Nature of Sunlight

Visible light is part of the spectrum of electromagnetic energy. It travels through space as waves.

Chloroplast Pigments

Pigment molecules absorb light energy of certain wavelengths and reflect other wavelengths. We see the reflected wavelengths as the color of the pigment. Several chloroplast pigments absorb light of various wavelengths, but it is the green pigment chlorophyll a that participates directly in the light reactions.

How Photosystems Harvest Light Energy; How the Light Reactions Generate ATP and NADPH

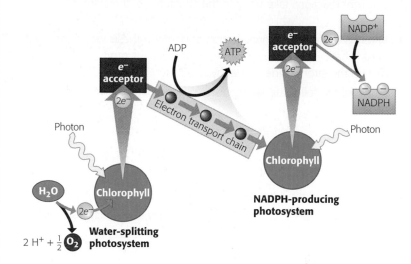

The Calvin Cycle: Making Sugar from Carbon Dioxide

Within the stroma (fluid) of the chloroplast, carbon dioxide from the air and ATP and NADPH produced during the light reactions are used to produce G3P, an energy-rich sugar molecule that can be used to make glucose and other organic molecules.

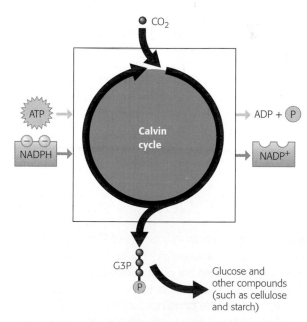

Evolution Connection: Solar-Driven Evolution

The photosynthetic adaptations of C_4 and CAM plants enable sugar production to continue even when stomata are closed, thereby reducing water loss in arid environments.

SELF-QUIZ

1. The light reactions take place in the structures of the chloroplast called the _____, while the Calvin cycle takes place in the _____.

2. In terms of the spatial organization of photosynthesis within the chloroplast, what is the advantage of the light reactions producing NADPH and ATP on the stroma side of the thylakoid membrane?

3. Which of the following are inputs to photosynthesis? Which are outputs?
   a. CO_2
   b. O_2
   c. sugar
   d. H_2O
   e. light

4. What color of light is the least effective in driving photosynthesis? Why?

5. When light strikes chlorophyll molecules, they lose electrons, which are ultimately replaced by splitting molecules of _____.

6. Which of the following are produced by reactions that take place in the thylakoids and are consumed by reactions in the stroma?
   a. CO_2 and H_2O
   b. $NADP^+$ and ADP
   c. ATP and NADPH
   d. glucose and O_2

7. The reactions of the Calvin cycle are not directly dependent on light, and yet they usually do not occur at night. Why?

8. Why is it difficult for most plants to carry out photosynthesis in very hot, dry environments, such as deserts?

9. What is the primary advantage offered by the C_4 and CAM pathways?

10. Of the following metabolic processes, which one is common to photosynthesis and cellular respiration?
    a. reactions that convert light energy to chemical energy
    b. reactions that split H_2O molecules and release O_2
    c. reactions that store energy by pumping H^+ across membranes
    d. reactions that convert CO_2 to sugar

Answers to the Self-Quiz questions can be found in Appendix D.

THE PROCESS OF SCIENCE

11. Tropical rain forests cover only about 3% of Earth's surface, but they are estimated to be responsible for more than 20% of global photosynthesis. For this reason, rain forests are often referred to as the "lungs" of the planet, providing O_2 for life all over Earth. However, most experts believe that rain forests make little or no net contribution to global O_2 production. From your knowledge of photosynthesis and cellular respiration, can you explain why they might think this? (*Hint*: What happens to the energy stored as sugars in the body of a plant when that plant dies or parts of it are eaten by animals?)

12. Suppose you wanted to discover whether the oxygen atoms in the glucose produced by photosynthesis come from H_2O or CO_2. Explain how you could use a radioactive isotope to find out.

BIOLOGY AND SOCIETY

13. There is strong evidence that Earth is getting warmer because of an intensified greenhouse effect resulting from increased CO_2 emissions from industry, vehicles, and the burning of forests. Global warming could influence agriculture, melt polar ice, and flood coastal regions. In response to these threats, 178 countries have accepted the Kyoto agreement, which calls for mandatory reductions of greenhouse gas emissions in 30 industrialized nations by 2012. As of the start of 2009, the United States had not signed the agreement, instead proposing a more modest set of voluntary goals allowing businesses to decide whether they wish to participate and providing tax incentives to encourage them to do so. The reasons given for rejecting the agreement are that it might hurt the American economy and that some less industrialized countries (such as India) are exempted from it, even though they produce a lot of pollution. Do you agree with this decision? In what ways might efforts to reduce greenhouse gases hurt the economy? How can those costs be weighed against the costs of global warming? Should poorer nations carry an equal burden to reduce their emissions?

14. As discussed in the Biology and Society section, burning biomass to produce electricity avoids many of the problems associated with gathering, refining, transporting, and burning fossil fuels. Yet the use of biomass as fuel is not without its own set of problems. What challenges might arise from a large-scale conversion to biomass energy? How do these challenges compare with those encountered with fossil fuels? Which set of challenges do you think is more likely to be overcome? Does one energy source have more benefits and fewer costs than the others? Explain.

Unit 2
Genetics

Chapter 8: **Cellular Reproduction: Cells from Cells**

Chapter Thread: **Life with and without Sex**

Chapter 9: **Patterns of Inheritance**

Chapter Thread: **Dog Breeding**

Chapter 10: **The Structure and Function of DNA**

Chapter Thread: **The Flu**

Chapter 11: **How Genes Are Controlled**

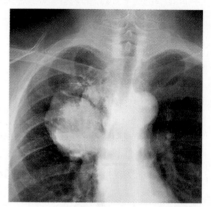

Chapter Thread: **Cancer**

Chapter 12: **DNA Technology**

Chapter Thread: **DNA Profiling**

8 Cellular Reproduction: Cells from Cells

Reproduction in the rain forest.
Botanists attempted to save the Hawaiian bellflower (visible in foreground) by means of sexual and asexual reproduction.

CHAPTER CONTENTS

BIOLOGY AND SOCIETY: Life with and without Sex

Rain Forest Rescue

Deep in a Hawaiian rain forest, the thrumming of a helicopter cut through the dense tropical air. After touching down in a clearing, a pair of rescuers jumped out, one of them carrying a precious bundle. A 20-minute hike through dense foliage brought them to their "patient": the wild *Cyanea kuhihewa*, a plant in the bellflower family. The rescuers were in the middle of a life-or-death battle, armed with the pollen of a species on the very brink of extinction. The stakes couldn't have been higher: The small bellflower before them was the last surviving member of its species in the wild.

The rescuers, scientists from the National Tropical Botanical Garden, were trying to promote sexual reproduction, the reproductive process that involves fertilization, the union of a sperm and an egg. Using a fine brush, the botanists transferred sperm-carrying pollen from a garden-grown plant onto the egg-containing wild bloom. Their hope was to initiate the development of a new seed, and from it, a new plant.

Unfortunately, the bellflower fertilization attempt failed. Even worse, the last remaining wild specimen died in 2003. But hope remains for *C. kuhihewa*. The garden's botanists have also propagated the plant by asexual reproduction, the production of offspring by a single parent. Many plants that normally reproduce sexually can be induced to reproduce asexually in the laboratory. Indeed, work at the botanical garden has produced several new bellflower plants in this way, using stem tissue snipped from wild plants in the 1990s.

Cell division is at the heart of organismal reproduction, whether by sexual or asexual means. The ability of organisms to reproduce their own kind is the one characteristic that best distinguishes living things from nonliving matter. And the perpetuation of life depends on the production of new cells. Plants created from cuttings result from repeated cell divisions, as does the development of a multicellular organism from a fertilized egg. And eggs and sperm themselves result from cell division of a special kind. In this chapter, we'll look at how individual cells reproduce and then see how cell reproduction underlies the process of sexual reproduction. During our discussion, we'll consider examples of asexual and sexual reproduction among both plants and animals.

What Cell Reproduction Accomplishes

When you hear the word *reproduction*, you probably think of the birth of new organisms. But reproduction actually occurs much more often at the cellular level. Consider the skin on your arm. Skin cells are constantly reproducing themselves and moving outward toward the surface, replacing dead cells that have rubbed off. This renewal of your skin goes on throughout your life. And when your skin is injured, additional cell reproduction helps heal the wound.

When a cell undergoes reproduction, or **cell division**, the two "daughter" cells that result are genetically identical to each other and to the original "parent" cell. (Biologists traditionally use the word *daughter* in this context; it does not imply gender.) Before the parent cell splits into two, it duplicates its **chromosomes**, the structures that contain most of the organism's DNA. Then, during the division process, one set of chromosomes is distributed to each daughter cell. As a rule, the daughter cells receive identical sets of chromosomes.

As summarized in **Figure 8.1**, cell division plays several important roles in the lives of organisms. For example, within your body, millions of cells must divide every second to replace damaged or lost cells. Another function of cell division is growth. All of the trillions of cells in your body are the result of repeated cell divisions that began in your mother's body with a single fertilized egg cell.

Another vital function of cell division is reproduction of the organism. Single-celled organisms, such as amoebas,

reproduce by dividing in half, and the offspring are genetic replicas of the parent. Because it does not involve fertilization of an egg by a sperm, this type of reproduction is called **asexual reproduction**. Offspring produced by asexual reproduction inherit all of their chromosomes from a single parent. Many multicellular organisms can reproduce asexually as well. For example, some sea star species have the ability to grow new individuals from fragmented pieces. And if you've ever grown a houseplant from a clipping, you've observed asexual reproduction in plants. In asexual reproduction, there is one simple principle of inheritance: The lone parent and each of its offspring have identical genes. The type of cell division responsible for asexual reproduction and for the growth and maintenance of multicellular organisms is called mitosis.

Sexual reproduction is different; it requires fertilization of an egg by a sperm. The production of egg and sperm cells involves a special type of cell division called meiosis, which occurs only in reproductive organs (such as testes and ovaries in humans). As we'll discuss later, a sperm or egg cell has only half as many chromosomes as the parent cell that gave rise to it.

Note that two kinds of cell division are involved in the lives of sexually reproducing organisms: meiosis for reproduction and mitosis for growth and maintenance. The remainder of the chapter is divided into two main sections, one for each type of cell division. ☑

☑CHECKPOINT

Ordinary cell division produces two daughter cells that are genetically identical. Name three functions of this type of cell division.

Answer: cell replacement, growth of an organism, asexual reproduction of an organism

---

► Figure 8.1 **Three functions of cell division.**

FUNCTIONS OF CELL DIVISION

Cell Replacement	Growth via Cell Division

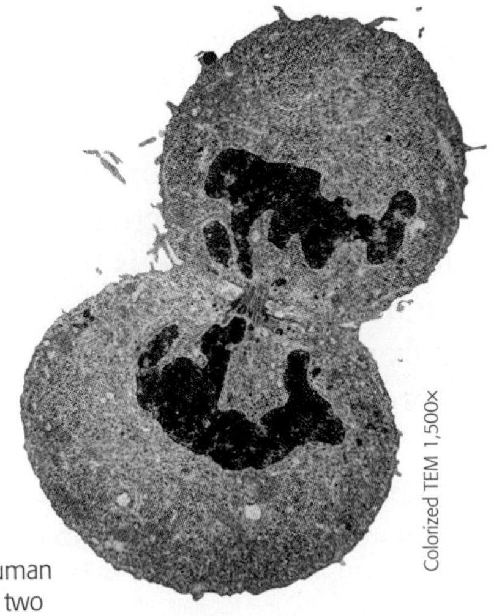

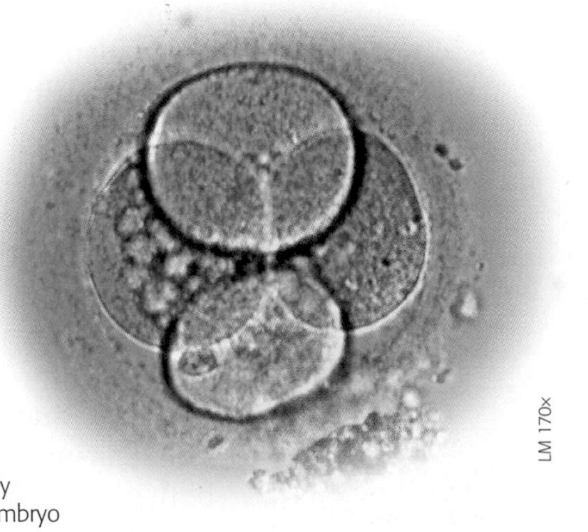

Division of a human kidney cell into two

Colorized TEM 1,500x

The cells of an early human embryo

LM 170x

The Cell Cycle and Mitosis

Almost all of the genes of a eukaryotic cell—around 21,000 in humans—are located on chromosomes in the cell nucleus. (The main exceptions are genes on small DNA molecules found in mitochondria and chloroplasts.) Because chromosomes are the lead players in cell division, let's focus on them before turning our attention to the cell as a whole.

Mitosis

Mitosis

Eukaryotic Chromosomes

Each eukaryotic chromosome contains one very long DNA molecule, typically bearing thousands of genes. The number of chromosomes in a eukaryotic cell depends on the species (**Figure 8.2**). Chromosomes are made up of a material called **chromatin**, a combination of DNA and protein molecules. The protein molecules help organize the chromatin and help control the activity of its genes.

Most of the time, the chromosomes exist as a diffuse mass of fibers that are much longer than the nucleus they are stored in. In fact, if stretched out, the DNA in just one of your cells would be taller than you! As a cell prepares to divide, its chromatin

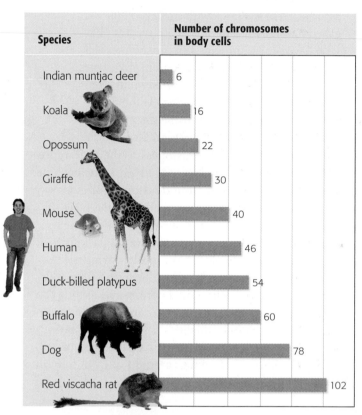

Species	Number of chromosomes in body cells
Indian muntjac deer	6
Koala	16
Opossum	22
Giraffe	30
Mouse	40
Human	46
Duck-billed platypus	54
Buffalo	60
Dog	78
Red viscacha rat	102

▶ Figure 8.2 **The number of chromosomes in the cells of selected mammals.** Notice that humans have 46 chromosomes and that the number of chromosomes does not correspond to the size or complexity of an organism.

Asexual Reproduction

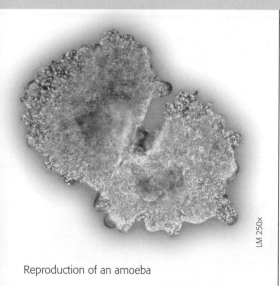

LM 250x

Reproduction of an amoeba

Fragmentation and regeneration of a sea star. The sea star on the right lost and replaced an arm. The severed arm grew into the new sea star on the left.

Reproduction of an African violet from a clipping (large leaf)

fibers coil up, forming compact chromosomes. When they are in this state, chromosomes are clearly visible under the light microscope, as shown in the plant cell in **Figure 8.3.** When a cell is not dividing, the chromosomes are too thin to be seen in a light micrograph.

Such long molecules of DNA can fit into the tiny nucleus because within each chromosome the DNA is packed into an elaborate, multilevel system of coiling and folding. A crucial aspect of DNA packing is the association of the DNA with small proteins called **histones**, found only in eukaryotes. (Bacteria have similar proteins, but prokaryotes lack the degree of DNA packing found in eukaryotes.)

Figure 8.4 presents a simplified model for the main levels of DNA packing. At the first level of packing, shown near the top, histones attach to the DNA. In electron micrographs, the combination of DNA and histones has the appearance of beads on a string. Each "bead," called a **nucleosome**, consists of DNA wound around histone molecules. When not dividing, the DNA of active genes takes on this lightly packed, "beads on a string" arrangement. When preparing to divide, chromosomes pack further. At the next level of packing, the beaded string is wrapped into a tight helical fiber. Then this fiber coils further into a thick supercoil. Looping and folding can further compact the DNA, as you can see in the chromosome at the bottom of the figure. Viewed as a whole, Figure 8.4 gives a sense of how successive levels of coiling and folding enable a huge amount of DNA to fit into a cell's tiny nucleus.

Before a cell begins the division process, it duplicates all of its chromosomes. The DNA molecule of each chromosome is copied through the process of DNA replication (see Chapter 10), and new protein molecules

▼ Figure 8.4 **DNA packing in a eukaryotic chromosome.** Successive levels of coiling of DNA and associated proteins ultimately results in highly compacted chromosomes.

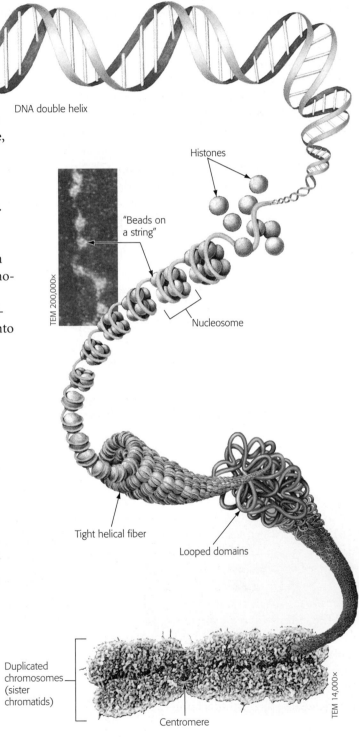

DNA double helix

Histones

"Beads on a string"

TEM 200,000x

Nucleosome

Tight helical fiber

Looped domains

Duplicated chromosomes (sister chromatids)

Centromere

TEM 14,000x

▼ Figure 8.3 **A plant cell just before division (colored by stains).**

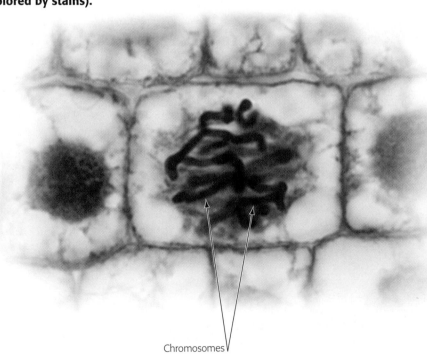

LM 1,400x

Chromosomes

attach as needed. The result is that each chromosome now consists of two copies called **sister chromatids**, which contain identical genes. At the bottom of Figure 8.4, you can see an electron micrograph of a human chromosome that has duplicated. The fuzzy appearance comes from the intricate twists and folds of the chromatin fibers. The two sister chromatids are joined together tightly at a narrow "waist" called the **centromere**.

When the cell divides, the sister chromatids of a duplicated chromosome separate from each other, as shown in the simple diagram in **Figure 8.5**. Once separated from its sister, each chromatid is considered a full-fledged chromosome, and it is identical to the original chromosome. One of the new chromosomes goes to one daughter cell, and the other goes to the other daughter cell. In this way, each daughter cell receives a complete and identical set of chromosomes. A dividing human skin cell, for example, has 46 duplicated chromosomes, and each of the two daughter cells that result from it has 46 single chromosomes.

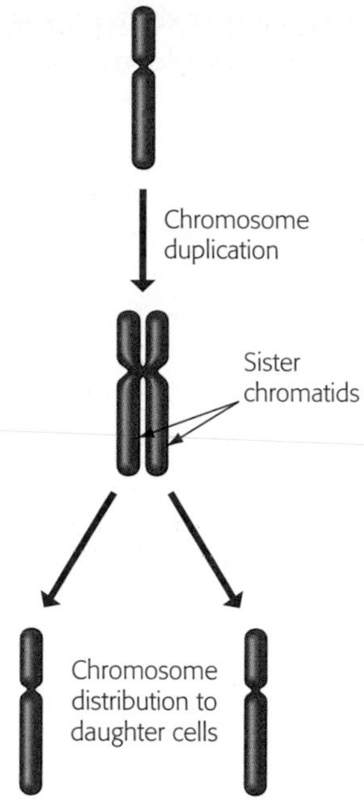

The Cell Cycle

How do chromosome duplication and cell division fit into the life of a cell? The rate at which a cell divides depends on its role within the organism's body. Some cells divide once a day, others less often, and highly specialized cells, such as mature muscle cells, not at all.

The ordered sequence of events that extends from the time a cell is first formed from a dividing parent cell until its own division into two cells is called the **cell cycle**. As **Figure 8.6** shows, most of the cell cycle is spent in **interphase**. This is a time when a cell performs its normal functions within the organism. For example, a cell in your stomach lining might make and release enzyme molecules that aid in digestion. During interphase, a cell roughly doubles everything in its cytoplasm. It increases its supply of proteins, increases the number of many of its organelles (such as mitochondria and ribosomes),

Chromosome duplication

Sister chromatids

Chromosome distribution to daughter cells

▲ **Figure 8.5 Duplication and distribution of a single chromosome.** During cell reproduction, the cell duplicates each chromosome and distributes the two copies to the daughter cells.

and grows in size. Typically, interphase lasts for at least 90% of the cell cycle.

From the standpoint of cell reproduction, the most important event of interphase is chromosome duplication, when the DNA in the nucleus is precisely doubled. This occurs approximately in the middle of interphase, and the period when it is occurring is called the S phase (for DNA *synthesis*). The interphase periods before and after the S phase are called the G_1 and G_2 phases, respectively (G stands for *gap*). During G_2, each chromosome in the cell consists of two identical sister chromatids, and the cell is preparing to divide.

The part of the cell cycle when the cell is actually dividing is called the **mitotic (M) phase**. It includes two overlapping processes, mitosis and cytokinesis. In **mitosis**, the nucleus and its contents, notably the duplicated chromosomes, divide and are evenly distributed, forming two daughter nuclei. In **cytokinesis**, the cytoplasm is divided in two. Cytokinesis usually begins before mitosis is completed. The combination of mitosis and cytokinesis produces two genetically identical daughter cells, each with a single nucleus, the surrounding cytoplasm with organelles, and a plasma membrane. ✓

▼ **Figure 8.6 The eukaryotic cell cycle.** The cell cycle extends from the "birth" of a cell (just after the dark blue arrow at the bottom of the cycle), resulting from cell reproduction, to the time the cell itself divides in two. (During interphase, the chromosomes are diffuse masses of thin fibers; they do not actually appear in the rodlike form you see here.)

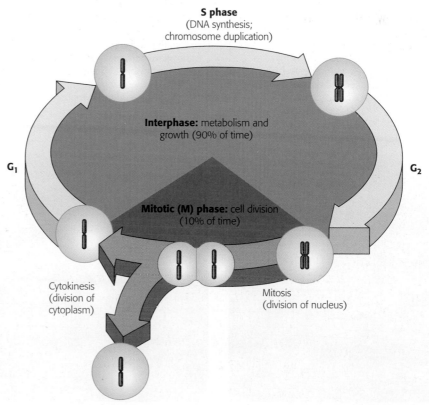

S phase
(DNA synthesis; chromosome duplication)

G_1

G_2

Interphase: metabolism and growth (90% of time)

Mitotic (M) phase: cell division (10% of time)

Cytokinesis (division of cytoplasm)

Mitosis (division of nucleus)

Mitosis and Cytokinesis

Figure 8.7 illustrates the cell cycle for an animal cell using drawings, texts, and photos. The micrographs running along the bottom of the page show dividing cells from a salamander, with chromosomes appearing in blue. The drawings in the top row (which are simplified to include just four chromosomes) include details that are not visible in the micrographs. The text within the figure describes the events occurring at each stage. As you study the figure, you'll see striking changes in the nucleus and other cellular structures.

▼ **Figure 8.7 Cell reproduction: A dance of the chromosomes.** After the chromatin doubles during interphase, the elaborately choreographed stages of mitosis—prophase, metaphase, anaphase, and telophase—distribute the duplicate sets of chromosomes to two separate nuclei. Cytokinesis then divides the cytoplasm, yielding two genetically identical daughter cells.

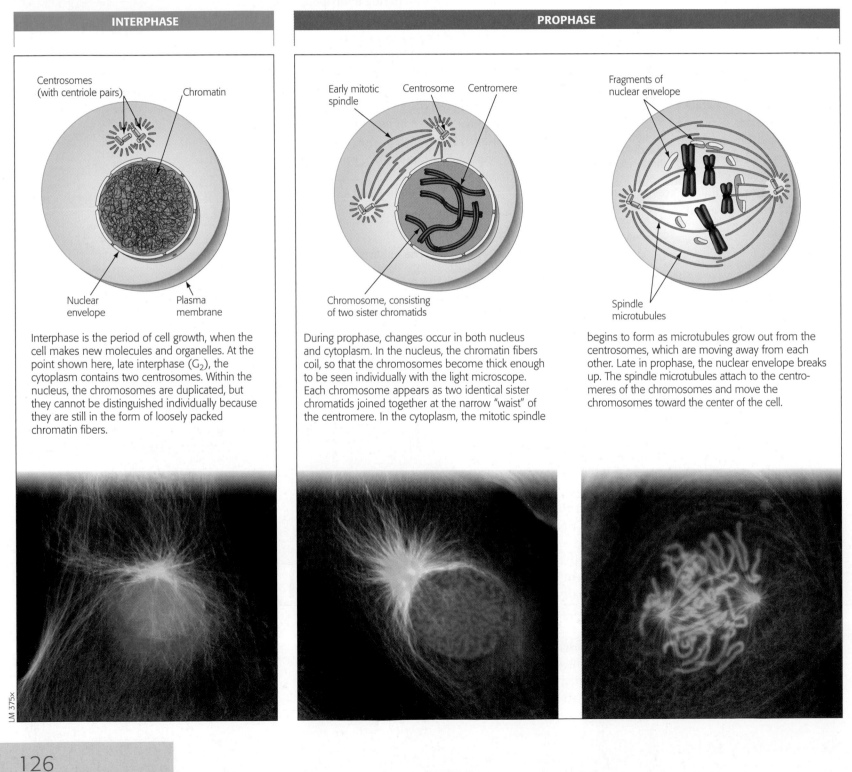

INTERPHASE

Centrosomes
(with centriole pairs)

Chromatin

Nuclear
envelope

Plasma
membrane

Interphase is the period of cell growth, when the cell makes new molecules and organelles. At the point shown here, late interphase (G_2), the cytoplasm contains two centrosomes. Within the nucleus, the chromosomes are duplicated, but they cannot be distinguished individually because they are still in the form of loosely packed chromatin fibers.

PROPHASE

Early mitotic
spindle

Centrosome

Centromere

Chromosome, consisting
of two sister chromatids

Fragments of
nuclear envelope

Spindle
microtubules

During prophase, changes occur in both nucleus and cytoplasm. In the nucleus, the chromatin fibers coil, so that the chromosomes become thick enough to be seen individually with the light microscope. Each chromosome appears as two identical sister chromatids joined together at the narrow "waist" of the centromere. In the cytoplasm, the mitotic spindle begins to form as microtubules grow out from the centrosomes, which are moving away from each other. Late in prophase, the nuclear envelope breaks up. The spindle microtubules attach to the centromeres of the chromosomes and move the chromosomes toward the center of the cell.

LM 375x

Mitosis is a continuum, but biologists distinguish four main stages: **prophase**, **metaphase**, **anaphase**, and **telophase**. The chromosomes are the stars of the mitotic drama, and their movements depend on the **mitotic spindle**, a football-shaped structure of microtubules that guides the separation of the two sets of daughter chromosomes (see Figure 4.21a to review microtubules; in the micrographs of Figure 8.7, they appear green). The spindle microtubules grow from two **centrosomes**, clouds of cytoplasmic material that in animal cells contain centrioles. (Centrioles are can-shaped structures made of microtubules; see Figure 4.5.)

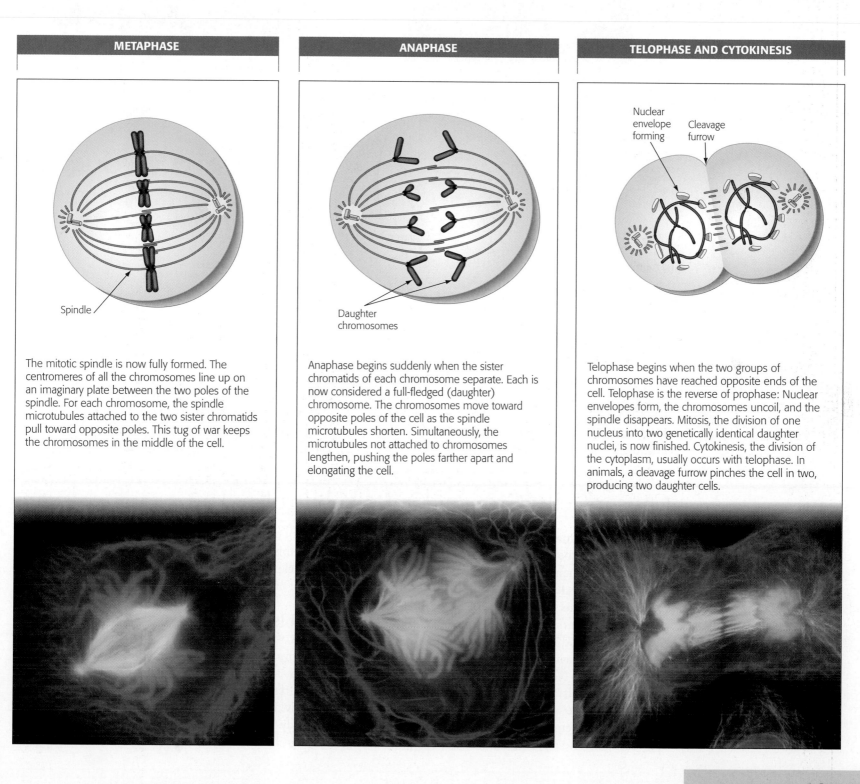

METAPHASE

Spindle

The mitotic spindle is now fully formed. The centromeres of all the chromosomes line up on an imaginary plate between the two poles of the spindle. For each chromosome, the spindle microtubules attached to the two sister chromatids pull toward opposite poles. This tug of war keeps the chromosomes in the middle of the cell.

ANAPHASE

Daughter chromosomes

Anaphase begins suddenly when the sister chromatids of each chromosome separate. Each is now considered a full-fledged (daughter) chromosome. The chromosomes move toward opposite poles of the cell as the spindle microtubules shorten. Simultaneously, the microtubules not attached to chromosomes lengthen, pushing the poles farther apart and elongating the cell.

TELOPHASE AND CYTOKINESIS

Nuclear envelope forming   Cleavage furrow

Telophase begins when the two groups of chromosomes have reached opposite ends of the cell. Telophase is the reverse of prophase: Nuclear envelopes form, the chromosomes uncoil, and the spindle disappears. Mitosis, the division of one nucleus into two genetically identical daughter nuclei, is now finished. Cytokinesis, the division of the cytoplasm, usually occurs with telophase. In animals, a cleavage furrow pinches the cell in two, producing two daughter cells.

▼ Figure 8.8 **Cytokinesis in animal and plant cells.**

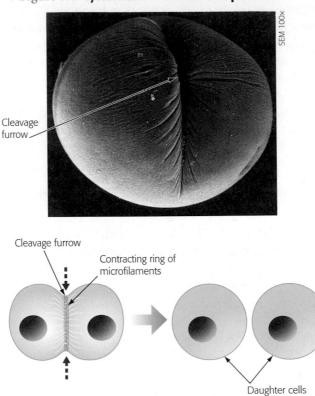

(a) **Animal cell cytokinesis**

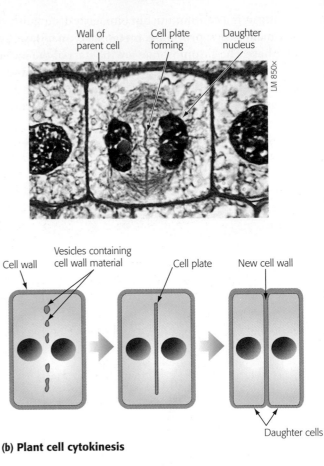

(b) **Plant cell cytokinesis**

Cytokinesis, the division of the cytoplasm into two genetically identical daughter cells, typically occurs during telophase. In animal cells, the cytokinesis process is known as cleavage. The first sign of cleavage is the appearance of a **cleavage furrow**, an indentation at the equator of the cell. A ring of microfilaments in the cytoplasm just under the plasma membrane contracts, like the pulling of a drawstring, deepening the furrow and pinching the parent cell in two (**Figure 8.8a**).

Cytokinesis in a plant cell occurs differently. Membranous vesicles containing cell wall material collect at the middle of the cell. The vesicles gradually fuse, forming a membranous disk called the **cell plate**. The cell plate grows outward, accumulating more cell wall material as more vesicles join it. Eventually, the membrane of the cell plate fuses with the plasma membrane, and the cell plate's contents join the parental cell wall. The result is two daughter cells (**Figure 8.8b**). ✓

Cancer Cells: Growing Out of Control

For a plant or animal to grow and maintain its tissues normally, it must be able to control the timing of cell division. The sequential events of the cell cycle are directed by a **cell cycle control system** that consists of specialized proteins within the cell. These proteins integrate information from the environment and from other body cells and send "stop" and "go-ahead" signals at certain key points during the cell cycle using signal transduction pathways (see Figure 5.19). For example, the cell cycle normally halts within the G_1 phase of interphase unless the cell receives a go-ahead signal via certain cell cycle control proteins. If that signal never arrives, the cell will switch into a permanently nondividing state.

Some of our nerve and muscle cells, for example, are arrested this way. If the go-ahead signal is received and the G_1 checkpoint is passed, the cell will usually complete the rest of the cycle.

What Is Cancer?

Cancer, which currently claims the lives of one out of every five people in the United States and other industrialized nations, is a disease of the cell cycle. Cancer cells do not respond normally to the cell cycle control system; they divide excessively and can invade other tissues of the body. If unchecked, cancer cells may continue to divide until they kill the host.

The abnormal behavior of cancer cells begins when a single cell undergoes transformation, a process whereby a normal cell converts to a cancer cell. Transformation occurs after a genetic change (mutation) in one or more genes that encode for proteins in the cell cycle control system. Because a transformed cell grows abnormally, the immune system normally recognizes and destroys it. However, if the cell evades destruction, it may proliferate to form a **tumor**, an abnormally growing mass of body cells. If the abnormal cells remain at the original site, the lump is called a **benign tumor**. Benign tumors can cause problems if they grow large and disrupt certain organs, such as the brain, but often they can be completely removed by surgery.

In contrast, a **malignant tumor** can spread into neighboring tissues and other parts of the body, displacing normal tissue and interrupting organ function (Figure 8.9). An individual with a malignant tumor is said to have **cancer**. Cancer cells may separate from the original tumor or secrete signal molecules that cause blood vessels to grow toward the tumor. A few tumor cells may then enter the circulatory or lymphatic system (blood or lymph vessels) and move to other parts of the body, where they may proliferate and form new tumors. The spread of cancer cells beyond their original site is called **metastasis**. Cancers are named according to where they originate. Liver cancer, for example, always begins in liver tissue and may spread from there.

Cancer Treatment

Once a tumor starts growing in the body, how can it be treated? The three main types of cancer treatment are sometimes referred to as "slash, burn, and poison." Surgery to remove a tumor ("slash") is usually the first step. "Burn" and "poison" refer to treatments that attempt to stop cancer cells from dividing. In **radiation therapy** ("burn"), parts of the body that have cancerous tumors are exposed to concentrated beams of high-energy radiation, which can often destroy cancer cells without seriously injuring the normal cells of the body. However, there is sometimes enough damage to normal body cells to produce unwanted side effects, such as nausea and hair loss.

Chemotherapy ("poison") uses the same basic strategy as radiation; in this case, drugs are administered that disrupt cell division. These drugs work in a variety of ways. Some prevent cell division by interfering with the mitotic spindle. For example, paclitaxel (trade name Taxol) freezes the spindle after it forms, keeping it from functioning. Paclitaxel is made from a chemical found in the bark of the Pacific yew, a tree found mainly in the northwestern United States. It has fewer side effects than many other anticancer drugs and seems to be effective against some hard-to-treat cancers of the ovary and breast. Another drug, vinblastine, prevents the mitotic spindle from forming in the first place. Vinblastine was first obtained from the periwinkle plant, which is native to the tropical rain forests of Madagascar.

Cancer Prevention and Survival

Although cancer can strike anyone, there are certain lifestyle changes you can make to reduce your chances of developing cancer or increase your chances of surviving it. Not smoking, exercising adequately, avoiding overexposure to the sun, and eating a high-fiber, low-fat diet can all help reduce the likelihood of getting cancer. Seven types of cancer can be easily detected: skin and oral (via physical exam), breast (via self-exams and mammograms for higher-risk women), prostate (via rectal exam), cervical (via Pap smear), testicular (via self-exam), and colon (via colonoscopy). Regular visits to the doctor can help identify tumors early, which is the best way to increase the chance of successful treatment. ✔

▼ Figure 8.9 **Growth and metastasis of a malignant tumor of the breast.**

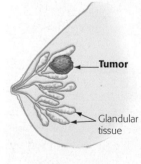

A tumor grows from a single cancer cell.

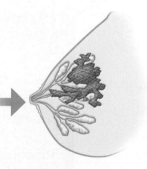

Cancer cells invade neighboring tissue.

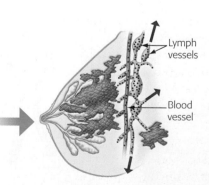

Metastasis: Cancer cells spread through lymph and blood vessels to other parts of the body.

Meiosis, the Basis of Sexual Reproduction

BioFlix ™

Meiosis

MP3 tutor sessions

Meiosis

Only maple trees produce more maple trees, only goldfish make more goldfish, and only people make more people. These simple facts of life have been recognized for thousands of years and are reflected in the age-old saying "Like begets like." But in a strict sense, "Like begets like" applies only to asexual reproduction, such as the reproduction of the African violet in Figure 8.1. Because the offspring of asexual reproduction inherit all their DNA from a single parent, they are exact genetic replicas of that one parent and of each other, and their appearances are very similar.

The family photo in **Figure 8.10** makes the point that in a sexually reproducing species, like does not exactly beget like. You probably resemble your parents more closely than you resemble a stranger, but you do not look exactly like your parents or your siblings. Each offspring of sexual reproduction inherits a unique combination of genes from its two parents, and this combined set of genes programs a unique combination of traits. As a result, sexual reproduction can produce tremendous variety among offspring.

Sexual reproduction depends on the cellular processes of meiosis and fertilization. But before discussing these processes, we need to return to chromosomes and their role in the life cycles of sexually reproducing organisms.

▼ **Figure 8.10 The varied products of sexual reproduction.** Every child inherits a unique combination of genes from his or her parents and displays a unique combination of traits.

Homologous Chromosomes

If we examine cells from different individuals of a single species—sticking to one gender, for now—we find that they have the same number and types of chromosomes. Viewed with a microscope, your chromosomes would look just like those of Angelina Jolie (if you're a woman) or Brad Pitt (if you're a man).

In humans, a typical body cell, called a **somatic cell**, has 46 chromosomes. If we break open a human cell in metaphase of mitosis, stain the chromosomes with dyes, take a picture with a microscope, and arrange them in matching pairs, we produce a display called a **karyotype** (Figure 8.11). Notice that each chromosome is duplicated, consisting of two sister chromatids joined at the centromere. Notice also that every (or almost every) chromosome has a twin that resembles it in length and centromere position. The two chromosomes of such a matching pair, called **homologous chromosomes**, carry genes controlling the same inherited characteristics. If a gene influencing eye color is located at a particular place on one chromosome—for example, within the yellow band in the drawing in Figure 8.11—then the homologous chromosome has that same gene in the same location. However, the two homologous chromosomes may have different versions of the same gene. (Let's restate this concept, since it can be confusing: A pair of homologous chromosomes has two nearly identical chromo-

▼ **Figure 8.11 Pairs of homologous chromosomes in a male karyotype.** This karyotype shows 22 completely homologous pairs (autosomes) and a 23rd pair that consists of an X chromosome and a Y chromosome (sex chromosomes). With the exception of X and Y, the homologous chromosomes of each pair match in size, centromere position, and staining pattern.

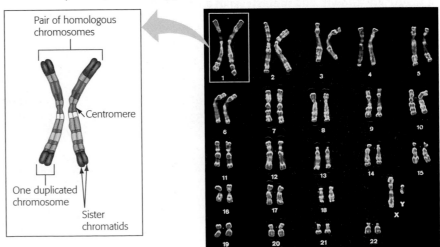

somes, each of which consists of two identical sister chromatids after chromosome duplication; see the drawing in Figure 8.11.) Altogether, we humans have 23 pairs of homologous chromosomes. Other species have different numbers of chromosomes (see Figure 8.2), but those, too, usually match in pairs.

For a human female, the 46 chromosomes fall neatly into 23 homologous pairs, with the members of each pair essentially identical in appearance. For a male, however, the chromosomes in one pair do not look alike. This nonmatching pair, only partly homologous, is the male's sex chromosomes. **Sex chromosomes** determine a person's sex (male versus female). As in all mammals, human males have one X chromosome and one Y chromosome. Females have two X chromosomes. (Other organisms have different systems; in this chapter, we focus on humans.) The remaining chromosomes, found in both males and females, are called **autosomes**. For both autosomes and sex chromosomes, we inherit one chromosome of each pair from our mother and the other from our father.

Gametes and the Life Cycle of a Sexual Organism

The **life cycle** of a multicellular organism is the sequence of stages leading from the adults of one generation to the adults of the next. Having two sets of chromosomes, one inherited from each parent, is a key factor in the life cycle of humans and all other species that reproduce sexually. **Figure 8.12** shows the human life cycle, emphasizing the number of chromosomes.

Humans (as well as most other animals and many plants) are said to be **diploid** organisms because all body cells contain pairs of homologous chromosomes. The total number of chromosomes, 46 in humans, is the diploid number, represented as $2n$. The exceptions are the egg and sperm cells, known as **gametes**. Made by meiosis in an ovary or testis, each gamete has a single set of chromosomes: 22 autosomes plus a sex chromosome, X or Y. A cell with a single chromosome set is called a **haploid** cell; it has only one member of each homologous pair. For humans, the haploid number, n, is 23.

In the human life cycle, a haploid sperm cell from the father fuses with a haploid egg cell from the mother in a process called **fertilization**. The resulting fertilized egg, called a **zygote**, is diploid. It has two sets of homologous chromosomes, one set from each parent. The life cycle is completed as a sexually mature adult develops from the zygote. Mitotic cell division ensures that all somatic

cells of the human body receive a copy of all of the zygote's 46 chromosomes. Thus, every one of the trillions of cells in your body can trace its ancestry back through mitotic divisions to the single zygote produced when your father's sperm and mother's egg fused about nine months before you were born.

All sexual life cycles involve an alternation of diploid and haploid stages. Producing haploid gametes by meiosis keeps the chromosome number from doubling in every generation. To illustrate, **Figure 8.13** tracks one pair of homologous chromosomes. ❶ Each of the chromosomes is duplicated during interphase (before mitosis). ❷ The first division, meiosis I, segregates the two chromosomes of the homologous pair, packaging them in separate (haploid) daughter cells. But each chromosome is still doubled. ❸ Meiosis II separates the sister chromatids. Each of the four daughter cells is haploid and contains only a single chromosome from the homologous pair.

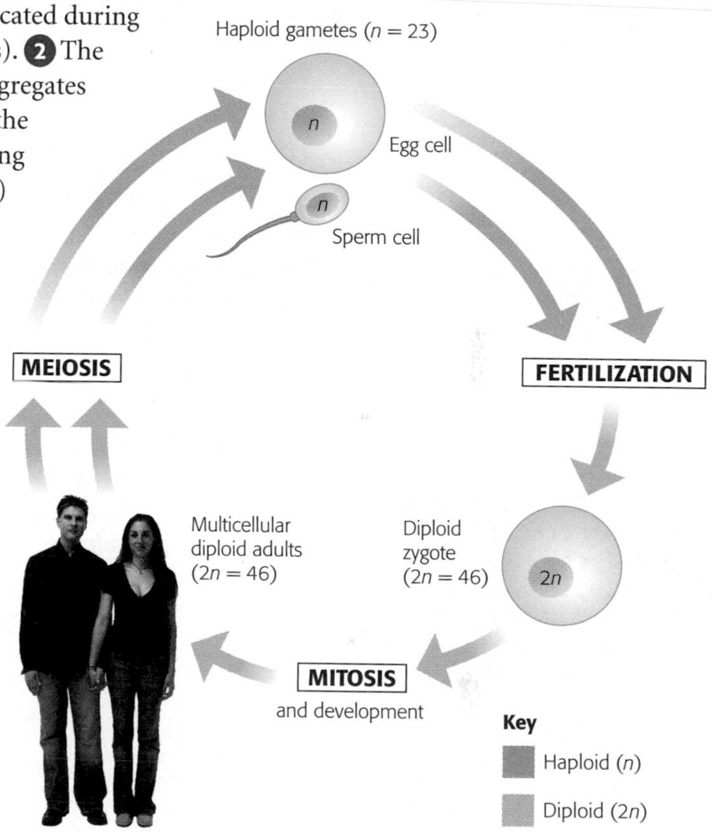

► Figure 8.12 **The human life cycle.** In each generation, the doubling of chromosome number that results from fertilization is offset by the halving of chromosome number during meiosis.

Haploid gametes ($n = 23$)

Egg cell

Sperm cell

MEIOSIS

FERTILIZATION

Multicellular diploid adults ($2n = 46$)

Diploid zygote ($2n = 46$)

MITOSIS
and development

Key
- Haploid (n)
- Diploid ($2n$)

▼ Figure 8.13 **How meiosis halves chromosome number.**

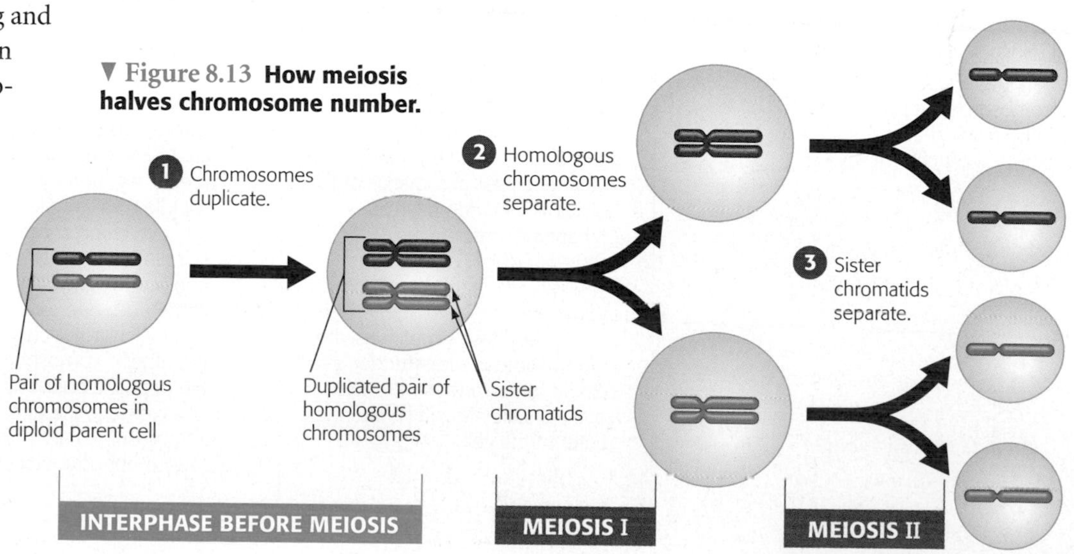

❶ Chromosomes duplicate.

❷ Homologous chromosomes separate.

❸ Sister chromatids separate.

Pair of homologous chromosomes in diploid parent cell

Duplicated pair of homologous chromosomes

Sister chromatids

INTERPHASE BEFORE MEIOSIS

MEIOSIS I

MEIOSIS II

131

The Process of Meiosis

Meiosis, the process that produces haploid daughter cells in diploid organisms, resembles mitosis, but with two special features. The first is that the number of chromosomes is reduced to half. In meiosis, a cell that has duplicated its chromosomes undergoes two consecutive divisions, called meiosis I and meiosis II. Because one duplication of the chromosomes is followed by two divisions, each of the four daughter cells resulting from meiosis has a haploid set of chromosomes—only half as many chromosomes as the starting cell.

The second special feature of meiosis is an exchange of genetic material—pieces of chromosomes—between homologous chromosomes. This exchange, called crossing over, occurs during the first prophase of meiosis. We'll look more closely at crossing over later. For now, study **Figure 8.14**, including the text below it, which describes the stages of meiosis in detail for an animal cell containing four chromosomes.

As you go through Figure 8.14, keep in mind the difference between homologous chromosomes and sister chromatids: The two chromosomes of a homologous pair are individual chromosomes that were inherited

▼ Figure 8.14 **The stages of meiosis.**

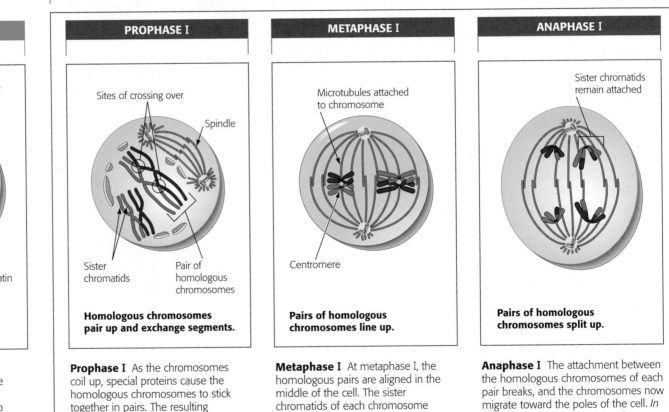

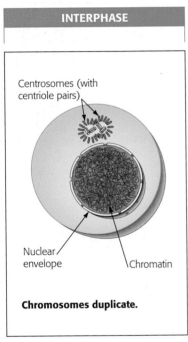

MEIOSIS I: HOMOLOGOUS CHROMOSOMES SEPARATE

INTERPHASE

Centrosomes (with centriole pairs)

Nuclear envelope

Chromatin

Chromosomes duplicate.

Like mitosis, meiosis is preceded by an interphase during which the chromosomes duplicate. Each chromosome then consists of two identical sister chromatids.

PROPHASE I

Sites of crossing over

Spindle

Sister chromatids

Pair of homologous chromosomes

Homologous chromosomes pair up and exchange segments.

Prophase I As the chromosomes coil up, special proteins cause the homologous chromosomes to stick together in pairs. The resulting structure has four chromatids. Within each set, chromatids of the homologous chromosomes exchange corresponding segments—they "cross over." Crossing over rearranges genetic information.

As prophase I continues, the chromosomes coil up further, a spindle forms, and the homologous pairs are moved toward the center of the cell.

METAPHASE I

Microtubules attached to chromosome

Centromere

Pairs of homologous chromosomes line up.

Metaphase I At metaphase I, the homologous pairs are aligned in the middle of the cell. The sister chromatids of each chromosome are still attached at their centromeres, where they are anchored to spindle microtubules. Notice that for each chromosome pair, the spindle microtubules attached to one homologous chromosome come from one pole of the cell, and the microtubules attached to the other chromosome come from the opposite pole. With this arrangement, the homologous chromosomes are poised to move toward opposite poles of the cell.

ANAPHASE I

Sister chromatids remain attached

Pairs of homologous chromosomes split up.

Anaphase I The attachment between the homologous chromosomes of each pair breaks, and the chromosomes now migrate toward the poles of the cell. *In contrast to mitosis, the sister chromatids migrate as a pair instead of splitting up.* They are separated not from each other, but from their homologous partners.

from different parents, one from the mother and one from the father. The members of a homologous pair of chromosomes in Figure 8.14 (and later figures) are colored red and blue to remind you that they differ in this way. In the interphase just before meiosis, each chromosome duplicates to form sister chromatids that remain together until anaphase of meiosis II. Before crossing over occurs, sister chromatids are identical and carry the same versions of all their genes. ☑

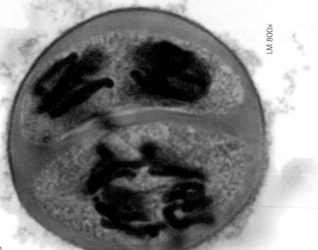

LM 800×

Metaphase II in a lily cell

☑**CHECKPOINT**

If a single diploid cell with 18 chromosomes undergoes meiosis and produces sperm, the result will be _____ sperm, each with _____ chromosomes. (Provide two numbers.)

Answer: four, nine

MEIOSIS II: SISTER CHROMATIDS SEPARATE

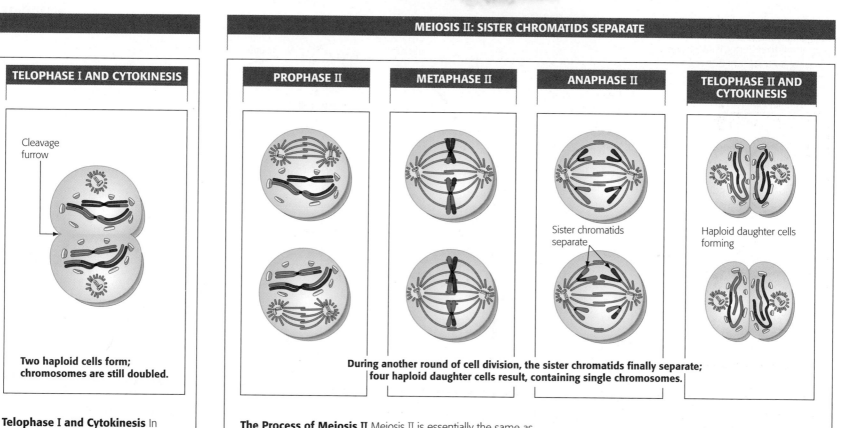

TELOPHASE I AND CYTOKINESIS

Cleavage furrow

Two haploid cells form; chromosomes are still doubled.

PROPHASE II

METAPHASE II

ANAPHASE II

Sister chromatids separate

TELOPHASE II AND CYTOKINESIS

Haploid daughter cells forming

During another round of cell division, the sister chromatids finally separate; four haploid daughter cells result, containing single chromosomes.

Telophase I and Cytokinesis In telophase I, the chromosomes arrive at the poles of the cell. When they finish their journey, each pole has a haploid chromosome set, although each chromosome is still in duplicate form. Usually, cytokinesis occurs along with telophase I, and two haploid daughter cells are formed.

The Process of Meiosis II Meiosis II is essentially the same as mitosis. The important difference is that meiosis II starts with a haploid cell that has *not* undergone a chromosome duplication during the preceding interphase.

During prophase II, a spindle forms and moves the chromosomes toward the middle of the cell. During metaphase II, the chromosomes are aligned as they are in mitosis, with the microtubules attached to the sister chromatids of each chromosome coming from opposite poles. In anaphase II, the centromeres of sister chromatids separate, and the sister chromatids of each pair move toward opposite poles of the cell. In telophase II, nuclei form at the cell poles, and cytokinesis occurs at the same time. There are now four daughter cells, each with the haploid number of single chromosomes.

Review: Comparing Mitosis and Meiosis

We have now described the two ways that cells of eukaryotic organisms divide. Mitosis, which provides for growth, tissue repair, and asexual reproduction, produces daughter cells that are genetically identical to the parent cell. Meiosis, needed for sexual reproduction, yields genetically unique haploid daughter cells—cells with only one member of each homologous chromosome pair.

For both mitosis and meiosis, the chromosomes duplicate only once, in the preceding interphase. Mitosis involves one division of the nucleus and cytoplasm, producing two diploid cells. Meiosis entails two nuclear and cytoplasmic divisions, yielding four haploid cells.

Figure 8.15 compares mitosis and meiosis, tracing these two processes for a diploid parent cell with four

▶ Figure 8.15 **Comparing mitosis and meiosis.** The events unique to meiosis occur during meiosis I: In prophase I, duplicated homologous chromosomes pair along their lengths, and crossing over occurs between homologous (nonsister) chromatids. In metaphase I, pairs of homologous chromosomes (rather than individual chromosomes) are aligned at the center of the cell. During anaphase I, sister chromatids of each chromosome stay together and go to the same pole of the cell as homologous chromosomes separate. At the end of meiosis I, there are two haploid cells, but each chromosome still has two sister chromatids.

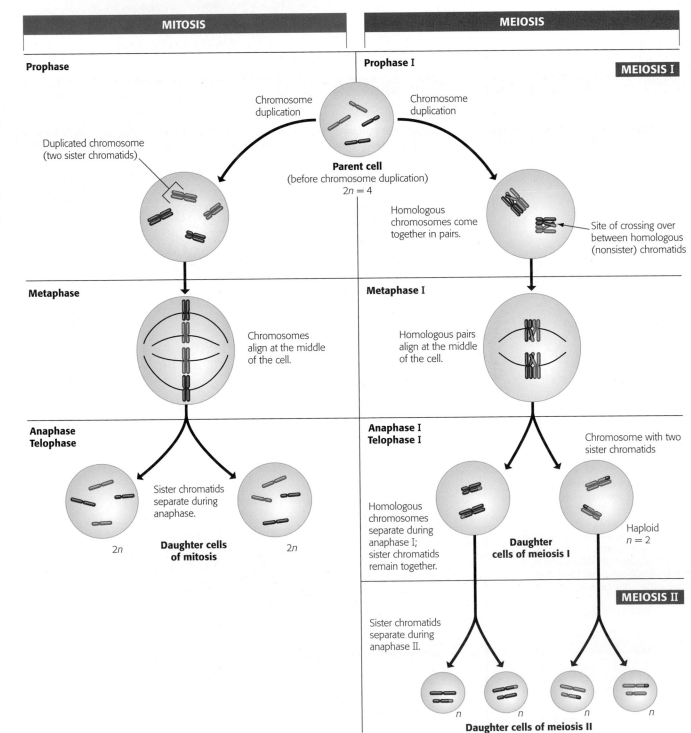

chromosomes. As before, homologous chromosomes are those matching in size. (Imagine that the red chromosomes were inherited from the mother and the blue chromosomes from the father.) Notice that all the events unique to meiosis occur during meiosis I. Meiosis II is virtually identical to mitosis in that it separates sister chromatids. But unlike mitosis, meiosis II yields daughter cells with a haploid set of chromosomes. ✓

The Origins of Genetic Variation

As we discussed earlier, offspring that result from sexual reproduction are genetically different from their parents and from one another. When we discuss evolution in Unit Three, we'll see that this genetic variety in offspring is the raw material for natural selection. For now, let's take another look at meiosis and fertilization to see how genetic variety arises.

Independent Assortment of Chromosomes

Figure 8.16 illustrates one way in which meiosis contributes to genetic variety. The figure shows how the arrangement of homologous chromosome pairs at metaphase of meiosis I affects the resulting gametes.

Once again, our example is from a diploid organism with four chromosomes (two homologous pairs), with colors used to differentiate homologous chromosomes (red for chromosomes inherited from the mother and blue for those from the father).

When aligned during metaphase I, the side-by-side orientation of each homologous pair of chromosomes (consisting of two sister chromatids) is a matter of chance—either the red or blue chromosome may be on the left or right. Thus, in this example, there are two possible ways that the chromosome pairs can align during metaphase I. In possibility 1, the chromosome pairs are oriented with both red chromosomes on the same side (blue/red and blue/red). In this case, each of the gametes produced at the end of meiosis II has only red or only blue chromosomes (combinations a and b). In possibility 2, the chromosome pairs are oriented differently (blue/red and red/blue). This arrangement produces gametes with one red and one blue chromosome (combinations c and d). Thus, with the two possible arrangements shown in this example, the organism will produce gametes with four different combinations of chromosomes. For a species with more than two pairs of chromosomes, such as the human, every chromosome pair orients independently of all the others at metaphase I. (Chromosomes X and Y behave as a homologous pair in meiosis.)

☑CHECKPOINT

True or false: Both mitosis and meiosis are preceded by chromosome duplication.

Answer: true

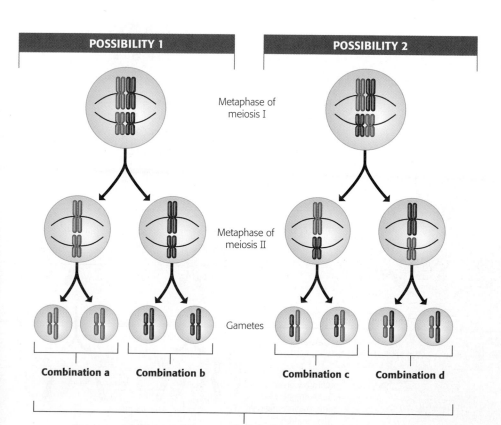

◀ Figure 8.16 **Results of alternative arrangements of chromosomes at metaphase of meiosis I.** The arrangement of chromosomes at metaphase I determines which chromosomes will be packaged together in the haploid gametes.

POSSIBILITY 1		POSSIBILITY 2

Metaphase of meiosis I

Metaphase of meiosis II

Gametes

Combination a   **Combination b**   **Combination c**   **Combination d**

Because possibilities 1 and 2 are equally likely, the four possible types of gametes will be made in approximately equal numbers.

For any species, the total number of chromosome combinations that can appear in gametes is 2^n, where n is the haploid number. For the organism in Figure 8.16, $n = 2$, so the number of chromosome combinations is 2^2, or 4. For a human ($n = 23$), there are 2^{23}, or about 8 million, possible chromosome combinations! This means that every gamete a human produces contains one of about 8 million possible combinations of maternal and paternal chromosomes.

Random Fertilization

How many possibilities are there when a gamete from one individual unites with a gamete from another individual during fertilization (Figure 8.17)? A human egg cell, representing one of about 8 million possibilities, is fertilized at random by one sperm cell, representing one of about 8 million other possibilities. By multiplying 8 million by 8 million, we find that a man and a woman can produce a diploid zygote with any of 64 trillion combinations of chromosomes! So we see that the random nature of fertilization adds a huge amount of potential variability to the offspring of sexual reproduction.

Crossing Over

So far, we have focused on genetic variety in gametes and zygotes at the whole-chromosome level. We'll now take a closer look at **crossing over**, the exchange of cor-

responding segments between chromatids of homologous chromosomes, which occurs during prophase I of meiosis. **Figure 8.18** shows crossing over between two homologous chromosomes and the resulting gametes. At the time that crossing over begins, homologous chromosomes are closely paired all along their lengths, with a precise gene-by-gene alignment. The sites of crossing over appear as X-shaped regions; each is called a **chiasma** (plural, *chiasmata*). The homologous chromatids remain attached to each other at chiasmata until anaphase I.

▼ Figure 8.17 **The process of fertilization: a close-up view.**

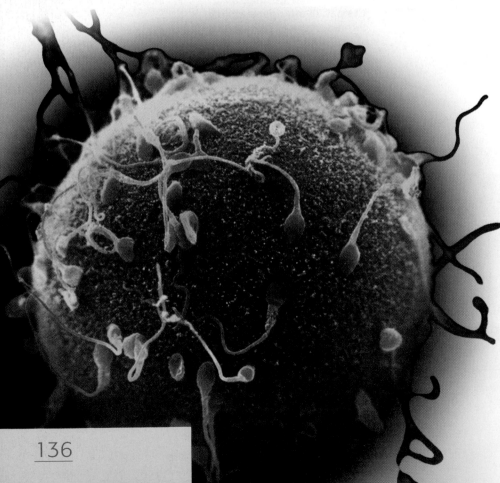

▼ Figure 8.18 **The results of crossing over during meiosis for a single pair of homologous chromosomes.** A real cell has multiple pairs of homologous chromosomes that produce a huge variety of recombinant chromosomes in the gametes.

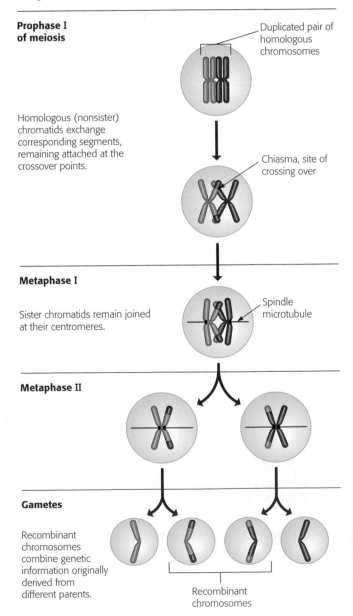

Prophase I of meiosis

Duplicated pair of homologous chromosomes

Homologous (nonsister) chromatids exchange corresponding segments, remaining attached at the crossover points.

Chiasma, site of crossing over

Metaphase I

Sister chromatids remain joined at their centromeres.

Spindle microtubule

Metaphase II

Gametes

Recombinant chromosomes combine genetic information originally derived from different parents.

Recombinant chromosomes

The exchange of segments between nonsister chromatids—one maternal chromatid and one paternal chromatid of a homologous pair—adds to the genetic variety resulting from sexual reproduction. In Figure 8.18, if there were no crossing over, meiosis could produce only two types of gametes. These would be the ones ending up with the "parental" types of chromosomes, either all blue or all red (as in Figure 8.16). With crossing over, gametes arise that have chromosomes that are part red and part blue. These chromosomes are called "recombinant" because they result from **genetic recombination**, the production of gene combinations different from those carried by the parental chromosomes.

Because most chromosomes contain thousands of genes, a single crossover event can affect many genes. When we also consider that multiple crossovers can occur in each pair of homologous chromosomes, it's not surprising that gametes and the offspring that result from them can be so varied. ☑

☑**CHECKPOINT**

Name two events during meiosis that contribute to genetic variety among gametes. During what stages of meiosis does each occur?

Answer: crossing over between homologous chromosomes during prophase I and independent orientation/assortment of the pairs of homologous chromosomes at metaphase I

THE PROCESS OF SCIENCE: Life with and without Sex

Do All Animals Have Sex?

As mentioned in the Biology and Society section, many plants can reproduce via both sexual and asexual routes. But while some animal species can also reproduce asexually, very few animals reproduce *only* asexually. In fact, evolutionary biologists have traditionally considered asexual reproduction an evolutionary dead end (for reasons we'll discuss in the Evolution Connection section).

To investigate a case where asexual reproduction seemed to be the norm, researchers from Harvard University studied a group of animals called the bdelloid rotifers (**Figure 8.19**). This class of nearly microscopic freshwater invertebrates includes over 300 known species. Despite hundreds of years of **observations**, no one had ever found bdelloid rotifer males or evidence of sexual reproduction. But the possibility remained that bdelloids had sex very infrequently or that the males were impossible to recognize by appearance. Thus, the Harvard research team posed the **question**, Does this entire class of animals reproduce solely by asexual means?

The researchers formed the **hypothesis** that bdelloid rotifers have indeed thrived for millions of years despite a lack of sexual reproduction. But how to prove it? In most species, the two versions of a gene in a pair of homologous chromosomes are very similar due to the constant trading of genes during sexual reproduction. If a species has survived without sex for millions of years, the researchers reasoned, then changes in the DNA sequences of homologous genes should accumulate independently, and the two versions of the genes should have significantly diverged from each other over time. This led to the **prediction** that bdelloid rotifers would display much more variation in their pairs of homologous genes than most organisms.

In a simple but elegant **experiment**, the researchers compared the sequences of a particular gene in bdelloid and non-bdelloid rotifers. Their **results** were striking. Among non-bdelloid rotifers that reproduce sexually, the two homologous versions of the gene were nearly identical, differing by only 0.5% on average. In contrast, the two versions of the same gene in bdelloid rotifers differed by 3.5–54%. These data provided strong evidence that bdelloid rotifers have evolved for millions of years without any sexual reproduction.

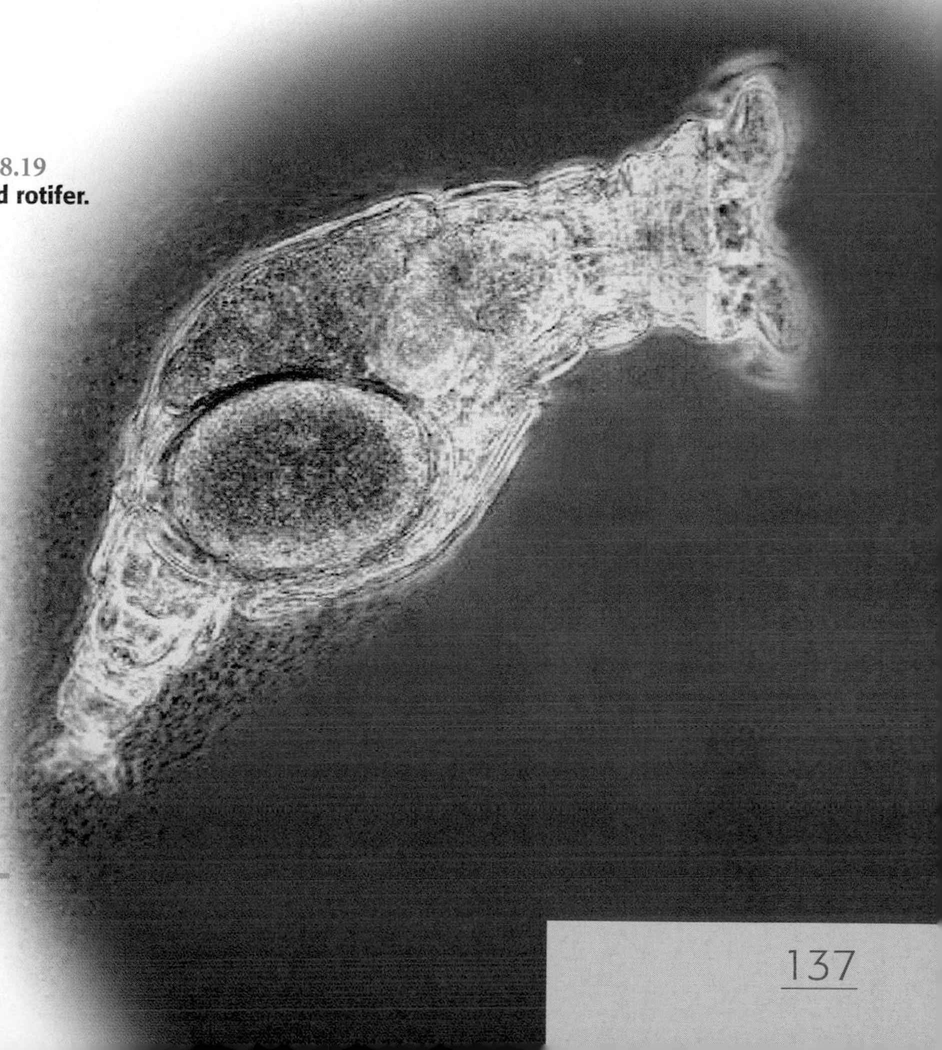

▶ **Figure 8.19**
A bdelloid rotifer.

LM 300×

When Meiosis Goes Awry

So far, our discussion of meiosis has focused on the process as it normally and correctly occurs. But what happens when an error occurs in the process? Such a mistake can result in genetic abnormalities that range from mild to severe to fatal.

How Accidents during Meiosis Can Alter Chromosome Number

Within the human body, meiosis occurs repeatedly as the testes or ovaries produce gametes. Almost always, the meiotic spindle distributes chromosomes to daughter cells without error. But occasionally there is a mishap, called a **nondisjunction**, in which the members of a chromosome pair fail to separate at anaphase. Nondisjunction can occur during meiosis I or II (Figure 8.20). In either case, gametes with abnormal numbers of chromosomes are the result.

Figure 8.21 shows what can happen when an abnormal gamete produced by nondisjunction unites with a normal gamete during fertilization. When a normal sperm fertilizes an egg cell with an extra chromosome,

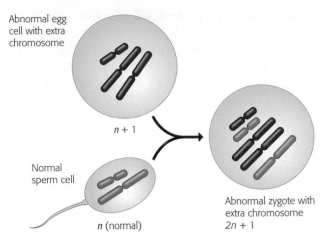

▲ Figure 8.21 **Fertilization after nondisjunction in the mother.**

the result is a zygote with a total of $2n + 1$ chromosomes. Mitosis then transmits the abnormality to all embryonic cells. If the organism survives, it will have an abnormal karyotype and probably a syndrome of disorders caused by the abnormal number of genes. ☑

▼ Figure 8.20 **Two types of nondisjunction.** In both examples in the figure, the cell at the top is diploid ($2n$), with two pairs of homologous chromosomes.

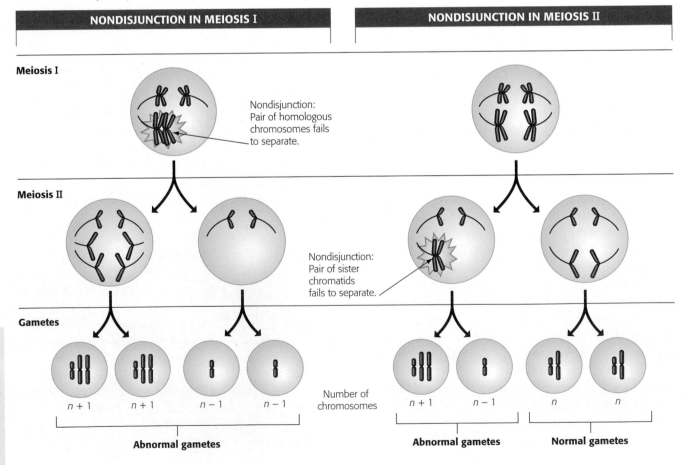

Down Syndrome: An Extra Chromosome 21

Figure 8.11 showed a normal human complement of 23 pairs of chromosomes. Compare it with Figure 8.22; besides having two X chromosomes (because it's from a female), the karyotype in Figure 8.22 has three number 21 chromosomes, making 47 chromosomes in total. This condition is called **trisomy 21**.

In most cases, a human embryo with an atypical number of chromosomes develops so abnormally that it is spontaneously aborted (miscarried) long before birth.

▼ **Figure 8.22 Trisomy 21 and Down syndrome.** This child displays the characteristic facial features of Down syndrome. The karyotype (bottom) shows trisomy 21; notice the three copies of chromosome 21.

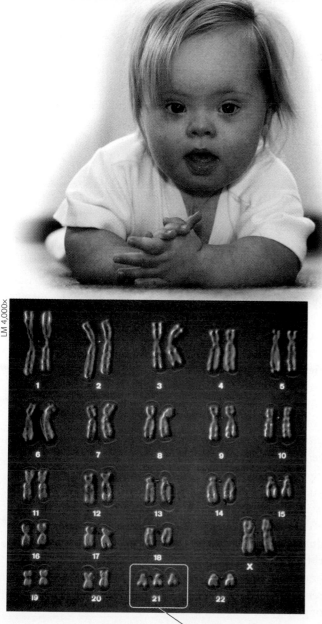

Chromosome 21

However, some aberrations in chromosome number seem to upset the genetic balance less drastically, and individuals with such abnormalities can survive. These people usually have a characteristic set of symptoms, called a syndrome. A person with trisomy 21, for instance, is said to have **Down syndrome** (named after John Langdon Down, who described it in 1866).

Affecting about 1 out of every 700 children, trisomy 21 is the most common chromosome number abnormality and the most common serious birth defect in the United States. Down syndrome includes characteristic facial features—frequently a fold of skin at the inner corner of the eye, a round face, and a flattened nose—as well as short stature, heart defects, and susceptibility to leukemia and Alzheimer's disease. People with Down syndrome usually have a life span shorter than normal. They also exhibit varying degrees of mental retardation. However, individuals with the syndrome may live to middle age or beyond, and many are socially adept and can function well within society.

As indicated in **Figure 8.23**, the incidence of Down syndrome in the offspring of normal parents increases markedly with the age of the mother. Down syndrome affects less than 0.05% of children (fewer than 1 in 2,000) born to women under age 30. The risk climbs to 1% (10 in 1,000) for mothers at age 40 and is even higher for older mothers. Because of this relatively high risk, pregnant women over 35 are candidates for fetal testing for trisomy 21 and other chromosomal abnormalities (see Chapter 9). ☑

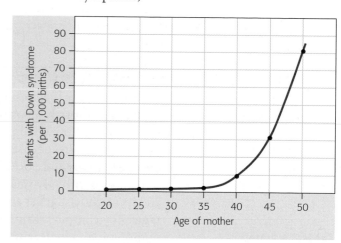

▲ **Figure 8.23 Maternal age and Down syndrome.** The chance of having a baby with Down syndrome rises with the age of the mother.

Abnormal Numbers of Sex Chromosomes

Nondisjunction in meiosis does not affect just autosomes, such as chromosome 21. It can also lead to abnormal numbers of sex chromosomes (X and Y). Unusual numbers of sex chromosomes seem to upset the genetic balance less than unusual numbers of autosomes. This may be because the Y chromosome is very small and carries relatively few genes. Furthermore, mammalian cells normally operate with only one functioning X chromosome because other copies of the chromosome become inactivated in each cell (see Chapter 11).

☑CHECKPOINT

According to Figure 8.23, how much more likely is a 45-year-old woman to have a child with Down syndrome than a 40-year-old mother?

Answer: three times (30 per 1,000 versus 10 per 1,000)

Table 8.1	Abnormalities of Sex Chromosome Number in Humans		
Sex Chromosomes	**Syndrome**	**Origins of Nondisjunction**	**Frequency in Population**
XXY	Klinefelter syndrome (male)	Meiosis in egg or sperm formation	$\frac{1}{2,000}$
XYY	None (normal male)	Meiosis in sperm formation	$\frac{1}{2,000}$
XXX	None (normal female)	Meiosis in egg or sperm formation	$\frac{1}{1,000}$
XO	Turner syndrome (female)	Meiosis in egg or sperm formation	$\frac{1}{5,000}$

Table 8.1 lists the most common human sex chromosome abnormalities. An extra X chromosome in a male, making him XXY, produces a condition called Klinefelter syndrome. Men with this disorder have male sex organs and normal intelligence, but the testes are abnormally small, the individual is sterile, and he often has breast enlargement and other feminine body contours. Klinefelter syndrome is also found in individuals with more than three sex chromosomes, such as XXYY, XXXY, or XXXXY. These abnormal numbers of sex chromosomes result from multiple nondisjunctions.

Human males with a single extra Y chromosome (XYY) do not have any well-defined syndrome, although they tend to be taller than average. Females with an extra X chromosome (XXX) cannot be distinguished from XX females except by karyotype.

Females who are lacking an X chromosome are designated XO; the O indicates the absence of a second sex chromosome. These women have Turner syndrome. They have a characteristic appearance, including short stature and often a web of skin extending between the neck and shoulders. Women with Turner syndrome are sterile because their sex organs do not fully mature at adolescence, and they have poor development of breasts and other secondary sex characteristics. However, they are usually of normal intelligence. The XO condition is the sole known case where having only 45 chromosomes is not fatal in humans.

The sex chromosome abnormalities described here illustrate the crucial role of the Y chromosome in determining a person's sex. In general, a single Y chromosome is enough to produce "maleness," regardless of the number of X chromosomes. The absence of a Y chromosome results in "femaleness." ✓

✓ CHECKPOINT

Why is an individual more likely to survive with an abnormal number of sex chromosomes than an abnormal number of autosomes?

Answer: because the Y chromosome is very small and extra X chromosomes are inactivated

EVOLUTION CONNECTION: Life with and without Sex

The Advantages of Sex

Throughout this chapter, we've examined cell division within the context of reproduction. Like the Hawaiian bellflower in the Biology and Society section, many plants can reproduce both sexually and asexually (Figure 8.24). An important advantage of asexual reproduction is that there is no need for a partner. Asexual reproduction may thus confer an evolutionary advantage when plants are sparsely distributed and unlikely to be able to exchange pollen. Furthermore, if a plant is superbly suited to a stable environment, asexual reproduction has the advantage of passing on its entire genetic legacy intact.

In contrast to plants, the vast majority of animals reproduce by sexual means. There are exceptions, such as organisms that can regenerate after fragmentation (see Figure 8.1) and the bdelloid rotifers discussed in the Process of Science section. But many animals reproduce only through sex. Therefore, sex must enhance evolutionary fitness. But how? The answer remains elusive. Most hypotheses focus on the unique combinations of genes formed during meiosis and fertilization. By producing offspring of varied genetic makeup, sexual reproduction may enhance survival by speeding adaptation to a changing environment, such as one that contains evolving pathogens. Another idea is that shuffling genes during sexual reproduction might allow a population to rid itself of harmful genes more rapidly. But for now, one of biology's most basic questions—Why have sex?—remains a hotly debated topic that is the focus of much ongoing research and discussion.

◄ **Figure 8.24 Sexual and asexual reproduction.** Many plants, such as this strawberry, have the ability to reproduce both sexually (via flowers that produce fruit) and asexually (via runners).

Chapter Review

SUMMARY OF KEY CONCEPTS

 Go to the Study Area at **www.masteringbiology.com** for practice quizzes, my**e**Book, BioFlix™ 3-D animations, MP3 Tutor Sessions, videos, current events, and more.

What Cell Reproduction Accomplishes

Cell reproduction, also called cell division, produces genetically identical daughter cells:

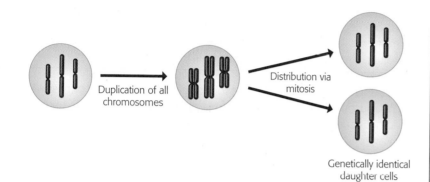

Duplication of all chromosomes

Distribution via mitosis

Genetically identical daughter cells

Some organisms use mitosis (ordinary cell division) to reproduce. This is called asexual reproduction, and it results in offspring that are genetically identical to the lone parent and to each other. Mitosis also enables multicellular organisms to grow and develop and to replace damaged or lost cells. Organisms that reproduce sexually, by the union of a sperm with an egg cell, carry out meiosis, a type of cell division that yields gametes with only half as many chromosomes as body (somatic) cells.

The Cell Cycle and Mitosis

Eukaryotic Chromosomes

The genes of a eukaryotic genome are grouped into multiple chromosomes in the nucleus. Each chromosome contains one very long DNA molecule, with many genes, that is wrapped around histone proteins. Individual chromosomes are coiled up and therefore visible with a light microscope only when the cell is in the process of dividing; otherwise, they are in the form of thin, loosely packed chromatin fibers. Before a cell starts dividing, the chromosomes duplicate, producing sister chromatids (containing identical DNA) joined together at the centromere.

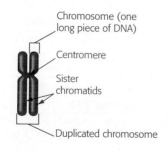

Chromosome (one long piece of DNA)

Centromere

Sister chromatids

Duplicated chromosome

The Cell Cycle

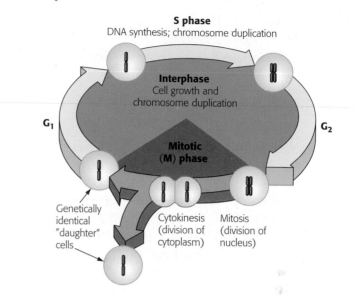

S phase
DNA synthesis; chromosome duplication

Interphase
Cell growth and chromosome duplication

G_1

G_2

Mitotic (M) phase

Genetically identical "daughter" cells

Cytokinesis (division of cytoplasm)

Mitosis (division of nucleus)

Mitosis and Cytokinesis

Mitosis is divided into four phases: prophase, metaphase, anaphase, and telophase. At the start of mitosis, the chromosomes coil up and the nuclear envelope breaks down (prophase). Then a mitotic spindle made of microtubules moves the chromosomes to the middle of the cell (metaphase). The sister chromatids then separate and are moved to opposite poles of the cell (anaphase), where two new nuclei form (telophase). Cytokinesis overlaps the end of mitosis. In animals, cytokinesis occurs by cleavage, which pinches the cell in two. In plants, a membranous cell plate divides the cell in two. Mitosis and cytokinesis produce genetically identical cells.

Cancer Cells: Growing Out of Control

When the cell cycle control system malfunctions, a cell may divide excessively and form a tumor. Cancer cells may grow to form malignant tumors, invade other tissues (metastasize), and even kill the organism. Surgery can remove tumors, and radiation and chemotherapy are effective as treatments because they interfere with cell division. You can increase the likelihood of surviving some forms of cancer through lifestyle changes and regular screenings.

Meiosis, the Basis of Sexual Reproduction

Homologous Chromosomes

The somatic cells (body cells) of each species contain a specific number of chromosomes; human cells have 46, made up of 23 pairs of homologous chromosomes. The chromosomes of a homologous pair carry genes for the same characteristics at the same places. Mammalian males have X and Y sex chromosomes (only partly homologous), while females have two X chromosomes.

Gametes and the Life Cycle of a Sexual Organism

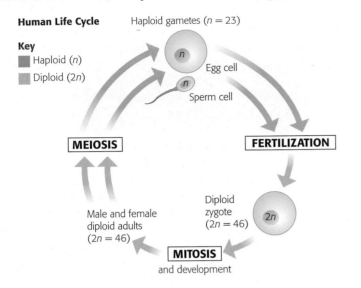

Human Life Cycle

Key
- Haploid (*n*)
- Diploid (2*n*)

Haploid gametes (*n* = 23)

n

Egg cell

n

Sperm cell

MEIOSIS

FERTILIZATION

Male and female
diploid adults
(2*n* = 46)

Diploid
zygote
(2*n* = 46)

2*n*

MITOSIS
and development

The Process of Meiosis

Meiosis, like mitosis, is preceded by chromosome duplication. But in meiosis, the cell divides twice to form four daughter cells. The first division, meiosis I, starts with the pairing of homologous chromosomes. In crossing over, homologous chromosomes exchange corresponding segments. Meiosis I separates the members of the homologous pairs and produces two daughter cells, each with one set of (duplicated) chromosomes. Meiosis II is essentially the same as mitosis; in each of the cells, the sister chromatids of each chromosome separate.

Review: Comparing Mitosis and Meiosis

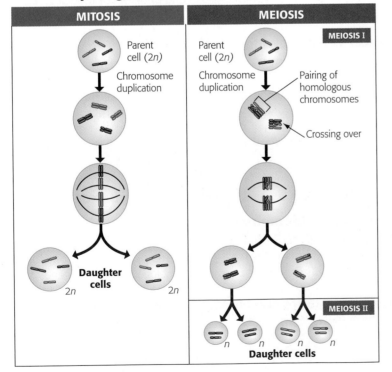

MITOSIS

Parent
cell (2*n*)

Chromosome
duplication

**Daughter
cells**
2*n*   2*n*

MEIOSIS

MEIOSIS I

Parent
cell (2*n*)

Chromosome
duplication

Pairing of
homologous
chromosomes

Crossing over

MEIOSIS II

n   *n*   *n*   *n*
Daughter cells

The Origins of Genetic Variation

Because the chromosomes of a homologous pair come from different parents, they carry different versions of many of their genes. The large number of possible arrangements of chromosome pairs at metaphase of meiosis I leads to many different combinations of chromosomes in eggs and sperm. Random fertilization of eggs by sperm greatly increases the variation. Crossing over during prophase of meiosis I increases variation still further.

When Meiosis Goes Awry

Sometimes a person has an abnormal number of chromosomes, which causes problems. Down syndrome is caused by an extra copy of chromosome 21. The abnormal chromosome count is a product of nondisjunction, the failure of a homologous pair of chromosomes to separate during meiosis I or of sister chromatids to separate during meiosis II. Nondisjunction can also produce gametes with extra or missing sex chromosomes, which lead to varying degrees of malfunction in humans but do not usually affect survival.

SELF-QUIZ

1. Which of the following is not a function of mitosis in humans?
   a. repair of wounds
   b. growth
   c. production of gametes from diploid cells
   d. replacement of lost or damaged cells

2. In what sense are the two daughter cells produced by mitosis identical?

3. Why is it difficult to observe individual chromosomes during interphase?

4. A biochemist measures the amount of DNA in cells growing in the laboratory. The quantity of DNA in a cell would be found to double
   a. between prophase and anaphase of mitosis.
   b. between the G_1 and G_2 phases of the cell cycle.
   c. during the M phase of the cell cycle.
   d. between prophase I and prophase II of meiosis.

5. Which two phases of mitosis are essentially opposites in terms of changes in the nucleus?

6. Complete the following table to compare mitosis and meiosis:

	Mitosis	Meiosis
a. Number of chromosomal duplications		
b. Number of cell divisions		
c. Number of daughter cells produced		
d. Number of chromosomes in daughter cells		
e. How chromosomes line up during metaphase		
f. Genetic relationship of daughter cells to parent cells		
g. Functions performed in the human body		

7. A chemical that disrupts microfilament formation would interfere with
  a. DNA replication.
  b. formation of the mitotic spindle.
  c. cleavage.
  d. crossing over.

8. If an intestinal cell in a dog contains 78 chromosomes, a dog sperm cell would contain _____ chromosomes.

9. A micrograph of a dividing cell from a mouse shows 19 chromosomes, each consisting of two sister chromatids. During which stage of meiosis could this picture have been taken? (Explain your answer.)

10. Tumors that remain at their site of origin are called _____, while tumors from which cells migrate to other body tissues are called _____.

11. A fruit fly somatic cell contains eight chromosomes. This means that _____ different combinations of chromosomes are possible in its gametes.

12. Although nondisjunction is a random event, there are many more individuals with an extra chromosome 21, which causes Down syndrome, than individuals with an extra chromosome 3 or chromosome 16. Propose an explanation for this.

Answers to the Self-Quiz questions can be found in Appendix D.

THE PROCESS OF SCIENCE

13. A mule is the offspring of a horse and a donkey. A donkey sperm contains 31 chromosomes and a horse egg 32 chromosomes, so the zygote contains a total of 63 chromosomes. The zygote develops normally. The combined set of chromosomes is not a problem in mitosis, and the mule combines some of the best characteristics of horses and donkeys. However, a mule is sterile; meiosis cannot occur normally in its testes or ovaries. Explain why mitosis is normal in cells containing both horse and donkey chromosomes but the mixed set of chromosomes interferes with meiosis.

14. You prepare a slide with a thin slice of an onion root tip. You see the following view in a light microscope. Identify the stage of mitosis for each of the outlined cells, a–d.

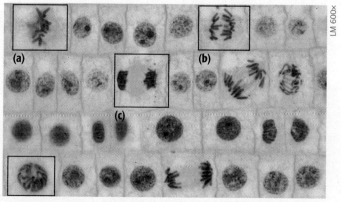

LM 600×

(a)   (b)   (c)   (d)

BIOLOGY AND SOCIETY

15. Every year, about a million Americans are diagnosed with cancer. This means that about 75 million Americans now living will eventually have cancer, and one in five will die of the disease. There are many kinds of cancers and many causes of the disease. For example, smoking causes most lung cancers. Overexposure to ultraviolet rays in sunlight causes most skin cancers. There is evidence that a high-fat, low-fiber diet is a factor in breast, colon, and prostate cancers. And agents in the workplace, such as asbestos and vinyl chloride, are also implicated as causes of cancer. Hundreds of millions of dollars are spent each year in the search for effective treatments for cancer, yet far less money is spent on preventing cancer. Why might this be true? What kinds of lifestyle changes could we make to help prevent cancer? What kinds of prevention programs could be initiated or strengthened to encourage these changes? What factors might impede such changes and programs? Should we devote more of our resources to treating cancer or preventing it? Defend your position.

16. The practice of buying and selling gametes, particularly eggs from fertile women, is becoming increasingly common in the United States and other industrialized countries. Do you have any objections to this type of transaction? Would you be willing to sell your gametes? At any price? Whether you are willing to do so or not, do you think that other people should be restricted from doing so?

9 Patterns of Inheritance

Canine genetics.
Dogs, such as these Siberian huskies, are one of humankind's longest-running genetic experiments.

CHAPTER CONTENTS

■■■ Chapter Thread: **DOG BREEDING**

BIOLOGY AND SOCIETY: Dog Breeding

A Matter of Breeding

The hard-working dogs shown on the facing page are Siberian huskies. Each member of the team looks like the others because they share a similar pedigree (genetic history). When these dogs are bred, they are expected to produce others that inherit the desirable traits that distinguish the breed, such as a thick double coat of fur and triangular well-insulated ears. This is a reasonable expectation because a purebred dog's well-documented pedigree includes parents, grandparents, and great-grandparents with similar genetic makeup and appearance.

But whether dogs are purebreds or mutts, we can ultimately explain their inborn traits through genetics, the scientific study of heredity. Because their patterns of inheritance are relatively simple, purebred dogs are important in genetic research. However, after generations of inbreeding, purebreds often suffer from serious genetic defects. For instance, a type of hereditary blindness called progressive retinal atrophy (PRA) is common among Siberian huskies and several other breeds. Studies of such defects may be useful for veterinary and human medicine because many genetic disorders are similar in dogs and humans.

Geneticists are also studying dogs to help shed light on the relationship between genetic makeup and behavior. Certain breeds have behavioral tendencies—perhaps you've witnessed the way border collies and sheepdogs will herd any animal!—though genetics clearly isn't everything. A dog's behavior is influenced not only by its genes but also by its environment and care.

In this chapter, you will learn the basic rules of how genetic traits are passed from generation to generation and how the behavior of chromosomes (the topic of Chapter 8) accounts for these rules. We will look at several inheritance patterns and see how to predict the ratios of offspring with particular traits. At several points in the chapter, we will return to the subject of dog breeding to help illustrate genetic principles.

Heritable Variation and Patterns of Inheritance

Heredity is the transmission of traits from one generation to the next. **Genetics**, the scientific study of heredity, began in the 1860s, when an Augustinian monk named Gregor Mendel deduced its fundamental principles by breeding garden peas (**Figure 9.1**). Mendel lived and worked in an abbey in Brunn, Austria (now Brno, in the Czech Republic). Strongly influenced by his study of physics, mathematics, and chemistry at the University of Vienna, his research was both experimentally and mathematically rigorous, and these qualities were largely responsible for his success.

▲ Figure 9.1
Gregor Mendel.

In a paper published in 1866, Mendel correctly argued that parents pass on to their offspring discrete "heritable factors" that are responsible for inherited traits, such as purple flowers or round seeds in pea plants. (It is interesting to note that Mendel's publication came just seven years after Darwin's 1859 publication of *The Origin of Species*, making the 1860s a banner decade in the development of modern biology.) In his paper, Mendel stressed that the heritable factors (today called genes) retain their individual identities generation after generation, no matter how they are mixed up or temporarily masked.

reach the carpel. When he wanted cross-fertilization (fertilization of one plant by pollen from a different plant), he pollinated the plants by hand, as shown in **Figure 9.3**. Thus, whether Mendel let a pea plant self-fertilize or cross-fertilized it with a known source of pollen, he could always be sure of the parentage of his new plants.

Each of the characters Mendel chose to study, such as flower color, occurred in two distinct forms. Mendel worked with his plants until he was sure he had **true-breeding** (purebred) varieties—that is, varieties for which self-fertilization produced offspring all identical to the parent. For instance, he identified a purple-flowered variety

In an Abbey Garden

Mendel probably chose to study garden peas because they were easy to grow and they came in many readily distinguishable varieties. For example, one variety has purple flowers, and another variety has white flowers. A heritable feature that varies among individuals, such as flower color, is called a **character**. A variant of a character, such as purple or white flowers, is called a **trait**.

Perhaps the most important advantage of pea plants as an experimental model was that Mendel could strictly control their reproduction. The petals of the pea flower almost completely enclose the egg- and sperm-producing parts—the carpel and stamens, respectively (**Figure 9.2**). Consequently, in nature, pea plants usually self-fertilize because sperm-carrying pollen grains released from the stamens land on the tip of the egg-containing carpel of the same flower. Mendel could ensure self-fertilization by covering a flower with a small bag so that no pollen from another plant could

▼ Figure 9.2 **The structure of a pea flower.** To reveal the reproductive organs—the stamens and carpel—one of the petals has been removed in this drawing.

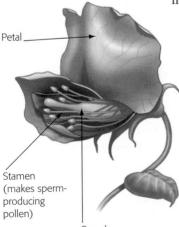

Petal

Stamen
(makes sperm-producing pollen)

Carpel
(produces eggs)

▼ Figure 9.3 **Mendel's technique for cross-fertilizing pea plants.**

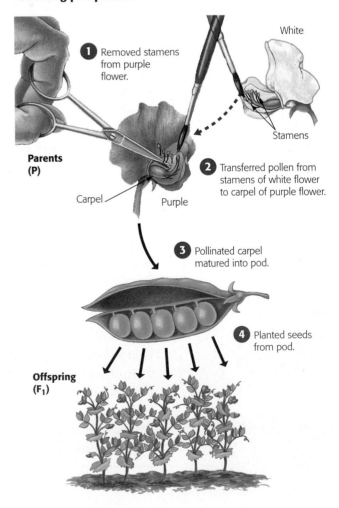

Parents (P)

White

1 Removed stamens from purple flower.

Stamens

2 Transferred pollen from stamens of white flower to carpel of purple flower.

Carpel   Purple

3 Pollinated carpel matured into pod.

4 Planted seeds from pod.

Offspring (F₁)

that, when self-fertilized, always produced offspring plants that had all purple flowers.

Now Mendel was ready to ask what would happen when he crossed different true-breeding varieties with each other. For example, what offspring would result if plants with purple flowers and plants with white flowers were cross-fertilized as shown in Figure 9.3? In the language of breeders and geneticists, the offspring of two different true-breeding varieties are called **hybrids**, and the cross-fertilization itself is referred to as a genetic **cross**. The parental plants are called the **P generation**, and their hybrid offspring are the **F₁ generation** (F for *filial*, from the Latin for "son" or "daughter"). When F_1 plants self-fertilize or fertilize each other, their offspring are the **F₂ generation**. ✔

Mendel's Law of Segregation

Mendel performed many experiments in which he tracked the inheritance of characters, such as flower color, that occur as two alternative traits (**Figure 9.4**). The results led him to formulate several hypotheses about inheritance. Let's look at some of his experiments and follow the reasoning that led to his hypotheses.

▼ Figure 9.4 **The seven characters of pea plants studied by Mendel.** Each character comes in the two alternative traits shown here.

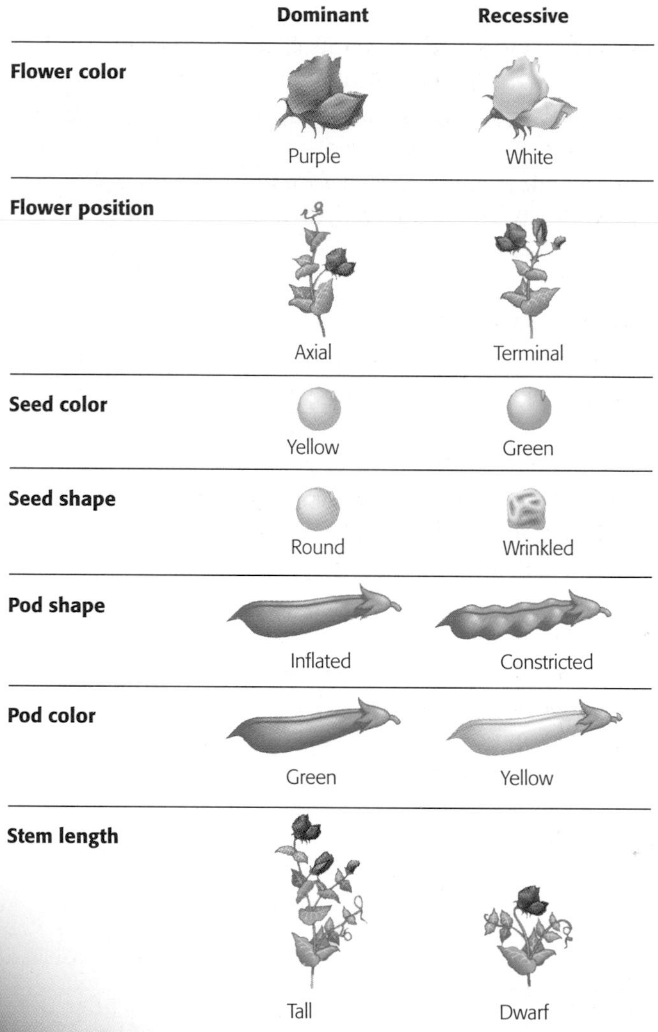

	Dominant	Recessive
Flower color	Purple	White
Flower position	Axial	Terminal
Seed color	Yellow	Green
Seed shape	Round	Wrinkled
Pod shape	Inflated	Constricted
Pod color	Green	Yellow
Stem length	Tall	Dwarf

✔CHECKPOINT

Why was the development of true-breeding pea plant varieties critical to Mendel's work?

Answer: True-breeding varieties allowed Mendel to predict the outcome of specific crosses and therefore to run controlled experiments.

Monohybrid Crosses

Figure 9.5 shows a cross between a true-breeding pea plant with purple flowers and a true-breeding pea plant with white flowers. This is called a **monohybrid cross** because the parent plants differ in only one character. Mendel saw that the F₁ plants all had purple flowers. Was the heritable factor for white flowers now lost as a result of the cross? By mating the F₁ plants with each other, Mendel found the answer to be no. Of the 929 F₂ plants, about three-fourths (705) had purple flowers and one-fourth (224) had white flowers; that is, there were about three purple F₂ plants for every white plant, or a 3:1 ratio of purple to white. Mendel concluded that the heritable factor for white flowers did not disappear in the F₁ plants, but was somehow hidden or masked when the purple-flower factor was present. He also deduced that the F₁ plants must have carried two factors for the flower-color character, one for purple and one for white. From these results and others, Mendel developed four hypotheses. Using modern terminology (including "gene" instead of "heritable factor"), here are his hypotheses:

1. *There are alternative versions of genes, the units that determine heritable traits.* For example, the gene for flower color in pea plants exists in one form for purple and another for white. The alternative versions of a gene are called **alleles**.

▼ **Figure 9.5 Mendel's cross tracking one character (flower color).** Note the 3:1 ratio of purple flowers to white flowers in the F₂ generation.

Mendel attending his peas

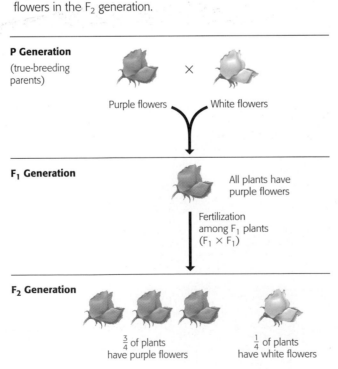

P Generation
(true-breeding parents)

Purple flowers   White flowers

F₁ Generation

All plants have purple flowers

Fertilization among F₁ plants
(F₁ × F₁)

F₂ Generation

$\frac{3}{4}$ of plants have purple flowers $\frac{1}{4}$ of plants have white flowers

2. *For each inherited character, an organism inherits two alleles, one from each parent.* These alleles may be the same or different. An organism that has two identical alleles for a gene is said to be **homozygous** for that gene (and is called a homozygote). An organism that has two different alleles for a gene is said to be **heterozygous** for that gene (and is called a heterozygote).

3. *If the two alleles of an inherited pair differ, then one determines the organism's appearance and is called the* **dominant allele**; *the other has no noticeable effect on the organism's appearance and is called the* **recessive allele**. We use uppercase italic letters to represent dominant alleles and lowercase italic letters to represent recessive alleles.

4. *A sperm or egg carries only one allele for each inherited character because the two members of an allele pair segregate (separate) from each other during the production of gametes.* This statement is now known as the **law of segregation**. When sperm and egg unite at fertilization, each contributes its alleles, restoring the paired condition in the offspring.

Do Mendel's hypotheses account for the 3:1 ratio he observed in the F₂ generation? **Figure 9.6** illustrates Mendel's law of segregation, which explains the inheritance pattern shown in Figure 9.5. His hypotheses predict that when alleles segregate during gamete formation in the F₁ plants, half the gametes will receive a purple-flower allele (*P*) and the other half a white-flower allele (*p*). During pollination among the F₁ plants, the gametes unite randomly. An egg with a purple-flower allele has an equal chance of being fertilized by a sperm with a purple-flower allele or one with a white-flower allele (that is, a *P* egg may fuse with a *P* sperm or a *p* sperm). Because the same is true for an egg with a white-flower allele (a *p* egg with a *P* sperm or *p* sperm), there are a total of four equally likely combinations of sperm and egg.

The diagram at the bottom of Figure 9.6, called a **Punnett square**, repeats the cross shown in Figure 9.5 in a way that highlights the four possible combinations of gametes and the offspring that result from each. Each square represents an equally probable product of fertilization. For example, the box in the upper right corner of the Punnett square shows the genetic combination resulting from a *p* sperm fertilizing a *P* egg.

According to the Punnett square, what will be the physical appearance of these F₂ offspring? One-fourth of the plants have two alleles specifying purple flowers (*PP*); clearly, these plants will have purple flowers. One-half (two-fourths) of the F₂ offspring have

▼ Figure 9.6 **The law of segregation.**

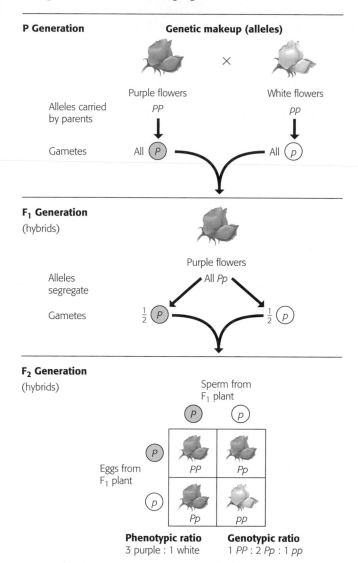

P Generation

Genetic makeup (alleles)

Purple flowers
PP

White flowers
pp

Alleles carried
by parents

Gametes   All ⓟ   All ⓟ

F₁ Generation
(hybrids)

Purple flowers
All *Pp*

Alleles
segregate

Gametes   ½ ⓟ   ½ ⓟ

F₂ Generation
(hybrids)

Sperm from
F₁ plant

Eggs from
F₁ plant

	P	*p*
P	*PP*	*Pp*
p	*Pp*	*pp*

Phenotypic ratio
3 purple : 1 white

Genotypic ratio
1 *PP* : 2 *Pp* : 1 *pp*

Mendel found that each of the seven characters he studied had the same inheritance pattern: One parental trait disappeared in the F₁ generation, only to reappear in one-fourth of the F₂ offspring. The underlying mechanism is stated by Mendel's law of segregation: Pairs of alleles segregate (separate) during gamete formation; the fusion of gametes at fertilization creates allele pairs again. Research since Mendel's day has established that the law of segregation applies to all sexually reproducing organisms, including dogs and humans.

Genetic Alleles and Homologous Chromosomes

Before continuing with Mendel's experiments, let's see how some of the concepts we discussed in Chapter 8 fit with what we've said about genetics so far. The diagram in **Figure 9.7** shows a pair of homologous chromosomes—chromosomes that carry alleles of the same genes. Recall from Chapter 8 that every diploid individual, whether pea plant or human, has chromosomes in homologous pairs. One member of each pair comes from the organism's female parent; the other member of each pair comes from the male parent. The labeled bands on the chromosomes in the figure represent three gene **loci** (singular, *locus*), specific locations of genes along the chromosome. You can see the connection between Mendel's law of segregation and homologous chromosomes: Alleles (alternative versions) of a gene reside at the same locus on homologous chromosomes. However, the two chromosomes may bear either identical alleles or different ones at any one locus. In other words, the organisms may be homozygous or heterozygous for the gene at that locus. We will return to the chromosomal basis of Mendel's law later in the chapter. ☑

☑CHECKPOINT

1. Genes come in different versions called _____. If both of these are the same, the individual is _____. If they are different, the individual is _____ and the version that is expressed is called _____.
2. How can two plants that have different genotypes for flower color be identical in phenotype?
3. You carry two alleles for every trait. Where did these alleles come from?

Answers: 1. alleles; homozygous; heterozygous; dominant. 2. One could be homozygous for the dominant allele, while the other is heterozygous. 3. One is from your father via his sperm, and one is from your mother via her egg.

inherited one allele for purple flowers and one allele for white flowers (*Pp*); like the F₁ plants, these plants will also have purple flowers, the dominant trait. (Note that *Pp* and *pP* are equivalent and usually written as the former.) Finally, one-fourth of the F₂ plants have inherited two alleles specifying white flowers (*pp*) and will express this recessive trait. Thus, Mendel's model accounts for the 3:1 ratio that he observed in the F₂ generation.

Geneticists distinguish between an organism's physical traits, called its **phenotype** (such as purple or white flowers), and its genetic makeup, called its **genotype** (in our example, *PP*, *Pp*, or *pp*). Now we can see that Figure 9.5 shows only the phenotypes and Figure 9.6 both the genotypes and phenotypes in our sample cross. For the F₂ plants, the ratio of plants with purple flowers to those with white flowers (3:1) is called the phenotypic ratio. The genotypic ratio is 1(*PP*):2(*Pp*):1(*pp*).

▼ Figure 9.7 **The relationship between alleles and homologous chromosomes.**
The matching colors of corresponding loci highlight the fact that homologous chromosomes carry alleles for the same genes at the same positions along their lengths.

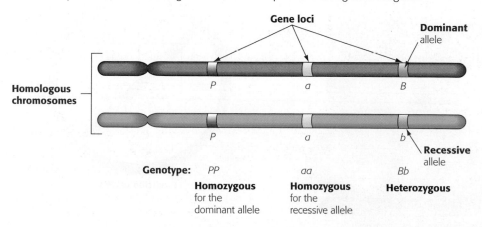

Gene loci

Dominant
allele

Homologous
chromosomes

P   *a*   *B*

P   *a*   *b*

Recessive
allele

Genotype: *PP*   *aa*   *Bb*

Homozygous
for the
dominant allele

Homozygous
for the
recessive allele

Heterozygous

Mendel's Law of Independent Assortment

Two other pea plant characters Mendel studied were seed shape and seed color. Mendel's seeds were either round or wrinkled in shape and either yellow or green in color. From tracking these characters one at a time in monohybrid crosses, Mendel knew that the allele for round shape (designated R) was dominant to the allele for wrinkled shape (r) and that the allele for yellow seed color (Y) was dominant to the allele for green seed color (y). What would result from a **dihybrid cross**, the crossing of parental varieties differing in two characters? Mendel crossed homozygous plants having round-yellow seeds (genotype $RRYY$) with plants having wrinkled-green seeds ($rryy$). As shown in **Figure 9.8**, the union of RY and ry gametes from the P generation yielded

hybrids heterozygous for both characters ($RrYy$)—that is, dihybrids. As we would expect, all of these offspring, the F_1 generation, had round-yellow seeds (the two dominant traits). But were the two characters transmitted from parents to offspring as a package, or was each character inherited independently of the other?

The question was answered when Mendel allowed fertilization to occur among the F_1 plants. If the genes for the two characters were inherited together (Figure 9.8a), then the F_1 hybrids would produce only the same two kinds of gametes that they received from their parents. In that case, the F_2 generation would show a 3:1 phenotypic ratio (three plants with round-yellow seeds for every one with wrinkled-green seeds), as in the Punnett square in Figure 9.8a. If, however, the two seed characters sorted independently, then the F_1 generation would produce four gamete genotypes—RY, rY, Ry, and ry—in equal quantities. The Punnett square in

▼ Figure 9.8 **Testing alternative hypotheses for gene assortment in a dihybrid cross.**

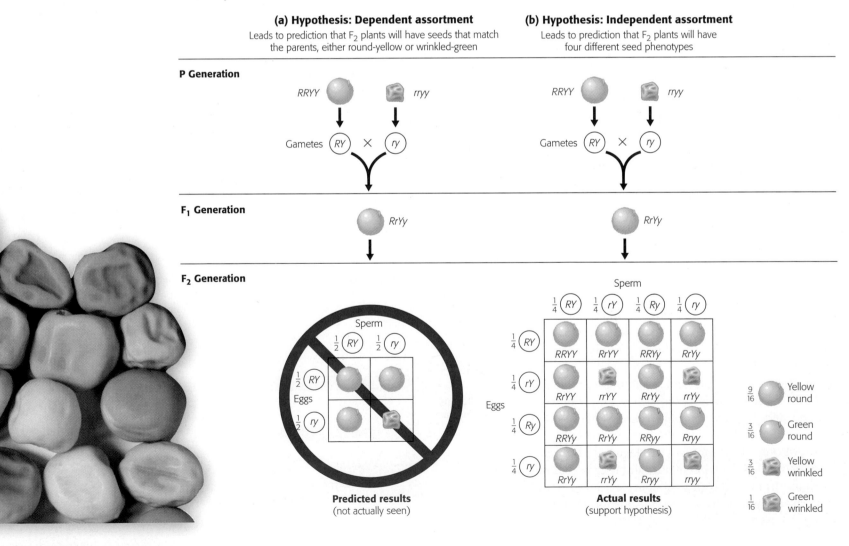

(a) Hypothesis: Dependent assortment
Leads to prediction that F_2 plants will have seeds that match the parents, either round-yellow or wrinkled-green

(b) Hypothesis: Independent assortment
Leads to prediction that F_2 plants will have four different seed phenotypes

Predicted results
(not actually seen)

Actual results
(support hypothesis)

Figure 9.8b shows all possible combinations of alleles that can result in the F_2 generation from the union of four kinds of sperm with four kinds of eggs. If you study the Punnett square, you'll see that it predicts nine different genotypes in the F_2 generation. These nine genotypes will produce four different phenotypes in a ratio of 9:3:3:1.

The Punnett square in Figure 9.8b also reveals that a dihybrid cross is equivalent to two monohybrid crosses occurring simultaneously. From the 9:3:3:1 ratio, we can see that the ratio of plants with round seeds to those with wrinkled seeds is 12:4, as is the ratio of yellow-seeded plants to green-seeded ones. These 12:4 ratios each reduce to 3:1, which is the F_2 ratio for a monohybrid cross. Mendel tried his seven pea characters in various dihybrid combinations and always observed a 9:3:3:1 ratio (or two simultaneous 3:1 ratios) of phenotypes in the F_2 generation. These results supported the hypothesis that *each pair of alleles assorts independently of the other pairs of alleles during gamete formation.* In other words, the inheritance of one character has no effect on the inheritance of another. This is called the **law of independent assortment.**

For another application of the law of independent assortment, examine Figure 9.9. The inheritance of two hereditary characters in Labrador retrievers is controlled by separate genes: black versus chocolate coat color, and normal vision versus the eye disorder progressive retinal atrophy (PRA). Black Labs have at least one copy of an allele called *B*, which gives their hairs densely packed granules of a dark pigment. The *B* allele is dominant to *b*, which leads to a less tightly packed distribution of pigment granules. As a result, the coats of dogs with genotype *bb* are chocolate in color. The allele that causes PRA, called *n*, is recessive to allele *N*, which is necessary for normal vision. Thus, only dogs of genotype *nn* become blind from PRA. If you mate two doubly heterozygous (*BbNn*) Labs (bottom of Figure 9.9), the phenotypic ratio of the offspring (F_2) is 9:3:3:1. These results resemble the F_2 results in Figure 9.8, demonstrating that the coat color and PRA genes are inherited independently. ☑

☑**CHECKPOINT**

Looking at the Punnett square in Figure 9.8b, what is the ratio of yellow seeds to green seeds? Of round seeds to wrinkled seeds?

Answer: 3:1; 3:1

▼ **Figure 9.9 Independent assortment of genes in Labrador retrievers.**
Blanks in the genotypes indicate alleles that can be either dominant or recessive.

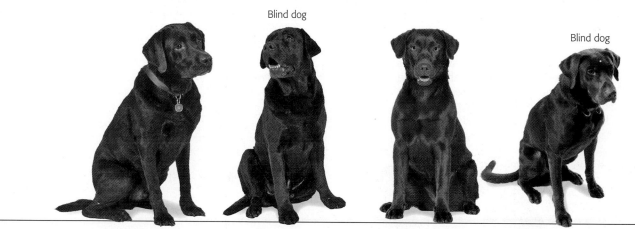

Blind dog

Blind dog

Phenotypes	Black coat, normal vision	Black coat, blind (PRA)	Chocolate coat, normal vision	Chocolate coat, blind (PRA)
Genotypes	*B_N_*	*B_nn*	*bbN_*	*bbnn*

(a) Possible phenotypes of Labrador retrievers

Mating of double heterozygotes
(black coat, normal vision)
BbNn ─── X ─── *BbNn*

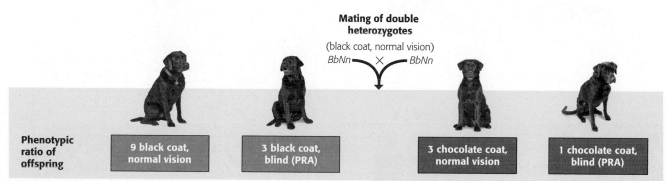

Phenotypic ratio of offspring	9 black coat, normal vision	3 black coat, blind (PRA)	3 chocolate coat, normal vision	1 chocolate coat, blind (PRA)

(b) A Labrador dihybrid cross

Using a Testcross to Determine an Unknown Genotype

Suppose you have a Labrador retriever with a chocolate coat. Consulting Figure 9.9, you can tell that its genotype must be *bb*, the only combination of alleles that produces the chocolate-coat phenotype. But what if you have a black Lab? It could have one of two possible genotypes—*BB* or *Bb*—and there is no way to tell which is correct by looking at the dog. To determine your dog's genotype, you could perform a **testcross**, a mating between an individual of dominant phenotype but unknown genotype (your black Lab) and a homozygous recessive individual—in this case, a *bb* chocolate Lab.

Figure 9.10 shows the offspring that could result from such a mating. If, as shown on the left, the black parent's genotype is *BB*, we would expect all the offspring to be black, because a cross between genotypes *BB* and *bb* can produce only *Bb* offspring. On the other hand, if the black parent is *Bb*, we would expect both black (*Bb*) and chocolate (*bb*) offspring in a 1:1 phenotypic ratio. Thus, the appearance of the offspring may reveal the original black dog's genotype. ☑

▼ **Figure 9.10 A Labrador retriever testcross.** To determine the genotype of a black Lab, it can be crossed with a chocolate Lab (homozygous recessive, *bb*). If all the offspring are black, the black parent most likely had genotype *BB*. If half the offspring are chocolate, the black parent must be heterozygous (*Bb*).

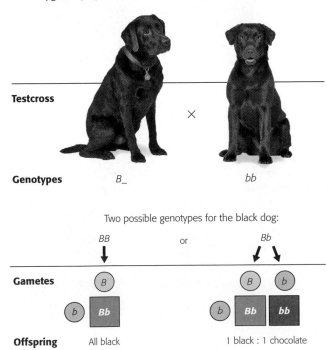

The Rules of Probability

Mendel's strong background in mathematics served him well in his studies of inheritance. For instance, he understood that genetic crosses obey the rules of probability—the same rules that apply to the tossing of coins, the rolling of dice, and the drawing of cards. An important lesson we can learn from coin tossing is that for each and every toss of the coin, the probability of heads is $\frac{1}{2}$. Even if heads has landed five times in a row, the probability of the next toss coming up heads is still $\frac{1}{2}$. In other words, the outcome of any particular toss is unaffected by what has happened on previous attempts. Each toss is an independent event.

If two coins are tossed simultaneously, the outcome for each coin is an independent event, unaffected by the other coin. What is the chance that both coins will land heads-up? The probability of such a dual event is the product of the separate probabilities of the independent events—for the coins, $\frac{1}{2} \times \frac{1}{2} = \frac{1}{4}$. This is called the **rule of multiplication**, and it holds true for independent events that occur in genetics as well as coin tosses, as shown in **Figure 9.11**. In our dihybrid cross of Labradors (see

▼ **Figure 9.11 Segregation of alleles and fertilization as chance events.** When a heterozygote (*Bb*) forms gametes, segregation of alleles during sperm and egg formation is like two separately tossed coins (that is, two independent events).

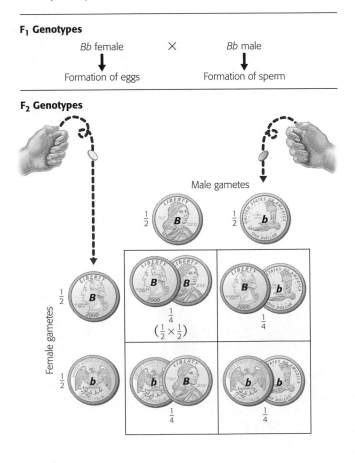

Figure 9.9), the genotype of the F₁ dogs for coat color was *Bb*. What is the probability that a particular F₂ dog will have the *bb* genotype? To produce a *bb* offspring, both egg and sperm must carry the *b* allele. The probability that an egg from a *Bb* dog will have the *b* allele is $\frac{1}{2}$, and the probability that a sperm will have the *b* allele is also $\frac{1}{2}$. By the rule of multiplication, the probability that two *b* alleles will come together at fertilization is $\frac{1}{2} \times \frac{1}{2} = \frac{1}{4}$. This is exactly the answer given by the Punnett square in Figure 9.11. If we know the genotypes of the parents, we can predict the probability for any genotype among the offspring. By applying the rules of probability to segregation and independent assortment, we can solve some rather complex genetics problems. ☑

Family Pedigrees

Mendel's laws apply to the inheritance of many human traits. **Figure 9.12** illustrates alternative forms of three human characters that are each thought to be determined by simple dominant-recessive inheritance at one gene locus. (The genetic basis of many other human characters—such as eye and hair color—are much more complex and poorly understood.) If we call the dominant allele of any such gene *A*, the dominant

phenotype results from either the homozygous genotype *AA* or the heterozygous genotype *Aa*. Recessive phenotypes always result from the homozygous genotype *aa*. In genetics, the word *dominant* does not imply that a phenotype is either normal or more common than a recessive phenotype; **wild-type traits** (those seen most often in nature) are not necessarily specified by dominant alleles. In genetics, dominance means that a heterozygote (*Aa*), carrying only a single copy of a dominant allele, displays the dominant phenotype. By contrast, the phenotype of the corresponding recessive allele is seen only in a homozygote (*aa*). Recessive traits can be more common in the population than dominant ones. For example, the absence of freckles is more common than their presence.

How do we know how particular human traits are inherited? Researchers working with pea plants or Labrador retrievers can perform testcrosses. But geneticists who study humans obviously cannot control the mating of their subjects. Instead, they must analyze the results of matings that have already occurred. First, the geneticist collects as much information as possible about a family's history for the trait. Then the researcher assembles this information into a family tree—the family **pedigree**. (You may associate pedigrees with purebred animals such as racehorses and champion dogs, but they

DOMINANT TRAITS

Freckles

Widow's peak

Free earlobe

RECESSIVE TRAITS

No freckles

Straight hairline

Attached earlobe

☑**CHECKPOINT**

Using a standard 52-card deck, what is the probability of being dealt an ace? What about being dealt an ace or a king? What about being dealt an ace and then another ace?

Answer: $\frac{1}{13}$ (4 aces / 52 cards); $\frac{2}{13}$ ($\frac{4}{52} + \frac{4}{52}$); ($\frac{4}{52} \times \frac{3}{51}$) (since there are 3 aces left in a deck with 51 cards remaining) = 0.0045, or $\frac{1}{221}$

◄ Figure 9.12 **Examples of inherited traits in humans thought to be controlled by a single gene.**

can represent human matings just as well.) To analyze a pedigree, the geneticist uses Mendel's concept of dominant and recessive alleles and his law of segregation.

Let's apply this approach to the example in **Figure 9.13**, which shows a pedigree tracing the incidence of free versus attached earlobes. The letter *F* stands for the dominant allele for free earlobes, and *f* symbolizes the recessive allele for attached earlobes. In the pedigree, □ represent males, ○ represent females, and colored symbols (■ and ●) indicate that the person has the trait under investigation (in this case, attached earlobes). The earliest generation studied is at the top of the pedigree. By applying Mendel's laws, we can deduce that the attached allele is recessive because that is the only way that one of the third-generation children (at the bottom of the pedigree) could have attached earlobes when both parents did not. We can therefore label all the individuals with attached earlobes in the pedigree (that is, all those with colored circles or squares) as homozygous recessive (*ff*). Mendel's laws also enable us to deduce the genotypes for most of the people in the pedigree. For example, both of the second-generation parents must have carried the *f* allele (which they passed on to the affected child) along with the *F* allele that gave them free earlobes. The same must be true of the set of grandparents on the left because they both had free earlobes but two of their sons had attached earlobes. Notice that we cannot deduce the genotype of every member of the pedigree. For example, the sister with the free earlobes must

have at least one *F* allele, but it is possible that she could be *FF* or *Ff*. We cannot distinguish between these two possibilities using the available data. ☑

Human Disorders Controlled by a Single Gene

The human genetic disorders listed in **Table 9.1** are known to be inherited as dominant or recessive traits controlled by a single gene. These disorders therefore show simple inheritance patterns like the ones Mendel studied in pea plants. The genes involved are all located on autosomes, chromosomes other than the sex chromosomes X and Y.

Recessive Disorders

Most human genetic disorders are recessive. They range in severity from relatively harmless to life-threatening. Most people who have recessive disorders are born to normal parents who are both heterozygotes—that is, who are **carriers** of the recessive allele for the disorder but appear normal themselves.

Using Mendel's laws, we can predict the fraction of affected offspring that is likely to result from a marriage between two carriers. Consider a form of inherited deafness caused by a recessive allele. Suppose two heterozygous carriers (*Dd*) have a child. What is the probability that the child will be deaf? As the Punnett square in

☑CHECKPOINT

What is a wild-type trait?

Answer: a trait that is the prevailing one in nature

► Figure 9.13 **A family pedigree showing inheritance of free versus attached earlobes.**

First generation
(grandparents)

Ff Ff ff Ff

Second generation
(parents, aunts, and uncles)

FF or Ff ff ff Ff Ff ff

Third generation
(brother and sister)

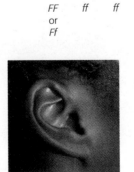

ff FF or Ff

Female   Male

● ■ Attached

○ □ Free

Table 9.1	Some Autosomal Disorders in Humans	
Disorder	Major Symptoms	Incidence
Recessive Disorders		
Albinism	Lack of pigment in skin, hair, and eyes	$\frac{1}{22,000}$
Cystic fibrosis	Excess mucus in lungs, digestive tract, liver; increased susceptibility to infections; death in early childhood unless treated	$\frac{1}{1,800}$ European-Americans
Phenylketonuria (PKU)	Accumulation of phenylalanine in blood; lack of normal skin pigment; mental retardation unless treated	$\frac{1}{10,000}$ in U.S. and Europe
Sickle-cell disease	Sickled red blood cells; damage to many tissues	$\frac{1}{500}$ African-Americans
Tay Sachs disease	Lipid accumulation in brain cells; mental deficiency; blindness; death in childhood	$\frac{1}{3,500}$ European Jews
Dominant Disorders		
Achondroplasia	Dwarfism	$\frac{1}{25,000}$
Alzheimer's disease (one type)	Mental deterioration; usually strikes late in life	Not known
Huntington's disease	Mental deterioration and uncontrollable movements; strikes in middle age	$\frac{1}{25,000}$
Hypercholesterolemia	Excess cholesterol in blood; heart disease	$\frac{1}{500}$

Figure 9.14 shows, each child of two carriers has a $\frac{1}{4}$ chance of inheriting two recessive alleles. Thus, we can say that about one-fourth of the children of this marriage are likely to be deaf. We can also say that a hearing child from such a marriage has a $\frac{2}{3}$ chance of being a carrier (that is, on average, two out of three of the offspring with the hearing phenotype will be *Dd*). We can apply this same method of pedigree analysis and prediction to any genetic trait controlled by a single gene locus.

The most common lethal genetic disease in the United States is **cystic fibrosis**. Affecting about 30,000 people in the United States and about 70,000 people worldwide, the cystic fibrosis allele is recessive and

carried by about 1 in 25 people of European ancestry. A person with two copies of this allele has cystic fibrosis, which is characterized by an excessive secretion of very thick mucus from the lungs, pancreas, and other organs. This mucus can interfere with breathing, digestion, and liver function and makes the person vulnerable to recurrent bacterial infections. Although there is no cure for this fatal disease, a special diet, antibiotics to prevent infection, frequent pounding of the chest and back to clear the lungs, and other treatments can greatly extend life. Once invariably fatal in childhood, advances in cystic fibrosis treatment have raised the median survival age of Americans with cystic fibrosis to 37.

Like cystic fibrosis, most genetic disorders are not evenly distributed across all ethnic groups. Such uneven distribution is the result of prolonged geographic isolation of certain populations. For example, the isolated lives of the early inhabitants of Martha's Vineyard (an island off the coast of Massachusetts) led to frequent marriages between close relatives. Consequently, the frequency of an allele that caused deafness was high, and the deafness allele was rarely transmitted to outsiders.

With the increased mobility in most societies today, it is relatively unlikely that two carriers of a rare, harmful allele will meet and mate. However, the probability increases greatly if close blood relatives marry and have children. People with recent common ancestors are more likely to carry the same recessive alleles than are unrelated people. Therefore, a mating between close blood relatives, called **inbreeding**, is more likely to produce offspring homozygous for a harmful recessive trait.

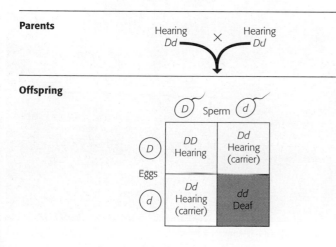

▼ **Figure 9.14 Predicted offspring when both parents are carriers for a recessive disorder.**

Geneticists have observed increased incidence of harmful recessive traits among many types of inbred animals. Most lines of purebred dogs have known genetic defects. The Siberian huskies discussed in the Biology and Society section, for example, are susceptible to the same form of blindness (PRA) as the retrievers shown in Figure 9.9. The detrimental effects of inbreeding are also seen in some endangered species, such as cheetahs (see Chapter 13).

Dominant Disorders

A number of human disorders are caused by dominant alleles. Some are nonlethal conditions, such as extra fingers and toes or fingers and toes that are webbed. A serious but nonlethal disorder caused by a dominant allele is **achondroplasia**, a form of dwarfism in which the head and torso develop normally, but the arms and legs are short (**Figure 9.15**). The homozygous dominant genotype causes death of the embryo, and therefore only heterozygotes, individuals with a single copy of the defective allele, have this disorder. This also means that a person with achondroplasia has a 50% chance of passing the condition on to any children (**Figure 9.16**). Therefore, all those who do not have achondroplasia, more than 99.99% of the population, are homozygous for the recessive allele. This example makes it clear that a dominant allele is not necessarily more plentiful in a population than the corresponding recessive allele.

Dominant alleles that are lethal are much less common than lethal recessive alleles. One example is the allele that causes **Huntington's disease**, a degeneration of the nervous system that usually does not begin until middle age. Once the deterioration of the nervous system begins, it is irreversible and inevitably fatal. Because the allele for Huntington's disease is dominant, any child born to a parent with the allele has a 50% chance of inheriting the allele and the disorder. This example makes it clear that a dominant allele is not necessarily "better" than the corresponding recessive allele.

The allele that causes Huntington's disease is dominant but lethal when inherited in two copies. Let's return to the chapter thread—dog breeding—to investigate a striking dog trait with a similar inheritance pattern.

▼ Figure 9.15
Achondroplasia, a dominant trait. Dr. Michael C. Ain, a pediatric orthopedic surgeon at the Johns Hopkins Children's Center, specializes in the repair of bone defects caused by achondroplasia and related disorders.

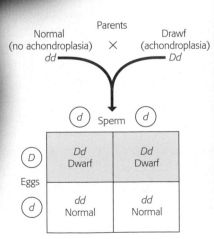

▼ Figure 9.16 **A family with and without achondroplasia.** Amy Roloff (who has achondroplasia) and her husband Matt (who has an unrelated form of dwarfism) have four children. Zachary received Amy's achondroplasia gene, while twin brother Jeremy, sister Molly Jo, and brother Jake did not. The hypothetical Punnett square shows the offspring expected from a mating of a wild-type male with a dwarf female.

Molly Jo Matt Amy Zachary Jake Jeremy

What Is the Genetic Basis of Hairless Dogs?

You've probably made the **observation** that compared to most mammals, dogs come in a wide variety of physical types. For instance, some dogs come in both hairless and thick-coated varieties (**Figure 9.17**). Breeders know from experience that the hairless phenotype is inherited as a dominant trait and that, as in achondroplasia, the homozygous dominant condition is lethal.

In 2005, the complete genome of a dog—the sequence of all the DNA of a female boxer named Tasha—was published. Since that time, canine geneticists have added a wealth of data from other breeds to that first genome. In 2008, an international group of researchers set out to investigate the **question** of the genetic basis for the hairless phenotype. The research team proposed the **hypothesis** that a comparison of genes of coated and hairless dogs would identify the gene or genes responsible. They made the **prediction** that a mutation in a single gene could account for the hairless appearance. Their **experiment** compared DNA sequences in 140 hairless dogs from 3 breeds with similar sequences in 87 coated

◀ **Figure 9.17 Hairless versus coated Chinese crested dogs.** The Chinese crested dog comes in both naturally hairless (left) and thick-coated (right) varieties.

dogs from 22 breeds. The **results** were clear: Every one of the hairless dogs, but none of the coated dogs, had a change in a single gene on chromosome 17. Due to the insertion of seven extra DNA bases, the chromosome 17 mutation knocks out a single protein that is involved in embryonic development. This experiment shows how the extreme range of phenotypes in dogs combined with the availability of genome sequences can be used to provide insight into interesting genetic questions.

Genetic Testing

Until relatively recently, the onset of symptoms was the only way to know if a person had inherited an allele that might lead to disease. Today, there are many tests that can detect the presence of disease-causing alleles in an individual's genome.

The most common forms of genetic testing are performed during pregnancy. Some prospective parents are aware that they have an increased risk of having a baby with a genetic disease. Genetic testing before birth usually requires the collection of fetal cells. In amniocentesis, a physician uses a needle to extract about 2 teaspoonfuls of the fluid that bathes the developing fetus. In chorionic villus sampling, a physician inserts a narrow, flexible tube through the mother's vagina and into her uterus, removing some placental tissue. Once cells are obtained, they can be screened for genetic diseases.

Because amniocentesis and chorionic villus sampling have risks of complications, these techniques are usually reserved for situations in which the possibility of a genetic disease is significantly higher than average. Newer

genetic screening procedures involve isolating tiny amounts of fetal cells or DNA released into the mother's bloodstream. Although few reliable tests are yet available using this method, this promising and complication-free technology may soon replace more invasive procedures.

As genetic testing becomes more routine, geneticists are working to make sure that the tests do not cause more problems than they solve. Geneticists stress that patients seeking genetic testing should receive counseling both before and after to explain the test and to help them cope with the results. But will sufficient numbers of genetic counselors be available to help individuals understand their test results? If fetal tests reveal a serious disorder, the parents must choose between terminating the pregnancy and preparing themselves for a baby with severe problems. Identifying a genetic disease early can give families time to prepare—emotionally, medically, and financially. Advances in biotechnology offer possibilities for reducing human suffering, but not before key ethical issues are resolved. The dilemmas posed by human genetics reinforce one of this book's themes: the immense social implications of biology. ☑

Variations on Mendel's Laws

Mendel's two laws explain inheritance in terms of discrete factors—genes—that are passed along from generation to generation according to simple rules of probability. These laws are valid for all sexually reproducing organisms, including garden peas, Siberian huskies, and human beings. But just as the basic rules of musical harmony cannot account for all the rich sounds of a symphony, Mendel's laws stop short of explaining some patterns of genetic inheritance. In particular, they do not explain the many cases of inherited characters that exist in more than two clear-cut variants—such as flower color that can be red, white, or pink or human skin color in all its range of shades. In fact, for most sexually reproducing organisms, cases where Mendel's rules can strictly account for the patterns of inheritance are relatively rare. More often, the observed inheritance patterns are more complex. We will now add several extensions to Mendel's laws that help account for this complexity.

Incomplete Dominance in Plants and People

The F_1 offspring of Mendel's pea crosses always looked like one of the two parent plants. In such situations, the dominant allele has the same effect on the phenotype whether present in one or two copies. But for some characters, the F_1 hybrids have an appearance in between the phenotypes of the two parents, an effect called **incomplete dominance**. For instance, when red snapdragons are crossed with white snapdragons, all the F_1 hybrids have pink flowers (**Figure 9.18**). And in the F_2 generation, the genotypic ratio and the phenotypic ratio are the same: 1:2:1.

We also see examples of incomplete dominance in humans. One case involves a recessive allele (*h*) that causes **hypercholesterolemia**, a condition characterized by dangerously high levels of cholesterol in the blood. Normal individuals are homozygous dominant, *HH*. Heterozygotes (*Hh*) have blood cholesterol levels about twice normal. They are unusually prone to cholesterol buildup in artery walls and may have heart attacks from blocked heart arteries by their mid-30s. Hypercholesterolemia is even more serious in homozygous individuals (*hh*). These homozygotes have about five times the normal amount of blood cholesterol and may have heart attacks as early as age 2. If we look at the molecular basis for hypercholesterolemia, we can understand the intermediate phenotype of heterozygotes (**Figure 9.19**). The *H* allele specifies a cell-surface receptor protein that

▼ **Figure 9.18 Incomplete dominance in snapdragons.**
Compare this diagram with Figure 9.6, where one of the alleles displays complete dominance.

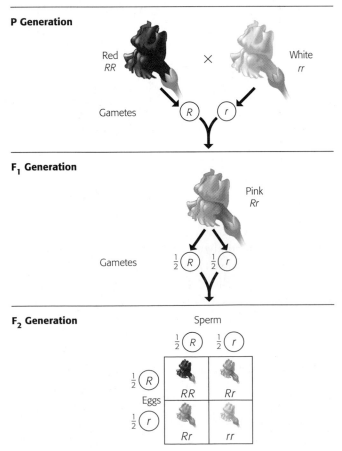

P Generation

Red
RR   ×   White
rr

Gametes R r

F₁ Generation

Pink
Rr

Gametes $\frac{1}{2}$ R $\frac{1}{2}$ r

F₂ Generation   Sperm

$\frac{1}{2}$ R $\frac{1}{2}$ r

Eggs $\frac{1}{2}$ R *RR*   *Rr*

$\frac{1}{2}$ r *Rr*   *rr*

▼ **Figure 9.19 Incomplete dominance in human hypercholesterolemia.**
LDL receptors promote the breakdown of cholesterol carried in the bloodstream by LDL (low-density lipoproteins). This process helps prevent the accumulation of cholesterol in the arteries. Having too few receptors allows dangerous levels of LDL to build up in the blood.

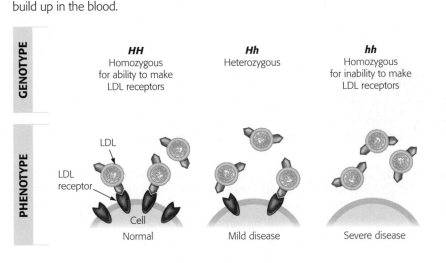

GENOTYPE

HH
Homozygous
for ability to make
LDL receptors

Hh
Heterozygous

hh
Homozygous
for inability to make
LDL receptors

PHENOTYPE

LDL

LDL
receptor

Cell

Normal

Mild disease

Severe disease

certain cells use to mop up excess low-density lipoprotein (LDL, or "bad cholesterol") from the blood. With only half as many receptors as *HH* individuals, heterozygotes can remove much less excess cholesterol.

ABO Blood Groups: An Example of Multiple Alleles and Codominance

So far, we have discussed inheritance patterns involving only two alleles per gene (*H* versus *h*, for example). But most genes occur in more than two forms, known as multiple alleles. Although each individual carries, at most, two different alleles for a particular gene, in cases of multiple alleles, more than two possible alleles exist in the population.

The **ABO blood groups** in humans are an example of multiple alleles. There are three common alleles for the character of ABO blood type, which in various combinations produce four phenotypes: A person's blood type may be A, B, AB, or O. These letters refer to two carbohydrates, designated A and B, that may be found on the surface of red blood cells (**Figure 9.20**). A person's red blood cells may be coated with carbohydrate A (giving them type A blood), carbohydrate B (type B), both (type AB), or neither (type O). Matching compatible blood groups is critical for safe blood transfusions. Our immune system produces blood proteins called antibodies that can bind specifically to the blood cell carbohydrates we lack. If a donor's blood cells have a carbohydrate that is foreign to the recipient, then the recipient's antibodies will cause the donated blood cells to clump together. This clumping can kill the recipient. The clumping reaction is also the basis of a blood-typing lab test.

The four blood groups result from various combinations of the three different alleles: I^A (for the ability to make substance A), I^B (for B), and i (for neither A nor B). Each person inherits one of these alleles from each parent. Because there are three alleles, there are six possible genotypes, as listed in Figure 9.20. Both the I^A and I^B alleles are dominant to the i allele. Thus, $I^A I^A$ and $I^A i$ people have type A blood, and $I^B I^B$ and $I^B i$ people have type B. Recessive homozygotes (ii) have type O blood; they make neither the A nor the B carbohydrate. Finally, people of genotype $I^A I^B$ make *both* carbohydrates. In other words, the I^A and I^B alleles exhibit **codominance**, meaning that both alleles are expressed in heterozygous individuals ($I^A I^B$), who have type AB blood. Be careful to distinguish codominance (the expression of both alleles) from incomplete dominance (the expression of one intermediate trait). ☑

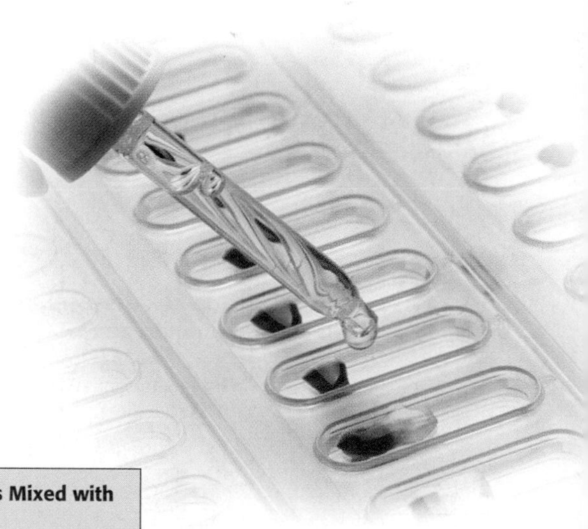

▼ Figure 9.20 **Multiple alleles for the ABO blood groups.** The three versions of the gene responsible for blood type may produce carbohydrate A (allele I^A), carbohydrate B (allele I^B), or neither carbohydrate (allele i). Because each person carries two alleles, six genotypes are possible that result in four different phenotypes. The clumping reaction that occurs between antibodies and foreign blood cells is the basis of blood-typing (shown in the photograph at right) and of the adverse reaction that occurs when someone receives a transfusion of incompatible blood.

Blood Group (Phenotype)	Genotypes	Red Blood Cells	Antibodies Present in Blood	Reactions When Blood from Groups Below Is Mixed with Antibodies from Groups at Left			
				O	A	B	AB
A	$I^A I^A$ or $I^A i$	Carbohydrate A	Anti-B				
B	$I^B I^B$ or $I^B i$	Carbohydrate B	Anti-A				
AB	$I^A I^B$		—				
O	ii		Anti-A Anti-B				

☑CHECKPOINT

1. Why is a testcross unnecessary to determine whether a snapdragon with red flowers is homozygous or heterozygous?
2. Maria has type O blood, and her sister has type AB blood. What are the genotypes of the girls' parents?

Answer: 1. Only plants homozygous for the dominant allele have red flowers; heterozygotes have pink flowers 2. One parent is $I^A i$ and the other parent is $I^B i$.

Pleiotropy and Sickle-Cell Disease

Our genetic examples to this point have been cases in which each gene specifies only one hereditary character. But in many cases, one gene influences several characters. The impact of a single gene on more than one character is called **pleiotropy**.

An example of pleiotropy in humans is **sickle-cell disease** (also sometimes called sick-cell anemia), a disorder characterized by a diverse set of symptoms. The direct effect of the sickle-cell allele is to make red blood cells produce abnormal hemoglobin proteins (see Figure 3.19). These abnormal molecules tend to link together and crystallize, especially when the oxygen content of the blood is lower than usual because of high altitude, overexertion, or respiratory ailments. As the hemoglobin crystallizes, the normally disk-shaped red blood cells deform to a sickle shape with jagged edges **(Figure 9.21)**. Sickled cells are destroyed rapidly by the body, and their destruction may cause anemia and general weakening of the body. Also, because of their angular shape, sickled cells do not flow smoothly in the blood and tend to accumulate and clog tiny blood vessels. Blood flow to body parts is reduced, resulting in

multiple symptoms, such as periodic fever, pain, and damage to various organs, including the heart, brain, and kidneys. Blood transfusions and drugs may relieve some of the symptoms, but there is no cure, and sickle-cell disease kills about 100,000 people in the world annually.

In most cases, only people who are homozygous for the sickle-cell allele have sickle-cell disease. Heterozygotes, who have one sickle-cell allele and one normal allele, are usually healthy; hence, the disease is considered recessive. However, at the molecular level, the two alleles are actually codominant: Both alleles are expressed in heterozygous individuals, and their red blood cells contain both normal and abnormal hemoglobin. A simple blood test can distinguish homozygotes from heterozygotes. (See the Evolution Connection section in Chapter 13 for further discussion of sickle-cell disease.) ☑

Polygenic Inheritance

Mendel studied genetic characters that could be classified on an either-or basis, such as purple or white flower color. However, many characters, such as human skin color and height, vary along a continuum in a population. Many such features result from **polygenic inheritance**, the additive effects of two or more genes on a single phenotypic character. (This is the converse of pleiotropy, in which one gene affects several characters.)

Let's consider a hypothetical model of how polygenic inheritance might work. Assume that skin pigmentation in humans is controlled by three genes that are inherited separately, like Mendel's pea genes. (Actually, genetic evidence indicates that *at least* three genes control this character.) The dark-skin allele for each gene (*A*, *B*, and *C*) contributes one "unit" of darkness to the phenotype and is incompletely dominant to the other alleles (*a*, *b*, and *c*). A person who is *AABBCC* would be very dark, while an *aabbcc* individual would be very light. An

☑CHECKPOINT

How does sickle-cell disease exemplify the concept of pleiotropy?

Answer: Homozygotes for the sickle-cell allele have abnormal hemoglobin, and its effect on the shape of red blood cells leads to a cascade of traits affecting many organs of the body.

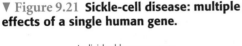

▼ Figure 9.21 **Sickle-cell disease: multiple effects of a single human gene.**

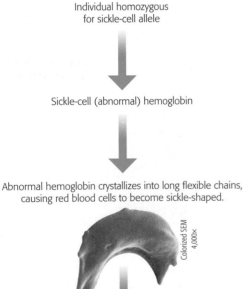

Individual homozygous
for sickle-cell allele

Sickle-cell (abnormal) hemoglobin

Abnormal hemoglobin crystallizes into long flexible chains,
causing red blood cells to become sickle-shaped.

Colorized SEM
4,000×

Sickled cells can lead to a cascade of symptoms, such as
weakness, pain, organ damage, and paralysis.

AaBbCc person would have skin of an intermediate shade. Because the alleles have an additive effect, the genotype *AaBbCc* would produce the same skin color as any other genotype with just three dark-skin alleles, such as *AABbcc*. The Punnett square in **Figure 9.22**

shows all possible genotypes of F₂ offspring. The row of squares below the Punnett square shows the seven skin pigmentation phenotypes that would theoretically result. This hypothetical example shows how inheritance of three genes could lead to seven levels of pigmentation at the frequencies indicated by the bars in the graph. ☑

☑CHECKPOINT

Based on the skin-tone model in Figure 9.22, put the following individuals in order by skin tone, from lightest to darkest: *AAbbCC*, *aaBBcc*, *AabBCc*, *Aabbcc*, *AaBBCC*.

Answer: Aabbcc, aaBBcc, AabBCc, AAbbCC, AaBBCC.

▼ **Figure 9.22** **A model for polygenic inheritance of skin color.** The seven bars in the graph at the bottom of the figure depict the relative numbers of each of the phenotypes in the F₂ generation. The bell-shaped curve indicates the distribution of an even greater variety of skin shades in the population that might result from the combination of heredity and environmental effects, such as sun-tanning.

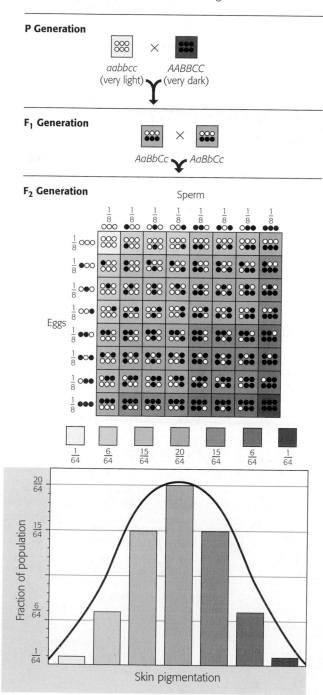

The Role of Environment

If we examine a real human population for the skin-color phenotype, we would see more shades than just seven. The true range might be similar to the entire spectrum of color under the bell-shaped curve in Figure 9.22. In fact, no matter how carefully we characterize the genes for skin color, a purely genetic description will always be incomplete. This is because some intermediate shades of skin color result from the effects of environmental factors, such as exposure to the sun.

Many phenotypic characters result from a combination of heredity and environmental factors. For example, although a single tree is locked into its inherited genotype, its leaves vary in size, shape, and color, depending on exposure to wind and sun and the tree's nutritional state. For humans, exercise alters build; experience improves performance on intelligence tests; and social and cultural factors can greatly affect appearance. As geneticists learn more and more about our genes, it's becoming clear that many human characters—such as a person's susceptibility to heart disease, cancer, alcoholism, and schizophrenia—are influenced by both genes and environment.

Whether human characters are more influenced by genes or by the environment—nature or nurture—is a very old and hotly contested issue. For some characters, such as the ABO blood group, a given genotype mandates a very specific phenotype. In contrast, a person's blood count of red and white cells varies quite a bit, depending on such factors as the altitude, the customary level of physical activity, and the presence of infectious agents.

Simply spending time with identical twins will convince anyone that environment, and not just genes, affects a person's traits (**Figure 9.23**). However, there is an important difference between these two sources of variation: Only genetic influences are inherited. Any effects of the environment are not passed on to the next generation.

▼ **Figure 9.23** **As a result of environmental influences, even identical twins can look different.**

Chromosomal Basis
of Inheritance

The Chromosomal Basis of Inheritance

Mendel published his results in 1866, but not until long after he died did biologists understand the significance of his work. Cell biologists worked out the processes of mitosis and meiosis (see Chapter 8) in the late 1800s. Then, around 1900, researchers began to notice parallels between the behavior of chromosomes and the behavior of Mendel's "heritable factors" (what we now call genes). One of biology's most important concepts—the chromosome theory of inheritance—began to emerge.

The **chromosome theory of inheritance** states that genes are located at specific positions on chromosomes

and that the behavior of chromosomes during meiosis and fertilization accounts for inheritance patterns. Indeed, it is chromosomes that undergo segregation and independent assortment during meiosis and thus account for Mendel's laws. **Figure 9.24** correlates the results of the dihybrid cross in Figure 9.8b with the movement of chromosomes through meiosis. Starting with two true-breeding parental plants, the diagram follows two genes on different chromosomes—one for seed shape (alleles *R* and *r*) and one for seed color (alleles *Y* and *y*)—through the F$_1$ and F$_2$ generations. ✔

► **Figure 9.24 The chromosomal basis of Mendel's laws.**

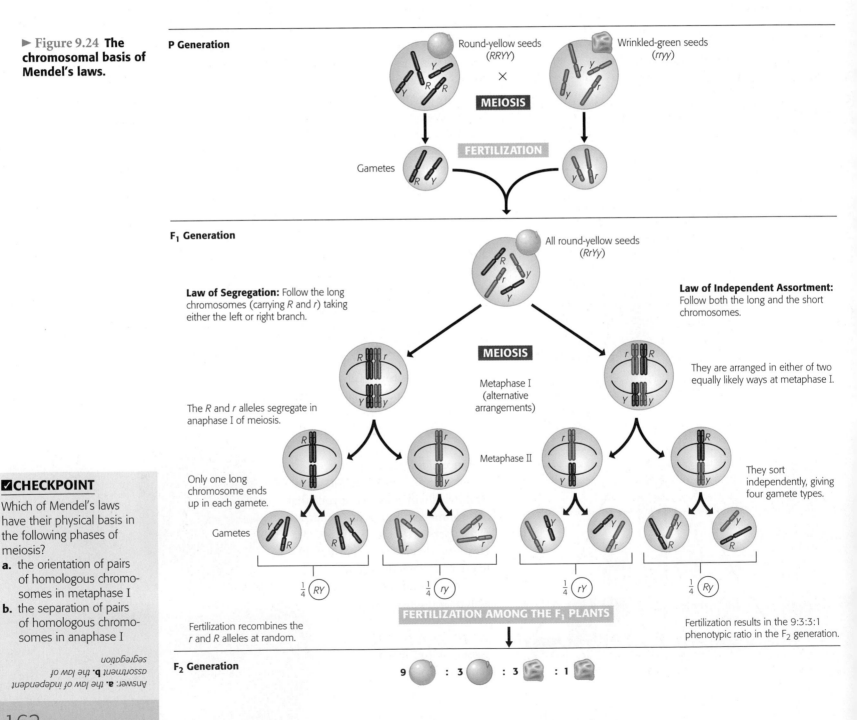

✔CHECKPOINT

Which of Mendel's laws have their physical basis in the following phases of meiosis?

a. the orientation of pairs of homologous chromosomes in metaphase I

b. the separation of pairs of homologous chromosomes in anaphase I

*Answer: **a.** the law of independent assortment **b.** the law of segregation*

Linked Genes

Realizing that genes are *on chromosomes* when they segregate enables us to extend Mendel's work in several important ways. For example, alleles that start out together on the same chromosome would be expected to travel together during meiosis and fertilization. In general, genes that are located close together on a chromosome, called **linked genes**, tend to be inherited as a set and therefore do not follow Mendel's law of independent assortment.

The most important early studies into linked genes took place in the New York City laboratory of American embryologist Thomas Hunt Morgan. In the early 1900s, Morgan and his colleagues studied the fruit fly *Drosophila melanogaster* (**Figure 9.25**). Often seen flying around overripe fruit, *Drosophila* is a good research animal for studies of inheritance because it is easily and inexpensively grown and can produce several generations in a matter of months.

Normal, or wild-type, *Drosophila* flies have gray bodies (genotype *GG*) and long wings (*LL*). Morgan cultivated true-breeding mutant (non-wild-type) fruit flies that had black bodies (*gg*) and short, underdeveloped wings (*ll*). He then performed a dihybrid testcross, crossing doubly heterozygous flies (*GgLl*) with double mutants (*ggll*). Morgan expected that the offspring would show equal numbers of all four possible phenotypes ($\frac{1}{4}$ gray-long, $\frac{1}{4}$ gray-short, $\frac{1}{4}$ black-long, $\frac{1}{4}$ black-short).

In fact, Morgan's testcross produced equal numbers of two of the four expected phenotypes but far smaller numbers of the other two phenotypes. He found that 83% of the flies displayed the phenotypes of the parents (gray-long and black-short), while only 17% showed nonparental phenotypes (gray-short and black-long). Morgan reasoned that the genes for body color and wing shape were linked—in this case, *G* with *L* and *g* with *l*—causing them to be inherited together. Such linkage would mean that meiosis in the heterozygous fruit flies would yield gametes with only two genotypes (*GL* and *gl*, but not *Gl* or *gL*). The large numbers of flies with gray-long and black-short traits in the experiment resulted from fertilization among the *GL* and *gl* gametes. ✔

Genetic Recombination: Crossing Over

But what of the smaller numbers of Morgan's gray-short and black-long flies? How could they be produced if the two genes involved are linked? In Chapter 8, we saw that

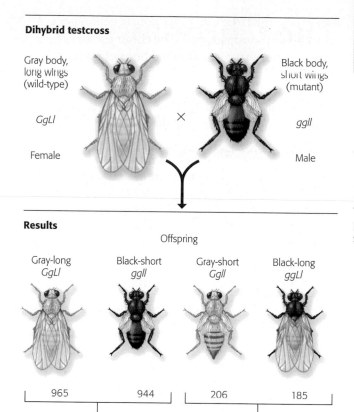

Dihybrid testcross

Gray body, long wings (wild-type) — *GgLl* — Female

×

Black body, short wings (mutant) — *ggll* — Male

Results

Offspring

Gray-long *GgLl*	Black-short *ggll*	Gray-short *Ggll*	Black-long *ggLl*
965	944	206	185

Parental phenotypes 83%

Recombinant phenotypes 17%

$$\text{Recombination frequency} = \frac{391 \text{ recombinants}}{2{,}300 \text{ total offspring}} = 0.17, \text{ or } 17\%$$

◄ **Figure 9.25 Thomas Morgan's experiment and results.** Among the 2,300 offspring of this dihybrid testcross, Morgan observed more than the expected parental phenotypes (gray-long and black-short) and fewer than expected nonparental phenotypes (gray-short and black-long).

during meiosis, crossing over between homologous chromosomes results in shuffled chromosome segments in the haploid daughter cells (see Figure 8.18), thereby producing new combinations of alleles. **Figure 9.26** reviews crossing over, showing how two linked genes can give rise to four different gamete genotypes. Two of the gamete genotypes reflect the presence of parental-type chromosomes, which have not been altered by crossing

▼ Figure 9.26 **Review: Crossing over can produce recombinant gametes.**

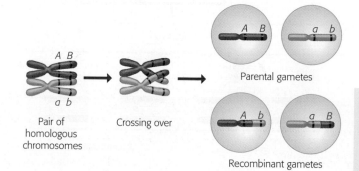

A B / *a b* — Pair of homologous chromosomes

Crossing over

A B / *a b* — Parental gametes

A b / *a B* — Recombinant gametes

☑CHECKPOINT

What are linked genes?

Answer: genes located near each other on the same chromosome that tend to be inherited together

over. In contrast, the other two gamete genotypes are recombinant (nonparental). The chromosomes of these gametes carry new combinations of alleles that result from the exchange of chromosome segments in crossing over.

Morgan hypothesized that crossing over accounted for the observed nonparental offspring, those with the recombinant phenotypes (gray-short and black-long). Furthermore, the 17% of offspring with recombinant phenotypes must have resulted from fertilization involving recombinant gametes (**Figure 9.27**). The percentage of recombinant offspring among the total is called the **recombination frequency** (see bottom of Figure 9.25).

▼ Figure 9.27 **Explaining the unexpected results from the dihybrid testcross in Figure 9.25.** The disproportionately large numbers of the parental phenotypes among the offspring is explained by gene linkage, the fact that the gene loci for the two characters are located nearby on the same chromosome and tend to remain together during meiosis and fertilization (*G* with *L* and *g* with *l*). As a result of crossing over, some of the gametes end up with recombinant chromosomes, carrying new combinations of alleles, either *Gl* or *gL*. When the recombinant gametes participate in fertilization, recombinant offspring can result.

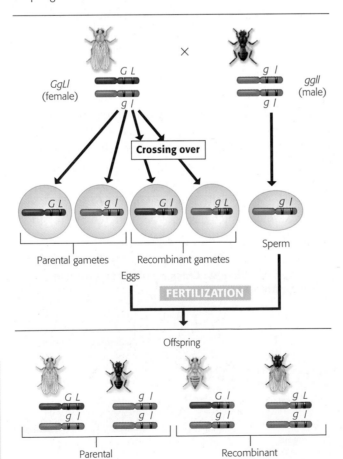

Linkage Maps

While working with *Drosophila*, Alfred H. Sturtevant, one of Morgan's students, developed a way to use crossover data to map gene loci. This technique is based on the assumption that the chance of crossing over is approximately equal at all points along a chromosome. He hypothesized that the farther apart two genes are on a chromosome, the higher the probability that a crossover will occur between them. His reasoning was elegantly simple: The greater the distance between two genes, the more points there are between them where crossing over can occur. (This assumption is not entirely accurate, but it is good enough to provide useful data.)

By applying Sturtevant's reasoning, geneticists can use recombination data to assign genes to relative positions on chromosomes—that is, to map genes. **Figure 9.28** shows some of the data used to map three linked genes (*g*, *c*, and *l*) that reside on one of the *Drosophila* chromosomes. Under the chromosome are the actual recombination frequencies (crossover frequencies) between these genes, taken two at a time: 17% between *g* and *l*, 9% between *g* and *c*, and 9.5% between *l* and *c*. Geneticists reasoned that these values represent the relative distances between the genes. Because the recombination frequencies between *g* and *c* and between *l* and *c* are approximately half the frequency between *g* and *l*, gene *c* must lie roughly midway between *g* and *l*. Thus, the sequence of these genes on their chromosome must be *g-c-l* (or the equivalent *l-c-g*). Such a diagram of relative gene locations is called a **linkage map**.

The linkage-mapping method has proved extremely valuable in establishing the relative positions of many genes in many organisms. The real beauty of the technique is that a wealth of information about genes can be learned simply by breeding and observing the organisms; no fancy equipment is required. ☑

▼ Figure 9.28 **Using crossover data to map genes.** The three recessive alleles specify black body (*g*), cinnabar eyes (*c*), and short wings (*l*). The corresponding dominant alleles specify the wild-type traits of gray body, red eyes, and long wings.

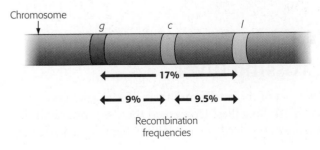

☑CHECKPOINT

If the order of three genes on a chromosome is *A-B-C*, which two genes will have the highest recombination frequency?

Answer: *A and C*

164

Sex Chromosomes and Sex-Linked Genes

The patterns of inheritance we've discussed so far have involved genes located only on autosomes, not on the sex chromosomes. We're now ready to look at the role of sex chromosomes in humans and the inheritance patterns exhibited by the characters they control.

Sex Determination in Humans

A number of organisms, including nearly all mammals, have a pair of sex chromosomes—designated X and Y—that determine an individual's sex (**Figure 9.29**). Individuals with one X chromosome and one Y chromosome are males; XX individuals are females. Human males and females both have 44 autosomes (chromosomes other than sex chromosomes). As a result of chromosome segregation during meiosis, each gamete contains one sex chromosome and a haploid set of autosomes (22 in humans). All eggs contain a single X chromosome. Of the sperm cells, half contain an X chromosome and half contain a Y chromosome. An offspring's sex depends on whether the sperm cell that fertilizes the egg bears an X or a Y. ☑

Sex-Linked Genes

Besides bearing genes that determine sex, the so-called sex chromosomes also contain genes for characters unrelated to maleness or femaleness. Any gene located on a sex chromosome is called a **sex-linked gene**. The X chromosome contains many more genes than the Y; therefore, most sex-linked genes are found on the X chromosome.

A number of human conditions, including red-green colorblindness, hemophilia, and a type of muscular dystrophy, result from sex-linked recessive alleles. **Red-green colorblindness** is a common sex-linked disorder characterized by a malfunction of light-sensitive cells in the eyes. (It is actually a class of disorders involving several sex-linked genes.) A person with normal color vision can see more than 150 colors. In contrast, someone with red-green colorblindness can see fewer than 25. For some affected people, red hues appear gray; others see gray instead of green; still others are green-weak or red-weak, tending to confuse shades of these colors. Figure 9.30 shows a simple test for red-green colorblindness.

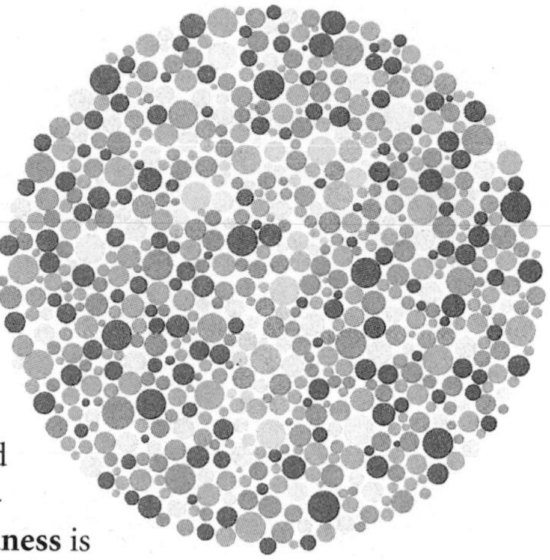

▲ **Figure 9.30 A test for red-green colorblindness.** Can you see a green numeral 7 against the reddish background? If not, you probably have some form of red-green colorblindness, a sex-linked trait.

▼ Figure 9.29 **The chromosomal basis of sex determination in humans.** The micrograph at right shows human X and Y chromosomes in a duplicated state.

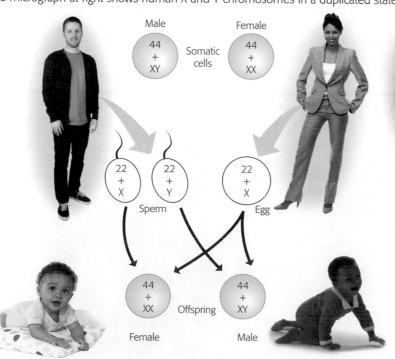

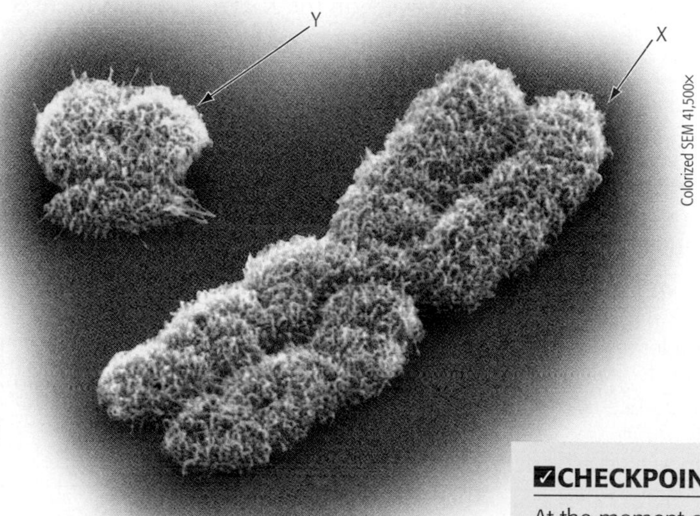

Colorized SEM 41,500×

▼ **Figure 9.31 Inheritance of colorblindness, a sex-linked recessive trait.** We use the uppercase letter *N* for the dominant, normal color vision allele and *n* for the recessive, colorblind allele. To indicate that these alleles are on the X chromosome, we show them as superscripts to the letter X. The Y chromosome does not have a gene locus for vision; therefore, the male's phenotype results entirely from the sex-linked gene on his single X chromosome.

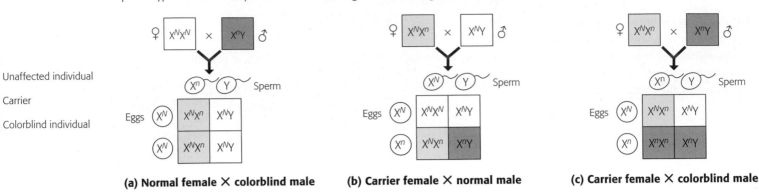

Key

☐ Unaffected individual

▨ Carrier

▨ Colorblind individual

(a) Normal female × colorblind male

(b) Carrier female × normal male

(c) Carrier female × colorblind male

Because they are located on the sex chromosomes, sex-linked genes exhibit unusual inheritance patterns. **Figure 9.31a** illustrates what happens when a colorblind male has offspring with a homozygous female with normal color vision. All the children have normal color vision, suggesting that the allele for wild type (normal color vision) is dominant. If a female carrier mates with a male who has normal color vision, the classic 3:1 phenotypic ratio of normal color vision to colorblindness appears among the children (**Figure 9.31b**). However, there is a surprising twist: The colorblind trait shows up only in males. All the females have normal vision, while half the males are colorblind and half are normal. This is because the gene involved in this inheritance pattern is located exclusively on the X chromosome; there is no corresponding locus on the Y. Thus, females (XX) carry two copies of the gene for this character, while males (XY) carry only one. Because the colorblindness allele is recessive, a female will be colorblind only if she receives that allele on both X chromosomes (**Figure 9.31c**). For a male, however, a single copy of the recessive allele confers colorblindness. For this reason, recessive sex-linked traits are expressed much more frequently in men than in women. For example, colorblindness is about 20-fold more common among males than among females.

Hemophilia is a sex-linked recessive trait with a long, well-documented history. Hemophiliacs bleed excessively when injured because they have inherited an abnormal allele for a factor involved in blood clotting. The most seriously affected individuals may bleed to death after relatively minor bruises or cuts. A high incidence of hemophilia has plagued the royal families of Europe. The first royal hemophiliac seems to have been a son of Queen Victoria (1819–1901) of England. It is likely that the hemophilia allele arose through a mutation in one of the gametes of Victoria's mother or father, making Victoria a carrier of the deadly allele. Victoria later passed the allele on to her daughter and granddaughter, both carriers. Hemophilia was eventually introduced into the royal families of Prussia, Russia, and Spain through the marriages of two of Victoria's daughters who were carriers. In this way, the age-old practice of strengthening international alliances by marriage effectively spread hemophilia through the royal families of several nations (**Figure 9.32**). ✔

☑ **CHECKPOINT**

1. What is meant by a sex-linked gene?
2. White eye color is a recessive sex-linked trait in fruit flies. If a white-eyed *Drosophila* female is mated with a red-eyed (wild-type) male, what do you predict for the numerous offspring?

Answers: 1. a gene that is located on a sex chromosome, usually the X chromosome 2. All female offspring will be heterozygous (X^{R}X^r), with red eyes; all male offspring will be white-eyed (X^{r}Y).

▼ **Figure 9.32 Hemophilia in the royal family of Russia.** The photograph shows Queen Victoria's granddaughter Alexandra, her husband Nicholas, who was the last czar of Russia, their son Alexis, and their daughters. In the pedigree, half-colored symbols represent heterozygous carriers of the hemophilia allele and fully colored symbols represent a person with hemophilia.

Queen Victoria — Albert

Alice — Louis

Alexandra — Czar Nicholas II of Russia

Alexis

Barking Up the Evolutionary Tree

As we've seen throughout this chapter, dogs are more than man's best friend—they are also one of our longest-running genetic experiments. About 15,000 years ago, in East Asia, humans began to cohabit with ancestral canines that were predecessors of both modern wolves and dogs. As people settled into permanent, geographically isolated settlements, populations of canines were separated from one another and eventually became inbred.

Different groups of people chose dogs with different traits, depending on their needs. Herders selected dogs that were good at controlling flocks of animals. Hunters chose dogs that were good at retrieving prey. Continued over millennia, such genetic tinkering has resulted in a diverse array of dog body types and behaviors, from tiny, feisty Chihuahuas to huge, docile St. Bernards. In each breed, a distinct genetic makeup results in a distinct set of physical and behavioral traits.

As discussed in the Process of Science section, our understanding of canine evolution took a big leap forward when researchers sequenced the complete genome of a dog. Using the genome sequence and a wealth of other data, canine geneticists produced an evolutionary tree of dog breeds (**Figure 9.33**). Published in 2006, the tree is based on genetic analysis of 414 dogs from 85 recognized breeds. The analysis shows that the canine family tree includes a series of well-defined branch points. Each fork represents artificial selection that produced a genetically distinct subpopulation with specific desired traits.

The genetic tree shows that the most ancient breeds, those most closely related to the wolf, are Asian species such as the shar-pei and Akita. Subsequent genetic splits created distinct breeds in Africa (basenji), the Arctic (Alaskan malamute and Siberian husky), and the Middle East (Afghan hound and saluki). The remaining breeds, primarily of European ancestry, can be grouped by genetic makeup into those bred for guarding (for example, the rottweiler), herding (such as various sheepdogs), and hunting (including the golden retriever and beagle).

The formulation of an evolutionary tree for the domestic dog shows that new technologies can provide answers to important genetic and evolutionary questions without ever performing controlled matings (as Mendel had to do with his peas). It also shows that the study of genetics can provide important insights into the evolutionary history of life on Earth.

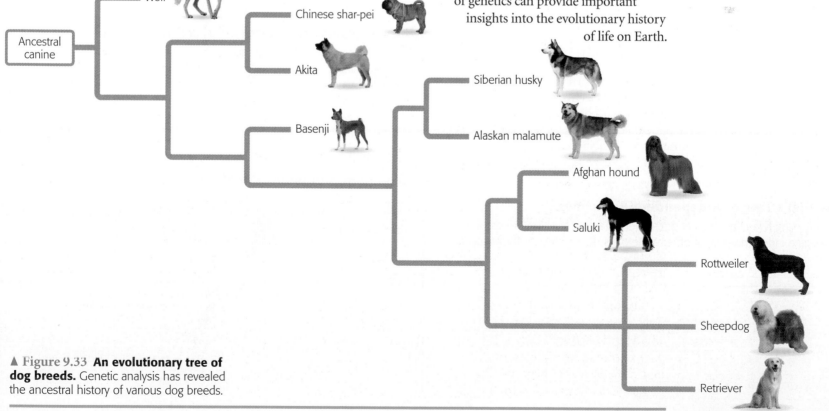

▲ Figure 9.33 **An evolutionary tree of dog breeds.** Genetic analysis has revealed the ancestral history of various dog breeds.

Chapter Review

SUMMARY OF KEY CONCEPTS

Heritable Variation and Patterns of Inheritance

Gregor Mendel was the first to study genetics, the science of heredity, by analyzing patterns of inheritance. He emphasized that heritable factors (genes) retain permanent identities.

In an Abbey Garden

Mendel started with true-breeding varieties of pea plants representing two alternative variants of a hereditary character, such as flower color. He then crossed the different varieties and traced the inheritance of traits from generation to generation.

Mendel's Law of Segregation

Pairs of alleles separate during gamete formation; fertilization restores the pairs.

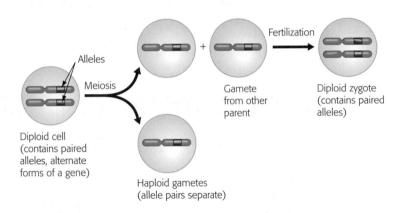

If an individual's genotype (genetic makeup) has two different alleles for a gene and only one influences the organism's phenotype (appearance), that allele is said to be dominant and the other allele recessive. Alleles of a gene reside at the same locus, or position, on homologous chromosomes. When the allele pair match, the organism is homozygous; when they're different, the organism is heterozygous.

Mendel's Law of Independent Assortment

By following two characters at once, Mendel found that the alleles of a pair segregate independently of other allele pairs during gamete formation.

Using a Testcross to Determine an Unknown Genotype

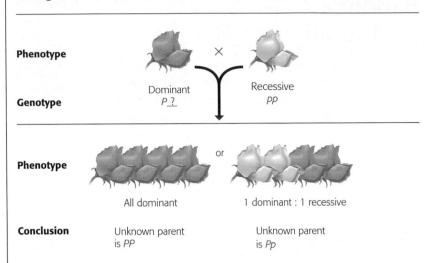

The Rules of Probability

Inheritance follows the rules of probability. The chance of inheriting a recessive allele from a heterozygous parent is $\frac{1}{2}$. The chance of inheriting it from both of two heterozygous parents is $\frac{1}{2} \times \frac{1}{2} = \frac{1}{4}$, illustrating the rule of multiplication for calculating the probability of two independent events.

Family Pedigrees

The inheritance of many human traits, from freckles to genetic diseases, follows Mendel's laws and the rules of probability. Geneticists can use family pedigrees to determine patterns of inheritance and individual genotypes among humans.

Human Disorders Controlled by a Single Gene

For traits that vary within a population, the one most commonly found in nature is called the wild type. Many inherited disorders in humans are controlled by a single gene (represented by two alleles). Most of these disorders, such as cystic fibrosis, are caused by autosomal recessive alleles. A few, such as Huntington's disease, are caused by dominant alleles.

Variations on Mendel's Laws

Incomplete Dominance in Plants and People

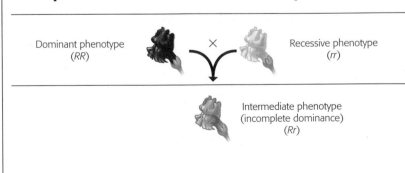

ABO Blood Groups: An Example of Multiple Alleles and Codominance

Within a population, there are often multiple kinds of alleles for a character, such as the three alleles for the ABO blood groups. The alleles determining the A and B blood factors are codominant; that is, both are expressed in a heterozygote.

Pleiotropy and Sickle-Cell Disease

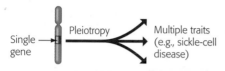

The presence of two copies of the sickle-cell allele at a single gene locus brings about the many symptoms of sickle-cell disease.

Polygenic Inheritance

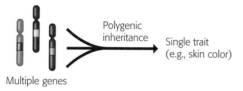

The Role of Environment

Many phenotypic characters result from a combination of genetic and environmental effects, but only genetic influences are biologically heritable.

The Chromosomal Basis of Inheritance

Genes are located on chromosomes. The behavior of chromosomes during meiosis and fertilization accounts for inheritance patterns.

Linked Genes

Certain genes are linked: They tend to be inherited together because they lie close together on the same chromosome.

Genetic Recombination: Crossing Over

Crossing over can separate linked alleles, producing gametes with recombinant chromosomes and offspring with recombinant phenotypes.

Linkage Maps

The fact that crossing over between linked genes is more likely to occur between genes that are farther apart enables geneticists to map the relative positions of genes on chromosomes.

Sex Chromosomes and Sex-Linked Genes

Sex Determination in Humans

In humans, sex is determined by whether a Y chromosome is present. A person who inherits two X chromosomes develops as a female. A person who inherits one X and one Y chromosome develops as a male.

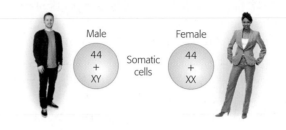

Sex-Linked Genes

Genes on the sex chromosomes (most often the X) are said to be sex-linked. Their inheritance pattern reflects the fact that females have two homologous X chromosomes, but males have only one. Most sex-linked human disorders, such as red-green colorblindness and hemophilia, are due to recessive alleles and are seen mostly in males. A male receiving a single sex-linked recessive allele from his mother will have the disorder; a female has to receive the allele from both parents to be affected.

Sex-Linked Traits				
Female: Two alleles	Genotype	$X^N X^N$	$X^N X^n$	$X^n X^n$
	Phenotype	Normal female	Carrier female	Affected female (rare)
Male: One allele	Genotype	$X^N Y$		$X^n Y$
	Phenotype	Normal male		Affected male

SELF-QUIZ

1. The genetic makeup of an organism is called its _____, while the physical traits of an organism are called its _____.

2. Which of Mendel's laws is represented by each statement?
   a. Alleles of each homologous pair separate independently during gamete formation.
   b. Alleles segregate during gamete formation; fertilization creates pairs of alleles once again.

3. Edward was found to be heterozygous (Ss) for the sickle-cell trait. The alleles represented by the letters S and s are
   a. on the X and Y chromosomes.
   b. linked.
   c. on homologous chromosomes.
   d. both present in each of Edward's sperm cells.

4. Whether an allele is dominant or recessive depends on
   a. how common the allele is, relative to other alleles.
   b. whether it is inherited from the mother or the father.
   c. whether it or another allele determines the phenotype when both are present.
   d. whether or not it is linked to other genes.

5. Two fruit flies with eyes of the usual red color are crossed, and their offspring are as follows: 77 red-eyed males, 71 ruby-eyed males, 152 red-eyed females. The gene that controls whether eyes are red or ruby is _____, and the allele for ruby eyes is _____.
   a. autosomal (carried on an autosome); dominant
   b. autosomal; recessive
   c. sex-linked; dominant
   d. sex-linked; recessive

6. All the offspring of a white hen and a black rooster are gray. The simplest explanation for this pattern of inheritance is
   a. pleiotropy.
   b. sex linkage.
   c. codominance.
   d. incomplete dominance.

7. A man who has type B blood and a woman who has type A blood could have children of which of the following phenotypes? (*Hint*: Review Figure 9.20.)
   a. A, B, or O
   b. AB only
   c. AB or O
   d. A, B, AB, or O

8. Duchenne muscular dystrophy is a sex-linked recessive disorder characterized by a progressive loss of muscle tissue. Neither Rudy nor Carla has Duchenne muscular dystrophy, but their first son does have it. If the couple has a second child, what is the probability that he or she will also have the disease?

9. In fruit flies, the genes for wing shape and body stripes are linked. In a fly whose genotype is $WwSs$, W is linked to S, and w is linked to s. Show how this fly can produce gametes containing four different combinations of alleles. Which are parental-type gametes? Which are recombinant gametes? What process produces recombinant gametes?

10. Adult height in humans is at least partially hereditary; tall parents tend to have tall children. But humans come in a range of sizes, not just tall or short. What extension of Mendel's model could produce this variation in height?

11. A true-breeding brown mouse is repeatedly mated with a true-breeding white mouse, and all their offspring are brown. If two of these brown offspring are mated, what fraction of the F_2 mice will be brown?

12. How could you determine the genotype of one of the brown F_2 mice in problem 11? How would you know whether a brown mouse is homozygous? Heterozygous?

13. Tim and Jan both have freckles (a dominant trait), but their son Michael does not. Show with a Punnett square how this is possible. If Tim and Jan have two more children, what is the probability that *both* of them will have freckles?

14. Incomplete dominance is seen in the inheritance of hyper-cholesterolemia. Mack and Toni are both heterozygous for this character, and both have elevated levels of cholesterol. Their daughter Katerina has a cholesterol level six times normal; she is apparently homozygous, hh. What fraction of Mack and Toni's children are likely to have elevated but not extreme levels of cholesterol, like their parents? If Mack and Toni have one more child, what is the probability that the child will suffer from the more serious form of hypercholesterolemia seen in Katerina?

15. A female fruit fly with forked bristles on her body is mated with a male fly with normal bristles. Their offspring are 121 females with normal bristles and 138 males with forked bristles. Explain the inheritance pattern for this trait.

16. Both parents of a boy are phenotypically normal, but their son suffers from hemophilia, a sex-linked recessive disorder. Draw a pedigree that shows the genotypes of the three individuals. What fraction of the couple's children are likely to suffer from hemophilia? What fraction are likely to be carriers?

17. Heather was surprised to discover that she suffered from red-green colorblindness. She told her biology professor, who said, "Your father is colorblind too, right?" How did her professor know this? Why did her professor not say the same thing to the colorblind males in the class?

Answers to Self-Quiz questions can be found in Appendix D.

THE PROCESS OF SCIENCE

18. In 1981, a stray cat with unusual curled-back ears was adopted by a family in Lakewood, California. Hundreds of descendants of this cat have since been born, and cat fanciers hope to develop the "curl" cat into a show breed. The curl allele is apparently dominant and carried on an autosome. Suppose you owned the first curl cat and wanted to develop a true-breeding variety. Describe tests that would determine whether the curl gene is dominant or recessive and whether it is autosomal or sex-linked.

19. Imagine that you have a large collection of fruit flies divided into ten different strains. Each strain is true-breeding (homozygous) and differs from wild-type flies in just one character. The only special equipment available is a magnifying glass that lets you determine the sex and readily observable traits of a fly, a large number of bottles to perform controlled matings, and an anesthetic liquid that enables you to examine and sort live flies. Using only this equipment, how much could you learn about the genetic makeup of the flies? Describe a series of experiments that would give you knowledge about fruit fly genetics.

BIOLOGY AND SOCIETY

20. Gregor Mendel never saw a gene, yet he concluded that "heritable factors" were responsible for the patterns of inheritance he observed in peas. Similarly, maps of *Drosophila* chromosomes (and the very idea that genes are carried on chromosomes) were conceived by observing the patterns of inheritance of linked genes, not by observing the genes directly. Is it legitimate for biologists to claim the existence of objects and processes they cannot actually see? How do scientists know whether an explanation is correct?

21. Many infertile couples turn to in vitro fertilization to try to have a baby. In this technique, sperm and ova are collected and used to create eight-cell embryos for implantation into a woman's uterus. At the eight-cell stage, one of the fetal cells can be removed without causing harm to the developing fetus. Once removed, the cell can be genetically tested. Some couples may know that a particular genetic disease runs in their family. They might wish to avoid implanting any embryos with the disease-causing genes. Do you think this is an acceptable use of genetic testing? What if a couple wanted to use genetic testing to select embryos for traits unrelated to disease, such as freckles? Do you think that couples undergoing in vitro fertilization should be allowed to perform whatever genetic tests they wish? Or do you think that there should be limits on what tests can be performed? How do you draw the line between genetic tests that are acceptable and those that are not?

10 The Structure and Function of DNA

The influenza virus. Although it can be deadly, influenza is caused by a relatively simple virus consisting of little more than protein and nucleic acid.

CHAPTER CONTENTS
■■■ Chapter Thread: **THE FLU**

BIOLOGY AND SOCIETY: The Flu

Tracking a Killer

On February 18, 2009, a group of elite investigators met in a conference room outside Washington, DC. Their mission was to track a serial killer and, by anticipating its next moves, prevent future deaths. The decisions they would make would affect hundreds of millions of people and save, or lose, tens of thousands of lives.

On that day, the Vaccines and Related Biological Products Advisory Committee of the U.S. Food and Drug Administration selected the influenza virus strains to be used in the upcoming 2009–2010 U.S. flu vaccine program. The committee evaluated the patterns of recent infections and predicted the three most dangerous strains of flu for the coming year. Based on their recommendations, pharmaceutical manufacturers produced hundreds of millions of doses of flu vaccine to be distributed during the fall and winter of 2009–2010. The precise formulation of the vaccine was nothing less than a matter of life and death.

While you may think of the flu as merely a seasonal inconvenience, the influenza virus is actually one of the deadliest pathogens known to science. Each year in the United States, over 20,000 people die from influenza infection, mostly the elderly or people with chronic diseases. And that is in a good year. Once every few decades, a new strain of flu explodes on the scene that causes global epidemics and widespread death. The worst of these was the flu of 1918–1919. In just 18 months, about 40 million died worldwide—that's more people than have died of AIDS since it was discovered over 25 years ago. Health officials are always on the lookout for new flu strains that have the potential to cause similar devastation. They are concerned that we may be overdue—the last major flu outbreak was the Hong Kong flu of 1968–1969—and that the new avian flu (discussed in the Evolution Connection section) could be particularly deadly.

The flu virus, like all viruses, consists of a relatively simple structure of nucleic acid (RNA in this case) and protein. The search for an effective flu vaccine depends on a detailed understanding of life at the molecular level. In this chapter, you will explore the structure of DNA, how it replicates and mutates, and how it controls the cell by directing the synthesis of RNA and protein.

DNA: Structure and Replication

DNA was known to be a chemical in cells by the late 1800s, but Mendel and other early geneticists did all their work without any knowledge of DNA's role in heredity. By the late 1930s, experimental studies had convinced most biologists that a specific kind of molecule, rather than some complex chemical mixture, was the basis of inheritance. Attention focused on chromosomes, which were already known to carry genes. By the 1940s, scientists knew that chromosomes consisted of two types of chemicals: DNA and protein. And by the early 1950s, a series of discoveries had convinced the scientific world that DNA acts as the hereditary material. This breakthrough ushered in the field of **molecular biology**, the study of heredity at the molecular level.

What came next was one of the most celebrated quests in the history of science: the effort to figure out the structure of DNA. A good deal was already known about DNA. Scientists had identified all its atoms and knew how they were bonded to one another. What was not understood was the specific three-dimensional arrangement of atoms that gave DNA its unique properties—the capacity to store genetic information, copy it, and pass it from generation to generation. A race was on to discover how the structure of this molecule could account for its role in heredity. We will describe that momentous discovery shortly. First, let's review the underlying chemical structure of DNA and its chemical cousin RNA.

DNA and RNA Structure

Recall from Chapter 3 that both DNA and RNA are nucleic acids, which consist of long chains (polymers) of chemical units (monomers) called **nucleotides**. (For an in-depth refresher, see the information on nucleic acids in Chapter 3, particularly Figures 3.22–3.26.) A diagram of a nucleotide polymer, or **polynucleotide**, is shown in **Figure 10.1**. This sample polynucleotide chain shows only one of many possible arrangements of the four different types of nucleotides (abbreviated A, C, T, and G) that make up DNA. Polynucleotides tend to be very long and can have any sequence of nucleotides, so a large number of polynucleotide chains are possible.

Nucleotides are joined together by covalent bonds between the sugar of one nucleotide and the phosphate of the next. This results in a **sugar-phosphate backbone**, a repeating pattern of sugar-phosphate-sugar-phosphate.

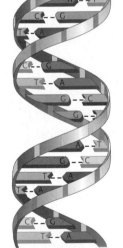

DNA double helix

▶ **Figure 10.1 The chemical structure of a DNA polynucleotide.** A molecule of DNA contains two polynucleotides, each a chain of nucleotides. Each nucleotide consists of a nitrogenous base, a sugar (blue), and a phosphate group (gold).

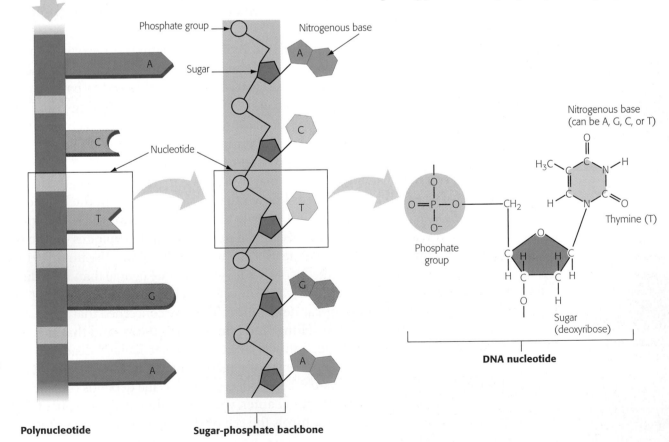

Phosphate group

Nitrogenous base

Sugar

Nucleotide

Polynucleotide

Sugar-phosphate backbone

Nitrogenous base (can be A, G, C, or T)

Thymine (T)

Phosphate group

Sugar (deoxyribose)

DNA nucleotide

The nitrogenous bases are arranged like ribs that project from this backbone. Moving from left to right across Figure 10.1, we can zoom in to see that each nucleotide consists of three components: a nitrogenous base, a sugar (blue), and a phosphate group (gold). Examining a single nucleotide even more closely, we see the chemical structure of its three components. The phosphate group, with a phosphorus atom (P) at its center, is the source of the *acid* in nucleic acid. (The phosphate has given up a hydrogen ion, H^+, leaving a negative charge on one of its oxygen atoms.) The sugar has five carbon atoms (shown in red): four in its ring and one extending above the ring. The ring also includes an oxygen atom. The sugar is called *deoxyribose* because, compared to the sugar ribose, it is missing an oxygen atom. The full name for **DNA** is *deoxyribonucleic acid*, with *nucleic* referring to DNA's location in the nuclei of eukaryotic cells. The nitrogenous base (thymine, in our example) has a ring of nitrogen and carbon atoms with various functional groups attached. In contrast to the acidic phosphate group, nitrogenous bases are basic; hence their name.

The four nucleotides found in DNA differ only in their nitrogenous bases (see Figure 3.24 for a review). At this point, the structural details are not as important as the fact that the bases are of two types. **Thymine (T)** and **cytosine (C)** are single-ring structures. **Adenine (A)** and **guanine (G)** are larger, double-ring structures. Instead of thymine, RNA has a similar base called **uracil (U)**. And RNA contains a slightly different sugar than DNA (ribose instead of deoxyribose). Other than that, RNA and DNA polynucleotides have the same chemical structure. Figure 10.2 is a computer graphic of a piece of RNA polynucleotide about 20 nucleotides long. ☑

▶ Figure 10.2 **An RNA polynucleotide.** The yellow used for the phosphorus atoms and the blue of the sugar atoms make it easy to spot the sugar-phosphate backbone.

Phosphate

Sugar (ribose)

☑**CHECKPOINT**

Compare and contrast the chemical components of DNA and RNA.

Answer: Both are polymers of nucleotides (a sugar + a nitrogenous base + a phosphate group). In RNA, the sugar is ribose; in DNA, it is deoxyribose. Both RNA and DNA have the bases A, G, and C, but DNA has T and RNA has U.

Watson and Crick's Discovery of the Double Helix

The celebrated partnership that resulted in the determination of the three-dimensional structure of DNA began soon after a 23-year-old American scientist named James D. Watson journeyed to Cambridge University, where Englishman Francis Crick was studying protein structure with a technique called X-ray crystallography. While visiting the laboratory of Maurice Wilkins at King's College in London, Watson saw an X-ray image of DNA, produced by Wilkins's colleague Rosalind Franklin. To Watson's trained eye, the photograph clearly revealed the basic shape of DNA to be a helix (spiral). On the basis of Watson's later recollection of the photo, he and Crick deduced that the diameter of the helix was uniform. The thickness of the helix suggested that it was made up of two polynucleotide strands—in other words, a **double helix**.

Using wire models, Watson and Crick began trying to construct a double helix that would conform both to Franklin's data and to what was then known about the chemistry of DNA (**Figure 10.3**). Watson placed the backbones on the outside of the model, forcing the nitrogenous bases to swivel to the interior of the molecule. It then occurred to him that the four kinds of bases might pair in a specific way. This idea of *specific base pairing* was a flash of inspiration that enabled Watson and Crick to solve the DNA puzzle.

▼ Figure 10.3 **Discoverers of the double helix.**

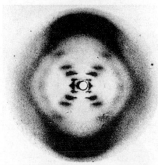

James Watson (left) and **Francis Crick.** The discoverers of the structure of DNA are shown in 1953 with their model of the double helix.

Rosalind Franklin (top). By generating X-ray images of DNA (bottom), Franklin provided Watson and Crick with some key data about the structure of DNA.

At first, Watson imagined that the bases paired like with like—for example, A with A, C with C. But that kind of pairing did not fit with the fact that the DNA molecule has a uniform diameter. An AA pair (made of two double-ringed bases) would be almost twice as wide as a CC pair (made of two single-ringed bases), causing bulges in the molecule. It soon became apparent that a double-ringed base on one strand must always be paired with a single-ringed base on the opposite strand.

Moreover, Watson and Crick realized that the individual structures of the bases dictated the pairings even more specifically. Each base has chemical side groups that can best form hydrogen bonds with one appropriate partner (to review hydrogen bonds, see Figure 2.8). Adenine can best form hydrogen bonds with thymine, and guanine with cytosine. In the biologist's shorthand, A pairs with T, and G pairs with C. A is also said to be "complementary" to T, and G to C.

You can picture the model of the DNA double helix proposed by Watson and Crick as a rope ladder having rigid, wooden rungs, with the ladder twisted into a spiral (Figure 10.4). Figure 10.5 shows three more detailed representations of the double helix. The ribbonlike diagram in **Figure 10.5a** symbolizes the bases with shapes that emphasize their complementarity. **Figure 10.5b** is a more chemically precise version showing only four base pairs, with the helix untwisted and the individual hydrogen bonds specified by dashed lines. **Figure 10.5c** is a computer model showing part of a double helix in atomic detail.

Although the base-pairing rules dictate the side-by-side combinations of nitrogenous bases that form the rungs of the double helix, they place no restrictions on the sequence of nucleotides along the length of a DNA strand. In fact, the sequence of bases can vary in countless ways.

▶ Figure 10.4 **A rope-ladder model of a double helix.** The ropes at the sides represent the sugar-phosphate backbones. Each wooden rung stands for a pair of bases connected by hydrogen bonds.

Twist

▼ Figure 10.5 **Three representations of DNA.**

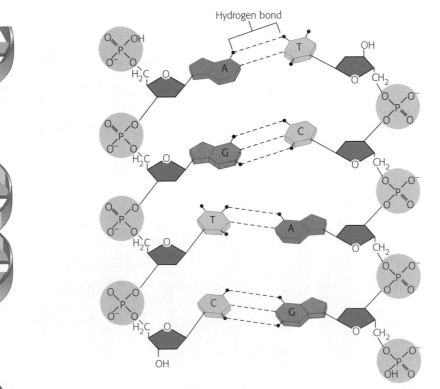

Hydrogen bond

(a) Ribbon model. The sugar-phosphate backbones are blue ribbons, and the bases are complementary shapes in shades of green and orange.

(b) Atomic model. In this more chemically detailed structure, you can see the individual hydrogen bonds (dashed lines). You can also see that the strands run in opposite directions: Notice that the sugars on the two strands are upside down with respect to each other.

(c) Computer model. Each atom is shown as a sphere, creating a space-filling model.

In April 1953, Watson and Crick rocked the scientific world with a succinct, two-page paper proposing their molecular model for DNA in the British scientific journal *Nature*. Few milestones in the history of biology have had as broad an impact as their double helix, with its AT and CG base pairing. In 1962, Watson, Crick, and Wilkins received the Nobel Prize for their work. (Franklin deserved the prize as well, but she had died from cancer in 1958.)

In their 1953 paper, Watson and Crick wrote that the structure they proposed "immediately suggests a possible copying mechanism for the genetic material." In other words, the structure of DNA also points toward a molecular explanation for life's unique properties of reproduction and inheritance, as we see next.

DNA Replication

When a cell or a whole organism reproduces, a complete set of genetic instructions must pass from one generation to the next. For this to occur, there must be a means of copying the instructions. Watson and Crick's model for DNA structure immediately suggested to them that each DNA strand serves as a mold, or template, to guide reproduction of the other strand. The logic behind the Watson-Crick proposal for how DNA is copied is quite simple. If you know the sequence of bases in one strand of the double helix, you can very easily determine the sequence of bases in the other strand by applying the base-pairing rules: A pairs with T (and T with A), and G pairs with C (and C with G). For example, if one polynucleotide has the sequence AGTC, then the complementary polynucleotide in that DNA molecule must have the sequence TCAG.

Figure 10.6 shows how this model can account for the direct copying of a piece of DNA. The two strands of parental DNA separate, and each becomes a template for the assembly of a complementary strand from a supply of free nucleotides. The nucleotides are lined up one at a time along the template strand in accordance with the base-pairing rules. Enzymes link the nucleotides to form the new DNA strands. The completed new molecules, identical to the parental molecule, are known as daughter DNA molecules (no gender should be inferred from this name).

Although the general mechanism of DNA replication is conceptually simple, the actual process is complex and requires the cooperation of more than a dozen enzymes and other proteins. The enzymes that make the covalent bonds between the nucleotides of a new DNA strand are called **DNA polymerases**. As an incoming nucleotide base-pairs with its complement on the template strand, a DNA polymerase adds it to the end of the growing daughter strand (polymer). The process is both fast and amazingly accurate; typically, DNA replication proceeds at a rate of 50 nucleotides per second, with fewer than one in a billion incorrectly paired. In addition to their roles in DNA replication, DNA polymerases and some of the associated proteins are also involved in repairing damaged DNA. DNA can be harmed by toxic chemicals in the environment or by high-energy radiation, such as X-rays and ultraviolet light.

DNA replication begins at specific sites on a double helix, called origins of replication. It then proceeds in both directions, creating what are called replication "bubbles" **(Figure 10.7)**. The parental DNA strands open up as daughter strands elongate on both sides of each bubble. The DNA molecule of a eukaryotic chromosome has many origins where replication can start simultaneously, shortening the total time needed for the process. Eventually, all the bubbles merge, yielding two completed double-stranded daughter DNA molecules.

DNA replication ensures that all the body cells in a multicellular organism carry the same genetic information. It is also the means by which genetic information is passed along to offspring. ☑

*Bio**Flix***™
DNA Replication

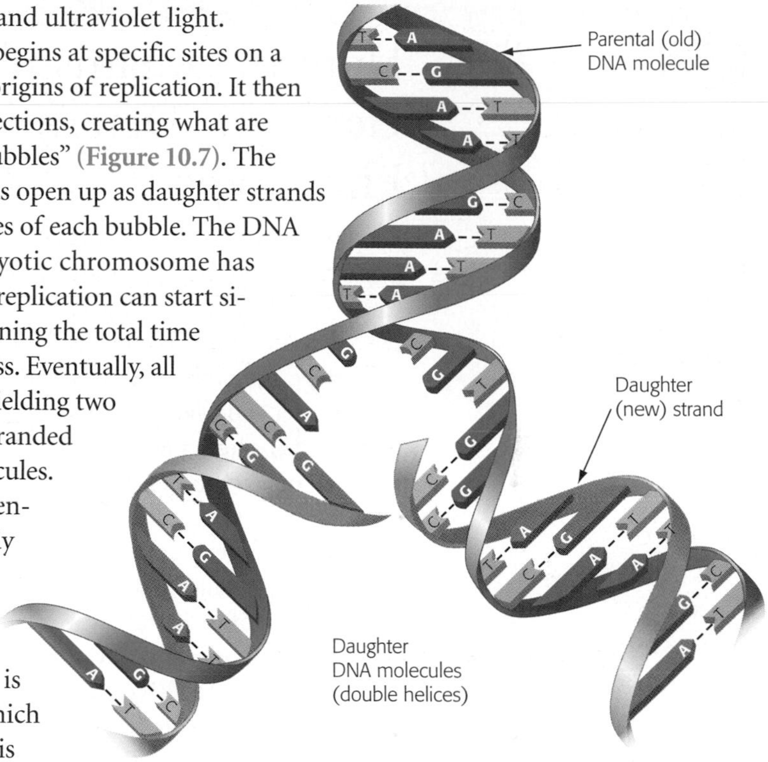

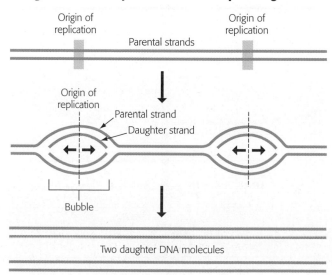

Parental (old) DNA molecule

Daughter (new) strand

Daughter DNA molecules (double helices)

▲ Figure 10.6 **DNA replication.** Replication results in two daughter DNA molecules, each consisting of one old strand and one new strand. The parental DNA untwists as its strands separate, and the daughter DNA rewinds as it forms.

▼ Figure 10.7 **Multiple "bubbles" in replicating DNA.**

Origin of replication

Origin of replication

Parental strands

Origin of replication

Parental strand

Daughter strand

Bubble

Two daughter DNA molecules

✓CHECKPOINT

1. How does complementary base pairing make DNA replication possible?
2. What enzymes connect nucleotides together during DNA replication?

*Answers: **1.** When the two strands of the double helix separate, each serves as a template on which nucleotides can be arranged by specific base pairing into new complementary strands. **2.** DNA polymerases*

BioFlix ™

Protein Synthesis

MP3
tutor sessions

DNA to RNA to Protein

The Flow of Genetic Information from DNA to RNA to Protein

Now that we've seen how the structure of DNA allows it to be copied, we can explore how DNA provides instructions to a cell and to an organism as a whole.

How an Organism's Genotype Determines Its Phenotype

Keeping in mind the structure of DNA, we can define genotype and phenotype more precisely than we did in Chapter 9. An organism's *genotype*, its genetic makeup, is the sequence of nucleotide bases in its DNA. The *phenotype*, the organism's physical traits, arises from the actions of a wide variety of proteins. For example, structural proteins help make up the body of an organism, and enzymes catalyze its metabolic activities.

What is the connection between the genotype and the protein molecules that more directly determine the phenotype? Recall from Chapter 4 that DNA specifies the synthesis of proteins. However, a gene does not build a protein directly, but rather dispatches instructions in the form of RNA, which in turn programs protein synthesis. This central concept in biology (termed the "central dogma" by Francis Crick) is summarized in **Figure 10.8**. The molecular "chain of command" is from DNA in the nucleus (purple area in the figure) to RNA

to protein synthesis in the cytoplasm (blue-green area). The two main stages are **transcription**, the transfer of genetic information from DNA into an RNA molecule, and **translation**, the transfer of the information from RNA into a protein.

The relationship between genes and proteins was first proposed in 1909, when English physician Archibald Garrod suggested that genes dictate phenotypes through enzymes, the proteins that catalyze chemical processes. Garrod hypothesized that inherited diseases reflect a person's inability to make a particular enzyme. He gave as one example the hereditary condition called alkaptonuria, in which the urine appears dark red because it contains a chemical called alkapton. Garrod reasoned that normal individuals have an enzyme that breaks down alkapton, whereas alkaptonuric individuals lack the enzyme. Garrod's hypothesis was ahead of its time, but research conducted decades later proved him right.

The major breakthrough in demonstrating the relationship between genes and enzymes came in the 1940s from the work of American geneticists George Beadle and Edward Tatum with the bread mold *Neurospora crassa* (**Figure 10.9**). Beadle and Tatum studied strains of the mold that were unable to grow on the usual growth medium. Each of these strains turned out to lack an enzyme in a metabolic pathway that synthesized an amino acid called arginine. Beadle and Tatum also showed that each mutant was defective in a single gene. Accordingly, they hypothesized that the function of an individual gene is to dictate the production of a specific enzyme.

▲ Figure 10.9 **A petri dish covered with the bread mold *Neurospora crassa.***

Beadle and Tatum's "one gene–one enzyme" hypothesis has since been modified. First it was extended beyond enzymes to include all types of proteins. For example, alpha-keratin, the structural protein of your hair, is the product of a gene. Then it was discovered that many proteins have two or more different polypeptide chains (see Figure 3.20); in such cases, each polypeptide is specified by its own gene. Thus, Beadle and Tatum's hypothesis is now stated as follows: the function of a gene is to dictate the production of a *polypeptide*. ✔

► Figure 10.8 **The flow of genetic information in a eukaryotic cell.** A sequence of nucleotides in the DNA is transcribed into a molecule of RNA in the cell's nucleus. The RNA travels to the cytoplasm, where it is translated into the specific amino acid sequence of a protein.

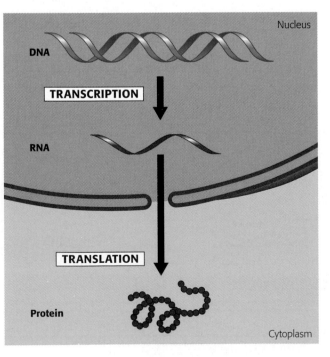

Nucleus

DNA

TRANSCRIPTION

RNA

TRANSLATION

Protein

Cytoplasm

☑**CHECKPOINT**

What are transcription and translation?

Answer: Transcription is the transfer of genetic information from DNA to RNA. Translation is the use of the information in an RNA molecule for the synthesis of a polypeptide.

From Nucleotides to Amino Acids: An Overview

Genetic information in DNA is transcribed into RNA and then translated into polypeptides. But how do these processes occur? Transcription and translation are linguistic terms, and it is useful to think of nucleic acids and polypeptides as having languages. To understand how genetic information passes from genotype to phenotype, we need to see how the chemical language of DNA is translated into the different chemical language of polypeptides.

What exactly is the language of nucleic acids? Both DNA and RNA are polymers made of monomers strung together in specific sequences that convey information, much as specific sequences of letters convey information in English. In DNA, the monomers are the four types of nucleotides, which differ in their nitrogenous bases (A, T, C, and G). The same is true for RNA, although it has the base U instead of T.

The language of DNA is written as a linear sequence of nucleotide bases, a sequence such as the blue sequence you see on the enlarged DNA strand in **Figure 10.10**. Specific sequences of bases, each with a beginning and an end, make up the genes on a DNA strand. A typical gene consists of thousands of nucleotides, and a single DNA molecule may contain thousands of genes.

When a segment of DNA is transcribed, the result is an RNA molecule. The process is called transcription because the nucleic acid language of DNA has simply been rewritten (transcribed) as a sequence of bases of RNA; the language is still that of nucleic acids. The nucleotide bases of the RNA molecule are complementary to those on the DNA strand. As you will soon see, this is because the RNA was synthesized using the DNA as a template.

Translation is the conversion of the nucleic acid language to the polypeptide language. Like nucleic acids, polypeptides are polymers, but the monomers that make them up—the letters of the polypeptide alphabet—are the 20 amino acids common to all organisms (represented as purple shapes in Figure 10.10). Again, the language is written in a linear sequence. In this case, the sequence of nucleotides of the RNA molecule dictates the sequence of amino acids of the polypeptide. (But remember, RNA is only a messenger; the genetic information that dictates the amino acid sequence originates in DNA.)

What are the rules for translating the RNA message into a polypeptide? In other words, what is the correspondence between the nucleotides of an RNA molecule and the amino acids of a polypeptide? Keep in mind that there are only four different kinds of nucleotides in DNA (A, G, C, T) and RNA (A, G, C, U). In translation, these four must somehow specify 20 amino acids. If each nucleotide base coded for one amino acid, only 4 of the 20 amino acids could be accounted for. In fact, triplets of bases are the smallest "words" of uniform length that can specify all the amino acids. There can be 64 (that is, 4^3) possible code words of this type—more than enough to specify the 20 amino acids. Indeed, there are enough triplets to allow more than one coding for each amino acid. For example, the base triplets AAA and AAG both code for the same amino acid.

Experiments have verified that the flow of information from gene to protein is based on a triplet code. The genetic instructions for the amino acid sequence of a polypeptide chain are written in DNA and RNA as a series of three-base words called **codons**. Three-base codons in the DNA are transcribed into complementary three-base codons in the RNA, and then the RNA codons are translated into amino acids that form a polypeptide. As summarized in Figure 10.10, one DNA codon (three nucleotides) → one RNA codon (three nucleotides) → one amino acid. Next we turn to the codons themselves. ☑

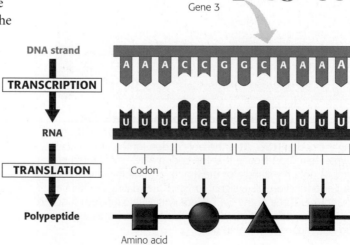

▼ Figure 10.10 **Transcription of DNA and translation of codons.** This figure focuses on a small region of one of the genes carried by a DNA molecule. The enlarged segment from one strand of gene 3 shows its specific sequence of bases. The red strand and purple chain represent the results of transcription and translation, respectively.

☑ CHECKPOINT

How many nucleotides are necessary to code for a polypeptide that is 100 amino acids long?

The Genetic Code

During the 1960s, scientists cracked the **genetic code**, the set of rules relating nucleotide sequence to amino acid sequence. As **Figure 10.11** shows, 61 of the 64 triplets code for amino acids. The triplet AUG has a dual function: It not only codes for the amino acid methionine (Met), but can also provide a signal for the start of a polypeptide chain. Three of the other codons do not designate amino acids. They are the stop codons that instruct the ribosomes to end the polypeptide.

Notice in Figure 10.11 that there is redundancy in the code but no ambiguity. For example, although codons UUU and UUC both specify phenylalanine (redundancy), neither of them ever represents any other amino acid (no ambiguity). The codons in the figure are the triplets found in RNA. They have a straightforward, complementary relationship to the codons in DNA. The nucleotides making up the codons occur in a linear order along the DNA and RNA, with no gaps separating the codons.

Almost all of the genetic code is shared by all organisms, from the simplest bacteria to the most complex plants and animals. The universality of the genetic vocabulary suggests that it arose very early in evolution and was passed on over the eons to all the organisms living on Earth today. As you will learn in Chapter 12, such universality is extremely important to modern DNA technologies. Because diverse organisms share a common genetic code, it is possible to program one species to produce a protein from another species by transplanting DNA (**Figure 10.12**). This allows scientists to mix and match genes from various species—a procedure with many useful applications. Besides having practical purposes, a shared genetic vocabulary also reminds us of the evolutionary kinship that connects all life on Earth. ☑

☑CHECKPOINT

An RNA molecule contains the nucleotide sequence CCAUUUACG. Using Figure 10.11, translate this sequence into the corresponding amino acid sequence.

Answer: Pro-Phe-Thr

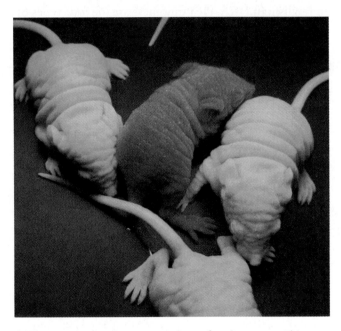

▲ Figure 10.12 **Mice expressing a foreign gene.** This photo shows the results of an experiment in which researchers incorporated a jelly (jellyfish) gene for a protein called green fluorescent protein (GFP) into the DNA of mice. The mouse in the middle lacks the gene for GFP; the other mice have it.

Second base of RNA codon

		U	C	A	G	
U		UUU ⎤ Phenylalanine (Phe) UUC ⎦ UUA ⎤ Leucine (Leu) UUG ⎦	UCU ⎤ UCC ⎥ Serine (Ser) UCA ⎥ UCG ⎦	UAU ⎤ Tyrosine (Tyr) UAC ⎦ UAA Stop UAG Stop	UGU ⎤ Cysteine (Cys) UGC ⎦ UGA Stop UGG Tryptophan (Trp)	U C A G
C		CUU ⎤ CUC ⎥ Leucine (Leu) CUA ⎥ CUG ⎦	CCU ⎤ CCC ⎥ Proline (Pro) CCA ⎥ CCG ⎦	CAU ⎤ Histidine (His) CAC ⎦ CAA ⎤ Glutamine (Gln) CAG ⎦	CGU ⎤ CGC ⎥ Arginine (Arg) CGA ⎥ CGG ⎦	U C A G
A		AUU ⎤ AUC ⎥ Isoleucine (Ile) AUA ⎦ AUG Met or start	ACU ⎤ ACC ⎥ Threonine (Thr) ACA ⎥ ACG ⎦	AAU ⎤ Asparagine (Asn) AAC ⎦ AAA ⎤ Lysine (Lys) AAG ⎦	AGU ⎤ Serine (Ser) AGC ⎦ AGA ⎤ Arginine (Arg) AGG ⎦	U C A G
G		GUU ⎤ GUC ⎥ Valine (Val) GUA ⎥ GUG ⎦	GCU ⎤ GCC ⎥ Alanine (Ala) GCA ⎥ GCG ⎦	GAU ⎤ Aspartic acid (Asp) GAC ⎦ GAA ⎤ Glutamic acid (Glu) GAG ⎦	GGU ⎤ GGC ⎥ Glycine (Gly) GGA ⎥ GGG ⎦	U C A G

First base of RNA codon (U, C, A, G — left side)
Third base of RNA codon (right side)

▲ Figure 10.11 **The dictionary of the genetic code, listed by RNA codons.** Practice using this dictionary by finding the codon UGG. (It is the only codon for the amino acid tryptophan, Trp.) Notice that the codon AUG (highlighted in green) not only stands for the amino acid methionine (Met), but also functions as a signal to "start" translating the RNA at that place. Three of the 64 codons (highlighted in red) function as "stop" signals that mark the end of a genetic message, but do not encode any amino acids.

Transcription: From DNA to RNA

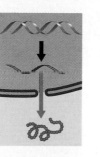

Let's look more closely at transcription, the transfer of genetic information from DNA to RNA. An RNA molecule is transcribed from a DNA template by a process that resembles the synthesis of a DNA strand during DNA replication. **Figure 10.13a** is a close-up view of this process. As with replication, the two DNA strands must first separate at the place where the process will start. In transcription, however, only one of the DNA strands serves as a template for the newly forming molecule. The nucleotides that make up the new RNA molecule take their places one at a time along the DNA template strand by forming hydrogen bonds with the nucleotide bases there. Notice that the RNA nucleotides follow the same base-pairing rules that govern DNA replication, except that U, rather than T, pairs with A. The RNA nucleotides are linked by the transcription enzyme **RNA polymerase**.

Figure 10.13b is an overview of the transcription of an entire gene. Special sequences of DNA nucleotides tell the RNA polymerase where to start and where to stop the transcribing process.

❶ Initiation of Transcription

The "start transcribing" signal is a nucleotide sequence called a **promoter**, which is located in the DNA at the beginning of the gene. A promoter is a specific place where RNA polymerase attaches. The first phase of transcription, called initiation, is the attachment of RNA polymerase to the promoter and the start of RNA

synthesis. For any gene, the promoter dictates which of the two DNA strands is to be transcribed (the particular strand varies from gene to gene).

❷ RNA Elongation

During the second phase of transcription, elongation, the RNA grows longer. As RNA synthesis continues, the RNA strand peels away from its DNA template, allowing the two separated DNA strands to come back together in the region already transcribed.

❸ Termination of Transcription

In the third phase, termination, the RNA polymerase reaches a special sequence of bases in the DNA template called a **terminator**. This sequence signals the end of the gene. At this point, the polymerase molecule detaches from the RNA molecule and the gene, and the DNA strands rejoin.

In addition to producing RNA that encodes amino acid sequences, transcription makes two other kinds of RNA that are involved in building polypeptides. We discuss these kinds of RNA a little later. ☑

☑CHECKPOINT

How does RNA polymerase "know" where to start transcribing a gene?

Answer: It recognizes the gene's promoter, a specific nucleotide sequence.

▼ Figure 10.13 **Transcription.**

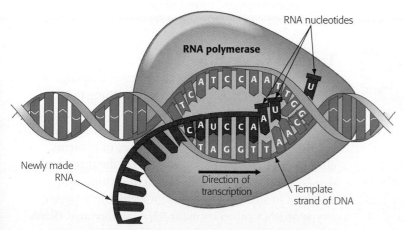

(a) A close-up view of transcription. As RNA nucleotides base-pair one by one with DNA bases on one DNA strand (called the template strand), the enzyme RNA polymerase (orange) links the RNA nucleotides into an RNA chain.

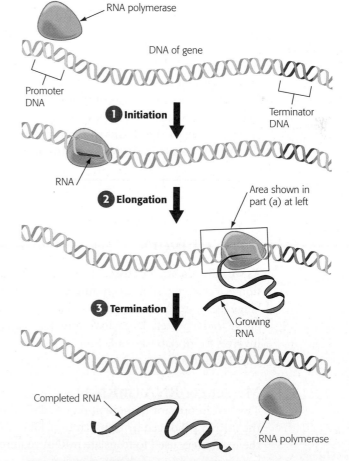

(b) Transcription of a gene. The transcription of an entire gene occurs in three phases: initiation, elongation, and termination of the RNA. The section of DNA where the RNA polymerase starts is called the promoter; the place where it stops is called the terminator.

The Processing of Eukaryotic RNA

In the cells of prokaryotes, which lack nuclei, the RNA transcribed from a gene immediately functions as **messenger RNA (mRNA)**, the messenger molecule that is translated. But this is not the case in eukaryotic cells. The eukaryotic cell not only localizes transcription in the nucleus but also modifies, or processes, the RNA transcripts there before they move to the cytoplasm for translation by the ribosomes.

One kind of RNA processing is the addition of extra nucleotides to the ends of the RNA transcript. These additions, called the **cap** and **tail**, protect the RNA from attack by cellular enzymes and help ribosomes recognize the RNA as mRNA.

Another type of RNA processing is made necessary in eukaryotes by noncoding stretches of nucleotides that interrupt the nucleotides that actually code for amino acids. It is as if unintelligible sequences of letters were randomly interspersed in an otherwise intelligible document. Most genes of plants and animals, it turns out, include such internal noncoding regions, which are called **introns**. The coding regions—the parts of a gene that are expressed—are called **exons**. As Figure 10.14 illustrates, both exons and introns are transcribed from DNA into RNA. However, before the RNA leaves the nucleus, the introns are removed, and the exons are joined to produce an mRNA molecule with a continuous coding sequence. This process is called **RNA splicing**. RNA splicing is believed to play a significant role in humans in allowing our approximately 25,000 genes to produce many

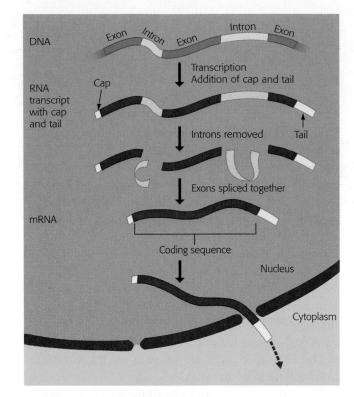

▲ Figure 10.14 **The production of messenger RNA (mRNA) in a eukaryotic cell.** Note that the molecule of mRNA that leaves the nucleus is substantially different from the molecule of RNA that was first transcribed from the gene. In the cytoplasm, the coding sequence of the final mRNA will be translated.

thousands more polypeptides. This is accomplished by varying the exons that are included in the final mRNA.

With capping, tailing, and splicing completed, the "final draft" of eukaryotic mRNA is ready for translation. ☑

Translation: The Players

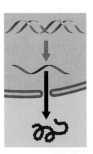

As we have already discussed, translation is a conversion between different languages—from the nucleic acid language to the protein language—and it involves more elaborate machinery than transcription.

Messenger RNA (mRNA)

The first important ingredient required for translation is the mRNA produced by transcription. Once it is present, the machinery used to translate mRNA requires enzymes and sources of chemical energy, such as ATP. In

addition, translation requires two heavy-duty components: ribosomes and a kind of RNA called transfer RNA.

Transfer RNA (tRNA)

Translation of any language into another language requires an interpreter, someone or something that can recognize the words of one language and convert them to the other. Translation of the genetic message carried in mRNA into the amino acid language of proteins also requires an interpreter. To convert the three-letter words (codons) of nucleic acids to the amino acid words of proteins, a cell uses a molecular interpreter, a type of RNA called **transfer RNA**, abbreviated **tRNA** (Figure 10.15).

▼ Figure 10.15 **The structure of tRNA.** At one end of the tRNA is the site where an amino acid will attach (purple) and at the other end is the three-nucleotide anticodon where the mRNA will attach (light green).

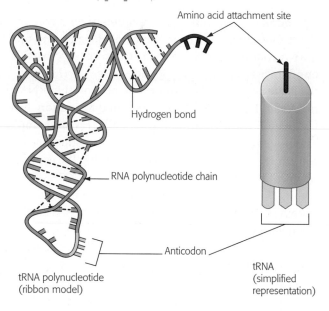

Amino acid attachment site

Hydrogen bond

RNA polynucleotide chain

Anticodon

tRNA polynucleotide
(ribbon model)

tRNA
(simplified
representation)

A cell that is producing proteins has in its cytoplasm a supply of amino acids. But amino acids themselves cannot recognize the codons arranged in sequence along messenger RNA. It is up to the cell's molecular interpreters, tRNA molecules, to match amino acids to the appropriate codons to form the new polypeptide. To perform this task, tRNA molecules must carry out two distinct functions: (1) pick up the appropriate amino acids and (2) recognize the appropriate codons in the mRNA. The unique structure of tRNA molecules enables them to perform both tasks.

As shown on the left in Figure 10.15, a tRNA molecule is made of a single strand of RNA—one polynucleotide chain—consisting of about 80 nucleotides. The chain twists and folds upon itself, forming several double-stranded regions in which short stretches of RNA base-pair with other stretches. At one end of the folded molecule is a special triplet of bases called an **anticodon**. The anticodon triplet is complementary to a codon triplet on mRNA. During translation, the anticodon on the tRNA recognizes a particular codon on the mRNA by using base-pairing rules. At the other end of the tRNA molecule is a site where one specific amino acid attaches. Although all tRNA molecules are similar, there are slightly different versions of tRNA for each amino acid.

Ribosomes

Ribosomes are the organelles that coordinate the functioning of the mRNA and tRNA and actually make polypeptides. As you can see in Figure 10.16a, a ribosome consists of two subunits. Each subunit is made up of proteins and a considerable amount of yet another kind of RNA, **ribosomal RNA (rRNA)**. A fully assembled ribosome has a binding site for mRNA on its small subunit and binding sites for tRNA on its large subunit. Figure 10.16b shows how two tRNA molecules get together with an mRNA molecule on a ribosome. One of the tRNA binding sites, the P site, holds the tRNA carrying the growing polypeptide chain, while another, the A site, holds a tRNA carrying the next amino acid to be added to the chain. The anticodon on each tRNA base-pairs with a codon on the mRNA. The subunits of the ribosome act like a vise, holding the tRNA and mRNA molecules close together. The ribosome can then connect the amino acid from the tRNA in the A site to the growing polypeptide. ☑

▼ Figure 10.16 **The ribosome.**

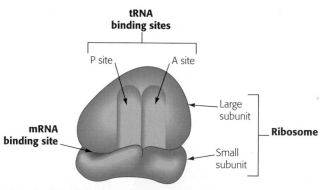

tRNA
binding sites

P site   A site

mRNA
binding site

Large
subunit

Ribosome

Small
subunit

(a) A simplified diagram of a ribosome. Notice the two subunits and sites where mRNA and tRNA molecules bind.

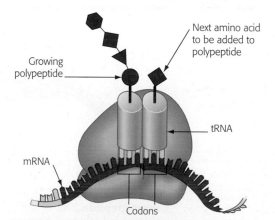

Growing
polypeptide

Next amino acid
to be added to
polypeptide

tRNA

mRNA

Codons

(b) The "players" of translation. When functioning in polypeptide synthesis, a ribosome holds one molecule of mRNA and two molecules of tRNA. The growing polypeptide is attached to one of the tRNAs.

☑**CHECKPOINT**

What is an anticodon?

Answer: An anticodon is the base triplet of a tRNA molecule that couples the tRNA to a complementary codon in the mRNA. The base pairing of anticodon to codon is a key step in translating mRNA to a polypeptide.

Translation: The Process

Translation is divided into the same three phases as transcription: initiation, elongation, and termination.

Initiation

This first phase brings together the mRNA, the first amino acid with its attached tRNA, and the two subunits of a ribosome. An mRNA molecule, even after splicing, is longer than the genetic message it carries (**Figure 10.17**). Nucleotide sequences at either end of the molecule (pink) are not part of the message, but along with the cap and tail in eukaryotes, they help the mRNA bind to the ribosome. The initiation process determines exactly where translation will begin so that the mRNA codons will be translated into the correct sequence of amino acids. Initiation occurs in two steps, as shown in **Figure 10.18**. ❶ An mRNA molecule binds to a small ribosomal subunit. A special initiator tRNA then binds to the **start codon**, where translation is to begin on the mRNA. The initiator tRNA carries the amino acid methionine (Met); its anticodon, UAC, binds to the start codon, AUG. ❷ A large ribosomal subunit binds to the small one, creating a functional ribosome. The initiator tRNA fits into the P site on the ribosome.

Elongation

Once initiation is complete, amino acids are added one by one to the first amino acid. Each addition occurs in the three-step elongation process shown in **Figure 10.19**. ❶ **Codon recognition**. The anticodon of an incoming tRNA molecule, carrying its amino acid, pairs with the mRNA codon in the A site of the ribosome. ❷ **Peptide bond formation**. The polypeptide leaves the tRNA in the P site and attaches to the amino acid on the tRNA in the A site. The ribosome catalyzes bond formation. Now the chain has one more amino acid. ❸ **Translocation**. The P site tRNA now leaves the ribosome, and the ribosome moves the remaining tRNA, carrying the growing polypeptide, to the P site. The mRNA and tRNA move as a unit. This movement brings into the A site the next mRNA codon to be translated, and the process can start again with step 1.

Termination

Elongation continues until a **stop codon** reaches the ribosome's A site. Stop codons—UAA, UAG, and UGA—do not code for amino acids but instead tell translation to stop. The completed polypeptide, typically several hundred amino acids long, is freed, and the ribosome splits into its subunits. ☑

▼ **Figure 10.17 A molecule of mRNA.**

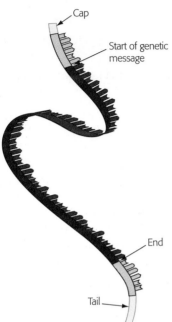

▼ **Figure 10.18 The initiation of translation.**

▶ **Figure 10.19 The elongation of a polypeptide.** The dashed red arrows indicate movement.

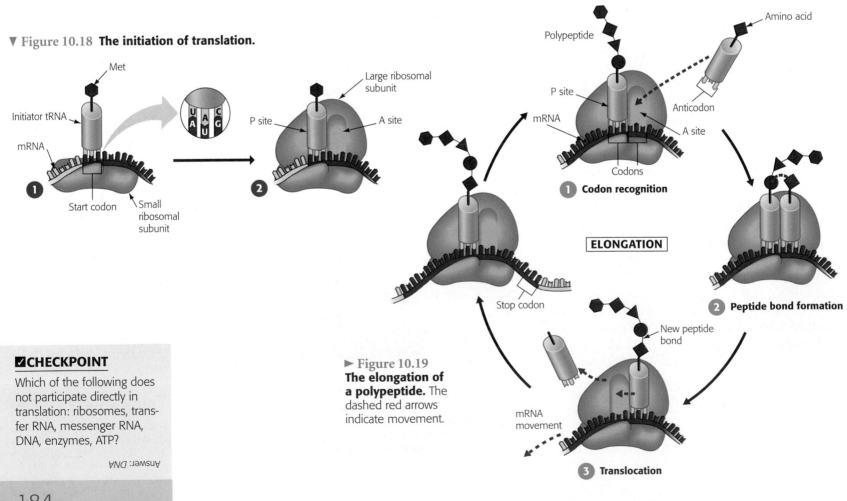

Review:
DNA → RNA → Protein

Figure 10.20 reviews the flow of genetic information in the cell, from DNA to RNA to protein. In eukaryotic cells, transcription—the stage from DNA to RNA—occurs in the nucleus, and the RNA is processed before it enters the cytoplasm. Translation is rapid; a single ribosome can make an average-sized polypeptide in less than a minute. As it is made, a polypeptide coils and folds, assuming a three-dimensional shape, its tertiary structure. Several polypeptides may come together, forming a protein with quaternary structure (see Figure 3.20).

What is the overall significance of transcription and translation? These are the processes whereby genes control the structures and activities of cells—or, more broadly, the way the genotype produces the phenotype. The chain of command originates with the information in a gene, a specific linear sequence of nucleotides in DNA. The gene dictates the transcription of a complementary sequence of nucleotides in mRNA. In turn, mRNA specifies the linear sequence of amino acids in a polypeptide. Finally, the proteins that form from the polypeptides determine the appearance and capabilities of the cell and organism.

For decades, the DNA → RNA → protein pathway was believed to be the sole means by which genetic information controls traits. In recent years, however, this notion has been challenged by discoveries that point to more complex roles for RNA. We will explore some of these special properties of RNA in Chapter 11. ☑

☑**CHECKPOINT**

Transcription is the synthesis of _____ , using _____ as a template. Translation is the synthesis of _____ , with one _____ determining each amino acid in the sequence. Translation is coordinated by _____ .

Answer: mRNA; DNA; protein (polypeptides); codon; ribosomes

▼ Figure 10.20 **A summary of transcription and translation.** This figure summarizes the main stages in the flow of genetic information from DNA to protein in a eukaryotic cell.

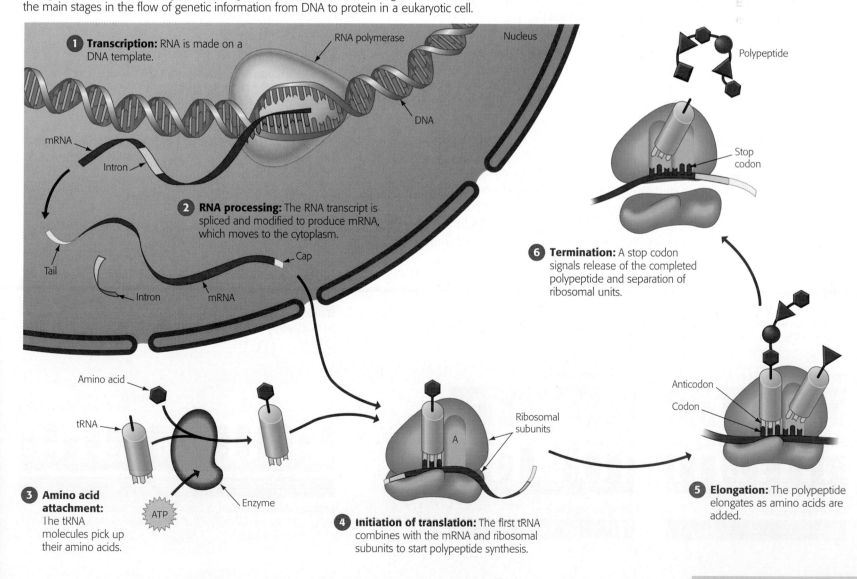

1 **Transcription:** RNA is made on a DNA template.

RNA polymerase

Nucleus

DNA

mRNA

Intron

Tail

2 **RNA processing:** The RNA transcript is spliced and modified to produce mRNA, which moves to the cytoplasm.

Cap

Intron

mRNA

Polypeptide

Stop codon

6 **Termination:** A stop codon signals release of the completed polypeptide and separation of ribosomal units.

Amino acid

tRNA

3 **Amino acid attachment:** The tRNA molecules pick up their amino acids.

ATP

Enzyme

4 **Initiation of translation:** The first tRNA combines with the mRNA and ribosomal subunits to start polypeptide synthesis.

A

Ribosomal subunits

Anticodon

Codon

5 **Elongation:** The polypeptide elongates as amino acids are added.

Mutations

Since discovering how genes are translated into proteins, scientists have been able to describe many heritable differences in molecular terms. For instance, sickle-cell disease can be traced to a change in a single amino acid in one of the polypeptides in the hemoglobin protein (see Figure 3.19). This difference is caused by a single nucleotide difference in the coding strand of DNA (**Figure 10.21**).

Any change in the nucleotide sequence of DNA is called a **mutation**. Mutations can involve large regions of a chromosome or just a single nucleotide pair, as in the sickle-cell allele. Occasionally, a base substitution leads to an improved protein or one with new capabilities that enhance the success of the mutant organism and its descendants. Much more often, though, mutations are harmful. Let's consider how mutations involving only one or a few nucleotide pairs can affect gene translation.

Types of Mutations

Mutations within a gene can occur as a result of base substitutions, deletions, or insertions (**Figure 10.22**). A base substitution is the replacement of one base, or nucleotide, by another. Depending on how a base substitution is translated, it can result in no change in the protein, in an insignificant change, or in a change that might be crucial to the life of the organism.

Because of the redundancy of the genetic code, some substitution mutations have no effect. For example, if a mutation causes an mRNA codon to change from GAA to GAG, no change in the protein product would result, because GAA and GAG both code for the same amino acid (Glu). Such a change is called a silent mutation.

Other changes of a single nucleotide do change the amino acid coding. Such mutations are called missense mutations. For example, if a mutation causes an mRNA codon to change from GGC to AGC, the resulting protein will have a serine (Ser) instead of a glycine (Gly) at this position (see Figure 10.22a). Some missense mutations have little or no effect on the shape or function of the resulting protein, but others, as we saw in the sickle-cell case, cause changes in the protein that prevent it from performing normally.

▼ Figure 10.22 **Three types of mutations and their effects.** Mutations are changes in DNA, but they are shown here in mRNA and the polypeptide product.

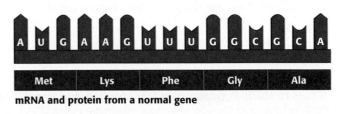

mRNA and protein from a normal gene

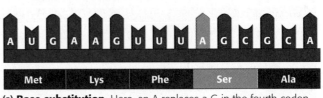

(a) Base substitution. Here, an A replaces a G in the fourth codon of the mRNA. The result in the polypeptide is a serine (Ser) instead of a glycine (Gly). This amino acid substitution may or may not affect the protein's function.

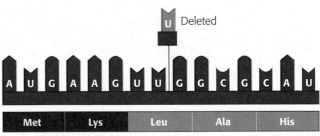

(b) Nucleotide deletion. When a nucleotide is deleted, all the codons from that point on are misread. The resulting polypeptide is likely to be completely nonfunctional.

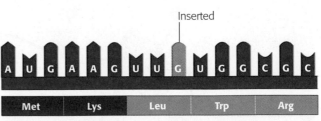

(c) Nucleotide insertion. As with a deletion, inserting one nucleotide disrupts all codons that follow, most likely producing a nonfunctional polypeptide.

▼ Figure 10.21 **The molecular basis of sickle-cell disease.** The sickle-cell allele differs from its normal counterpart, a gene for hemoglobin, by only one nucleotide (orange). This difference changes the mRNA codon from one that codes for the amino acid glutamic acid (Glu) to one that codes for valine (Val).

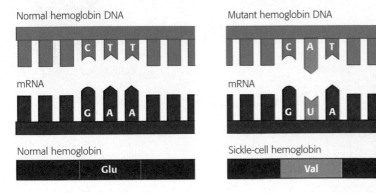

Some base substitutions, called nonsense mutations, change an amino acid codon into a stop codon. For example, if an AGA (Arg) codon is mutated to a UGA (stop) codon, the result will be a prematurely terminated protein, which probably will not function properly.

Mutations involving the deletion or insertion of one or more nucleotides in a gene often have disastrous effects (see Figure 10.22b and c). Because mRNA is read as a series of nucleotide triplets during translation, adding or subtracting nucleotides may alter the triplet grouping of the genetic message. All the nucleotides that are "downstream" of the insertion or deletion will be regrouped into different codons. For example, consider an mRNA molecule containing the sequence AAG-UUU-GGC-GCA; this codes for Lys-Phe-Gly-Ala. If a U is missing in the second codon (as a result of a deletion in the DNA), the resulting sequence will be AAG-UUG-GCG-CA, which codes for Lys-Leu-Ala (see Figure 10.22b). The altered polypeptide is likely to be nonfunctional. Inserting one or two nucleotides would have a similarly profound effect.

Mutagens

What causes mutations? Mutagenesis, the creation of mutations, can occur in a number of ways. Mutations resulting from errors during DNA replication or recombination are known as spontaneous mutations, as are other mutations of unknown cause. Other sources of mutation are physical and chemical agents called **mutagens**. The most common physical mutagen is high-energy radiation, such as X-rays and ultraviolet (UV) light. Chemical mutagens are of various types. One type, for example, consists of chemicals that are similar to normal DNA bases but that base-pair incorrectly when incorporated into DNA.

Many mutagens can act as carcinogens, agents that cause cancer. What can you do to avoid exposure to mutagens? Several lifestyle practices can help, including wearing protective clothing and sunscreen to minimize direct exposure to the sun's UV rays and not smoking. But such precautions are not foolproof, and it is not possible to avoid mutagens (such as UV radiation and secondhand smoke) entirely.

Although mutations are often harmful, they can also be extremely useful, both in nature and in the laboratory. Mutations are one source of the rich diversity of genes in the living world, a diversity that makes evolution by natural selection possible (Figure 10.23). Mutations are also essential tools for geneticists. Whether naturally occurring or created in the laboratory, mutations are responsible for the different alleles needed for genetic research. ☑

▼ Figure 10.23 **Mutations and diversity.** Mutations are one source of the diversity of life visible in this scene from the big island of Hawaii.

☑**CHECKPOINT**

1. What would happen if a mutation changed a start codon to some other codon?
2. What happens when one nucleotide is lost from the middle of a gene?

Answers: **1.** mRNA transcribed from the mutated gene would be non-functional because ribosomes would not initiate translation. **2.** In the mRNA, the reading of the triplets downstream from the deletion is shifted, leading to a long string of incorrect amino acids in the polypeptide.

Viruses and Other Noncellular Infectious Agents

Viruses share some of the characteristics of living organisms, such as having genetic material in the form of nucleic acid packaged within a highly organized structure. A virus is generally not considered alive, however, because it is not cellular and cannot reproduce on its own. In many cases, a **virus** is nothing more than "genes in a box": a bit of nucleic acid wrapped in a protein coat (**Figure 10.24**). A virus can survive only by infecting a living cell with genetic material that directs the cell's molecular machinery to make more viruses. In this section, we'll look at viruses that infect different types of host organisms, starting with bacteria.

Bacteriophages

Viruses that attack bacteria are called **bacteriophages** ("bacteria-eaters"), or **phages** for short. **Figure 10.25** shows a micrograph of a bacteriophage called T4 infecting an *E. coli* bacterium. The phage consists of a molecule of DNA enclosed within an elaborate structure made of proteins. The "legs" of the phage (called tail fibers) bend when they touch the cell surface. The tail is a hollow rod enclosed in a springlike sheath. As the legs bend, the spring compresses, the bottom of the rod punctures the cell membrane, and the viral DNA passes from inside the head of the virus into the cell.

Once they infect a bacterium, most phages enter a reproductive cycle called the **lytic cycle**. The lytic cycle gets its name from the fact that, after many copies of the phage are produced within the bacterial cell, the bacterium lyses (breaks open). Some viruses can also reproduce by an alternative route—the **lysogenic cycle**.

▼ Figure 10.24 **Adenovirus.** A virus that infects the human respiratory system, an adenovirus consists of DNA enclosed in a protein coat shaped like a 20-sided polyhedron, shown here in a computer-generated model that is magnified approximately 500,000 times the actual size. At each vertex of the polyhedron is a protein spike, which helps the virus attach to a susceptible cell.

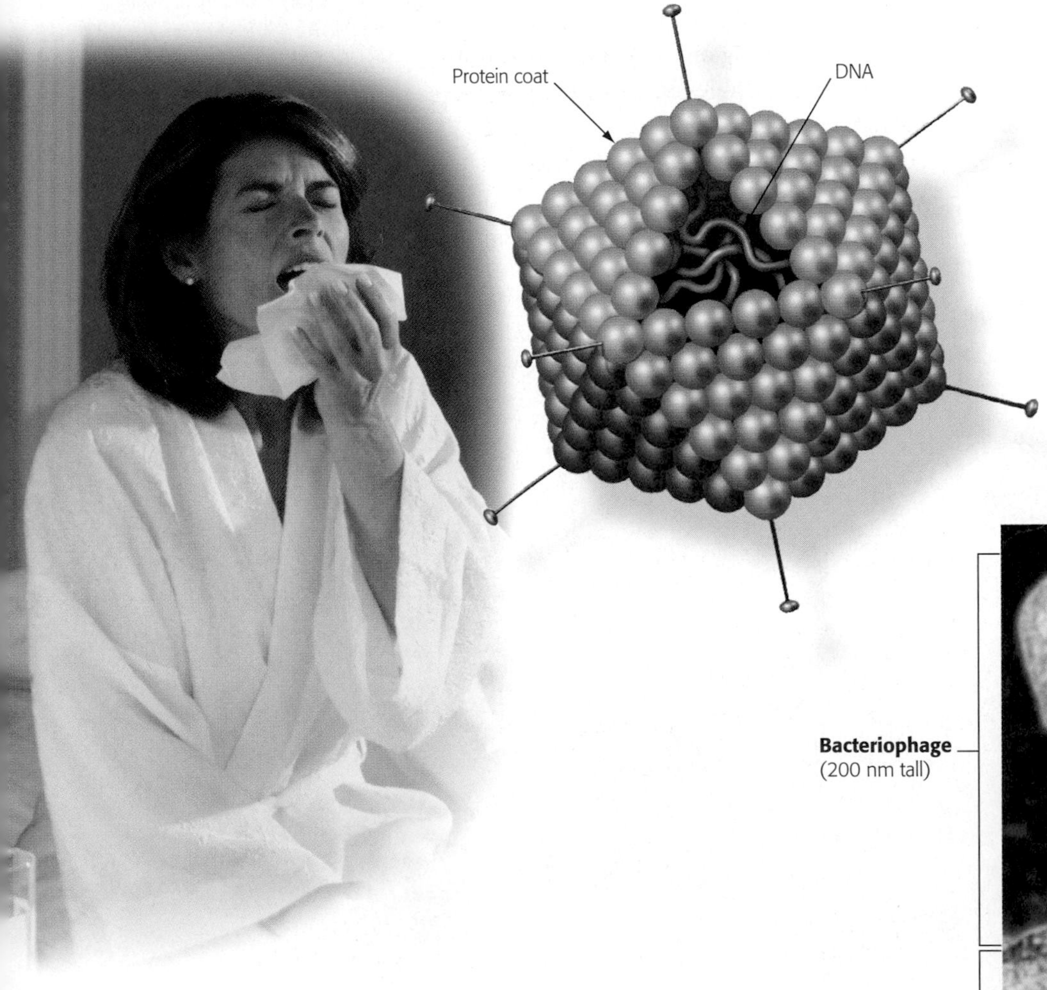

Protein coat

DNA

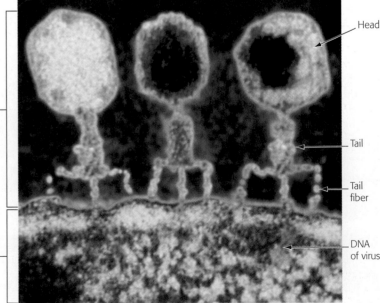

▼ Figure 10.25 **Bacteriophages (viruses) infecting a bacterial cell.**

Head

Bacteriophage (200 nm tall)

Tail

Tail fiber

Bacterial cell

DNA of virus

Colorized TEM 225,000×

During a lysogenic cycle, viral DNA replication occurs without phage production or the death of the cell.

Figure 10.26 illustrates the two kinds of cycles for a phage named lambda that can infect *E. coli* bacteria. Lambda has a head (containing DNA) and a tail. Before embarking on one of the two cycles, **1** lambda binds to the outside of a bacterium and injects its DNA inside. **2** The injected lambda DNA forms a circle. In the lytic cycle, this DNA immediately turns the cell into a virus-producing factory. **3** The cell's own machinery for DNA replication, transcription, and translation is hijacked by the virus and used to produce copies of the virus. **4** The cell lyses, releasing the new phages.

In the lysogenic cycle, **5** the viral DNA is inserted into the bacterial chromosome. Once there, the phage DNA is referred to as a **prophage**, and most of its genes are inactive. Survival of the prophage depends on the reproduction of the cell where it resides. **6** The host cell replicates the prophage DNA along with its cellular DNA and then, upon dividing, passes on both the prophage and the cellular DNA to its two daughter cells. A single infected bacterium can quickly give rise to a large population of bacteria that all carry prophages. The prophages may remain in the bacterial cells indefinitely. **7** Occasionally, however, a prophage leaves its chromosome; this event may be triggered by environmental conditions such as exposure to a mutagen. Once separate, the lambda DNA usually switches to the lytic cycle, which results in the production of many copies of the virus and lysing of the host cell.

Sometimes the few prophage genes active in a lysogenic bacterial cell can cause medical problems. For example, the bacteria that cause diphtheria, botulism, and scarlet fever would be harmless to humans if it were not for the prophage genes they carry. Certain of these genes direct the bacteria to produce toxins that make people ill. ☑

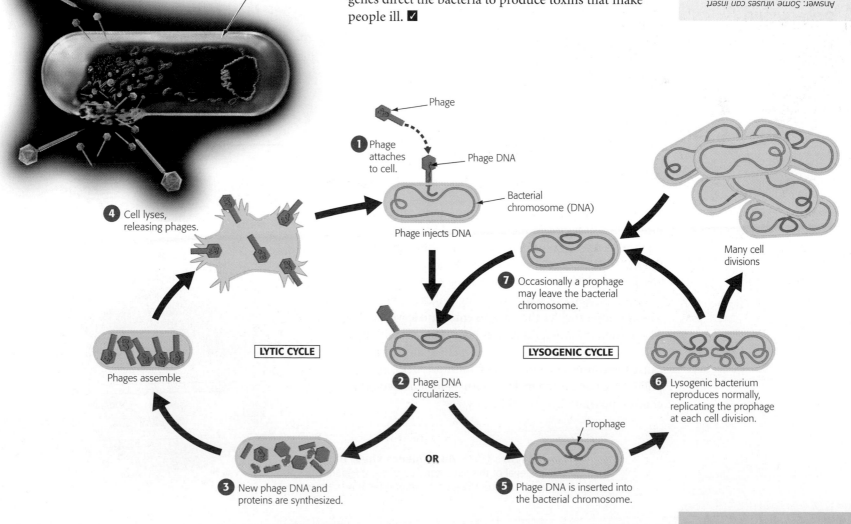

▼ Figure 10.26 **Alternative phage reproductive cycles.** Certain phages can undergo alternative reproductive cycles. After entering the bacterial cell, the phage DNA can either integrate into the bacterial chromosome (lysogenic cycle) or immediately start the production of progeny phages (lytic cycle), destroying the cell. In most cases, the phage follows the lytic pathway, but once it enters a lysogenic cycle, the phage's DNA may be carried in the host cell's chromosome for many generations.

Phage lambda

E. coli

Phage

1 Phage attaches to cell.

Phage DNA

Bacterial chromosome (DNA)

Phage injects DNA

4 Cell lyses, releasing phages.

7 Occasionally a prophage may leave the bacterial chromosome.

Many cell divisions

LYTIC CYCLE

LYSOGENIC CYCLE

Phages assemble

2 Phage DNA circularizes.

6 Lysogenic bacterium reproduces normally, replicating the prophage at each cell division.

Prophage

3 New phage DNA and proteins are synthesized.

OR

5 Phage DNA is inserted into the bacterial chromosome.

☑**CHECKPOINT**

Describe one way some viruses can perpetuate their genes without destroying the cells they infect.

Answer: Some viruses can insert their DNA into the DNA of the cell they infect. The viral DNA is replicated along with the cell's DNA every time the cell divides.

Plant Viruses

Viruses that infect plant cells can stunt plant growth and diminish crop yields. Most known plant viruses have RNA rather than DNA as their genetic material. Many of them, like the tobacco mosaic virus shown in **Figure 10.27**, are rod-shaped with a spiral arrangement of proteins surrounding the nucleic acid.

▼ **Figure 10.27 Tobacco mosaic virus.** The photo shows the mottling of leaves in tobacco mosaic disease. The rod-shaped virus causing the disease has RNA as its genetic material.

To infect a plant, a virus must first get past the plant's outer protective layer of cells (the epidermis). For this reason, a plant damaged by wind, chilling, injury, or insects is more susceptible to infection than a healthy plant. Some insects carry and transmit plant viruses, and farmers and gardeners may spread plant viruses through the use of pruning shears and other tools. Also, infected plants may pass viruses to their offspring.

There is no cure for most viral plant diseases, and agricultural scientists focus on preventing infection and on breeding or genetically engineering varieties of crop plants that resist viral infection. In Hawaii, for example, the spread of papaya ringspot potyvirus (PRSV) by aphids wiped out the papaya (Hawaii's second largest crop) in certain island regions. But since 1998, farmers have been able to plant a genetically engineered PRSV-resistant strain of papaya, and papayas have been reintroduced into their old habitats. ☑

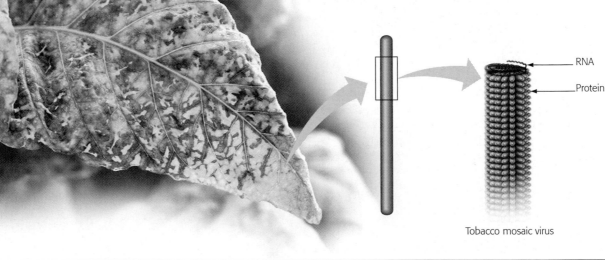

Tobacco mosaic virus

RNA

Protein

Animal Viruses

Viruses that infect animal cells are common causes of disease. As discussed in the Biology and Society section, no virus is a greater human health threat than the influenza (flu) virus (**Figure 10.28**). Like many animal viruses, this one has an outer envelope made of phospholipid membrane, with projecting spikes of protein. The envelope enables the virus to enter and leave a cell. As mentioned earlier, flu viruses have RNA as their genetic material. Other RNA viruses include those that cause the common cold, measles, mumps, AIDS, and polio. Diseases caused by DNA viruses include hepatitis, chicken pox, and herpes infections.

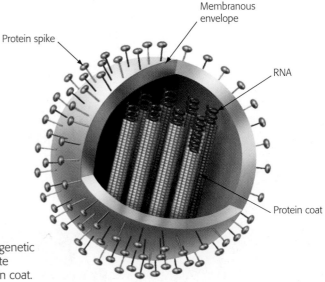

Protein spike

Membranous envelope

RNA

Protein coat

► **Figure 10.28 An influenza virus.** The genetic material of this virus consists of eight separate molecules of RNA, each wrapped in a protein coat.

Figure 10.29 shows the reproductive cycle of an enveloped RNA virus: the mumps virus, which causes a once-common childhood disease that has been almost eliminated by vaccines. When the virus contacts a susceptible cell, protein spikes on its outer surface attach to receptor proteins on the cell's plasma membrane. **1** The viral envelope fuses with the cell's membrane, allowing the protein-coated RNA to enter the cytoplasm. **2** Enzymes then remove the protein coat. **3** An enzyme that entered the cell as part of the virus uses the virus's RNA genome as a template for making complementary strands of RNA. The new strands have two functions: **4** They serve as mRNA for the synthesis of new viral proteins, and **5** they serve as templates for synthesizing new viral genome RNA. **6** The new coat proteins assemble around the new viral RNA. **7** Finally, the viruses leave the cell by cloaking themselves in plasma membrane. In other words, the virus obtains its envelope from the cell, leaving the cell without necessarily lysing it.

Not all animal viruses reproduce in the cytoplasm. For example, the viruses called herpesviruses—which cause chicken pox, shingles, cold sores, and genital herpes—are enveloped DNA viruses that reproduce in a cell's nucleus, and they get their envelopes from the cell's nuclear membrane. Copies of the herpesvirus DNA usually remain behind as mini-chromosomes in the nuclei of certain nerve cells. There they remain latent until some sort of physical stress, such as a cold or sunburn, or emotional stress triggers virus production, resulting in unpleasant symptoms. Once acquired, herpes infections may flare up repeatedly throughout a person's life. Over 75% of American adults are thought to carry herpes simplex 1 (which causes cold sores), and over 20% carry herpes simplex 2 (which causes genital herpes).

The amount of damage a virus causes the body depends partly on how quickly the immune system responds to fight the infection and partly on the ability of the infected tissue to repair itself. We usually recover completely from colds because our respiratory tract tissue can efficiently replace damaged cells by mitosis. In contrast, the poliovirus attacks nerve cells, which are not usually replaceable. The damage to such cells by polio, unfortunately, is permanent. In such cases, the only medical option is to prevent the disease with vaccines.

How effective are vaccines? Next, we'll look at the example of the flu vaccine. ☑

☑CHECKPOINT

Why is infection by herpesvirus permanent?

Answer: because herpesvirus leaves DNA in the nuclei of nerve cells

▼ Figure 10.29 **The reproductive cycle of an enveloped virus.** This virus is the one that causes mumps. Like the flu virus, it has a membranous envelope with protein spikes, but its genome is a single molecule of RNA.

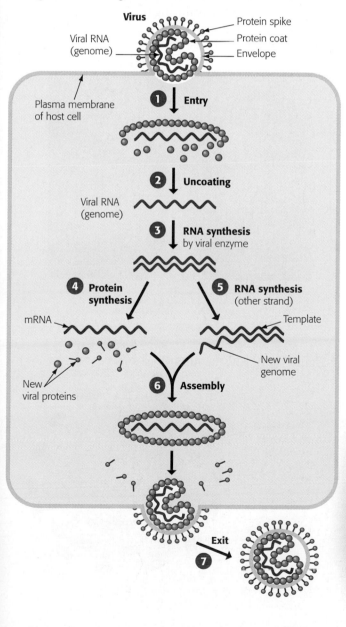

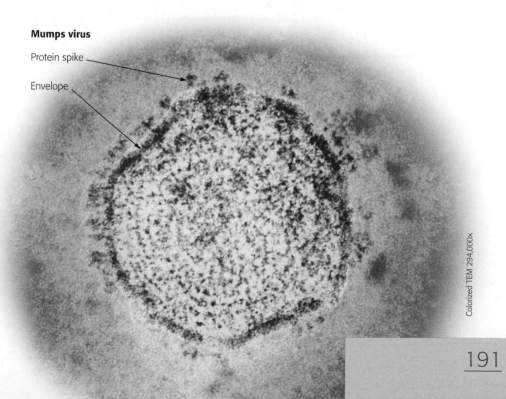

Mumps virus

Protein spike

Envelope

Colorized TEM 294,000x

Do Flu Vaccines Protect the Elderly?

Yearly flu vaccinations are recommended for people under the age of 19 and over 50. But how can we be sure they are effective? Epidemiologists (who study the distribution, causes, and control of diseases in populations) have made the **observation** that vaccination rates among the elderly rose from 15% in 1980 to 65% in 1996. This observation has led them to ask the **question**: Do flu vaccines decrease the mortality rate among those elderly people who receive them? To find out, researchers investigated data from the general population. Their main **hypothesis** was that elderly people who were immunized would have fewer hospital stays and deaths during the winter after vaccination. Their **experiment** followed tens of thousands of people over the age of 65 during the ten flu seasons of the 1990s. The **results** are summarized in **Figure 10.30**. People who were vaccinated had a 27% less chance of being hospitalized during the next flu season and a 48% less chance of dying. But could some factor other than flu shots be at play? For example, maybe people who choose to be vaccinated are healthier for other reasons. As a control, the researchers examined health data for the summer (when flu is not a factor). During these months, there was no difference in the hospitalization rates and only 16% fewer deaths for the immunized, suggesting that flu vaccines provide a significant health benefit among the elderly during the flu season.

▶ **Figure 10.30 The effect of flu vaccines on the elderly.** Receiving a flu vaccine greatly reduced the risk of hospitalization and death in the flu season following the shot. The reduction was much smaller or nonexistent in later summer months.

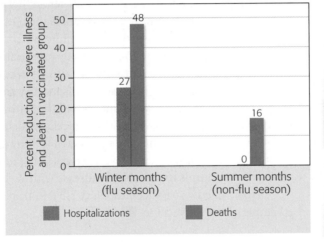

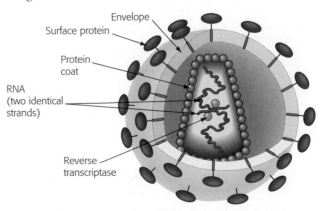

HIV, the AIDS Virus

The devastating disease **AIDS** (acquired immunodeficiency syndrome) is caused by **HIV** (human immunodeficiency virus), a type of RNA virus with some special twists. In outward appearance, the AIDS virus (**Figure 10.31**) somewhat resembles the flu or mumps virus. Its envelope enables HIV to enter and leave a cell much the way the mumps virus does. But HIV has a different mode of reproduction. It is a **retrovirus**, an RNA virus that reproduces by means of a DNA molecule. Retroviruses are so named because they reverse the usual DNA → RNA flow of genetic information. These viruses carry molecules of an enzyme called **reverse transcriptase**, which catalyzes reverse transcription, the synthesis of DNA on an RNA template.

▼ **Figure 10.31 HIV, the AIDS virus.**

Figure 10.32 illustrates what happens after HIV RNA is uncoated in the cytoplasm of a cell. The reverse transcriptase (green) **1** uses the RNA as a template to make a DNA strand and then **2** adds a second, complementary DNA strand. **3** The resulting double-stranded viral DNA then enters the cell nucleus and inserts itself into the chromosomal DNA, becoming a **provirus**. Occasionally, the provirus is **4** transcribed into RNA and **5** translated into viral proteins. **6** New viruses assembled from these components eventually leave the cell and can then infect other cells. This is the standard reproductive cycle for retroviruses.

HIV infects and eventually kills several kinds of white blood cells that are important in the body's immune system. The loss of such cells causes the body to become susceptible to other infections that it would normally be able to fight off. Such secondary infections cause the

syndrome (a collection of symptoms) that eventually kills AIDS patients. Since it was first recognized in 1981, HIV has infected tens of millions of people worldwide, resulting in millions of deaths.

While there is no cure for AIDS, its progression can be slowed by two categories of anti-HIV drugs. Both types of medicine interfere with the reproduction of the virus. The first type inhibits the action of enzymes called proteases, which help produce the final versions of HIV proteins. The second type, which includes the drug AZT, inhibits the action of the HIV enzyme reverse transcriptase. The key to AZT's effectiveness is its shape. The shape of a molecule of AZT is very similar to the shape of part of the T (thymine) nucleotide (**Figure 10.33**). In fact, AZT's shape is so similar to the T nucleotide that AZT can bind to reverse transcriptase instead of T. But unlike thymine, AZT cannot be incorporated into a growing DNA chain. Thus, AZT "gums up the works," interfering with the synthesis of HIV DNA. Because this synthesis is an essential step in the reproductive cycle of HIV, AZT may block the spread of the virus within the body.

Many HIV-infected people in the United States and other industrialized countries take a "drug cocktail" that contains both reverse transcriptase inhibitors and protease inhibitors, and the combination seems to be much more effective than the individual drugs in keeping the virus at bay and extending patients' lives. However, even in combination, the drugs do not completely rid the body of the virus. Typically, HIV reproduction and the symptoms of AIDS return if a patient discontinues the medications. Because AIDS has no cure yet, prevention (namely, the avoidance of unprotected sex and needle sharing) is the only healthy option. ✔

▼ Figure 10.32 **The behavior of HIV nucleic acid in an infected cell.**

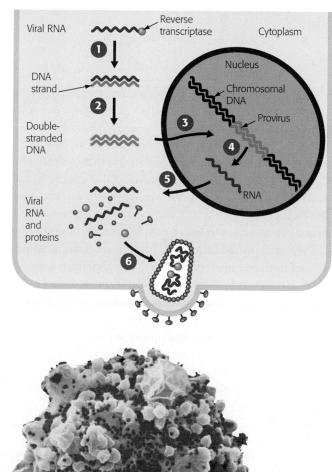

Viral RNA — Reverse transcriptase — Cytoplasm

1

DNA strand

2

Nucleus

Chromosomal DNA

Double-stranded DNA

3 Provirus

4

5 RNA

Viral RNA and proteins

6

HIV (red dots) infecting a white blood cell

SEM 5,500×

▼ Figure 10.33 **AZT and the T nucleotide.** The anti-HIV drug AZT (right) has a chemical shape very similar to part of the T (thymine) nucleotide of DNA.

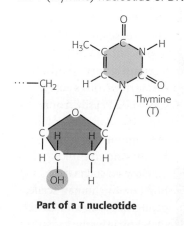

Part of a T nucleotide

Thymine (T)

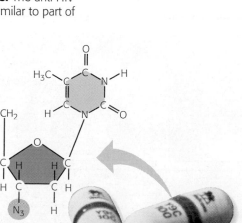

AZT

Viroids and Prions

Viruses may be small and simple, but they dwarf two other classes of pathogens: viroids and prions. Viroids are small circular RNA molecules that infect plants. Viroids do not encode proteins but can nonetheless replicate in host plant cells, apparently using cellular enzymes. These small RNA molecules seem to cause disease by interfering with the regulatory systems that control plant growth.

Even stranger are infectious proteins called **prions**. Prions are believed to cause a number of degenerative brain diseases in various animal species, including scrapie in sheep and goats, chronic wasting disease in deer and elk, Creutzfeldt-Jakob disease in humans, and mad cow disease (which infected over 2 million cattle in the United Kingdom in the 1980s; **Figure 10.34**). A prion is thought to be a misfolded form of a protein normally present in brain cells. When the prion enters a cell containing the normal form of protein, the prion

somehow converts the normal protein molecules to the misfolded prion version. To date, there is no known cure for prion diseases, so the only hope for avoiding future illnesses lies in understanding and preventing the process of infection. ☑

▼ Figure 10.34 **Mad cow disease.** An outbreak of mad cow disease in England required the culling of hundreds of thousands of cows to prevent the spread of the disease.

EVOLUTION CONNECTION: The Flu

Emerging Viruses

Emerging viruses are viruses that have appeared suddenly or have only recently come to the attention of medical scientists. Consider the avian flu. In 1997, at least 18 people in Hong Kong were infected with a strain of flu virus previously seen only in birds. A mass culling of all of Hong Kong's 1.5 million domestic birds appeared to stop that outbreak. Beginning in 2002, however, new cases of human infection by this bird strain began to crop up around southeast Asia. As of 2009, the disease caused by this virus has spread to Europe and Africa, infecting over 400 people and killing over 250 of them. Over 100 million birds have either died from the disease or been killed to prevent the spread of infection (**Figure 10.35**). If this virus evolves so that it can spread easily from person to person, the potential for a major human outbreak is significant.

How do such viruses burst on the human scene, giving rise to new diseases? One way is by the mutation of existing viruses. RNA

▼ Figure 10.35 **Rounding up ducks in Vietnam to help prevent the spread of the Avian flu virus.**

viruses tend to have unusually high rates of mutation because errors in replicating their RNA genomes are not subject to certain proofreading mechanisms that help reduce errors during DNA replication. Some mutations enable existing viruses to evolve into new strains that can cause disease in individuals who have developed resistance to the ancestral virus. This is why we need yearly flu vaccines (see the Biology and Society section): Mutations create new influenza virus strains to which people have no immunity.

New viral diseases also arise from the spread of existing viruses from one host species to another. Scientists estimate that about three-quarters of new human diseases have originated in other animals. The avian flu is a prime example. The spread of a viral disease from a small, isolated population can also lead to widespread epidemics. For instance, AIDS went unnamed and virtually unnoticed for decades before it began to spread around the world. In this case, technological and social factors, including affordable international travel, blood transfusions, sexual promiscuity, and the abuse of intravenous drugs, allowed a previously rare human disease to become a global scourge.

Acknowledging the persistent threat that viruses pose to human health, geneticist and Nobel Prize winner Joshua Lederberg once warned: "We live in evolutionary competition with microbes. There is no guarantee that we will be the survivors." If we ever do manage to control HIV, influenza, and other emerging viruses, this success will likely arise from our understanding of molecular biology.

Chapter Review

SUMMARY OF KEY CONCEPTS

(MB) Go to the Study Area at **www.masteringbiology.com** for practice quizzes, my**e**Book, BioFlix™ 3-D animations, MP3 Tutor Sessions, videos, current events, and more.

DNA: Structure and Replication

DNA and RNA Structure

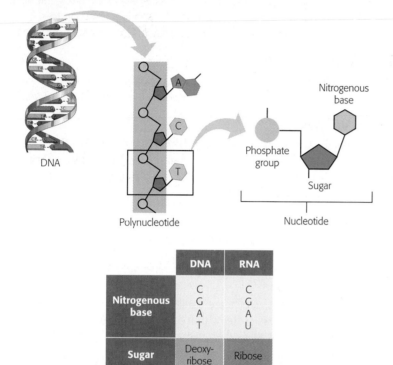

	DNA	RNA
Nitrogenous base	C G A T	C G A U
Sugar	Deoxyribose	Ribose
Number of strands	2	1

Watson and Crick's Discovery of the Double Helix

Watson and Crick worked out the three-dimensional structure of DNA: two polynucleotide strands wrapped around each other in a double helix. Hydrogen bonds between bases hold the strands together. Each base pairs with a complementary partner: A with T, and G with C.

DNA Replication

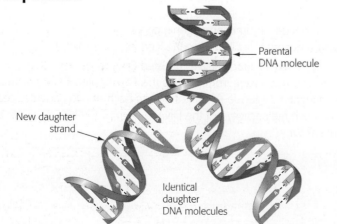

The Flow of Genetic Information from DNA to RNA to Protein

How an Organism's Genotype Determines Its Phenotype

The information constituting an organism's genotype is carried in the sequence of its DNA bases. Studies of inherited metabolic defects first suggested that phenotype is expressed through proteins.

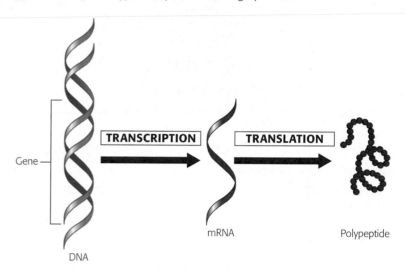

From Nucleotides to Amino Acids: An Overview

The DNA of a gene is transcribed into RNA using the usual base-pairing rules, except that an A in DNA pairs with U in RNA. In the translation of a genetic message, each triplet of nucleotide bases in the RNA, called a codon, specifies one amino acid in the polypeptide.

The Genetic Code

In addition to codons that specify amino acids, the genetic code has one codon that is a start signal and three that are stop signals for translation. The genetic code is redundant: There is more than one codon for most amino acids.

Transcription: From DNA to RNA

In transcription, RNA polymerase binds to the promoter of a gene, opens the DNA double helix there, and catalyzes the synthesis of an RNA molecule using one DNA strand as a template. As the single-stranded RNA transcript peels away from the gene, the DNA strands rejoin.

The Processing of Eukaryotic RNA

The RNA transcribed from a eukaryotic gene is processed before leaving the nucleus to serve as messenger RNA (mRNA). Introns are spliced out, and a cap and tail are added.

Translation: The Players

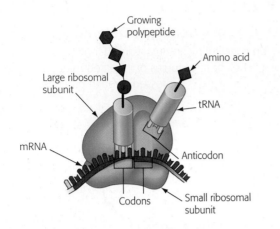

Growing polypeptide

Amino acid

Large ribosomal subunit

tRNA

mRNA

Anticodon

Codons

Small ribosomal subunit

Translation: The Process

In initiation, a ribosome assembles with the mRNA and the initiator tRNA bearing the first amino acid. Beginning at the start codon, the codons of the mRNA are recognized one by one by tRNAs bearing succeeding amino acids. The ribosome bonds the amino acids together. With each addition, the mRNA translocates by one codon through the ribosome. When a stop codon is reached, the completed polypeptide is released.

Review: DNA → RNA → Protein

Figure 10.20 summarizes transcription, RNA processing, and translation. The sequence of codons in DNA, via the sequence of codons in mRNA, spells out the primary structure of a polypeptide.

Mutations

Mutations are changes in the DNA base sequence, caused by errors in DNA replication or recombination or by mutagens. Substituting, deleting, or inserting nucleotides in a gene has varying effects on the polypeptide and organism.

Type of Mutation	Effect
Substitution of one DNA base for another	**Silent** mutations result in no change to amino acids.
	Missense mutations swap one amino acid for another.
	Nonsense mutations change an amino acid codon to a stop codon.
Insertions or **deletions** of DNA nucleotides	These mutations can alter the triplet grouping of codons and greatly change the amino acid sequence.

Viruses and Other Noncellular Infectious Agents

Viruses can be regarded as genes packaged in protein.

Bacteriophages

When phage DNA enters a lytic cycle inside a bacterium, it is replicated, transcribed, and translated. The new viral DNA and protein molecules then assemble into new phages, which burst from the cell. In the lysogenic cycle, phage DNA inserts into the cell's chromosome and is passed on to generations of daughter cells. Much later, it may initiate phage production.

Plant Viruses

Viruses that infect plants can be a serious agricultural problem. Most have RNA genomes. Viruses enter plants via breaks in the plant's outer layers.

Animal Viruses

Many animal viruses, such as flu viruses, have RNA genomes; others, such as hepatitis viruses, have DNA. Some animal viruses "steal" a bit of cell membrane as a protective envelope. Some, such as the herpesvirus, can remain latent inside cells for long periods.

HIV, the AIDS Virus

HIV is a retrovirus. Inside a cell it uses its RNA as a template for making DNA, which is then inserted into a chromosome.

Viroids and Prions

Even smaller than viruses, viroids are small molecules of RNA that can infect plants. Prions are infectious proteins that cause a number of degenerative brain diseases in humans and other animals.

SELF-QUIZ

1. A molecule of DNA contains two polymer strands called _____, made by bonding together many monomers called _____. Each monomer contains three parts: a _____, a _____, and a _____.

2. Which of the following correctly ranks nucleic acid structures in order of size, from largest to smallest?
   a. gene, chromosome, nucleotide, codon
   b. chromosome, gene, codon, nucleotide
   c. nucleotide, chromosome, gene, codon
   d. chromosome, nucleotide, gene, codon

3. A scientist inserts a radioactively labeled DNA molecule into a bacterium. The bacterium replicates this DNA molecule and distributes one daughter molecule (double helix) to each of two daughter cells. How much radioactivity will the DNA in each of the two daughter cells contain? Why?

4. The nucleotide sequence of a DNA codon is GTA. In an mRNA molecule transcribed from this DNA, the codon has the sequence _____. In the process of protein synthesis, a tRNA pairs with the mRNA codon. The nucleotide sequence of the tRNA anticodon is _____. The amino acid attached to the tRNA is _____. (See Figure 10.11.)

5. Describe the process by which the information in a gene is transcribed and translated into a protein. Correctly use these terms in your description: tRNA, amino acid, start codon, transcription, mRNA, gene, codon, RNA polymerase, ribosome, translation, anticodon, peptide bond, stop codon.

6. Match the following molecules with the cellular process or processes in which they are primarily involved.
   a. ribosomes
   b. tRNA
   c. DNA polymerases
   d. RNA polymerase
   e. mRNA

   1. DNA replication
   2. transcription
   3. translation

7. A geneticist found that a particular mutation had no effect on the polypeptide encoded by the gene. This mutation probably involved
   a. deletion of one nucleotide.
   b. alteration of the start codon.
   c. insertion of one nucleotide.
   d. substitution of one nucleotide.

8. Scientists have discovered how to put together a bacteriophage with the protein coat of phage A and the DNA of phage B. If this composite phage were allowed to infect a bacterium, the phages produced in the cell would have
   a. the protein of A and the DNA of B.
   b. the protein of B and the DNA of A.
   c. the protein and DNA of A.
   d. the protein and DNA of B.

9. How do some viruses reproduce without ever having DNA?

10. HIV requires an enzyme called _____ to convert its RNA genome to a DNA version.

11. Why is reverse transcriptase a particularly good target for anti-AIDS drugs? (*Hint*: Would you expect such a drug to harm the human host?)

Answers to the Self-Quiz questions can be found in Appendix D.

THE PROCESS OF SCIENCE

12. A cell containing a single chromosome is placed in a medium containing radioactive phosphate, making any new DNA strands formed by DNA replication radioactive. The cell replicates its DNA and divides. Then the daughter cells (still in the radioactive medium) replicate their DNA and divide, resulting in a total of four cells. Sketch the DNA molecules in all four cells, showing a normal (nonradioactive) DNA strand as a solid line and a radioactive DNA strand as a dashed line.

13. In a classic 1952 experiment, biologists Alfred Hershey and Martha Chase labeled two batches of bacteriophages, one with radioactive sulfur (which only tags protein) and the other with radioactive phosphorus (which only tags DNA). In separate test tubes, they allowed each batch of phages to bind to nonradioactive bacteria and inject its DNA. After a few minutes, they separated the bacterial cells from the viral parts that remained outside the bacterial cells and measured the radioactivity of both portions. What results do you think they obtained? How would these results help them to determine which viral component—DNA or protein—was the infectious portion?

BIOLOGY AND SOCIETY

14. Scientists at the National Institutes of Health (NIH) have worked out thousands of sequences of genes and the proteins they encode, and similar analysis is being carried out at universities and private companies. Knowledge of the nucleotide sequences of genes might be used to treat genetic defects or produce lifesaving medicines. The NIH and some U.S. biotechnology companies have applied for patents on their discoveries. In Britain, the courts have ruled that a naturally occurring gene cannot be patented. Do you think individuals and companies should be able to patent genes and gene products? Before answering, consider the following: What are the purposes of a patent? How might the discoverer of a gene benefit from a patent? How might the public benefit? What negative effects might result from patenting genes?

15. Your college roommate seeks to improve her appearance by visiting a tanning salon. How would you explain the dangers of this to her?

How Genes Are Controlled

Smoking and cancer. Cancer, like the orange-tinted lung cancer shown in this X-ray, is caused by a breakdown of gene control.

CHAPTER CONTENTS

■■■ Chapter Thread: **CANCER**

BIOLOGY AND SOCIETY: Cancer

Tobacco's Smoking Gun

When European explorers returned from their first voyages to the Americas, they brought back tobacco, a common trade item among Native Americans. It quickly became popular in Europe, and the southern United States developed into a major tobacco producer. By the 1950s, about half of all Americans smoked over a pack of cigarettes each day. Little credence was paid to health risks; in fact, cigarette advertising often touted the "health benefits" of tobacco. Smoking was pervasive in American society.

But doctors began to notice a disturbing trend: As tobacco use increased, so did the rate of lung cancer. This disease was rare in 1930, but by 1955 it had become the deadliest form of cancer among American men. In fact, by 1990, lung cancer was killing over twice as many men each year as any other type of cancer. But a few skeptics doubted the link between smoking and cancer. They pointed out that the evidence was purely statistical or based on animal studies; no direct proof had been found that tobacco smoke caused cancer in humans.

The "smoking gun" of proof was found in 1996 when researchers added one component of tobacco smoke, called BPDE, to human lung cells growing in the lab. In these cells, the researchers showed, BPDE binds to DNA within a gene called *p53*, which codes for a protein that normally helps suppress the formation of tumors. The researchers proved that BDPE causes mutations in the *p53* gene that deactivate the protein and lead to tumors. This work directly linked a chemical in tobacco smoke with the formation of human lung tumors.

How can a mutation in a gene lead to cancer? It turns out that many cancer-associated genes encode proteins that turn other genes on or off within a cell. When these proteins malfunction, the cell may become cancerous. In fact, the ability to properly control which genes are active at any given time is crucial to normal cell function. How genes are controlled and how the regulation of genes affects cells and organisms are the subjects of this chapter. As you will see, these issues touch on some of the most interesting topics in all of biology.

Control of Gene Expression

☑CHECKPOINT

If your nerve cells and skin cells have the same genes, how can they be so different?

Answer: Each cell type expresses genes that are not expressed in the other cell type.

How and Why Genes Are Regulated

In a multicellular organism, every somatic cell (every cell except the gametes) is produced by repeated rounds of mitosis, starting from a zygote. Since a cell's genes are duplicated with each round of the cell cycle, each somatic cell has the same DNA as the zygote. Simply put, every somatic cell contains every gene.

If every somatic cell contains identical genetic instructions, how do cells become different from one another? Cells with the same genetic information can develop into cells with different structures and functions only if gene activity is regulated. Control mechanisms must turn *on* certain genes, while other genes remain turned *off* in a particular cell. In other words, individual cells must undergo **cellular differentiation**—that is, they must become specialized in structure and function. It is **gene regulation**, the turning on and off of genes, that leads to this specialization.

Patterns of Gene Expression in Differentiated Cells

What does it mean to say that genes are active or inactive, turned on or off? As discussed in Chapter 10, genes determine the nucleotide sequence of specific mRNA molecules, and mRNA in turn determines the sequence of amino acids in proteins (DNA→RNA→protein). A gene that is turned on is being transcribed into mRNA, and that message is being translated into specific proteins. The overall process by which genetic information flows from genes to proteins—that is, from genotype to phenotype—is called **gene expression**.

Because all the differentiated cells in an individual organism contain the same genes, the differences among the cells must result from the selective expression of genes—that is, from the pattern of genes turned on in a given cell at a given time. Such regulation of gene expression plays a central role in the development of a unicellular zygote into a multicellular organism. During embryonic growth, groups of cells follow diverging developmental pathways, and each group becomes a particular kind of tissue. In the mature organism, each cell type—nerve or pancreas, for instance—has a different pattern of turned-on genes.

Figure 11.1 shows the patterns of gene expression for four genes in three different specialized cells of an adult human. Note that the genes for "housekeeping" enzymes, such as those that provide energy via glycolysis, are "on" in all the cells. In contrast, the genes for some proteins, such as insulin and hemoglobin, are expressed only by particular kinds of cells. One protein, hemoglobin, is not expressed in any of the cell types shown in the figure. ☑

Gene Regulation in Bacteria

To understand how a cell can regulate gene expression, consider the relatively simple case of bacteria. In the course of their lives, bacteria must regulate their genes in response to environmental changes. For example,

► Figure 11.1 **Patterns of gene expression in three types of human cells.** Different types of cells express different combinations of genes. The specialized proteins whose genes are represented here are a glycolysis enzyme; an antibody, which aids in fighting infection; insulin, a hormone made in the pancreas; and the oxygen transport protein hemoglobin, which is expressed only in red blood cells.

Key

✓
Active gene

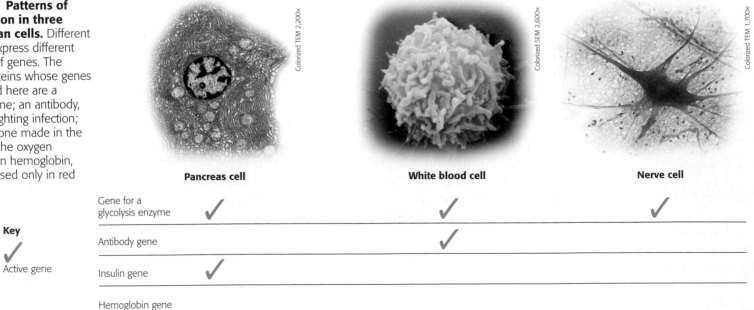

	Pancreas cell	White blood cell	Nerve cell
Gene for a glycolysis enzyme	✓	✓	✓
Antibody gene		✓	
Insulin gene	✓		
Hemoglobin gene			

when a nutrient is plentiful, bacteria do not squander resources to make it from scratch. Bacterial cells that can conserve resources and energy have a survival advantage over cells that are unable to do so. Thus, natural selection has favored bacteria that express only the genes whose products are needed by the cell.

Imagine an *Escherichia coli* bacterium living in your intestines. It will be bathed in various nutrients, depending on what you eat. If you drink a milk shake, for example, there will be a sudden rush of the sugar lactose. In response, *E. coli* will express three genes for enzymes that enable the bacterium to absorb and digest this sugar. After the lactose is gone, *E. coli* turns the genes off; it does not waste its energy continuing to produce these enzymes. Thus, a bacterium can adjust its gene expression to changes in the environment.

How does the presence or absence of lactose influence the activity of the genes that code for lactose enzymes? The key is the way the three genes are organized: They are adjacent in the DNA and regulated (turned on and off) as a single unit. This regulation is achieved through control sequences, short stretches of DNA that help turn all three genes on and off, coordinating their expression. Such a cluster of genes with related functions, along with the control sequences, is called an **operon** (Figure 11.2). The operon considered here, the *lac* (short for lactose) operon, was first described in the 1960s by French biologists François Jacob and Jacques Monod. The *lac* operon illustrates principles of gene regulation that apply to a wide variety of prokaryotic genes.

How do DNA control sequences turn genes on or off? One control sequence, called a **promoter** (green in the figure), is the site where the enzyme RNA polymerase attaches and initiates transcription—in our example, transcription of the three genes for lactose enzymes. Between the promoter and the enzyme genes, a DNA segment called an **operator** (yellow) acts as a switch that is turned on or off, depending on whether a specific protein is bound there. The operator and protein together determine whether RNA polymerase can attach to the promoter and start transcribing the genes (light blue). In the *lac* operon, when the operator switch is turned on, all the enzymes needed to metabolize lactose are made at once.

The top half of Figure 11.2 shows the *lac* operon in "off" mode, its status when there is no lactose around. Transcription is turned off because a protein called a **repressor** ( ) ❶ binds to the operator ( ) and ❷ physically blocks the attachment of RNA polymerase ( ) to the promoter ( ).

The bottom half of Figure 11.2 shows the operon in "on" mode, when lactose is present. The lactose ( )

interferes with attachment of the *lac* repressor to the operator by ❶ binding to the repressor and ❷ changing the repressor's shape. In its new shape ( ), the repressor cannot bind to the operator, and the operator switch remains on. ❸ RNA polymerase is no longer blocked, so it can now bind to the promoter and from there ❹ transcribe the genes for the lactose enzymes into mRNA. ❺ Translation produces all three lactose enzymes (purple).

Many operons have been identified in bacteria. Some are quite similar to the *lac* operon, while others have somewhat different mechanisms of control. For example, operons that control amino acid synthesis cause bacteria to stop making these molecules when they are already present in the environment, saving materials and energy for the cells. In these cases, the amino acid *activates* the repressor. Armed with a variety of operons, *E. coli* and other prokaryotes can thrive in frequently changing environments. ✔

✔CHECKPOINT

A mutation in *E. coli* makes the *lac* operator unable to bind the active repressor. How would this mutation affect the cell? Why would this effect be a disadvantage?

Answer: The cell would wastefully produce the enzymes for lactose metabolism continuously, even in the absence of lactose.

▼ Figure 11.2 **The *lac* operon of *E. coli*.**

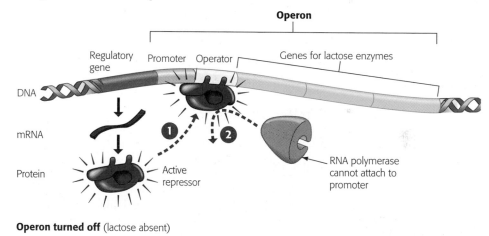

Operon turned off (lactose absent)

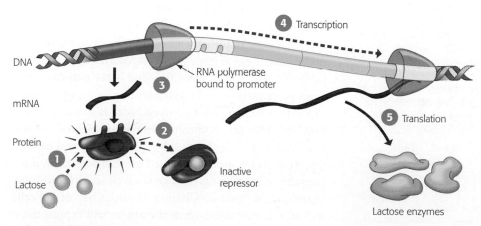

Operon turned on (lactose inactivates repressor)

Gene Regulation in Eukaryotic Cells

Eukaryotes, especially multicellular ones, have more sophisticated mechanisms than bacteria for regulating the expression of their genes. This is not surprising because a prokaryote, being a single cell, does not require the elaborate regulation of gene expression that leads to cell specialization in multicellular eukaryotic organisms.

The pathway from gene to functioning protein in eukaryotic cells is a long one, providing a number of points where the process can be regulated—turned on or off, speeded up or slowed down. Picture the series of pipes that carry water from your local water supply to a faucet in your home. At various points, valves control the flow of water. We use this analogy in **Figure 11.3** to illustrate the flow of genetic information from a eukaryotic chromosome—a reservoir of genetic information—to an active protein that has been made in the cell's cytoplasm. The multiple mechanisms that control gene expression are analogous to the control valves in your water pipes. In the figure, each control knob indicates a gene expression "valve." All these knobs represent possible control points, although only one or a few control points are likely to be important for a typical protein.

Using a reduced version of Figure 11.3 as a guide to the flow of genetic information through the cell, we will explore several ways that eukaryotes can control gene expression, starting within the cell nucleus.

The Regulation of DNA Packing

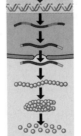

Recall from Chapter 8 that eukaryotic chromosomes may be in a more or less condensed state, with the DNA and accompanying proteins more or less tightly wrapped together. DNA packing tends to prevent gene expression by preventing RNA polymerase and other transcription proteins from binding to the DNA.

Cells may use DNA packing for the long-term inactivation of genes. One intriguing case is seen in female mammals, where one X chromosome in each somatic cell is highly compacted and almost entirely inactive. This **X chromosome inactivation** first takes place early in embryonic development, when one of the two X chromosomes in each cell is inactivated at random. After one X chromosome is inactivated in each embryonic cell, all of that cell's descendants will have the same X chromosome turned off. Consequently, if a female is heterozygous for a gene on the X chromosome (a sex-linked gene; see Chapter 9), about half of her cells will express one allele, while the others will express the alternate allele (**Figure 11.4**). ☑

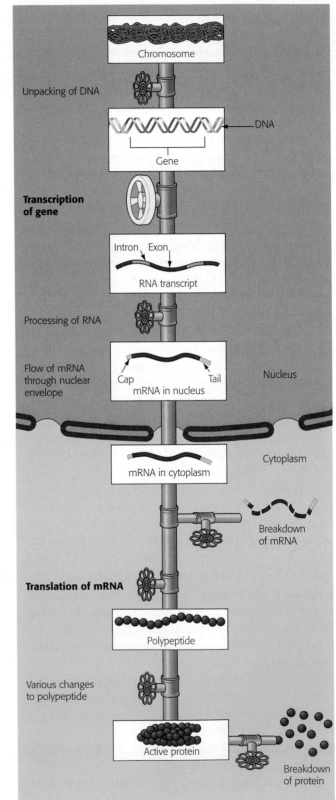

▼ **Figure 11.3 The gene expression "pipeline" in a eukaryotic cell.** Each valve in the pipeline represents a stage at which the pathway from chromosome to functioning protein can be regulated.

Chromosome

Unpacking of DNA

DNA

Gene

Transcription of gene

Intron   Exon

RNA transcript

Processing of RNA

Flow of mRNA through nuclear envelope

Cap   Tail
mRNA in nucleus

Nucleus

mRNA in cytoplasm

Cytoplasm

Breakdown of mRNA

Translation of mRNA

Polypeptide

Various changes to polypeptide

Active protein

Breakdown of protein

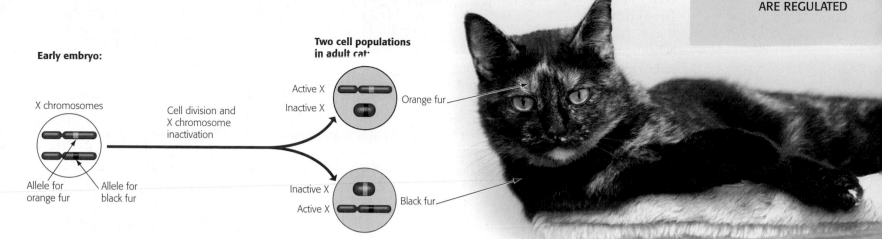

▲ Figure 11.4 **X chromosome inactivation: the tortoiseshell pattern on a cat.** The tortoiseshell gene is on the X chromosome, and the tortoiseshell phenotype requires the presence of two different alleles, one for orange fur and one for non-orange (black) fur. If a female is heterozygous for the tortoiseshell gene, orange patches are formed by populations of cells in which the X chromosome with the orange allele is active; black patches have cells in which the X chromosome with the non-orange allele is active.

The Initiation of Transcription

The initiation of transcription (whether transcription starts or not) is the most important stage for regulating gene expression. In both prokaryotes and eukaryotes, regulatory proteins bind to DNA and turn the transcription of genes on and off. Unlike prokaryotic genes, however, most eukaryotic genes have individual promoters and other control sequences. That is, eukaryotic genes are generally not organized into groups as operons.

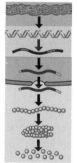

As seen in **Figure 11.5**, transcriptional regulation in eukaryotes is complex, typically involving many proteins (collectively called **transcription factors**, purple in the figure) acting in concert to bind to DNA sequences called **enhancers** (yellow) and to the promoter (green). The DNA-protein assembly promotes the binding of RNA polymerase (orange) to the promoter. Genes coding for related enzymes, such as those in a metabolic pathway, may share a specific kind of enhancer (or collection of enhancers), allowing these genes to be activated at the same time. Not shown in the figure are repressor proteins, which may bind to DNA sequences called **silencers**, inhibiting the start of transcription.

In fact, repressor proteins that turn genes off are less common in eukaryotes than **activators**, proteins that turn genes on by binding to DNA. Activators act by making it easier for RNA polymerase to bind to the promoter. The use of activators is efficient because a typical animal or plant cell needs to turn on (transcribe) only a small percentage of its genes, those required for the cell's specialized structure and function. The "default" state for most genes in multicellular eukaryotes seems to be "off," with the exception of "housekeeping" genes for routine activities such as glucose metabolism. ✔

☑**CHECKPOINT**

Of all the control points of DNA expression shown in Figure 11.3, which is under the tightest regulation?

Answer: the initiation of transcription

▼ Figure 11.5 **A model for turning on a eukaryotic gene.** A large assembly of proteins and several control sequences in the DNA are involved in initiating the transcription of a eukaryotic gene.

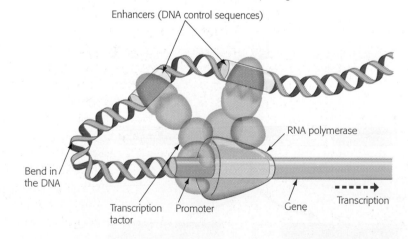

Enhancers (DNA control sequences)

RNA polymerase

Bend in the DNA

Transcription factor

Promoter

Gene

Transcription

RNA Processing and Breakdown

Within a eukaryotic cell, transcription occurs in the nucleus, where RNA transcripts are processed into mRNA before moving to the cytoplasm for translation by the ribosomes (see Figure 10.20). RNA processing includes the addition of a cap and a tail, as well as the removal of any introns—noncoding DNA segments that interrupt the genetic message—and the splicing together of the remaining exons.

Sometimes a cell can carry out exon splicing in more than one way, generating different mRNA molecules from the same starting RNA molecule. Notice in **Figure 11.6**, for example, that one mRNA ends up with the green exon and the other with the brown exon. With this sort of **alternative RNA splicing**, an organism can get more than one type of polypeptide from a single gene. Alternative RNA splicing is very common in humans. In one case, the RNA transcript from a gene can be spliced to encode seven different versions of a cellular protein.

After an mRNA is produced in its final form, its "lifetime" can be highly variable, from hours to weeks to months. Controlling the timing of mRNA breakdown provides another opportunity for control. But all mRNAs are eventually broken down and their parts recycled.

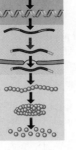

☑CHECKPOINT

After a gene is transcribed in the nucleus, how is the transcript converted to mRNA? After the mRNA reaches the cytoplasm, what are four control mechanisms that can regulate the amount of active protein in the cell?

Answer: by RNA processing, including the addition of cap and tail, and RNA splicing; control by microRNAs, initiation of translation, activation of the protein, and breakdown of the protein

microRNAs

Recent research has established an important role for a variety of small single-stranded RNA molecules, called microRNAs (miRNAs), that can bind to complementary sequences on mRNA molecules in the cytoplasm. After binding, some miRNAs trigger breakdown of their target mRNA, whereas others block translation. It has been estimated that miRNAs may regulate the expression of up to one-third of all human genes, a striking figure given that miRNAs were unknown 20 years ago. These discoveries hint at a large, diverse population of RNA molecules that help regulate gene expression.

The Initiation of Translation

The process of translation offers additional opportunities for regulation. Among the molecules involved in translation are many regulatory proteins. Red blood cells, for instance, have a protein that prevents the translation of hemoglobin mRNA unless the cell has a supply of heme, an iron-containing chemical group essential for hemoglobin function.

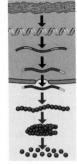

Protein Activation and Breakdown

The final opportunities for regulating gene expression occur after translation. Post-translational control mechanisms in eukaryotes often involve cutting polypeptides into smaller, active final products. The hormone insulin, for example, is synthesized as one long, inactive polypeptide (**Figure 11.7**). After translation, an enzyme removes an interior section, leaving two shorter chains that constitute the active insulin molecule.

Another control mechanism operating after translation is the selective breakdown of proteins. Some proteins that trigger metabolic changes in cells are broken down within a few minutes or hours. This regulation allows a cell to adjust the kinds and amounts of its proteins in response to changes in its environment. ☑

▼ Figure 11.6 **Alternative RNA splicing: producing two different mRNAs from the same gene.** Two different cells can use a given DNA gene to synthesize somewhat different mRNAs and proteins. The cells might be from different tissues or the same type of cell at different times in the organism's life. In this example, one mRNA has ended up with exon 3 (brown) and the other with exon 4 (green). These two mRNAs can then be translated into different but related proteins.

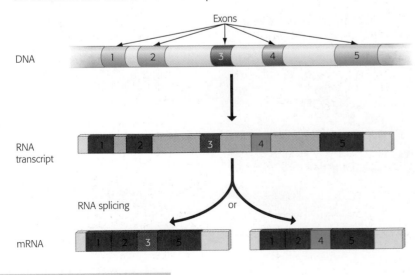

▼ Figure 11.7 **The formation of an active insulin molecule.** Only in its final form does insulin act as a hormone.

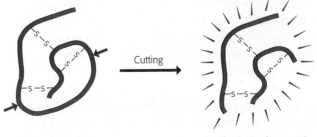

Initial polypeptide   Insulin (active hormone)

Cell Signaling

So far, we have considered gene regulation only within a single cell. In a multicellular organism, the process can cross cell boundaries. A cell can produce and secrete chemicals, such as hormones, that affect gene regulation in another cell.

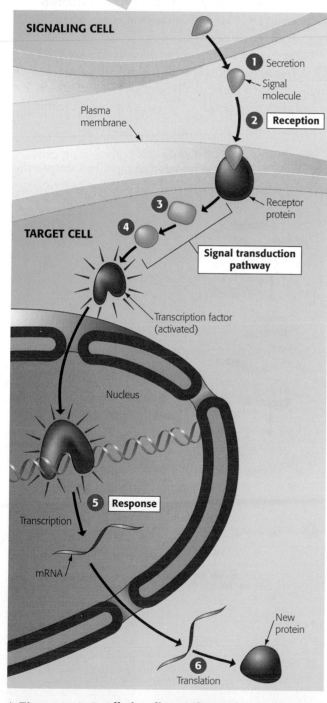

▲ Figure 11.8 **A cell-signaling pathway that turns on a gene.** The coordination of cellular activities in a multicellular organism depends on cell-to-cell signaling that helps regulate genes.

As you saw in Figure 5.19, a signal molecule can act by binding to a receptor protein and initiating a signal transduction pathway, a series of molecular changes that converts a signal to a specific response inside the target cell. **Figure 11.8** shows an example of cell-to-cell signaling in which the target cell's response is the transcription (turning on) of a gene. **1** First, the signaling cell secretes the signal molecule (⬤). **2** This molecule binds to a specific receptor protein (⬤) embedded in the target cell's plasma membrane. **3** The binding activates a signal transduction pathway consisting of a series of relay proteins (green) within the target cell. Each relay molecule activates another. **4** The last relay molecule in the pathway activates a transcription factor (⬤) that **5** triggers the transcription of a specific gene. **6** Translation of the mRNA produces a protein. ✓

Homeotic Genes

Cell-to-cell signaling and the control of gene expression are most important during early embryonic development, when a zygote develops into a multicellular organism. Master control genes called **homeotic genes** regulate groups of other genes that determine what body parts will develop in which locations. For example, one set of homeotic genes in fruit flies instructs cells in the embryonic head and thorax (midbody) to form antennae and legs, respectively. Elsewhere, these homeotic genes remain turned off, while others are turned on. Mutations in homeotic genes can produce bizarre effects. For example, fruit flies with mutations in homeotic genes may have extra sets of wings or have legs growing from their head (**Figure 11.9**).

► Figure 11.9 **The effect of homeotic genes.** The strange mutant fruit flies shown at the bottom result from mutations in homeotic (master control) genes.

Normal fruit fly

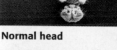

Normal head

Mutant fly with extra wings

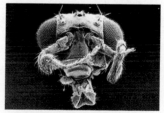

Mutant fly with extra legs growing from head

☑**CHECKPOINT**

How can one homeotic
gene mutation dramatically
affect an organism's physical
appearance?

*Answer: Because homeotic genes
control many other genes, a single
change can affect the expression of
many of the proteins that control
appearance.*

One of the exciting biological discoveries in recent years is that similar homeotic genes help direct embryonic development in nearly every eukaryotic organism examined so far, including yeasts, plants, earthworms, frogs, chickens, mice, and humans. **Figure 11.10** highlights some striking similarities in the chromosomal locations and developmental roles of homeotic genes in two quite different animals: a fruit fly and a mouse. The colored segments represent homeotic genes that are very similar in both animals. Notice that the fruit fly and mouse chromosomes have the same order of genes, corresponding to analogous body regions in both animals. These similarities suggest that the original version of these homeotic genes arose very early in the history of life and that the genes have remained remarkably unchanged over eons of animal evolution. ☑

DNA Microarrays: Visualizing Gene Expression

Scientists who study gene regulation can use a DNA microarray to visualize patterns of gene expression. A **DNA microarray** is a glass slide with thousands of single-stranded DNA fragments attached to wells in a tightly spaced array (grid). Each DNA fragment is obtained from a particular gene; a single microarray thus carries DNA from thousands of genes, perhaps even all the genes of an organism.

Figure 11.11 outlines how microarrays are used.
❶ A researcher collects all of the mRNA transcribed from genes in a particular type of cell. This mRNA is mixed with reverse transcriptase, a viral enzyme that

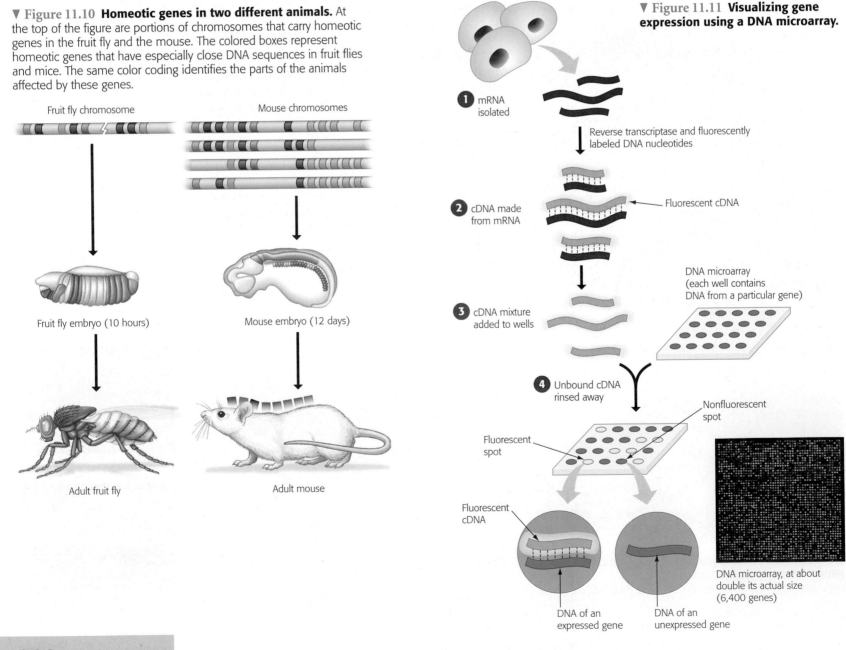

▼ Figure 11.10 **Homeotic genes in two different animals.** At the top of the figure are portions of chromosomes that carry homeotic genes in the fruit fly and the mouse. The colored boxes represent homeotic genes that have especially close DNA sequences in fruit flies and mice. The same color coding identifies the parts of the animals affected by these genes.

Fruit fly chromosome

Mouse chromosomes

Fruit fly embryo (10 hours)

Mouse embryo (12 days)

Adult fruit fly

Adult mouse

▼ Figure 11.11 **Visualizing gene expression using a DNA microarray.**

❶ mRNA isolated

Reverse transcriptase and fluorescently labeled DNA nucleotides

❷ cDNA made from mRNA

Fluorescent cDNA

❸ cDNA mixture added to wells

DNA microarray (each well contains DNA from a particular gene)

❹ Unbound cDNA rinsed away

Nonfluorescent spot

Fluorescent spot

Fluorescent cDNA

DNA of an expressed gene

DNA of an unexpressed gene

DNA microarray, at about double its actual size (6,400 genes)

2 catalyzes synthesis of DNA that is complementary to each mRNA sequence. This **complementary DNA (cDNA)** is synthesized using nucleotides that have been modified to fluoresce (glow). The fluorescent cDNA collection thus represents all of the genes being actively transcribed in the cell. **3** A small amount of the fluorescently labeled cDNA mixture is added to the DNA fragment in each well of the microarray. If a molecule in the cDNA mixture is complementary to a DNA fragment at a particular location on the microarray, the cDNA molecule binds to it, becoming fixed there. **4** After unbound cDNA is rinsed away, the remaining cDNA glows in the microarray. The pattern of glowing spots enables the researcher to determine which genes were being transcribed in the starting cells. Researchers can thus learn which genes are active in different tissues or in tissues from individuals in different states of health. ✓

Cloning Plants and Animals

Now that we have examined how gene expression is regulated, we will devote the rest of this chapter to how gene regulation affects two important processes: cloning and cancer.

The Genetic Potential of Cells

One of the most important "take-home lessons" from this chapter is that differentiated cells express only a small percentage of their genes. So how do we know that all the genes are still present? And if all the genes are still there, do differentiated cells retain the potential to express them?

One way to approach these questions is to see if a differentiated cell can be dedifferentiated and stimulated to generate a whole new organism. This sort of genetic potential can be readily demonstrated in plants. For example, if you have ever grown a plant from a small cutting, you've seen evidence that a differentiated plant cell can undergo cell division and give rise to all the tissues of an entire plant. On a larger scale, the technique described in **Figure 11.12** can be used to produce hundreds or thousands of genetically identical organisms—clones—from the cells of a single plant.

Plant cloning is now used extensively in agriculture. For some plants, such as orchids, cloning is the only commercially practical means of reproducing plants. In other cases, cloning has been used to reproduce a plant with specific desirable traits, such as high fruit yield or resistance to disease. The success of plant cloning shows that cell differentiation in plants does not result in irreversible changes in the DNA.

A similar, naturally occurring process in animals is **regeneration**, the regrowth of lost body parts. When a salamander loses a leg, for example, certain cells in the leg stump reverse their differentiated state, divide, and then differentiate again to give rise to a new leg. Many other animals, especially among the invertebrates, can regenerate lost parts, and isolated pieces of a few relatively simple animals can dedifferentiate and then develop into an entirely new organism (see Figure 8.1).

▶ **Figure 11.12 Test-tube cloning of a carrot plant.** A single cell removed from the root of a carrot plant and placed in growth medium may begin dividing and eventually grow into an adult plant. The new plant is a genetic duplicate of the parent plant. This process proves that mature plant cells can reverse their differentiation and develop into all the specialized cells of an adult plant.

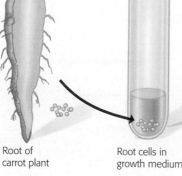

Single
cell

Root of
carrot plant

Root cells in
growth medium

Cell division
in culture

Young plant

Adult plant

Reproductive Cloning of Animals

Animal cloning is achieved through a procedure called **nuclear transplantation** (Figure 11.13). First performed in the 1950s on frog embryos, nuclear transplantation involves replacing the nucleus of an egg cell or a zygote with a nucleus removed from an adult body cell. The recipient cell may then begin to divide. Repeated cell divisions form a blastocyst, a hollow ball of about 100 cells that is an early stage in normal animal development. At this point, the blastocyst may be used for different purposes, as indicated by the two branches in Figure 11.13.

If the animal to be cloned is a mammal, further development requires implanting the blastocyst into the uterus of a surrogate mother (Figure 11.13, upper branch). The resulting animal will be genetically identical to the donor of the nucleus—a "clone" of the donor. This type of cloning is called **reproductive cloning** because it results in the birth of a new animal.

Scottish researcher Ian Wilmut and his colleagues used reproductive cloning to produce the celebrated sheep Dolly in 1997. The researchers used an electric shock to fuse specially treated adult sheep udder cells with 277 eggs from which they had removed the nuclei. After several days of growth, 29 of the resulting embryos were implanted in the uteruses of surrogate mothers. One of the embryos developed into the world-famous Dolly. As expected, Dolly resembled her genetic parent, the nucleus donor, not the egg donor or the surrogate mother.

Practical Applications of Reproductive Cloning

In the years since Dolly's landmark birth, researchers have cloned over a dozen other species of mammals, including mice, horses, dogs, mules, cows, pigs, rabbits, ferrets, and cats (Figure 11.14a). Why would anyone want to do this? In agriculture, farm animals with specific sets of desirable traits might be cloned to produce herds with these traits. In research, genetically identical animals may provide perfect "control animals" for experiments. The pharmaceutical industry is experimenting with cloning animals for potential medical use. For example, the pigs in Figure 11.14b are clones that lack a gene for a protein that can cause immune system rejection in humans. Organs from such pigs may one day be used in human patients who need transplants.

Perhaps the most intriguing use of reproductive cloning is to restock populations of endangered animals. Among the rare animals that have been cloned are a wild mouflon (a small European sheep), a banteng (a Javanese cow), a gaur (an Asian ox), and gray wolves (Figure 11.14c). The 2003 cloning of a banteng, whose numbers have dwindled to just a few in the wild, was a remarkable case. Using frozen cells from a zoo-raised

▼ Figure 11.13 **Cloning by nuclear transplantation.** In nuclear transplantation, an adult somatic nucleus is injected into a nucleus-free egg cell. The resulting embryo may then be used to produce a new organism (reproductive cloning, shown in the upper branch) or to provide stem cells (therapeutic cloning, lower branch). The lamb in the photograph is the famous Dolly, shown with her surrogate mother.

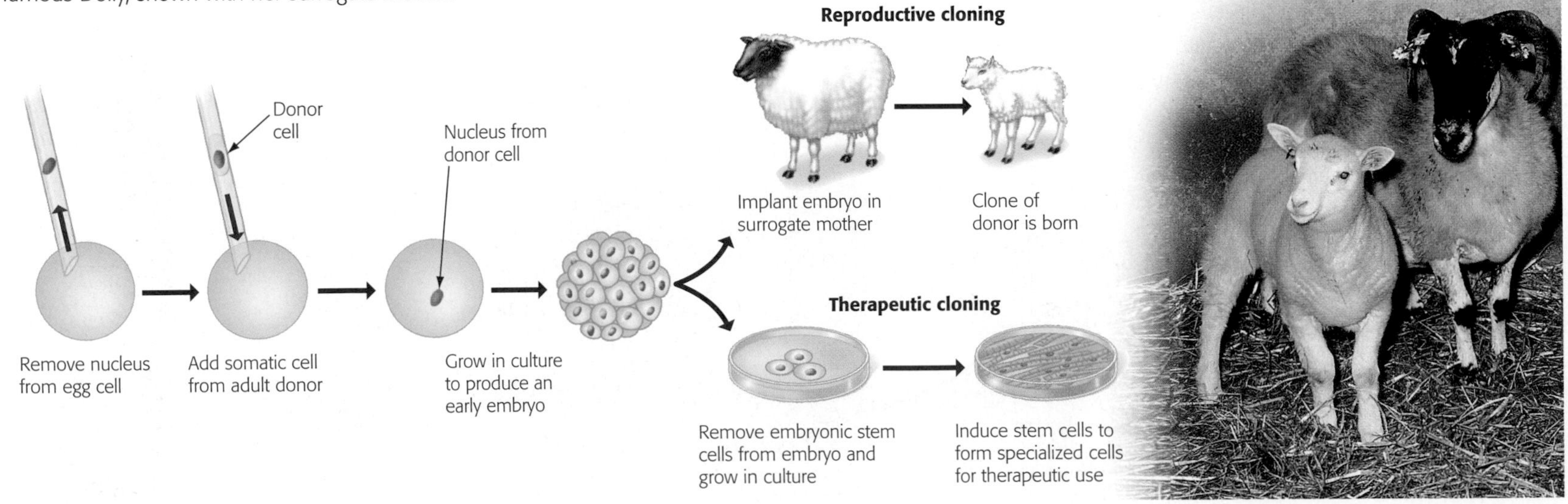

banteng that had died 23 years prior, scientists transplanted nuclei from the frozen cells into nucleus-free eggs from dairy cows. The resulting embryos were implanted into surrogate cows, leading to the birth of a healthy baby banteng. This success shows that it is possible to produce a baby even when a female of the donor species is unavailable. Scientists may someday be able to use similar cross-species methods to clone an animal from a recently extinct species.

The use of cloning to repopulate endangered species holds tremendous promise. However, cloning may also create new problems. Conservationists object that cloning may detract from efforts to preserve natural habitats. They correctly point out that cloning does not increase genetic diversity and is therefore not as beneficial to endangered species as natural reproduction. Cloned animals are also less healthy than those arising from a fertilized egg. In 2003, Dolly was euthanized after suffering complications from a lung disease that is normally seen only in much older sheep. Other cloned animals have exhibited defects such as susceptibility to obesity, pneumonia, liver failure, and premature death.

Human Cloning

The cloning of various mammals has heightened speculation that humans could be cloned. Critics point out practical and ethical objections to human cloning. Practically, cloning of mammals is extremely difficult and inefficient. Only a small percentage of cloned embryos develop normally, and they appear less healthy than naturally born kin. Ethically, creating human embryos for cloning raises troubling questions about the status of the human blastocyst. Consensus on this topic is unlikely anytime soon. Meanwhile, the research and the debate continue. ☑

☑ CHECKPOINT

Suppose a nucleus from an adult body cell of a black mouse is injected into an egg removed from a white mouse, and then the embryo is implanted into a brown mouse. What would be the color of the resulting cloned mice?

Answer: black, the color of the nucleus donor

▼ Figure 11.14 **Reproductive cloning of mammals.**

(a) The first cloned cat. CC ("Copy Cat") and her lone parent (left).

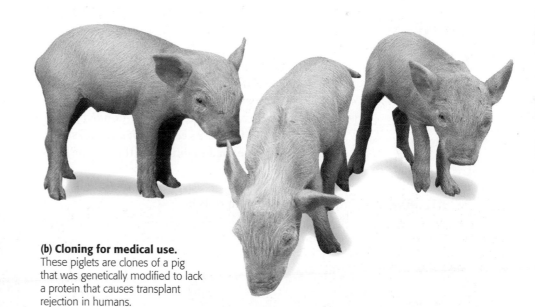

(b) Cloning for medical use. These piglets are clones of a pig that was genetically modified to lack a protein that causes transplant rejection in humans.

(c) Clones of endangered animals

Mouflon calf with mother   Banteng   Gaur   Gray wolf

Therapeutic Cloning and Stem Cells

The lower branch of Figure 11.13 (p. 208) depicts an alternative cloning outcome called therapeutic cloning. The purpose of **therapeutic cloning** is not to produce a viable organism but to produce embryonic stem cells.

Embryonic Stem Cells

In mammals, **embryonic stem cells (ES cells)** are derived from blastocysts. During development, embryonic stem cells in the blastocyst differentiate, giving rise to all the specialized cells in the body. When removed from an early embryo and grown in laboratory culture, embryonic stem cells can divide indefinitely. The right conditions—such as the presence of certain growth-stimulating proteins—can induce changes in gene expression that cause the cells to develop into a particular cell type **(Figure 11.15)**. If scientists can discover the right conditions, they may be able to grow cells for the repair of injured or diseased organs. The use of embryonic stem cells in therapeutic cloning is controversial, however, because the removal of ES cells destroys the embryo.

Adult Stem Cells

Embryonic stem cells are not the only stem cells available to researchers. **Adult stem cells** are cells in adult tissues that generate replacements for nondividing differentiated cells. Unlike ES cells, adult stem cells are partway along the road to differentiation; in the body,

they usually give rise to only a few related types of specialized cells. For example, stem cells in bone marrow can generate the different kinds of blood cells (see Figure 11.15). Adult stem cells are much more difficult to grow in culture than ES cells, and they are not as versatile, but researchers have had some success. Because no embryonic tissue is involved in their harvest, adult stem cells may provide an ethically less problematic route for human tissue and organ replacement than ES cells. However, some researchers think that ES cells are the only cells likely to lead to groundbreaking advances in human health.

Umbilical Cord Blood Banking

Another source of stem cells is blood collected from the umbilical cord and placenta at birth. Such stem cells appear to be partially differentiated, less so than adult stem cells, more so than embryonic stem cells. To obtain these cells, a physician inserts a needle into the umbilical cord and extracts 1/4 to 1/2 cup of blood **(Figure 11.16)**. The cells are then frozen and kept in a blood bank until needed for later medical treatment. In 2005, doctors reported that an infusion of umbilical cord blood stem cells from a compatible (but unrelated) donor appeared to cure some babies of Krabbe's disease, a usually fatal inherited disorder of the nervous system. Other people have received cord blood as a treatment for leukemia. To date, however, most attempts at umbilical cord blood therapy have not been successful. At present, the American Academy of Pediatrics recommends cord blood banking only for babies born into families with a known genetic risk. So far, the promise of cord blood banking vastly exceeds the accomplishments. ☑

✓ CHECKPOINT

How do the results of reproductive cloning and therapeutic cloning differ?

Answer: Reproductive cloning results in the production of a live individual; therapeutic cloning produces stem cells.

▼ Figure 11.15 **Differentiation of embryonic stem cells in culture.** Scientists hope to discover growth conditions that will stimulate cultured stem cells to differentiate into specialized cells.

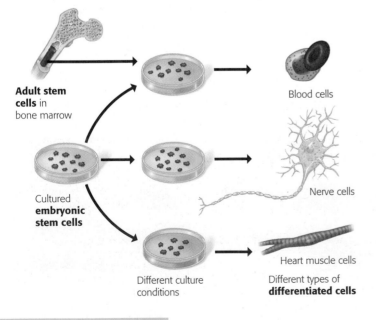

Adult stem cells in bone marrow

Cultured **embryonic stem cells**

Blood cells

Nerve cells

Heart muscle cells

Different culture conditions

Different types of **differentiated cells**

▼ Figure 11.16 **Umbilical cord blood banking.** Just after birth, a doctor may collect blood from a newborn's umbilical cord. The umbilical cord blood (inset), rich in stem cells, is stored for possible future use.

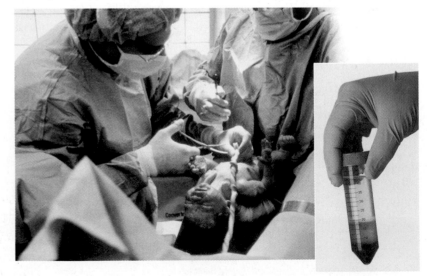

The Genetic Basis of Cancer

Chapter 8 introduced cancer as a variety of diseases in which cells escape from the control mechanisms that normally limit their growth and division. This escape involves changes in gene expression.

Genes That Cause Cancer

One of the earliest clues to the role of genes in cancer was the discovery in 1911 of a virus that causes cancer in chickens. Recall that viruses are simply "genes in a box," molecules of DNA or RNA surrounded by protein and in some cases a membrane. Viruses that cause cancer can become permanent residents in host cells by inserting their nucleic acid into the DNA of host chromosomes. Researchers have identified a number of viruses that harbor cancer-causing genes. When inserted into a host cell, these genes can make the cell cancerous. A gene that causes cancer is called an **oncogene** ("tumor gene").

Oncogenes and Tumor-Suppressor Genes

In 1976, American molecular biologists J. Michael Bishop, Harold Varmus, and their colleagues made a startling discovery. They found that the virus that causes cancer in chickens contains an oncogene that is an altered version of a normal chicken gene. Subsequent research has shown that the chromosomes of many animals, including humans, contain genes that can be converted to oncogenes. A normal gene with the potential to become an oncogene is called a **proto-oncogene**. (These terms can be confusing, so they bear repeating: a *proto-oncogene* is a normal gene that, if changed, can become a cancer-causing *oncogene*.) A cell can acquire an oncogene from a virus or from the conversion of one of its own proto-oncogenes.

How can a change in a gene cause cancer? Searching for the normal roles of proto-oncogenes in the cell, researchers found that many of these genes code for **growth factors**—proteins that stimulate cell division—or for other proteins that affect the cell cycle. When all these proteins are functioning normally, in the right amounts at the right times, they help keep the rate of cell division at an appropriate level. When they malfunction—if a growth factor becomes hyperactive, for example—cancer (uncontrolled cell growth) may result.

For a proto-oncogene to become an oncogene, a mutation must occur in the cell's DNA. **Figure 11.17** illustrates three kinds of changes in DNA that can produce active oncogenes. In all three cases, abnormal gene expression stimulates the cell to divide excessively.

Changes in genes whose products inhibit cell division are also involved in cancer. These genes are called **tumor-suppressor genes** because the proteins they encode normally help prevent uncontrolled cell growth (**Figure 11.18**). Any mutation that keeps a normal growth-inhibiting protein from being made or from functioning may contribute to development of cancer. Researchers have identified many mutations in both tumor-suppressor and growth factor genes that are associated with cancer, as we'll discuss next.

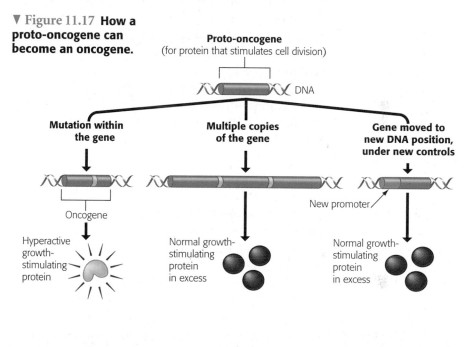

▼ Figure 11.17 **How a proto-oncogene can become an oncogene.**

Proto-oncogene
(for protein that stimulates cell division)

DNA

Mutation within the gene — Oncogene — Hyperactive growth-stimulating protein

Multiple copies of the gene — Normal growth-stimulating protein in excess

Gene moved to new DNA position, under new controls — New promoter — Normal growth-stimulating protein in excess

▼ Figure 11.18 **Tumor-suppressor genes.**

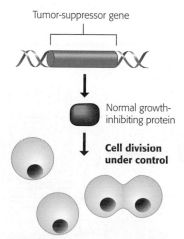

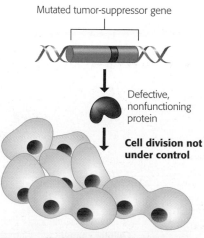

Tumor-suppressor gene — Normal growth-inhibiting protein — **Cell division under control**

Mutated tumor-suppressor gene — Defective, nonfunctioning protein — **Cell division not under control**

(a) Normal cell growth. A tumor-suppressor gene normally codes for a protein that inhibits cell growth and division. In this way, the gene helps prevent cancerous tumors from arising.

(b) Uncontrolled cell growth (cancer). When a mutation in a tumor-suppressor gene makes its protein defective, cells that are usually under the control of the normal protein may divide excessively, forming a tumor.

Can Cancer Therapy Be Personalized?

Medical researchers have made many **observations** of specific mutations that can lead to cancer. But it is reasonable to **question** whether this information can be used to help patients. A research team from the Dana Farber Cancer Institute in Boston formed the **hypothesis** that DNA-sequencing technology can be used to test tumors and identify which cancer-causing mutations they carry. They made the **prediction** that such information could be used to design customized therapies for patients based on the particular mutations involved.

Their **experiment** involved screening for 238 possible mutations in 1,000 human tumors from 18 different body tissues. A subset of their **results**, presented in Table 11.1, shows that no single mutation is present in every tumor and that each tumor involves different mutations. (In this table, a ✓ indicates that a given mutation was found in at least one tumor of that type.) These results reveal that it is possible to cheaply and accurately determine which mutations are present in a given cancer patient. Indeed, there are now a number of cancer types (including leukemia, breast cancer, and lung cancer) for which personalized therapies—based on genetic screening—are available. Personalized cancer treatment remains an active area of research.

Table 11.1	Cancer-Causing Mutations			
Tumor location	*A*	*B*	*C*	*D*
Pancreas		✓		
Ovary	✓	✓		✓
Lung	✓		✓	✓
Colon	✓	✓	✓	✓
Breast	✓			✓

The Progression of a Cancer

Over 150,000 Americans will be stricken by cancer of the colon (the main part of the large intestine) or the rectum this year. One of the best-understood types of human cancer, colon cancer illustrates an important principle about how cancer develops: More than one mutation is needed to produce a full-fledged cancer cell. As in many cancers, the development of colon cancer is a gradual process.

As shown in Figure 11.19, ❶ colon cancer begins when an oncogene arises or is activated through mutation, causing unusually frequent division of normal-looking cells in the colon lining. ❷ Later, additional DNA mutations (such as the inactivation of a tumor-suppressor gene) cause the growth of a small benign tumor (polyp) in the colon wall. ❸ Further mutations eventually lead to formation of a malignant tumor—a tumor that has the potential to metastasize (spread).

▼ Figure 11.19 **Stepwise development of a typical colon cancer.**

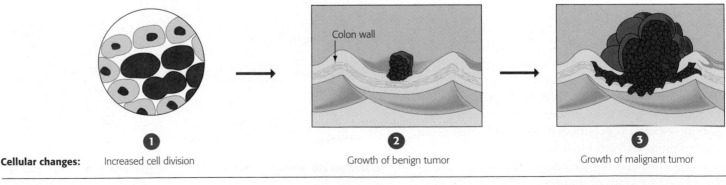

Cellular changes:	Increased cell division	Growth of benign tumor	Growth of malignant tumor
DNA changes:	Oncogene activated	Tumor-suppressor gene inactivated	Second tumor-suppressor gene inactivated

The development of a malignant tumor is accompanied by a gradual accumulation of mutations that convert proto-oncogenes to oncogenes and knock out tumor-suppressor genes (Figure 11.20). The requirement for several DNA mutations—usually four or more—explains why cancers can take a long time to develop. This requirement may also help explain why the incidence of cancer increases greatly with age; the longer we live, the more likely we are to accumulate mutations that cause cancer.

"Inherited" Cancer

Most mutations that lead to cancer arise in the organ where the cancer starts—the colon, for example. Because these mutations do not affect the cells that give rise to eggs or sperm, they are not passed from parent to child. Sometimes, however, a cancer-causing mutation occurs in a cell that gives rise to gametes and is therefore passed on from generation to generation. Such mutations predispose the people who inherit them to developing cancer. Such cancer is called familial or inherited. But even familial cancers don't appear unless the person acquires additional mutations in the susceptible tissue.

One well-studied inherited cancer gene is *BRCA1* (pronounced "braca-1"), whose normal allele encodes a tumor-suppressor protein. Certain alleles of *BRCA1* are associated with breast cancer, a disease that strikes one out of every ten American women (Figure 11.21). Some *BRCA1* mutations put a woman at high risk for both breast cancer and ovarian cancer—a more than 80% risk

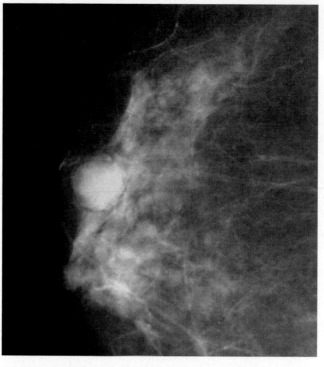

▲ Figure 11.21 **Breast cancer.** In this mammogram, normal breast tissue appears as gray and white, while the tumor is highlighted in yellow and red.

of developing cancer during her lifetime. In recent years, tests for *BRCA1* mutations have become available. Unfortunately, these tests are of limited use because surgical removal of the breasts and/or ovaries is the only preventive option currently available to women who carry the mutant genes. ☑

▼ Figure 11.20 **Accumulation of mutations in the development of a cancer cell.**
Mutations leading to cancer accumulate in a lineage of cells. In this figure, colors distinguish the normal cells from cells with one or more mutations, leading to increased cell division and cancer. Once a cancer-promoting mutation occurs (orange band on chromosome), it is passed to all the descendants of the cell carrying it.

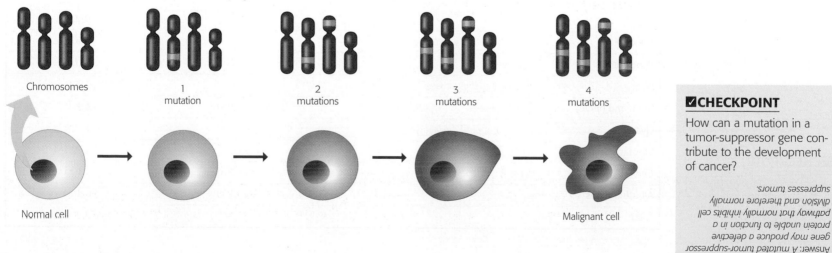

Chromosomes   1 mutation   2 mutations   3 mutations   4 mutations

Normal cell   Malignant cell

☑ **CHECKPOINT**

How can a mutation in a tumor-suppressor gene contribute to the development of cancer?

Answer: A mutated tumor-suppressor gene may produce a defective protein unable to function in a pathway that normally inhibits cell division and therefore normally suppresses tumors.

Cancer Risk and Prevention

Cancer is the second-leading cause of death in most industrialized countries (after heart disease). Death rates due to certain forms of cancer have decreased in recent years, but the overall cancer death rate is still on the rise, currently increasing at about 1% per decade.

Most cancers arise from mutations that are caused by **carcinogens**, cancer-causing agents found in the environment. Mutations often result from decades of exposure to carcinogens. One of the most potent carcinogens is ultraviolet (UV) radiation. Excessive exposure to UV radiation from the sun can cause skin cancer, including a deadly type called melanoma.

The one substance known to cause more cases and types of cancer than any other is tobacco. More people die from lung cancer (over 160,000 Americans in 2008) than from any other form of cancer. Most tobacco-related cancers are due to smoking, but the passive inhalation of secondhand smoke also poses a risk. As Table 11.2 indicates, tobacco use, sometimes in combination with alcohol consumption, causes some types of cancer. In nearly all cases, cigarettes are the main culprit, but smokeless tobacco products (snuff and chewing tobacco) are linked to cancer of the mouth and throat.

Exposure to some of the most lethal carcinogens is often a matter of individual choice: Tobacco use, the consumption of alcohol, and excessive time spent in the sun are all avoidable behaviors that affect cancer risk.

There is increasing evidence that some food choices significantly reduce a person's cancer risk. For instance, eating 20–30 g of plant fiber daily (about twice the amount the average American consumes) while eating less animal fat may help prevent colon cancer. Some studies suggest that certain substances in fruits and vegetables, including vitamins C and E and certain compounds related to vitamin A, may help protect against a variety of cancers. Cabbage and its relatives, such as broccoli and cauliflower, are thought to be especially rich in substances that help prevent cancer, although some of the specific substances have not yet been identified. Determining how diet influences cancer has become an important focus of nutrition research.

The battle against cancer is being waged on many fronts, and there is reason for optimism in the progress being made. It is especially encouraging that we can help reduce our risk of acquiring some of the most common forms of cancer by the choices we make in our daily lives. ☑

Table 11.2	Cancer in the United States (Ranked by Number of Cases)		
Cancer	**Known or Likely Carcinogens or Factors**	**Estimated Cases (2009)**	**Estimated Deaths (2009)**
Lung	Cigarette smoke	219,000	159,000
Breast	Estrogen; possibly dietary fat	194,000	40,600
Prostate	Testosterone; possibly dietary fat	192,000	27,000
Colon and rectum	High dietary fat; low dietary fiber	147,000	49,900
Skin	Ultraviolet light	74,600	11,600
Lymphomas	Viruses (for some types)	74,500	20,800
Bladder	Cigarette smoke	71,000	14,300
Kidney	Cigarette smoke	57,800	13,000
Uterus	Estrogen	53,400	11,800
Leukemias	X-rays; benzene; viruses (for some types)	44,800	21,900
Pancreas	Cigarette smoke	42,500	35,200
Liver	Alcohol; hepatitis viruses	22,600	18,200
Brain and nerve	Trauma; X-rays	22,100	12,900
Ovary	Large number of ovulation cycles	21,600	14,600
Stomach	Table salt; cigarette smoke	21,100	10,600
Cervix	Viruses; cigarette smoke	11,300	4,100
All other types		210,100	96,800
Total		1,479,400	562,300

Source: *Cancer Facts and Figures 2009* (American Cancer Society Inc.)

☑ **CHECKPOINT**

Of all known behavioral factors, which one causes the most cancer cases and deaths?

Answer: *tobacco use*

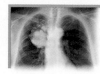

The Evolution of Cancer in the Body

The theory of evolution describes natural selection acting on populations of organisms. Recently, medical researchers have been using an evolutionary perspective to gain insight into the development of tumors, such as the bone tumor shown in **Figure 11.22**. Evolution drives the growth of a tumor—which can be thought of as a population of cancer cells—and also affects how those cells respond to cancer treatments.

Recall from Chapter 1 that there are several assumptions behind Darwin's theory of natural selection. Let's consider how each one can be applied to cancer. First, all evolving populations have the potential to produce more offspring than can be supported by the environment. Cancer cells, with their uncontrolled growth, clearly demonstrate such overproduction. Second, there must be variation among individuals of the population.

Studies of tumor cell DNA, like the one described in the Process of Science section, show genetic variability within tumors. Finally, variations in the population must affect survival and reproductive success. Indeed, the accumulation of mutations in cancer cells renders them less susceptible to normal mechanisms of reproductive control. Mutations that enhance survival of malignant cancer cells are passed on to that cell's descendants. In short, a tumor evolves, much like a population of organisms.

Viewing the progression of cancer through the lens of evolution helps explain why there is no easy "cure" for cancer but may also pave the way for novel therapies. For example, some researchers are attempting to "prime" tumors for treatment by increasing the reproductive success of only those cells that will be susceptible to a chemotherapy drug. Our understanding of cancer, like all other aspects of biology, benefits from an evolutionary perspective.

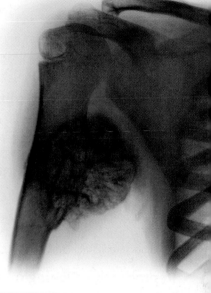

▼ **Figure 11.22 X-ray of shoulder and upper arm, revealing a large bone tumor.**

Chapter Review

SUMMARY OF KEY CONCEPTS

How and Why Genes Are Regulated

Patterns of Gene Expression in Differentiated Cells

The various cell types in a multicellular organism are a consequence of different combinations of genes being turned on and off via gene regulation.

Gene Regulation in Bacteria

An operon is a cluster of genes with related functions together with their promoter and other DNA sequences that control their transcription. The *lac* operon allows *E. coli* to produce enzymes for lactose use only when the sugar is present.

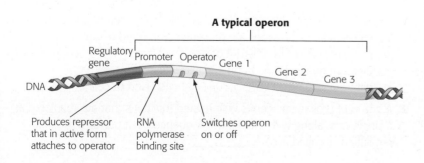

A typical operon

Regulatory gene | Promoter | Operator | Gene 1 | Gene 2 | Gene 3

DNA

Produces repressor that in active form attaches to operator

RNA polymerase binding site

Switches operon on or off

Gene Regulation in Eukaryotic Cells

In the nucleus of eukaryotic cells, there are several possible control points in the pathway of gene expression.

- DNA packing tends to block gene expression, presumably by preventing access of transcription proteins to the DNA. An extreme example is X chromosome inactivation in the cells of female mammals.

- The most important control point in both eukaryotes and prokaryotes is at gene transcription. Various regulatory proteins interact with DNA and with each other to turn the transcription of eukaryotic genes on or off.

- There are also opportunities for the control of eukaryotic gene expression after transcription, when introns are cut out of the RNA and a cap and tail are added to process RNA transcripts into mRNA.

- In the cytoplasm, presence of microRNAs may block the translation of an mRNA, and various proteins may regulate the start of translation.

- Finally, the cell may activate the finished protein in various ways (for instance, by cutting out portions). Eventually, the protein may be selectively broken down.

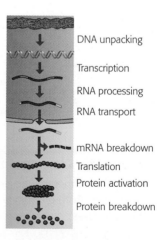

DNA unpacking

Transcription

RNA processing

RNA transport

mRNA breakdown

Translation

Protein activation

Protein breakdown

Cell Signaling

Cell-to-cell signaling is key to the development and functioning of multicellular organisms. Signal transduction pathways convert molecular messages to cell responses, often the transcription of particular genes.

Homeotic Genes

Evidence for the evolutionary importance of gene regulation is apparent in homeotic genes, master genes that regulate other genes that in turn control embryonic development.

DNA Microarrays: Visualizing Gene Expression

DNA microarrays can be used to determine which genes are turned on in a particular cell type.

Cloning Plants and Animals

The Genetic Potential of Cells

Most differentiated cells retain a complete set of genes, so a carrot plant, for example, can be made to grow from a single carrot cell. Under special conditions, animals can also be cloned.

Reproductive Cloning of Animals

Nuclear transplantation is a procedure whereby a donor cell nucleus is inserted into a nucleus-free egg. First demonstrated in frogs in the 1950s, reproductive cloning was used in 1997 to clone a sheep from an adult mammary cell and has since been used to create many other cloned animals.

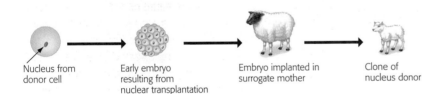

Nucleus from donor cell → Early embryo resulting from nuclear transplantation → Embryo implanted in surrogate mother → Clone of nucleus donor

Therapeutic Cloning and Stem Cells

The purpose of therapeutic cloning is to produce embryonic stem cells for medical uses. Both embryonic and adult stem cells show promise for future therapeutic uses.

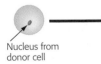

Nucleus from donor cell → Early embryo resulting from nuclear transplantation → Embryonic stem cells in culture → Specialized cells

The Genetic Basis of Cancer

Genes That Cause Cancer

Cancer cells, which divide uncontrollably, can result from mutations in genes whose protein products regulate the cell cycle.

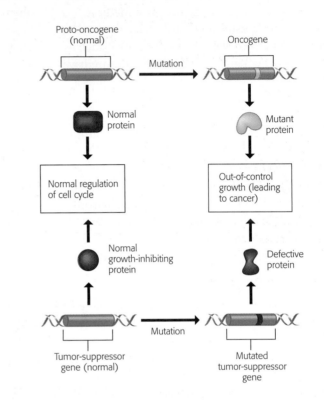

Many proto-oncogenes and tumor-suppressor genes code for proteins active in signal transduction pathways regulating cell division. Mutations of these genes cause malfunction of the pathways. Cancers result from a series of genetic changes in a cell lineage. Researchers have gained insight into the genetic basis of breast cancer by studying families in which a disease-predisposing mutation is inherited.

Cancer Risk and Prevention

Reducing exposure to carcinogens (which induce cancer-causing mutations) and making other healthful lifestyle choices can help reduce cancer risk.

SELF-QUIZ

1. Your bone cells, muscle cells, and skin cells look different because
   a. different kinds of genes are present in each kind of cell.
   b. they are present in different organs.
   c. different genes are active in each kind of cell.
   d. different mutations have occurred in each kind of cell.

2. A group of prokaryotic genes with related functions that are regulated as a single unit, along with the control sequences that perform this regulation, is called a(n) _____.

3. The regulation of gene expression must be more complex in multicellular eukaryotes than in prokaryotes because
  a. eukaryotic cells are much larger.
  b. in a multicellular eukaryote, different cells are specialized for different functions.
  c. prokaryotes are restricted to stable environments.
  d. eukaryotes have fewer genes, so each gene must do several jobs.

4. A eukaryotic gene was inserted into the DNA of a bacterium. The bacterium then transcribed this gene into mRNA and translated the mRNA into protein. The protein produced was useless and contained many more amino acids than the protein made by the eukaryotic cell. Why?
  a. The mRNA was not spliced as it is in eukaryotes.
  b. Eukaryotes and prokaryotes use different genetic codes.
  c. Repressor proteins interfered with transcription and translation.
  d. Ribosomes were not able to bind to tRNA.

5. How does dense packing of DNA in chromosomes prevent gene expression?

6. What evidence demonstrates that differentiated cells in a plant or animal retain their full genetic potential?

7. The most common procedure for cloning an animal is _____.

8. What is learned from a DNA microarray?

9. Which of the following is a valid difference between embryonic stem cells and the stem cells found in adult tissues?
  a. In laboratory culture, only adult stem cells are immortal.
  b. In nature, only embryonic stem cells give rise to all the different types of cells in the organism.
  c. Only adult stem cells can be made to differentiate in the laboratory.
  d. Only embryonic stem cells are in every tissue of the adult body.

10. Name three potential sources of stem cells.

11. What is the difference between oncogenes and proto-oncogenes? How can one turn into the other? What function do proto-oncogenes serve?

12. A mutation in a single gene may cause a major change in the body of a fruit fly, such as an extra pair of legs or wings. Yet it takes many genes to produce a wing or leg. How can a change in just one gene cause such a big change in the body? What are such genes called?

Answers to the Self-Quiz questions can be found in Appendix D.

THE PROCESS OF SCIENCE

13. Study the depiction of the *lac* operon in Figure 11.2. Normally, the genes are turned off when lactose is not present. Lactose activates the genes, which code for enzymes that enable the cell to use lactose. Mutations can alter the function of this operon; in fact, the effects of various mutations enabled Jacob and Monod to figure out how the operon works. Predict how the following mutations would affect the function of the operon in the presence and absence of lactose:
  a. mutation of regulatory gene; repressor will not bind to lactose
  b. mutation of operator; repressor will not bind to operator
  c. mutation of regulatory gene; repressor will not bind to operator
  d. mutation of promoter; RNA polymerase will not attach to promoter

14. The human body has a far greater variety of proteins than genes, a fact that seems to highlight the importance of alternative RNA splicing, which allows several different mRNAs to be made from a single gene. Suppose you have samples of two types of adult cells from one person. Design an experiment using microarrays to determine whether or not the different gene expression is due to alternative RNA splicing.

15. Because a cat must have both orange and non-orange alleles to be tortoiseshell (see Figure 11.4), we would expect only female cats, which have two X chromosomes, to be tortoiseshell. Normal male cats (XY) can carry only one of the two alleles. Male tortoiseshell cats are rare and usually sterile. What might you guess their genotype to be?

16. Design a DNA microarray experiment that measures the difference in gene expression between normal colon cells and cells from a colon tumor.

BIOLOGY AND SOCIETY

17. A chemical called dioxin is produced as a by-product of certain chemical manufacturing processes. Trace amounts of this substance were present in Agent Orange, a defoliant sprayed on vegetation during the Vietnam War. There has been a continuing controversy over its effects on soldiers exposed to Agent Orange during the war. Animal tests have suggested that dioxin can cause cancer, liver and thymus damage, immune system suppression, and birth defects; at high dosage it can be lethal. But such animal tests are inconclusive; a hamster is not affected by a dose that can kill a much larger guinea pig, for example. Researchers have discovered that dioxin enters a cell and binds to a protein that in turn attaches to the cell's DNA. How might this mechanism help explain the variety of dioxin's effects on different body systems and in different animals? How might you determine whether a particular individual became ill as a result of exposure to dioxin? Do you think this information is relevant in the lawsuits of soldiers suing over exposure to Agent Orange? Why or why not?

18. There are genetic tests available for several types of "inherited cancer." The results from these tests cannot usually predict that someone will get cancer within a particular amount of time. Rather, they indicate only that a person has an increased risk of developing cancer. For many of these cancers, lifestyle changes cannot decrease a person's risk. Therefore, some people consider the tests useless. If your close family had a history of cancer and a test were available, would you want to get screened? Why or why not? What would you do with this information? If a sibling decided to get screened, explain whether you would want to know the results.

12 DNA Technology

A DNA profile.
Even minuscule bits of
evidence can provide a
DNA profile.

BIOLOGY AND SOCIETY: DNA Profiling

DNA, Guilt, and Innocence

Early in the morning of May 3, 1992, Christine Jackson, a 3-year-old girl sleeping in her Mississippi home, was abducted, raped, murdered, and thrown into a creek. A tiny semen sample was recovered at the crime scene. Police arrested the mother's boyfriend, who had been babysitting the girl that night. He was convicted of the horrifying crime and sentenced to death.

The cells of every person (except identical twins) contain unique DNA. DNA profiling is the analysis of DNA samples to determine whether they come from the same individual. Although the crime scene semen sample was insufficient for DNA testing in 1992, new methods allowed a DNA profile to be determined in 2001. The results were conclusive: The boyfriend did not leave his semen at the crime scene.

DNA profiling can provide evidence of guilt as well as innocence. Several years later, the DNA profile obtained from the crime scene was matched to a man who confessed to the crime and to the nearly identical 1990 murder of 3-year-old Courtney Smith, for which a third man was serving jail time. After years of legal maneuvers, the two wrongly convicted men were freed in 2008, exonerated based on the DNA evidence. They were assisted by a nonprofit legal organization called the Innocence Project. To date, this organization has used DNA profiling data to exonerate 227 prisoners, including 17 death row inmates.

Because of its unbiased nature, DNA profiling has rapidly transformed crime investigation. Beyond the courtroom, DNA technology has led to some of the most remarkable scientific advances in recent years: Corn has been genetically modified to produce its own insecticide; human genes are being compared with those of other animals; and significant advances have been made toward curing fatal genetic diseases. This chapter will describe these and other uses of DNA technology and explain how various techniques are performed. We'll also examine some social, legal, and ethical issues that these new technologies raise.

DNA Technology

Recombinant DNA Technology

You may think of **biotechnology**, the manipulation of organisms or their components to make useful products, as a modern field, but it actually dates back to the dawn of civilization. Consider such ancient practices as the use of yeast to make bread and the selective breeding of livestock. But when people use the term *biotechnology* today, they are usually referring to **DNA technology**, methods for studying and manipulating genetic material. Using these techniques, scientists can modify specific genes and move them between organisms as different as bacteria, plants, and animals.

In the 1970s, the field of biotechnology exploded with the invention of methods for making recombinant DNA in the laboratory. **Recombinant DNA** results when scientists combine nucleotide sequences (pieces of DNA) from two different sources—often from different species—to form a single DNA molecule. Recombinant DNA technology is widely used in **genetic engineering**, the direct manipulation of genes for practical purposes. Scientists have genetically engineered bacteria to mass-produce a variety of useful chemicals, from cancer drugs to pesticides. Scientists have also transferred genes from bacteria to plants and from one animal species to another **(Figure 12.1)**. ☑

☑ CHECKPOINT

What is biotechnology? What is recombinant DNA?

Answer: the manipulation of organisms or their parts to produce a useful product; a molecule containing DNA from two different sources, often different species

▼ **Figure 12.1 Glowing fish.** Genetic engineers produced glowing fish by transferring a gene for a fluorescent protein originally obtained from jellies ("jellyfish").

Applications: From Humulin to Foods to "Pharm" Animals

By transferring the gene for a desired protein into a bacterium, yeast, or other kind of cell that is easy to grow in culture, scientists can produce large quantities of proteins that are present naturally in only small amounts. In this section, you'll learn about some applications of recombinant DNA technology.

Making Humulin

Humulin is human insulin produced by genetically modified bacteria **(Figure 12.2)**. In humans, insulin is a protein normally made by the pancreas. Functioning as a hormone, insulin helps regulate the level of glucose in the blood. If the body fails to produce enough insulin, the result is type 1 diabetes. There is no cure, so people with this disease must inject themselves with daily doses of insulin for the rest of their lives.

Since human insulin is not readily available, diabetes was historically treated using cow and pig insulins. This treatment was not without problems, however. Pig and cow insulins can cause allergic reactions because their chemical structures differ slightly from that of human insulin. In addition, by the 1970s, the supply of beef and pork pancreas available for insulin extraction could not keep up with the demand. A new source was needed.

In 1978, scientists working at the biotechnology company Genentech chemically synthesized two genes, one for each of the polypeptides of the active form of human insulin (see Figure 11.7). Since the amino acid sequences of the two insulin polypeptides were already known, it was easy to use the genetic code (see Figure 10.11) to determine nucleotide sequences that would encode for them. Researchers synthesized DNA fragments and linked them

► **Figure 12.2 Humulin, human insulin produced by genetically modified bacteria.**

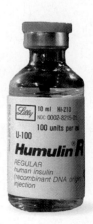

to form the insulin genes. In 1979, they succeeded in inserting these artificial genes into *E. coli* host cells. Under proper growing conditions, these bacteria cranked out large quantities of the human protein.

In 1982, Humulin hit the market as the world's first genetically engineered pharmaceutical product. Today, it is produced in gigantic fermentation vats filled with a liquid culture of bacteria; the vats operate around the clock. Each day, more than 4 million people with diabetes use the insulin collected, purified, and packaged at such facilities (Figure 12.3).

Insulin is just one of many human proteins produced by genetically modified bacteria. Another example is human growth hormone (HGH). Abnormally low levels of this hormone during childhood and adolescence can cause a form of dwarfism. Because growth hormones from other animals are not effective in humans, HGH was an early target of genetic engineers. Before genetically engineered HGH became available in 1985, children with an HGH deficiency could only be treated with scarce and expensive supplies of HGH obtained from human cadavers.

Besides bacteria, yeast and mammalian cells can also be used to produce medically valuable human proteins. For example, genetically modified mammalian cells growing in laboratory cultures are currently used to produce a hormone called EPO that stimulates production of red blood cells. EPO is used to treat anemia; unfortunately, some athletes abuse the drug to seek the advantage of artificially high levels of oxygen-carrying red blood cells (called "blood doping").

DNA technology is also helping medical researchers develop vaccines. A **vaccine** is a harmless variant or derivative of a disease-causing microbe—such as a bacterium or virus—that is used to prevent an infectious disease. When a person is inoculated, the vaccine stimulates the immune system to develop lasting defenses against the microbe. For many viral diseases, the only way to prevent serious harm from the illness is to use vaccination to prevent the illness in the first place. One approach to vaccine production is to use genetically engineered yeast cells to make large amounts of a protein found on the microbe's outer surface. The vaccine against hepatitis B, a disabling and sometimes fatal liver disease, is made in this way.

Genetically Modified (GM) Foods

Since ancient times, humans have selectively bred agricultural crops to make them more useful (see Figure 1.13a). Today, DNA technology is quickly replacing traditional breeding programs as scientists work to improve the productivity of agriculturally important plants and animals. Scientists have produced many types of **genetically modified (GM) organisms**, organisms that have acquired one or more genes by artificial means. If the newly acquired gene is from another organism, typically of another species, the recombinant organism is called a **transgenic organism**.

In the United States today, roughly half the corn crop and over three-quarters of the soybean and cotton crops are genetically modified. Figure 12.4 shows corn that has been genetically engineered to resist attack by an insect called the European corn borer. Growing insect-resistant plants reduces the need for chemical insecticides. In another example, modified strawberry plants produce bacterial proteins that act as a natural antifreeze, protecting the plants from cold weather, which can harm the delicate crop. Potatoes and rice have been experimentally modified to produce harmless proteins derived from the cholera bacterium; researchers hope that these modified foods will one day serve as an edible vaccine against cholera, a disease that kills thousands of children in less-industrialized nations every year. In India, the insertion of a natural but rare saltwater-resistance gene has enabled new varieties of rice to grow in water three times as salty as seawater.

▼ Figure 12.4 **Genetically modified corn.** The corn plants in this field carry a bacterial gene that helps prevent infestation by the European corn borer (inset).

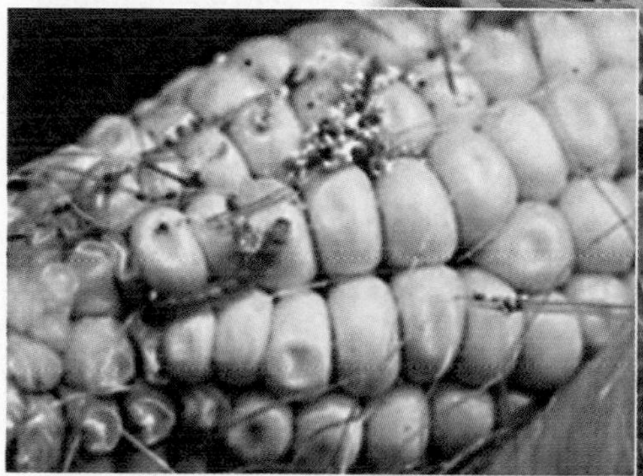

▼ Figure 12.3 **A factory that produces genetically engineered insulin.**

▲ **Figure 12.5 Genetically modified rice.** "Golden rice," shown here alongside ordinary rice, has been genetically modified to produce high levels of beta-carotene, a molecule that the body converts to vitamin A.

Scientists are also using genetic engineering to improve the nutritional value of crop plants. One example is "golden rice," a transgenic variety of rice that carries two daffodil genes. This rice produces yellow grains containing beta-carotene, which our body uses to make vitamin A (**Figure 12.5**). This rice could help prevent vitamin A deficiency and resulting blindness. However, controversy surrounds the use of GM foods, as we'll discuss at the end of the chapter. ☑

"Pharm" Animals

Figure 12.6 shows a transgenic pig that carries a gene for human hemoglobin. The pig-produced hemoglobin can be isolated and used in human blood transfusions. Because transgenic animals are difficult to produce, researchers may create a single transgenic animal and then breed or clone it. The resulting herd of transgenic animals, all carrying a recombinant human gene, could then serve as a grazing pharmaceutical factory— "pharm" animals.

DNA technology may eventually replace traditional animal breeding. Scientists might, for example, identify a gene that causes the development of larger muscles (which make up most of the meat we eat) in one variety of cattle and transfer it to other cattle or even to chickens. In 2006, University of Pittsburgh researchers genetically modified pigs to carry a roundworm gene whose protein converts less healthy fatty acids to omega-3 fatty acids. Meat from the modified pigs contains four to five times as much healthy omega-3 fat as regular pork. Unlike transgenic plants, however, transgenic animals are currently used only to produce potentially useful proteins; at present, no transgenic animals are sold as food.

Recombinant DNA technology serves many roles today and will certainly play an even larger part in our future. In the next section, you'll learn how scientists create and manipulate recombinant DNA.

Recombinant DNA Techniques

Although recombinant DNA techniques can use a variety of cell types, bacteria are the workhorses of modern biotechnology. To manipulate genes in the laboratory, biologists often use bacterial **plasmids**, which are small, circular DNA molecules that replicate (duplicate) separately from the much larger bacterial chromosome (**Figure 12.7**). Because plasmids can carry virtually any gene and are passed on from one generation of bacteria to the next, they are key tools for **gene cloning**, the production of multiple identical copies of a gene-carrying piece of DNA. Gene-cloning methods are central to the production of useful products from genetically engineered organisms. Consider a typical genetic engineering challenge: A molecular biologist identifies a gene of interest that codes for a valuable protein. The biologist wants to manufacture the protein on a large scale. **Figure 12.8** illustrates a way to accomplish this by using recombinant DNA techniques.

To start, the biologist isolates two kinds of DNA: ❶ bacterial plasmids that will serve as **vectors** (gene carriers) and ❷ DNA from another organism that includes the gene of interest, along with other genes. This other DNA may be from any type of organism, even a human. ❸ The researcher uses an enzyme to cut the two kinds of DNA. Each plasmid is cut in only one place; the other DNA is cut into many fragments, one of which carries the gene of interest. The figure shows the processing of just three of these DNA fragments and three plasmids, but actually millions of plasmids and

◀ **Figure 12.6 A genetically modified swine.** This pig has been genetically modified to carry a gene for human hemoglobin.

▼ **Figure 12.7 Bacterial plasmids.** The micrograph shows a bacterial cell that has been ruptured, revealing one long chromosome and several smaller plasmids. The inset is an enlarged view of a single plasmid.

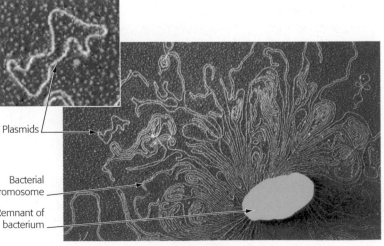

Plasmids

Bacterial chromosome

Remnant of bacterium

Colorized TEM 2,700x

▼ Figure 12.8 **Using recombinant DNA technology to produce useful products.**

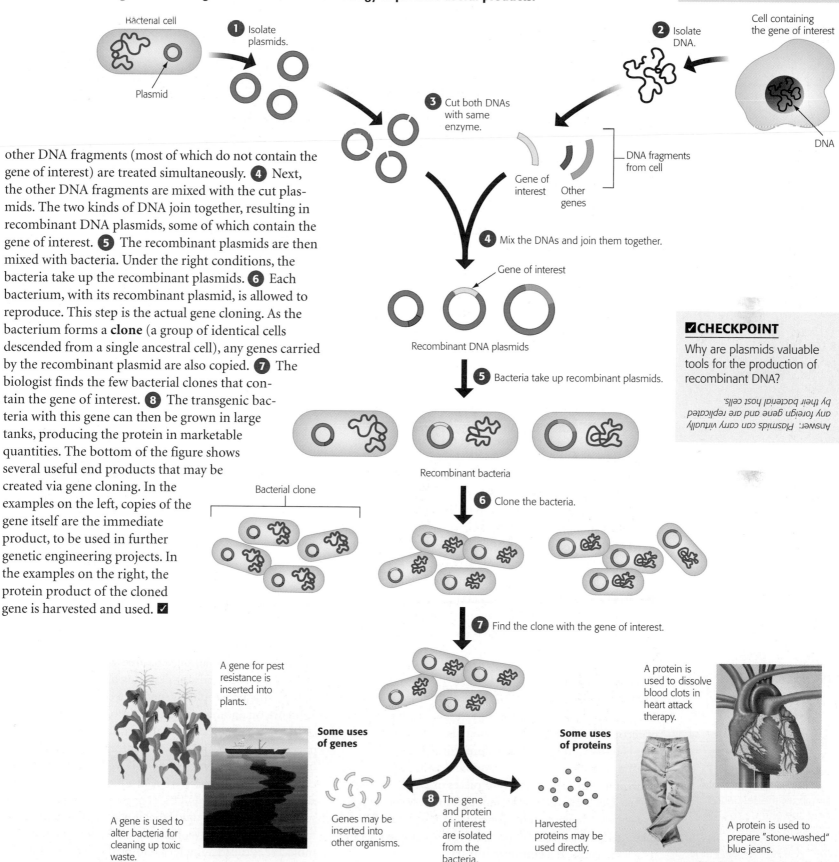

other DNA fragments (most of which do not contain the gene of interest) are treated simultaneously. ④ Next, the other DNA fragments are mixed with the cut plasmids. The two kinds of DNA join together, resulting in recombinant DNA plasmids, some of which contain the gene of interest. ⑤ The recombinant plasmids are then mixed with bacteria. Under the right conditions, the bacteria take up the recombinant plasmids. ⑥ Each bacterium, with its recombinant plasmid, is allowed to reproduce. This step is the actual gene cloning. As the bacterium forms a **clone** (a group of identical cells descended from a single ancestral cell), any genes carried by the recombinant plasmid are also copied. ⑦ The biologist finds the few bacterial clones that contain the gene of interest. ⑧ The transgenic bacteria with this gene can then be grown in large tanks, producing the protein in marketable quantities. The bottom of the figure shows several useful end products that may be created via gene cloning. In the examples on the left, copies of the gene itself are the immediate product, to be used in further genetic engineering projects. In the examples on the right, the protein product of the cloned gene is harvested and used. ✔

① Isolate plasmids.

Bacterial cell

Plasmid

② Isolate DNA.

Cell containing the gene of interest

DNA

③ Cut both DNAs with same enzyme.

Gene of interest

Other genes

DNA fragments from cell

④ Mix the DNAs and join them together.

Gene of interest

Recombinant DNA plasmids

⑤ Bacteria take up recombinant plasmids.

Recombinant bacteria

Bacterial clone

⑥ Clone the bacteria.

⑦ Find the clone with the gene of interest.

A gene for pest resistance is inserted into plants.

A gene is used to alter bacteria for cleaning up toxic waste.

Some uses of genes

Genes may be inserted into other organisms.

⑧ The gene and protein of interest are isolated from the bacteria.

Some uses of proteins

Harvested proteins may be used directly.

A protein is used to dissolve blood clots in heart attack therapy.

A protein is used to prepare "stone-washed" blue jeans.

A Closer Look: Cutting and Pasting DNA with Restriction Enzymes

As you saw in Figure 12.8, recombinant DNA is created by combining two ingredients: a bacterial plasmid and the gene of interest. To understand how these DNA molecules are spliced together, you need to learn how enzymes cut and paste DNA.

The cutting tools used for making recombinant DNA are bacterial enzymes called **restriction enzymes**. Biologists have identified hundreds of restriction enzymes, each recognizing a particular short DNA sequence (usually four to eight nucleotides long). For example, one restriction enzyme only recognizes the DNA sequence GAATTC, whereas another recognizes GGATCC. After a restriction enzyme binds to its recognition sequence on a molecule of DNA, it cuts the two strands of the DNA at specific points within the sequence.

The top of **Figure 12.9** shows a piece of DNA (blue) that contains one recognition sequence for a particular restriction enzyme. ❶ The restriction enzyme cuts the DNA strands between the bases A and G within the recognition sequence, producing pieces of DNA called **restriction fragments**. The staggered cuts yield two double-stranded DNA fragments with single-stranded ends, called "sticky ends." Sticky ends are the key to joining DNA restriction fragments originating from different sources. ❷ Next, a piece of DNA from another source (green) is added. Notice that the green DNA has single-stranded ends identical in base sequence to the sticky ends on the blue DNA because the same restriction enzyme was used to cut both types of DNA. ❸ The complementary ends on the blue and green fragments stick together by base pairing. The union between the blue and green fragments is then made permanent by the "pasting" enzyme **DNA ligase**. ❹ This enzyme, which is one of the proteins the cell normally uses in DNA replication, connects the DNA pieces into continuous strands by forming bonds between adjacent nucleotides. The final outcome is a single molecule of recombinant DNA. ✓

A Closer Look: Obtaining the Gene of Interest

The procedure shown in Figure 12.8 can yield millions of recombinant plasmids carrying many different segments of foreign DNA. Such a procedure is called a "shotgun" approach to gene cloning because it "hits" an enormous number of different pieces of DNA. A typical cloned DNA fragment is big enough to carry one or a few genes. A collection of cloned DNA fragments that includes an organism's entire genome (a complete set of its genes) is called a **genomic library**.

▼ Figure 12.9 **Cutting and pasting DNA.** The production of recombinant DNA requires two enzymes: a restriction enzyme, which cuts the original DNA molecules into pieces, and DNA ligase, which pastes the pieces together.

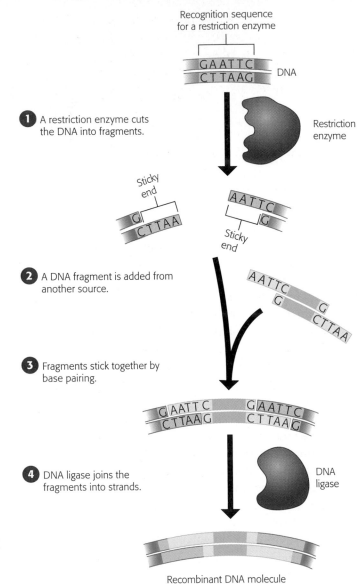

❶ A restriction enzyme cuts the DNA into fragments.

❷ A DNA fragment is added from another source.

❸ Fragments stick together by base pairing.

❹ DNA ligase joins the fragments into strands.

Recombinant DNA molecule

After you've created a genomic library, you have to find the right "book"—that is, you must identify the bacterial clone containing a desired gene (step 7 in Figure 12.8). Methods for detecting a gene depend on base pairing between the gene and a complementary sequence on another nucleic acid molecule, either DNA or RNA. When at least part of the nucleotide sequence of a gene is already known, this information can be used to advantage. For example, if we know that a gene contains the sequence TAGGCT, a biologist can synthesize a short single strand of DNA with a complementary sequence (ATCCGA) and label it with a radioactive isotope or fluorescent dye. This labeled complementary molecule is called a **nucleic acid probe** because it is used to find a specific gene or other nucleotide sequence within a mass of DNA. (In actual practice, probe molecules are considerably longer than six nucleotides.) When a radioactive DNA probe is added to the DNA of various clones, it tags the correct molecule—finds the right book in the library—by base-pairing to the complementary sequence in the gene of interest (**Figure 12.10**). After a probe detects the desired clone within a library, more of the tagged cells can be grown, resulting in the production of large quantities of the gene of interest.

Another approach to obtaining a gene of interest is to synthesize it. One method uses reverse transcriptase, a viral enzyme that can synthesize DNA by using an mRNA template (see Chapter 10). **Figure 12.11** shows the steps involved. A eukaryotic cell **1** transcribes the gene of interest and **2** processes the transcript, removing introns and splicing exons together to produce mRNA. A researcher then **3** isolates the mRNA in a test tube and **4** makes single-stranded DNA from it using reverse transcriptase. **5** DNA polymerase is then used to synthesize a second DNA strand.

When the researcher starts with an mRNA mixture from a particular cell type, the DNA that results from this procedure, called complementary DNA (cDNA), represents only those genes that were actually transcribed in the starting cells. Because cDNA molecules lack introns, they are shorter than the full version of the genes and therefore easier to work with.

A final approach is to synthesize a gene of interest from scratch. Automated DNA-synthesizing machines can accurately and rapidly produce customized DNA molecules of any sequence up to lengths of a few hundred nucleotides (**Figure 12.12**). ☑

▼ Figure 12.11 **Making a gene from eukaryotic mRNA.** Using the enzyme reverse transcriptase (green), a researcher can produce an artificial DNA gene (cDNA, in blue) from a molecule of messenger RNA (mRNA, in red).

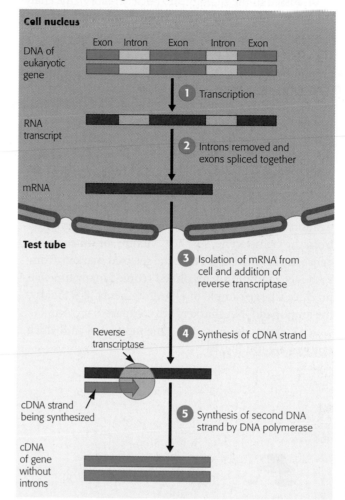

☑**CHECKPOINT**

Name three ways a gene of interest can be obtained.

Answer: *The gene can be isolated from a genomic library, produced from mRNA using reverse transcriptase, or synthesized from scratch.*

▼ Figure 12.10 **How a DNA probe tags a gene.** The probe is a short, radioactive, single-stranded molecule of DNA or RNA. When it is mixed with single-stranded DNA from a gene with a complementary sequence, it attaches by hydrogen bonds (shown in red), "labeling" the gene.

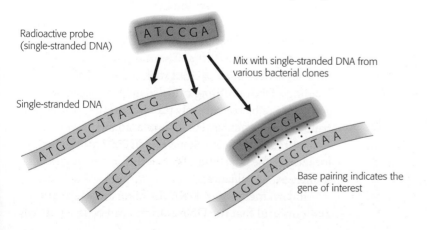

▶ Figure 12.12 **A DNA synthesizer.** This synthesizer constructs molecules of DNA from a programmed sequence, using solutions of the four DNA nucleotides (stored in the bottles beneath the machine).

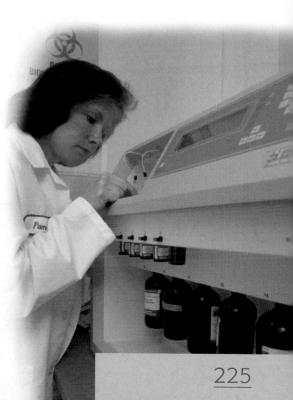

DNA Profiling and Forensic Science

As discussed in the Biology and Society section, the analysis of DNA samples to determine whether they come from the same individual, known as **DNA profiling**, can help in criminal investigations. Indeed, DNA profiling has rapidly transformed the field of **forensics**, the scientific analysis of evidence for crime scene investigations and other legal proceedings. To produce a DNA profile, scientists compare **genetic markers**, sequences in the genome that vary from person to person. Like a gene (which is a type of genetic marker), a noncoding genetic marker is more likely to be a match between relatives than between unrelated individuals.

Figure 12.13 presents an overview of a typical investigation using DNA profiling. **1** First, DNA samples from the crime scene, suspects, victims, or other evidence are isolated. **2** Next, the selected markers from each DNA sample are amplified (copied many times) to produce a large sample of DNA fragments. **3** Finally, the amplified DNA markers are compared, proving which samples are from the same individual and which samples are unique. ☑

CHECKPOINT

DNA profiling depends on analyzing genetic markers. What is a genetic marker?

Answer: A genetic marker is any sequence in the genome that varies from person to person.

► Figure 12.13 **Overview of DNA profiling.** In this example, DNA from suspect 1 does not match DNA found at the crime scene, but DNA from suspect 2 does match.

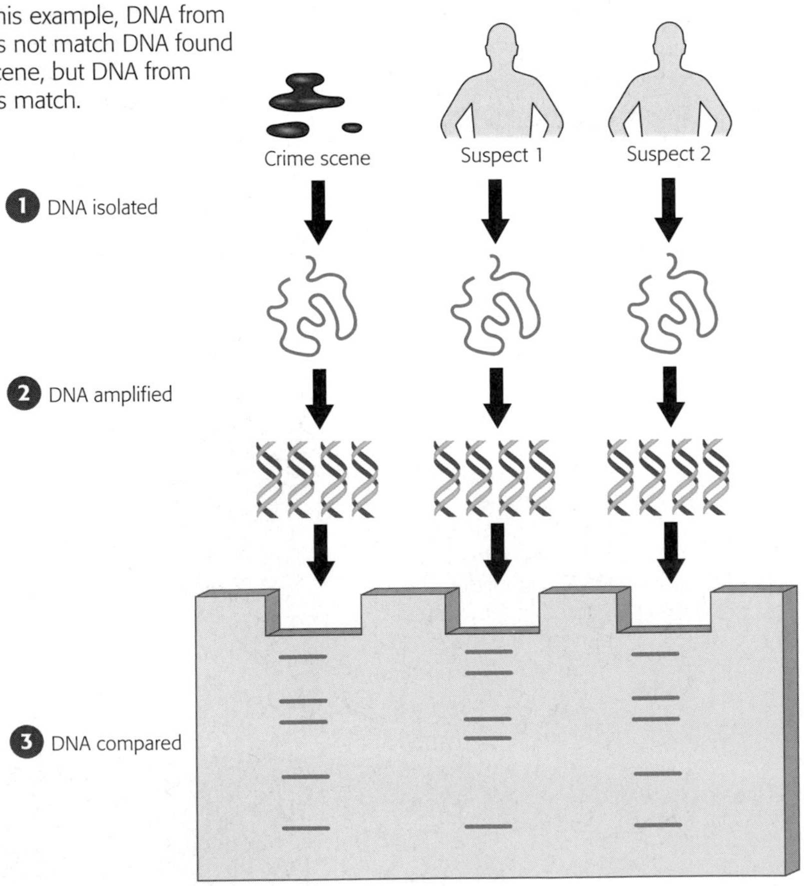

Crime scene   Suspect 1   Suspect 2

1 DNA isolated

2 DNA amplified

3 DNA compared

Investigating Murder, Paternity, and Ancient DNA

Since its introduction in 1986, DNA profiling has become a standard tool of forensics and has provided crucial evidence in many famous cases. The first well-known use of DNA profiling was in the O. J. Simpson murder trial. In this case, DNA analysis proved that blood in Simpson's car belonged to the victims and that blood at the crime scene belonged to Simpson. (The jury did not find the DNA evidence alone to be sufficient to convict the suspect, and Simpson was found not guilty.) During the investigation that led up to his impeachment, President Bill Clinton repeatedly denied that he had sexual relations with Monica Lewinsky—until DNA profiling proved that his semen was on her dress. Of course, DNA evidence can prove innocence as well as guilt. As discussed in the Biology and Society section, DNA profiling has helped lawyers at the Innocence Project exonerate over 240 convicted criminals, including some on death row. In more than a third of these cases, DNA profiling has also identified the true perpetrators.

DNA profiling can also be used to identify crime victims. The largest such effort in history took place after the World Trade Center attack on September 11, 2001. Forensic scientists in New York City worked for years to identify over 20,000 samples of victims' remains. DNA profiles of tissue samples from the disaster site were matched to DNA profiles from tissue known to be from the victims. If no sample of a victim's DNA was available, blood samples from close relatives were used to confirm identity through near matches. Over half of the victims identified at the World Trade Center site were recognized solely by DNA evidence, providing closure to many grieving families.

The use of DNA profiling extends beyond crimes. For instance, comparing the DNA of a mother, her child, and the purported father can settle a question of paternity. Sometimes paternity is of historical interest: DNA profiling proved that Thomas Jefferson or a close male relative fathered a child with his slave Sally Hemings. DNA profiling can also help protect endangered species by conclusively proving the origin of contraband animal products. For example, an analysis of elephant tusks seized in Singapore in 2007 pinpointed the location of the poaching, allowing enforcement officials to increase surveillance.

Modern methods of DNA profiling are so specific and powerful that the DNA samples can be in a partially degraded state. This allows DNA analysis to be applied in a great number of ways. In evolution research, the technique has been used to study DNA pieces recovered from an ancient mummified human and from a 30-million year-old plant fossil. A 2005 study determined that DNA extracted from a 27,000-year-old Siberian mammoth was 98.6% identical to DNA from modern African elephants. One of the strangest cases of DNA profiling is that of Cheddar Man, a 9,000-year-old skeleton found in a cave near Cheddar, England (Figure 12.14). DNA was extracted from his tooth and used to construct a DNA profile. The results suggested that Cheddar Man was a direct ancestor (through approximately 300 generations) of a present-day schoolteacher who lived only half a mile from the cave!

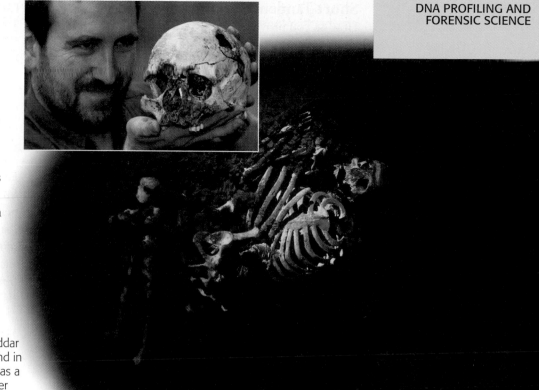

► Figure 12.14 **Cheddar Man.** Analysis of DNA extracted from "Cheddar Man"—a 9,000-year-old skeleton found in an English cave—suggested that he was a direct ancestor of a local schoolteacher (shown above).

DNA Profiling Techniques

In this section, you'll learn about techniques for making a DNA profile (steps 2 and 3 in Figure 12.13).

The Polymerase Chain Reaction (PCR)

The **polymerase chain reaction (PCR)** is a technique by which a specific segment of DNA can be targeted and copied quickly and precisely. Through PCR, a scientist can obtain enough DNA from even minute amounts of blood or other tissue to allow a DNA profile to be constructed.

In principle, PCR is simple. A DNA sample is mixed with nucleotides, the DNA replication enzyme DNA polymerase, and a few other ingredients. The solution is then exposed to cycles of heating (to separate the DNA strands) and cooling (to allow double-stranded DNA to re-form). During these cycles, specific regions of each molecule of DNA are replicated, doubling the amount of that DNA (Figure 12.15). The key to automated PCR is an unusually heat-stable DNA polymerase, first isolated from prokaryotes living in hot springs. Unlike most proteins, this enzyme can withstand the heat at the start of each cycle. Beginning with a single DNA molecule, automated PCR can generate hundreds of billions of copies in a few hours. ☑

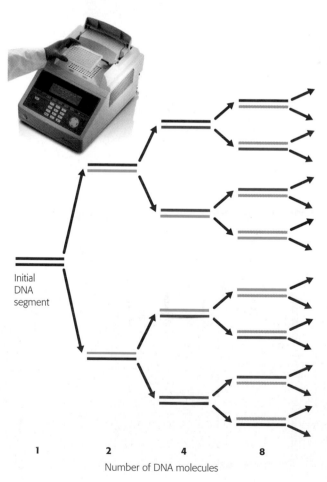

◄ Figure 12.15 **DNA amplification by PCR.** The polymerase chain reaction (PCR) is a method for making many copies of a specific segment of DNA. Each round of PCR, performed on a tabletop thermal cycler (shown at top), doubles the total quantity of DNA.

Initial
DNA
segment

| 1 | 2 | 4 | 8 |

Number of DNA molecules

☑CHECKPOINT

Why is only the slightest trace of DNA at a crime scene often sufficient for forensic analysis?

Answer: because PCR can be used to produce enough molecules for analysis

Short Tandem Repeat (STR) Analysis

How do you prove that two samples of DNA come from the same person? You could compare the entire genomes found in the two samples. But such an approach would be extremely impractical, requiring a lot of time and money. Instead, forensic scientists typically compare about a dozen short segments of noncoding repetitive DNA. **Repetitive DNA**, which makes up much of the DNA that lies between genes in humans, consists of nucleotide sequences that are present in multiple copies in the genome. Some of this DNA consists of short sequences repeated many times tandemly (one after another); such a series of repeats is called a **short tandem repeat** (STR). For example, one person might have the sequence AGAT repeated 12 times in a row at one place in the genome, the sequence GATA repeated 35 times at a second place, and so on; another

person is likely to have the same sequences at the same places but with a different number of repeats. These stretches of repetitive DNA, like any genetic marker, are more likely to be an exact match between relatives than between unrelated individuals.

STR analysis is a method of DNA profiling that compares the lengths of STR sequences at certain sites in the genome. Most commonly, STR analysis compares the number of repeats of specific four-nucleotide DNA sequences at 13 sites scattered throughout the genome. Each repeat site, which typically contains from 3 to 50 four-nucleotide repeats in a row, varies widely from person to person. In fact, some STRs used in the standard procedure have up to 80 variations in the number of repeats. In the United States, the number of repeats at each site is entered into a database called CODIS (Combined DNA Index System) administered by the Federal Bureau of Investigation.

Consider the two samples of DNA shown in **Figure 12.16**. Imagine that the top DNA segment was obtained at a crime scene and the bottom from a suspect's blood. The two segments have the same number of repeats at the first site: 7 repeats of the four-nucleotide DNA sequence AGAT (orange). Notice, however, that they differ in the number of repeats at the second site: 8 repeats of GATA (purple) in the crime scene DNA, compared with 13 repeats in the suspect's DNA. To create a DNA profile, a scientist uses PCR to specifically amplify the regions of DNA that include these STR sites. The resulting fragments are then compared. Next we'll look at how this comparison is made. ☑

▼ **Figure 12.16 Short tandem repeat (STR) sites.** Scattered throughout the genome, STR sites contain tandem repeats of four-nucleotide sequences. The number of repetitions at each site can vary from individual to individual. In this figure, both DNA samples have the same number of repeats (7) at the first STR site, but different numbers (8 versus 13) at the second.

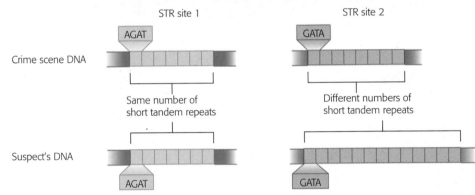

Gel Electrophoresis

The lengths of DNA fragments are compared using **gel electrophoresis**, a method for sorting macromolecules—usually proteins or nucleic acids—primarily by their electrical charge and size. **Figure 12.17** shows how gel electrophoresis separates DNA fragments obtained from

▼ **Figure 12.17 Gel electrophoresis of DNA molecules.**

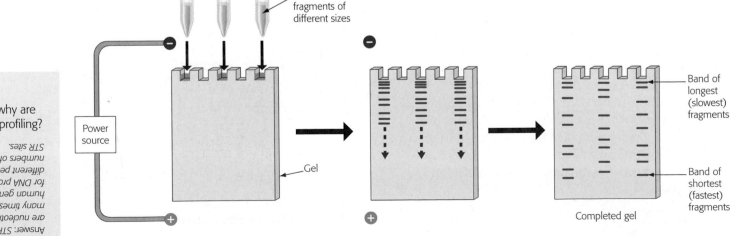

different sources. A DNA sample from each source is placed in a separate well (hole) at one end of a flat, rectangular gel, a thin slab of jellylike material that acts as a molecular sieve. A negatively charged electrode is then attached to the DNA-containing end of the gel and a positive electrode to the other end. Because the phosphate (PO_4^-) groups of nucleotides give DNA fragments a negative charge, the fragments move through the gel toward the positive pole. However, longer DNA fragments move more slowly through the thicket of polymer fibers in the gel than do shorter DNA fragments. Over time, shorter molecules move farther through the gel than longer molecules. Gel electrophoresis thus separates DNA fragments by length. When the current is turned off, a series of bands is left in each "lane" of the gel. Each band is a collection of DNA fragments of the same length. The bands can be made visible by staining, by exposure onto photographic film (if the DNA is radioactively labeled), or by measuring fluorescence (if the DNA is labeled with a fluorescent dye).

Figure 12.18 shows the gel that would result from using gel electrophoresis to separate the DNA fragments from the example in Figure 12.16. The differences in the locations of the bands reflect the different lengths of the DNA fragments. This gel would provide evidence that the crime scene DNA did not come from the suspect. Because gel electrophoresis reveals similarities and differences between DNA samples, DNA profiling can provide evidence of either guilt or innocence. ✓

RFLP Analysis

Gel electrophoresis has many uses besides STR analysis. One application is RFLP analysis. RFLP (pronounced "rif-lip") stands for restriction fragment length polymorphism. In this method, the DNA molecules to be compared are exposed to a restriction enzyme (Figure 12.19). The resulting restriction fragments are separated and made visible on a gel. In this case, both the number and location of bands indicate whether the original DNA samples had identical nucleotide sequences at the sites shown.

☑CHECKPOINT
You use a restriction enzyme to cut a DNA molecule that has three copies of the enzyme's recognition sequence clustered near one end. When you separate the restriction fragments by gel electrophoresis, how do you expect the bands to appear?

Answer: three bands near the positive pole at the bottom of the gel (small fragments) and one band near the negative pole at the top of the gel (large fragment)

▼ **Figure 12.19 RFLP analysis.** The DNA segments shown have nucleotide sequences that differ at one base pair (orange boxes). A particular restriction enzyme may therefore cut the segments at different places. In this case, the difference in DNA sequence results in three restriction fragments from the first DNA sample and two from the second. This difference, revealed by gel electrophoresis, indicates that the two DNA samples come from different individuals.

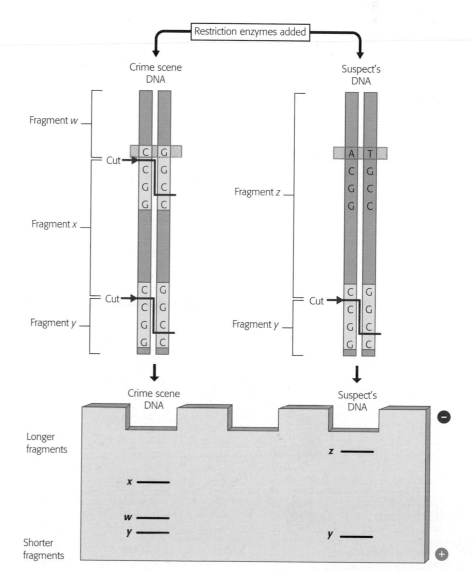

▼ **Figure 12.18 Visualizing STR fragment patterns.** This figure shows the bands that would result from gel electrophoresis of the STR sites illustrated in Figure 12.16. Notice that one of the bands from the crime scene DNA doesn't match one of the bands from the suspect's DNA. (A gel from an actual DNA profile would typically contain more than just two bands in each lane.)

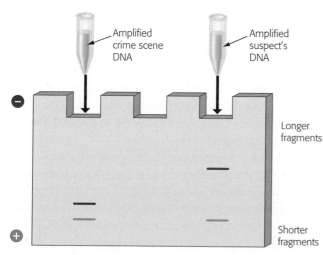

Genomics and Proteomics

In 1995, a team of scientists announced that they had determined the nucleotide sequence of the entire genome of *Haemophilus influenzae*, a bacterium that can cause several human diseases, including pneumonia and meningitis. **Genomics**, the study of complete sets of genes (genomes), was born.

The first targets of genomics research were bacteria, which have relatively little DNA. The *H. influenzae* genome, for example, contains 1.8 million nucleotides and 1,709 genes. But soon the attention of genomics researchers turned toward more complex organisms with much larger genomes (Table 12.1). As of 2009, the genomes of nearly 1,000 species have been published, and thousands more are in progress. The majority of organisms sequenced to date are prokaryotes, including *E. coli* and several hundred other bacteria (some of medical importance) and a few dozen archaea. Over 100 eukaryotic genomes have been completed.

Baker's yeast (*Saccharomyces cerevisiae*) was the first eukaryote to have its full sequence determined, and the roundworm *Caenorhabditis elegans* was the first multicellular organism. Other sequenced animals include the fruit fly (*Drosophila melanogaster*) and lab rat (*Ratus norvegicus*), model organisms for genetics. Among the sequenced plants are *Arabidopsis thaliana*, a type of mustard plant used as a model organism, and rice (*Oryza sativa*) and sorghum (*Sorghum bicolor*), two of the world's most economically important grain crops. Next we'll discuss a particularly notable example of a sequenced animal genome—our own.

The Human Genome Project

The **Human Genome Project** was a massive scientific endeavor to determine the nucleotide sequence of all the DNA in the human genome and to identify the location and sequence of every gene. The project began in 1990 as an effort by government-funded researchers from six countries. Several years into the project, private companies, chiefly Celera Genomics, joined the effort. At the completion of the project, over 99% of the genome had been determined to 99.999% accuracy. (There remain a few hundred gaps of unknown sequence that will require special methods to figure out.) This ambitious project has provided a wealth of data that may illuminate the genetic basis of what it means to be human.

The chromosomes in the human genome (22 autosomes plus the X and Y sex chromosomes) contain approximately 3.2 billion nucleotide pairs of DNA. To try to get a sense of this quantity of DNA, imagine that its nucleotide sequence is printed in letters (A, T, C, and G) in books the size of this textbook. The sequence would fill a stack of books 18 stories high! However, the biggest surprise from the Human Genome Project is the small number of human genes—only about 20,000— very close to the number found in a roundworm! This number is well below estimates made before the project began. In fact, a friendly betting pool set up in the year 2000 by the world's leading genome researchers included a wide range of guesses—from 26,000 to 150,000. Every single genome expert overestimated!

Our genome was a major challenge to sequence because, like the genomes of most complex eukaryotes, only a small amount of our total DNA consists of genes that code for proteins, tRNAs, or rRNAs. Most complex eukaryotes have a huge amount of noncoding DNA— about 98% of human DNA is of this type. Some of this

Table 12.1	Some Important Sequenced Genomes		
Organism	**Year Completed**	**Size of Genome (in base pairs)**	**Approximate Number of Genes**
Haemophilus influenzae (bacterium)	1995	1.8 million	1,700
Saccharomyces cerevisiae (yeast)	1996	12 million	5,800
Escherichia coli (bacterium)	1997	4.6 million	4,400
Caenorhabditis elegans (roundworm)	1998	97 million	19,100
Drosophila melanogaster (fruit fly)	2000	180 million	13,700
Arabidopsis thaliana (mustard plant)	2000	120 million	25,500
Oryza sativa (rice)	2002	430 million	40,000
Homo sapiens (human)	2003	3.2 billion	20,000
Rattus norvegicus (lab rat)	2004	2.8 billion	20,000
Pan troglodytes (chimpanzee)	2005	3.1 billion	20,000
Macaca mulatta (macaque)	2007	2.9 billion	22,000
Ornithorhynchus anatinus (duck-billed platypus)	2008	1.8 billion	18,500
Sorghum bicolor (sorghum)	2009	730 million	34,500

noncoding DNA is made up of gene control sequences such as promoters, enhancers, and microRNAs (see Chapter 11). Other noncoding regions include introns (whose total length in a gene may be ten times greater than the total length of the exons) and repetitive DNA (some of which is used in DNA profiling). Some noncoding DNA is important to our health, with certain regions known to carry several disease-causing mutations. But the function (if any) of most noncoding DNA remains unknown.

The potential benefits of having a complete map of the human genome are enormous. For instance, hundreds of disease-associated genes have already been identified. One example is the gene that is mutated in an inherited type of Parkinson's disease, a debilitating brain disorder that causes tremors of increasing severity. Half a million Americans suffer from this disorder (Figure 12.20). Until recently, Parkinson's disease was thought to have only an environmental basis; there was no evidence of a hereditary component. But data from the Human Genome Project mapped some cases of Parkinson's disease to a specific gene. Interestingly, an altered version of the protein encoded by this gene has also been tied to Alzheimer's disease, suggesting a link between these two brain disorders. Moreover, the same gene is also found in rats, where it plays a role in the sense of smell, and in zebra finches, where it is thought to be involved in song learning. Cross-species comparisons such as these may uncover clues about the role played by the normal version of the protein in the human brain. And such knowledge could eventually lead to treatment for Parkinson's disease. ✓

Tracking the Anthrax Killer

In October 2001, a 63-year-old Florida man died from inhalation anthrax, a disease caused by breathing spores of the bacterium *Bacillus anthracis*. As the first victim of this disease in the United States since 1976, his death was immediately suspicious. By the end of the year, four more people had died from anthrax. Law enforcement officials realized that someone was sending anthrax spores through the mail (Figure 12.21). The United States was facing an unprecedented bioterrorist attack.

In the investigation that followed, one of the most helpful clues turned out to be the anthrax spores themselves. Investigators compared the genomes of the mailed anthrax spores with several laboratory strains. They quickly established that all of the mailed spores were genetically identical, suggesting that a single perpetrator was behind all the attacks. Furthermore, they were able to match the deadly spores with a laboratory subtype isolated at the U.S. Army Medical Research Institute of Infectious Diseases in Fort Detrick, Maryland. A second, more comprehensive whole-genome analysis of the spores used in the attack was completed in 2008. This analysis found four unique mutations in the mailed anthrax and traced the mutations to a single flask at the army facility. Based in part on this evidence, the FBI named an army research scientist as a suspect in the case.

✓CHECKPOINT

1. Approximately how many nucleotides and genes are contained in the human genome?
2. Name three types of DNA that do not code for another molecule.

Answers: 1. about 3.2 billion nucleotides and 20,000 genes 2. introns, repetitive DNA, and gene control sequences

▼ Figure 12.20 **The fight against Parkinson's disease.** Actor Michael J. Fox and boxer Muhammad Ali—both of whom have Parkinson's disease—testify before the Senate on the status of federal funding for Parkinson's research.

▼ Figure 12.21 **The 2001 anthrax attacks.** In 2001, envelopes containing anthrax spores caused five deaths.

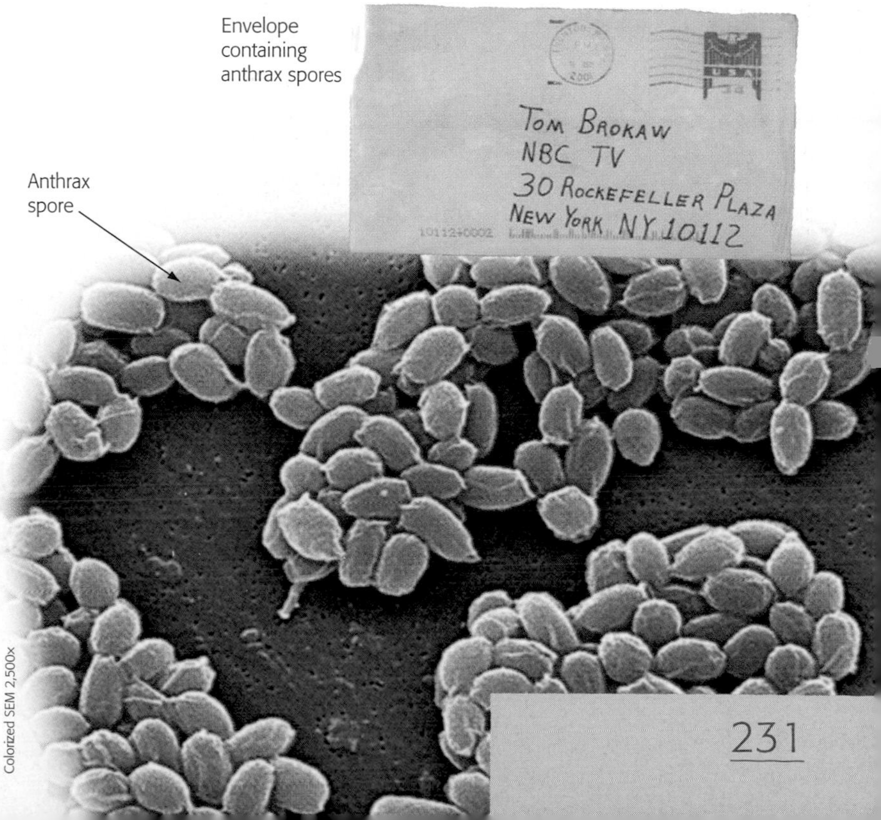

Envelope containing anthrax spores

Tom Brokaw
NBC TV
30 Rockefeller Plaza
New York, NY 10112

Anthrax spore

Colorized SEM 2,500x

The anthrax investigation is just one example of the new field of comparative genomics, the comparison of whole genomes. In 1991, sequence data provided strong evidence that a Florida dentist transmitted HIV to several patients. In 1993, after a cult released anthrax spores in downtown Tokyo, genomic analysis showed why their attack didn't kill anyone: They had used a harmless veterinary vaccine strain. And investigation of the West Nile virus outbreak in 1999 proved that a single natural strain of virus was infecting both birds and humans. Comparative genomics has even allowed geneticists to produce a family tree of dog breeds (see the Evolution Connection section that ends Chapter 9).

Comparative genomics can also reveal similarities and differences in organisms more closely related to humans. In 2005, researchers completed the genome sequence for our closest living relative on the evolutionary tree of life, the chimpanzee (*Pan troglodytes*). Comparisons with human DNA revealed that we share 96% of our genome with our closest animal relative. Genomic scientists are currently finding and studying the important differences, shedding scientific light on the age-old question of what makes us human.

Genome-Mapping Techniques

Genomes are most often sequenced using the *whole-genome shotgun method*. The first step in this method is to chop the entire genome into fragments using restriction enzymes. Next, the fragments are cloned and sequenced. Finally, computers running specialized mapping software reassemble the millions of overlapping short sequences into a single continuous sequence for every chromosome—an entire genome (**Figure 12.22**).

The DNA sequences determined by the Human Genome Project have been deposited in a database that is available via the Internet. (You can browse it yourself at the website for the National Center for Biotechnology Information.) Scientists use software to scan and analyze the sequences for genes, control elements, and other features. The result is a genetic map containing all the genes and their locations on chromosomes. Now comes the most exciting challenge: figuring out the functions of the genes and other sequences and how they work together to direct the structure and function of a living organism. This challenge and the applications of the new knowledge should keep scientists busy well into the twenty-first century.

One interesting question to ask about the Human Genome Project is whose genome was sequenced. The human genome sequenced by government-funded scientists was actually a reference genome compiled from a group of individuals. The genome sequenced by Celera consisted primarily of DNA from scientist

▼ Figure 12.22 **Genome sequencing.** In the photo at the bottom, a technician performs a step in the whole-genome shotgun method (depicted in the diagram).

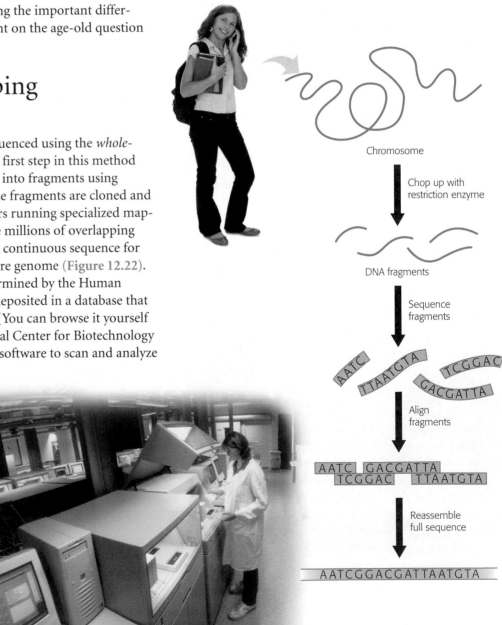

Chromosome

Chop up with restriction enzyme

DNA fragments

Sequence fragments

AATC TTAATGTA TCGGAC GACGATTA

Align fragments

AATC GACGATTA
TCGGAC TTAATGTA

Reassemble full sequence

AATCGGACGATTAATGTA

Craig Venter, the company's president. These representative sequences will serve as standards so that comparisons of individual differences and similarities can be made. Starting in 2007, the genomes of a few other individuals—the first was James Watson, codiscoverer of the structure of DNA—have also been sequenced. These sequences are part of a larger effort called the Human Variome Project (where "variome" refers to variations in the genome). Begun in 2006, this project seeks to collect information on all of the genetic variations that affect human health. As the amount of sequence data grows, the small differences that account for individual variation within our species will come to light. We may soon enter an age of "personal genomics," where individual genetic differences among people will be put to routine medical use. Some of these seemingly minuscule differences can actually be a matter of life or death, as we'll see next.

THE PROCESS OF SCIENCE: DNA Profiling

Can Genomics Cure Cancer?

Lung cancer, which kills more Americans every year than any other type of cancer, has long been the target of searches for effective chemotherapy drugs. One drug used to treat lung cancer, called gefitinib, targets the protein encoded by the *EGFR* gene. This protein is found on the surface of cells that line the lungs and is also found in lung cancer tumors.

Unfortunately, gefitinib is ineffective for many patients. While studying the effectiveness of gefitinib, researchers at the Dana-Farber Cancer Institute in Boston made the **observation** that a few patients actually responded quite positively to the drug. This posed a **question**: Are genetic differences among lung cancer patients responsible for the differences in gefitinib's effectiveness? The researchers' **hypothesis** was that mutations in the *EGFR* gene were causing the different responses to gefitinib. The team made the **prediction** that DNA profiles focusing on the *EGFR* gene would reveal different DNA sequences in the tumors of responsive patients compared with the tumors of unresponsive patients. The researchers' **experiment** involved sequencing the *EGFR* gene in cells extracted from the tumors of five patients who responded to the drug and four who did not.

The **results**, published in 2004, were quite striking: All five tumors from gefitinib-responsive patients had mutations in *EGFR*, whereas none of the other four tumors did (**Figure 12.23**). These results suggest that doctors can use DNA profiling techniques to screen lung cancer patients for those who are most likely to benefit from treatment with this drug. In broader terms, this work suggests that genomics may bring about a revolution in the treatment of disease by allowing therapies to be custom-tailored to the genetic makeup of each patient.

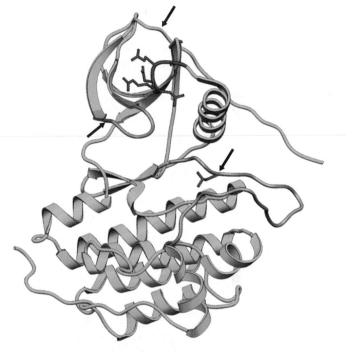

▲ Figure 12.23 **The *EGFR* protein: Fighting cancer with genomics.** Mutations (located at sites indicated by black arrows) in the *EFGR* protein can affect the ability of a cancer-fighting drug to destroy lung tumors. Here, the amino acid backbone of the protein is shown in green, with some important regions highlighted in orange, blue, and red.

Proteomics

The successes in the field of genomics have encouraged scientists to attempt similar systematic studies of the full protein sets (proteomes) that genomes encode, an approach called **proteomics**. The number of different proteins in humans far exceeds the number of different genes (about 100,000 proteins versus about 20,000 genes). And since proteins, not genes, actually carry out the activities of the cell, scientists must study when and where proteins are produced and how they interact in order to understand the functioning of cells and organisms.

Genomics and proteomics are enabling biologists to approach the study of life from an increasingly holistic perspective. Biologists are now in a position to compile catalogs of genes and proteins—that is, listings of all the "parts" that contribute to the operation of cells, tissues, and organisms. With such catalogs in hand, researchers are shifting their attention from the individual parts to how these parts function as a whole in biological systems. ✔

Human Gene Therapy

Human gene therapy is a recombinant DNA procedure intended to treat disease by altering an afflicted person's genes. In some cases, a mutant version of a gene may be replaced or supplemented with the normal allele. This could potentially correct a genetic disorder, perhaps permanently. In other cases, genes are inserted and expressed only long enough to treat a medical problem.

Figure 12.24 summarizes one approach to human gene therapy. **1** A gene from a normal individual is isolated and cloned by recombinant DNA techniques. **2** The gene is inserted into a vector, such as a harmless virus. **3** The virus is then injected into the patient. The virus inserts a copy of its genome, including the human gene, into the DNA of the patient's cells. The normal gene is then transcribed and translated within the patient's body, producing the desired protein. Ideally, the nonmutant version of the gene would be inserted into cells that multiply throughout a person's life. Bone marrow cells, which include the stem cells that give rise to all the types of blood cells, are prime candidates. If the procedure succeeds, the cells will multiply through-out the patient's life and express the normal gene. The engineered cells will supply the missing protein, and the patient will be cured.

Gene therapy has been used to treat severe combined immunodeficiency (SCID), a fatal inherited disease caused by a defective gene. Absence of the enzyme encoded by this gene prevents the development of the immune system, requiring patients to remain isolated within protective "bubbles." Unless treated with a bone marrow transplant (effective just 60% of the time), SCID patients quickly die from infections by microbes that most of us easily fend off.

Since the year 2000, gene therapy has cured 22 children with inborn SCID, providing the first strong evidence of the effectiveness of gene therapy. As part of this treatment, researchers periodically removed immune system cells from the patients' blood, infected them with a virus engineered to carry the normal allele of the defective gene, then reinjected the blood into the patient. The celebrations of this medical breakthrough were short-lived, however; four of the patients developed leukemia, and one died. Apparently, the retrovirus used as a vector activated an oncogene (see Chapter 11), creating cancerous blood cells. Thus, although gene therapy remains promising, there is very little evidence of safe and effective application. Active research continues, with new, tougher safety guidelines. ✔

✔CHECKPOINT

Why are bone marrow stem cells ideally suited as targets for gene therapy?

Answer: because bone marrow stem cells multiply throughout a person's life

▼ Figure 12.24 **One approach to human gene therapy.**

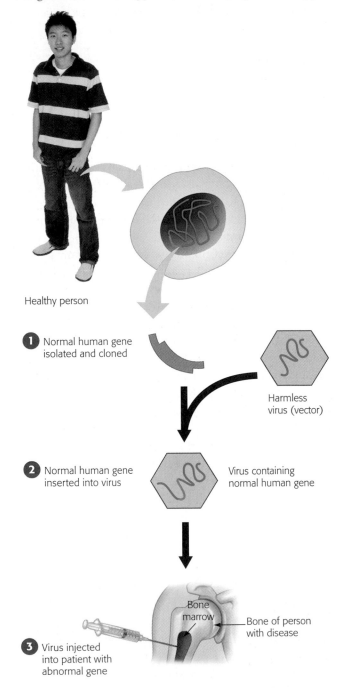

Healthy person

1 Normal human gene isolated and cloned

Harmless virus (vector)

2 Normal human gene inserted into virus

Virus containing normal human gene

3 Virus injected into patient with abnormal gene

Bone marrow

Bone of person with disease

Safety and Ethical Issues

As soon as scientists realized the power of DNA technology, they began to worry about potential dangers. Early concerns focused on the possibility of creating hazardous new disease-causing organisms. What might happen, for instance, if cancer-causing genes were transferred into infectious bacteria or viruses? To address such concerns, scientists developed a set of guidelines that have become formal government regulations in the United States and some other countries.

One safety measure is a set of strict laboratory procedures to protect researchers from infection by engineered microbes and to prevent microbes from accidentally leaving the laboratory (Figure 12.25). In addition, strains of microbes to be used in recombinant DNA experiments are genetically crippled to ensure that they cannot survive outside the laboratory. Finally, certain obviously dangerous experiments have been banned. Today, most public concern about possible hazards centers not on recombinant microbes but on genetically modified (GM) foods.

The Controversy over Genetically Modified Foods

GM strains account for a significant percentage of several crops. Controversy about the safety of these foods is an important political issue (Figure 12.26). For example, the European Union has suspended the introduction of new GM crops and considered banning the import of all GM foodstuffs. In the United States and other countries where the GM revolution has proceeded more quietly than in Europe, the labeling of GM foods is now being debated but has not yet become law.

Advocates of a cautious approach fear that crops carrying genes from other species might harm the environment or be hazardous to human health (by, for example, introducing new allergens, molecules that can cause allergic reactions, into foods). A major concern is that transgenic plants might pass their new genes to close relatives in nearby wild areas. We know that lawn and

▼ Figure 12.25 **Maximum-security laboratory.** A scientist in a high-containment laboratory wears a biohazard suit, used for working with dangerous microorganisms.

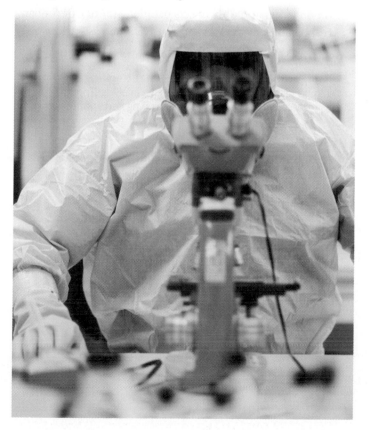

▶ Figure 12.26 **Opposition to genetically modified organisms (GMOs).** This photo shows tractors destroying a Belgian field of genetically modified rape, a forage crop used to feed sheep, birds, and hogs and also to make cooking oil. The Belgian Minister for Consumer Protection ordered the destruction as part of a widespread opposition to GM crops. The sign reads "GMO, no thanks! Yes to biodiversity."

crop grasses, for example, commonly exchange genes with wild relatives via pollen transfer. If domestic plants carrying genes for resistance to herbicides, diseases, or insect pests pollinated wild plants, the offspring might become "superweeds" that would be very difficult to control. However, researchers may be able to prevent the escape of such plant genes in various ways—for example, by engineering plants so they cannot hybridize. Concern has also been raised that the widespread use of GM seeds may reduce natural genetic diversity, leaving crops susceptible to catastrophic die-offs in the event of a sudden change to the environment or introduction of a new pest (see, for example, the Biology and Society section in Chapter 16).

The U.S. National Academy of Sciences released a study finding no scientific evidence that transgenic crops pose any special health or environmental risks. But the authors of the study also recommended more stringent long-term monitoring to watch for unanticipated environmental impacts.

In 2000, negotiators from 130 countries (including the United States) agreed on a Biosafety Protocol that requires exporters to identify GM organisms present in bulk food shipments and allows importing countries to decide whether the shipments pose environmental or health risks. This agreement has been hailed as a breakthrough by environmentalists.

Today, governments and regulatory agencies throughout the world are grappling with how to facilitate the use of biotechnology in agriculture, industry, and medicine while ensuring that new products and procedures are safe. In the United States, all projects are evaluated for potential risks by a number of regulatory agencies, including the Food and Drug Administration, the Environmental Protection Agency, the National Institutes of Health, and the Department of Agriculture. ✔

✔CHECKPOINT

What is the main concern about adding genes for herbicide resistance to crop plants?

Answer: the possibility that the genes could escape, via cross-pollination, to wild plants that are closely related to the crop species.

Ethical Questions Raised by DNA Technology

DNA technology raises legal and ethical questions—few of which have clear answers. Consider, for example, the treatment of dwarfism with injections of human growth hormone (HGH) produced by genetically engineered cells. Should parents of short but hormonally normal children be able to seek HGH treatment to make their kids taller? If not, who decides which children are "tall enough" to be excluded from treatment?

Genetic engineering of gametes (sperm or ova) and zygotes has been accomplished in lab animals. It has not been attempted in humans because such a procedure would raise very difficult ethical questions. Should we

try to eliminate genetic defects in our children and their descendants? Should we interfere with evolution in this way? From a long-term perspective, the elimination of unwanted versions of genes from the gene pool could backfire. Genetic variety is a necessary ingredient for the adaptation of a species as environmental conditions change with time. Genes that are damaging under some conditions may be advantageous under others (one example is the sickle-cell allele—see the Evolution Connection section at the end of Chapter 13). Are we willing to risk making genetic changes that could be detrimental to our species in the future? We may have to face such questions soon.

Advances in genetic profiling raise privacy issues. If we were to create a DNA profile of every person at birth, then we could theoretically match nearly every violent crime to a perpetrator because it is virtually impossible for someone to commit a violent crime without leaving behind DNA evidence. But are we, as a society, prepared to sacrifice our genetic privacy, even for such a worthwhile goal?

As more information becomes available about our personal genetic makeup, ethicists question whether greater access to this information is always beneficial. For example, mail-in kits have become available that can tell healthy people their relative risk of developing various diseases (such as Parkinson's and Crohn's) later in life (Figure 12.27). Some argue that such information helps families to prepare. Others say that the test preys on our fears and only upsets people without offering any real benefit because certain diseases, such as Parkinson's,

▼ Figure 12.27 **Personalized genetic testing.** This kit can be used to send saliva for genetic analysis. The results can indicate a person's risk of developing certain diseases.

are not currently preventable or treatable. Other tests (such as ones for breast cancer risk) may help a person make changes that can prevent disease. How can we identify truly useful tests?

There is a danger that information about disease-associated genes could be abused. One issue is the possibility of discrimination and stigmatization. Would you, as an employer, want to know if a potential employee had an increased risk of schizophrenia? Should you be able to find out? Should insurance companies have the right to screen applicants for disease genes? People might be coerced into taking a DNA test in order to be considered for a job or an insurance policy. How do we prevent genetic information from being used in a discriminatory manner?

A much broader ethical question is how do we really feel about wielding one of nature's powers—the evolution of new organisms? Some might ask if we have any right to alter an organism's genes—or to create new organisms.

The benefits to people and the environment must also be considered. For example, bacteria are being engineered to clean up mining wastes and other pollutants that threaten the soil, water, and air. Since these organisms may be the only feasible solutions to some of our most pressing environmental problems, many people think this type of genetic engineering should be encouraged.

DNA technologies raise many complex issues that have no easy answers. It is up to you, as a "citizen scientist," to make informed choices. ✔

☑ **CHECKPOINT**

Why does genetically modifying a human gamete raise different ethical questions than genetically modifying a human somatic (body) cell?

Answer: A genetically modified somatic cell will affect only the patient. Modifying a gamete will affect an unborn individual as well as all of his or her descendants.

EVOLUTION CONNECTION: DNA Profiling

Profiling the Y Chromosome

Barring mutations, the human Y chromosome passes essentially intact from father to son. By comparing Y DNA, researchers can learn about the ancestry of human males. DNA profiling can thus provide data about recent human evolution.

In 2003, geneticists discovered that about 8% of males currently living in central Asia have Y chromosomes of striking genetic similarity. Further analysis traced their common genetic heritage to a single man living about 1,000 years ago. In combination with historical records, the data led to the speculation that the Mongolian ruler Genghis Khan **(Figure 12.28)** may be responsible for the spread of the chromosome to nearly 16 million men living today. A similar study of Irish men in 2006 suggested that nearly 10% of them were descendants of Niall of the Nine Hostages, a warlord who lived during the 1400s. Another study of Y DNA seemed to confirm the claim by the Lemba people of southern Africa that they are descended from ancient Jews. Sequences of Y DNA distinctive of the Jewish priestly caste called Kohanim are found at high frequencies among the Lemba.

Comparison of Y chromosome DNA profiles is part of a larger effort to learn more about the human genome. Other research efforts are extending genomic studies to many more species. These studies will advance our understanding of all aspects of biology, including health and ecology as well as evolution. In fact, comparisons of the completed genome sequences of bacteria, archaea, and eukaryotes first supported the theory that these are the three fundamental domains of life—a topic we discuss further in the next unit, "Evolution and Diversity."

▲ Figure 12.28 **Genghis Khan.**

Chapter Review

SUMMARY OF KEY CONCEPTS

Recombinant DNA Technology

DNA technology, the manipulation of genetic material, is a new branch of biotechnology, the use of organisms to make helpful products. DNA technology often involves the use of recombinant DNA, the combination of nucleotide sequences from two different sources.

Applications: From Humulin to Foods to "Pharm" Animals

Recombinant DNA techniques have been used to create genetically modified organisms, ones that carry artificially introduced genes. Nonhuman cells have been engineered to produce human proteins, genetically modified food crops, and transgenic farm animals. A transgenic organism is one that carries artificially introduced genes, typically from a different species.

Recombinant DNA Techniques

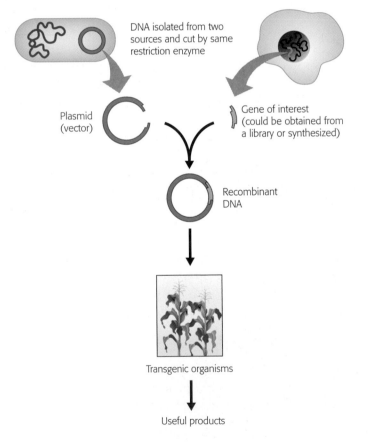

DNA isolated from two sources and cut by same restriction enzyme

Plasmid (vector)

Gene of interest (could be obtained from a library or synthesized)

Recombinant DNA

Transgenic organisms

Useful products

DNA Profiling and Forensic Science

Forensics, the scientific analysis of legal evidence, has been revolutionized by DNA technology. DNA profiling is used to determine whether two DNA samples come from the same individual.

Investigating Murder, Paternity, and Ancient DNA

DNA profiling can be used to establish innocence or guilt of a criminal suspect, identify victims, determine paternity, and contribute to basic research.

DNA Profiling Techniques

Short tandem repeat (STR) analysis compares DNA fragments using the polymerase chain reaction (PCR) and gel electrophoresis.

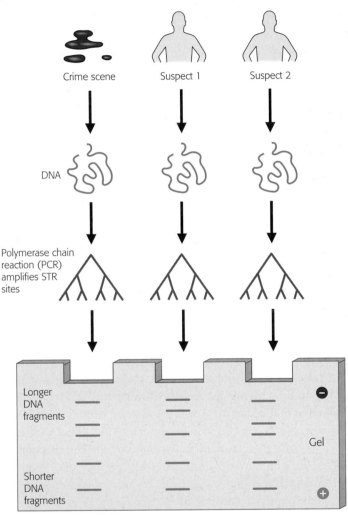

Crime scene   Suspect 1   Suspect 2

DNA

Polymerase chain reaction (PCR) amplifies STR sites

Longer DNA fragments

Shorter DNA fragments

Gel

DNA fragments compared by gel electrophoresis
(Bands of shorter fragments move faster toward the positive pole.)

Genomics and Proteomics

The Human Genome Project

The nucleotide sequence of the human genome is providing a wealth of useful data. The 24 different chromosomes of the human genome contain about 3.2 billion nucleotide pairs and 20,000 genes. The majority of the genome consists of noncoding DNA.

Tracking the Anthrax Killer

Comparing genomes can aid criminal investigations and basic research.

Genome-Mapping Techniques

The whole-genome shotgun method involves sequencing DNA fragments from an entire genome and then assembling the sequences.

Proteomics

Success in genomics has given rise to proteomics, the systematic study of the full set of proteins found in organisms.

Human Gene Therapy

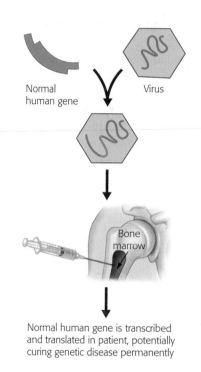

Normal human gene

Virus

Bone marrow

Normal human gene is transcribed and translated in patient, potentially curing genetic disease permanently

Gene therapy trials have focused on SCID, an inherited immune disease, with some success and some setbacks.

Safety and Ethical Issues

The Controversy over Genetically Modified Foods

The debate about genetically modified crops centers on whether they might harm humans or damage the environment by transferring genes through cross-pollination with other species.

Ethical Questions Raised by DNA Technology

We as a society and as individuals must become educated about DNA technologies to address the ethical questions raised by their use.

SELF-QUIZ

1. Suppose you wish to create a large batch of the protein lactase using recombinant DNA. Place the following steps in the order you would have to perform them.
   a. Find the clone with the gene for lactase.
   b. Insert the plasmids into bacteria and grow the bacteria into clones.
   c. Isolate the gene for lactase.
   d. Create recombinant plasmids, including one that carries the gene for lactase.

2. Why is an artificial gene that is made using reverse transcriptase often shorter than the natural form of the gene?

3. A carrier that moves DNA from one cell to another, such as a plasmid, is called a _____.

4. In making recombinant DNA, what is the benefit of using a restriction enzyme that cuts DNA in a staggered fashion?

5. A paleontologist has recovered a bit of organic material from the 400-year-old preserved skin of an extinct dodo. She would like to compare DNA from the sample with DNA from living birds. The most useful method for initially increasing the amount of dodo DNA available for testing is _____.

6. Why do DNA fragments containing STR sites from different people tend to migrate to different locations during gel electrophoresis?

7. What feature of a DNA fragment causes it to move through a gel during electrophoresis?
   a. the electrical charges of its phosphate groups
   b. its nucleotide sequence
   c. the hydrogen bonds between its base pairs
   d. its double helix shape

8. After a gel electrophoresis procedure is run, the pattern of bars in the gel shows
   a. the order of bases in a particular gene.
   b. the presence of various-sized fragments of DNA.
   c. the order of genes along particular chromosomes.
   d. the exact location of a specific gene in a genomic library.

9. Name the steps of the whole-genome shotgun method.

10. Put the following steps of human gene therapy in the correct order.
    a. Virus is injected into patient.
    b. Human gene is inserted into a virus.
    c. Normal human gene is isolated and cloned.
    d. Normal human gene is transcribed and translated in the patient.

Answers to the Self-Quiz questions can be found in Appendix D.

THE PROCESS OF SCIENCE

11. A biochemist hopes to find a gene in human liver cells that codes for a blood-clotting protein. The nucleotide sequence of a small part of the gene is CTGGACTGACA. Briefly explain how to obtain the desired gene.

12. Some scientists once joked that when the DNA sequence of the human genome was complete, "we can all go home" because there would be nothing left for genetic researchers to discover. Why haven't they all "gone home"?

BIOLOGY AND SOCIETY

13. In the not-too-distant future, gene therapy may be used to treat many inherited disorders. What do you think are the most serious ethical issues to face before human gene therapy is used on a large scale? Explain.

14. Today, it is fairly easy to make transgenic plants and animals. What are some safety and ethical issues raised by this use of recombinant DNA technology? What are some dangers of introducing genetically engineered organisms into the environment? What are some reasons for and against leaving such decisions to scientists? Who should make these decisions?

15. In October 2002, the government of the African nation of Zambia announced that it was refusing to distribute 15,000 tons of corn donated by the United States, enough corn to feed 2.5 million Zambians for three weeks. The government rejected the corn because it was likely to contain genetically modified kernels. The government made the decision after its scientific advisers concluded that the studies of the health risks posed by GM crops "are inconclusive." Do you agree with this assessment? Do you think that it is a good justification for refusing the donated corn? At the time of the government's decision, Zambia was facing food shortages, and 35,000 Zambians were expected to die from starvation over the next six months. In light of this, do you think it was morally acceptable for the government to refuse the food? How do the relative risks posed by GM crops compare with the relative risks posed by starvation?

16. In 1983, a 10-year-old girl was kidnapped from her home, raped, and murdered. A jury convicted a local teenager of the crimes and sentenced him to death for the brutal killing. In 1995, DNA analysis proved that semen found near the scene could not have come from the man accused. After 12 years on death row, he was exonerated and released from prison. His case, which took place in Illinois, was far from unique. From 1977 to 2000, 12 convicts were executed and 13 exonerated. In 2000, the governor of Illinois declared a moratorium on all executions in his state because the death penalty system was "fraught with errors." Do you support the Illinois governor's decision? What rights should death penalty inmates have with regard to DNA testing of old evidence? Who should pay for this additional testing?

Unit 3
Evolution and Diversity

Chapter 13: **How Populations Evolve**

Chapter Thread: **Evolution in Action**

Chapter 14: **How Biological Diversity Evolves**

Chapter Thread: **Mass Extinctions**

Chapter 15: **The Evolution of Microbial Life**

Chapter Thread: **The Origin of Life**

Chapter 16: **Plants, Fungi, and the Move onto Land**

Chapter Thread: **Plant-Fungus Symbiosis**

Chapter 17: **The Evolution of Animals**

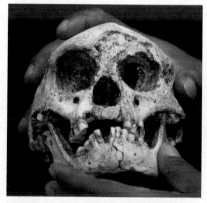

Chapter Thread: **Human Evolution**

13 How Populations Evolve

Evolutionary adaptations.
Through natural selection, species such as the leafy sea dragon can become highly adapted to their environments.

CHAPTER CONTENTS

■■■ Chapter Thread: **Evolution in Action**

BIOLOGY AND SOCIETY: Evolution in Action

Persistent Pests

In the 1960s, the World Health Organization (WHO) began a campaign to eradicate the mosquitoes that transmit the disease malaria. It was a noble goal because malaria kills an estimated 3 million people each year in tropical regions. The effort focused on spraying the mosquitoes' habitat with a pesticide called DDT. Early results were promising. Progress soon stalled, however, and the plan was dropped. How could a tiny mosquito thwart the best efforts of well-funded scientists?

Such failures, it turns out, are common. Pesticides often have encouraging early results: A relatively small amount dusted onto a crop may kill 99% of the insects. However, the few surviving insects have alleles (alternate forms of a gene) that enable them to resist the chemical attack. For example, the alleles may code for enzymes that destroy the pesticide. When the survivors reproduce, their offspring may inherit these alleles. In each generation, the proportion of resistant insects in the population increases, making subsequent sprayings less effective.

Pesticide-resistant insects are just one example of evolution in action. The sea dragon on the facing page demonstrates how well natural selection can drive the evolutionary adaptations of species to their surroundings.

An understanding of evolution informs all of biology, from exploring life's molecules to analyzing ecosystems. And applications of evolutionary biology are transforming medicine, agriculture, biotechnology, and conservation biology. This unit of chapters features mechanisms of evolution and traces the history of life on Earth. This chapter starts with the story of how Charles Darwin formulated his ideas. Next, it presents some evidence in support of evolution, followed by a closer look at Darwin's theory of natural selection. It then focuses on the genetic basis of evolution and the way evolution proceeds. Throughout the discussion, you'll read about evolution in action: verifiable, measurable examples of evolution that affect our world.

Charles Darwin and *The Origin of Species*

Biology came of age on November 24, 1859, the day Charles Darwin published *On the Origin of Species by Means of Natural Selection*. Darwin's book presented two main concepts. First, he argued convincingly from several lines of evidence that contemporary species arose from a succession of ancestors through a process of "descent with modification," his phrase for evolution. Darwin's second concept in *The Origin of Species* was a mechanism for how life evolves: natural selection.

Natural selection is a process in which organisms with certain inherited traits are more likely to survive and reproduce than are individuals with other traits (see Figure 1.12). As a result of natural selection, a **population**—a group of individuals of the same species living in the same place at the same time—can change over generations. Natural selection leads to **evolutionary adaptation**, a population's increase in the frequency of traits suited to the environment (**Figure 13.1**). (The term *adaptation* can also refer to the trait itself; for example, an insect's camouflage is an adaptation that helps it avoid predators.) In modern terms, we would say that the genetic composition of the population has changed over time, and that is one way of defining **evolution**. But we can also use the term *evolution* on a much grander scale to mean all of biological history, from the earliest microbes to the enormous diversity of organisms that live on the Earth today.

Darwin's book drew a cohesive picture of life by connecting the dots between a bewildering array of seemingly unrelated facts. *The Origin of Species* focused biologists' attention on the great diversity of organisms—their origins and relationships, their similarities and differences, their geographic distribution, and their adaptations to surrounding environments. ☑

Darwin's Cultural and Scientific Context

Before we examine how natural selection works and how Darwin derived the idea, let's place the Darwinian revolution in its historical context. As you read, you can refer to the time line in **Figure 13.2**.

The view of life developed in *The Origin of Species* contrasts sharply with the prevailing view during Darwin's lifetime. Many scientists of his day thought Earth was relatively young and populated by a huge number of unrelated species. *The Origin of Species* challenged that widely held notion. It was truly radical for its time—not only challenging the current scientific views but also shaking the deepest roots of Western culture.

☑CHECKPOINT

What were the two main concepts in Darwin's *The Origin of Species*?

Answer: descent of diverse species from common ancestors and natural selection as the mechanism of evolution

▶ Figure 13.1 **Camouflage as an example of evolutionary adaptation.** Related species of insects called mantids have diverse shapes and colors that evolved in different environments.

A Trinidad tree mantid that mimics dead leaves

A leaf mantid in Costa Rica

A flower mantid in Malaysia

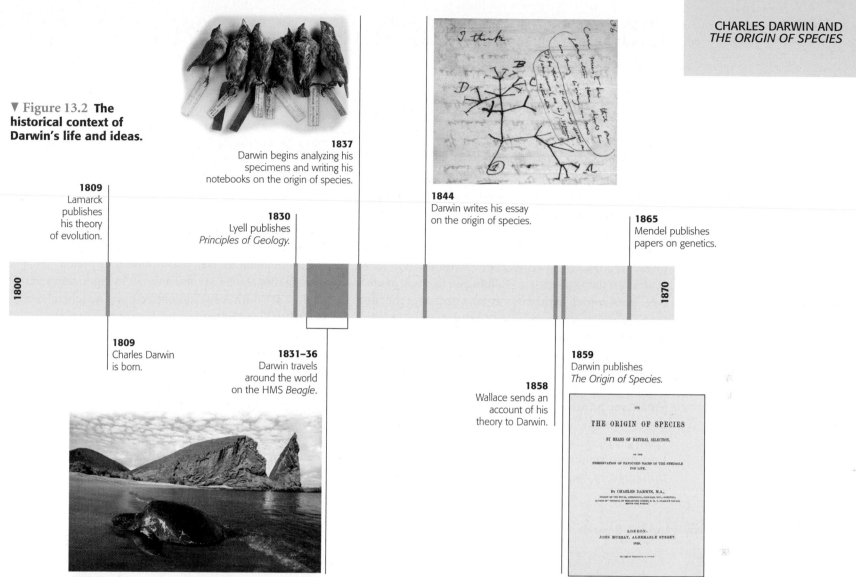

▼ Figure 13.2 **The historical context of Darwin's life and ideas.**

1837
Darwin begins analyzing his specimens and writing his notebooks on the origin of species.

1809
Lamarck publishes his theory of evolution.

1830
Lyell publishes *Principles of Geology.*

1844
Darwin writes his essay on the origin of species.

1865
Mendel publishes papers on genetics.

1800

1870

1809
Charles Darwin is born.

1831–36
Darwin travels around the world on the HMS *Beagle.*

1858
Wallace sends an account of his theory to Darwin.

1859
Darwin publishes *The Origin of Species.*

Green sea turtle in the Galápagos Islands

The Idea of Fixed Species

The Greek philosopher Aristotle, whose views had an enormous impact on Western culture, generally held that species are fixed, or permanent, and do not evolve. Judeo-Christian culture fortified this idea with a literal interpretation of the biblical book of Genesis, which tells the story of each form of life being individually created in its present-day form. The idea that all living species are unchanging in form and inhabit an Earth that is only about 6,000 years old dominated the intellectual climate of the Western world for centuries.

Lamarck and Evolutionary Adaptations

In the mid-1700s, the study of **fossils**—imprints or remains of organisms that lived in the past—led French naturalist Georges Buffon to suggest that Earth might be much older than 6,000 years. He also observed similarities between particular fossils and living animals. In 1766, Buffon proposed that certain fossil forms might

be ancient versions of similar living species. Then, in the early 1800s, French naturalist Jean-Baptiste de Lamarck suggested that the best explanation for this relationship of fossils to current organisms is that life evolves. Lamarck explained evolution as the refinement of traits that equip organisms to perform successfully in their environments. For example, some birds have powerful beaks that enable them to crack tough seeds.

We remember Lamarck mainly for his erroneous view of *how* species evolve. He proposed that by using or not using its body parts, an individual may develop certain traits that it passes on to its offspring. In other words, Lamarck proposed that acquired traits are inherited. He suggested, for example, that the strong beaks of seed-cracking birds are the cumulative result of ancestors exercising their beaks during feeding and passing that acquired beak power on to offspring. However, simple observations provide evidence against the inheritance of acquired traits: A carpenter who builds up strength and stamina through a lifetime of pounding nails with a heavy hammer will not pass enhanced biceps on to children. However, Lamarck's mistaken idea obscures the important fact that he helped set the stage for Darwin by proposing that species evolve as a result of interactions between organisms and their environments.

The Voyage of the *Beagle*

Charles Darwin was born in 1809, on the same day as Abraham Lincoln. Even as a boy, Darwin's consuming interest in nature was evident. When he was not reading nature books, he was in the fields and forests fishing, hunting, and collecting insects. His father, an eminent physician, could see no future for a naturalist and sent Charles to the University of Edinburgh to study medicine. But Charles, who was only 16 years old at the time, found medical school boring and distasteful. He left Edinburgh without a degree and then enrolled at Christ College at Cambridge University, intending to become a minister. Darwin received his B.A. degree, which included courses in biology, in 1831. Soon after, his botany professor recommended him to Captain Robert FitzRoy, who was preparing the survey ship HMS *Beagle* for a voyage around the world. It was a tour that would have a profound effect on Darwin's thinking and eventually on the thinking of the entire world.

Darwin was 22 years old when he sailed from Great Britain on the *Beagle* in December 1831 (Figure 13.3). The main mission of the voyage was to chart poorly known stretches of the South American coastline. Darwin spent most of his time on shore collecting thousands of specimens of fossils and living plants and animals. He noted the unique adaptations of organisms that inhabited such diverse environments as the Brazilian jungles, the grasslands of the Argentine pampas, and the desolate and frigid lands at the southern tip of South America.

In spite of their unique adaptations, the plants and animals throughout the continent all had a definite South American stamp, very distinct from the life-forms of Europe. That in itself may not have surprised Darwin. But the plants and animals living in temperate regions of South America seemed more closely related to species living in tropical regions of that continent than to species living in temperate regions of Europe. And the South American fossils Darwin found, though clearly different species from living ones, were distinctly South American in their resemblance to the living plants and animals of that continent. These observations led Darwin to wonder if contemporary South American species owed their features to descent from ancestral species on that continent.

Darwin was particularly intrigued by the geographic distribution of organisms on the Galápagos Islands.

▼ Figure 13.3 **The voyage of the *Beagle*.**

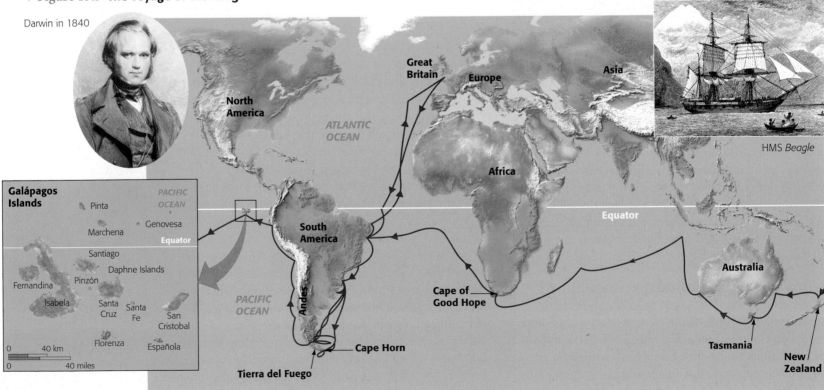

The Galápagos are relatively young volcanic islands about 900 kilometers (about 540 miles) off the Pacific coast of South America. Most of the animals that inhabit these remote islands are found nowhere else in the world, but they resemble species living on the South American mainland (**Figure 13.4**).

While on his voyage, Darwin was strongly influenced by the newly published *Principles of Geology*, by Scottish geologist Charles Lyell. The book presented the case for an ancient Earth sculpted by gradual geologic processes that continue today. Having witnessed an earthquake that raised part of the coastline of Chile almost a meter, Darwin realized that natural forces gradually changed Earth's surface and that these forces still operate. Thus, the growth of mountains as a result of earthquakes could account for the presence of marine snail fossils he collected on mountaintops in the Andes. Darwin would eventually apply this principle of *gradualism* to the evolution of Earth's life.

By the time Darwin returned to Great Britain five years after the *Beagle* set sail (a full three years longer than was originally planned), his experiences and reading had led him to seriously doubt that Earth and all its living organisms had been specially created only a few thousand years earlier. Darwin had come to realize that Earth was very old and constantly changing. He began to analyze his collections and to discuss them with colleagues. He continued to read, correspond with other scientists, and maintain extensive journals of his observations, studies, and thoughts.

Descent with Modification

By the early 1840s, Darwin had composed a long essay describing the major features of his theory of evolution. He realized that his ideas would cause a social furor, however, and therefore delayed publishing his essay. Then, in the mid-1850s, Alfred Wallace, a British naturalist doing fieldwork in Indonesia, developed a theory almost identical to Darwin's. When Wallace sent Darwin a manuscript describing his own ideas on natural selection, Darwin wrote, "All my originality ... will be smashed." However, in 1858, two of Darwin's colleagues presented Wallace's paper and excerpts from Darwin's earlier essay together to the scientific community. With the publication in 1859 of *The Origin of Species*, Darwin presented the world with an avalanche of evidence and a strong, logical argument for evolution.

As noted earlier, Darwin made two main points in *The Origin of Species*. First, he presented evidence that each living species descended from a succession of ancestral species. In the first edition of his book, he did not use the word *evolution*, referring instead to "descent with modification." Darwin hypothesized that as the descendants of the earliest organisms spread into various habitats over millions of years, they accumulated modifications, or adaptations, to diverse ways of life. Darwin's second main point was that natural selection is the mechanism for descent with modification. Next, we'll examine the evidence for evolution and then look more closely at Darwin's theory of natural selection. ☑

☑**CHECKPOINT**

Darwin's phrase for evolution, _____ with_____, captured the idea that an ancestral species could diversify into many descendant species by the accumulation of different _____ to various environments.

Answer: descent; modification; adaptations

▼ Figure 13.4 **A marine iguana (right), an example of the unique species inhabiting the Galápagos.** Darwin noticed that Galápagos marine iguanas—with webbed feet and flattened tail that aid in swimming—are similar to, but distinct from, land-dwelling iguanas on the islands and on the South American mainland (left).

Evidence of Evolution

Evolution leaves observable signs. Such clues to the past are essential to any historical science. Historians of human civilization can study written records from earlier times. But they can also piece together the evolution of societies by recognizing vestiges of the past in modern cultures. Even if we did not know from written documents that Spaniards colonized the Americas, we would deduce this from the Hispanic stamp on Latin American culture. Similarly, biological evolution has left evidence in today's organisms, as well as in fossils.

In this section, we will examine five of the many lines of evidence in support of evolution. One of them—fossils—is a historical record. The other four—biogeography, comparative anatomy, comparative embryology, and molecular biology—encompass historical vestiges of evolution evident in modern life.

The Fossil Record

Over millions of years, sand and silt that eroded from the land were carried by rivers and deposited in the oceans, piling up and compressing older deposits below into rock. Some dead organisms that settled along with the sediments left imprints in the rocks. Thus, each rock layer, or stratum (plural, *strata*), bears a unique set of fossils representing a local sampling of the organisms that lived and died when that sediment was deposited. Younger strata are on top of older ones, so the positions of fossils in the strata reveal their relative age. (The ages of fossils can be confirmed using radiometric dating—see Figure 14.15.) The **fossil record** is this ordered sequence of fossils as they appear in the rock layers, marking the passage of geologic time (**Figure 13.5**).

The fossil record reveals the appearance of organisms in a historical sequence. The oldest known fossils, dating from about 3.5 billion years ago, are prokaryotes. This fossil evidence fits with the molecular and cellular evidence that prokaryotes are the ancestors of all life. Fossils in younger layers of rock reveal the evolution of various groups of eukaryotic organisms.

Paleontologists (scientists who study fossils) have discovered many transitional forms that link past and present. For example, a series of transitional fossils provides evidence that birds descended from one branch of dinosaurs. Another example is a series of transitional whale fossils connecting these aquatic mammals to four-legged land mammals (**Figure 13.6**). Whales living today have forelegs in the form of flippers and small bones that may be remnants of ancestral hind legs and feet. In the past few decades, a series of remarkable fossils of extinct mammals have been discovered in Pakistan, Egypt, and North America that document the transition from life on land to life in the sea. Some whale ancestors had a type of anklebone that is otherwise unique to the group of land mammals that includes pigs, hippos, cows, camels, and deer. The anklebone similarity strongly suggests that whales (as well as dolphins and porpoises) are most closely related to this group of land mammals. ☑

☑CHECKPOINT

Why are older fossils generally in deeper rock layers than younger fossils?

Answer: *Sedimentation places younger rock layers on top of older ones.*

▼ Figure 13.5 **Strata of sedimentary rock at the Grand Canyon.** The Colorado River has cut through over 2,000 m of rock, exposing sedimentary strata that are like huge pages from the book of life. Each stratum entombs fossils that represent some of the organisms from that period of Earth's history.

▼ Figure 13.6 **A transitional fossil linking past and present.** The hypothesis that whales evolved from terrestrial (land-dwelling) ancestors predicts a four-limbed beginning for whales. Paleontologists digging in Egypt and Pakistan have identified extinct whales that had hind limbs. Shown here are fossilized leg bones of *Basilosaurus*, one of those ancient whales. These whales were already aquatic animals that no longer used their legs to support their weight.

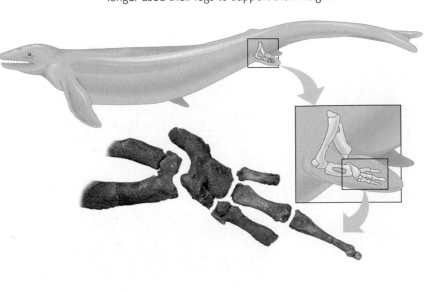

Biogeography

It was the geographic distribution of species, called **biogeography**, that first suggested to Darwin that today's organisms evolved from ancestral forms. Consider, for example, Darwin's visit to the Galápagos Islands. Darwin noted that the Galápagos animals resembled species of the South American mainland more than they resembled animals on similar but distant islands. The logical explanation was that the Galápagos species evolved from animals that had migrated from South America, each species adapting to its new environment.

Many other examples from biogeography seem baffling without an evolutionary perspective, such as the diversity of marsupials in Australia (**Figure 13.7**). Why is Australia home to so many kinds of marsupials (mammals that complete embryonic development outside the uterus, typically in a mother's pouch) but relatively few placental mammals (mammals that complete embryonic development in the uterus)? It is not because Australia is inhospitable to placental mammals. Humans have introduced rabbits, foxes, and many other placental mammals to Australia, where these introduced species have thrived to the point of becoming ecological and economic nuisances. The prevailing hypothesis is that the unique Australian wildlife evolved on that island continent in isolation from regions where early placental mammals diversified.

The geographic distribution of species makes little sense if we imagine that species were individually placed in suitable environments. In the Darwinian view, we find species where they are because they evolved from ancestors that inhabited those regions.

▼ Figure 13.7 **Biogeography.** The continent of Australia is home to many unique plants and animals, such as these marsupials, mammals that evolved in relative isolation from other continents where placental mammals diversified.

Common ringtail possum

Koala

Australia

Red kangaroo

Common wombat

Comparative Anatomy

The comparison of body structures in different species is called **comparative anatomy**. Certain anatomical similarities among species are signs of evolutionary history. For example, the same skeletal elements make up the forelimbs of humans, cats, whales, and bats, even though the functions of these forelimbs differ greatly; clearly, a whale's flipper does not do the same job as a bat's wing. If these limbs had been uniquely engineered in their current forms, we would expect a variety of basic designs that reflect their unique tasks. It is more logical that the arms, forelegs, flippers, and wings of these different mammals are variations on anatomical structures of an ancestral organism, structures that over millions of years have become adapted to different functions. These adaptations account for the observed similarities in structures despite different uses. Such similarity in structure due to common ancestry is called **homology**. The forelimbs of diverse mammals are therefore known as homologous structures (**Figure 13.8**).

Comparative anatomy attests that evolution is a remodeling process in which ancestral structures become modified as they take on new functions—the kind of process that Darwin referred to as descent with modification. The historical constraints of this modification are evident in anatomical imperfections. For example, the human spine and knee joint were derived from ancestral structures that supported four-legged mammals. Consequently, almost none of us will reach old age without experiencing knee or back problems. If these structures had first taken form specifically to support our bipedal posture, we would expect them to be less subject to sprains, spasms, and other common injuries.

Some of the most interesting homologous structures are "leftover" structures of marginal, if any, importance to the organism. These **vestigial structures** are remnants of features that served important functions in the organism's ancestors, such as the rear leg bones evident in ancient whale fossils (see Figure 13.6). In another example, the skeletons of some snakes retain vestiges of the pelvis and leg bones of walking ancestors. If limbs were a hindrance to ancient snakes' way of life, natural selection would favor snake descendants with successively smaller limbs. ☑

▼ Figure 13.8 **Homologous structures: anatomical signs of descent with modification.** The forelimbs of all mammals are constructed from the same skeletal elements. (Homologous bones in each of these four mammals are colored the same.) The hypothesis that all mammals descended from a common ancestor predicts that their forelimbs, though diversely adapted, would be variations on a common anatomical theme.

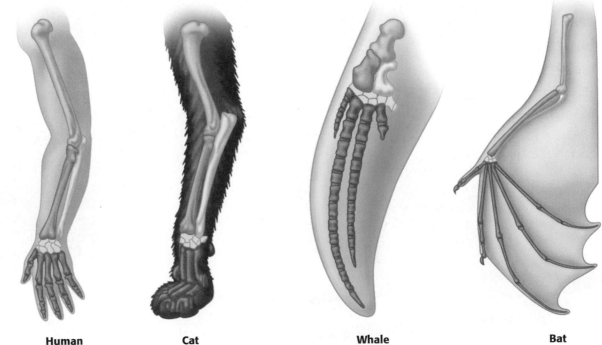

Human **Cat** **Whale** **Bat**

Comparative Embryology

Comparing early stages of development in different animal species reveals additional homologies not visible in adult organisms. For example, all vertebrate embryos have a developmental stage in which structures called pharyngeal pouches appear on the sides of the throat. At this stage, the embryos of fishes, frogs, snakes, birds, and apes—indeed, all vertebrates—look more alike than different (Figure 13.9). The different classes of vertebrates take on more distinctive features as development progresses. For example, pharyngeal pouches develop into gills in fishes but into parts of the ear and throat in humans.

Molecular Biology

As you saw in Chapter 10, the hereditary background of an organism is documented in its DNA and in the proteins encoded by the DNA. If two species have genes with nucleotide sequences that match closely (and thus proteins with amino acid sequences that match closely), biologists conclude that these sequences are homologous and must have been inherited from a relatively recent common ancestor (Figure 13.10). In contrast, the greater the number of sequence differences between species, the less likely they share a close common ancestor. Molecular comparisons between diverse organisms have allowed biologists to develop hypotheses about the evolutionary divergence of branches on the tree of life. For example, genetic analyses first suggested that the domain Archaea is more closely related to the eukaryotes than it is to the domain Bacteria.

Darwin's boldest hypothesis was that *all* forms of life are related to some extent through branching evolution from the earliest organisms. About 100 years after Darwin made his claim, molecular biology began providing strong evidence for evolution: All forms of life use the same genetic language of DNA and RNA, and the genetic code (how RNA triplets are translated into amino acids) is nearly universal. This genetic language has been passed along through all the branches of evolution since its beginnings in an early form of life. ✓

▼ Figure 13.9 **Evolutionary signs from comparative embryology.** At the early stage of development shown here, the kinship of vertebrates is unmistakable. Notice, for example, the pharyngeal pouches and tails in both the chicken embryo and the human embryo.

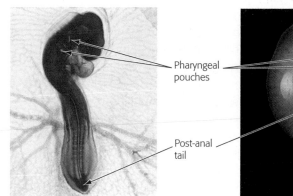
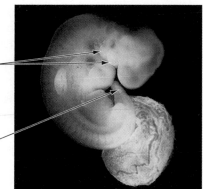

Pharyngeal pouches

Post-anal tail

Chicken embryo   **Human embryo**

▼ Figure 13.10 **Genetic relationships among some primates.** The bars in this diagram show the percent of selected DNA sequences that match between a chimpanzee and other primates, the animal group that includes monkeys, apes, and humans. For example, note that the selected DNA sequences of chimps and humans are better than a 98% match. In contrast, the DNA sequences of chimps and Old World monkeys (such as macaques, mandrills, baboons, and rhesus monkeys) have less than a 93% match.

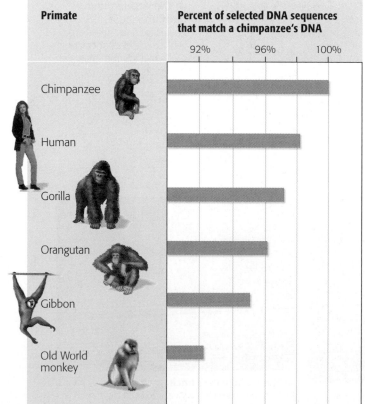

Primate | **Percent of selected DNA sequences that match a chimpanzee's DNA**

92%   96%   100%

Chimpanzee

Human

Gorilla

Orangutan

Gibbon

Old World monkey

✓CHECKPOINT

Name the five lines of evidence for evolution presented in this section.

Answer: the fossil record, biogeography, comparative anatomy, comparative embryology, molecular biology

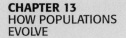

MP3 tutor sessions

Natural Selection

Natural Selection

Darwin perceived adaptation to the environment and the origin of new species as closely related processes. Imagine, for example, that an animal species from a mainland colonizes a chain of distant, relatively isolated islands. In the Darwinian view, populations on the different islands may diverge more and more in appearance as each population adapts to its local environment. Over many generations, the populations on different islands could become dissimilar enough to be designated as separate species. Evolution of the finches on the Galápagos Islands is an example. It is a reasonable hypothesis that the islands were colonized by finches that strayed from elsewhere and then diversified on the different islands. Among the differences between the Galápagos finches are their beaks, which are adapted to the specific foods available on each species' home island (Figure 13.11). Darwin anticipated that explaining how such adaptations arise is the key to understanding evolution. And his theory of natural selection remains our best explanation for the formation of new adaptations. ☑

☑CHECKPOINT

Explain how natural selection acts on new adaptations.

Answer: Natural selection is the process by which organisms with certain inherited adaptations are more likely to survive and reproduce than organisms that lack those adaptations.

Darwin's Theory of Natural Selection

Darwin based his theory of natural selection on two key observations, both of which had already been noted by scientists. First, he recognized that all species tend to produce excessive numbers of offspring (Figure 13.12). Darwin deduced that because natural resources are limited, the production of more individuals than the environment can support leads to a struggle for existence among the individuals of a population. In most cases, only a small percentage of offspring will survive in each generation. Many eggs are laid, young born, and seeds spread, but only a tiny fraction complete their development and leave offspring of their own. The rest are starved, eaten, frozen, diseased, unmated, or unable to reproduce for other reasons.

The second key observation that led Darwin to natural selection was his awareness of variation among individuals of a population. Just as no two people in a

▼ Figure 13.11 **Galápagos finches with beaks adapted for specific diets.**

(a) The large ground finch. This species of Galápagos finch has a large beak specialized for cracking seeds that fall from plants to the ground.

(b) The small tree finch. The smaller beak of the small tree finch is used to grasp insects.

(c) The woodpecker finch. The long, narrow beak of the woodpecker finch allows it to hold tools such as cactus spines to probe for wood-boring insects.

▼ Figure 13.12 **Overproduction of offspring.** A cloud of millions of spores is exploding from these puffballs, a type of fungus. (Each puffball in this photo is about 2 cm across.) Only a tiny fraction of the spores will actually give rise to offspring that survive and reproduce.

Spore cloud

▲ Figure 13.13 **Color variations within a single species of Asian lady beetles.**

human population are alike, individual variation abounds in all species (Figure 13.13). Much of this variation is heritable. Siblings share more traits with each other and with their parents than they do with less closely related members of the population.

From these two observations, Darwin arrived at the conclusion that defines natural selection: Individuals whose inherited traits are best suited to the local environment are more likely than less fit individuals to survive and reproduce. In other words, the individuals that function best should leave the most surviving offspring. Darwin's genius was in connecting two observations that anyone could make and drawing a conclusion that could explain how adaptations evolve.

Observation 1: **Overproduction and competition.**
Populations of all species have the potential to produce many more offspring than the environment can possibly support with food, space, and other resources. This overproduction makes a struggle for existence among individuals inevitable.

Observation 2: **Individual variation.**
Individuals in a population vary in many heritable traits.

Conclusion: **Unequal reproductive success.**
Those individuals with heritable traits best suited to the local environment generally leave a larger share of surviving, fertile offspring.

Darwin's insight was both simple and profound. The environment screens a population's inherent variability. Unequal success in reproduction (natural selection) leads to an accumulation of the favored traits in the population over generations (evolution). In other words, natural selection promotes evolutionary adaptations.

It is important to emphasize three key but subtle points about evolution by natural selection. The first point is that individuals do not evolve, even though natural selection occurs through interactions between individual organisms and their environment. Evolution refers to generation-to-generation changes in populations (as you'll see in the next section).

The second key point is that natural selection can amplify or diminish only heritable traits. An organism may become modified through interactions with the environment during its lifetime, and those acquired traits may help it survive. But unless these traits are coded for in the genes of the organism's gametes (sperm or egg), they cannot be passed on to offspring and therefore will not affect the reproductive success of the next generation.

The third key point is that evolution is not goal directed; it does not lead to perfectly adapted organisms. Natural selection is the result of environmental factors that vary from place to place and from time to time. A trait that is favorable in one situation may be useless—or even detrimental—in different circumstances. For example, some genetic mutations that happen to endow mosquitoes with resistance to the pesticide DDT also reduce a mosquito's growth rate. Before DDT was introduced, the gene for resistance was a handicap. But after DDT became part of the environment, the mutant alleles were advantageous, and natural selection increased their frequency in mosquito populations. This example shows that significant evolutionary change can occur in a short time—something we'll stress again in the next section. ☑

Natural Selection in Action

Natural selection and evolution are observable. You learned about one unsettling example in the Biology and Society section: the evolution of pesticide resistance within insect species. Pesticides control insects and prevent them from eating crops, transmitting diseases such as malaria, or just annoying us. But widespread use has led to the unintended evolution of pesticide-resistant insect populations (Figure 13.14). Chapter 1 described another example of natural selection in action: the evolution of antibiotic-resistant bacteria. More recently, doctors have documented an increase in drug-resistant strains of HIV, the virus that causes AIDS. The unifying thread sewn throughout this chapter is that evolution and natural selection did not operate only in the distant past, but are observable today. Let's look at another example of evolution in action.

☑CHECKPOINT

Explain why the following statement is incorrect: "Pesticides cause pesticide resistance in insects."

Answer: An environmental factor does not create new traits such as pesticide resistance, but favors traits that are already represented in the population.

▼ Figure 13.14 **Evolution of pesticide resistance in insect populations.** By spraying crops with poisons to kill insect pests, people have unwittingly favored the reproductive success of insects with inherent resistance to the poisons.

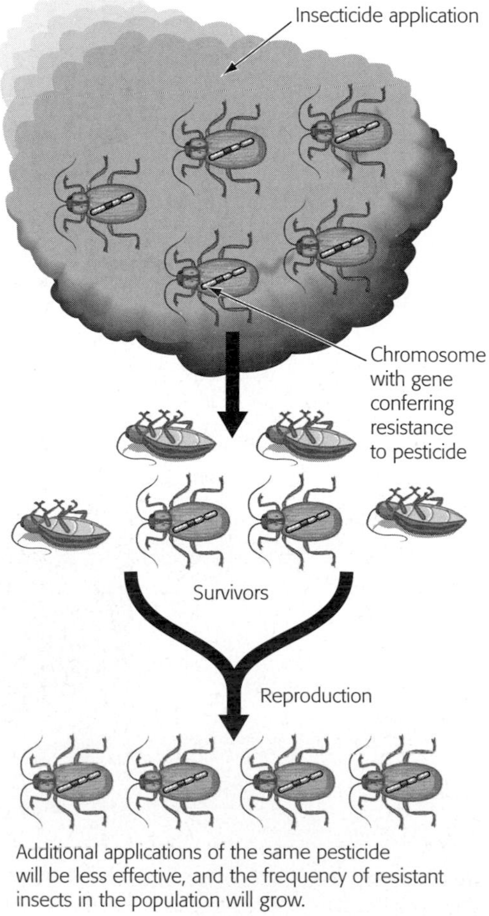

Insecticide application

Chromosome with gene conferring resistance to pesticide

Survivors

Reproduction

Additional applications of the same pesticide will be less effective, and the frequency of resistant insects in the population will grow.

Does Predation Drive the Evolution of Lizard Horn Length?

One recent and particularly elegant demonstration of evolution in action involved the flat-tailed horned lizard *Phrynosoma mcalli*, a desert inhabitant of the American Southwest. The lizard's main predator, a bird called the shrike, attacks by biting a lizard's neck just behind the skull, severing the spine. The shrike then carries the dead prey to a convenient place, such as a fence or branch, impales it, and eats it (Figure 13.15).

Evolutionary biologists at Utah State University and Indiana University made the **observation** that the flat-tailed horned lizard defends against attack by thrusting its head backward, stabbing the shrike with the spiked horns that protrude from the rear and sides of the skull. This led the researchers to **question** whether longer horn length and spread represented a survival advantage. Their **hypothesis** was that it did, so they formed the **prediction** that live horned lizards would have longer and more widely spread horns than killed ones.

The **experiment** to test this hypothesis was simple: The researchers measured the length of rear horns and the tip-to-tip spread distance of side horns from the skulls of 29 killed lizards (found where they had been impaled) and 155 live lizards. Their **results**, shown in the graph in Figure 13.15, indicate that the average horn length and spread of live lizards is about 10% greater than that of killed lizards (note that each blue bar is about 10% taller than the corresponding orange bar). The researchers concluded that defensive behavior against predators is one factor driving natural selection of horn length among this lizard species. Thus, evolution—in this case of the length of lizard horns—is observable and measurable in the world around us today.

(a) A flat-tailed horned lizard. The lizards use the spiky horns that protrude from the back and sides of the skull to ward off attacks.

◀ Figure 13.15 **The effect of predation on the evolution of lizard horn length.**
Researchers measured the horns of lizards killed by birds and compared them with the horns of live lizards. The fact that the length of the rear horns and the spread of the side horns are both significantly greater in live lizards than in killed lizards suggests that horn length is an adaptation that is evolving in response to predation by birds.

(b) The remains of a lizard impaled by a shrike. After killing a lizard, the bird often impales the lizard on a fence or branch.

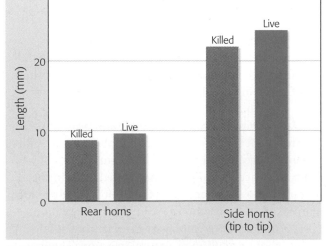

(c) Results of measurement of lizard horns.

Evolutionary Trees

In Darwin's view, the history of life is analogous to a tree. Patterns of descent branch off from a common trunk, the first organism, to the tips of millions of twigs representing the species living today. At each fork of the evolutionary tree is an ancestor common to all evolutionary branches extending from that fork. Closely related species share many traits because their lineage of common descent traces to a recent fork of the tree of life. Biologists represent these patterns of descent with an **evolutionary tree**, although today they usually turn the trees sideways as in Figure 13.16.

Homologous structures, both anatomical and molecular, can be used to determine the branching sequence of an evolutionary tree. Some homologous structures, such as the genetic code, are shared by all species because they date to the deep ancestral past. In contrast, traits that evolved more recently are shared by smaller groups of organisms. For example, all tetrapods (from the Greek *tetra*, four, and *pod*, foot) have the same basic limb bone structure illustrated in Figure 13.8, but their ancestors do not.

Figure 13.16 is an evolutionary tree of tetrapods (amphibians, mammals, and reptiles, including birds) and their closest living relatives, the lungfishes. In this diagram, each branch point represents the common ancestor of all species that descended from it. For example, lungfishes and tetrapods descended from ancestor ❶. Three homologies are shown by the blue dots on the tree— tetrapod limbs, the amnion (a protective embryonic membrane), and feathers. Tetrapod limbs were present in common ancestor ❷ and hence are found in its descendants (the tetrapods). The amnion was present in ancestor ❸ and thus is shared only by mammals and reptiles, which are known as amniotes. Feathers were present only in ancestor ❻ and hence are found only in birds.

Evolutionary trees are hypotheses reflecting our current understanding of patterns of evolutionary descent. Some trees are more speculative because sufficient data are not yet available. Other trees are based on a convincing combination of fossil, anatomical, and DNA data. ✓

✓CHECKPOINT

In Figure 13.16, which number represents the most recent common ancestor of humans and canaries?

Answer: *The most recent ancestor of humans (mammals) and canaries (birds) is* ❸.

▼ Figure 13.16 **An evolutionary tree of tetrapods (four-limbed animals).**

Each branch point represents the common ancestor of the lineages beginning there or to the right of it.

A blue dot represents a homologous trait shared by all the groups to the right of the mark.

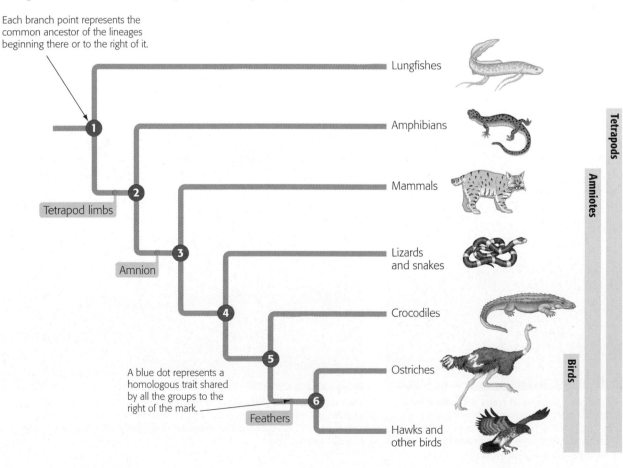

The Modern Synthesis: Darwinism Meets Genetics

Natural selection requires hereditary processes that Darwin could not explain. How do the variations that are the raw material for natural selection arise in a population? And how are these variations passed along from parents to offspring? Darwin and Gregor Mendel lived and worked at the same time. In fact, by breeding peas in his abbey garden, Mendel illuminated the very hereditary processes required for natural selection to work. However, Mendel's discoveries went unappreciated by the scientific community during his lifetime. Mendelism and Darwinism finally came together in the mid-1900s, decades after both scientists had died. This fusion of genetics with evolutionary biology came to be known as the **modern synthesis** (here, the term *synthesis* means "combination"). One of its key elements is an emphasis on the biology of populations.

Populations as the Units of Evolution

As noted earlier, individual organisms do not evolve during their lifetimes. Although natural selection affects an organism's survival and reproductive success, the evolutionary impact is only apparent in the changes in a population over time.

One population may be isolated from other populations of the same species, with little interbreeding and thus little exchange of genes between them. Such isolation is common for populations confined to widely separated islands, unconnected lakes, or mountain ranges separated by lowlands. However, populations are not usually so isolated, and they rarely have sharp boundaries (**Figure 13.17a**). One population center may blur into another in a region of overlap, where members of both populations are present but less numerous. Nevertheless, individuals are more concentrated in the population centers and are more likely to breed with other locals (**Figure 13.17b**). Therefore, organisms of a population are generally more closely related to one another than to members of other populations.

A population is the smallest biological unit that can evolve. In the Biology and Society section and Figure 13.14, you saw that natural selection can affect pesticide resistance in populations of insects. That impact is measured by the change in the relative numbers of resistant insects over a span of generations, not by the survivability of any individual insect.

▼ Figure 13.17 **Populations.**

(a) Two dense populations of trees separated by a lake. Interbreeding occurs when wind blows pollen between the populations. Nevertheless, trees are more likely to breed with members of the same population than with trees on the other side of the lake.

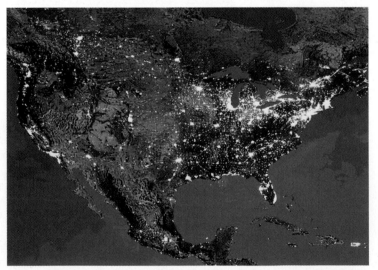

(b) A nighttime satellite view of North America. Notice the lights of human population centers, or cities. People move around, of course, and there are suburban and rural communities between cities, but people are more likely to choose mates locally.

In studying evolution at the population level, biologists focus on what is called the **gene pool**, the total collection of alleles in a population at any one time (that is, the total of all the alleles in all the individuals making up the population). For many genes, there are two or more alleles (versions of the gene) in the gene pool. For example, in an insect population, there may be one allele that codes for an enzyme that breaks down a certain pesticide and one for a version of the enzyme that does not. In fields sprayed with pesticide, the first allele will increase in frequency and the other allele will decrease in frequency. When the relative frequencies of alleles in a population change like this over a number of generations, evolution is occurring on its smallest scale. Later we'll look at the causes of evolution within a population. But first let's explore the sources of genetic variation. ☑

phenotypes; there are no in-between types. Next, we'll focus on the sources of different alleles.

Sources of Genetic Variation

Mutations and sexual reproduction, both involving random processes, produce genetic variation. Let's examine each of these processes in detail.

Mutations, random changes in the nucleotide sequence of DNA (see Figure 10.22), can result in new alleles. For example, a mutation in a gene may substitute one nucleotide for another. If such a change affects the protein's function, the mutation will probably be harmful. A random mutation is like a shot in the dark; it is not likely to improve a genome any more than shooting a bullet through the hood of a car is likely to improve engine performance.

☑**CHECKPOINT**

What is the smallest biological unit that can evolve?

Answer: a population

Genetic Variation in Populations

You have no trouble recognizing your friends in a crowd. People vary in appearance, reflecting the individual differences in their genomes. Individual variation abounds in all species (**Figure 13.18**). In addition to visible differences, most populations vary greatly at the molecular level. For example, you cannot tell someone's blood group (A, B, AB, or O) just by looking at that person, but biochemical analyses can reveal such variations.

Not all variation in a population is heritable. Phenotype results from a combination of the genotype, which is inherited, and many environmental influences. For instance, a strength-training program can build up your muscle mass beyond what would naturally occur from your genetic makeup. However, you would not pass this environmentally induced physique on to your offspring. Only the genetic component of variation is relevant to natural selection.

Many of the variable traits in a population result from the combined effect of several genes. This polygenic ("many genes") inheritance produces traits that vary more or less continuously—in human height, for instance, from very short individuals to very tall ones. By contrast, other features, such as human blood group, are determined by a single gene locus, with different alleles producing one of only a few distinct

▼ Figure 13.18 **Variation in a garter snake population.** These four garter snakes, which belong to the same species, were all captured in one Oregon field. The behavior of each physical type is correlated with its coloration. When approached, spotted snakes, which blend in with their background, generally freeze. In contrast, snakes with stripes, which make it difficult to judge the speed of motion, usually flee rapidly when approached.

On rare occasions, however, a mutant allele may actually enhance reproductive success. This kind of effect is more likely when the environment is changing in such a way that alleles that were once disadvantageous are favorable under the new conditions. We already considered one example, the changing fortunes of the DDT resistance allele in insect populations.

A new mutation that is transmitted in gametes can immediately change the gene pool of a population by replacing one allele with another. For any given gene locus, mutation alone has little effect on a large population in a single generation. This is because a mutation at any given gene locus is a very rare event. However, the cumulative impact of mutations across the genome can be significant because an individual has thousands of genes, and many populations have thousands or millions of individuals.

Organisms with very short generation spans, such as bacteria, can evolve rapidly with mutation as the only source of genetic variation. Bacteria multiply so quickly that natural selection can increase a population's frequency of a beneficial mutation in just hours or days. For most animals and plants, however, their long generation times prevent new mutations from significantly affecting overall genetic variation in the short term. Consequently, in sexually reproducing organisms, most of the genetic variation in a population results from the unique combination of alleles that each individual inherits. (Of course, the origin of that allele variation is past mutation.)

As you saw in Chapter 8, fresh assortments of existing alleles arise every generation from three random components of sexual reproduction: independent orientation of homologous chromosomes at metaphase I of meiosis (see Figure 8.16), random fertilization, and crossing over (see Figure 8.18). During prophase I of meiosis, pairs of homologous chromosomes, one set inherited from each parent, trade some of their genes by crossing over, and then each pair separate into gametes independently of other chromosome pairs. Gametes from one individual vary extensively in their genetic makeup, and each zygote made by a mating pair has a unique assortment of alleles resulting from the random union of sperm and egg.

While the processes that generate genetic variation—mutation and events during sexual reproduction—are random, natural selection (and hence evolution) is not. The environment selectively promotes the propagation of those genetic combinations that enhance survival and reproductive success. ☑

Analyzing Gene Pools

As described earlier, a gene pool consists of all the alleles in a population at any one time. The gene pool is the reservoir from which the next generation of organisms draws its genes.

Imagine a wildflower population with two varieties contrasting in flower color (**Figure 13.19**). An allele for red flowers, which we will symbolize by R, is dominant to an allele for white flowers, symbolized by r. These are the only two alleles for flower color in the gene pool of this hypothetical plant population. Now, let's say that 80%, or 0.8, of all flower-color loci in the gene pool have the R allele. We'll use the letter p to represent the relative frequency of the R allele in the population. Thus, $p = 0.8$. Because there are only two alleles in this example, the r allele must be present at the other 20% (0.2) of the gene pool's flower-color loci. Let's use the letter q for the frequency of the r allele in the population. For the wildflower population, $q = 0.2$. And since there are only two alleles for flower color, we know that we can express their frequencies as follows:

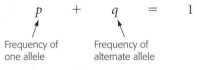

$$p + q = 1$$

Frequency of one allele   Frequency of alternate allele

Notice that if we know the frequency of either allele in the gene pool, we can subtract it from 1 to calculate the frequency of the other allele.

From the frequencies of alleles, we can also calculate the frequencies of different genotypes in the population if the gene pool is completely stable (not evolving). In the wildflower population, what is the probability of producing an RR individual by "drawing" two R alleles from the pool of gametes? Here we apply the rule of multiplication that you learned in Chapter 9 (review Figure 9.11). The probability of drawing an R sperm

▼ Figure 13.19 **A population of wildflowers with two varieties of color.**

multiplied by the probability of drawing an *R* egg is $p \times p = p^2$, or $0.8 \times 0.8 = 0.64$. In other words, 64% of the plants in the population will have the *RR* genotype. Applying the same math, we also know the frequency of *rr* individuals in the population: $q^2 = 0.2 \times 0.2 = 0.04$. Thus, 4% of the plants are *rr*, giving them white flowers. Calculating the frequency of heterozygous individuals, *Rr*, is trickier. That's because the heterozygous genotype can form in two ways, depending on whether the sperm or egg supplies the dominant allele. So the frequency of the *Rr* genotype is $2pq$, which is $2 \times 0.8 \times 0.2 = 0.32$. In our imaginary wildflower population, 32% of the plants are *Rr*, with red flowers. **Figure 13.20** reviews these calculations graphically.

Now we can write a general formula for calculating the frequencies of genotypes in a gene pool from the frequencies of alleles, and vice versa:

$$p^2 \;+\; 2pq \;+\; q^2 \;=\; 1$$

Frequency of homozygotes for one allele   Frequency of heterozygotes   Frequency of homozygotes for alternate allele

Notice that the frequencies of all genotypes in the gene pool must add up to 1. This formula is called the Hardy-Weinberg formula, named for the two scientists who derived it in 1908.

▼ **Figure 13.20 A mathematical swim in the gene pool.** Each of the four boxes in the grid corresponds to an equally probable "draw" of alleles from the gene pool.

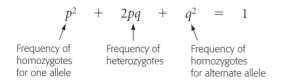

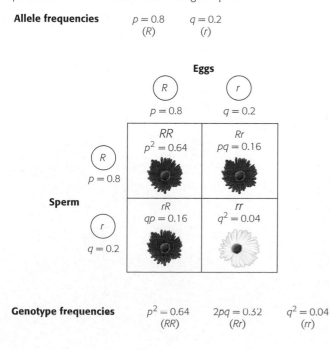

Allele frequencies $p = 0.8$ (*R*) $q = 0.2$ (*r*)

Eggs

	R $p = 0.8$	*r* $q = 0.2$
R $p = 0.8$	*RR* $p^2 = 0.64$	*Rr* $pq = 0.16$
r $q = 0.2$	*rR* $qp = 0.16$	*rr* $q^2 = 0.04$

Sperm

Genotype frequencies $p^2 = 0.64$ (*RR*) $2pq = 0.32$ (*Rr*) $q^2 = 0.04$ (*rr*)

Population Genetics and Health Science

Public health scientists use the Hardy-Weinberg formula to calculate the percentage of a human population that carries the allele for certain inherited diseases. Consider phenylketonuria (PKU), which is an inherited inability to break down the amino acid phenylalanine. If untreated, the disorder causes severe mental retardation. PKU occurs in about 1 out of 10,000 babies born in the United States. Newborn babies are now routinely tested for PKU, and symptoms can be prevented if individuals living with the disease follow a strict diet (**Figure 13.21**).

PKU is caused by a recessive allele (that is, one that must be present in two copies to produce the phenotype). Thus, we can represent the frequency of individuals in the U.S. population born with PKU with the q^2 term in the Hardy-Weinberg formula. For one PKU occurrence per 10,000 births, $q^2 = 0.0001$. Therefore, q, the frequency of the recessive allele in the population, equals the square root of 0.0001, or 0.01. And p, the frequency of the dominant allele, equals $1 - q$, or 0.99.

Now let's calculate the frequency of carriers, who are heterozygous individuals who carry the PKU allele in a single copy and may pass it on to offspring. Carriers are represented in the formula by $2pq$: $2 \times 0.99 \times 0.01$, or 0.0198. Thus, the Hardy-Weinberg formula tells us that about 2% of the U.S. population are carriers for the PKU allele. Estimating the frequency of a harmful allele is essential for any public health program dealing with genetic diseases.

Microevolution as Change in a Gene Pool

As stated earlier, evolution can be measured as changes in the genetic composition of a population over time. It helps, as a basis of comparison, to know what to expect if a population is *not* evolving. A nonevolving population is in genetic equilibrium, which is also known as **Hardy-Weinberg equilibrium**. The population's gene pool remains constant. From generation to generation, the frequencies of alleles (p and q) and genotypes (p^2, $2pq$, and q^2) are unchanged. Sexual shuffling of genes cannot by itself change a large gene pool.

One of the products of the modern synthesis was a definition of evolution that is based on the genetics of populations: Evolution is a generation-to-generation change in a population's frequencies of alleles. Because this is evolution viewed on the smallest scale, it is sometimes referred to as **microevolution**. ☑

INGREDIENTS: SORBITOL, MAGNESIUM STEARATE, ARTIFICIAL FLAVOR, **ASPARTAME†** (SWEETENER), ARTIFICIAL COLOR (YELLOW 5 LAKE, BLUE 1 LAKE), ZINC GLUCONATE. **†PHENYLKETONURICS: CONTAINS PHENYLALANINE.**

▲ **Figure 13.21 A warning to individuals with PKU.** People with PKU (phenylketonuria) must strictly regulate their dietary intake of the amino acid phenylalanine. In addition to natural sources, phenylalanine is found in aspartame, a common artificial sweetener. The frequency of the PKU allele is high enough to warrant warnings on foods that contain phenylalanine.

☑ **CHECKPOINT**

1. Which term in the Hardy-Weinberg formula ($p^2 + 2pq + q^2 = 1$) corresponds to the frequency of individuals with *no* alleles for the recessive disease PKU?
2. Define microevolution.

Answers: 1. p^2 2. Microevolution is a change in a population's frequencies of alleles.

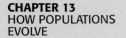

Mechanisms of Evolution

Now that we understand that evolution can be evident in changes in a population's genetic makeup from generation to generation (microevolution), we come to an obvious question: What mechanisms can change a gene pool? The three main causes of evolutionary change are genetic drift, gene flow, and natural selection.

Genetic Drift

Flip a coin 1,000 times, and a result of 700 heads and 300 tails would make you very suspicious about that coin. But flip a coin 10 times, and an outcome of 7 heads and 3 tails would seem within reason. With a smaller sample, there is a greater chance of deviation from an idealized result—in this case, an equal number of heads and tails.

Let's apply coin toss logic to a population's gene pool. If a new generation draws its alleles at random from the previous generation, then the larger the population (the sample size), the better the new generation will represent the gene pool of the previous generation. Thus, one requirement for a gene pool to maintain the status quo is a large population size. The gene pool of a small population may not be accurately represented in the next generation because of sampling error. The changed gene pool is analogous to the erratic outcome from a small sample of coin tosses.

Figure 13.22 applies this concept of sampling error to a small population of wildflowers. Chance causes the frequencies of the alleles for red (*R*) and white (*r*) flowers to change over the generations. And that fits our definition of microevolution. This evolutionary mechanism, a change in the gene pool of a population due to chance, is called **genetic drift**. But what would cause a population to shrink down to a size where there is genetic drift? Two ways this can occur are the bottleneck effect and the founder effect.

The Bottleneck Effect

Disasters such as earthquakes, floods, and fires may kill large numbers of individuals, producing a small surviving population that is unlikely to have the same genetic makeup as the original population. By chance, certain alleles may be overrepresented among the survivors. Other alleles may be underrepresented. And some alleles may be eliminated. Chance may continue to change the gene pool for many generations until the population is again large enough for sampling errors to be insignificant.

▼ Figure 13.22 **Genetic drift.** This hypothetical wildflower population consists of only ten plants. Due to random change over the generations, genetic drift can eliminate some alleles, as is the case for the *r* allele in generation 3 of this imaginary population.

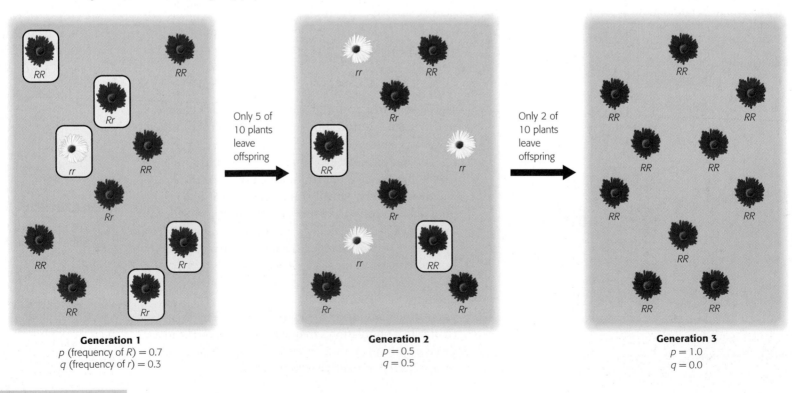

Generation 1
p (frequency of *R*) = 0.7
q (frequency of *r*) = 0.3

Generation 2
p = 0.5
q = 0.5

Generation 3
p = 1.0
q = 0.0

The analogy in **Figure 13.23** illustrates why genetic drift due to a drastic reduction in population size is called the **bottleneck effect**.

Bottlenecking usually reduces the overall genetic variability in a population because at least some alleles are likely to be lost from the gene pool. An important

▼ **Figure 13.23 The bottleneck effect.** The colored marbles in this analogy represent three alleles in an imaginary population. Shaking just a few of the marbles through the bottleneck is like drastically reducing the size of a population struck by some environmental disaster. By chance, purple marbles are overrepresented in the new population, green marbles are underrepresented, and orange marbles are absent. Similarly, bottlenecking a population of organisms tends to reduce variability.

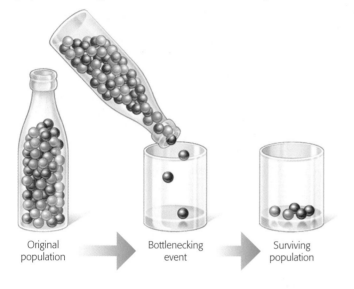

Original population → Bottlenecking event → Surviving population

application of this concept is the potential loss of individual variation, and hence adaptability, in bottlenecked populations of endangered species, such as the cheetah (**Figure 13.24**). The fastest of all running animals, cheetahs are magnificent cats that were once widespread in Africa and Asia. Like many African mammals, the number of cheetahs fell drastically during the last ice age (around 10,000 years ago). At that time, the species suffered a severe bottleneck, possibly as a result of disease, human hunting, and periodic droughts. Evidence suggests that the South African cheetah population suffered a second bottleneck during the 1800s, when farmers hunted the animals to near extinction. Today, only a few small populations of cheetahs exist in the wild. Genetic variability in these populations is very low compared with populations of other mammals. This lack of variability, coupled with an increasing loss of habitat, makes the cheetah's future precarious. The cheetahs remaining in Africa are being crowded into nature preserves and parks as human demands on the land increase. Along with crowding comes an increased potential for the spread of disease. With so little variability, the cheetah has a reduced capacity to adapt to such environmental challenges. Captive breeding programs are already under way and may be required for the cheetah's long-term survival. ☑

☑**CHECKPOINT**

Would you expect modern cheetahs to have more genetic variation or less genetic variation than cheetahs did 1,000 years ago?

Answer: less, because the bottleneck effect reduces genetic variability

▼ **Figure 13.24 Implications of the bottleneck effect in conservation biology.** Some endangered species, such as the cheetah, have low genetic variability. As a result, they are less adaptable to environmental changes, such as new diseases, than are species with a greater resource of genetic variation.

The Founder Effect

Genetic drift is also likely when a few individuals colonize an isolated island, lake, or other new habitat. The smaller the colony, the less its genetic makeup will match the gene pool of the larger population from which the colonists emigrated. If the colony succeeds, random drift will continue to affect the frequency of alleles until the population is large enough for genetic drift to be minimal.

The type of genetic drift resulting from the establishment of a small, new population whose gene pool differs from that of the parent population is called the **founder effect**. The founder effect undoubtedly contributed to the evolutionary divergence of the finches and other organisms that arrived as strays on the remote Galápagos Islands that Darwin visited.

The founder effect also explains the relatively high frequency of certain inherited disorders in some small human populations. In 1814, a group of 15 people founded a British colony on Tristan da Cunha, a cluster of small islands in the middle of the Atlantic Ocean (**Figure 13.25**). Apparently, one of the colonists carried a recessive allele for retinitis pigmentosa, a progressive form of blindness. Of the 240 descendants who still lived on the islands in the 1960s, 4 had retinitis pigmentosa, and 9 others were heterozygous carriers with one copy of the recessive allele. The frequency of the retinitis pigmentosa allele was ten times higher on Tristan da Cunha than in the British population from which the founders came.

▼ Figure 13.25 **Residents of Tristan da Cunha in the early 1900s.** The island of Tristan da Cunha, located in the middle of the Atlantic Ocean, is listed in the *Guinness World Records* as the world's most remote inhabited island. The genetic isolation of the island's residents resulted in a disproportionately high rate of hereditary blindness.

▲ Figure 13.26 **Human gene flow.** Human migration is transferring alleles between populations that were once isolated. This magazine cover celebrates our changing gene pools and culture with a computer-generated image blending facial features from several races.

Gene Flow

Another source of evolutionary change, separate from genetic drift, is **gene flow**, where a population may gain or lose alleles when fertile individuals move into or out of the population or when gametes (such as plant pollen) are transferred between populations. Gene flow tends to reduce differences between populations. For example, because humans today move more freely about the world than in the past, gene flow has become an important agent of evolutionary change in previously isolated human populations (**Figure 13.26**). ☑

Natural Selection: A Closer Look

Genetic drift, gene flow, and even mutation can cause evolution, but they do not necessarily lead to adaptation. In fact, random drift, migrant alleles, and shot-in-the-dark mutations are unlikely to improve a population's fit to its environment. Of all causes of evolution, only natural selection promotes adaptation. And such evolutionary adaptation, remember, is a blend of chance and sorting—chance in the random generation of genetic variability, and sorting in the unequal reproductive success among the varying individuals. Darwin explained the basics of natural selection. But it took the modern synthesis to fill in the details.

Hardy-Weinberg equilibrium, which defines a nonevolving population, demands that all individuals in a population be equal in their ability to survive and reproduce. This condition is probably never actually met. On average, individuals that function best in the environment leave the most offspring and therefore have a disproportionate impact on the gene pool. When farmers began spraying their fields with pesticides, resistant pests started outreproducing other members of the insect populations. This spraying increased the frequency of alleles for pesticide resistance in gene pools. As a result, evolution by natural selection occurred.

Darwinian Fitness

The phrases "struggle for existence" and "survival of the fittest" are misleading if we take them to mean direct competitive contests between individuals. There *are* animal species in which individuals lock horns or otherwise fight one another to determine mating privilege. But

▲ **Figure 13.27 Darwinian fitness of some flowering plants depends in part on competition in attracting pollinators.**

is most common when the local environment changes or when organisms migrate to a new environment. An actual example is the shift of insect populations toward a greater frequency of pesticide-resistant individuals.

Disruptive selection can lead to a balance between two or more contrasting phenotypes in a population (Figure 13.28b). A patchy environment, which favors different phenotypes in different patches, is one situation associated with disruptive selection. The variations seen in the snake population in Figure 13.18 result from disruptive selection.

Stabilizing selection favors intermediate phenotypes (Figure 13.28c). It typically occurs in relatively stable environments, where conditions tend to reduce physical variation. This evolutionary conservatism works by selecting against the more extreme phenotypes. For example, stabilizing selection keeps the majority of human birth weights between 3 and 4 kg (approximately 6.5 to 9 pounds). For babies much lighter or heavier than this, infant mortality is greater.

reproductive success is generally more subtle and passive. Plants in a wildflower population, for example, may differ in reproductive success because some attract more pollinators—perhaps the result of slight differences in flower color, shape, or fragrance (Figure 13.27). A frog may produce more eggs than her neighbors because she is better at catching insects for food. These examples point to a biological definition of **fitness**: the contribution an individual makes to the gene pool of the next generation relative to the contributions of other individuals. Thus, the fittest individuals in the context of evolution are those that produce the largest numbers of viable, fertile offspring and thus pass on the most genes to the next generation. ☑

Three General Outcomes of Natural Selection

Imagine a population of mice with individuals ranging in fur color from very light to very dark gray. If we graph the number of mice in each color category, we get a bell-shaped curve, like the curve some of your instructors draw after grading an exam. If natural selection favors certain fur-color phenotypes over others, the population of mice will change over the generations. Three general outcomes are possible, depending on which phenotypes are favored. These three modes of natural selection are called directional selection, disruptive selection, and stabilizing selection.

Directional selection shifts the overall makeup of a population by selecting in favor of one extreme phenotype—the darkest mice, for example (Figure 13.28a). Directional selection

☑**CHECKPOINT**

What is the best measure of Darwinian fitness?

Answer: the number of fertile offspring an individual leaves

◀ **Figure 13.28 Three possible outcomes for selection working on fur color in imaginary populations of mice.** The large downward arrows symbolize the pressure of natural selection working against certain phenotypes.

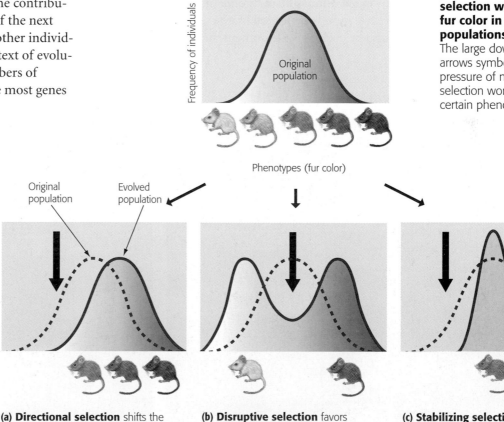

(a) **Directional selection** shifts the overall makeup of the population by favoring variants at one extreme. In this case, the trend is toward darker color, perhaps because the landscape has been shaded by the growth of trees, making darker mice less noticeable to predators.

(b) **Disruptive selection** favors variants at opposite extremes over intermediate individuals. Here, the relative frequencies of very light and very dark mice have increased. Perhaps the mice have colonized a patchy habitat where a background of light soil is studded with dark rocks.

(c) **Stabilizing selection** culls extreme variants from the population, in this case eliminating individuals that are unusually light or dark. The trend is toward reduced phenotypic variation and maintenance of the status quo.

Of the three selection modes, stabilizing selection prevails most of the time, resisting change in well-adapted populations. Evolutionary spurts occur when a population is stressed by a change in the environment or by migration to a new place. When challenged with a new set of environmental problems, a population either adapts through natural selection or dies off in that locale. The fossil record tells us that extinction is the most common result. Those populations that do survive crises may change enough to be designated new species, as we will see in Chapter 14. ☑

Sexual Selection

Darwin was the first to explore the implications of **sexual selection**, a form of natural selection in which individuals with certain traits are more likely than other individuals to obtain mates. The males and females of an animal species obviously have different reproductive organs. But they may also have secondary sexual traits, noticeable differences not directly associated with reproduction or survival. This distinction in appearance, called **sexual dimorphism**, is often manifested in a size difference. Among male vertebrates, sexual dimorphism may also be evident in adornment, such as manes on

lions, antlers on deer, or colorful plumage on peacocks and other birds (**Figure 13.29a**).

In some species, secondary sex structures may be used to compete with members of the same sex (usually males) for mates. Contests may involve physical combat, but are more often ritualized displays (**Figure 13.29b**). Such selection is common in species where the winning individual acquires a harem of mates (an obvious boost to that male's evolutionary fitness).

In a more common type of sexual selection, individuals of one sex (usually females) are choosy in selecting their mates. Males with the largest or most colorful adornments are often the most attractive to females. The extraordinary feathers of a peacock's tail are an example of this sort of "choose me!" statement. Every time a female chooses a mate based on a certain appearance or behavior, she perpetuates the alleles that caused her to make that choice and allows a male with that particular phenotype to perpetuate his alleles.

What is the advantage to females of being choosy? One hypothesis is that females prefer male traits that are correlated with "good" genes. In several bird species, research has shown that traits preferred by females, such as bright beaks or long tails, are related to overall male health. This link—between genes and health—brings us to this chapter's final example of evolution in action.

☑CHECKPOINT

The thickness of fur in a bear population increases over several generations as the climate in the region becomes colder. This is an example of which type of selection: directional, disruptive, or stabilizing?

Answer: *directional*

▼ Figure 13.29 **Sexual dimorphism.**

(a) Sexual dimorphism in a finch species. Among vertebrates, including this pair of green-winged Pytilia (native to Africa), males (right) are usually the showier sex.

(b) Competing for mates. Male Spanish ibex engage in nonlethal combat for the right to mate with females.

The Genetics of the Sickle-Cell Allele

▼ Figure 13.30 **Mapping malaria and the sickle-cell allele.** The inset shows sickled red blood cells.

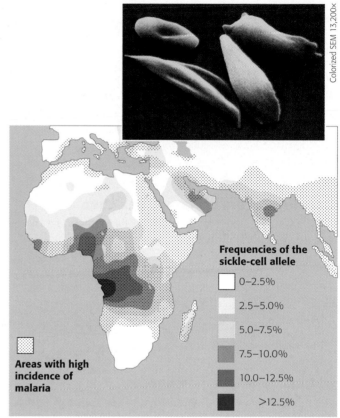

Colorized SEM 13,200x

Frequencies of the sickle-cell allele

- 0–2.5%
- 2.5–5.0%
- 5.0–7.5%
- 7.5–10.0%
- 10.0–12.5%
- >12.5%

Areas with high incidence of malaria

Throughout this book, we emphasize evolution as biology's unifying theme. Within this chapter, we have highlighted examples of evolution in action, observable instances of evolution by natural selection in our world today. Let's continue that thread by considering one illness in an evolutionary context: sickle-cell disease.

About 1 out of every 400 African-Americans has sickle-cell disease, a genetic disorder in which oxygen delivery by the blood is impaired due to abnormally shaped red blood cells. Red blood cell sickling causes periodic painful episodes as well as some potentially life-threatening complications, such as stroke. Sickle-cell disease is caused by a recessive allele. Only homozygous individuals, who inherit the recessive allele from both parents, have the disorder. About one in ten African-Americans has a single copy of the sickle-cell allele (as do a very small number from other ethnic groups). These heterozygous individuals do not have sickle-cell disease but can pass the allele for the disorder on to their children.

Why is the sickle-cell allele so much more common in African-Americans than in the general U.S. population? And how can we explain such a high frequency among African-Americans for an allele with the potential to shorten life (and hence reduce reproductive success)? Evolutionary biology holds the answers.

In the African tropics, the sickle-cell allele is both boon and bane. It is true that when two copies are inherited, the allele causes a life-threatening disease. But heterozygous individuals, who have just one copy of the sickle-cell allele, are relatively resistant to the symptoms of malaria. This is an important advantage in tropical regions where malaria, caused by a parasitic microorganism, is a major cause of death. The frequency of the sickle-cell allele in Africa is generally highest in areas where the malaria parasite is most common (Figure 13.30). The representation of the allele among black Americans is thus a vestige of African roots.

This example of the intersection of evolutionary biology and health science is a reminder that biology is the foundation of all medicine. And evolution is the foundation of all biology.

Chapter Review

SUMMARY OF KEY CONCEPTS

(MB) Go to the Study Area at **www.masteringbiology.com** for practice quizzes, myⓔBook, BioFlix™ 3-D animations, MP3 Tutor Sessions, videos, current events, and more.

Charles Darwin and *The Origin of Species*

Charles Darwin established the ideas of evolution and natural selection in his 1859 publication *On the Origin of Species by Means of Natural Selection*.

Darwin's Cultural and Scientific Context

During his around-the-world voyage on the *Beagle*, Darwin observed adaptations of organisms that inhabited diverse environments. In particular, Darwin was struck by the geographic distribution of organisms on the Galápagos Islands, off the South American coast. When Darwin considered his observations in light of new evidence for a very old Earth that changed slowly, he arrived at ideas that were at odds with the long-held notion of a young Earth populated by unrelated and unchanging species.

Descent with Modification

Darwin made two proposals in *The Origin of Species*:
(1) Existing species descended from ancestral species, and
(2) natural selection is the mechanism of evolution.

Evidence of Evolution

The Fossil Record

The fossil record shows that organisms have appeared in a historical sequence, and many fossils link ancestral species with those living today.

Biogeography

Biogeography, the study of the geographic distribution of species, suggests that species evolved from ancestors that inhabited the same region.

Comparative Anatomy

Homologous structures among species and vestigial organs provide evidence of evolutionary history.

Comparative Embryology

Closely related species often have similar stages in their embryonic development.

Molecular Biology

All species share a common genetic code, suggesting that all forms of life are related through branching evolution from the earliest organisms. Comparisons of DNA and proteins provide evidence of evolutionary relationships.

Natural Selection

Darwin's Theory of Natural Selection

Individuals best suited for a particular environment are more likely to survive and reproduce than less fit individuals.

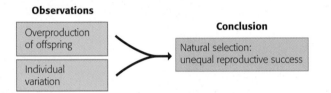

Natural Selection in Action

Natural selection can be observed in the evolution of pesticide-resistant insects, drug-resistant microbes, and horned lizards, among many other organisms.

Evolutionary Trees

An evolutionary tree represents a succession of related species, with the most recent at the tips of the branches. Each branch point represents a common ancestor of all species that radiate from it.

The Modern Synthesis: Darwinism Meets Genetics

The modern synthesis fused genetics (Mendelism) and evolutionary biology (Darwinism) in the mid-1900s.

Populations as the Units of Evolution

A population, members of the same species living in the same time and place, is the smallest biological unit that can evolve. Population genetics emphasizes the extensive genetic variation within populations and tracks the genetic makeup of populations over time.

Genetic Variation in Populations

Polygenic ("many genes") inheritance produces traits that vary continuously, whereas traits determined by one genetic locus may be present in two or more distinct forms. Mutation and sexual reproduction produce genetic variation. Individual mutations have little short-term effect on a large gene pool, but in the long term, mutation is the source of genetic variation.

Analyzing Gene Pools

A gene pool consists of all the alleles in all the individuals making up a population. The Hardy-Weinberg formula can be used to calculate the frequencies of genotypes in a gene pool from the frequencies of alleles, and vice versa:

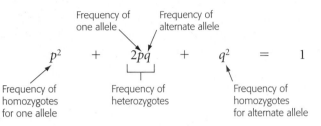

$$p^2 + 2pq + q^2 = 1$$

Population Genetics and Health Science

The Hardy-Weinberg formula can be used to estimate the frequency of a harmful allele, which is useful information for public health programs dealing with genetic diseases.

Microevolution as Change in a Gene Pool

Microevolution is generation-to-generation change in a population's frequencies of alleles.

Mechanisms of Evolution

Genetic Drift

Genetic drift is a change in the gene pool of a small population due to chance. Bottlenecking (a drastic reduction in population size) and the founder effect (occurring in a new population started by a few individuals) are two situations leading to genetic drift.

Gene Flow

A population may gain or lose alleles by gene flow, which is genetic exchange with another population.

Natural Selection: A Closer Look

Of all causes of evolution, only natural selection promotes evolutionary adaptations. Darwinian fitness is the contribution an individual makes to the gene pool of the next generation relative to the contributions of other individuals. The outcome of natural selection may be directional, disruptive, or stabilizing.

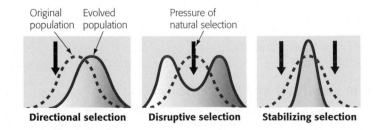

Sexual Selection

Secondary sexual traits (such as sex-specific plumage or behaviors) can promote sexual selection, a type of natural selection where inherited traits determine mating preferences.

SELF-QUIZ

1. Which of the following is *not* an observation or inference on which Darwin's theory of natural selection is based?
   a. There is heritable variation among individuals.
   b. Poorly adapted individuals never produce offspring.
   c. Because excessive numbers of offspring are produced, there is a competition for limited resources.
   d. Individuals whose inherited traits best fit them to the environment will generally produce more offspring.

2. Which of the following is a true statement about Charles Darwin?
   a. He was the first to discover that living things can change, or evolve.
   b. He based his theory on the inheritance of acquired traits.
   c. He proposed natural selection as the mechanism of evolution.
   d. He was the first to realize that Earth is more than 6,000 years old.

3. What is gradualism? How did Darwin apply that idea to the evolution of life?

4. In a population with two alleles for a particular genetic locus, *B* and *b*, the allele frequency of *B* is 0.7. If this population is in Hardy-Weinberg equilibrium, the frequency of heterozygotes is _____, the frequency of homozygous dominants is _____, and the frequency of homozygous recessives is _____.

5. Define fitness from an evolutionary perspective.

6. The processes of _____ and _____ generate variation, and _____ produces adaptation to the environment.
   a. sexual reproduction . . . natural selection . . . mutation
   b. mutation . . . sexual reproduction . . . genetic drift
   c. genetic drift . . . mutation . . . sexual reproduction
   d. mutation . . . natural selection . . . sexual reproduction
   e. mutation . . . sexual reproduction . . . natural selection

7. As a mechanism of evolution, natural selection can be most closely equated with
   a. random mating.
   b. genetic drift.
   c. unequal reproductive success.
   d. gene flow.

8. Compare and contrast how bottlenecking and a founder event can lead to genetic drift.

9. In a particular bird species, individuals with average-sized wings survive severe storms more successfully than other birds in the same population with longer or shorter wings. Of the three general outcomes of natural selection (directional, disruptive, or stabilizing), this example illustrates _____.

10. Which of the following statements is (are) true about a population in Hardy-Weinberg equilibrium? (More than one may be true.)
    a. The population is quite small.
    b. The population is not evolving.
    c. Gene flow between the population and surrounding populations does not occur.
    d. Natural selection is not occurring.

11. What environmental factor accounts for the relatively high frequency of the sickle-cell allele in tropical Africa?

Answers to the Self-Quiz questions can be found in Appendix D.

THE PROCESS OF SCIENCE

12. A population of snails has recently become established in a new region. The snails are preyed on by birds that break the snails open on rocks, eat the soft bodies, and leave the shells. The snails occur in both striped and unstriped forms. In one area, researchers counted both live snails and broken shells. Their data are summarized here:

	Striped shells	Unstriped shells
Number of live snails	264	296
Number of broken snail shells	486	377
Total	750	673

Based on these data, which snail form is more subject to predation by birds? Predict how the frequencies of striped and unstriped individuals might change over the generations.

13. Imagine that the presence or absence of stripes on the snails from the previous question is determined by a single gene locus, with the dominant allele (*S*) producing striped snails and the recessive allele (*s*) producing unstriped snails. Combining the data from both the living snails and broken shells, calculate the following: the frequency of the dominant allele, the frequency of the recessive allele, and the number of heterozygotes in the observed groups.

BIOLOGY AND SOCIETY

14. To what extent are humans in a technological society exempt from natural selection? Explain your answer.

14 How Biological Diversity Evolves

An endangered species.
These mountain gorillas,
photographed in Rwanda,
are just one of many
species threatened with
extinction.

CHAPTER CONTENTS

■■■ Chapter Thread: **MASS EXTINCTIONS**

BIOLOGY AND SOCIETY: Mass Extinctions

The Sixth Mass Extinction

The fossil record reveals that the evolutionary history of life on Earth has been episodic, with long, relatively stable periods punctuated by brief, cataclysmic ones. During these upheavals, new species formed and others died out in great numbers.

Extinctions are inevitable in a changing world, but the fossil record reveals a few instances of great change, times when the majority of life on Earth—between 50% and 90% of living species—suddenly died out, disappearing forever. Scientists have documented five such mass extinctions during the last 500 million years. We'll discuss the causes and effects of these ancient mass extinctions later in this chapter.

Today, many biologists believe that we are witnessing a sixth mass extinction. Human activities are modifying the global environment to such an extent that many species are currently threatened with extinction. In the past 400 years—a very short time on a geologic scale—more than a thousand species are known to have become extinct. Scientists estimate that this is 100 to 1,000 times the extinction rate seen in most of the fossil record.

Do these events represent the beginning of a mass extinction? This question is difficult to answer, partly because it is hard to document both the total number of species on Earth (estimates range from 10 million to 200 million) and the number of extinctions. Although losses have not reached the levels of the other "big five" extinctions, the populations of many species—such as the mountain gorillas pictured on the facing page—are declining at an alarming rate, suggesting that a sixth (human-caused) mass extinction could occur within the next few centuries. And, as with prior mass extinctions, life on Earth may take millions of years to recover.

But the fossil record also shows a creative side to the destruction. Mass extinctions can pave the way for the evolution of many diverse species from a common ancestor. Accordingly, we'll begin this chapter by discussing the birth of new species and then examine how biologists trace the evolution of biological diversity and classify living organisms.

Macroevolution and the Diversity of Life

When Darwin traveled to the Galápagos Islands, he realized that he was visiting a place of origins. Though the volcanic islands were geologically young, they were already home to many plants and animals known nowhere else in the world. Among these unique inhabitants were giant tortoises, for which the Galápagos are named (*galápago* means "tortoise" in Spanish). After visiting the islands, Darwin wrote in his diary: "Both in space and time, we seem to be brought somewhat near to that great fact—that mystery of mysteries—the first appearance of new beings on this Earth."

To understand the "appearance of new beings," it is not enough to explain microevolution, the generation-to-generation change of allele frequencies within a population. If that were all that ever happened, then Earth would be populated only by a highly adapted version of one species. The evolutionary mechanisms you read about in Chapter 13 must also explain **macroevolution**, the major changes in the history of life, which are usually evident in the fossil record. Macroevolution includes the origin of new species, which generates biological diversity; the origin of evolutionary novelty, such as the wings and feathers of birds and the large brains of humans; the explosive diversification following some evolutionary breakthroughs, such as the origin of thousands of plant species after the flower evolved; and mass extinctions, which clear the way for new adaptive explosions, such as the diversification of mammals following the disappearance of most of the dinosaurs.

The discussion of macroevolution in this chapter focuses on the origin of species. Sometimes a whole population may change significantly through adaptation to a changing environment. Such linear, nonbranching evolution does not create a new species (**Figure 14.1**, left). The formation of new species, called **speciation**, occurs when one species evolves into two or more species (see Figure 14.1, right).

Each time speciation occurs, the diversity of life increases. Beginning 3.6 billion years ago, an ancestral species gave rise to two or more species, which then gave rise to new lineages, which in turn branched again and again, populating Earth with the millions of species that have ever existed and that exist today. A new species often closely resembles the species that gave rise to it. Occasionally, however, a new species has properties novel enough to define a major branch in the tree of life, such as the flowers of flowering plants. But how is a species recognized and defined? We'll explore that question next. ☑

▶ Figure 14.1
Two patterns of evolution.

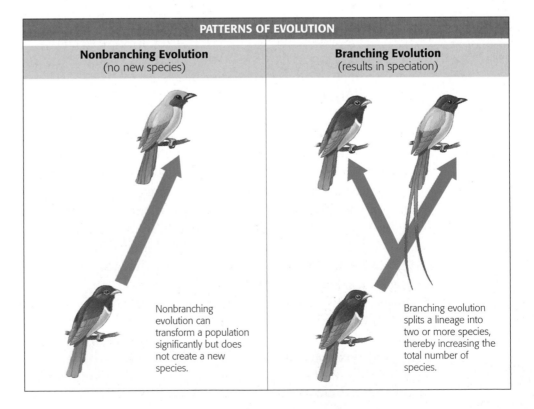

PATTERNS OF EVOLUTION

Nonbranching Evolution (no new species)	**Branching Evolution** (results in speciation)
Nonbranching evolution can transform a population significantly but does not create a new species.	Branching evolution splits a lineage into two or more species, thereby increasing the total number of species.

The Origin of Species

Species is a Latin word meaning "kind" or "appearance." Indeed, we learn to distinguish between the kinds of plants and animals—between dogs and cats, for example, or between roses and dandelions—from differences in their appearance. Although the basic idea of species as distinct life-forms seems intuitive, devising a more formal definition is not so easy.

What Is a Species?

One way of defining a species (and the main definition used in this book) is the **biological species concept**. It defines a **species** as a group of populations whose members have the potential to interbreed with one another in nature to produce fertile offspring (offspring that can reproduce) (**Figure 14.2**). Geography and culture may conspire to keep a Manhattan business-woman and a Mongolian dairyman apart. But if the two did meet and mate, they could have viable babies that develop into fertile adults because all humans belong to the same species. In contrast, humans and chimpanzees, despite having a shared evolutionary history, remain distinct species because they do not interbreed.

We cannot apply the biological species concept to all situations. For example, basing the definition of species on reproductive compatibility excludes organisms that only reproduce asexually (producing offspring from a single parent). Fossils aren't doing any sexual reproduction either, so they cannot be evaluated by the biological species concept.

In response to such challenges, biologists have developed several other ways to define species. For example, most of the 1.8 million species named to date have been classified based on measurable physical traits. Another approach identifies species in terms of ecological niches, focusing on unique adaptations to particular roles in a biological community. Yet another classification scheme defines a species as the smallest group of individuals sharing a common ancestor and forming one branch on the tree of life.

Each species concept is useful, depending on the situation and the questions being asked. The biological species concept, however, is particularly useful when focusing on how discrete groups of organisms may arise and be maintained by reproductive isolation. Next, we'll look at reproductive isolation more closely because it is such an essential factor in the evolution of so many species. ✔

Speciation

✔**CHECKPOINT**

By defining a species by its reproductive _____ from other populations, the biological species concept can only be applied to organisms that reproduce _____.

Answer: isolation; sexually

▼ Figure 14.2 **The biological species concept is based on reproductive compatibility rather than physical similarity.**

Similarity between different species. The eastern meadowlark (left) and the western meadowlark (right) are very similar in appearance, but they are separate species and do not interbreed.

Diversity within one species. Humans, as diverse in appearance as we are, belong to a single species (*Homo sapiens*) and can interbreed.

Reproductive Barriers between Species

Clearly, a fly will not mate with a frog or a fern. But what prevents closely related species from interbreeding? What, for example, maintains the species boundary between the eastern meadowlark and the western meadowlark (shown in Figure 14.2)? Their geographic ranges overlap in the Great Plains region, and they are so similar that only expert birders can tell them apart. And yet, these two bird species do not interbreed.

A **reproductive barrier** is anything that prevents individuals of closely related species from interbreeding. Let's examine the different kinds of reproductive barriers that isolate the gene pools of species (Figure 14.3). We can classify reproductive barriers as either prezygotic or postzygotic, depending on whether they block interbreeding before or after the formation of zygotes (fertilized eggs).

Prezygotic barriers prevent mating or fertilization between species (Figure 14.4). The barrier may be time based (temporal isolation). For example, western spotted skunks breed in the fall, but the eastern species breeds in late winter. Temporal isolation keeps the species from mating even where they coexist on the Great Plains. In other cases, species live in the same region but not in the same habitats (habitat isolation). For example, one species of North American garter snake lives mainly in water, while a closely related species lives on land. Traits that enable individuals to recognize potential mates, such as a particular odor, coloration, or courtship ritual, can also function as reproductive barriers (behavioral isolation). In many bird species, for example, courtship

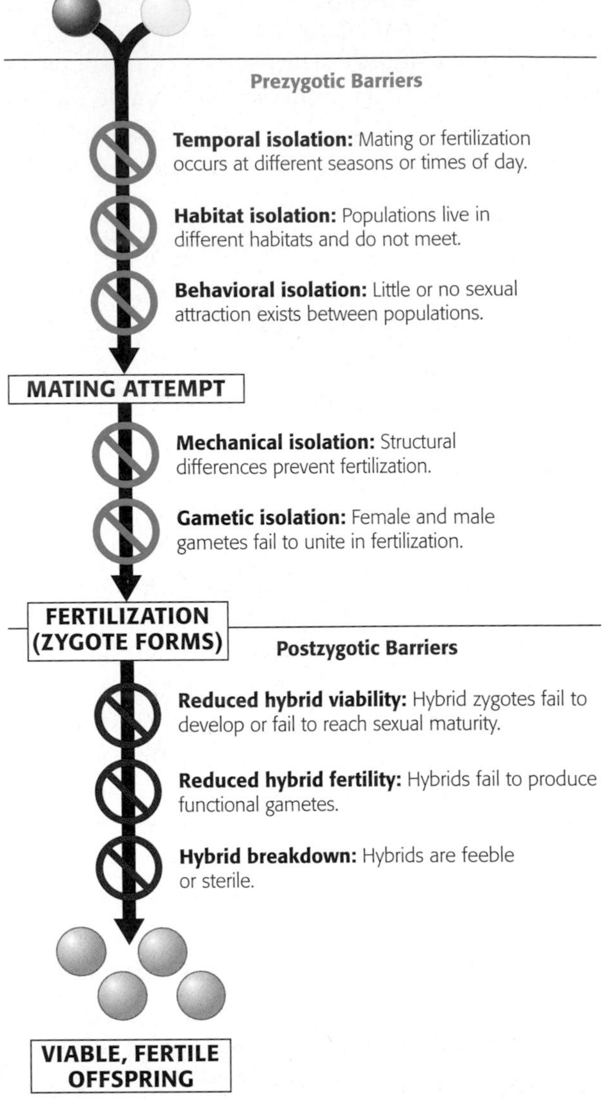

▼ Figure 14.3 **Reproductive barriers between closely related species.**

INDIVIDUALS OF DIFFERENT SPECIES

Prezygotic Barriers

Temporal isolation: Mating or fertilization occurs at different seasons or times of day.

Habitat isolation: Populations live in different habitats and do not meet.

Behavioral isolation: Little or no sexual attraction exists between populations.

MATING ATTEMPT

Mechanical isolation: Structural differences prevent fertilization.

Gametic isolation: Female and male gametes fail to unite in fertilization.

FERTILIZATION (ZYGOTE FORMS)

Postzygotic Barriers

Reduced hybrid viability: Hybrid zygotes fail to develop or fail to reach sexual maturity.

Reduced hybrid fertility: Hybrids fail to produce functional gametes.

Hybrid breakdown: Hybrids are feeble or sterile.

VIABLE, FERTILE OFFSPRING

► Figure 14.4 **Prezygotic barriers.** Prezygotic barriers prevent mating or fertilization.

PREZYGOTIC BARRIERS

Temporal Isolation	Habitat Isolation
These two closely related species of skunks mate at different times of the year.	These two closely related species of garter snakes do not mate because one lives in the water while the other lives on land.

behavior is so elaborate that individuals are unlikely to mistake a bird of a different species as one of their kind. In still other cases, the egg-producing and sperm-producing structures of different species are anatomically incompatible (mechanical isolation). For example, the two closely related species of snails in Figure 14.4 cannot join their male and female sex organs because their shells spiral in opposite directions. In still other cases, gametes (eggs and sperm) of different species are incompatible, preventing fertilization (gametic isolation). Gametic isolation is very important when fertilization is external. Male and female sea urchins of many species release eggs and sperm into the sea, but fertilization occurs only if species-specific molecules on the surface of egg and sperm attach to each other.

Postzygotic barriers operate if interspecies mating actually occurs and results in hybrid zygotes (Figure 14.5). (In this context, "hybrid" means that the egg comes from one species and the sperm from another species.) In some cases, hybrid offspring die before reaching reproductive maturity (reduced hybrid viability). For example, although certain closely related salamander species will hybridize, the offspring fail to develop normally because of genetic incompatibilities between the two species. In other cases of hybridization, offspring may become vigorous adults, but are infertile (reduced hybrid fertility). A mule, for example, is the hybrid offspring of a female horse and a male donkey. Because mules are sterile, there is no avenue for gene transfer between the two parental species, horse and donkey. In other cases, the first-generation hybrids are viable and fertile, but when these hybrids mate with one another or with either parent species, the offspring are feeble or

sterile (hybrid breakdown). For example, different species of cotton plants can produce fertile hybrids, but the offspring of the hybrids do not survive.

In summary, reproductive barriers form the boundaries around closely related species. In most cases, it is not a single reproductive barrier but some combination of two or more that keeps species isolated. Next, we examine situations that make reproductive isolation and speciation possible. ✔

▼ **Figure 14.5 Postzygotic barriers.** Postzygotic barriers prevent development of fertile adults.

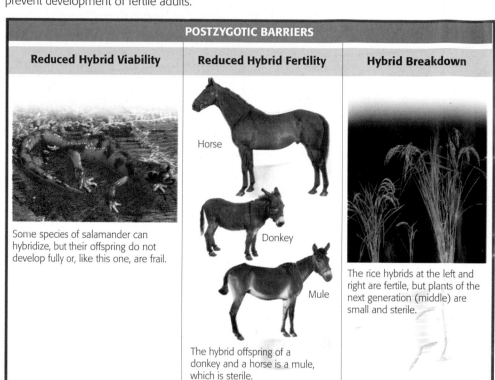

POSTZYGOTIC BARRIERS		
Reduced Hybrid Viability	**Reduced Hybrid Fertility**	**Hybrid Breakdown**
Some species of salamander can hybridize, but their offspring do not develop fully or, like this one, are frail.	Horse / Donkey / Mule — The hybrid offspring of a donkey and a horse is a mule, which is sterile.	The rice hybrids at the left and right are fertile, but plants of the next generation (middle) are small and sterile.

Behavioral Isolation

Galápagos blue-footed boobies mate only after a specific ritual of high stepping that advertises a mate's bright blue feet.

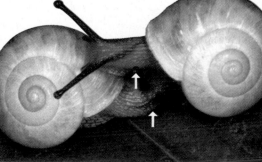

Mechanical Isolation

Because these snails' shells spiral in opposite directions, their genital openings (indicated by arrows) cannot be aligned and mating cannot occur.

Gametic Isolation

Gametes of these red and purple urchins are unable to fuse because proteins on the surface of the eggs and sperm cannot bind to one another.

Mechanisms of Speciation

A key event in the origin of many species occurs when a population is somehow severed from other populations of the parent species. With its gene pool isolated, the splinter population can follow its own evolutionary course. Changes in its allele frequencies caused by genetic drift and natural selection will not be diluted by alleles entering from other populations (gene flow). Such reproductive isolation can result from two general scenarios: allopatric ("different country") speciation and sympatric ("same country") speciation (**Figure 14.6**). In **allopatric speciation**, the initial block to gene flow is a geographic barrier that physically isolates the splinter population. In contrast, **sympatric speciation** is the origin of a new species without geographic isolation. The splinter population becomes reproductively isolated even though it is right in the midst of the parent population.

Allopatric Speciation

Several geologic processes can fragment a population into two or more isolated populations. A mountain range may emerge and gradually split a population of organisms that can inhabit only lowlands. A land bridge, such as the Isthmus of Panama, may form and separate the marine life on either side. A large lake may subside and form several smaller lakes, with their populations now isolated. Allopatric speciation can also occur if individuals colonize a new, geographically remote area and become isolated from the parent population. An example is the speciation that occurred on the Galápagos Islands following initial colonization by immigrant organisms.

How formidable must a geographic barrier be to keep allopatric populations apart? The answer depends partly on the ability of the organisms to move about. Birds, mountain lions, and coyotes can cross mountain ranges, rivers, and canyons. Nor do such barriers hinder the windblown pollen of pine trees or the spread of seeds carried by animals. In contrast, small rodents may find a deep canyon or a wide river an impassable barrier (**Figure 14.7**).

Allopatric speciation is more common for a small, isolated population because it is more likely than a large population to have its gene pool changed substantially by both genetic drift and natural selection. For example, in less than 2 million years, the few foreign animals and plants that first colonized the Galápagos Islands gave rise to all the species now found there. But for each

▼ Figure 14.6 **Two modes of speciation.**

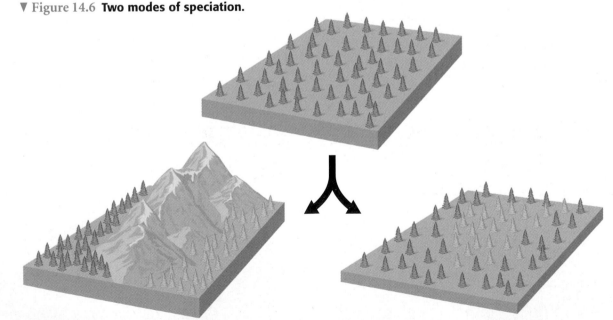

Allopatric speciation occurs when a population forms a new species after being geographically isolated from its parent population.

Sympatric speciation occurs when part of a population becomes a new species while in the midst of its parent population.

Ammospermophilus harrisii

Ammospermophilus leucurus

▲ **Figure 14.7 Allopatric speciation of antelope squirrels on opposite rims of the Grand Canyon.** Harris's antelope squirrel (*Ammospermophilus harrisii*) is found on the south rim of the Grand Canyon. Just a few miles away on the north rim is the closely related white-tailed antelope squirrel (*Ammospermophilus leucurus*). Birds and other organisms that can disperse easily across the canyon have not diverged into different species on opposite rims.

small, isolated population that becomes a new species, many more simply perish in their new environment. Life on the frontier is harsh, and most pioneer populations become extinct.

Even if a small, isolated population survives, it does not necessarily evolve into a new species. The population may adapt to its local environment and begin to look very different from the ancestral population, but that doesn't necessarily make it a new species. Speciation occurs with the evolution of reproductive barriers between the isolated population and its parent population. In other words, if speciation occurs during geographic separation, the new species cannot breed with its ancestral population, even if the two populations should come back into contact at some later time (**Figure 14.8**). ☑

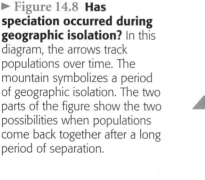

► Figure 14.8 **Has speciation occurred during geographic isolation?** In this diagram, the arrows track populations over time. The mountain symbolizes a period of geographic isolation. The two parts of the figure show the two possibilities when populations come back together after a long period of separation.

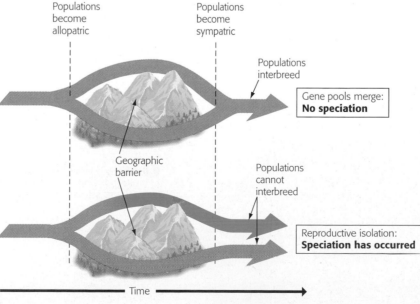

Populations become allopatric

Populations become sympatric

Populations interbreed

Gene pools merge:
No speciation

Geographic barrier

Populations cannot interbreed

Reproductive isolation:
Speciation has occurred

Time

Sympatric Speciation

In sympatric speciation, a new species emerges from an old one, even though they live in the same time and place. How can a subpopulation become reproductively isolated while in the midst of its parent population? In some cases, subgroups of a population evolve adaptations for exploiting food sources in different habitats (such as the shallow versus deep habitats of a lake). In other cases, a type of sexual selection in which females choose mates based on color can also contribute to rapid reproductive isolation. But the most frequently observed mechanism of sympatric speciation involves large-scale genetic changes that occur in a single generation.

Biologists have documented many examples of plant species that originated from accidents during cell division. Such accidents can produce organisms with extra sets of chromosomes. A new species that evolves this way has polyploid cells, meaning that each cell has more than two sets of chromosomes. The new species cannot produce fertile hybrids with its parent species. This is a good example of sympatric speciation because reproductive isolation has occurred in a single generation, without geographic isolation.

Polyploids do not always come from a single parent species. In fact, most polyploid species arise from the hybridization of two parent species. This mechanism of sympatric speciation accounts for many of the plant species we grow for food, including oats, potatoes, bananas, peanuts, barley, plums, apples, sugarcane, coffee, and wheat. Wheat makes a good case study. The most widely cultivated plant in the world, what we call wheat is actually represented by 20 species. Humans began domesticating wheat from wild grasses at least 11,000 years ago in the Middle East. **Figure 14.9** traces the evolutionary path from that first cultivated wheat species to bread wheat, the most important wheat species today. ✓

✓**CHECKPOINT**

Each speciation episode in the evolution of bread wheat (see Figure 14.9) is an example of _____ speciation, which is the origin of a new species without geographic isolation from the parent species.

Answer: sympatric

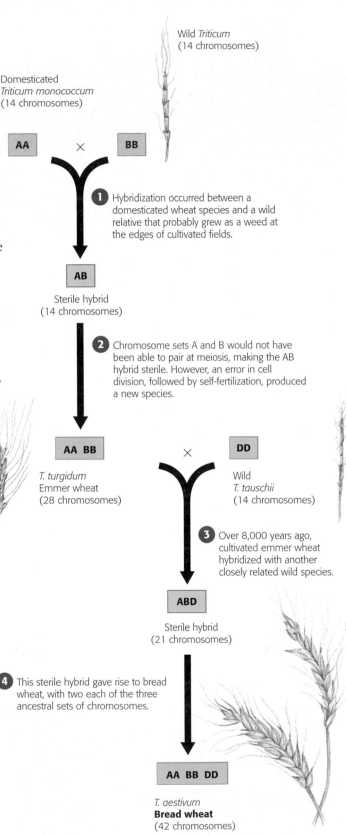

▼ **Figure 14.9 The evolution of wheat.** The uppercase letters in this diagram represent not alleles but sets of chromosomes that can be traced in the evolution of wheat, or *Triticum*. Interestingly, the AABB hybrid, emmer wheat (*Triticum turgidum*), is still grown widely in Eurasia and western North America and is used mainly for making pasta.

Wild *Triticum*
(14 chromosomes)

Domesticated
Triticum monococcum
(14 chromosomes)

AA × **BB**

1 Hybridization occurred between a domesticated wheat species and a wild relative that probably grew as a weed at the edges of cultivated fields.

AB

Sterile hybrid
(14 chromosomes)

2 Chromosome sets A and B would not have been able to pair at meiosis, making the AB hybrid sterile. However, an error in cell division, followed by self-fertilization, produced a new species.

AA BB × **DD**

T. turgidum
Emmer wheat
(28 chromosomes)

Wild
T. tauschii
(14 chromosomes)

3 Over 8,000 years ago, cultivated emmer wheat hybridized with another closely related wild species.

ABD

Sterile hybrid
(21 chromosomes)

4 This sterile hybrid gave rise to bread wheat, with two each of the three ancestral sets of chromosomes.

AA BB DD

T. aestivum
Bread wheat
(42 chromosomes)

What Is the Tempo of Speciation?

Biologists continue to make field observations and devise experiments to study evolution in progress. However, much of the evidence of evolution comes from the fossil record, the chronicle of extinct organisms engraved in layers of rock over millions of years of geologic time. What does this record say about the process of speciation?

Many fossil species appear suddenly in a layer of rock and persist essentially unchanged through several layers (strata) until disappearing as suddenly as they appeared. The term **punctuated equilibria** describes these long periods of little change, or equilibrium, punctuated by abrupt episodes of speciation.

Figure 14.10 shows two models for the tempo of speciation: the punctuated model and the graduated model. In the punctuated model, there are no transitional stages in the lineages. The new species change little, if at all, once they appear. In the graduated model, differences evolve little by little in populations as the species become adapted to their local environments, and new species (represented by the two butterflies at the far right) evolve gradually from the ancestral population. According to the graduated model, big changes (speciations) occur by the steady accumulation of many small changes.

What do punctuated and graduated patterns tell us about how long it takes new species to form? Suppose that a species survived for 5 million years, but most of the changes in its body features occurred during the first 50,000 years of its existence, just 1% of its overall history. Time periods this short often cannot be distinguished in fossil strata. Thus, based on its fossils, the species would seem to have appeared suddenly and continued with little change before becoming extinct. Even though such a species may have originated more slowly than its fossils suggest, a punctuated pattern indicates that speciation occurred relatively rapidly. For species whose fossils show much more gradual change, we cannot tell exactly when a new biological species formed. It is likely that speciation in such groups occurred relatively slowly, perhaps taking millions of years.

How often do species evolve abruptly and then remain essentially unchanged for most of their existence? Rapid speciation is certainly known to occur. As we saw earlier, abrupt speciation can occur by polyploidy in plants and even occasionally in a few animals. And genetic drift and natural selection can significantly alter the gene pool of a small, isolated population in just a few hundred generations.

Regardless of the tempo of speciation, as species diverge, differences accumulate and become more pronounced, eventually leading to the formation of new groups of organisms that differ greatly from their ancestors. Furthermore, as one group produces many new species, another group may lose species to extinction. The cumulative effects of multiple speciation and extinction events have helped to shape the dramatic changes documented in the fossil record. This type of macroevolution is the subject of the next section. ☑

▼ Figure 14.10 **Two models for the tempo of evolution.**

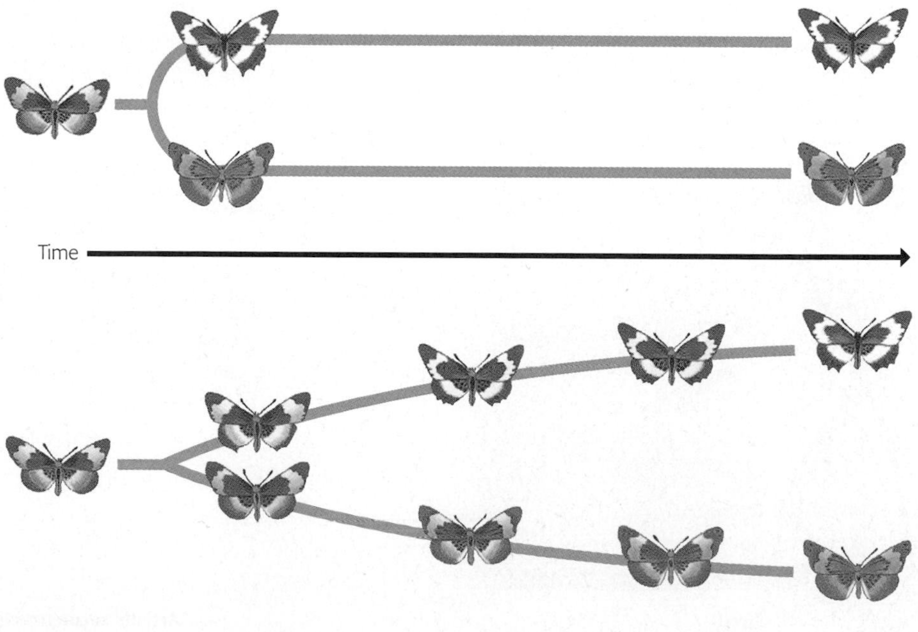

Punctuated model. A new species changes most as it first branches from a parent species. After this speciation episode, there is little change for the rest of the species' existence.

Time ⟶

Graduated model. Species that are descended from a common ancestor diverge gradually in form as they acquire unique adaptations.

☑CHECKPOINT

How does the punctuated model account for the relative rarity of transitional fossils linking newer species to older ones?

Answer: According to the punctuated model, the time required for speciation in most cases is relatively short compared with the overall duration of the species' existence. Thus, on the vast geologic time scale of the fossil record, the transition of one species to another seems abrupt.

The Evolution of Biological Novelty

The two squirrels in Figure 14.7 are different species, but they are very similar animals that live very much the same way. When most people think of evolution, they envision much more dramatic transformation. How can we account for such evolutionary products as flight in birds and braininess in humans?

Adaptation of Old Structures for New Functions

Birds are derived from a lineage of earthbound reptiles (Figure 14.11). How could flying animals evolve from flightless ancestors? More generally, how do major novelties of biological structure and function evolve? One way evolutionary novelty arises is when structures that originally played one role gradually acquire a different role. Structures that evolve in one context but become adapted for another function are called exaptations. This term suggests that a structure can become adapted to alternative functions; it does not mean that a structure somehow evolves in anticipation of future use. New features can arise gradually via intermediate stages, each functioning in the organism's current context.

Exaptation can account for the evolution of novel structures. Consider the evolution of birds from a dinosaur ancestor. Birds have lightweight skeletons with honeycombed bones, a feature found in their dinosaur ancestors. Since the fossil record clearly indicates that light bones predated flight, they must have had some function on the ground. The ancestors of birds were small, agile, bipedal dinosaurs that would have benefited from a light frame. Moreover, a winglike form with feathers would have increased the surface area of the forelimbs. These enlarged forelimbs were adapted for flight after functioning in some other capacity, such as thermal regulation, courtship displays, or camouflage (roles that wings still play today). The first flights may have been only glides or extended hops as the animal pursued prey or fled from a predator. As flight evolved, animals with better-adapted feathers and wings would have had a selective advantage.

According to Harvard zoologist Karel Liem, "Evolution is like modifying a machine while it's running." The concept that biological novelties can evolve by the remodeling of old structures is in the Darwinian tradition of large changes being an accumulation of many small changes crafted by natural selection. ☑

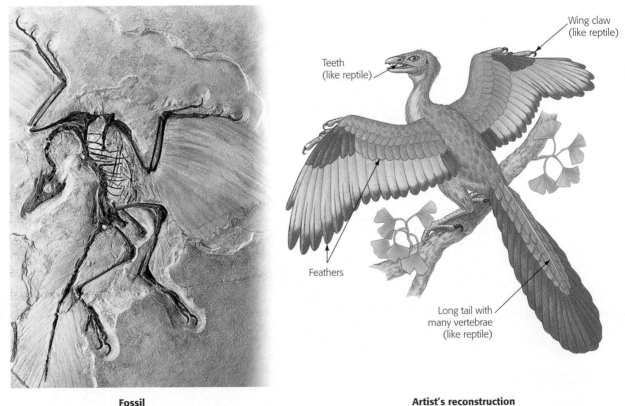

► Figure 14.11 **An extinct bird.** Called *Archaeopteryx* ("ancient wing"), this animal lived near tropical lagoons in central Europe about 150 million years ago. Like birds today, it had flight feathers, but otherwise it was more like some small bipedal dinosaurs of its era. *Archaeopteryx* probably relied mainly on gliding from trees. Despite its feathers, *Archaeopteryx* is not considered an ancestor of today's birds. Instead, it probably represents an extinct side branch of the bird lineage.

Fossil

Wing claw (like reptile)

Teeth (like reptile)

Feathers

Long tail with many vertebrae (like reptile)

Artist's reconstruction

Evo-Devo: Development and Evolutionary Novelty

Gradual evolutionary remodeling, such as the accumulation of flight adaptations in birds, probably involves a large number of genetic changes in populations. In other cases of macroevolution, just a few genetic changes can become magnified into major structural differences between organisms. Such cases are of particular interest to scientists working in the field of **evo-devo**, or evolutionary developmental biology, which studies the evolution of developmental processes in multicellular organisms.

Genes that program development control the rate, timing, and spatial pattern of changes in an organism's form as it develops from a zygote into an adult. (Chapter 11 provides a closer look at how genes control development.) A subtle change in a species' developmental program can have profound effects. The same or very similar genes are involved in the development of form across multiple lineages. Changes in the number, nucleotide sequence, and regulation of these genes have led to the huge diversity in body forms. Many striking evolutionary transformations are the result of a change in the rate or timing of developmental events.

For example, the animal in **Figure 14.12**, a salamander called an axolotl, illustrates a phenomenon called **paedomorphosis**, which is the retention into adulthood of features that were solely juvenile in ancestral species. The axolotl grows to full size and reproduces without losing its external gills, a juvenile feature in most species of salamanders.

Paedomorphosis has also been important in human evolution. Humans and chimpanzees are even more alike in body form as fetuses than they are as adults. In the fetuses of both species, the skulls are rounded and the jaws are small, making the face rather flat (**Figure 14.13**). As development proceeds, uneven bone growth makes the chimpanzee skull sharply angular, with heavy browridges and massive jaws. The adult chimpanzee has much greater jaw strength than we have, and its teeth are proportionately larger. In contrast, the adult human has a skull with decidedly rounded, more fetus-like contours. Put another way, our skull is paedomorphic; it retains fetal features even after we are mature. Our large skull is one of our most distinctive features. Our large, complex brain, which fills that bulbous skull, is another. The human brain is proportionately larger than the chimpanzee brain because growth of the organ is switched off much later in human development. Compared with the brain of a chimpanzee, our brain continues to grow for several more years, which can be interpreted as the prolonging of a juvenile process.

Homeotic genes, the master control genes described in Chapter 11, determine such basic features as where a pair of wings or legs will develop on a fruit fly. Changes in such genes or in where such genes are expressed can profoundly affect body form. Consider, for example, the evolution of tetrapods (animals with backbones and four limbs) from fishes. The location within a developing limb where certain homeotic genes are expressed is initially the same in both fishes and tetrapods. A second region of expression in the developing tetrapod limb, however, produces the extra skeletal elements that develop into foot bones. Thus, changes in the expression of these genes appear to have led to the evolution of walking legs from the paired fins of fishes.

Duplications of homeotic genes probably facilitated the origin of new body shapes in animals, including the evolution of vertebrates (animals with backbones) from invertebrates (animals without backbones). For example, consider the homeotic genes of a fruit fly (an invertebrate) and a mouse (a vertebrate). A fruit fly has a single cluster of homeotic genes that direct the development of major body parts, whereas a mouse has four clusters of these genes. Two duplications of these gene clusters appear to have occurred in the evolution of vertebrates from invertebrates. Mutations in these duplicated genes may then have led to the origin of novel vertebrate characteristics. For instance, some genes may have taken on new roles, such as directing the development of a backbone, jaws, and limbs.

Gills

▲ Figure 14.12
Paedomorphosis. The axolotl, a salamander, becomes an adult (shown here) and reproduces while retaining certain tadpole characteristics, including gills.

▼ Figure 14.13 **Comparison of human and chimpanzee skull development.** Starting with fetal skulls that are very similar (left), the differential growth rates of the bones making up the skulls produce adult heads with very different proportions. The grid lines will help you relate the fetal skulls to the adult skulls.

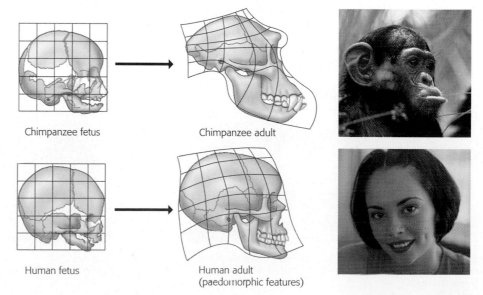

Chimpanzee fetus

Chimpanzee adult

Human fetus

Human adult (paedomorphic features)

Earth History and Macroevolution

Having examined how new species and evolutionary adaptations arise, we are ready to turn our attention to the history of biological diversity. Macroevolution is closely tied to the history of Earth.

Geologic Time and the Fossil Record

The fossil record, the sequence in which fossils appear in rock strata, is an archive of macroevolution. **Figure 14.14** surveys some of the diverse ways that organisms can fossilize. Sedimentary rocks are the richest sources

of fossils and provide a record of life on Earth in their strata (see Chapter 13). The fossils in each stratum of sedimentary rock are a local sample of the organisms that existed at the time the sediment was deposited.

By studying many sites over the past two centuries, geologists have established a **geologic time scale**, reflecting a consistent sequence of geologic periods (**Table 14.1**). The time line presented in Table 14.1 is separated into four broad divisions: the Precambrian (a general term for the time before about 540 million years ago), followed by the Paleozoic, Mesozoic, and Cenozoic eras. Each of these divisions represents a distinct age in the history of Earth and its life.

▼ Figure 14.14 **A gallery of fossils.**

Sedimentary fossils are formed when minerals seep into and replace organic matter. These petrified (stone) trees in Arizona's Petrified Forest National Park are about 190 million years old.

Sedimentary rocks are the richest hunting grounds for paleontologists, scientists who study fossils. This researcher is excavating a fossilized dinosaur skeleton from sandstone in Dinosaur National Monument, located in Utah and Colorado.

This 30-million-year-old insect is embedded in amber (hardened resin from a tree).

Trace fossils are footprints, burrows, and other remnants of an ancient organism's behavior. A dinosaur left these footprints in a creek bed in what is now Oklahoma.

These tusks belong to a whole 23,000-year-old mammoth, which scientists discovered in Siberian ice in 1999.

Table 14.1	The Geologic Time Scale				

Geologic Time	Period	Epoch	Age (millions of years ago)	Some Important Events in the History of Life	
Cenozoic era	Quaternary	Recent		Historical time	
			0.01		
		Pleistocene		Ice ages; humans appear	
			1.8		
	Tertiary	Pliocene		Origin of genus *Homo*	
			5		
		Miocene		Continued speciation of mammals and angiosperms	
			23		
		Oligocene		Origins of many primate groups, including apes	
			34		
		Eocene		Angiosperm dominance increases; origins of most living mammalian orders	
			56		
		Paleocene		Major speciation of mammals, birds, and pollinating insects	
			65		
Mesozoic era	Cretaceous			Flowering plants (angiosperms) appear; many groups of organisms, including most dinosaur lineages, become extinct at end of period (Cretaceous extinctions)	
			145		
	Jurassic			Gymnosperms continue as dominant plants; dinosaurs become dominant	
			200		
	Triassic			Cone-bearing plants (gymnosperms) dominate landscape; speciation of dinosaurs, early mammals, and birds	
			251		
Paleozoic era	Permian			Extinction of many marine and terrestrial organisms (Permian extinctions); speciation of reptiles; origins of mammal-like reptiles and most living orders of insects	
			299		
	Carboniferous			Extensive forests of vascular plants; first seed plants; origin of reptiles; amphibians become dominant	
			359		
	Devonian			Diversification of bony fishes; first amphibians and insects	
			416		
	Silurian			Early vascular plants dominate land	
			444		
	Ordovician			Marine algae are abundant; colonization of land by diverse fungi, plants, and animals	
			488		
	Cambrian			Origin of most living animal phyla (Cambrian explosion)	
			542		
Precambrian			600	Diverse algae and soft-bodied invertebrate animals appear	
			635	Oldest animal fossils	
			2,100	Oldest eukaryotic fossils	
			2,700	Oxygen begins accumulating in atmosphere	
			3,500	Oldest fossils known (prokaryotes)	
			4,600	Approximate time of origin of Earth	

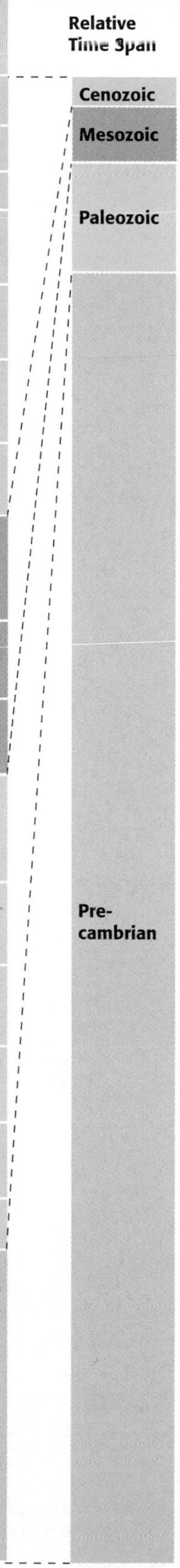

Relative Time Span

Cenozoic

Mesozoic

Paleozoic

Pre-cambrian

☑**CHECKPOINT**

Use Table 14.1 to estimate
how long prokaryotes inhabited Earth before eukaryotes
evolved.

*Answer: about 1,800 million years,
or 1.8 billion years.*

The boundaries between eras are marked by mass extinctions, when many forms of life disappeared from the fossil record and were replaced by species that diversified from the survivors. For example, the beginning of the Cambrian period is delineated by a great diversity of fossilized animals that are absent in rocks of the late Precambrian. And most of the animals that lived during the late Precambrian became extinct at the end of that era.

Fossils are reliable chronological records only if we can determine their ages. The record of the rocks reveals only the *relative* ages of fossils and therefore the order in which groups of species evolved. However, the series of sedimentary layers alone does not tell us the *absolute* ages of the embedded fossils. The rock strata are analogous to the layers of wallpaper you might peel from the walls of a very old house. You could determine the sequence in which the wallpapers had been applied, but not the year that each layer was added. Geologists use a variety of methods to determine the ages of rocks and the fossils they contain. The most common method is **radiometric dating** (Figure 14.15), which is based on the decay of radioactive isotopes (recall that isotopes are alternative forms of elements; see Table 2.1). The dates in the geologic time scale in Table 14.1 were established by radiometric dating. ☑

▼ Figure 14.15 **Radiometric dating.** Amounts of radioactive isotopes can be measured by the radiation they emit as they decompose to more stable atoms (see Chapter 2). Paleontologists use this clocklike decay to date fossils.

Radioactive decay of carbon-14.
Carbon-14 is a radioactive isotope with a half-life of 5,600 years. From the time an organism dies, it takes 5,600 years for half of the radioactive carbon-14 to decay; half of the remainder is present after another 5,600 years; and so on.

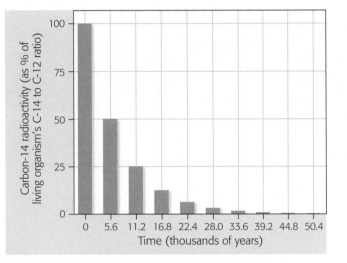

How carbon-14 dating is used to determine the vintage of a fossilized clam shell.

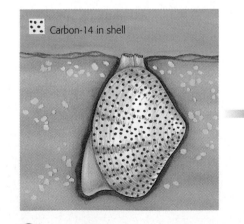

❶ While an organism, in this case a clam, is alive, it assimilates the different isotopes of each element in proportions determined by their relative abundances in the environment. Carbon-14 is taken up in trace quantities, along with much larger quantities of the more common carbon-12.

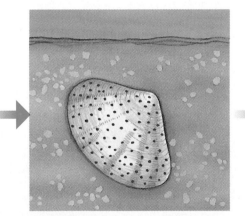

❷ After the clam dies, it is covered with sediment, and its shell eventually becomes consolidated into a layer of rock as the sediment is compressed. From the time the clam dies and ceases to assimilate carbon, the amount of carbon-14 relative to carbon-12 in the fossil declines due to radioactive decay.

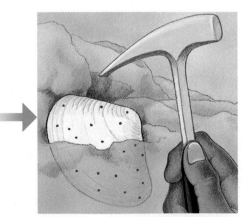

❸ After the clam fossil is found, its age can be determined by measuring the ratio of the two isotopes to learn how many half-life reductions have occurred since it died. For example, if the ratio of carbon-14 to carbon-12 in this fossil clam was found to be 25% of that of a living organism, this fossil would be about 11,200 years old (see graph above).

Plate Tectonics and Macroevolution

The continents are not locked in place. They drift about Earth's surface like passengers on huge, irregularly shaped plates of crust floating on a flexible layer of hot, underlying material called the mantle. When two landmasses are embedded in different plates, their positions relative to each other change. For example, North America and Europe are presently drifting apart at a rate of about 2 cm per year. Many important geologic processes, including mountain building, volcanic activity, and earthquakes, occur at plate boundaries. California's infamous San Andreas Fault is at a border where two plates slide past each other **(Figure 14.16)**.

Plate movements rearrange geography constantly, but two chapters in the continuing saga of continental drift had an especially strong influence on life. About 250 million years ago, near the end of the Paleozoic era, plate movements brought all the previously separated landmasses together into a supercontinent called Pangaea ("all land") **(Figure 14.17)**. Imagine some of the possible effects on life. Species that had been evolving in isolation came together and competed. As the landmasses joined over millions of years, the total amount of shoreline was reduced. There is also evidence that the ocean basins increased in depth, lowering sea level and draining the shallow coastal seas. Then, as now, most marine species inhabited shallow waters, and the formation of Pangaea destroyed a considerable amount of that habitat. It was probably a long, traumatic period for terrestrial life as well. The continental interior, which had a drier and more erratic climate than coastal regions, increased in area substantially when the land came together. Changing ocean currents also undoubtedly affected land life as well as sea life. Thus,

the formation of Pangaea had a tremendous environmental impact that reshaped biological diversity by causing extinctions and providing new opportunities for the survivors, which diversified as a result of branching evolution.

The second dramatic chapter in the history of continental drift began about 180 million years ago, during the Mesozoic era. Pangaea started to break up, causing geographic isolation of colossal proportions. As the continents drifted apart, each became a separate evolutionary arena as climates changed and the organisms of the different biogeographic realms diverged.

The pattern of continental separations is the solution to many puzzles of biogeography. For example, paleontologists have discovered matching fossils of Mesozoic reptiles in Ghana (West Africa) and Brazil. These two parts of the world, now separated by 3,000 km of ocean, were contiguous during the early Mesozoic era. Plate tectonics also explains much about the current distribution of organisms. For example, the mammals of Australia contrast so sharply with those of the rest of the world because the plate on which Australia rests drifted away from other plates early in the evolutionary history of mammals, leaving marsupials free to evolve in isolation on that continent (see Figure 13.7). ✔

◀ **Figure 14.16 California's San Andreas fault.** This fissure is on the San Andreas fault in California's Santa Cruz Mountains, near the epicenter of the 1989 Loma Prieta earthquake. The earthquake caused extensive damage to the surrounding area as well as to the city of San Francisco.

▶ **Figure 14.17 The history of plate tectonics.** The continents continue to drift, though not at a rate that's likely to cause any motion sickness for their passengers.

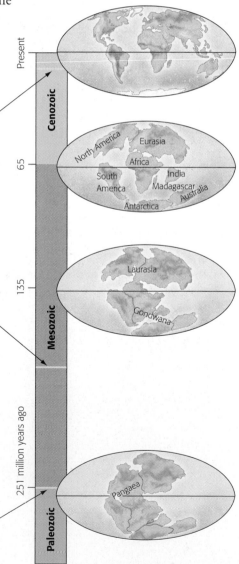

India collided with Eurasia just 10 million years ago, forming the Himalayas, the tallest and youngest of Earth's mountain ranges.

About 180 million years ago, Pangaea began to split into northern (Laurasia) and southern (Gondwana) landmasses, which later separated into the modern continents.

Pangaea formed about 250 million years ago.

Mass Extinctions and Explosive Diversifications of Life

As discussed in the Biology and Society section, the fossil record reveals that five mass extinctions have occurred over the last 500 million years. In each of these events, 50% or more of Earth's species died out. Of all the mass extinctions, those marking the ends of the Permian and Cretaceous periods have been the most intensively studied.

The Permian mass extinction, at about the time the merging continents formed Pangaea, claimed about 96% of marine species and took a tremendous toll on terrestrial life as well. Equally notable is the mass extinction at the end of the Cretaceous period. For 150 million years prior, dinosaurs dominated Earth's land and air, whereas mammals were few and small, resembling today's rodents. Then, about 65 million years ago, most of the dinosaurs became extinct, leaving behind only the descendants of one lineage, the birds. Remarkably, the massive die-off (which also included half of all other species) occurred in less than 10 million years—a brief period in geologic time.

But there is a flip side to the destruction. Each massive dip in species diversity has been followed by explosive diversification of certain survivors. Extinctions seem to have provided the surviving organisms with new environmental opportunities. For example, mammals existed for at least 75 million years before undergoing an explosive increase in diversity just after the Cretaceous period. Their rise to prominence was undoubtedly associated with the void left by the extinction of the dinosaurs. The world would be a very different place today if many dinosaur lineages had escaped the Cretaceous extinctions or if none of the mammals had survived.

THE PROCESS OF SCIENCE: Mass Extinctions

Did a Meteor Kill the Dinosaurs?

For decades, scientists have been debating the cause of the rapid dinosaur die-off that occurred 65 million years ago. Many **observations** provide clues. The fossil record shows that the climate had cooled and that shallow seas were receding from continental lowlands. It also shows that many plant species died out. Perhaps the most telling evidence was discovered by physicist Luis Alvarez and his geologist son Walter Alvarez, both of the University of California. In 1980, they found that rock deposited around 65 million years ago contains a thin layer of clay rich in iridium, an element very rare on Earth but common in meteors and other extraterrestrial material that occasionally falls to Earth. This discovery led the Alvarez team to ask the following **question**: Is

▼ Figure 14.18 **Trauma for planet Earth and its Cretaceous life.**

An artist's representation of the impact of an asteroid or comet.

The impact's immediate effect was most likely a cloud of hot vapor and debris that could have killed many of the plants and animals in North America within hours.

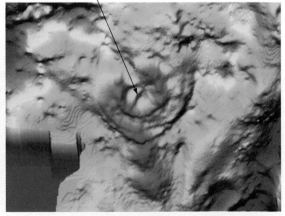

Chicxulub crater

The 65-million-year-old Chicxulub impact crater is located in the Caribbean Sea near the Yucatán Peninsula of Mexico. The horseshoe shape of the crater and the pattern of debris in sedimentary rocks indicate that an asteroid or comet struck at a low angle from the southeast.

the iridium layer the result of fallout from a huge cloud of dust that billowed into the atmosphere when a large meteor or asteroid hit Earth?

The father and son formed the **hypothesis** that the mass extinction 65 million years ago was caused by the impact of an extraterrestrial object. This hypothesis makes a clear **prediction**: A huge impact crater of the right age should be found somewhere on Earth's surface. (This is a good example of discovery science, which relies on verifiable observations rather than a direct **experiment**; see Chapter 1.) In 1981, two petroleum geologists found the **results** predicted by Alvarezes' hypothesis: the Chicxulub crater, located near Mexico's Yucatán Peninsula in the Caribbean Sea **(Figure 14.18)**. This impact site, about 180 km wide (about 112 miles)

and dating from the predicted time, was created when a meteor or asteroid about 10 km in diameter (about 6 miles) slammed into Earth, releasing thousands of times more energy than is stored in the world's combined stockpile of nuclear weapons. Such a cloud could have blocked sunlight and disturbed climate severely for months, perhaps killing off many plant species and, later, the animals that depended on those plants for food.

Debate continues about whether this impact alone caused the dinosaurs to die out or whether other factors—such as continental movements or volcanic activity—also contributed. Most scientists agree, however, that the collision that created the Chicxulub crater could indeed have been a major factor in global climatic changes and mass extinctions.

Classifying the Diversity of Life

Systematics is a discipline of biology that focuses on classifying organisms and determining their evolutionary relationships. Systematics includes **taxonomy**, which is the identification, naming, and classification of species.

Some Basics of Taxonomy

Assigning scientific names to species is an essential part of systematics. Common names such as monkey, fly, and pea may work well in everyday communication, but they can be ambiguous because there are many species of each of these organisms. And some common names are misleading. For example, consider the following so-called "fishes": jellyfish (a cnidarian), crayfish (a crustacean), and silverfish (an insect).

Using an agreed-upon formal naming system eases communication among scientists, allows researchers to unambiguously identify an organism, and makes it easier to recognize when a new species is discovered. The formal taxonomic system used by biologists today dates back to Carolus Linnaeus (1707–1778), a Swedish physician and botanist (plant specialist). Linnaeus's system has two main characteristics: a two-part name for each species and a hierarchical classification of species into broader groups of organisms.

Naming Species

Linnaeus's system assigns to each species a two-part latinized name, or **binomial**. The first part of a binomial is the **genus** (plural, *genera*) to which the species belongs. The second part of a binomial is unique for each species within the genus. The two parts must be used together

to name a species. For example, the scientific name for the leopard is *Panthera pardus*. Notice that the first letter of the genus is capitalized and that the whole binomial is italicized and latinized. (You can name a bug you discover after a friend, but you must add the appropriate Latin ending.) The scientific name of our own species, *Homo sapiens*, which Linnaeus assigned in a show of optimism, means "wise man."

Hierarchical Classification

In addition to naming species, a major objective of systematics is to group species into an ordered hierarchy of categories. The first step of such a hierarchical classification is built into the binomial. We group species that are closely related into the same genus. For example, the leopard (*Panthera pardus*) belongs to a genus that also contains three other cats: the lion (*Panthera leo*), the tiger (*Panthera tigris*), and the jaguar (*Panthera onca*) **(Figure 14.19)**. Grouping species is natural for us, at least in concept. We lump together several trees we know as oaks and distinguish them from several other species of trees we call maples. Indeed, oaks and maples belong to separate genera. Biology's taxonomic scheme formalizes our tendency to group related objects as a way of structuring our view of the world.

▼ **Figure 14.19 The four species within the genus *Panthera*.**

Leopard (*Panthera pardus*)

Tiger (*Panthera tigris*)

Lion (*Panthera leo*)

Jaguar (*Panthera onca*)

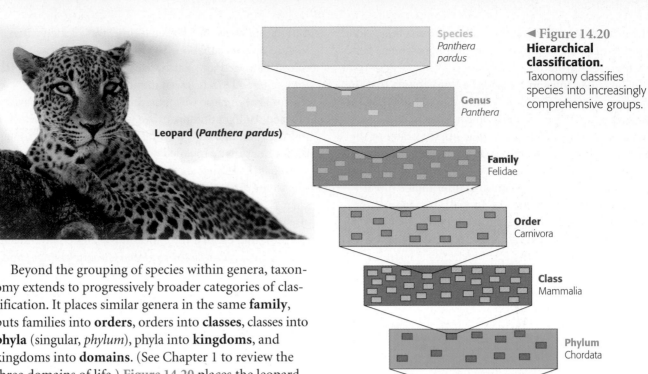

Leopard (*Panthera pardus*)

◀ Figure 14.20
Hierarchical classification. Taxonomy classifies species into increasingly comprehensive groups.

Species
Panthera pardus

Genus
Panthera

Family
Felidae

Order
Carnivora

Class
Mammalia

Phylum
Chordata

Kingdom
Animalia

Domain
Eukarya

Beyond the grouping of species within genera, taxonomy extends to progressively broader categories of classification. It places similar genera in the same **family**, puts families into **orders**, orders into **classes**, classes into **phyla** (singular, *phylum*), phyla into **kingdoms**, and kingdoms into **domains**. (See Chapter 1 to review the three domains of life.) **Figure 14.20** places the leopard in this taxonomic scheme of groups within groups. The resulting classification of a particular organism is somewhat like a postal address identifying a person in a particular apartment, in a building with many apartments, on a street with many apartment buildings, in a city with many streets, and so on. ✔

☑CHECKPOINT

How much of the classification in Figure 14.20 do humans share with the leopard?

Answer: We are classified the same down to the class level: mammals. We do not belong to the same order.

Classification and Phylogeny

Ever since Darwin, systematics has had a goal beyond simple organization: to have classification reflect evolutionary relationships. In other words, how an organism is named and classified should reflect its place within the evolutionary tree of life. Biologists use **phylogenetic trees** to depict hypotheses about the evolutionary history of species. These branching diagrams reflect the hierarchical classification of groups nested within more inclusive groups. The tree in **Figure 14.21** shows the classification of some carnivores and their probable evolutionary relationships. Note that each branch point represents the divergence of two lineages from a common ancestor.

Sorting Homology from Analogy

Homologous structures are one of the best sources of information about phylogenetic relationships. Recall from Chapter 13 that homologous structures in different species may vary in form and function but exhibit fundamental similarities because they evolved from the

▼ Figure 14.21 **The relationship of classification and phylogeny for some members of the order Carnivora.** The hierarchical classification is reflected in the finer and finer branching of the phylogenetic tree. Each branch point in the tree represents an ancestor common to species to the right of that branch point.

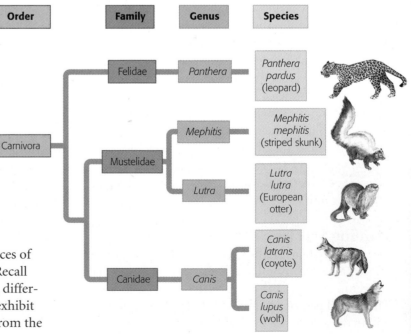

Order	Family	Genus	Species
	Felidae	*Panthera*	*Panthera pardus* (leopard)
Carnivora	Mustelidae	*Mephitis*	*Mephitis mephitis* (striped skunk)
		Lutra	*Lutra lutra* (European otter)
	Canidae	*Canis*	*Canis latrans* (coyote)
			Canis lupus (wolf)

same structure in a common ancestor. Among the vertebrates, for instance, the whale forelimb is adapted for steering in the water, whereas the bat wing is adapted for flight. Nonetheless, there are many basic similarities in the bones supporting these two structures, as you saw in Figure 13.8. The greater the number of homologous structures between two species, the more closely the species are related.

There are pitfalls in the search for homology: Not all likeness is inherited from a common ancestor. Species from different evolutionary branches may have certain structures that are superficially similar if natural selection has shaped analogous adaptations. This is called **convergent evolution**. Similarity due to convergence is called **analogy**, not homology. For example, the wings of insects and those of birds are analogous flight equipment: They evolved independently and are built from entirely different structures.

To develop phylogenetic trees and classify organisms according to evolutionary history, we must use only homologous similarities. This guideline is generally straightforward, but there can be complications. Adaptation can obscure homologies, and convergence can create misleading similarities. As we saw in Chapter 13, comparing the embryonic development of two species can often expose homology that is not apparent in the mature structures.

There is another clue to distinguishing homology from analogy: The more complex two similar structures are, the less likely it is they evolved independently. For example, compare the skulls of a human and a chimpanzee (see Figure 14.13). Although each is a fusion of many bones, they match almost perfectly, bone for bone. It is highly improbable that such complex structures matching in so many details could have separate origins. Most likely, the genes required to build these skulls were inherited from a common ancestor. ☑

Molecular Biology as a Tool in Systematics

If homology reflects common ancestry, then comparing the genes and gene products (proteins) of organisms gets right to the heart of their evolutionary relationships. Sequences of nucleotides in DNA are inherited, and they program corresponding sequences of amino acids in proteins. At the molecular level, the evolutionary divergence of species parallels the accumulation of differences in their genomes. The more recently two species have branched from a common ancestor, the more similar their DNA and amino acid sequences should be.

Today, the amino acid sequences for many proteins and the nucleotide sequences for a rapidly increasing number of genomes are in databases that are available

on the Internet (see Chapter 12). These databases have fostered a boom in systematics as researchers use them to compare the hereditary information of different species in search of homology at its most basic level.

Molecular systematics provides a new way to test hypotheses about the phylogeny of species. The strongest support for any such hypothesis is agreement between molecular data and other means of tracing phylogeny, such as evaluating anatomical homology and analyzing the fossil record. And some fossils are preserved in such a way that DNA fragments can be extracted for comparison with living organisms (Figure 14.22).

The Cladistic Revolution

Systematics entered a vigorous new era in the 1960s. Just as molecular methods became readily available for comparing species, computer technology that could crunch a new wealth of data helped usher in a new approach called cladistics.

Cladistics is the scientific search for clades (from the Greek word for "branch"). A **clade** consists of an ancestral species and all its descendants—a distinct branch in the tree of life. Identifying clades involves identifying homologies unique to a species or to a higher taxonomic cluster, such as a class or phylum (Figure 14.23).

Cladistics requires comparing an ingroup with an outgroup. The ingroup (for example, the three mammals in Figure 14.23) is the taxonomic cluster that is actually being analyzed. The outgroup (in Figure 14.23, the iguana, representing reptiles) is a species or group of species known to have diverged before the lineage that contains the groups being studied. By comparing members of the ingroup with each other and with the outgroup, we can determine what distinguishes the ingroup

▲ **Figure 14.22 Studying ancient DNA.** Some sedimentary fossils, such as this 40-million-year-old leaf, retain DNA and other organic material that scientists can analyze.

☑CHECKPOINT

Our forearms and a bat's wings are derived from the same ancestral prototype; thus, they are _____. In contrast, the wings of a bat and the wings of a bee are derived from totally unrelated structures; thus, they are _____.

Answer: homologous; analogous

► **Figure 14.23 A simplified example of cladistics.**

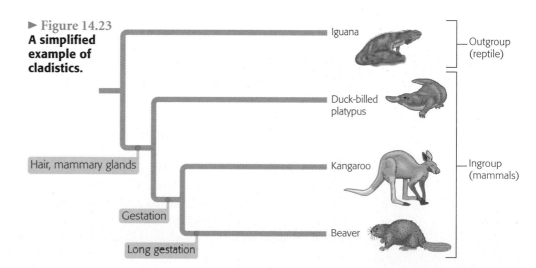

Iguana — Outgroup (reptile)

Duck-billed platypus

Hair, mammary glands

Kangaroo — Ingroup (mammals)

Gestation

Beaver

Long gestation

from the outgroup. All the mammals in the ingroup have hair and mammary glands. These were present in the ancestral mammal, but not in the outgroup. But gestation, the carrying of offspring in the uterus within the female parent, is absent from the duck-billed platypus (which lays eggs with a shell). From this absence we might infer that the duck-billed platypus represents an early branch point in the mammalian clade. Proceeding in this manner, we can construct a phylogenetic tree. Each branch represents the divergence of two groups from a common ancestor, with the emergence of a lineage possessing one or more new features. The sequence of branching represents the order in which they evolved and when groups last shared a common ancestor. In other words, cladistics focuses on the changes that define the branch points in evolution. Identifying clades makes it possible to construct classification schemes that reflect the branching pattern of evolution.

Cladistics has become the most widely used method in systematics. This approach clarifies evolutionary relationships that were not always apparent in other taxonomic classifications. For instance, biologists traditionally placed birds and reptiles in separate classes of vertebrates (class Aves and class Reptilia, respectively). This classification, however, is inconsistent with cladistics. An inventory of homologies indicates that birds and

crocodiles make up one clade, and lizards and snakes form another. If we go back as far as the ancestor that crocodiles share with lizards and snakes to make up a clade, then the class Reptilia must also include birds. The tree in **Figure 14.24** is thus more consistent with cladistics than with traditional classifications. ☑

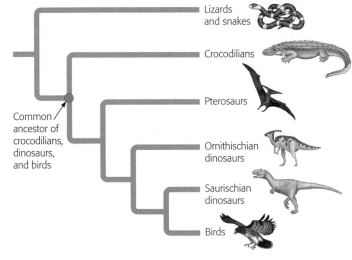

▼ **Figure 14.24 How cladistics is shaking phylogenetic trees.** Strict application of cladistics sometimes produces phylogenetic trees that conflict with classical taxonomy.

Lizards and snakes

Crocodilians

Pterosaurs

Common ancestor of crocodilians, dinosaurs, and birds

Ornithischian dinosaurs

Saurischian dinosaurs

Birds

Classification: A Work in Progress

Phylogenetic trees are hypotheses about evolutionary history. Like all hypotheses, they are revised (or in some cases rejected) based on new evidence. Molecular systematics and cladistics are combining to remodel phylogenetic trees and challenge traditional classifications.

Linnaeus divided all known forms of life between the plant and animal kingdoms, and the two-kingdom system prevailed in biology for over 200 years. In the mid-1900s, the two-kingdom system was replaced by a five-kingdom system that placed all prokaryotes in one kingdom and divided the eukaryotes among four other kingdoms.

In the late 1900s, molecular studies and cladistics led to the development of a **three-domain system** (**Figure 14.25**). This current scheme recognizes three basic groups: two domains of prokaryotes—Bacteria and Archaea—and one domain of eukaryotes, called Eukarya. The domains Bacteria and Archaea differ in a number of important structural, biochemical, and functional features, which we will discuss in Chapter 15.

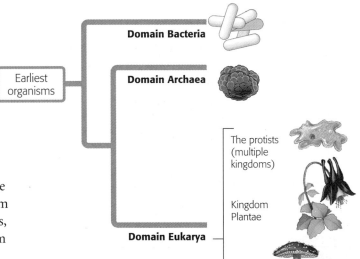

Domain Bacteria

Earliest organisms

Domain Archaea

The protists (multiple kingdoms)

Kingdom Plantae

Domain Eukarya

Kingdom Fungi

Kingdom Animalia

▲ **Figure 14.25 The three-domain classification system.** Molecular and cellular evidence supports the phylogenetic hypothesis that two lineages of prokaryotes, the domains Bacteria and Archaea, diverged very early in the history of life. Molecular evidence also suggests that domain Archaea is more closely related to domain Eukarya than to domain Bacteria.

The domain Eukarya is currently divided into kingdoms, but the exact number of kingdoms is still under debate. Biologists generally agree on the kingdoms Plantae, Fungi, and Animalia. These kingdoms consist of multicellular eukaryotes that differ in structure, development, and modes of nutrition. Plants make their own food by photosynthesis. Fungi live by decomposing the remains of other organisms and absorbing small organic molecules. Most animals live by ingesting food and digesting it within their bodies.

The remaining eukaryotes, the protists, include all those that do not fit the definition of plant, fungus, or animal—effectively, a taxonomic grab bag. Most protists are unicellular (amoebas, for example). But the protists also include certain large, multicellular organisms that are believed to be direct descendants of unicellular protists. For example, many biologists classify the seaweeds as protists because they are more closely related to some single-celled algae than they are to true plants.

It is important to understand that classifying Earth's diverse species is a work in progress as we learn more about organisms and their evolution. Charles Darwin envisioned the goals of modern systematics when he wrote in *The Origin of Species*, "Our classifications will come to be, as far as they can be so made, genealogies." ☑

☑**CHECKPOINT**

The current classification scheme favored by most biologists places life into _____ domains. The domain _____ includes humans in the kingdom called _____.

Answer: three; Eukarya; Animalia

EVOLUTION CONNECTION: Mass Extinctions

Rise of the Mammals

In this chapter, you've read about mass extinctions and their effects on the evolution of life on Earth. In the fossil record, each mass extinction was followed by a period of evolutionary change in which many new species arose whose adaptations allowed them to fill new habitats or community roles. The appearance of many new species followed each mass extinction as survivors became adapted to the many vacant ecological niches.

For example, fossil evidence indicates that the number of mammal species increased dramatically after the extinction of most of the dinosaurs around 65 million years ago (Figure 14.26). Although mammals originated 180 million years ago, fossils older than 65 million years indicate that they were mostly small and not very diverse. Early mammals may have been eaten or outcompeted by the larger and more diverse dinosaurs. With the disappearance of most of the dinosaurs, mammals expanded greatly in both diversity and size, filling the ecological roles once occupied by dinosaurs. Had it not been for the dinosaur extinction, mammals may never have expanded their territories and become the predominant land animals. Therefore, we humans may owe our existence to the demise of older species. Through the process of evolution by natural selection, this pattern of death and renewal is repeated throughout the history of life on Earth.

▼ Figure 14.26 **The increase in mammalian species after the extinction of dinosaurs.** Although mammals originated nearly 150 million years ago, they did not begin to widely diverge until after the demise of the dinosaurs.

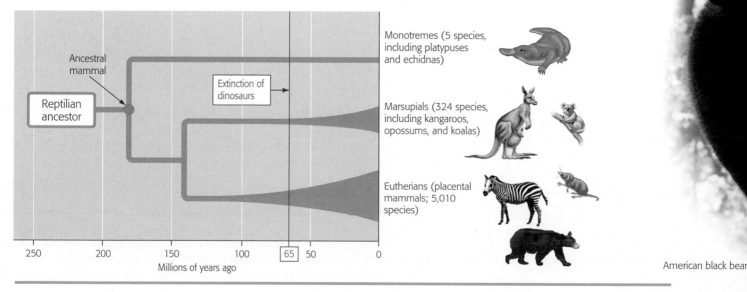

American black bear

Chapter Review

SUMMARY OF KEY CONCEPTS

(MB) Go to the Study Area at **www.masteringbiology.com** for practice quizzes, my**e**Book, BioFlix™ 3-D animations, MP3 Tutor Sessions, videos, current events, and more.

Macroevolution and the Diversity of Life

Microevolution refers to evolutionary change in the gene pool of a population within a species. Macroevolution refers to evolutionary changes evident at or above the species level, which are often seen in the fossil record. It includes the appearance of new species, the origins of evolutionary novelties, and the explosive diversification that follows some evolutionary breakthroughs and mass extinctions.

The Origin of Species

What Is a Species?

According to the biological species concept, a species is a group of populations whose members have the potential to interbreed in nature to produce fertile offspring. The biological species concept is just one of several possible ways to define species.

Reproductive Barriers between Species

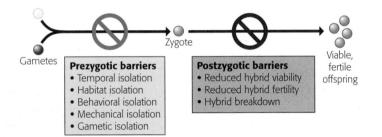

Gametes

Prezygotic barriers
- Temporal isolation
- Habitat isolation
- Behavioral isolation
- Mechanical isolation
- Gametic isolation

Zygote

Postzygotic barriers
- Reduced hybrid viability
- Reduced hybrid fertility
- Hybrid breakdown

Viable, fertile offspring

Mechanisms of Speciation

When the gene pool of a population is severed from other gene pools of the parent species, the splinter population can follow its own evolutionary course.

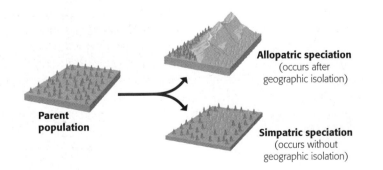

Parent population

Allopatric speciation
(occurs after geographic isolation)

Simpatric speciation
(occurs without geographic isolation)

Hybridization leading to polyploids is a common mechanism of sympatric speciation in plants.

What Is the Tempo of Speciation?

According to the punctuated equilibria model, the time required for speciation is relatively short compared with the overall duration of the species' existence. This model accounts for the relative rarity of transitional fossils linking newer species to older ones.

The Evolution of Biological Novelty

Adaptation of Old Structures for New Functions

An exaptation is a structure that evolves in one context and gradually becomes adapted for other functions.

Evo-Devo: Development and Evolutionary Novelty

A subtle change in the genes that control a species' development can have profound effects. In paedomorphosis, for example, the adult retains body features that were strictly juvenile in ancestral species.

Earth History and Macroevolution

Geologic Time and the Fossil Record

Geologists have established a geologic time scale with four broad divisions: Precambrian, Paleozoic, Mesozoic, and Cenozoic. The most common method for determining the ages of fossils is radiometric dating.

Plate Tectonics and Macroevolution

About 250 million years ago, plate movements brought all the landmasses together into the supercontinent Pangaea, causing extinctions and providing new opportunities for the survivors to diversify. About 180 million years ago, Pangaea began to break up, causing geographic isolation.

Mass Extinctions and Explosive Diversifications of Life

The fossil record reveals long, relatively stable periods punctuated by mass extinctions followed by explosive diversification of certain survivors. For example, during the Cretaceous extinctions, about 65 million years ago, the world lost an enormous number of species, including most of the dinosaurs. Mammals greatly increased in diversity after the Cretaceous period.

Classifying the Diversity of Life

Systematics, the study of biological diversity, includes taxonomy, which is the identification, naming, and classification of species.

Some Basics of Taxonomy

Each species is assigned a two-part name. The first part is the genus, and the second part is unique for each species within the genus. In the taxonomic hierarchy, domain > kingdom > phylum > class > order > family > genus > species.

Classification and Phylogeny

The goal of classification is to reflect phylogeny, which is the evolutionary history of a species. Classification is based on the fossil record, homologous structures, and comparisons of DNA and amino acid sequences. Homology (similarity based on shared ancestry) must be distinguished from analogy (similarity based on convergent evolution). Cladistics uses shared characteristics to group related organisms into clades—distinctive branches in the tree of life.

Classification: A Work in Progress

Biologists currently classify life into a three-domain system:

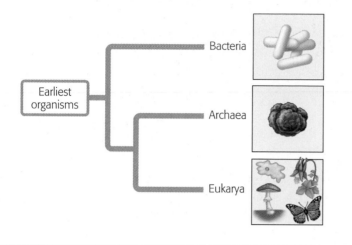

Bacteria

Earliest organisms

Archaea

Eukarya

SELF-QUIZ

1. Contrast microevolution with macroevolution.

2. Bird guides once listed the myrtle warbler and Audubon's warbler as distinct species that lived side by side in parts of their ranges. However, recent books describe them as the eastern and western forms of a single species, the yellow-rumped warbler. Apparently, the two kinds of warblers
   a. live in the same areas.
   b. successfully interbreed.
   c. are almost identical in appearance.
   d. are merging to form a single species.

3. Identify each of the following reproductive barriers as prezygotic or postzygotic.
   a. One lilac species lives on acidic soil, another on basic soil.
   b. Mallard and pintail ducks mate at different times of year.
   c. Two species of leopard frogs have different mating calls.
   d. Hybrid offspring of two species of jimsonweed always die before reproducing.
   e. Pollen of one kind of pine tree cannot fertilize another kind.

4. Why is a small, isolated population more likely to undergo speciation than a large one?

5. Many species of plants and animals adapted to desert conditions probably did not arise there. Their success in living in deserts could be due to _____, structures that evolved in one context but became adapted for different functions.

6. Mass extinctions
   a. cut the number of species to the few survivors left today.
   b. resulted mainly from the separation of the continents.
   c. occurred regularly, about every million years.
   d. were followed by diversification of the survivors.

7. The animals and plants of India are almost completely different from the species in nearby Southeast Asia. Why might this be true?
   a. They have become separated by convergent evolution.
   b. The climates of the two regions are completely different.
   c. India is in the process of separating from the rest of Asia.
   d. India was a separate continent until relatively recently.

8. Place these levels of classification in order from least inclusive to most inclusive: class, domain, family, genus, kingdom, order, phylum, species.

9. A paleontologist estimates that when a particular rock formed, it contained 12 mg of the radioactive isotope potassium-40. The rock now contains 3 mg of potassium-40. The half-life of potassium-40 is 1.3 billion years. From this information, you can conclude that the rock is approximately _____ billion years old.

10. In the three-domain system, which two domains contain prokaryotic organisms?

Answers to the Self-Quiz questions can be found in Appendix D.

THE PROCESS OF SCIENCE

11. Imagine you are conducting fieldwork and discover two groups of mice living on opposite sides of a river. Assuming that you will not disturb the mice, design a study to determine whether these two groups belong to the same species. If you could capture some of the mice and bring them to the lab, how might that affect your experimental design?

BIOLOGY AND SOCIETY

12. Experts estimate that human activities cause the extinction of hundreds or thousands of species every year. The natural rate of extinction is thought to be a few species per year. As we continue to alter the global environment, especially by cutting down tropical rain forests, the resulting extinction will probably rival that at the end of the Cretaceous period. Most biologists are alarmed at this prospect. What are some reasons for their concern? Considering that life has endured numerous mass extinctions and has always bounced back, how is the present mass extinction different? What might be some consequences for the surviving species?

15 The Evolution of Microbial Life

Signs of life.
These present-day rocky outcrops closely resemble fossils of ancient prokaryotes.

CHAPTER CONTENTS
■■■ Chapter Thread THE ORIGIN OF LIFE

BIOLOGY AND SOCIETY: The Origin of Life

Can Life Be Created in the Lab?

In *The Origin of Species*, Charles Darwin described the emergence of new species as a consequence of natural selection acting on existing species. But if you take his argument back to its logical start, you arrive at the ultimate mystery: How did life first arise on Earth?

Some biologists seek insight into this question through experiments on artificial systems. While artificial life has not yet been created in the lab, several important steps toward that goal have been completed. In 2008, a team led by Craig Venter (famous for his role in sequencing the human genome—see Chapter 12) synthesized from scratch the entire genome of *Mycoplasma genitalium*, a species of bacteria found naturally in the human urinary tract. The researchers constructed DNA strands, about 6,000 bases each, and then connected them to make a complete genome. In earlier experiments, Venter's team had proved that they could transplant the complete genome of one species of *Mycoplasma* bacteria into another. In one fell swoop, the recipient organism took up the new genome and expressed its genes. By combining these two approaches, Venter's team hopes to create an artificial genome and then transplant it into a genome-free host cell, essentially "rebooting" that cell to express the artificial genome. If successful, a new, artificial life-form could result.

Why pursue this line of research? One reason is that it may help answer fundamental questions about the origins of life in nature. Additionally, Venter hopes to engineer artificial organisms that can be completely controlled—a bacterial strain that might, for example, clean up toxic wastes or generate biofuels but be unable to survive outside rigidly controlled conditions (so that it would not spread uncontrollably). But could the creation of artificial life have unintended consequences that might ultimately prove harmful? Are there certain experiments that just should not be performed?

The question, How did life begin? continues to fascinate scientists, though it may never be answered with certainty. Accordingly, this chapter starts with a discussion of some of the key events in the history of life on Earth. We'll then discuss the origins, structures, and diversity of microbes (a term generally used to describe microscopic organisms), starting with the prokaryotes and then moving on to the protists.

Major Episodes in the History of Life

This chapter is the first of three that survey Earth's life-forms and their evolution. This section helps to set the stage by giving a brief overview of the major events in the history of life on Earth.

Life began when Earth was young. The planet formed about 4.6 billion years ago, and its crust began to solidify about 4 billion years ago. A few hundred million years later, by 3.5 billion years ago, Earth was already inhabited by a diversity of organisms (Figure 15.1). Those earliest organisms were all **prokaryotes**, their cells lacking true nuclei. Within the next billion years, two distinct groups of prokaryotes—bacteria and archaea—diverged.

An oxygen revolution began about 2.7 billion years ago. Photosynthetic prokaryotes that split water molecules released oxygen gas, profoundly changing Earth's atmosphere (see the Evolution Connection section at the end of Chapter 6). The accumulation of O_2 in the atmosphere doomed many prokaryotic groups. Among the survivors, a diversity of metabolic modes evolved, including cellular respiration, which uses O_2 in extracting energy from food (see Chapter 6). All of this metabolic evolution occurred during the almost 2 billion years that prokaryotes had Earth to themselves.

The oldest eukaryotic fossils are about 2.1 billion years old. Recall from Chapter 4 that **eukaryotes** are composed of one or more cells that contain nuclei and many other organelles absent in prokaryotic cells. The eukaryotic cell evolved from a prokaryotic community, a host cell containing even smaller prokaryotes. The mitochondria of our cells and those of every other eukaryote are descendants of those smaller prokaryotes, as are the chloroplasts of plants and algae.

The origin of more complex cells launched a period of tremendous diversification of eukaryotic forms. These new organisms were the protists. Represented today by a great diversity of species, protists are mostly microscopic and unicellular. Examples you may recognize include algae, amoebas, and *Paramecium*.

The next great event in the evolution of life was multicellularity. As discussed in this chapter's Evolution Connection section, the first multicellular eukaryotes evolved at least 1.2 billion years ago as colonies of single-celled organisms. Their modern descendants

▼ Figure 15.1 **Some major episodes in the history of life.** The timing of events shown on this phylogenetic tree is based on fossil evidence and molecular analysis. (Review Figure 14.23 for a reminder of how to read and interpret phylogenetic trees.)

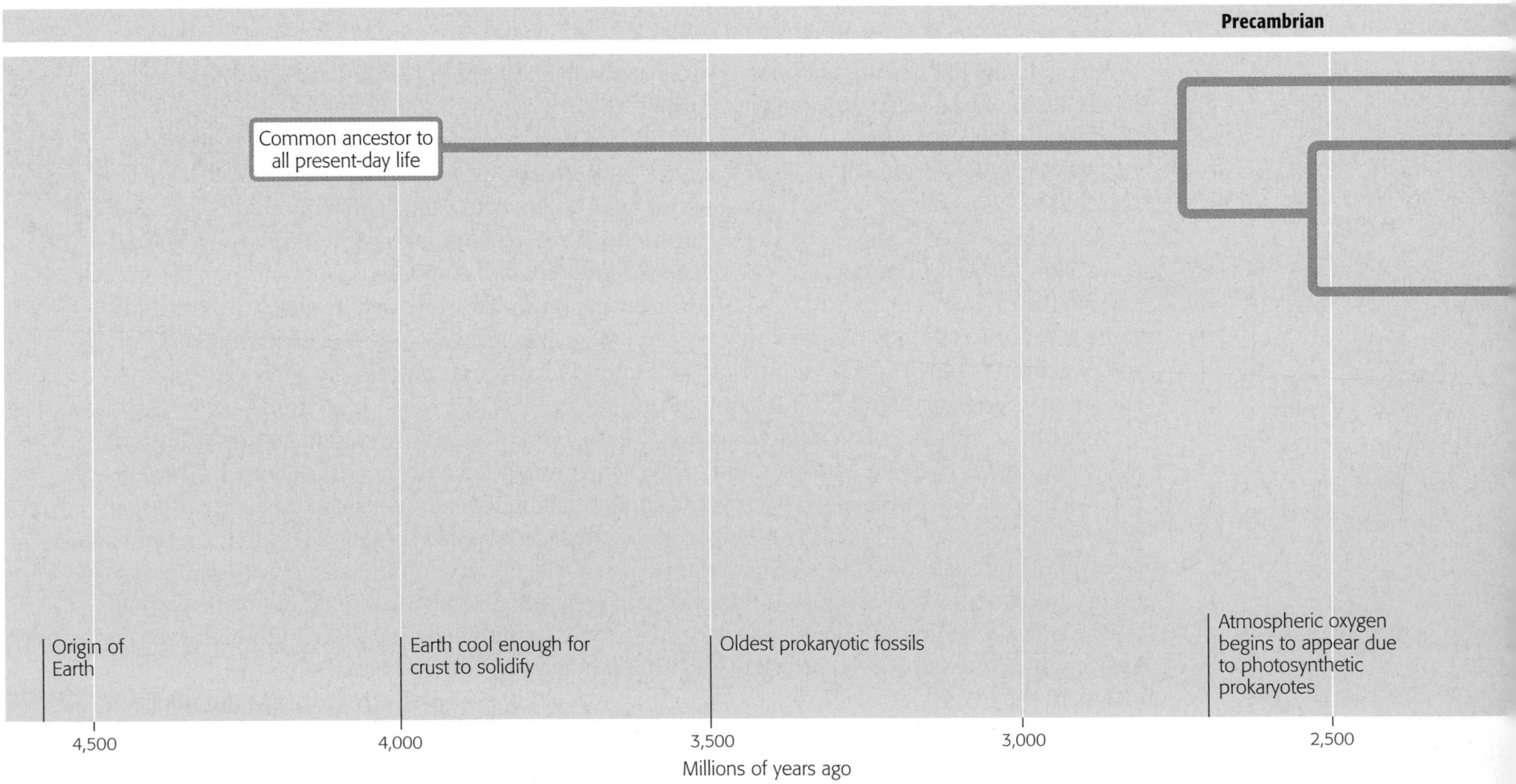

Precambrian

Common ancestor to all present-day life

Origin of Earth

Earth cool enough for crust to solidify

Oldest prokaryotic fossils

Atmospheric oxygen begins to appear due to photosynthetic prokaryotes

4,500   4,000   3,500   3,000   2,500

Millions of years ago

include multicellular protists, such as seaweeds. Other evolutionary branches stemming from the ancient protists gave rise to animals, fungi, and plants.

The greatest diversification of animals was the so-called Cambrian explosion. The Cambrian was the first period of the Paleozoic era, which began about 540 million years ago (see Table 14.1). The earliest animals lived in late-Precambrian seas, but they diversified extensively over a span of just 10 million years during the early Cambrian. In fact, all the major animal body plans—as well as all the major groups—had evolved by the end of that evolutionary eruption.

For over 85% of biological history—life's first 3 billion years—life was confined mostly to aquatic habitats. The colonization of land was a major milestone in the history of life. Plants and fungi together led the way about 500 million years ago. Plants transformed the landscape, creating new opportunities for all life-forms, especially herbivorous (plant-eating) animals and their predators.

The evolutionary venture onto land included vertebrate animals in the form of the first amphibians. These prototypes of today's frogs and salamanders descended from air-breathing fish with fleshy fins that could support the animal's weight on land. Further evolution by natural selection led to the appearance of reptiles and mammals. Among the mammals are the primates, the animal group that includes humans and their closest relatives, apes and monkeys.

Figure 15.2 uses the analogy of a clock ticking down from the origin of Earth 4.6 billion years ago to the present to summarize the major events in the history of life. As the figure makes clear, to trace the history of life on Earth, we must go back to the origin and diversification of microbes, starting with the prokaryotes.

▶ Figure 15.2 **A clock analogy for the major events in the history of life on Earth.**

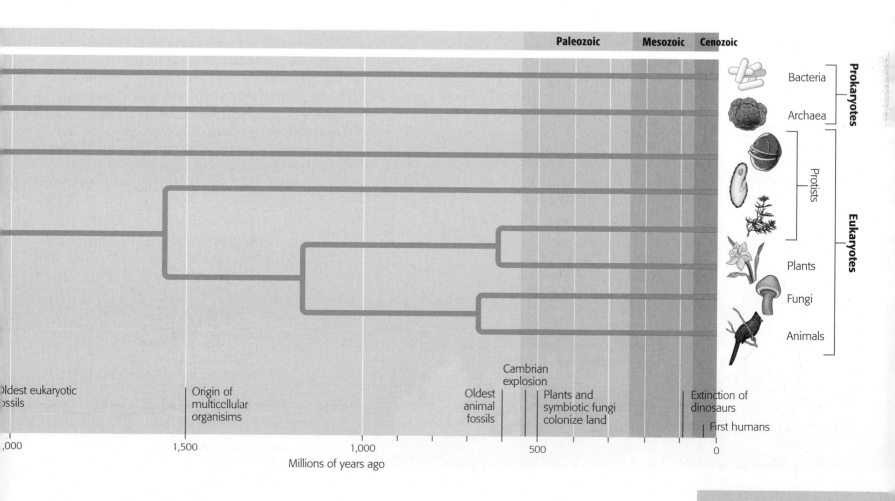

The Origin of Life

Although we'll never know for sure how life on Earth began, in this section we'll discuss hypotheses for the formation of the first life. Then we'll talk about its subsequent early evolution.

Resolving the Biogenesis Paradox

From the time of the ancient Greeks until well into the 1800s, it was commonly believed that life could regularly arise from nonliving matter, an idea called **spontaneous generation**. Many people believed, for instance, that flies could arise from rotting meat and fish from ocean mud. Then, in 1862, experiments by Louis Pasteur confirmed what many others had suspected: All life today, including microbes, arises only by the reproduction of preexisting life. This "life-from-life" principle is called **biogenesis**.

But wait! If life always arises from previous life, then how could the first organisms arise? Although there is no evidence that spontaneous generation occurs today,

it could have early in Earth's history, when conditions were very different. For instance, the O_2 in today's atmosphere is a corrosive agent that tends to disrupt chemical bonds, thereby preventing the formation of complex molecules—but O_2 wasn't present in the ancient atmosphere. And such energy sources as lightning, volcanic activity, and ultraviolet sunlight were all more intense on the early Earth than they are today. So life did not begin on a planet anything like the modern Earth, but on a young Earth that was a very different world (**Figure 15.3**). Most biologists now think that chemical and physical processes in Earth's primordial environment could have led to very simple cells through a sequence of stages. Debate abounds about the nature of those stages. ☑

A Four-Stage Hypothesis for the Origin of Life

According to one hypothesis for the origin of life, the first organisms were products of chemical evolution in four stages: (1) the abiotic (nonliving) synthesis of small organic molecules, such as amino acid and nucleotide monomers; (2) the joining of these small molecules into macromolecules, including proteins and nucleic acids; (3) the packaging of all these molecules into pre-cells, droplets with membranes that maintained an internal chemistry different from the surroundings; and (4) the origin of self-replicating molecules that eventually made inheritance possible. This is all speculative, of course, but what makes it a valid scientific hypothesis is that it leads to predictions that can be tested in the laboratory. Let's take a closer look at each of these four stages.

Stage 1: Abiotic Synthesis of Organic Monomers
This stage was the first to be extensively studied in the laboratory. We'll begin our investigation of chemical evolution with a breakthrough experiment that first brought this idea to the forefront of scientific thinking.

☑ CHECKPOINT
One reason why the spontaneous generation of life on Earth could not occur today is the abundance of
_____ in our modern atmosphere.

Answer: oxygen (O_2)

◀ Figure 15.3 **An artist's rendition of Earth about 3 billion years ago.** The pad-like objects in the scene represent colonies of prokaryotes known from the fossil record.

THE PROCESS OF SCIENCE: The Origin of Life

Can Biological Monomers Form Spontaneously?

In 1953, University of Chicago scientist Harold Urey and his 23-year-old graduate student Stanley Miller conducted what is now considered a classic experiment. They began with the **observation** that modern biological macromolecules (DNA, protein, carbohydrates, and so on) are all composed of elements (primarily oxygen, hydrogen, carbon, and nitrogen) that were present in abundance on the early Earth. This led to the **question**, Could biological molecules arise spontaneously under conditions like those on the early Earth? Miller and Urey began with the **hypothesis** that a closed system designed to simulate such conditions in the laboratory could produce biologically important organic molecules from inorganic ingredients.

Figure 15.4 shows the apparatus they created to test their hypothesis. A flask of warmed water simulated the primordial sea. An "atmosphere"—in the form of gases added to a reaction chamber—contained hydrogen gas (H_2), methane (CH_4), ammonia (NH_3), and water vapor (H_2O). To mimic the prevalent lightning of the early Earth, sparks were discharged into the chamber. A condenser cooled the atmosphere, causing water and any dissolved compounds to "rain" into the miniature "sea." Miller and Urey's **prediction** was that organic molecules would form and accumulate during this **experiment**.

Miller and Urey's **results** made front-page news. After the apparatus had run for a week, an abundance of organic molecules essential for life, including amino acids, the monomers of proteins, had collected in the "sea."

Since Miller and Urey's classic experiment, other scientists have repeated and extended the research, varying such conditions as the composition of the ancient "atmosphere" and "sea" (since Miller's "early atmosphere" was almost certainly incorrect in some ways). Laboratory analogs of the primeval Earth have produced all 20 amino acids and several sugars. These laboratory results support the concept of the abiotic synthesis of organic molecules on the early Earth.

▼ Figure 15.4 **The abiotic production of organic molecules: A laboratory simulation of early-Earth chemistry.**

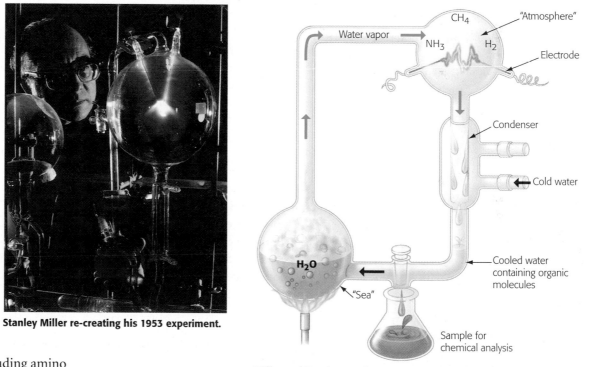

Stanley Miller re-creating his 1953 experiment.

Miller and Urey's experiment

Stage 2: Abiotic Synthesis of Polymers

If the hypothesis of an abiotic origin of life is correct, then it should be possible to link organic monomers to form polymers such as proteins and nucleic acids without the help of enzymes and other cellular equipment. Researchers have brought about such polymerization by dripping solutions of organic monomers onto hot sand, clay, or rock. The heat vaporizes the water in the solutions and concentrates the monomers on the underlying material. Some of the monomers then spontaneously bond together to form polymers. On the early Earth, raindrops or waves may have splashed dilute solutions of organic monomers onto fresh lava or other hot rocks and then washed polypeptides and other polymers back into the sea. There, they could accumulate in great quantities (since no life existed to consume them). ☑

Stage 3: Formation of Pre-Cells

According to the four-stage hypothesis, a key step in the origin of life was the isolation of a collection of abiotically created molecules within a membrane. We'll call these molecular aggregates pre-cells—not really cells, but molecular packages with some of the properties of life. Within a confined space, certain combinations of molecules could be concentrated and interact more efficiently. Furthermore, the internal environment of a pre-cell could differ from its surroundings.

Laboratory experiments demonstrate that pre-cells could have formed spontaneously from abiotically produced organic compounds (see Figure 5.20). Such pre-cells produced in the laboratory display some lifelike properties: They have a selectively permeable surface, can grow by absorbing molecules from their surroundings, and swell or shrink when placed in solutions of different salt concentrations.

Stage 4: Origin of Self-Replicating Molecules

Life is defined partly by the process of inheritance, which is based on self-replicating molecules. Today's cells store their genetic information as DNA. They transcribe the information into RNA and then translate RNA messages into specific enzymes and other proteins (see Chapter 10). This mechanism of information flow probably emerged gradually through a series of small changes to much simpler processes.

What were the first genes like? One hypothesis is that they were short strands of RNA that replicated themselves without the assistance of proteins. In laboratory experiments, short RNA molecules can assemble spontaneously from nucleotide monomers in the absence of enzymes (**Figure 15.5**). The result is a population of RNA molecules, each with a random sequence of monomers.

Some of the molecules self-replicate, but their success at this reproduction varies. What happens can be described as molecular evolution: The RNA varieties that replicate fastest increase their frequency in the population.

In addition to the experimental evidence, there is another reason the idea of RNA genes in the primordial world is plausible. Cells actually have RNAs that can act as enzymes; they are called ribozymes. Perhaps early ribozymes catalyzed their own replication. That would help with the "chicken and egg" paradox of which came first, enzymes or genes. Maybe the "chicken and egg" came together in the same RNA molecules. The molecular biology of today may have been preceded by an ancient "RNA world." ☑

From Chemical Evolution to Darwinian Evolution

If pre-cells with self-replicating RNA (and later DNA) did form on the young Earth, they would be refined by natural selection. Mutations, errors in the copying of the "genes," would result in variation among the pre-cells. And the most successful of these pre-cells would grow, divide (reproduce), and continue to evolve. Of course, the gap between such pre-cells and even the simplest of modern cells is enormous. But with millions of years of incremental changes through natural selection, these molecular cooperatives could have become more and more cell-like. The point at which we stop calling them pre-cells and start calling them living cells is as fuzzy as our understanding of how life originated. But we do know that prokaryotes were already flourishing at least 3.5 billion years ago and that all branches of life arose from those ancient prokaryotes.

☑**CHECKPOINT**

What are ribozymes? Why are they a logical step in the formation of life?

Answer: A ribozyme is an RNA molecule that functions as an enzyme. Ribozymes can perform some of the functions of both DNA and protein.

▼ Figure 15.5 **Self-replication of RNA "genes."**

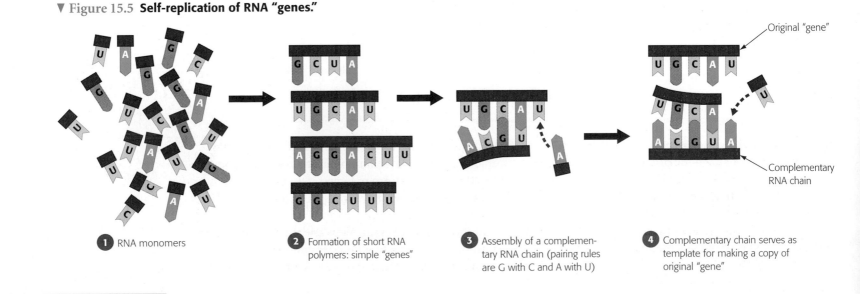

1 RNA monomers

2 Formation of short RNA polymers: simple "genes"

3 Assembly of a complementary RNA chain (pairing rules are G with C and A with U)

4 Complementary chain serves as template for making a copy of original "gene"

Original "gene"

Complementary RNA chain

Prokaryotes

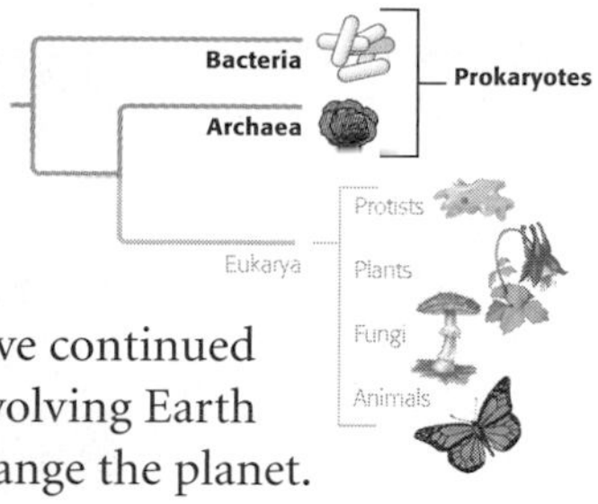

The history of prokaryotic life is a success story spanning billions of years. Prokaryotes lived and evolved all alone on Earth for about 2 billion years. They have continued to adapt and flourish on an evolving Earth and in turn have helped to change the planet. In this section, you will study prokaryotic structure and function, diversity, effects on humans, and ecological significance.

They're Everywhere!

Today, prokaryotes are found wherever there is life. They thrive in habitats too cold, too hot, too salty, too acidic, or too alkaline for any eukaryote (**Figure 15.6**). Biologists have even discovered prokaryotes living on the walls of a gold mine 2 miles below Earth's surface. In addition, prokaryotes far outnumber eukaryotes. In fact, more prokaryotes are living in your mouth right now than the total number of people who have ever lived! The collective biological mass (biomass) of prokaryotes is at least ten times that of all eukaryotes.

Though individual prokaryotes are small organisms (**Figure 15.7**), they are giants in their collective impact on Earth and its life. In terms of their impact on humans, we hear most about the relatively few species that can cause serious illness. Bacterial infections are responsible for about half of all human diseases, including tuberculosis, cholera, many sexually transmissible diseases, and certain types of food poisoning. During the 1300s, Black Death—bubonic plague, a bacterial disease—spread across Europe, killing an estimated 25% of the human population.

However, prokaryotic life is much more than just a rogues' gallery. Benign or beneficial prokaryotes are far more common than harmful ones. For example, humans have long used bacteria to produce various types of food, converting milk to cheese, yogurt, and sour cream, for example. In addition, each of us harbors several hundred species of bacteria in and on our body. Bacteria in our intestines provide us with important vitamins, and others living in our mouth prevent harmful fungi from growing there.

The ecological significance of prokaryotes is hard to overstate. Prokaryotes recycle carbon and other vital chemical elements back and forth between organic matter and the soil and atmosphere. For example, some prokaryotes decompose dead organisms. Found in soil and at the bottom of lakes, rivers, and oceans, these decomposers return chemical elements to the environment in the form of inorganic compounds that can be used by plants, which in turn feed animals. If prokaryotic decomposers were to disappear, the chemical cycles that sustain life would come to a halt. All forms of eukaryotic life would also be doomed. In contrast, prokaryotic life would undoubtedly persist in the absence of eukaryotes, as it once did for 2 billion years.

Microbial Life

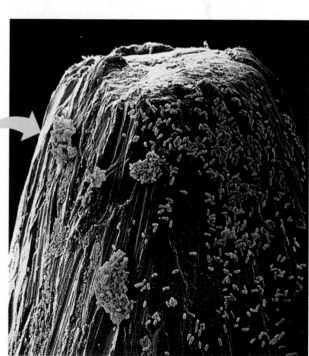

◀ **Figure 15.6 A window to early life?** An instrument on the research submarine *Alvin* samples the water around a hydrothermal vent more than 1.5 km below the ocean's surface. Prokaryotes that live near the vent use the emitted gases as an energy source. This environment, which is very dark, hot, and under high pressure, is among the most extreme in which life exists today.

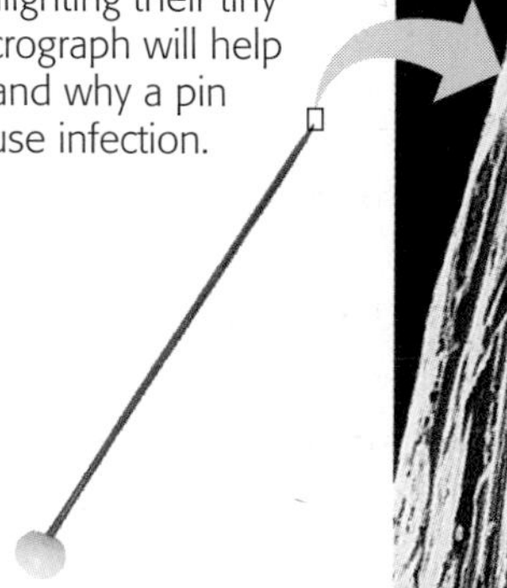

▶ **Figure 15.7 Bacteria on the head of a pin.** The orange rods are individual bacteria, each about 5 μm long, on the head of a pin. Besides highlighting their tiny size, this micrograph will help you understand why a pin prick can cause infection.

Colorized SEM 605×

The Structure and Function of Prokaryotes

Prokaryotes have a cellular organization fundamentally different from that of eukaryotes. Whereas eukaryotic cells have a membrane-enclosed nucleus and numerous other membrane-enclosed organelles, prokaryotic cells lack these structural features (see Chapter 4, especially Figures 4.4 and 4.5). And nearly all species of prokaryotes have cell walls exterior to their plasma membranes. In this section, we will discuss other features that are common to all prokaryotes, including aspects of their form, reproduction, and nutrition.

Prokaryotic Forms

Determining cell shape by microscopic examination is an important step in identifying a prokaryote (**Figure 15.8**). Spherical species are called **cocci** (singular, *coccus*). Cocci that occur in clusters are called staphylococci. Other cocci occur in chains; they are called streptococci. For example, the bacterium that causes strep throat in humans is a streptococcus. Rod-shaped prokaryotes are called **bacilli** (singular, *bacillus*). A third prokaryotic cell shape is spiral or curved. These include spirochetes, different species of which cause syphilis and Lyme disease.

Most prokaryotes are unicellular and very small, but there are exceptions to both of these generalizations. The cells of some species usually exist as groups of two or more cells, such as the streptococci already mentioned. Others form true colonies, which are permanent aggregates of identical cells (**Figure 15.9a**). And some species even exhibit a simple multicellular organization, with a division of labor among specialized types of cells (**Figure 15.9b**). Among unicellular species, moreover, there are some giants that actually dwarf most eukaryotic cells (**Figure 15.9c**).

▼ Figure 15.9 **A diversity of prokaryotic shapes and sizes.**

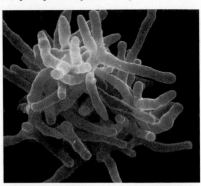

(a) Actinomycete. An actinomycete is a mass of branching chains of rod-shaped cells. These bacteria are common in soil, where they secrete antibiotics that inhibit the growth of other bacteria. Various antibiotic drugs, such as streptomycin, are obtained from actinomycetes.

Colorized SEM 4,800x

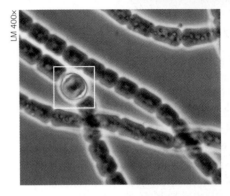

LM 400x

(b) Cyanobacteria. These photosynthetic cyanobacteria divide labor among their cells. The box highlights a cell that converts atmospheric nitrogen to ammonia, which can then be incorporated into amino acids and other organic compounds.

LM 18x

(c) Giant bacterium. The larger white blob in this photo is the marine bacterium *Thiomargarita namibiensis*. This prokaryotic cell is over 0.5 mm in diameter, about the size of the fruit fly's head below it.

▼ Figure 15.8 **Three common shapes of prokaryotic cells.**

SHAPES OF PROKARYOTIC CELLS		
Spherical (cocci)	**Rod-shaped (bacilli)**	**Spiral**

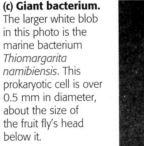

Colorized SEM 10,000x

Colorized SEM 9,000x

Colorized TEM 30,000x

About half of all prokaryotic species are mobile. Many of those that travel have one or more flagella that propel the cells away from unfavorable places or toward more favorable places, such as nutrient-rich locales (Figure 15.10).

Prokaryotic Reproduction

Many prokaryotes can reproduce at a phenomenal rate if conditions are favorable. The cells copy their DNA almost continuously and divide again and again by the process called **binary fission**. Dividing by binary fission, a single cell becomes 2 cells, which then become 4, 8, 16, and so on. Some species can produce a new generation in only 20 minutes under optimal conditions. If reproduction continued unchecked at this rate, a single prokaryote could give rise to a colony outweighing Earth in only three days!

Fortunately, few prokaryotic populations can sustain this kind of exponential growth for long. Environments are usually limiting in resources such as food and space. Prokaryotes also produce metabolic waste products that may eventually pollute the colony's environment. Still, you can understand why certain bacteria can make you sick so soon after infection or why food can spoil so rapidly. Refrigeration retards food spoilage not because the cold kills bacteria, but because most microorganisms reproduce very slowly at such low temperatures.

Some prokaryotes can survive during very harsh conditions and for extended periods, even centuries, by forming specialized cells called endospores. An **endospore** is a thick-coated, protective cell produced within the prokaryotic cell when the prokaryote is exposed to unfavorable conditions (Figure 15.11). The endospore can survive all sorts of trauma, including lack of water and nutrients, extreme heat or cold, and most poisons. Not even boiling water kills most of these resistant cells. And when the environment becomes more hospitable, the endospore can absorb water and resume growth. To ensure that all cells, including endospores, are killed when laboratory equipment is sterilized, microbiologists use an appliance called an autoclave, a pressure cooker that uses high-pressure steam at a temperature of 121°C (250°F). The food-canning industry employs similar methods to kill endospores of dangerous soil bacteria such as *Clostridium botulinum*, which produces a toxin that causes the potentially fatal disease botulism. ☑

▼ Figure 15.10 **Prokaryotic flagella.** At the base of the prokaryotic flagellum is a motor and set of rings embedded in the plasma membrane and cell wall. This machinery actually spins like a wheel, rotating the filament of the flagellum.

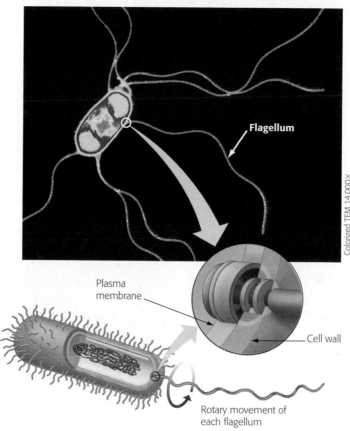

Flagellum

Plasma membrane

Cell wall

Rotary movement of each flagellum

Colorized TEM 14,000×

▼ Figure 15.11 **An endospore in an anthrax bacterium.** This prokaryote is *Bacillus anthracis*, the notorious bacterium that produces the deadly disease called anthrax in cattle, sheep, and humans. There are actually two cells here, one inside the other. The outer cell produced the specialized dormant inner cell, the endospore.

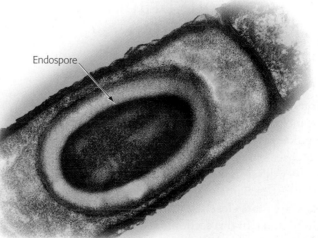

Endospore

Colorized SEM 45,000×

☑**CHECKPOINT**

1. Using a microscope, how could you distinguish the cocci that cause staph infections from those that cause strep throat?
2. Why do microbiologists autoclave their laboratory instruments and glassware rather than just washing them in very hot water?

Answers: 1. by the arrangement of the cell aggregates: grapelike clusters for staphylococcus and chains of cells for streptococcus 2. to kill bacterial endospores, which can survive boiling water

Prokaryotic Nutrition

When classifying diverse organisms, biologists often use the phrase "mode of nutrition" to describe how an organism obtains energy and carbon, the two main resources needed for synthesizing organic compounds. Species that obtain energy from light are called phototrophs, while species that obtain energy from environmental chemicals are called chemotrophs. Species that obtain carbon from the inorganic compound carbon dioxide (CO_2) are called autotrophs, while species that obtain carbon from at least one organic nutrient—the sugar glucose, for instance—are called heterotrophs. We can combine energy source (phototroph versus chemotroph) and carbon source (autotroph versus heterotroph) to group all organisms according to the four major modes of nutrition shown in **Figure 15.12**. Two of these modes—photoautotrophs (such as most plants), and chemoheterotrophs (such as most animals)—are dominant among multicellular organisms. The other two modes are used only by certain prokaryotes. ☑

► Figure 15.12
Modes of nutrition.

☑**CHECKPOINT**

A bacterium requires only water and the amino acid methionine to grow and lives in very deep caves where no light penetrates. Based on its mode of nutrition, this species would be classified as a _____.

Answer: *chemoheterotroph*

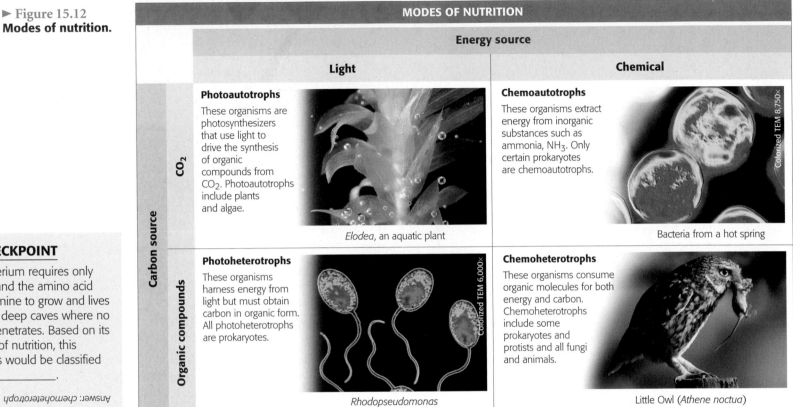

MODES OF NUTRITION

Energy source — **Light** | **Chemical**

Carbon source

CO_2

Photoautotrophs
These organisms are photosynthesizers that use light to drive the synthesis of organic compounds from CO_2. Photoautotrophs include plants and algae.

Elodea, an aquatic plant

Chemoautotrophs
These organisms extract energy from inorganic substances such as ammonia, NH_3. Only certain prokaryotes are chemoautotrophs.

Colorized TEM 8,750×

Bacteria from a hot spring

Organic compounds

Photoheterotrophs
These organisms harness energy from light but must obtain carbon in organic form. All photoheterotrophs are prokaryotes.

Colorized TEM 6,000×

Rhodopseudomonas

Chemoheterotrophs
These organisms consume organic molecules for both energy and carbon. Chemoheterotrophs include some prokaryotes and protists and all fungi and animals.

Little Owl (*Athene noctua*)

The Two Main Branches of Prokaryotic Evolution: Bacteria and Archaea

By comparing diverse prokaryotes at the molecular level, biologists have identified two major branches of prokaryotic evolution: **bacteria** and **archaea**. Thus, life is organized into three domains—**Bacteria, Archaea**, and **Eukarya** (review Figure 14.25). Although bacteria and archaea have prokaryotic cell organization in common, they differ in many structural and physiological characteristics. Some of these differences suggest that archaea are more closely related to eukaryotes than they are to bacteria. In this section, we'll focus on the special characteristics of archaea before turning our attention to bacteria.

The term *archaea* ("ancient") refers to the antiquity of this group's origin from the earliest cells. Even today, many species of archaea inhabit extreme environments, such as hot springs and salt ponds. Few other modern organisms (if any) can survive in some of these environments, which may resemble habitats on the early Earth.

Biologists refer to some archaea as "extremophiles," meaning "lovers of the extreme." There are extreme halophiles ("salt lovers"), archaea that thrive in such environments as Utah's Great Salt Lake, the Dead Sea, and seawater-evaporating ponds used to produce salt

(Figure 15.13a). There are also extreme thermophiles ("heat lovers") that live in very hot water (Figure 15.13b); some archaea even populate the deep-ocean vents that gush water hotter than 100°C (212°F), such as the one shown in Figure 15.6. Also among the archaea are the methanogens, which live in anaerobic (oxygen-free) environments and give off methane as a waste product. They are abundant in the mud at the bottom of lakes and swamps. You may have seen methane, also called marsh gas, bubbling up from a swamp. Great numbers of methanogens also inhabit the digestive tracts of animals. In humans, intestinal gas is largely the result of their metabolism. More importantly, methanogens aid digestion in cattle, deer, and other animals that depend heavily on cellulose for their nutrition. Normally, bloating does not occur in these animals because they regularly expel large volumes of gas produced by the methanogens. (And that may be more than you wanted to know about these gas-producing microbes!)

▼ Figure 15.13 **Archaeal "extremophiles."**

(a) Salt-loving archaea. This is an aerial photo of commercial salt-producing ponds at the edge of San Francisco Bay. The colors of the ponds result from dense growth of the harmless prokaryotes that thrive when the salinity of the water reaches five to eight times that of seawater.

(b) Heat-loving archaea. In this photo, you can see orange and yellow colonies of heat-loving prokaryotes growing near a Nevada geyser.

Bacteria and Humans

In this section, we'll consider some of the ways that bacteria interact with humans. These interactions can be for both good and ill.

Bacteria That Cause Disease

We are constantly exposed to bacteria, some of which are potentially harmful (Figure 15.14). Bacteria and other organisms that cause disease are called **pathogens**. We're healthy most of the time because our body's defenses check the growth of pathogens. Occasionally, the balance shifts in favor of a pathogen, and we become ill. Even some of the bacteria that are normal residents of the human body can make us sick when our defenses have been weakened by poor nutrition, medical treatment (especially cancer therapy), or a viral infection.

Most pathogenic bacteria cause disease by producing poisons. There are two classes of these poisons: exotoxins and endotoxins. **Exotoxins** are poisonous proteins secreted by bacterial cells. A single gram of the exotoxin that causes botulism could kill a million people. Another exotoxin producer is *Staphylococcus aureus* (abbreviated *S. aureus*). It is a common, usually harmless resident of our skin surface. However, if *S. aureus* enters the body through a cut or other wound or is swallowed in contaminated food, it can cause serious diseases collectively called staph infections. One type of *S. aureus* produces exotoxins that cause layers of skin to

slough off ("flesh-eating disease"); another can cause vomiting and severe diarrhea; yet another can produce a potentially deadly disease called toxic shock syndrome.

In contrast to exotoxins, **endotoxins** are not cell secretions but are chemical components of the outer membrane of certain bacteria. All endotoxins induce the same general symptoms: fever, aches, and sometimes shock (a dangerous drop in blood pressure). The severity of symptoms varies with the host's condition and with the bacterium. Different species of *Salmonella*, for example, produce endotoxins that cause food poisoning and typhoid fever.

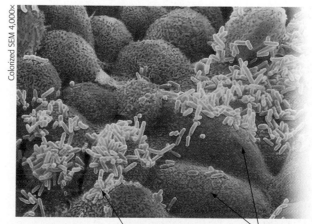

Colorized SEM 4,000×

Haemophilus influenzae

Cells of nasal lining

◀ Figure 15.14 **Bacteria that cause pneumonia.** *Haemophilus influenzae*, shown here on cells lining the interior of a human nose, causes pneumonia and other lung infections, which kill about 4 million people worldwide per year. Most victims are children in less industrialized countries, where malnutrition lowers resistance to all pathogens.

During the last century, following the discovery that "germs" cause disease, the incidence of bacterial infections declined, particularly in more industrialized nations. Sanitation is generally the most effective way to prevent bacterial disease. The installation of water treatment and sewage systems continues to be a public health priority throughout the world. Antibiotics have been discovered that can cure most bacterial diseases. However, resistance to widely used antibiotics has evolved in many of these pathogens (see Chapter 1).

In addition to sanitation and antibiotics, a third defense against bacterial disease is education. A case in point is Lyme disease, the most widespread pest-carried disease in the United States. The disease is caused by a spirochete bacterium carried by ticks (**Figure 15.15**). Disease-carrying ticks live on deer and field mice but also bite humans. Lyme disease usually starts as a red rash shaped like a bull's-eye around a tick bite. Antibiotics can cure the disease if administered within a month of exposure. If untreated, Lyme disease can cause debilitating arthritis, heart disease, and nervous system disorders. So far, the best defense is public education about avoiding tick bites and the importance of seeking treatment if a rash develops. The Centers for Disease Control and Prevention recommends that you avoid vegetation in tick-infested areas, wear light-colored clothing when walking through brush so that you can easily see any ticks, and use an appropriate insect repellent. ☑

Bioterrorism

During the fall of 2001, five Americans died from the disease anthrax in a presumed terrorist attack (**Figure 15.16**). Unfortunately, this act of bioterrorism was hardly unprecedented; there is a long and ugly history of humans using organisms as weapons.

During the Middle Ages, the bacterium *Yersinia pestis* (the cause of bubonic plague) played a role in battle when armies hurled the bodies of plague victims into enemy ranks. Early conquerors, settlers, and warring armies in South and North America gave native peoples items purposely contaminated with infectious bacteria. In the 1930s, the Japanese government instituted a biowarfare program that killed tens of thousands of Chinese soldiers and civilians. Their weapon? The bacteria that cause plague, anthrax, and cholera. In 1984, members of a cult in Oregon contaminated restaurant salad bars with *Salmonella* bacteria; over 700 people became sick, and 45 were hospitalized. During the 1990s, another cult tried to start an anthrax epidemic in Tokyo; and in Iraq, the Iraqi army loaded missiles filled with harmful bacteria. Luckily, neither of these attempts resulted in casualties.

The United States opened its first biological weapons research facility in 1943. There, the military bred new strains of bacteria that cause such illnesses as anthrax and botulism. To "weaponize" these pathogens, researchers selected highly virulent strains and developed formulations for effective dispersal. But the practical difficulties of controlling such weapons—and a measure of moral repugnance—led President Richard Nixon to end the U.S. bioweapons program in 1969 and to order its products destroyed. In 1975, the United States signed the Biological Weapons Convention, pledging never to develop or store biological weapons. Eventually, 103 nations joined the ban, although not every signing nation has honored it.

▼ **Figure 15.15 Lyme disease, a bacterial disease transmitted by ticks.** The bacterium that causes Lyme disease (shown in micrograph at right) is carried from deer to humans by ticks.

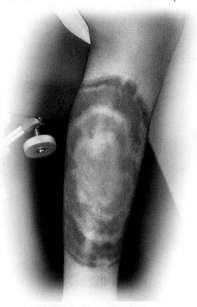

"Bull's-eye" rash

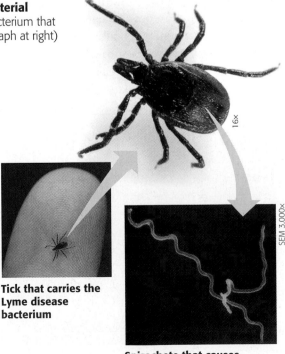

Tick that carries the Lyme disease bacterium

Spirochete that causes Lyme disease

▼ **Figure 15.16 Cleaning up after a bioterrorist attack.** This 2001 photo shows hazardous-material workers spraying themselves after they had searched the Senate Office Building in Washington, DC, for anthrax spores.

The Ecological Impact of Prokaryotes

Pathogenic bacteria are in the minority among prokaryotes. Far more common are species that are essential to our well-being, either directly or indirectly. Let's turn our attention now to the vital role that prokaryotes play in sustaining the biosphere.

Prokaryotes and Chemical Recycling

Not too long ago, the atoms making up the organic molecules in your body were part of the inorganic compounds of soil, air, and water, as they will be again. Life depends on the recycling of chemical elements between the biological and physical components of ecosystems. Prokaryotes play essential roles in these chemical cycles. For example, all the nitrogen that plants use to make proteins and nucleic acids comes from prokaryotic metabolism in the soil. In turn, animals get their nitrogen compounds from plants.

Another vital function of prokaryotes, mentioned earlier in the chapter, is the breakdown of organic wastes and dead organisms. Prokaryotes decompose organic matter and, in the process, return elements to the environment in inorganic forms that can be used by other organisms. If it were not for such decomposers, carbon, nitrogen, and other elements essential to life would become locked in the organic molecules of corpses and waste products. You'll learn more about the role that prokaryotes play in chemical cycling in Chapter 20.

Prokaryotes and Bioremediation

People have put the metabolically diverse prokaryotes to work in cleaning up the environment. The use of organisms to remove pollutants from water, air, or soil is called **bioremediation**. One example of bioremediation is the use of prokaryotic decomposers to treat our sewage. Raw sewage is first passed through a series of screens and shredders, and solid matter settles out from the liquid waste. This solid matter, called sludge, is then gradually added to a culture of anaerobic prokaryotes, including both bacteria and archaea. The microbes decompose the organic matter in the sludge, converting it to material that can be used as landfill or fertilizer. The liquid wastes may then be passed through a trickling filter system consisting of a long horizontal bar that slowly rotates, spraying liquid wastes onto a bed of rocks (Figure 15.17). Aerobic prokaryotes and fungi growing on the rocks remove much of the organic matter from the liquid. The outflow from the rock bed is then sterilized and released back into the environment.

We are just beginning to explore the great potential that prokaryotes offer for bioremediation. Certain

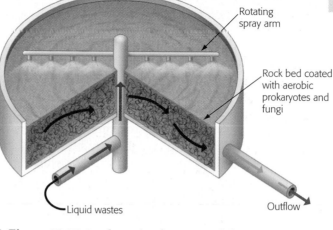

▲ Figure 15.17 **Putting microbes to work in sewage treatment facilities.** This is a trickling filter system, which uses bacteria, archaea, and fungi to treat liquid wastes after sludge is removed.

Labels: Rotating spray arm; Rock bed coated with aerobic prokaryotes and fungi; Liquid wastes; Outflow

bacteria that occur naturally on ocean beaches can decompose petroleum and are useful in cleaning up oil spills (Figure 15.18). Genetically engineered bacteria may be able to degrade oil more rapidly than naturally occurring oil-eaters. Bacteria may also help us clean up old mining sites. The water that drains from mines is highly acidic and is also laced with heavy metals and other poisons. Bacteria called *Thiobacillus* thrive in the acidic waters that drain from mines. While obtaining energy by oxidizing sulfur or sulfur-containing compounds, the bacteria also accumulate metals from the mine waters. Unfortunately, their use in cleaning up mine wastes is limited because their metabolism also adds sulfuric acid to the water. If this problem is solved, perhaps through genetic engineering, *Thiobacillus* and other prokaryotes may help us overcome some environmental dilemmas that seem unsolvable today. One current research focus is a bacterium that tolerates radiation doses thousands of times stronger than those that would kill a person. This species may help clean up toxic dump sites that include radioactive waste.

The various modes of nutrition and metabolic pathways we find in organisms living today are all variations on themes that evolved in prokaryotes during their long reign as Earth's exclusive inhabitants. The subsequent breakthroughs in evolution were mostly structural, including the origin of the eukaryotic cell and the diversification of the protists. ✓

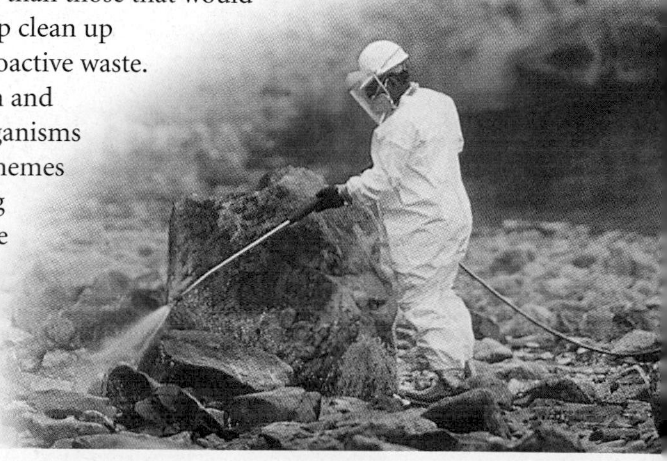

▼ Figure 15.18 **Treatment of an oil spill in Alaska.** These workers are spraying fertilizers onto an oil-soaked beach. The fertilizers stimulate growth of naturally occurring bacteria that initiate the breakdown of the oil. This technique is the fastest and least expensive way yet devised to clean up spills on beaches.

Protists

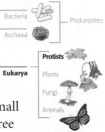

Anton van Leeuwenhoek, an early Dutch microscopist, was the first person to describe the microbial world: "No more pleasant sight has met my eye than this of so many thousands of living creatures in one small drop of water," he wrote more than three centuries ago. It is a world every biology student should have the opportunity to rediscover by peering through a microscope into a droplet of pond water filled with diverse creatures called protists (**Figure 15.19**).

The term **protists** is a bit of a catch-all category that includes all eukaryotes that are not fungi, animals, or plants. It should therefore not be surprising that the protists are highly diverse. Most, but not all, are unicellular. Because their cells are eukaryotic, even the simplest protists are much more complex than any prokaryote.

The first eukaryotes to evolve from prokaryotic ancestors were protists. These primal eukaryotes were not only the predecessors of the great variety of modern protists; they were also ancestral to all other eukaryotes—plants, fungi, and animals. Two of the most significant chapters in the history of life—the origin of the eukaryotic cell and the subsequent emergence of multicellular eukaryotes—both occurred during the evolution of protists.

▼ Figure 15.19 **A diversity of protists in a drop of pond water.**

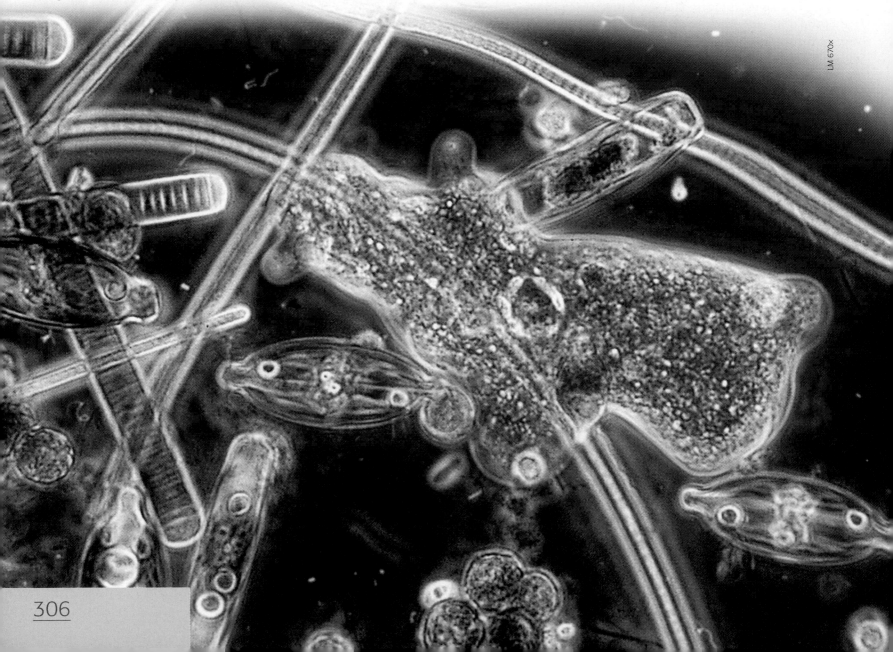

The Origin of Eukaryotic Cells

The fossil record indicates that eukaryotes evolved from prokaryotes around 2 billion years ago. One of biology's most engaging questions is how this happened—in particular, how the membrane enclosed organelles of eukaryotic cells arose. A widely accepted theory is that eukaryotic cells evolved through a combination of two processes. In one process, the eukaryotic cell's endomembrane system—all of the membrane-enclosed organelles except mitochondria and chloroplasts (review Figure 4.18)—evolved from inward folds of the plasma membrane of a prokaryotic cell (Figure 15.20a).

A second, very different process, called endosymbiosis, generated mitochondria and chloroplasts. **Symbiosis** ("living together") is a close association between organisms of two or more species, and **endosymbiosis** refers to one species living inside another host species. Chloroplasts and mitochondria seem to have evolved from small symbiotic prokaryotes (symbionts) that established residence within other, larger host prokaryotes (Figure 15.20b). The ancestors of mitochondria may have been aerobic bacteria that were able to use oxygen to release large amounts of energy from organic molecules by cellular respiration. At some point, such a prokaryote might have been an internal parasite of a larger heterotroph, or an ancestral host cell may have ingested some of these aerobic cells for food. If some of the smaller cells were indigestible, they might have remained alive and continued to perform respiration in the host cell. In a similar way, photosynthetic bacteria ancestral to chloroplasts may have come to live inside a larger host cell. Because almost all eukaryotes have mitochondria but only some have chloroplasts, it is logical to suppose that mitochondria evolved first.

However the relationships began, it is not hard to imagine how a symbiosis between engulfed aerobic or photosynthetic cells and a larger host cell might have become mutually beneficial. In a world that was becoming increasingly aerobic, a cell that was itself an anaerobe would have benefited from aerobic endosymbionts that turned the oxygen to advantage. And a heterotrophic host could derive nourishment from photosynthetic endosymbionts. In the process of becoming more interdependent, the host and endosymbionts would have become a single organism, its parts inseparable.

Developed primarily by Lynn Margulis, of the University of Massachusetts, the endosymbiosis model is supported by extensive evidence. Present-day mitochondria and chloroplasts are similar to prokaryotic cells in a number of ways. For example, both types of organelles contain small amounts of DNA, RNA, and ribosomes that resemble prokaryotic versions more than eukaryotic ones. These components enable chloroplasts and mitochondria to exhibit some autonomy in their activities. The organelles transcribe and translate their DNA into polypeptides, contributing some of their own enzymes to the cell. They also replicate their own DNA and reproduce within the cell by a process resembling the binary fission of prokaryotes.

The origin of the eukaryotic cell made more complex organisms possible, and a vast variety of protists evolved. ☑

☑CHECKPOINT

Which organelles of eukaryotic cells probably descended from endosymbiotic bacteria?

Answer: mitochondria and chloroplasts

▼ Figure 15.20 **A two-stage hypothesis for the evolution of eukaryotes through endosymbiosis.**

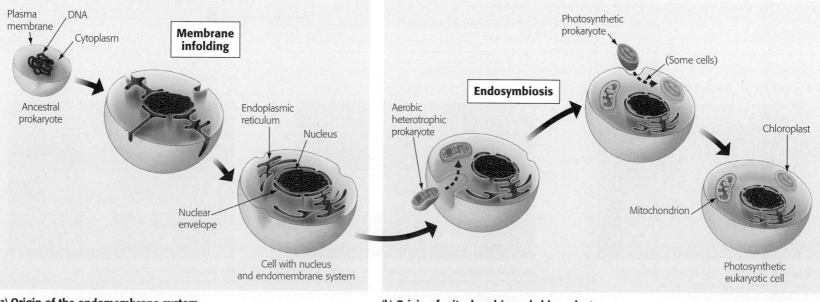

(a) Origin of the endomembrane system

(b) Origin of mitochondria and chloroplasts

The Diversity of Protists

All protists are eukaryotes, but they are so diverse that it is hard to name any other general characteristics about them. Biologists used to classify all protists in a kingdom called Protista but now think that these organisms constitute several kingdoms within domain Eukarya. While our knowledge of the evolutionary relationships between these diverse groups remains incomplete, the term *protist* is still a convenient term to refer to eukaryotes that are not plants, animals, or fungi. This catch-all category will almost certainly be reorganized in the coming years.

Given their diversity, it is not surprising that protists vary in structure and function more than any other group of organisms. Because most protists are unicellular (although there are some colonial and multicellular species), they are justifiably considered the simplest eukaryotic organisms. But at the cellular level, many protists are exceedingly complex—among the most elaborate of all cells. We should expect this of organisms that must carry out, within the boundaries of a single cell, all the basic functions performed by the collective of specialized cells that make up the bodies of plants and animals. Each unicellular protist is not analogous to a single cell from a human, but is itself an organism as complete as any whole animal or plant.

The classification of protists remains a work in progress. Therefore, rather than discuss any hypothesis about protist phylogeny, let's instead look at four major categories of protists, grouped by lifestyle: protozoans, slime molds, unicellular algae, and seaweeds.

Protozoans

Protists that live primarily by ingesting food are called **protozoans** (Figure 15.21). Protozoans thrive in all types of aquatic environments. Most species eat bacteria or other protozoans, but some can absorb nutrients dissolved in the water. Protozoans that live as parasites in animals, though in the minority, cause some of the world's most harmful human diseases.

Flagellates are protozoans that move by means of one or more flagella. Most species are free-living (nonparasitic). However, this group also includes some nasty parasites that make humans sick. An example is *Giardia*, a flagellate that infects the human intestine and can cause abdominal cramps and severe diarrhea. People become infected mainly by drinking water contaminated with feces from infected animals. Another group of dangerous flagellates are the trypanosomes, including a species that causes sleeping sickness, a serious illness prevalent in tropical Africa and transmitted by the tsetse fly.

Amoebas are characterized by great flexibility in their body shape and the absence of permanent organelles for locomotion. Most species move and feed by means of **pseudopodia** (singular, *pseudopodium*), temporary extensions of the cell. Amoebas can assume virtually any shape as they creep over rocks, sticks, or mud at the bottom of a pond or ocean. One species of parasitic amoeba causes amoebic dysentery, responsible for up to 100,000 deaths worldwide every year. Other protozoans with pseudopodia include the **forams**, which also have shells.

Apicomplexans are all parasitic, and some cause serious human diseases. They are named for a structure at their apex (tip) that is specialized for penetrating host cells and tissues. This group of protozoans includes *Plasmodium*, the parasite that causes malaria. As part of the effort to combat malaria, scientists determined the complete sequence of the *Plasmodium* genome in 2002.

Ciliates are protozoans that use structures called cilia to move and feed. Nearly all ciliates are free-living (nonparasitic). The best-known example is the freshwater ciliate *Paramecium*. ✓

☑ **CHECKPOINT**

What three modes of locomotion occur among protozoans?

Answer: movement using flagella, cilia, and pseudopodia

▼ Figure 15.21 **A diversity of protozoans.**

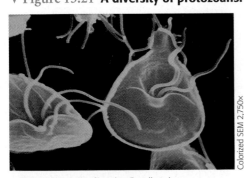

A flagellate: *Giardia*. This flagellated protozoan parasite can colonize and reproduce within the human intestine, causing disease.

Colorized SEM 2,750×

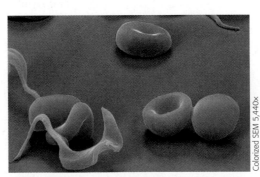

Another flagellate: trypanosomes. The squiggly forms among these human red blood cells are trypanosomes, parasitic flagellates that cause sleeping sickness, a debilitating disease common in parts of Africa.

Colorized SEM 5,440×

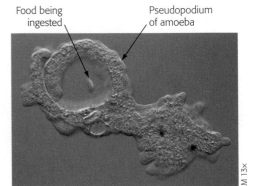

Food being ingested

Pseudopodium of amoeba

An amoeba. This amoeba is ingesting a smaller protozoan as food. The amoeba's pseudopodium arches around the prey and engulfs it into a food vacuole.

LM 13×

Slime Molds

These protists are more attractive than their name. Slime molds resemble fungi in appearance and lifestyle, but the similarities are due to convergent evolution; slime molds and fungi are not at all closely related. The weblike body of a slime mold, like that of a fungus, is an adaptation that increases exposure to the environment. This suits the role of these organisms as decomposers. The two main groups of these protists are plasmodial slime molds and cellular slime molds.

Plasmodial slime molds are named for the feeding stage in their life cycle, an amoeboid mass called a plasmodium (**Figure 15.22**)—not to be confused with *Plasmodium*, the apicomplexan parasite that causes malaria. You can find plasmodial slime molds among the leaf litter and other decaying material on a forest floor, and you won't need a microscope to see them. A plasmodium can measure several centimeters across, with its network of fine filaments taking in bacteria and bits of dead organic matter amoeboid style. Large as it is, the plasmodium is actually a single cell with many nuclei.

The study of **cellular slime molds** raise a question about what it means to be an individual organism, given the different forms that they can take during their successive life stages (**Figure 15.23**). The feeding stage of a cellular slime mold consists of ❶ solitary amoeboid cells. They function individually, using their pseudopodia to creep through organic matter and engulf bacteria. But when food is in short supply, the amoeboid cells swarm together to form ❷ a slug-like colony that moves and functions as a single unit. After a brief period of mobility, the colony ❸ extends a stalk and develops into a multicellular reproductive structure. ☑

▼ Figure 15.22 **A plasmodial slime mold.** The weblike form of the slime mold's feeding stage is an adaptation that enlarges the organism's surface area, increasing its contact with food, water, and oxygen.

▼ Figure 15.23 **Life stages of a cellular slime mold.**

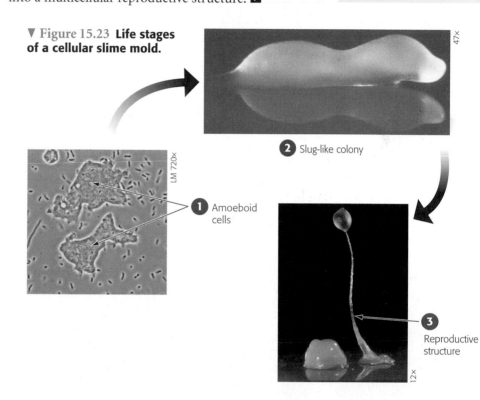

❷ Slug-like colony

❶ Amoeboid cells

❸ Reproductive structure

A foram. A foram cell secretes a shell made of organic material hardened with calcium carbonate. The shells of fossilized forms are a major component of limestone.

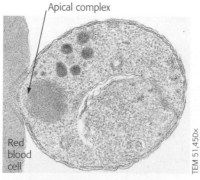

Apical complex

Red blood cell

An apicomplexan. *Plasmodium*, which causes malaria, enters red blood cells of its human host. The parasite feeds on the host cell from within, eventually destroying it.

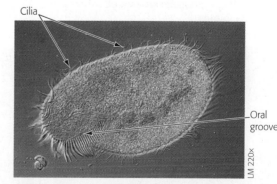

Cilia

Oral groove

A ciliate. The ciliate *Paramecium* uses its cilia to move through pond water. Cilia lining the oral groove keep a current of water containing food moving toward the cell "mouth."

Unicellular and Colonial Algae

Photosynthetic protists belong to an informal category called **algae** (singular, *alga*). Algae come in unicellular, colonial, and multicellular forms. Their chloroplasts support food chains in freshwater and marine ecosystems. Many unicellular algae are components of **plankton**, the communities of organisms, mostly microscopic, that drift or swim weakly near the surfaces of ponds, lakes, and oceans. We'll look at three groups of unicellular algae—dinoflagellates, diatoms, and green algae—and one type of colonial algae.

Dinoflagellates are abundant in the vast aquatic pastures of plankton. Each **dinoflagellate** species has a characteristic shape reinforced by external plates made of cellulose (**Figure 15.24a**). The beating of two flagella in perpendicular grooves produces a spinning movement. Dinoflagellate blooms—population explosions—sometimes cause warm coastal waters to turn pinkish orange, a phenomenon known as a red tide. Toxins produced by some red-tide dinoflagellates have caused massive fish kills, especially in the tropics, and are poisonous to humans as well.

Diatoms have glassy cell walls containing silica, the mineral used to make glass (**Figure 15.24b**). The cell wall consists of two halves that fit together like the bottom and lid of a shoe box. Diatoms store their food reserves in the form of an oil that provides buoyancy, keeping diatoms floating as plankton near the sunlit surface. Massive accumulations of fossilized diatoms make up thick sediments known as diatomaceous earth, which is mined for its use as a filtering material, an abrasive, and a natural insecticide.

Green algae are named for their grass-green chloroplasts. Unicellular green algae flourish in most freshwater lakes and ponds, as well as many home pools and aquariums. Some species are flagellated, such as *Chlamydomonas* (**Figure 15.24c**). The green algal group also includes colonial forms, such as *Volvox*, shown in **Figure 15.24d**. Each *Volvox* colony is a hollow ball of flagellated cells (the small green dots in the photo) that are very similar to certain unicellular green algae. The balls within the balls in Figure 15.24d are daughter colonies that will be released when the parent colonies rupture. Of all photosynthetic protists, green algae are the most closely related to plants. (We'll examine the evidence of this evolutionary relationship in the next chapter.)

Seaweeds

Defined as large, multicellular marine algae, **seaweeds** grow on rocky shores and just offshore beyond the zone of the pounding surf. Their cell walls have slimy and rubbery substances that cushion their bodies against the agitation of the waves. Some seaweeds are as large and complex as many plants. And though the word *seaweed* implies a plantlike appearance, the similarities between these algae and plants are a consequence of convergent evolution. In fact, the closest relatives of seaweeds are certain unicellular algae, which is why many biologists include seaweeds with the protists. Seaweeds are classified into three diferent groups, based partly on the types of pigments present in their chloroplasts: green algae, red algae, and brown algae (some of which are known as kelp) (**Figure 15.25**).

Many coastal people, particularly in Asia, harvest seaweeds for food. For example, in Japan and Korea, some seaweed species (such as brown algae called kombu) are ingredients in soups. Other seaweeds (such as red algae called nori) are used to wrap sushi. Marine algae are rich in iodine and other essential minerals. However, much of their organic material consists of unusual polysaccharides that humans cannot digest, which prevents seaweeds from becoming staple foods. They are ingested mostly for their rich tastes and unusual textures. The gel-forming substances in the cell walls of seaweeds are widely used as thickeners for such processed foods as puddings, ice cream, and salad dressing. The seaweed extract called agar provides the gel-forming base for the media microbiologists use to culture bacteria in petri dishes and for a type of gel commonly used to perform gel electrophoresis (as shown in Figure 12.17). ☑

▼ **Figure 15.24 Unicellular and colonial algae.**

(a) A dinoflagellate, with its wall of protective plates

SEM 1,065×

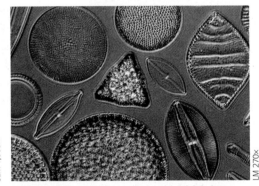

(b) A sample of diverse diatoms, which have glassy walls

LM 270×

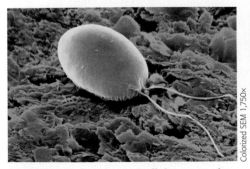

(c) *Chlamydomonas*, a unicellular green alga with a pair of flagella

Colorized SEM 1,750×

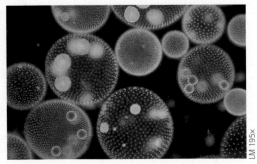

(d) *Volvox*, a colonial green alga

LM 195×

▼ Figure 15.25 **The three major groups of seaweeds.**

Green algae. This sea lettuce is an edible species that inhabits the intertidal zone, where the land meets the ocean. In addition to seaweeds, green algae include unicellular and colonial species.

Red algae. These seaweeds are most abundant in the warm coastal waters of the tropics. Because their chloroplasts have special pigments that absorb the blue and green light that penetrates best through water, red algae can generally live in the deepest water.

Brown algae. This group includes the largest seaweeds, known as kelp, which grow as marine "forests" in relatively deep water beyond the intertidal zone.

EVOLUTION CONNECTION: The Origin of Life

The Origin of Multicellular Life

An orchestra can play a greater variety of musical compositions than a violin soloist can. Put simply, increased complexity makes more variations possible. Thus, the origin of the eukaryotic cell led to an evolutionary radiation of new forms of life. Unicellular protists, which are organized on the complex eukaryotic plan, are much more diverse in form than the simpler prokaryotes. The evolution of multicellular bodies crossed another threshold in structural organization.

Multicellular organisms are fundamentally different from unicellular ones. In a unicellular organism, all of life's activities are carried out by a single cell. In contrast, a multicellular organism has various specialized cells that perform different functions—such as feeding, waste disposal, gas exchange, and protection—and are dependent on each other.

The evolutionary links between unicellular and multicellular life were probably colonial forms, in which unicellular protists stuck together as loose federations of independent cells (Figure 15.26). The gradual transition from colonies to truly multicellular organisms involved the cells becoming increasingly interdependent as a division of labor evolved. We can scc one level of specialization and cooperation in the colonial green alga *Volvox* (see Figure 15.24d). *Volvox* produces gametes (sperm and ova), which depend on nonreproductive (somatic) cells while developing. Cells in truly multicellular organisms are specialized for many more nonreproductive functions.

Multicellularity evolved many times among the ancestral stock of protists, leading to new waves of biological diversification. The diverse seaweeds are examples of their descendants, and so are plants, fungi, and animals. In the next chapter, we'll trace the long evolutionary movement of plants and fungi onto land.

▼ Figure 15.26 **A model for the evolution of multicellular organisms from unicellular protists.**

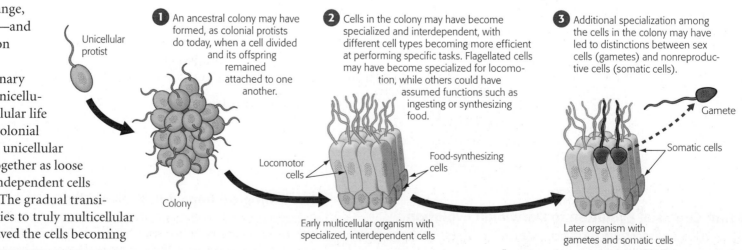

Unicellular protist

1 An ancestral colony may have formed, as colonial protists do today, when a cell divided and its offspring remained attached to one another.

Colony

2 Cells in the colony may have become specialized and interdependent, with different cell types becoming more efficient at performing specific tasks. Flagellated cells may have become specialized for locomotion, while others could have assumed functions such as ingesting or synthesizing food.

Locomotor cells

Food-synthesizing cells

Early multicellular organism with specialized, interdependent cells

3 Additional specialization among the cells in the colony may have led to distinctions between sex cells (gametes) and nonreproductive cells (somatic cells).

Gamete

Somatic cells

Later organism with gametes and somatic cells

Chapter Review

SUMMARY OF KEY CONCEPTS

Major Episodes in the History of Life

Major episode	Millions of years ago
Plants and fungi colonize land	500
All major animal phyla established	530
First multicellular organisms	1,200
Oldest eukaryotic fossils	1,800
Accumulation of O_2 in atmosphere	2,400
Oldest prokaryotic fossils	3,500
Origin of Earth	4,600

The Origin of Life

Resolving the Biogenesis Paradox

All life today arises only by the reproduction of preexisting life. However, most biologists think it is possible that chemical and physical processes in Earth's primordial environment produced the first cells through a series of stages.

A Four-Stage Hypothesis for the Origin of Life

One scenario suggests that the first organisms were products of chemical evolution in four stages:

Inorganic compounds

1 **Abiotic synthesis of organic monomers**

Organic monomers

2 **Abiotic synthesis of polymers**

Polymer

3 **Formation of pre-cells**

Membrane-enclosed compartment

4 **Self-replicating molecules**

Complementary chain

From Chemical Evolution to Darwinian Evolution

Over millions of years, natural selection favored the most efficient pre-cells, which evolved into the first prokaryotic cells.

Prokaryotes

They're Everywhere!

Prokaryotes are found wherever there is life and greatly outnumber eukaryotes. Prokaryotes thrive in habitats where eukaryotes cannot live. A few prokaryotic species cause serious diseases, but most are either benign or beneficial to other forms of life.

The Structure and Function of Prokaryotes

Prokaryotic cells lack nuclei and other membrane-enclosed organelles. Most have cell walls. Some of the most common shapes of prokaryotes are:

Spherical   **Rod-shaped**   **Spiral**

About half of all prokaryotic species are mobile, most of these using flagella to move. Some prokaryotes can survive extended periods of harsh conditions by forming endospores. Many prokaryotes can reproduce by binary fission at high rates if conditions are favorable, but growth is usually restricted by limited resources.

Prokaryotes exhibit all four major modes of nutrition:

Nutritional Mode	Energy Source	Carbon Source
Photoautotroph	Sunlight	CO_2
Chemoautotroph	Inorganic chemicals	
Photoheterotroph	Sunlight	Organic compounds
Chemoheterotroph	Organic compounds	

The Two Main Branches of Prokaryotic Evolution: Bacteria and Archaea

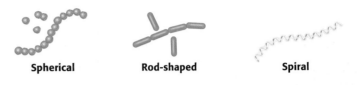

The prokaryotic lineage includes domains Bacteria and Archaea. Many archaea are "extremophiles" capable of surviving under conditions (such as high heat or salt concentrations) that would kill other forms of life.

Bacteria and Humans

Most pathogenic bacteria cause disease by producing exotoxins or endotoxins. Sanitation, antibiotics, and education are the best defenses against bacterial disease. Humans have a long history of employing bacteria as bioterrorist agents.

The Ecological Impact of Prokaryotes

Prokaryotes help recycle chemical elements between the biological and physical components of ecosystems. Humans can use prokaryotes to remove pollutants from water, air, and soil in the process called bioremediation.

Protists

The Origin of Eukaryotic Cells

The nucleus and endomembrane system of eukaryotes probably evolved from infoldings of the plasma membrane of ancestral prokaryotes. Mitochondria and chloroplasts probably evolved from symbiotic prokaryotes that took up residence inside larger cells, a process called endosymbiosis.

The Diversity of Protists

Protists are unicellular eukaryotes and their closest multicellular relatives.

- Protozoans (including flagellates, amoebas, apicomplexans, and ciliates) primarily live in aquatic environments and ingest their food.
- Slime molds (including plasmodial slime molds and cellular slime molds) resemble fungi in appearance and lifestyle as decomposers, but are not at all closely related.
- Unicellular algae (including dinoflagellates, diatoms, and unicellular green algae) are photosynthetic protists that support food chains in freshwater and marine ecosystems.
- Seaweeds—which include green, red, and brown algae—are large, multicellular marine algae that grow on and near rocky shores.

SELF-QUIZ

1. Place these events in the history of life on Earth in the order that they occurred.
   a. colonization of land by animals
   b. colonization of land by plants and fungi
   c. diversification of animals (Cambrian explosion)
   d. origin of eukaryotes
   e. origin of humans
   f. origin of multicellular organisms
   g. origin of prokaryotes

2. Place the following steps in the origin of life in the order that they are hypothesized to have occurred.
   a. integration of abiotically produced molecules into membrane-enclosed pre-cells
   b. origin of the first molecules capable of self-replication
   c. abiotic joining of organic monomers into polymers
   d. abiotic synthesis of organic monomers
   e. natural selection among pre-cells

3. DNA replication relies on the enzyme DNA polymerase. Why does this suggest that the earliest genes were made from RNA?

4. Contrast exotoxins with endotoxins.

5. What is the difference between autotrophs and heterotrophs in terms of the source of their organic compounds?

6. The bacteria that cause tetanus can be killed only by prolonged heating at temperatures considerably above boiling. What does this suggest about tetanus bacteria?

7. To what nutritional classification do you belong? To what nutritional classification does a mushroom belong? (*Hint*: Review Figure 15.12.)

8. Why are protists especially important to biologists investigating the evolution of eukaryotic life?

9. Of the following, which describes protists most inclusively?
   a. multicellular eukaryotes
   b. protozoans
   c. eukaryotes that are not plants, fungi, or animals
   d. single-celled organisms closely related to bacteria

10. Which algal group is most closely related to plants?
    a. diatoms
    b. green algae
    c. dinoflagellates
    d. seaweeds

Answers to the Self-Quiz questions can be found in Appendix D.

THE PROCESS OF SCIENCE

11. Imagine you are on a team designing a moon base that will be self-contained and self-sustaining. Once supplied with building materials, equipment, and organisms from Earth, the base will be expected to function indefinitely. One of the members of your team has suggested that everything sent to the base be chemically treated or irradiated so that no bacteria of any kind are present. Do you think this is a good idea? Predict some of the consequences of eliminating all bacteria from an environment.

12. Your classmate says that organisms that require oxygen existed before photosynthetic organisms. Do you support this idea? Explain why or why not.

BIOLOGY AND SOCIETY

13. Many local newspapers publish a weekly list of restaurants that have been cited by inspectors for poor sanitation. Locate such a report and highlight the cases that are probably associated with food contamination by pathogenic prokaryotes.

14. What do you think should be done to prevent bioterrorism?

Plants, Fungi, and the Move onto Land

Will the blight end the chestnut?
The farmers rather guess not.
It keeps smoldering at the roots
And sending up new shoots
Till another parasite
Shall come to end the blight.

—ROBERT FROST, "EVIL TENDENCIES" (1930)

An American chestnut tree, circa 1920.
The blight that devastated the chestnut is an example of a harmful interaction between plants and fungi.

CHAPTER CONTENTS

BIOLOGY AND SOCIETY: Plant-Fungus Symbiosis

Will the Blight End the Chestnut?

The forests of the eastern United States, from Maine to Georgia, were once dominated by the American chestnut tree (*Castanea dentate*). Prized for their rapid growth, huge size, rot-resistant wood (which made them ideal for log cabin foundations), and bountiful harvest of edible nuts (traditionally "roasting on an open fire" during the Thanksgiving-to-Christmas holiday season), American chestnuts were a mainstay of rural life.

Tragically, all this changed in just a few decades. Around 1900, an Asian fungus called *Cryphonectria parasitica* was accidentally introduced from China into North America. While many Asian trees had evolved defenses against the fungus, American trees had not. In just 25 years, blight caused by the fungus killed virtually every one of the estimated 3.5 billion adult American chestnut trees.

Despite the decimation of trees by the blight, the American chestnut is not extinct. You can still find small chestnut trees sprouting from old roots or stumps in many forests. Unfortunately, the blight fungus kills practically all the trees before they reach sexual maturity, so the trees are not propagating and restoring the population. However, forest researchers are working with the wealth of genetic material contained in the many surviving young sprouts to try to develop a blight-resistant strain through breeding and genetic engineering. Perhaps this magnificent tree will once again be a presence in American forests. But the question posed by Robert Frost in the opening line of his poem remains unanswered.

The chestnut blight is an example of **symbiosis**, a close association of two or more species. The harmful interaction between plant and fungus exemplified by the chestnut blight is an unusual case. As we explore the diversity of plants and fungi in this chapter, you will see that it is much more common for the members of these two kingdoms to benefit from each other's presence. Indeed, as we'll discuss first, aquatic plants probably never could have adapted to land without the aid of fungi.

Colonizing Land

Plants are terrestrial (land-dwelling) organisms. It is true that some, such as water lilies, have returned to the water, but they evolved from terrestrial ancestors (just as several species of aquatic mammals, such as whales, evolved from terrestrial mammals). In this section, we'll discuss some of the adaptations that allowed plants to move onto land.

Terrestrial Adaptations of Plants

What exactly is a plant? A **plant** is a multicellular eukaryote that makes organic molecules by photosynthesis (in other words, plants are photoautotrophs—see Figure 15.12). Photosynthesis distinguishes plants from the animal and fungal kingdoms. But what about large algae, including seaweeds, which we classified as protists in the preceding chapter? They, too, are multicellular, eukaryotic, and photosynthetic. What distinguishes plants from algae is a set of structural and reproductive terrestrial adaptations.

Structural Adaptations

Living on land poses different problems than living in water (**Figure 16.1**). In terrestrial habitats, the resources that a photosynthetic organism needs are found in two very different places. Light and carbon dioxide are mainly available in the air, while water and mineral nutrients are found mainly in the soil. Thus, the complex bodies of plants are specialized to take advantage of these two environments by having both aerial leaf-bearing organs called **shoots** and subterranean organs called **roots**.

Most plants have symbiotic fungi associated with their roots. These root-fungus combinations are called **mycorrhizae** ("fungus roots") (**Figure 16.2**). For their part, the fungi absorb water and essential minerals from the soil and provide these materials to the plant. The sugars produced by the plant nourish the fungi. Mycorrhizae are evident on some of the oldest plant fossils, suggesting that they are key adaptations that made it possible for plants to live on land.

Shoots also show structural adaptations to the terrestrial environment. Leaves are the main photosynthetic organs of most plants. Exchange of carbon dioxide (CO_2) and oxygen (O_2) between the atmosphere and the

▼ Figure 16.1 **Structural adaptations of algae and plants.**

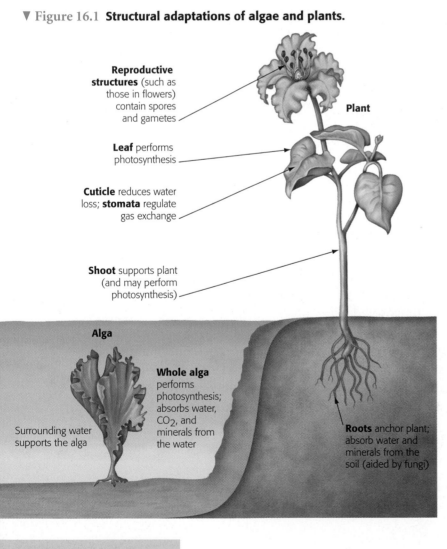

Reproductive structures (such as those in flowers) contain spores and gametes

Leaf performs photosynthesis

Cuticle reduces water loss; **stomata** regulate gas exchange

Shoot supports plant (and may perform photosynthesis)

Plant

Alga

Whole alga performs photosynthesis; absorbs water, CO_2, and minerals from the water

Surrounding water supports the alga

Roots anchor plant; absorb water and minerals from the soil (aided by fungi)

▼ Figure 16.2 **Mycorrhizae: symbiotic associations of fungi and roots.** The finely branched filaments of the fungus (white in the photo) provide an extensive surface area for absorption of water and minerals from the soil.

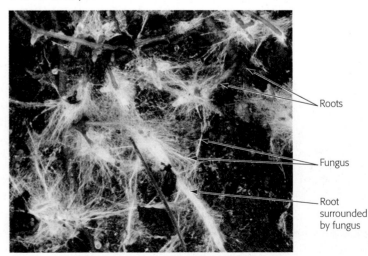

Roots

Fungus

Root surrounded by fungus

photosynthetic interior of a leaf occurs via **stomata** (singular, *stoma*), the microscopic pores found on a leaf's surface (see Figure 7.2). A waxy layer called the **cuticle** coats the leaves and other aerial parts of most plants, helping the plant body retain water. (You've probably noticed the waxy surface of some houseplant leaves.)

Differentiation of the plant body into shoot and root systems solved one problem but created new ones. For the shoot system to stand up straight in the air, it must have support. This is not a major problem in the water: Huge seaweeds do not need skeletons because the surrounding water buoys them. An important terrestrial adaptation of plants is **lignin**, a chemical that hardens cell walls. Imagine what would happen to you if your skeleton were to suddenly turn mushy. A tree would also collapse if it were not for its "skeleton," its framework of lignin-rich cell walls.

Specialization of the plant body into shoots and roots also introduced the problem of transporting vital materials between the distant organs. The terrestrial equipment of most plants includes **vascular tissue**, a system of tube-shaped cells that branch throughout the plant (**Figure 16.3**). The vascular tissue actually has two types of tissues specialized for transport: **xylem**, consisting of dead cells with tubular cavities for transporting water and minerals from roots to leaves; and **phloem**, consisting of living cells that distribute sugars from the leaves to the roots and other nonphotosynthetic parts of the plant.

Reproductive Adaptations

Adapting to land also required a new mode of reproduction. For algae, the surrounding water ensures that gametes (sperm and eggs) and developing offspring stay moist. The aquatic environment also provides a means of dispersing the gametes and offspring. Plants, however, must keep their gametes and developing offspring from drying out in the air. Plants (and some algae) produce their gametes in protective structures that are called **gametangia** (singular, *gametangium*). A gametangium has a jacket of protective cells surrounding a moist chamber where gametes can develop without dehydrating.

For most plants, sperm reach the eggs by traveling inside pollen grains, which are carried by wind or animals. The egg remains within tissues of the mother plant and is fertilized there. In plants, but not algae, the zygote (fertilized egg) develops into an embryo while still contained within the female parent, which protects the embryo and keeps it from dehydrating (**Figure 16.4**). Most plants rely on wind or animals, such as fruit-eating birds or mammals, to disperse their offspring, which are in the form of embryos contained in seeds. ☑

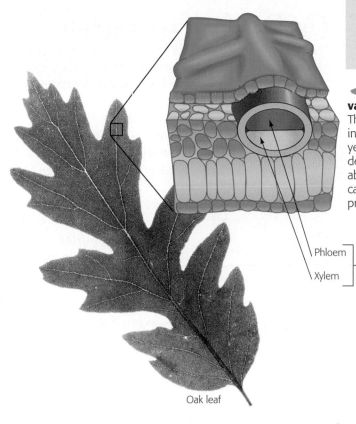

The vascular tissue of the leaf in the photograph is visible as yellow veins. This tissue delivers water and minerals absorbed by the roots and carries away the sugars produced in the leaves.

Phloem

Xylem

Vascular tissue

Oak leaf

▼ Figure 16.4 **The protected embryo of a plant.** Internal fertilization, with sperm and egg combining within a moist chamber on the mother plant, is an adaptation for living on land. The female parent continues to nurture and protect the plant embryo, which develops from the zygote.

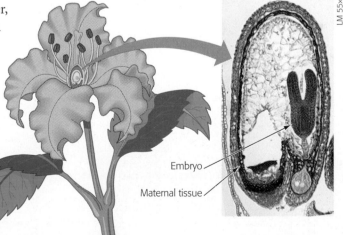

LM 55×

Embryo

Maternal tissue

☑CHECKPOINT

Name some adaptations of plants for living on land.

Answer: any of the following: cuticle; stomata; vascular tissue; lignin-hardened cell walls; gametangia, which protect gametes; protected embryos; and differentiation of the body into aerial shoots and subterranean roots

The Origin of Plants from Green Algae

The algal ancestors of plants carpeted moist fringes of lakes or coastal salt marshes over 500 million years ago. These shallow-water habitats were subject to occasional drying, and natural selection would have favored algae that could survive periodic droughts. Some species accumulated adaptations that enabled them to live permanently above the water line. A modern-day lineage of green algae, the **charophytes** (Figure 16.5), may resemble one of these early plant ancestors. Plants and present-day charophytes probably evolved from a common ancestor.

Adaptations making life on dry land possible had accumulated by about 475 million years ago, the age of the oldest known plant fossils. The evolutionary novelties of these first land plants opened the new frontier of a terrestrial habitat. Early plant life would have thrived in the new environment. Bright sunlight was abundant on land, the atmosphere had a wealth of carbon dioxide, and at first there were relatively few pathogens and plant-eating animals. The stage was set for an explosive diversification of plant life.

▼ Figure 16.5 **Two species of charophytes, the closest algal relatives of plants.**

LM 85x

LM 265x

Evolution of Plants

Plant Diversity

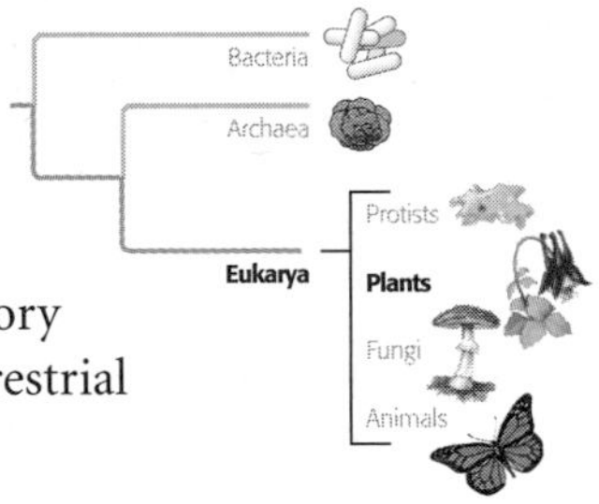

As we survey the diversity of modern plants, remember that the past is the key to the present. The history of the plant kingdom is a story of adaptation to diverse terrestrial habitats.

Bacteria

Archaea

Eukarya

Protists

Plants

Fungi

Animals

① After plants originated from an algal ancestor approximately 475 million years ago, early diversification gave rise to nonvascular plants, including mosses, liverworts, and hornworts. These plants, called **bryophytes**, lack true roots and leaves. Bryophytes also lack lignin, the wall-hardening material that enables other plants to stand tall. Without lignified cell walls, bryophytes have weak upright support. The most familiar bryophytes are **mosses**. A mat of moss actually consists of many plants growing in a tight pack, holding one another up. Gametangia, which protect the gametes and embryos, are a terrestrial adaptation that originated in bryophytes.

Highlights of Plant Evolution

The fossil record chronicles four major periods of plant evolution, which are also evident in the diversity of modern plants (Figure 16.6). Each stage is marked by the evolution of structures that opened new opportunities on land.

② The second period of plant evolution, begun about 425 million years ago, was the diversification of

▼ Figure 16.6 **Highlights of plant evolution.** This phylogenetic tree highlights the evolution of structures that allowed plants to move onto land; these structures still exist in modern plants. As we survey the diversity of plants, miniature versions of this tree will help you place each plant group in its evolutionary context.

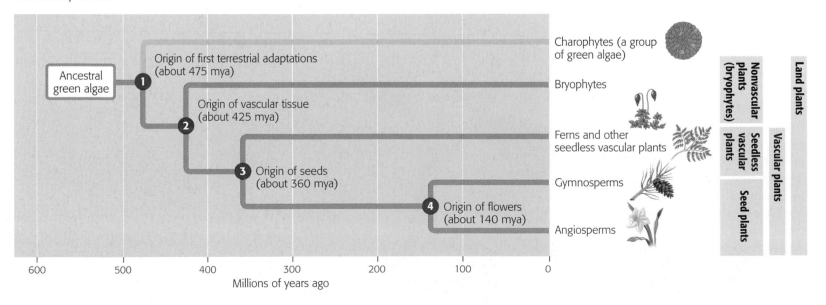

Ancestral green algae

① Origin of first terrestrial adaptations (about 475 mya)

② Origin of vascular tissue (about 425 mya)

③ Origin of seeds (about 360 mya)

④ Origin of flowers (about 140 mya)

Charophytes (a group of green algae)

Bryophytes

Ferns and other seedless vascular plants

Gymnosperms

Angiosperms

Nonvascular plants (bryophytes)

Seedless vascular plants

Seed plants

Vascular plants

Land plants

600   500   400   300   200   100   0

Millions of years ago

plants with vascular tissue. The presence of conducting tissues hardened with lignin allowed vascular plants to grow much taller, rising above the ground to achieve significant height. The earliest vascular plants lacked seeds. Today, this seedless condition is retained by **ferns** and a few other groups of vascular plants.

❸ The third major period of plant evolution began with the origin of the seed about 360 million years ago. Seeds advanced the colonization of land by further protecting plant embryos from drying and other hazards. A **seed** consists of an embryo packaged along with a store of food within a protective covering. The seeds of early seed plants were not enclosed in any specialized chambers. These plants gave rise to the **gymnosperms** ("naked seeds"). Today, the most widespread and

diverse gymnosperms are the **conifers**, consisting mainly of cone-bearing trees, such as pines.

❹ The fourth major episode in the evolutionary history of plants was the emergence of flowering plants, or **angiosperms** ("contained seeds"), at least 140 million years ago. The **flower** is a complex reproductive structure that bears seeds within protective chambers called ovaries. This contrasts with the naked seeds of gymnosperms. The great majority of living plants—some 250,000 species—are angiosperms, including all our fruit and vegetable crops, grains, grasses, and most trees.

With these highlights as our framework, we are now ready to survey the four major groups of modern plants: bryophytes, ferns, gymnosperms, and angiosperms (Figure 16.7). ☑

▼ Figure 16.7 **The major groups of plants.**

PLANT DIVERSITY			
Bryophytes (nonvascular plants)	**Ferns** (seedless vascular plants)	**Gymnosperms** (naked-seed plants)	**Angiosperms** (flowering plants)

Bryophytes

Mosses, which are bryophytes, may sprawl as low mats over acres of land (Figure 16.8). Mosses display two of the key terrestrial adaptations that made the move onto land possible: (1) a waxy cuticle that helps prevent dehydration and (2) the retention of developing embryos within the mother plant's gametangium. However, mosses are not totally liberated from their ancestral aquatic habitat. Mosses need water to reproduce because their sperm need to swim to reach eggs within the female gametangium. (A film of rainwater or dew is enough moisture for the sperm to travel.) In addition,

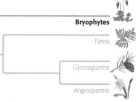

▼ Figure 16.8 **A peat moss bog in Scotland.** Mosses are bryophytes, which are nonvascular plants. Peat mosses, or sphagnum, carpet at least 3% of Earth's terrestrial surface. They are most commonly found in high northern latitudes. The ability of peat moss to absorb and retain water makes it an excellent addition to garden soil.

because most mosses have no vascular tissue to carry water from soil to aerial parts of the plant, they need to live in damp, shady places.

If you examine a mat of moss closely, you can see two distinct forms of the plant. The green, sponge-like plant that is the more obvious is called the **gametophyte**. Careful examination will reveal the other form of the moss, called a **sporophyte**, growing out of a gametophyte as a stalk with a capsule at its tip (Figure 16.9). The cells of the gametophyte are haploid (they have one set of chromosomes; see Chapter 8). In contrast, the sporophyte is made up of diploid cells (with two chromosome sets). These two different stages of the plant life cycle are named for the types of reproductive cells they produce. Gametophytes produce gametes (sperm and eggs), while sporophytes produce spores. As reproductive cells, **spores** differ from gametes in two ways: A spore can develop into a new organism without fusing with another cell (two gametes must fuse to form a zygote); and spores usually have tough coats that enable them to resist harsh environments (whereas gametes must stay moist).

The gametophyte and sporophyte are alternating generations that take turns producing each other. Gametophytes produce gametes that unite to form zygotes, which develop into new sporophytes. And sporophytes produce spores that give rise to new gametophytes. This type of life cycle, called **alternation of generations**, occurs only in plants and multicellular green algae (Figure 16.10). Among plants, mosses and other bryophytes are unique in having the gametophyte as the larger, more obvious plant. As we continue our survey of plants, we'll see an increasing dominance of the sporophyte as the more highly developed generation. ☑

▼ Figure 16.9 **The two forms of a moss.** The feathery plant we generally know as a moss is the gametophyte. The stalk with the capsule at its tip is the sporophyte. This photo shows the capsule releasing its tiny spores.

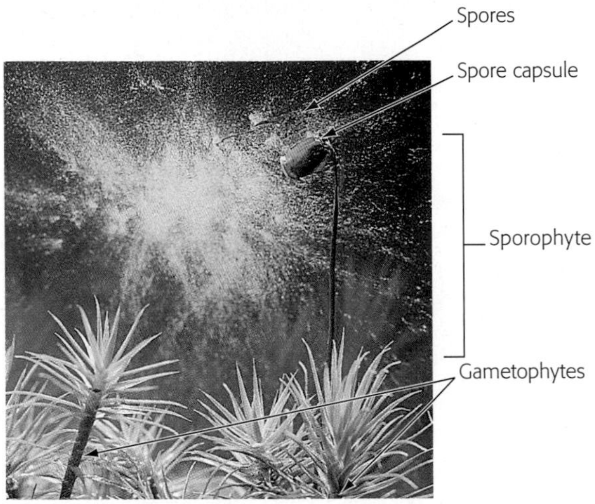

MP3 tutor sessions
Alternation of Generations

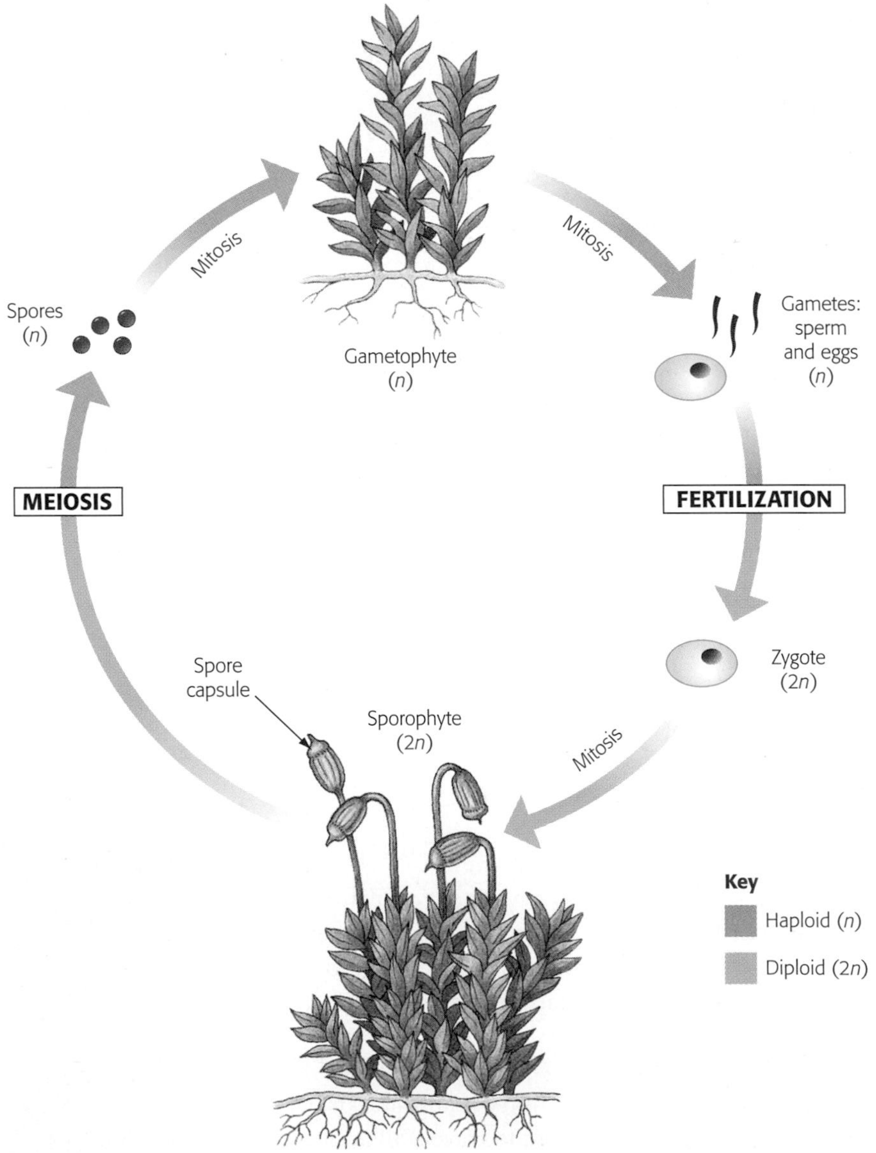

► Figure 16.10 **Alternation of generations.** Plants have life cycles very different from ours. Each of us is a diploid individual; the only haploid stages in the human life cycle, as for nearly all animals, are sperm and eggs. By contrast, plants have alternating generations: Diploid (2*n*) individuals (sporophytes) and haploid (*n*) individuals (gametophytes) generate each other in the life cycle.

☑CHECKPOINT

Bryophytes, like all plants, have a life cycle that involves an alternation of generations. What are the two generations called? Which generation dominates in bryophytes?

Answer: gametophyte, sporophyte; gametophyte

Ferns

Ferns took terrestrial adaptation to the next level with the evolution of vascular tissue. Ferns are by far the most diverse seedless vascular plants, with more than 12,000 known species. However, the sperm of ferns, like those of mosses, have flagella and must swim through a film of water to fertilize eggs. Most ferns inhabit the tropics, although many species are found in temperate forests, such as many woodlands in the United States (Figure 16.11).

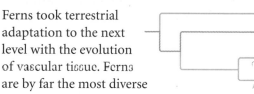

During the Carboniferous period, from about 360 to 300 million years ago, ancient ferns were part of a much greater diversity of seedless plants that formed vast, swampy tropical forests over much of what is now Eurasia and North America (Figure 16.12). As the plants died, they fell into stagnant wetlands and did not decay completely. Their remains formed thick organic deposits called peat. Later, seawater flooded the swamps, marine sediments covered the peat, and pressure and heat gradually converted the peat to coal. Coal is black sedimentary rock made up of fossilized plant material. Like coal, oil and natural gas also formed from the remains of long-dead organisms; thus, all three are known as **fossil fuels**. Since the Industrial Revolution, coal has been a crucial source of energy for human society. However, burning these fossil fuels releases CO_2 and other greenhouse gases into the atmosphere, contributing to global climate change. ✔

▼ Figure 16.11 **Ferns (seedless vascular plants).** The fern species in the foreground grows on the forest floor in the eastern United States. The fern generation familiar to us is the sporophyte generation. You would have to crawl on the forest floor and explore with careful hands and sharp eyes to find fern gametophytes (upper right), tiny plants growing on or just below the soil surface.

Spore capsule

"Fiddlehead" (young leaves ready to unfurl)

◄ Figure 16.12 **A "coal forest" of the Carboniferous period.** This painting, based on fossil evidence, reconstructs one of the great seedless forests. Most of the large trees with straight trunks are seedless vascular plants called lycophytes. On the left, the tree with numerous feathery branches is another type of seedless vascular plant called a horsetail. The plants near the base of the trees are ferns.

✔CHECKPOINT

Why are ferns able to grow taller than mosses?

Answer: Vascular tissue hardened with lignin allows ferns to stand taller and transport nutrients farther.

Gymnosperms

"Coal forests" dominated
the North American and
Eurasian landscapes until
near the end of the
Carboniferous period. At that
time, the global climate turned drier
and colder, and the vast swamps began to disappear.
This climatic change provided an opportunity for seed
plants, which can complete their life cycles on dry land
and withstand long, harsh winters. Of the earliest seed
plants, the most successful were the gymnosperms, and
several kinds grew along with the seedless plants in the
Carboniferous swamps. Their descendants include the
conifers, or cone-bearing plants. Let's take a closer look
at the conifers before discussing the adaptations that
allowed seed plants to dominate the land.

Conifers

Perhaps you have had the fun of hiking or skiing
through a forest of conifers, the most common gymnosperms. Pines, firs, spruces, junipers, cedars, and redwoods are all conifers. A broad band of coniferous
forests covers much of northern Eurasia and North
America and extends southward in mountainous
regions (**Figure 16.13**). Today, about 190 million
acres of coniferous forests in the United States,
mostly in the western states and Alaska, are designated national forests.

Conifers are among the tallest, largest, and
oldest organisms on Earth. Coastal redwoods,
native to the northern California coast, are the
world's tallest trees. Three coastal redwoods discovered in 2006 are more than 110 m (370 ft)
tall. Some of the most massive organisms alive are the
giant sequoias, relatives of redwoods that grow in the
Sierra Nevada mountains of California. One, known as
the General Sherman tree, is about 84 m (275 ft) high
and weighs more than the combined weight of a dozen
space shuttles. Bristlecone pines, another species of
California conifer, are among the oldest organisms alive.
One bristlecone, named Methuselah, is more than 4,600
years old; it was a young tree when humans invented
writing.

Nearly all conifers are evergreens, meaning they
retain leaves throughout the year. Even during winter,
they perform a limited amount of photosynthesis on
sunny days. And when spring comes, conifers already
have fully developed leaves that can take advantage of
the sunnier days. The needle-shaped leaves of pines and
firs are also adapted to survive dry seasons. A thick cuticle covers the leaf, and the stomata are located in pits,
further reducing water loss.

Coniferous forests are highly productive; you probably use products harvested from them every day. For
example, conifers provide much of our lumber for
building and wood pulp for paper production. What we
call wood is actually an accumulation of vascular tissue
with lignin, which gives the tree structural support.

Terrestrial Adaptations of Seed Plants

Compared to ferns, most gymnosperms have three
additional adaptations that make survival in diverse terrestrial habitats possible: (1) further reduction of the
gametophyte, (2) pollen, and (3) seeds.

The first adaptation is an even greater development
of the diploid sporophyte compared to the haploid
gametophyte generation (**Figure 16.14**). A pine tree
or other conifer is actually a sporophyte with tiny

▼ Figure 16.13 **A coniferous forest in Banff National Park, in Alberta, Canada.** Coniferous forests are widespread in northern North America and Eurasia; conifers also grow in the Southern Hemisphere, though they are less numerous there.

▶ Figure 16.14 **Three variations on alternation of generations in plants.**

Key

Haploid (*n*)

Diploid (2*n*)

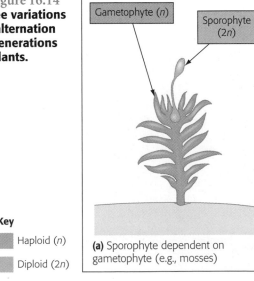

(a) Sporophyte dependent on gametophyte (e.g., mosses)

Gametophyte (*n*)

Sporophyte (2*n*)

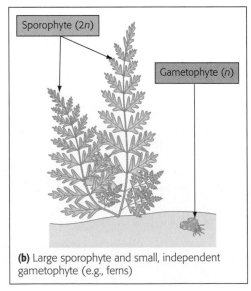

(b) Large sporophyte and small, independent gametophyte (e.g., ferns)

Sporophyte (2*n*)

Gametophyte (*n*)

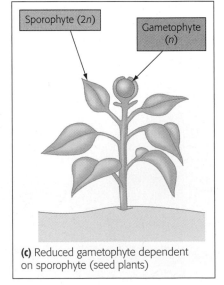

(c) Reduced gametophyte dependent on sporophyte (seed plants)

Sporophyte (2*n*)

Gametophyte (*n*)

gametophytes living in its cones (**Figure 16.15**). In contrast to bryophytes and ferns, gymnosperm gametophytes are totally dependent on and protected by the tissues of the parent sporophyte.

A second adaptation of seed plants to dry land came with the evolution of pollen. A **pollen grain** is actually the much-reduced male gametophyte; it houses cells that will develop into sperm. In the case of conifers, wind carries the pollen from male to female cones, where eggs develop within female gametophytes. This mechanism for sperm transfer contrasts with the swimming sperm of mosses and ferns. In seed plants, this use of tough, airborne pollen that carries sperm to egg is a terrestrial adaptation that led to even greater success and diversity of plants on land.

The third important terrestrial adaptation of seed plants is the seed itself. A seed consists of a plant embryo packaged along with a food supply within a protective coat. Seeds develop from structures called **ovules** (**Figure 16.16**). In conifers, the ovules are located on the scales of female cones. Conifers and other gymnosperms, lacking ovaries, bear their seeds "naked" on

the cone scales (though the seeds do have protective coats). Once released from the parent plant, the seed can remain dormant for days, months, or even years. Under favorable conditions, the seed can then **germinate**: Its embryo emerges through the seed coat as a seedling. Some seeds drop close to their parents, while others are carried far by the wind or animals. ☑

☑CHECKPOINT

Contrast the modes of sperm delivery in ferns and conifers.

Answer: The flagellated sperm of ferns must swim through water to reach eggs. In contrast, the airborne pollen of conifers delivers sperm to eggs in ovules without the need to go through water.

▼ **Figure 16.16** **From ovule to seed.**

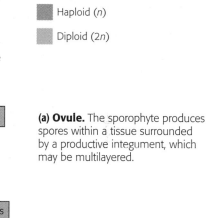

Female cone, cross section

Cross section of scale

Key

Haploid (*n*)

Diploid (2*n*)

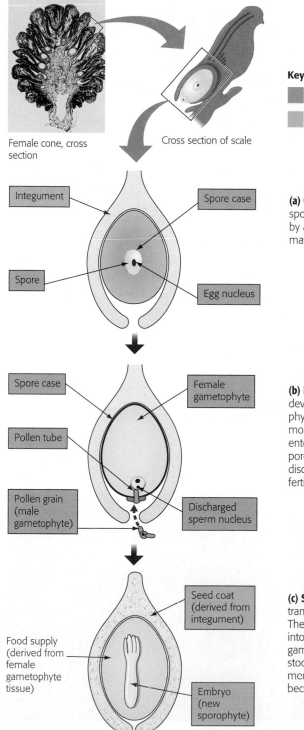

Integument

Spore case

Spore

Egg nucleus

(a) Ovule. The sporophyte produces spores within a tissue surrounded by a productive integument, which may be multilayered.

Spore case

Female gametophyte

Pollen tube

Pollen grain (male gametophyte)

Discharged sperm nucleus

(b) Fertilized ovule. The spore develops into a female gametophyte, which produces one or more eggs. When a pollen grain enters the ovule through a special pore in the integument, it discharges sperm cells that fertilize eggs.

Seed coat (derived from integument)

Food supply (derived from female gametophyte tissue)

Embryo (new sporophyte)

(c) Seed. Fertilization initiates the transformation of ovule to seed. The fertilized egg (zygote) develops into an embryo; the rest of the gametophyte forms a tissue that stockpiles food; and the integument of the ovule hardens to become the seed coat.

▼ **Figure 16.15** **A pine tree, the sporophyte, bearing two types of cones containing gametophytes.** Each scale of the female cone is actually a modified leaf that bears a structure called an ovule containing a female gametophyte. Male cones release clouds of millions of pollen grains, the male gametophytes. Some of these pollen grains land on female cones on trees of the same species. The sperm can fertilize eggs in the ovules of the female cones. The ovules eventually develop into seeds.

Scale

Ovule-producing cones; the scales contain female gametophytes

Pollen-producing cones; they produce male gametophytes

Ponderosa pine

Angiosperms

Angiosperms dominate the modern landscape. There are about 250,000 angiosperm species versus about 700 species of gymnosperms. Several unique adaptations account for the success of angiosperms. For example, refinements in vascular tissue make water transport even more efficient in angiosperms than in gymnosperms. Of all terrestrial adaptations, however, it is the flower that accounts for the unparalleled success of the angiosperms.

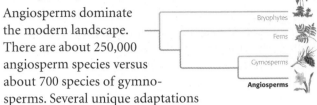

Flowers, Fruits, and the Angiosperm Life Cycle

No organism makes a showier display of its sex life than the angiosperm. From roses to dandelions, flowers display a plant's sex organs. For many angiosperms, this showiness helps to attract insects and other animals that transfer pollen from the sperm-bearing organs of one flower to the egg-bearing organs of another. This dependence on animals

for pollen transfer targets the pollen to other plants of the same species, rather than relying on uncertain winds to blow the pollen around.

A flower is actually a short stem with four whorls of modified leaves: sepals, petals, stamens, and carpels (**Figure 16.17**). At the bottom of the flower are the **sepals**, which are usually green. They enclose the flower before it opens (think of the green "wrapping" on a rosebud). Above the sepals are the **petals**, which are often colorful and help to attract insects and other pollinators. The actual reproductive structures are multiple stamens and one or more carpels. Each **stamen** consists of a stalk—the **filament**—bearing a sac called an **anther**, in which the pollen grains develop. The **carpel** consists of a stalk—the **style**—with an ovary at the base and a sticky tip known as the **stigma**, which traps pollen. The **ovary** is a protective chamber containing one or more ovules, in which the eggs develop. As you can see in **Figure 16.18**, the basic structure of a flower can exist in many beautiful variations.

◄ Figure 16.17 **Structure of a flower.**

▼ Figure 16.18 **A diversity of flowers.**

Pansy

Bleeding heart

California poppy

Water lily

▼ Figure 16.19 **The angiosperm life cycle.**

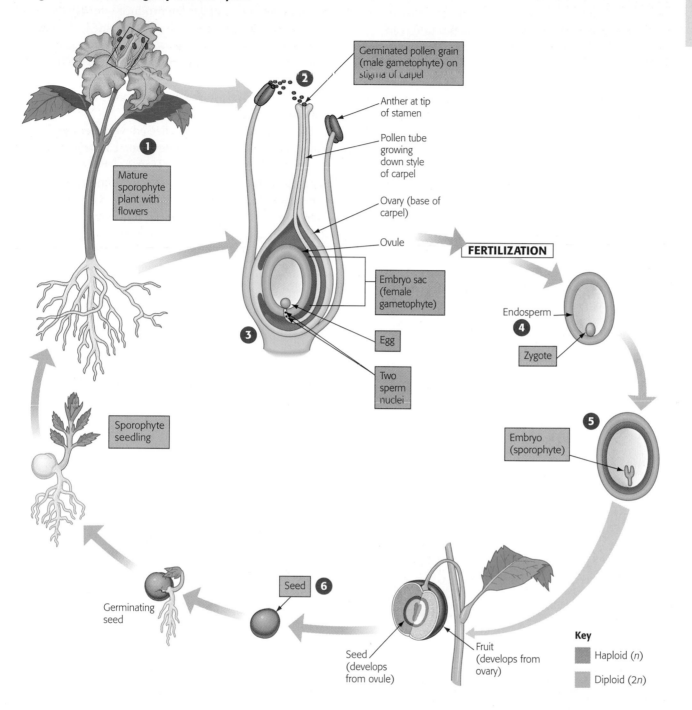

Germinated pollen grain (male gametophyte) on stigma of carpel

Anther at tip of stamen

Pollen tube growing down style of carpel

Ovary (base of carpel)

Ovule

Mature sporophyte plant with flowers

FERTILIZATION

Embryo sac (female gametophyte)

Endosperm

Zygote

Egg

Two sperm nuclei

Sporophyte seedling

Embryo (sporophyte)

Germinating seed

Seed

Seed (develops from ovule)

Fruit (develops from ovary)

Key

Haploid (*n*)

Diploid (2*n*)

Figure 16.19 highlights key stages in the angiosperm life cycle. ❶ The flower we see is part of the sporophyte plant. As in gymnosperms, the pollen grain is the male gametophyte of angiosperms. The female gametophyte—the embryo sac—is located within an ovule, which in turn resides within a chamber of the ovary. ❷ A pollen grain that lands on the sticky stigma of a carpel extends a tube down to the ovule and ❸ deposits two sperm nuclei within the embryo sac. This **double fertilization** is an angiosperm characteristic. One sperm cell fertilizes an egg in the embryo sac. ❹ This produces a zygote, which ❺ develops into an embryo. The second sperm cell "fertilizes" another female gametophyte cell, which then develops into a nutrient-storing tissue called **endosperm**, which nourishes the embryo. Double fertilization therefore synchronizes the development of the embryo and food reserves within an ovule. ❻ The whole ovule develops into a seed. The seed's enclosure within an ovary is what distinguishes angiosperms from gymnosperms, which have a naked seed.

A **fruit** is the ripened ovary of a flower. As seeds are developing from ovules, the ovary wall thickens, forming the fruit that encloses the seeds. A pea pod is an example of a fruit, with seeds (mature ovules, the peas) encased in the ripened ovary (the pod). Fruits protect and help disperse seeds. As **Figure 16.20** demonstrates, many angiosperms depend on animals to disperse seeds. Conversely, most land animals, including humans, rely on angiosperms as a food source, directly or indirectly. ✓

Angiosperms and Agriculture

Whereas gymnosperms supply most of our lumber and paper, angiosperms supply nearly all of our food—as well as the food eaten by domesticated animals, such as cows and chickens. Over 90% of the plant kingdom is made up of angiosperms, including cereal grains such as wheat and corn, citrus and other fruit trees, garden vegetables, and cotton. Many types of garden produce—tomatoes, squash, strawberries, and oranges, to name just a few— are the edible fruits of plants we have domesticated. Fine hardwoods from flowering plants such as oak, cherry, and walnut trees supplement the lumber we get from conifers. We also grow angiosperms for fiber, medications, perfumes, and decoration.

Early humans probably collected wild seeds and fruits. Agriculture gradually developed as people began sowing seeds and cultivating plants to have a more dependable food source. And as they domesticated certain plants, people began to select those with improved yield and quality. Agriculture can thus be seen as yet another facet of the evolutionary relationship between plants and animals.

▼ Figure 16.20 **Fruits and seed dispersal.** Different types of fruits are adapted for different methods of dispersal.

Wind dispersal. Some angiosperms depend on wind for seed dispersal. Here, you can see seeds blowing off a plant called great reedmace.

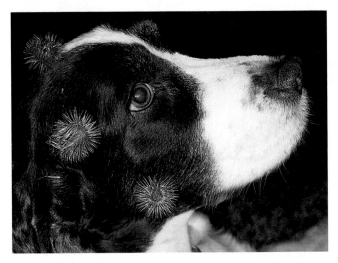

Animal transportation. Some fruits are adapted to hitch free rides on animals. The cockleburs attached to the fur of this dog may be carried miles before opening and releasing seeds.

Animal ingestion. Many angiosperms produce fleshy, edible fruits that are attractive to animals as food. When a weasel eats a berry, it digests the fleshy part of the fruit, but most of the tough seeds pass unharmed through the animal's digestive tract. The weasel later deposits the seeds, along with a fertilizer supply, some distance from where it ate the fruit.

Plant Diversity as a Nonrenewable Resource

The exploding human population, with its demand for space and natural resources, is extinguishing plant species at an unprecedented rate. The problem is especially critical in the tropics, where more than half the human population lives and population growth is fastest. Tropical rain forests are being destroyed at a frightening pace (Figure 16.21). The most common cause of this destruction is large-scale slash-and-burn clearing of forest for agricultural use. Fifty million acres, an area about the size of the state of Washington, are cleared each year, a rate that could completely eliminate Earth's tropical forests within 25 years. As the forest disappears, so do thousands of plant species. Insects and other rain forest animals that depend on these plants are also vanishing. In all, researchers estimate that the destruction of habitat in the rain forest and other ecosystems is claiming hundreds of species each year. The toll is greatest in the tropics because that is where most species live; but environmental assault seems to be a general human tendency. Europeans eliminated most of their forests centuries ago, and habitat destruction is now endangering many species in North America. What is lost is irreplaceable—entire ecosystems that provide medicinal plants, food, timber, and clean water and air.

Many people have ethical concerns about contributing to the extinction of living forms. But there are also practical reasons to be concerned about the loss of plant diversity. As already mentioned, we depend on plants for thousands of products, including food, building materials, and medicines (Table 16.1). More than 120 prescription drugs are extracted from plants. However, researchers have investigated fewer than 5,000 of the 300,000 known plant species as potential sources of medicine. Pharmaceutical companies were led to most of these species by local peoples who use the plants in preparing their traditional medicines.

Scientists are now rallying to slow the loss of plant diversity, in part by offering less destructive ways for people to benefit from forests. The goal of such efforts is to encourage management practices that use forests as resources without damaging them. The solutions we propose must be economically realistic; people who live where there are tropical rain forests must be able to make a living. But if the only goal is profit for the short term, then we will continue to slash and burn until the forests are gone. We need to appreciate the rain forests and other ecosystems as living treasures that can regenerate only slowly. Only then will we learn to work with them in ways that preserve their biological diversity for the future.

Throughout our survey of plants in this chapter, we have seen how entangled the botanical world is with other terrestrial life. We switch our attention now to that other group of organisms that moved onto land with plants: the kingdom Fungi. ☑

☑ CHECKPOINT

In what way are forests renewable resources? In what way are they not?

Answer: Forests are renewable in the sense that new trees can grow where old growth has been removed by logging. Habitats that are permanently destroyed cannot be replaced, however, so forests must be harvested in a sustainable manner.

▼ Figure 16.21 **Clear-cutting of a tropical forest in Brazil.**

Table 16.1	A Sampling of Medicines Derived from Plants		
Compound	**Source**		**Example of Use**
Atropine	Belladonna plant		Pupil dilator in eye exams
Digitalin	Foxglove		Heart medication
Menthol	Eucalyptus tree		Ingredient in cough medicines
Morphine	Opium poppy		Pain reliever
Quinine	Quinine tree		Malaria preventive
Paclitaxel (Taxol)	Pacific yew		Ovarian cancer drug
Tubocurarine	Curare tree		Muscle relaxant during surgery
Vinblastine	Periwinkle		Leukemia drug

Source: Adapted from Randy Moore et al., Botany, 2nd ed. Dubuque, IA: Brown, 1998. Table 2.2, p. 37.

Fungi

The word *fungus* often evokes unpleasant images. Fungi rot timbers, spoil food, and afflict humans with athlete's foot and worse. However, ecosystems would collapse without fungi to decompose dead organisms, fallen leaves, feces, and other organic materials. Fungi recycle vital chemical elements back to the environment in forms that other organisms can assimilate. And you have already learned that nearly all plants have mycorrhizae, fungus-root associations that help plants absorb minerals and water from the soil. In addition to these ecological roles, fungi have been used by people in various ways for centuries. We eat fungi (mushrooms and truffles, for instance), culture fungi to produce antibiotics and other drugs, add them to dough to make bread rise, culture them in milk to produce a variety of cheeses, and use them to ferment beer and wine.

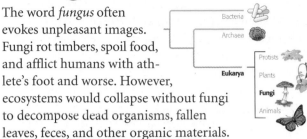

Fungi are eukaryotes, and most are multicellular, but many have body structures and modes of reproduction unlike those of any other organism (**Figure 16.22**). Molecular studies indicate that fungi and animals arose from a common ancestor around 1.5 billion years ago. The oldest undisputed fossils of fungi, however, are only about 460 million years old, perhaps because the ancestors of terrestrial fungi were microscopic and fossilized poorly. Despite appearances, a mushroom is more closely related to you than it is to any plant!

Biologists who study fungi have described over 100,000 species, and there may be as many as 1.5 million. Classifying fungi is an ongoing area of research. (One widely accepted phylogenetic tree divides the kingdom Fungi into five groups.) You are probably familiar with several kinds of fungi, including mushrooms, mold, and yeast. In this section, we'll discuss the characteristics common to all fungi and then survey their wide-ranging ecological impact. ☑

☑CHECKPOINT

Name three ways that we benefit from fungi in our environment.

Answer: Fungi help recycle nutrients by decomposing dead organisms; mycorrhizae help plants absorb water and nutrients; and some fungi serve us as food.

▼ Figure 16.22 **A gallery of diverse fungi.**

Orange fungi. These are the reproductive structures of a fungus that absorbs nutrients as it decomposes a fallen tree on a forest floor.

A "fairy ring." Some mushroom-producing fungi poke up "fairy rings," which can appear on a lawn overnight. A ring develops at the edge of the main body of the fungus, which consists of an underground mass of tiny filaments (hyphae) within the ring. As the underground fungal mass grows outward from its center, the diameter of the fairy ring produced at its expanding perimeter increases annually.

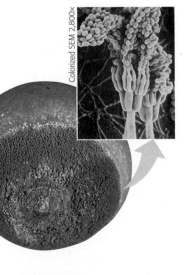

Colorized SEM 2,800x

Mold. Molds grow rapidly on their food sources, which are often our food sources as well. The mold on this orange reproduces asexually by producing chains of microscopic spores (inset) that are dispersed via air currents.

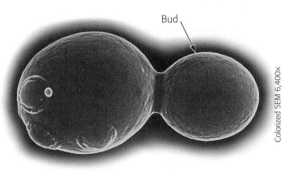

Bud

Colorized SEM 6,400x

Budding yeast. Yeasts are unicellular fungi. This yeast cell is reproducing asexually by a process called budding.

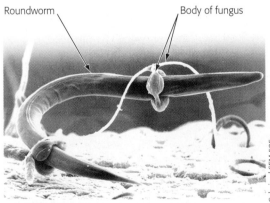

Roundworm   Body of fungus

Colorized SEM 525x

Predatory fungus. This predatory fungus traps and feeds on tiny roundworms in the soil. The fungus is equipped with hoops that can constrict around a worm in less than a second.

Characteristics of Fungi

We'll begin our look at the structure and function of fungi with an overview of how fungi obtain nutrients.

Fungal Nutrition

Fungi are chemoheterotrophs (see Figure 15.12) that acquire their nutrients by **absorption**. In this mode of nutrition, small organic molecules are absorbed from the surrounding medium. A fungus digests food outside its body by secreting powerful digestive enzymes into the food. The enzymes decompose complex molecules to simpler compounds that the fungus can absorb. Fungi absorb nutrients from such nonliving organic material as fallen logs, animal corpses, and the wastes of live organisms.

Fungal Structure

The bodies of most fungi are constructed of threadlike filaments called **hyphae** (singular, *hypha*). Fungal hyphae are minute threads of cytoplasm surrounded by a plasma membrane and cell wall. The cell walls of fungi differ from the cellulose walls of plants. Fungal cell walls are usually built mainly of chitin, a strong but flexible polysaccharide that is also found in the external skeletons of insects. Most fungi have multicellular hyphae, which consist of chains of cells separated by cross-walls with pores. In many fungi, cell-to-cell channels allow ribosomes, mitochondria, and even nuclei to flow between cells.

Fungal hyphae branch repeatedly, forming an interwoven network called a **mycelium** (plural, *mycelia*), the feeding structure of the fungus (**Figure 16.23**). Fungal mycelia usually escape our notice because they are often subterranean, but they can be huge. In fact, scientists have discovered that the mycelium of one humongous fungus in Oregon is 5.5 kilometers (km)—that's 3.4 miles!—in diameter and spreads through 2,200 acres of forest. This fungus is at least 2,600 years old and weighs hundreds of tons, qualifying it as one of Earth's oldest and largest organisms.

A mycelium maximizes contact with its food source by mingling with the organic matter it is decomposing and absorbing. A bucketful of rich organic soil may contain as much as a kilometer of hyphae. A fungal mycelium grows rapidly, adding hyphae as it branches within its food. The great majority of fungi are non-motile; they cannot run, swim, or fly in search of food. But the mycelium makes up for the lack of mobility by swiftly extending the tips of its hyphae into new territory. ☑

Fungal Reproduction

The mushroom in Figure 16.23 is actually made up of tightly packed hyphae. Mushrooms arise from an underground mycelium. While the mycelium obtains food from organic material via absorption, the function of the mushroom is reproduction. It must be above ground to disperse its spores on air currents.

Fungi typically reproduce by releasing haploid spores that are produced either sexually or asexually. The output of spores is mind-boggling. For example, puffballs, which are the reproductive structures of certain fungi, can spew clouds containing trillions of spores (see Figure 13.12). Easily carried by wind or water, spores germinate to produce mycelia if they land in a moist place where there is food. Spores thus function in dispersal and account for the wide geographic distribution of many species of fungi. The airborne spores of fungi have been found more than 160 km (100 miles) above Earth. Closer to home, try leaving a slice of bread out for a week and you will observe the furry mycelia that grow from the invisible spores raining down from the surrounding air.

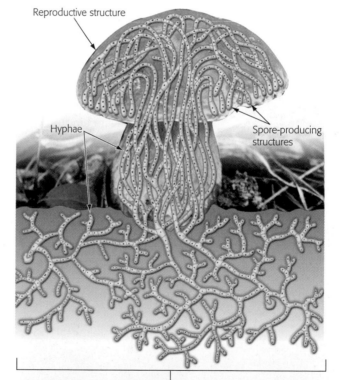

Reproductive structure

Hyphae

Spore-producing structures

Mycelium

◀ **Figure 16.23 The fungal mycelium.** A mushroom consists of tightly packed hyphae that extend upward from a much more massive mycelium of hyphae growing underground. The photo shows a mycelium made up of the cottony threads that decompose organic litter.

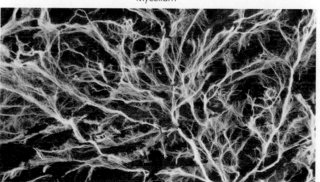

☑ CHECKPOINT

Describe how the structure of a fungal mycelium reflects its function.

Answer: The extensive network of hyphae puts a large surface area in contact with the food source.

The Ecological Impact of Fungi

Fungi have been major players in terrestrial communities ever since plants and fungi together moved onto land. Let's examine a few examples of how fungi continue to have an enormous ecological impact, including numerous interactions with humans.

Fungi as Decomposers

Fungi and bacteria are the principal decomposers that keep ecosystems stocked with the inorganic nutrients essential for plant growth. Without decomposers, carbon, nitrogen, and other elements would accumulate in non-living organic matter. Plants and the animals they feed would starve because elements taken from the soil would not be returned (see Chapter 20).

Fungi are well adapted as decomposers of organic refuse. Their invasive hyphae enter the tissues and cells of dead organisms and digest polymers, including the cellulose of plant cell walls. A succession of fungi, in concert with bacteria and, in some environments, invertebrate animals, is responsible for the complete breakdown of organic litter. The air is so loaded with fungal spores that as soon as a leaf falls or an insect dies, it is covered with spores and soon after infiltrated by fungal hyphae.

We may applaud fungi that decompose forest litter or dung, but it's a different story when molds attack our fruit or our shower curtains. A significant amount of the world's fruit harvest is lost each year to fungal attack. And a wood-digesting fungus does not distinguish between a fallen oak limb and the oak planks of a boat. During the Revolutionary War, the British lost more ships to fungal rot than to enemy attack. What's more, soldiers stationed in the tropics during World War II watched as their tents, clothing, boots, and binoculars were destroyed by molds.

Parasitic Fungi

Parasitism is a relationship in which two species live in contact and one organism benefits while the other is harmed. Of the 100,000 known species of fungi, about 30% make their living as parasites. Parasitic fungi absorb nutrients from the cells or body fluids of living hosts.

About 50 species of fungi are known to be parasitic in humans and other animals. Among the diseases that fungi cause in humans are yeast infections of the lungs, some of which can be fatal, and vaginal yeast infections. Other fungal parasites produce a skin disease called ringworm, so named because it appears as circular red areas on the skin. Most commonly, ringworm fungus attacks the feet and causes intense itching and sometimes blisters. This condition, known as athlete's foot, is highly contagious but can be treated with fungicidal lotions and powders.

The great majority of fungal parasites infect plants. In some cases, fungi that infect plants have literally changed landscapes. In the Biology and Society section, you learned how a fungal blight devastated the American chestnut. A related fungus killed most American elm trees (**Figure 16.24a**). Fungi are also serious agricultural pests, and some of the fungi that attack food crops are toxic to humans. The seed heads of many kinds of grain and grasses, including rye, wheat, and oats, are sometimes infected with fungal growths called ergots (**Figure 16.24b**). Consumption of flour made from ergot-infested grain can cause hallucinations, temporary insanity, and death. In fact, lysergic acid, the raw material from which the hallucinogenic drug LSD is made, has been isolated from ergots. This fact may help explain a centuries-old mystery, as we'll see next.

(a) American elm trees killed by Dutch elm disease fungus

(b) Ergots

◀ Figure 16.24 **Parasitic fungi that cause plant disease. (a)** The parasitic fungus that causes Dutch elm disease evolved with European species of elm trees, and it is relatively harmless to them. But it has been deadly to American elms since it was accidentally introduced after World War I. **(b)** Ergots, parasitic fungi, are the dark structures on these rye seed heads.

Did a Fungus Lead to the Salem Witch Hunt?

In January 1692, eight young girls in the town of Salem, Massachusetts, began to act bizarrely. The girls suffered from incomprehensible speech, odd skin sensations, convulsions, and hallucinations. The worried community blamed the girls' symptoms on witchcraft and began to accuse one another. By the time the hysteria ended that autumn, more than 150 villagers had been accused of witchcraft, and 20 of them had been hanged. Finding the cause behind the "Salem witch hunt" has long intrigued historians.

In 1976, a University of California psychology graduate student offered a new explanation. She began with the **observation** that the symptoms reported by the girls were consistent with ergot poisoning (Figure 16.25). This led her to **question** whether an ergot outbreak could have been behind the witch hunt. The researcher tested her **hypothesis** by examining the historical records, making the **prediction** that facts consistent with ergot poisoning would be uncovered.

Her **results** were suggestive, though not conclusive. Agricultural records confirm that rye—the principal host for ergot—grew abundantly around Salem at that time and that the growing season of 1691 had been particularly warm and wet, conditions under which ergot thrives. This suggests that the rye crop consumed during the winter of 1691–1692 could easily have been contaminated. The summer of 1692, when the accusations began to die down, was dry, consistent with an ergot die-off. Most importantly, the reported symptoms appear consistent with those of ergot poisoning. These clues suggest (but do not prove) that the girls, and perhaps others in Salem, were in the grips of ergot-induced illness. Some historians dispute this idea, and other hypotheses have been proposed. Conclusive evidence may never be found, but this story reinforces the unifying thread of this chapter—the importance of the interaction of plants and fungi—and illustrates how the scientific method can be applied in a wide variety of academic disciplines.

▼ **Figure 16.25 Ergot and the Salem witch hunt.** Ergot poisoning may have been the catalyst for the Salem witch hunt of 1692.

Commercial Uses of Fungi

It would not be fair to fungi to end our discussion with an account of diseases. In addition to their positive global impact as decomposers, fungi also have a number of practical uses for humans.

Most of us have eaten mushrooms, although we may not have realized that we were ingesting the reproductive extensions of subterranean fungi. Mushrooms are often cultivated commercially in artificial caves in which cow manure is piled—thus, you should be sure to wash your store-bought mushrooms thoroughly. Edible mushrooms also grow wild in fields, forests, and backyards, but so do poisonous ones. There are no simple rules to help the novice distinguish edible from deadly mushrooms. Only experts should dare to collect wild fungi for eating.

Mushrooms are only one of many fungi we eat (Figure 16.26). Truffles are fungi that grow in association with tree roots; they are highly prized by gourmets. And the distinctive flavors of certain kinds of cheeses (such as blue cheese) come from the fungi used to ripen them. Particularly important in food production are unicellular fungi, the yeasts. As discussed in Chapter 6, yeasts are used in baking, brewing, and winemaking.

▼ **Figure 16.26 Fungi eaten by humans.**

Truffles (the fungal kind, not the chocolates). These are the reproductive structures of certain fungi that grow with tree roots as mycorrhizae. Truffles release strong odors that attract certain mammals, which excavate the fungi and disperse their spores. Truffle hunters use pigs or dogs to locate their prizes, which may command prices of hundreds of dollars per pound.

Blue cheese. The turquoise streaks in blue cheese are the mycelia of a specific fungus.

Chanterelle mushrooms. Gourmet mushrooms are highly prized by chefs for their earthy flavors and interesting textures.

Fungi are medically valuable as well. Some fungi produce antibiotics that are used to treat bacterial diseases. In fact, the first antibiotic discovered was penicillin, made by the common mold *Penicillium* (Figure 16.27). In 2007, researchers discovered that several varieties of fungi that attack hazelnut trees produce a powerful anticancer drug.

As sources of medicines and food, as decomposers, and as partners with plants in mycorrhizae, fungi play vital roles in life on Earth. ☑

► Figure 16.27 **Fungal production of an antibiotic.** Penicillin is made by the common mold *Penicillium.* In this petri dish, the clear area between the mold and the growing *Staphylococcus* bacteria is where the antibiotic produced by the *Penicillium* inhibits the growth of the bacteria.

EVOLUTION CONNECTION: Plant-Fungus Symbiosis

Mutually Beneficial Symbiosis

Discussing fungi in the same chapter as plants may seem to indicate that these two kingdoms are close relatives. Actually, as we've discussed, fungi are more closely related to animals than to plants. But the success of plants on land and the great diversity of fungi are interconnected; neither could have populated the land without the other.

Evolution is not just about the origin and adaptation of individual species. Relationships between species are also an evolutionary product. Symbiosis is the term used to describe ecological relationships between organisms of different species that are in direct physical contact.

Although the Biology and Society and Process of Science sections presented examples of harmful symbiotic relationships, there are many other examples where

symbiosis is mutually beneficial. Eukaryotic cells evolved from symbiosis among prokaryotes. And today, bacteria living in the roots of certain plants provide nitrogen compounds to their host and receive food in exchange. We have our own symbiotic bacteria that help keep our skin healthy and produce certain vitamins in our intestines. Particularly relevant to this chapter is the symbiotic association of fungi and plant roots—mycorrhizae—that made life's move onto land possible.

Lichens, symbiotic associations of fungi and algae, are striking examples of how two species can become so merged that the cooperative is essentially a new life-form. At a distance, it is easy to mistake lichens for mosses or other simple plants growing on rocks, rotting logs, trees, roofs, or gravestones (Figure 16.28). In fact, lichens are not mosses or any other kind of plant, nor are they even individual organisms. A lichen is a symbiotic association of millions of tiny algal cells embraced by a mesh of fungal hyphae. The photosynthetic algae feed the fungi. The fungal mycelium, in turn, provides a suitable habitat for the algae, helping the algae absorb and retain water and minerals. The mutualistic merger is so complete that lichens are actually named as species, as though they were individual organisms.

After protists and plants, fungi is the third group of eukaryotes that we have surveyed so far. Strong evidence suggests that they evolved from protist ancestors that also gave rise to the fourth and most diverse group of eukaryotes: the animals, the topic of the next chapter.

▼ Figure 16.28 **Lichens: symbiotic associations of fungi and algae.** Lichens generally grow very slowly, sometimes less than a millimeter per year. Some lichens are thousands of years old, rivaling the oldest plants as Earth's elders. The close relationship between the fungal and algal partners is evident in the microscopic blowup of a lichen.

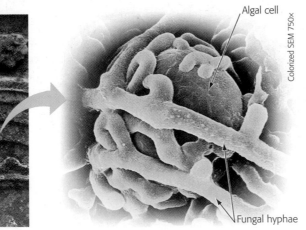

Chapter Review

SUMMARY OF KEY CONCEPTS

 Go to the Study Area at **www.masteringbiology.com** for practice quizzes, my**e**Book, BioFlix™ 3-D animations, MP3 Tutor Sessions, videos, current events, and more.

Colonizing Land

Terrestrial Adaptations of Plants

Plants are multicellular photosynthetic eukaryotes with adaptations for living on land.

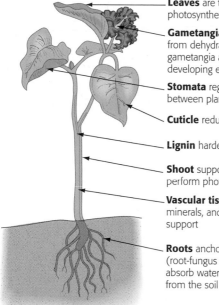

Leaves are the main photosynthetic organs

Gametangia protect gametes from dehydration; female gametangia also protect developing embryos

Stomata regulate gas exchange between plant and atmosphere

Cuticle reduces water loss

Lignin hardens cell walls

Shoot supports plant; may perform photosynthesis

Vascular tissues transport water, minerals, and sugars; provide support

Roots anchor plant; mycorrhizae (root-fungus associations) help absorb water and minerals from the soil

The Origin of Plants from Green Algae

Plants evolved from a group of multicellular green algae called charophytes.

Plant Diversity

Highlights of Plant Evolution

Four major periods of plant evolution are marked by terrestrial adaptations.

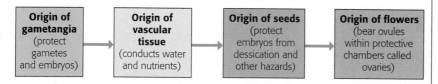

| **Origin of gametangia** (protect gametes and embryos) | → | **Origin of vascular tissue** (conducts water and nutrients) | → | **Origin of seeds** (protect embryos from dessication and other hazards) | → | **Origin of flowers** (bear ovules within protective chambers called ovaries) |

Bryophytes

 The most familiar bryophytes are mosses. Mosses display two key terrestrial adaptations: a waxy cuticle that prevents dehydration and the retention of developing embryos within the mother plant's gametangia. Mosses are most common in moist environments because their sperm must swim to the eggs and because they lack lignin in their cell walls and thus cannot stand tall. Bryophytes are unique among plants in having the gametophyte as the dominant generation in the life cycle.

Ferns

Ferns are seedless plants that have vascular tissues but still use flagellated sperm to fertilize eggs. During the Carboniferous period, giant ferns were among the plants that decayed to thick deposits of organic matter, which were gradually converted to coal.

Gymnosperms

 A drier and colder global climate near the end of the Carboniferous period favored the evolution of the first seed plants. The most successful were the gymnosperms, represented by conifers. Needle-shaped leaves with thick cuticles and sunken stomata are adaptations to dry conditions. Conifers and most other gymnosperms have three additional terrestrial adaptations: (1) further reduction of the haploid gametophyte and greater development of the diploid sporophyte, (2) sperm-bearing pollen, which doesn't require water for transport, and (3) seeds, which consist of a plant embryo packaged along with a food supply inside a protective coat.

Angiosperms

 Angiosperms supply nearly all our food and much of our fiber for textiles. The evolution of the flower and more efficient water transport help account for the success of the angiosperms. The dominant stage is a sporophyte with gametophytes in its flowers. The female gametophyte is located within an ovule, which in turn resides within a chamber of the ovary. Fertilization of an egg in the female gametophyte produces a zygote, which develops into an embryo. The whole ovule develops into a seed. The seed's enclosure within an ovary is what distinguishes angiosperms from gymnosperms, which have naked seeds. A fruit is the ripened ovary of a flower. Fruits protect and help disperse seeds. Angiosperms are a major food source for animals, while animals aid plants in pollination and seed dispersal. Agriculture constitutes a unique kind of evolutionary relationship among plants, humans, and other animals.

Plant Diversity as a Nonrenewable Resource

The exploding human population, with its demand for space and natural resources, is causing the extinction of plant species at an unprecedented rate. Scientists are researching how to slow this loss and help us learn to work with forests in ways that preserve their biological diversity.

Fungi

Characteristics of Fungi

Fungi are unicellular or multicellular eukaryotes; they are chemoheterotrophs that digest their food externally and absorb the nutrients from the environment. They are more closely related to animals than to plants. A fungus usually consists of a mass of threadlike hyphae, forming a mycelium. The cell walls of fungi are mainly composed of chitin. Although most fungi are nonmotile, a mycelium can grow very quickly, extending the tips of its hyphae into new territory. Mushrooms are reproductive structures that extend from the underground mycelium. Fungi reproduce and disperse by releasing spores that are produced either sexually or asexually.

The Ecological Impact of Fungi

Fungi and bacteria are the principal decomposers of ecosystems. Many molds destroy fruit, wood, and human-made materials. About 50 species of fungi are known to be parasites of humans and other animals. Fungi are also commercially important as food and in baking, beer and wine production, and the manufacture of antibiotics.

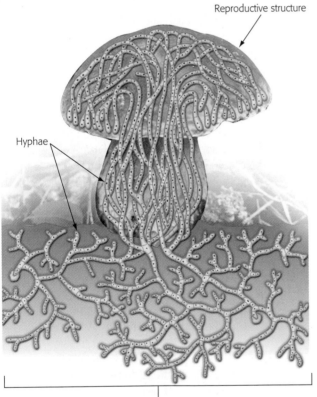

Reproductive structure

Hyphae

Mycelium

Evolution Connection: Mutually Beneficial Symbiosis

Lichens, in which algae are surrounded by fungal hyphae, are an example of a mutually beneficial symbiotic relationship.

SELF-QUIZ

1. Which of the following structures is common to all four major plant groups: vascular tissue, flowers, seeds, cuticle, pollen?

2. Angiosperms are distinguished from all other plants because only angiosperms have reproductive structures called _____.

3. Complete the following analogies:
   a. Gametophyte is to haploid as _____ is to diploid.
   b. _____ are to conifers as flowers are to _____.
   c. Ovule is to seed as ovary is to _____.

4. Under a microscope, a piece of a mushroom would look most like
   a. jelly.
   b. a tangle of string.
   c. grains of sand.
   d. a sponge.

5. During the Carboniferous period, the dominant plants, which later formed the great coal beds, were mainly
   a. mosses and other bryophytes.
   b. ferns and other seedless vascular plants.
   c. charophytes and other green algae.
   d. conifers and other gymnosperms.

6. You discover a new species of plant. Under the microscope, you find that it produces flagellated sperm. A genetic analysis shows that its dominant generation has diploid cells. What kind of plant do you have?

7. How does the evergreen nature of pines and other conifers adapt the plants for living where the growing season is very short?

8. Which of the following terms includes all others in the list? angiosperm, fern, vascular plant, gymnosperm, seed plant

9. Plant diversity is greatest in
   a. tropical forests.
   b. the temperate forests of Europe.
   c. deserts.
   d. the oceans.

10. What is a fruit?

11. Lichens are symbionts of photosynthetic _____ with _____.

12. Contrast the heterotrophic nutrition of a fungus with your own heterotrophic nutrition.

Answers to the Self-Quiz questions can be found in Appendix D.

THE PROCESS OF SCIENCE

13. In April 1986, an accident at a nuclear power plant in Chernobyl, Ukraine, scattered radioactive fallout for hundreds of miles. In assessing the biological effects of the radiation, researchers found mosses to be especially valuable as organisms for monitoring the damage. Radiation damages organisms by causing mutations. Explain why it is faster to observe the genetic effects of radiation on mosses than on other types of plants. Imagine that you are conducting tests shortly after a nuclear accident. Using potted moss plants as your experimental organisms, design an experiment to test the hypothesis that the frequency of mutations decreases with the organism's distance from the source of radiation.

14. You discover what you think may be one extremely large underground fungal mycelium living beneath your campus. How could you prove that it is, in fact, one individual organism spread across a very large area, as opposed to a group of separate organisms?

BIOLOGY AND SOCIETY

15. Why are tropical rain forests being destroyed at such an alarming rate? What kinds of social, technological, and economic factors are responsible? Most forests in more industrialized Northern Hemisphere countries have already been cut. Do the more industrialized nations have a right to pressure the less industrialized nations in the Southern Hemisphere to slow or stop the destruction of their forests? Defend your answer. What kinds of benefits, incentives, or programs might slow the assault on the rain forests?

16. Imagine you were charged with the task of managing a coniferous forest. How would you balance the need for productive use of the forest (to provide lumber, for example) with preservation of its diversity? What activities would you allow or prohibit in the forest (for example, snowmobiling, logging, hiking, mushroom harvesting, mining, camping, grazing, making campfires)? How would you defend your choices to the public?

The Evolution of Animals

A skull from one of Indonesia's "hobbit people."
Scientists are debating whether the skull is from an ancient human-like species.

CHAPTER CONTENTS

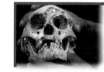

BIOLOGY AND SOCIETY: Human Evolution

Rise of the Hobbit People

In 2003, Australian anthropologists digging on the Indonesian island of Flores stumbled upon bones—including a nearly complete female skeleton—of some highly unusual people. The find, dated to 12,000 to 18,000 years ago (very recent in terms of human evolution), had some very odd features: The bones belonged to adults who stood just over 3 feet tall with a brain less than one-third the size of modern humans (about the size of a chimpanzee's). Since that initial find, bones from about 12 similarly sized individuals have also been unearthed.

While there are many known humans of small stature, including some pygmies living on Flores, the tiny head and brain were unprecedented. The unusual bones were attributed to a previously unknown species named *Homo floresiensis* and nicknamed "hobbits." The discoverers speculated that a small-bodied, small-brained band of ancestral humans arrived on Flores millions of years ago from Africa and, within the isolated environment of the island, evolved into the even smaller *Homo floresiensis*. There is precedent for animal evolution of this type: Biologists have discovered island-bound dwarf populations of deer, elephants, and hippos. One hypothesis is that a lack of predators favors the evolution of smaller, more energy-efficient forms.

Other scientists are highly skeptical, arguing that the bones are from diseased members of *Homo sapiens*. Several human disorders can cause malformations similar to those found in the Flores bones. Are these bones from an isolated human-like species with a tiny head, or from diseased members of our own species? The debate rages on.

Homo sapiens and *Homo floresiensis* are just two of the over 1 million species of animals that have been named and described by biologists. This amazing diversity arose through hundreds of millions of years of evolution as natural selection shaped animal adaptations to Earth's many environments. In this chapter, we'll look at the 9 most abundant and widespread of the roughly 35 phyla (major groups) in the kingdom Animalia. We'll give special attention to the major milestones in animal evolution and conclude by reconnecting with the fascinating subject of human evolution.

The Origins of Animal Diversity

Animal life began in Precambrian seas with the evolution of multicellular creatures that ate other organisms. We are among their descendants.

What Is an Animal?

Animals are eukaryotic, multicellular, heterotrophic organisms that obtain nutrients by eating. This mode of nutrition contrasts animals with plants and other organisms that construct organic molecules through photosynthesis. It also contrasts with fungi, which obtain nutrients by absorption after digesting the food outside the body (see Chapter 16). Most animals digest their food within their bodies after ingesting other organisms, dead or alive, whole or by the piece (**Figure 17.1**).

Animal cells lack the cell walls that provide strong support in the bodies of plants and fungi. And most animals have muscle cells for movement and nerve cells that control the muscles. The most complex animals can use their muscular and nervous systems for many functions other than eating. Some species even use massive networks of nerve cells called brains to think.

Most animals are diploid and reproduce sexually; eggs and sperm are the only haploid cells. The life cycle of a sea star (**Figure 17.2**) includes basic stages found in most animal life cycles. **1** Male and female adult animals make haploid gametes by meiosis, and **2** an egg and a sperm fuse, producing a zygote. **3** The zygote divides by mitosis, forming **4** an early embryonic stage called a **blastula**, which is usually a hollow ball of cells. **5** In most animals, one side of the blastula folds inward, forming a stage called a **gastrula**. **6** The gastrula develops into a saclike embryo with a two-layered wall and an opening at one end. After the gastrula stage, many animals develop directly into adults. **7** Others, such as the sea star, develop into a **larva**, an immature individual that looks different from the adult animal (a tadpole, for another example, is a larval frog). **8** The larva undergoes a major change of body form, called **metamorphosis**, in becoming an adult capable of reproducing sexually. ☑

CHECKPOINT

What mode of nutrition distinguishes animals from fungi, both of which are heterotrophs?

Answer: ingestion (eating)

▼ Figure 17.1 **Nutrition by ingestion, the animal way of life.** Few animals ingest a piece of food as large as the gazelle being eaten by this rock python. The snake will spend two weeks or more digesting its meal.

▼ Figure 17.2 **Life cycle of a sea star as an example of animal development.**

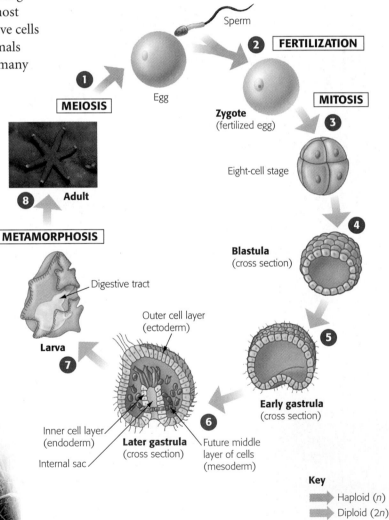

Sperm

1 MEIOSIS

Egg

2 FERTILIZATION

Zygote (fertilized egg)

MITOSIS

3

Eight-cell stage

4

Blastula (cross section)

5

Outer cell layer (ectoderm)

Early gastrula (cross section)

6

Inner cell layer (endoderm)

Later gastrula (cross section)

Future middle layer of cells (mesoderm)

Internal sac

7 Larva

Digestive tract

METAMORPHOSIS

8 Adult

Key
→ Haploid (*n*)
→ Diploid (2*n*)

Early Animals and the Cambrian Explosion

Animals probably evolved from a colonial, flagellated protist that lived in Precambrian seas about a billion years ago (Figure 17.3). By the late Precambrian, about 600–700 million years ago, a diversity of animals had evolved. At the beginning of the Cambrian period, 542 million years ago, animals underwent a relatively rapid diversification called "the Cambrian explosion." During a span of only about 15 million years, all the major animal body plans we see today evolved. Although many of the Cambrian animals seem bizarre, most zoologists now agree that Cambrian fossils can be classified as ancient representatives of modern animal phyla (Figure 17.4).

Why did the evolution of animal forms accelerate so dramatically at that time? Hypotheses abound. One hypothesis emphasizes increasingly complex predator-prey relationships that led to diverse adaptations for feeding, motility, and protection. This hypothesis would help explain why most Cambrian animals had shells or hard outer skeletons, in contrast with Precambrian animals, which were mostly soft-bodied. Another hypothesis focuses on the evolution of genes that control the development of animal form, such as the placement of body parts in embryos. The study of the relationship between evolution and development, called evo-devo, is a very active area of biological research. At least some "master control" genes are important in development of diverse animal phyla (see Chapter 11). Variations in how, when, and where these genes are expressed

in an embryo can produce some of the major differences in body form that distinguish the phyla. Perhaps such changes in gene expression were partly responsible for the relatively rapid diversification of animals during the early Cambrian period.

In the last half billion years, animal evolution has to a large degree merely generated variations of the animal forms that originated in the Cambrian seas. Continuing research will help test hypotheses about the Cambrian explosion. But even as the explosion becomes less mysterious, it will seem no less wondrous. ☑

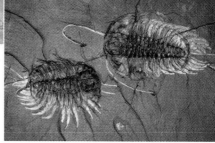

◀ Figure 17.4 **A Cambrian seascape.** This drawing is based on fossils (such as *Olenoides serratus*, a kind of trilobite, shown below) collected at a site called the Burgess Shale in British Columbia, Canada.

☑CHECKPOINT

Why is animal evolution during the early Cambrian referred to as an "explosion"?

Answer: because a great diversity of animals evolved in a relatively short time span

▼ Figure 17.3 **One hypothesis for a sequence of stages in the origin of animals from a colonial protist.** With its specialized cells and a simple digestive compartment, the proto-animal shown at the end of the process could have fed on organic matter on the seafloor.

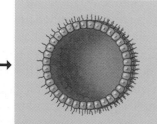

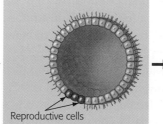

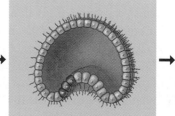

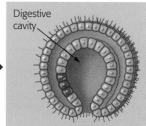

The earliest colonies may have consisted of only a few cells, all of which were flagellated and basically identical.

Some of the later colonies may have been hollow spheres of cells that ingested organic nutrients from the water.

Eventually, cells in the colony may have specialized, with some cells adapted for reproduction and others for locomotion and feeding. *(Reproductive cells)*

A simple multicellular organism with cell layers may have evolved from a hollow colony, with cells on one side of the colony cupping inward.

A layered body plan would have enabled further division of labor, with cells specialized for locomotion, protection, reproduction, or feeding. *(Digestive cavity)*

Animal Phylogeny

Historically, biologists have categorized animals by "body plan"—general features of body structure. Distinctions between body plans are used to help infer the evolutionary relationships between animal groups. More recently, a wealth of genetic data has allowed evolutionary biologists to modify and refine groups. **Figure 17.5** represents a set of hypotheses about the evolutionary relationships between nine major animal phyla based on structural and genetic similarities.

▶ Figure 17.5 **An overview of animal phylogeny.** Only 9 of the more than 30 animal phyla (the exact number is not agreed upon) are included in the tree and in the text. The branching takes into account the body plan of the organisms as well as genetic data.

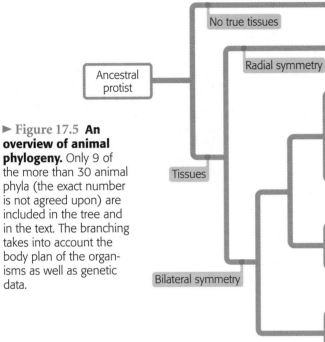

A major branch point in animal evolution distinguishes sponges from all other animals based on structural complexity. Unlike more complex animals, sponges lack true tissues, groups of similar cells that perform a function (such as nervous tissue). A second major evolutionary split is based on body symmetry: radial versus bilateral (**Figure 17.6**). To understand this difference, imagine a pail and shovel. The pail has **radial symmetry**, identical all around a central axis. The shovel has **bilateral symmetry**, which means there's only one way to split it into two equal halves—right down the midline. A bilateral animal has a definite "head end" that first encounters food, danger, and other stimuli when traveling. In most bilateral animals, a nerve center in the form of

▼ Figure 17.6 **Body symmetry.**

Radial symmetry. Parts radiate from the center, so any slice through the central axis divides into mirror images.

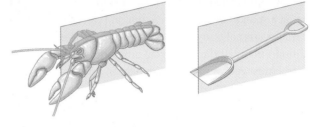

Bilateral symmetry. Only one slice can divide left and right sides into mirror-image halves.

a brain is at the head end, near a concentration of sense organs such as eyes. Thus, bilateral symmetry is an adaptation that aids movement, such as crawling, burrowing, or swimming. Indeed, many radial animals are stationary, whereas most bilateral animals are mobile.

The evolution of body cavities also helped lead to more complex animals. A **body cavity** is a fluid-filled space separating the digestive tract from the outer body wall. The cavity enables the internal organs to grow and move independently of the outer body wall, and the fluid cushions them from injury. In soft-bodied animals such as earthworms, the fluid is under pressure and functions as a hydrostatic skeleton. Of the phyla shown in Figure 17.5, only sponges, cnidarians, and flatworms lack a body cavity.

Among animals with a body cavity, there are differences in how the cavity develops (**Figure 17.7**). In all cases, the cavity is at least partly lined by a middle layer of tissue, called mesoderm, which develops between the inner (endoderm) and outer (ectoderm) layers of the gastrula embryo. If the body cavity is not completely lined by tissue derived from mesoderm, it is called a **pseudocoelom**. Among the animal phyla discussed in this chapter, only the roundworms (nematodes) have a pseudocoelom. A true **coelom**, the type of body cavity humans and many other animals have, is completely lined by tissue derived from mesoderm. (Despite its name, a pseudocoelom functions just like a coelom.)

With the overview of animal evolution in Figure 17.5 as our guide, we're ready to take a closer look at the nine most numerous animal phyla. ✓

▼ Figure 17.7 **Body plans of bilateral animals.** The various organ systems of these animals develop from the three tissue layers that form in the embryo.

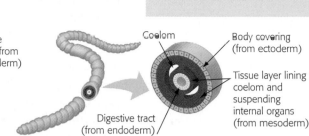

Body covering (from ectoderm)

Tissue-filled region (from mesoderm)

Digestive tract (from endoderm)

(a) No body cavity: for example, flatworm

Body covering (from ectoderm)

Muscle layer (from mesoderm)

Pseudocoelom

Digestive tract (from endoderm)

(b) Pseudocoelom: a body cavity only partially lined by the mesoderm, the middle tissue layer; for example, roundworm

Coelom

Body covering (from ectoderm)

Tissue layer lining coelom and suspending internal organs (from mesoderm)

Digestive tract (from endoderm)

(c) True coelom: a fluid-filled body cavity completely lined by mesoderm; for example, annelid

Major Invertebrate Phyla

Living as we do on land, our sense of animal diversity is biased in favor of vertebrates, animals with a backbone, such as amphibians, reptiles, and mammals. However, vertebrates make up less than 5% of all animal species. If we were to sample the animals in an aquatic habitat, such as a pond, tide pool, or coral reef, or if we were to consider the millions of insects that share our terrestrial world, we would find ourselves in the realm of **invertebrates**, animals without backbones. We give special attention to the vertebrates simply because we humans are among the backboned ones. However, by exploring the other 95% of the animal kingdom—the invertebrates—we'll discover an astonishing diversity of beautiful creatures that too often escape our notice.

cavity and then flows out of the sponge through a larger opening (**Figure 17.8**). Cells called choanocytes have flagella that sweep water through the sponge's porous body. Choanocytes trap bacteria and other food particles in mucus and then engulf the food by phagocytosis (see Chapter 4). Cells called amoebocytes pick up food from the choanocytes, digest it, and carry the nutrients to other cells. Amoebocytes also manufacture the fibers that make up a sponge's skeleton. In some sponges, these fibers are sharp and spur-like. Other sponges have softer, more flexible skeletons; these pliant, honeycombed skeletons are often used as natural sponges in the bath or to wash cars. ✔

☑CHECKPOINT

In what fundamental way does the structure of a sponge differ from that of all other animals?

Answer: A sponge has no true tissues.

Sponges

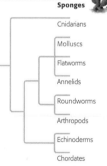

Sponges were once grouped into a single phylum Porifera. However, genetic evidence now suggests that there are multiple phyla. Sponges are stationary animals that appear so sedate that the ancient Greeks believed them to be plants. The simplest of all animals, sponges probably evolved very early from colonial protists. Sponges range in height from about 1 cm to 2 m. They have no nerves or muscles, but their individual cells can sense and react to changes in the environment. The cell layers of sponges are loose federations of cells, not really tissues, because the cells are relatively unspecialized. Of the 9,000 or so species of sponges, only about 100 live in fresh water; the rest are marine.

The body of a sponge resembles a sac perforated with holes. Water is drawn through the pores into a central

Sponges

Cnidarians

Molluscs

Flatworms

Annelids

Roundworms

Arthropods

Echinoderms

Chordates

▼ Figure 17.8 **Anatomy of a sponge.**

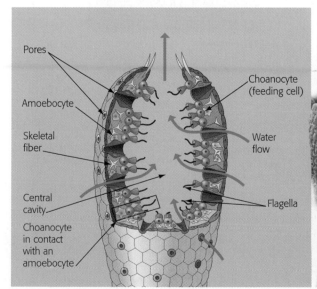

Pores

Amoebocyte

Skeletal fiber

Central cavity

Choanocyte in contact with an amoebocyte

Choanocyte (feeding cell)

Water flow

Flagella

Cnidarians

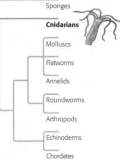

Cnidarians (phylum Cnidaria) are characterized by the presence of body tissues—as are all the remaining animals we will discuss—as well as by radial symmetry and tentacles with stinging cells. Cnidarians include sea anemones, hydras, corals, and jellies (sometimes called jellyfish, though they are not fish). Most of the 10,000 cnidarian species are marine.

The basic body plan of a cnidarian is a sac with a central digestive compartment, the **gastrovascular cavity**. A single opening to this cavity functions as both mouth and anus. This basic body plan has two variations: the stationary **polyp** and the floating **medusa (Figure 17.9)**. Polyps adhere to larger objects and extend their tentacles,

waiting for prey. Examples of the polyp body plan are hydras, sea anemones, and corals. A medusa (plural, *medusae*) is a flattened, mouth-down version of the polyp. It moves freely by a combination of passive drifting and contractions of its bell-shaped body. Some jellies are medusae with tentacles over 100 m long dangling from an umbrella up to 2 m in diameter. There are some species of cnidarians that live only as polyps, others only as medusae, and still others that pass through both a medusa stage and a polyp stage in their life cycle.

Cnidarians are carnivores that use tentacles arranged in a ring around the mouth to capture prey and push the food into the gastrovascular cavity, where digestion begins. The undigested remains are eliminated through the mouth/anus. The tentacles are armed with batteries of cnidocytes ("stinging cells") that function in defense and in the capture of prey **(Figure 17.10)**. The phylum Cnidaria is named for these stinging cells. ✔

☑ CHECKPOINT

In what fundamental way does the body plan of a cnidarian differ from that of all other animals?

Answer: The body of a cnidarian is radially symmetrical.

▼ **Figure 17.9 Polyp and medusa forms of cnidarians.** Note that cnidarians have two tissue layers, distinguished in the diagrams by blue and yellow. The gastrovascular cavity has only one opening, which functions as both mouth and anus.

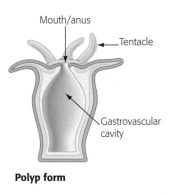

Mouth/anus

Tentacle

Gastrovascular cavity

Polyp form

Coral

Sea anemone

Hydra

▼ **Figure 17.10 Cnidocyte action.** When a trigger on a tentacle is stimulated by touch, a fine thread shoots out from a capsule. Some cnidocyte threads entangle prey, while others puncture the prey and inject a poison.

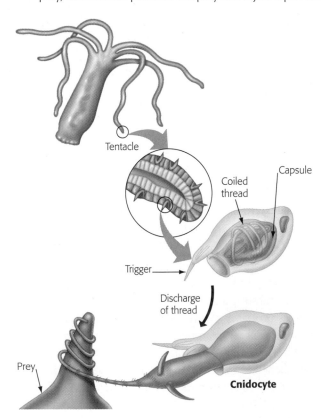

Tentacle

Coiled thread

Capsule

Trigger

Discharge of thread

Prey

Cnidocyte

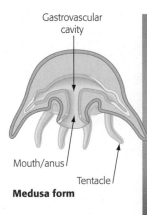

Gastrovascular cavity

Mouth/anus

Tentacle

Medusa form

Jelly

Molluscs

Snails and slugs, oysters and clams, and octopuses and squids are all **molluscs** (phylum Mollusca). Molluscs are soft-bodied animals, but most are protected by a hard shell. Slugs, squids, and octopuses have reduced shells, most of which are internal, or they have lost their shells completely during their evolution. Many molluscs feed by using a file-like organ called a **radula** to scrape up food. Garden snails use their radulas like tiny saws to cut pieces out of leaves.

All molluscs have a similar body plan (**Figure 17.11**). The body has three main parts: a muscular foot, usually used for movement; a visceral mass containing most of the internal organs; and a fold of tissue called the mantle. The **mantle** drapes over the visceral mass and secretes the shell if one is present.

There are 93,000 known species of molluscs, with most being marine animals. The three major groups of molluscs are gastropods (including snails and slugs), bivalves (including clams and oysters), and cephalopods (including squids and octopuses) (**Figure 17.12**).

Most **gastropods** are protected by a single spiraled shell into which the animal can retreat when threatened. Many have a distinct head with eyes at the tips of tentacles (think of a garden snail). Marine, freshwater, and terrestrial gastropods make up about three-quarters of the living mollusc species.

The **bivalves**, including clams, oysters, mussels, and scallops, have shells divided into two halves hinged together. None of the bivalves have a radula. There are both marine and freshwater species, with most being

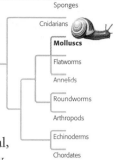

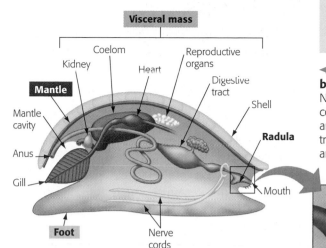

Visceral mass

Coelom — Reproductive organs

Kidney — Heart — Digestive tract

Mantle — Shell

Mantle cavity — **Radula**

Anus

Gill — Mouth

Foot — Nerve cords

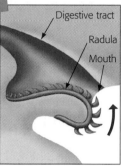

Digestive tract

Radula

Mouth

◄ Figure 17.11 **The general body plan of a mollusc.** Note the body cavity (a true coelom, though a small one) and the complete digestive tract, with both mouth and anus.

sedentary, using their muscular foot for digging and anchoring in sand or mud.

Cephalopods are all marine animals and generally differ from gastropods and sedentary bivalves in that their bodies are fast and agile. A few have large, heavy shells, but in most the shell is small and internal (as in squids) or missing (as in octopuses). Cephalopods have large brains and sophisticated sense organs, which contribute to their success as mobile predators. They use beak-like jaws and a radula to crush or rip prey apart. The mouth is at the base of the foot, which is drawn out into several long tentacles for catching and holding prey. Cephalopods include the largest invertebrates: the giant squid (averaging about 10 m in length) and the closely related colossal squid (up to 13 m). Very little is known about these majestic creatures. In fact, until 2004, none was ever seen alive. In that year, Japanese zoologists captured the first video footage of a giant squid in its natural habitat. ☑

☑ CHECKPOINT

Classify these molluscs: A garden snail is an example of a _____; a clam is an example of a _____; a squid is an example of a _____.

Answer: gastropod; bivalve; cephalopod

▼ Figure 17.12 **Mollusc diversity.**

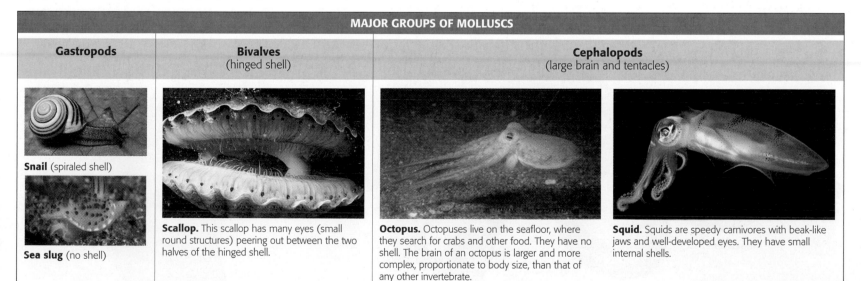

MAJOR GROUPS OF MOLLUSCS			
Gastropods	**Bivalves** (hinged shell)	**Cephalopods** (large brain and tentacles)	
Snail (spiraled shell) **Sea slug** (no shell)	**Scallop.** This scallop has many eyes (small round structures) peering out between the two halves of the hinged shell.	**Octopus.** Octopuses live on the seafloor, where they search for crabs and other food. They have no shell. The brain of an octopus is larger and more complex, proportionate to body size, than that of any other invertebrate.	**Squid.** Squids are speedy carnivores with beak-like jaws and well-developed eyes. They have small internal shells.

Flatworms

Flatworms (phylum Platyhelminthes) are the simplest animals with bilateral symmetry. True to their name, these worms are ribbonlike and range from about 1 mm to about 20 m in length. Most flatworms have a gastrovascular cavity with a single opening. There are about 20,000 species of flatworms living in marine, freshwater, and damp terrestrial habitats (Figure 17.13).

The gastrovascular cavity of free-living flatworms called planarians is highly branched, providing an extensive surface area for the absorption of nutrients. When the animal feeds, a muscular tube projects through the mouth and sucks food in. Planarians live on the undersurfaces of rocks in freshwater ponds and streams.

Parasitic flatworms called blood flukes (*Schistosoma mansoni*) are a major health problem in the tropics. These worms have suckers that attach to the inside of the blood vessels near the human host's intestines. Infection by these flatworms causes a long-lasting disease

called schistosomiasis or blood fluke disease, with such symptoms as severe abdominal pain, anemia, and dysentery. About 250 million people in 70 countries suffer from blood fluke disease each year.

Tapeworms parasitize many vertebrates, including humans. Most tapeworms have a very long, ribbonlike body with repeated parts. They differ from other flatworms in not having a gastrovascular cavity. Living in partially digested food in the intestines of their hosts, tapeworms simply absorb nutrients across their body surface. The head of a tapeworm is armed with suckers and hooks that lock the worm to the intestinal lining of the host. Behind the head is a long ribbon of units that are little more than sacs of sex organs. At the back of the worm, mature units containing thousands of eggs break off and leave the host's body with the feces. Humans can become infected with tapeworms by eating undercooked beef, pork, or fish containing tapeworm larvae. The larvae are microscopic, but the adults can reach lengths of 20 m in the human intestine. Such large tapeworms can cause intestinal blockage and rob enough nutrients from the human host to cause nutritional deficiencies. Fortunately, an orally administered drug can kill the adult worms. ✔

Sponges
Cnidarians
Molluscs
Flatworms
Annelids
Roundworms
Arthropods
Echinoderms
Chordates

☑**CHECKPOINT**

Flatworms are the simplest animals to display a body plan that is _____.

Answer: bilaterally symmetrical

▼ Figure 17.13 **Flatworm diversity.**

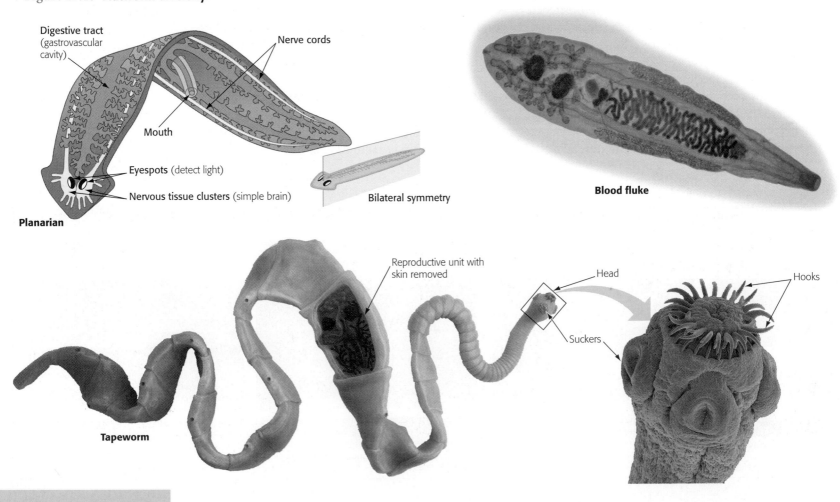

Digestive tract (gastrovascular cavity)

Nerve cords

Mouth

Eyespots (detect light)

Nervous tissue clusters (simple brain)

Planarian

Bilateral symmetry

Blood fluke

Reproductive unit with skin removed

Head

Hooks

Suckers

Tapeworm

Annelids

Annelids (phylum Annelida) are worms that have **body segmentation**, which is the subdivision of the body along its length into a series of repeated parts called segments. In annelids, the segments look like a set of fused rings. There are about 16,000 annelid species, ranging in length from less than 1 mm to the 3-m giant Australian earthworm. Annelids live in damp soil, the sea, and most freshwater habitats. There are three main groups: earthworms, polychaetes, and leeches (**Figure 17.14**).

Annelids exhibit two innovations shared by all other bilateral animals except flatworms. One is a **complete digestive tract**, which is a digestive tube with two openings: a mouth and an anus. A complete digestive tract can process food and absorb nutrients as a meal moves in one direction from one specialized digestive organ to the next. In humans, for example, the mouth, stomach, and intestines act as digestive organs. A second innovation is a body cavity, which in this case is a coelom (a body cavity completely lined by mesoderm-derived tissue; see Figure 17.7c).

Earthworms, like all annelids, are segmented externally and internally (**Figure 17.15**). Many of the internal structures are repeated, segment by segment. The coelom (body cavity) is partitioned by walls (only two segment walls are fully shown here). The nervous system (yellow in the figure) and waste-disposal organs (green) are repeated in each segment. The digestive tract, however, is not segmented; it passes through the

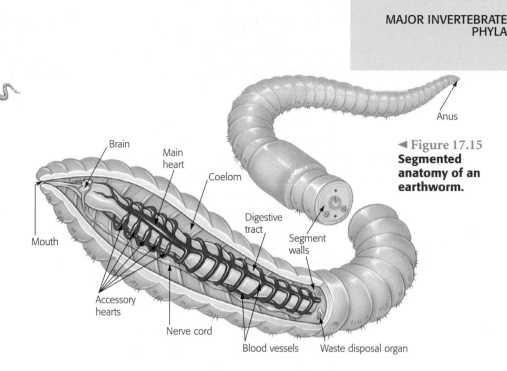

▲ Figure 17.15
Segmented anatomy of an earthworm.

Labels: Anus, Brain, Main heart, Coelom, Digestive tract, Segment walls, Mouth, Accessory hearts, Nerve cord, Blood vessels, Waste disposal organ

segment walls from the mouth to the anus. Segmental blood vessels include one main heart and five pairs of accessory hearts.

Earthworms eat their way through the soil, extracting nutrients as the soil passes through the digestive tract. Undigested material, mixed with mucus secreted into the digestive tract, is eliminated as castings through the anus. Farmers and gardeners value earthworms because the animals till the earth, and the castings improve the texture of the soil. Charles Darwin estimated that each acre of British farmland contained about 50,000 earthworms that produced 18 tons of castings per year.

▼ Figure 17.14 **Annelid diversity.**

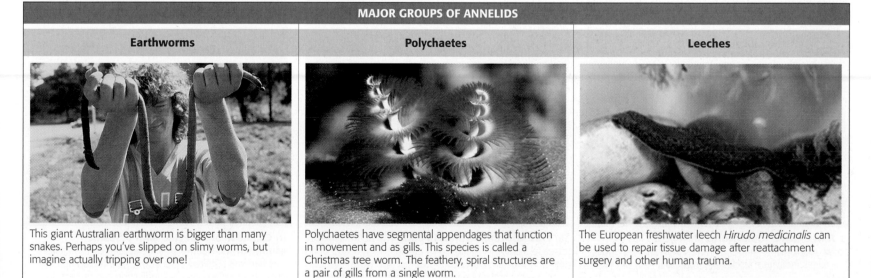

MAJOR GROUPS OF ANNELIDS		
Earthworms	**Polychaetes**	**Leeches**
This giant Australian earthworm is bigger than many snakes. Perhaps you've slipped on slimy worms, but imagine actually tripping over one!	Polychaetes have segmental appendages that function in movement and as gills. This species is called a Christmas tree worm. The feathery, spiral structures are a pair of gills from a single worm.	The European freshwater leech *Hirudo medicinalis* can be used to repair tissue damage after reattachment surgery and other human trauma.

Cladogram labels: Sponges, Cnidarians, Molluscs, Flatworms, **Annelids**, Roundworms, Arthropods, Echinoderms, Chordates

In contrast to earthworms, most **polychaetes** are marine, mainly crawling on or burrowing in the seafloor. Segmental appendages with hard bristles help the worm wriggle about in search of small invertebrates to eat. The appendages also increase the animal's surface area for taking up oxygen and disposing of metabolic wastes, including carbon dioxide.

The third group of annelids, **leeches**, are notorious for the bloodsucking habits of some species. However, most species are free-living carnivores that eat small invertebrates such as snails and insects. A few terrestrial species inhabit moist vegetation in the tropics, but the majority of leeches live in fresh water. A European freshwater

Medicinal leech

species called *Hirudo medicinalis* is used for the treatment of circulatory complications. Most commonly, leeches are applied after reconstructive microsurgery in which limbs or digits are reattached. Because arteries (which transport blood into a reattached area) are easier to reconnect than veins (which transport blood out), blood can pool in the reattached area and stagnate, starving the healing tissue of oxygen. Medicinal leeches have razor-like jaws with hundreds of tiny teeth that cut through the skin. They secrete saliva containing an anesthetic and an anticoagulant into the wound. The anesthetic makes the bite virtually painless, and the anticoagulant prevents clotting as the leech drains excess blood from the wound. ☑

☑**CHECKPOINT**

The body plan of an annelid displays _____, meaning that the body is divided into a series of repeated regions.

Answer: segmentation

Roundworms

Roundworms (also called **nematodes**, members of the phylum Nematoda) get their common name from their cylindrical body, which is usually tapered at both ends (**Figure 17.16**). Roundworms are among the most diverse (in species number) and widespread of all animals. About 25,000 species of roundworms are known, and perhaps ten times that

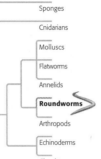

number actually exist. Roundworms range in length from about a millimeter to a meter. They are found in most aquatic habitats, in wet soil, and as parasites in the body fluids and tissues of plants and animals.

Free-living roundworms in the soil are important decomposers. A single rotting apple may contain 90,000 nematodes; an acre of topsoil contains billions. Other species of nematodes are major agricultural pests that attack the roots of plants. At least 50 parasitic roundworm species infect humans, including pinworms, hookworms, and the parasite that causes trichinosis. ☑

☑**CHECKPOINT**

Which phylum is most closely related to the roundworms? (*Hint*: Refer to the phylogenetic tree.)

Answer: arthropods

▼ Figure 17.16 **Roundworm diversity.**

(a) A free-living roundworm. This species has the classic roundworm shape: cylindrical with tapered ends. The ridges indicate muscles that run the length of the body.

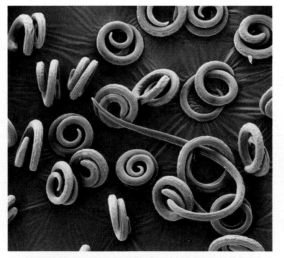

(b) Parasitic roundworms in pork. The potentially fatal disease trichinosis is caused by eating undercooked pork infected with *Trichinella* roundworms. The worms (shown here in pork tissue) burrow into the human intestine and then invade muscle tissue.

(c) Canine heart infected with parasitic roundworms. Heartworms are spread by mosquitoes that transmit fertilized eggs from one infected host to another. Regular doses of medication can protect dogs from heartworms.

Arthropods

Arthropods (phylum Arthropoda) are named for their jointed appendages. Crustaceans (such as crabs and lobsters), arachnids (such as spiders and scorpions), and insects (such as grasshoppers and moths) are examples of arthropods (**Figure 17.17**). Zoologists estimate that the total arthropod population numbers about a billion billion (10^{18}) individuals—that's about 150 million arthropods for each person! Researchers have identified over a million arthropod species, mostly insects. In fact, two out of every three species of life that have been scientifically described are arthropods. And arthropods are represented in nearly all habitats of the biosphere. In species diversity, distribution, and sheer numbers, arthropods must be regarded as the most successful animal phylum.

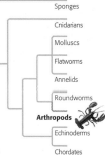

Sponges
Cnidarians
Molluscs
Flatworms
Annelids
Roundworms
Arthropods
Echinoderms
Chordates

◀ **Figure 17.17 Arthropod diversity.**

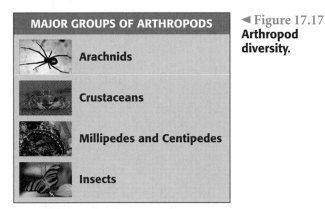

MAJOR GROUPS OF ARTHROPODS

Arachnids

Crustaceans

Millipedes and Centipedes

Insects

General Characteristics of Arthropods

Arthropods are segmented animals. In contrast with the repeating similar segments of annelids, however, arthropod segments and their appendages have become specialized for a great variety of functions. This evolutionary flexibility contributed to the great diversification of arthropods. Specialization of segments (or of fused groups of segments) provides for an efficient division of labor among body regions. For example, the appendages of different segments are variously adapted for walking, feeding, sensory reception, swimming, and defense (**Figure 17.18**).

The body of an arthropod is completely covered by an **exoskeleton**, an external skeleton. This coat is constructed from layers of protein and a polysaccharide called chitin. The exoskeleton can be a thick, hard armor over some parts of the body (such as the head), yet be paper-thin and flexible in other locations (such as the joints). The exoskeleton protects the animal and provides points of attachment for the muscles that move the appendages. There are, of course, advantages to wearing hard parts on the outside. Our own skeleton is interior to most of our soft tissues, an arrangement that doesn't provide much protection from injury. But our skeleton does offer the advantage of being able to grow along with the rest of our body. In contrast, a growing arthropod must occasionally shed its old exoskeleton and secrete a larger one. This process, called molting, leaves the animal temporarily vulnerable to predators and other dangers. The next five pages explore the major groups of arthropods. ☑

☑CHECKPOINT

What is the primary difference between your skeleton and a crab's skeleton?

Answer: Your skeleton is interior, whereas a crab has an exterior skeleton (an exoskeleton).

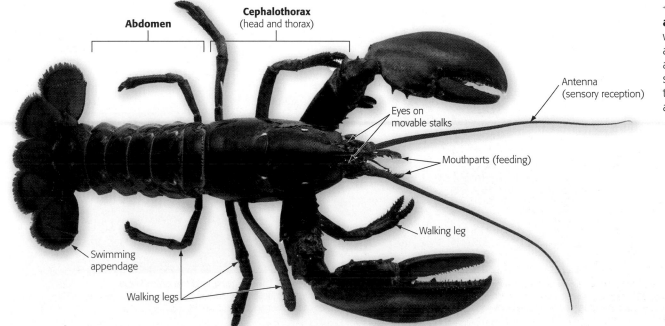

Abdomen

Cephalothorax (head and thorax)

Eyes on movable stalks

Antenna (sensory reception)

Mouthparts (feeding)

Walking leg

Swimming appendage

Walking legs

◀ **Figure 17.18 Anatomy of a lobster, a crustacean.** The whole body, including the appendages, is covered by an exoskeleton. The body is segmented, but this characteristic is obvious only in the abdomen.

Arachnids

Most **arachnids** live on land. Scorpions, spiders, ticks, and mites are examples (Figure 17.19). Arachnids usually have four pairs of walking legs and a specialized pair of feeding appendages. In spiders, these feeding appendages are fang-like and equipped with poison glands. As a spider uses these appendages to immobilize and dismantle its prey, it spills digestive juices onto the torn tissues and sucks up its liquid meal.

▼ Figure 17.19 **Arachnid characteristics and diversity.**

Two feeding appendages

Leg (four pairs)

Scorpion. Scorpions have a pair of large pincers that function in defense and food capture. The tip of the tail bears a poisonous stinger. Scorpions sting people only when prodded or stepped on.

Dust mite. This microscopic house dust mite is a ubiquitous scavenger in our homes. Dust mites are harmless except to people who are allergic to the mites' feces.

Spider. Like most spiders, including the tarantula in the large photo, this black widow spins a web of liquid silk, which solidifies as it comes out of specialized glands. A black widow's venom can kill small prey but is rarely fatal to humans.

Wood tick. Wood ticks carry bacteria that cause Rocky Mountain spotted fever. Lyme disease is caused by a separate species called deer ticks.

Crustaceans

Crustaceans are nearly all aquatic. Crabs, lobsters, crayfish, shrimps, and barnacles are all crustaceans (Figure 17.20). They all exhibit the arthropod hallmark of multiple pairs of specialized appendages. One group of crustaceans, the isopods, is represented on land by pill bugs, which you can find on the undersides of moist leaves and other organic debris.

▼ Figure 17.20 **Crustacean characteristics and diversity.**

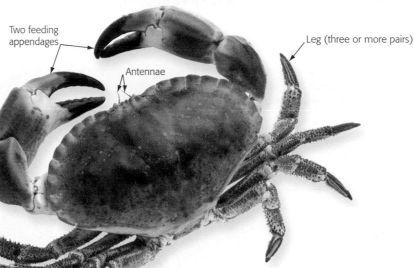

Two feeding appendages

Antennae

Leg (three or more pairs)

Crab. Ghost crabs are common along shorelines throughout the world. They scurry along the surf's edge and then quickly bury themselves in sand.

Shrimp. Naturally found in Pacific waters from Africa to Asia, giant prawns are widely cultivated as food.

Pill bug. Commonly found in moist locations with decaying leaves, such as under logs, pill bugs get their name from their tendency to roll up into a tight ball when they sense danger.

Crayfish. The Miami cave crayfish, an endangered crustacean, was only recently discovered living in underground Florida aquifers.

Barnacles. Barnacles are stationary crustaceans with exoskeletons hardened into shells by calcium carbonate (lime). The jointed appendages projecting from the shell capture small plankton.

Millipedes and Centipedes

Millipedes and **centipedes** have similar segments over most of the body and superficially resemble annelids, but their jointed legs reveal they are arthropods (**Figure 17.21**). Millipedes are landlubbers that eat decaying plant matter. They have two pairs of short legs per body segment. Centipedes are terrestrial carnivores, with a pair of poison claws used in defense and to paralyze prey, such as cockroaches and flies. Each of their body segments bears a single pair of legs.

► Figure 17.21 **Millipedes and centipedes.**

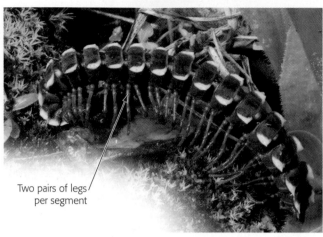

Two pairs of legs per segment

One pair of legs per segment

Millipede. Like most millipedes, this blue millipede has an elongated body with two pairs of legs per trunk segment.

Centipede. Centipedes can be found in dirt and leaf litter. Their venomous claws can harm cockroaches and spiders, but not humans.

Insect Anatomy

Like the grasshopper in **Figure 17.22**, most **insects** have a three-part body: head, thorax, and abdomen. The head usually bears a pair of sensory antennae and a pair of eyes. The mouthparts of insects are adapted for particular kinds of eating—for example, for biting and chewing plant material in grasshoppers; for lapping up fluids in houseflies; and for piercing skin and sucking blood in mosquitoes. Most adult insects have three pairs of legs and one or two pairs of wings, all extending from the thorax.

Flight is obviously one key to the great success of insects. An animal that can fly can escape many predators, find food and mates, and disperse to new habitats much faster than an animal that must crawl on the ground. Because their wings are extensions of the exoskeleton and not true appendages, insects can fly without sacrificing legs. By contrast, the flying vertebrates—birds and bats—have one of their two pairs of legs modified for wings, which explains why these vertebrates are generally not very swift on the ground.

▼ Figure 17.22 **Anatomy of a grasshopper.**

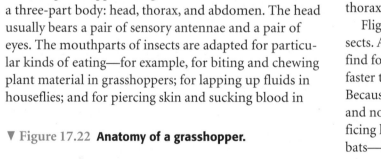

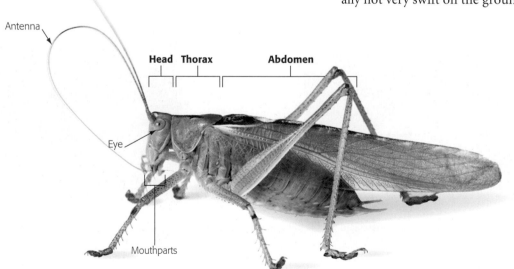

Antenna

Head Thorax Abdomen

Eye

Mouthparts

Insect Diversity

In species diversity, insects outnumber all other forms of life combined (Figure 17.23). They live in almost every terrestrial habitat and in fresh water, and flying insects fill the air. Insects are rare in the seas, where crustaceans are the dominant arthropods. The oldest insect fossils date back to about 400 million years ago. Later, the evolution of flight sparked an explosion in insect variety.

▼ Figure 17.23 **Insect diversity.**

Peacock katydid. The eyespots on the wings of this katydid may startle predators.

Leaf roller. Unlike most moths, the Costa Rican leaf roller moth is active during the daytime.

Banded Orange Heliconian. The range of this butterfly extends from Brazil to Kansas, where it can be seen feeding on the nectar of flowers in summertime.

Giraffe weevil. This species is found in Madagascar. Only males have an extended neck.

Praying mantis. There are over 2,000 species of praying mantises throughout the world.

Yellow jacket wasp. Common in North America, yellow jackets are important insect predators.

Leaf beetle. This beetle belongs to the family of leaf beetles, the most commonly encountered beetles in North America.

Longhorn beetle. The longhorn or Titan beetle is the largest known beetle in the Amazon rain forest and among the largest insects in the world.

Many insects undergo metamorphosis in their development. In the case of grasshoppers and some other insect groups, the young resemble adults but are smaller and have different body proportions. The animal goes through a series of molts, each time looking more like an adult, until it reaches full size. In other cases, insects have distinctive larval stages specialized for eating and growing that are known by such names as maggots (fly larvae) or caterpillars (larvae of moths and butterflies). The larval stage looks entirely different from the adult stage, which is specialized for dispersal and reproduction. Metamorphosis from the larva to the adult occurs during a pupal stage (**Figure 17.24**).

Doctors can use maggots—specifically, juvenile blowflies (*Phaenicia sericata*)—to cleanse infected wounds. This practice is becoming increasingly common for the treatment of bedsores, burns, surgical wounds, and traumatic injuries. It is safe and effective because the enzymes secreted by the immature flies break down only dead or dying tissue, leaving healthy tissue unharmed.

Animals so numerous, diverse, and widespread as insects are bound to affect the lives of all other terrestrial organisms, including humans, in many ways. On one hand, we depend on bees, flies, and other insects to pollinate our crops and orchards. On the other hand, insects are carriers of the microbes that cause many human diseases, such as malaria and West Nile disease. Insects also compete with humans for food by eating our field crops. Trying to minimize their losses, farmers in the United States spend billions of dollars each year on pesticides, spraying crops with massive doses of insecticide poisons. But try as they might, not even humans have significantly challenged the preeminence of insects and their arthropod kin. As Cornell University's Thomas Eisner puts it: "Bugs are not going to inherit the Earth. They own it now. So we might as well make peace with the landlord." ☑

▼ Figure 17.24 **Metamorphosis of a monarch butterfly.**

The **larva (caterpillar)** spends its time eating and growing, molting as it grows.

After several molts, the larva becomes a **pupa** encased in a cocoon.

Within the pupa, the larval organs break down and adult organs develop from cells that were dormant in the larva.

Finally, the **adult** emerges from the cocoon.

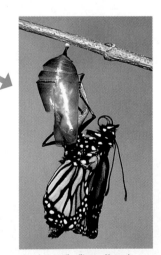

The butterfly flies off and reproduces, nourished mainly by calories stored when it was a caterpillar.

Echinoderms

The **echinoderms** (phylum Echinodermata) are named for their spiny surfaces (*echin* is Greek for "spiny"). Among the echinoderms are sea stars, sea urchins, sea cucumbers, and sand dollars (**Figure 17.25**).

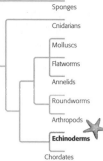

Echinoderms include about 7,000 species, all of them marine. Most move slowly (if at all). Echinoderms lack body segments, and most have radial symmetry as adults. Both the external and the internal parts of a sea star, for instance, radiate from the center like spokes of a wheel. In contrast to the adult, the larval stage of echinoderms is bilaterally symmetrical. This supports other evidence that echinoderms are not closely related to other radial animals, such as cnidarians, that never show bilateral symmetry. Most echinoderms have an **endoskeleton** (interior skeleton) constructed from hard plates just beneath the skin. Bumps and spines of this endoskeleton account for the animal's rough or prickly surface. Unique to echinoderms is the **water vascular system**, a network of water-filled canals that circulate water throughout the echinoderm's body, facilitating gas exchange (the entry of O_2 and the removal of CO_2) and waste disposal. The water vascular system also branches into extensions called tube feet. A sea star or sea urchin pulls itself slowly over the seafloor using its suction-cup-like tube feet. Sea stars also use their tube feet to grip prey during feeding.

Looking at sea stars and other adult echinoderms, you may think they have little in common with humans and other vertebrates. But as shown by the phylogenetic tree at the left, echinoderms share an evolutionary branch with chordates, the phylum that includes vertebrates. Analysis of embryonic development can differentiate the echinoderms and chordates from the evolutionary branch that includes molluscs, flatworms, annelids, roundworms, and arthropods. With this context in mind, we're now ready to make the transition from invertebrates to vertebrates. ☑

☑CHECKPOINT

Contrast the skeleton of an echinoderm with that of an arthropod.

Answer: An echinoderm has an endoskeleton; an arthropod has an exoskeleton.

▼ Figure 17.25 **Echinoderm diversity.**

Sea star. When a sea star encounters an oyster or clam, it grips the mollusc's shell with its tube feet (see inset) and positions its mouth next to the narrow opening between the two halves of the prey's shell. The sea star then pushes its stomach out through its mouth and the crack in the mollusc's shell.

Sea urchin. In contrast to sea stars, sea urchins are spherical and have no arms. If you look closely, you can see the long tube feet projecting among the spines. Unlike sea stars, which are mostly carnivorous, sea urchins mainly graze on seaweed and other algae.

Tube feet

Sea cucumber. On casual inspection, this California sea cucumber does not look much like other echinoderms. However, a closer look reveals many echinoderm traits, including five rows of tube feet.

Sand dollar. Live sand dollars have a skin of movable spines covering a rigid skeleton. A set of five pores (arranged in a star pattern) allows seawater to be drawn into the sand dollar's body.

Vertebrate Evolution and Diversity

Most of us are curious about our family ancestry. Biologists are also interested in the larger question of tracing human ancestry within the animal kingdom. In this section, we trace the evolution of the vertebrates, the group that includes humans and their closest relatives. All vertebrates have endoskeletons, a characteristic shared with most echinoderms. However, vertebrate endoskeletons are unique in having a cranium (skull) and a backbone, a series of bones called vertebrae (singular, vertebra), for which the group is named (**Figure 17.26**). Our first step in tracing the vertebrate lineage is to determine where vertebrates fit in the animal kingdom.

▼ Figure 17.26 **A vertebrate endoskeleton.** This snake skeleton, like those of all vertebrates, has a cranium (skull) and a backbone consisting of vertebrae.

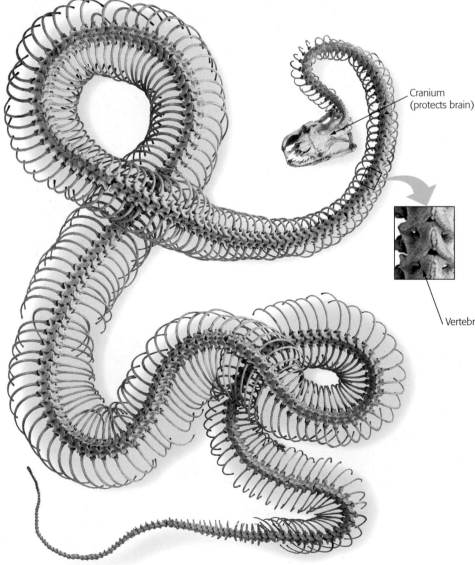

Cranium
(protects brain)

Vertebra

Characteristics of Chordates

The last phylum in our survey of the animal kingdom is the phylum Chordata. **Chordates** share four key features that appear in the embryo and sometimes in the adult (**Figure 17.27**). These four chordate hallmarks are (1) a **dorsal, hollow nerve cord**; (2) a **notochord**, which is a flexible, longitudinal rod located between the digestive tract and the nerve cord; (3) **pharyngeal slits**, which are grooves in the pharynx, the region of the digestive tube just behind the mouth; and (4) a **post-anal tail**, which is a tail to the rear of the anus. Though these chordate characteristics are often difficult to recognize in the adult animal, they are always present in chordate embryos. For example, the notochord, for which our phylum is named, persists in adult humans only in the form of the cartilage disks that function as cushions between the vertebrae. (Back injuries described as "ruptured disks" or "slipped disks" refer to these structures.)

Body segmentation is another chordate characteristic. Chordate segmentation is apparent in the backbone of vertebrates (see Figure 17.26) and is also evident in the segmental muscles of all chordates (see the chevron-shaped—>>>>—muscles in the lancelet in Figure 17.28). Segmental musculature is not so obvious in adult humans unless one is motivated enough to sculpt those "washboard abs."

Three groups of chordates are invertebrates. Two of these groups, **tunicates** and **lancelets**, have no cranium (**Figure 17.28**). The third group— hagfishes—have a cranium. (Hagfishes will be described shortly.) All other

Sponges

Cnidarians

Molluscs

Flatworms

Annelids

Roundworms

Arthropods

Echinoderms

Chordates

▼ Figure 17.27 **Chordate characteristics.**

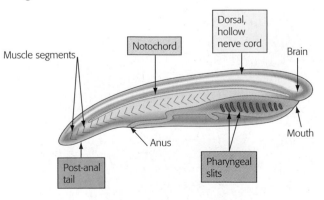

Muscle segments

Notochord

Dorsal, hollow nerve cord

Brain

Mouth

Pharyngeal slits

Anus

Post-anal tail

chordates are **vertebrates**, which retain the basic chordate characteristics but have additional features that are unique—including, of course, the backbone. **Figure 17.29** is an overview of chordate and vertebrate evolution that will provide a context for our survey. ☑

► Figure 17.28 **Chordates that have no cranium or vertebrae.**

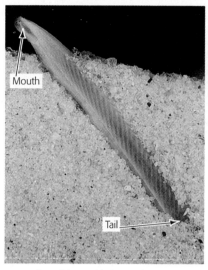

Mouth

Tail

Lancelet. This marine invertebrate owes its name to its bladelike shape. Only a few centimeters long, lancelets wiggle backward into the sand, leaving their mouth exposed, and filter tiny food particles from the seawater.

Tunicates. The tunicate, or sea squirt, is a stationary animal that filters food from the water. These pastel sea squirts get their nickname from their coloration and the fact that they can quickly expel water to startle intruders.

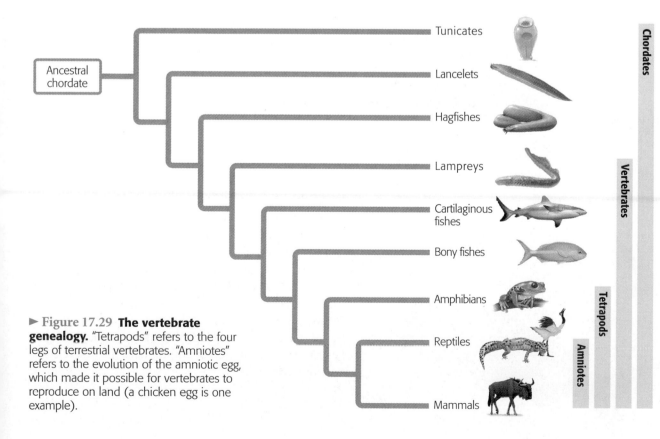

Ancestral chordate

Tunicates

Lancelets

Hagfishes

Lampreys

Cartilaginous fishes

Bony fishes

Amphibians

Reptiles

Mammals

Chordates

Vertebrates

Tetrapods

Amniotes

► Figure 17.29 **The vertebrate genealogy.** "Tetrapods" refers to the four legs of terrestrial vertebrates. "Amniotes" refers to the evolution of the amniotic egg, which made it possible for vertebrates to reproduce on land (a chicken egg is one example).

☑**CHECKPOINT**

During our early embryonic development, what four features do we share with invertebrate chordates such as lancelets?

Answer: (1) dorsal, hollow nerve cord; (2) notochord; (3) pharyngeal slits; (4) post-anal tail

355

Fishes

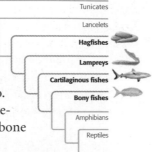

The first vertebrates were aquatic and probably evolved during the early Cambrian period about 542 million years ago. In contrast with most other vertebrates, they lacked jaws, hinged bone structures that work the mouth.

Two types of jawless fishes survive today: hagfishes and lampreys. As noted earlier, hagfishes are invertebrates but have a cranium. Present-day hagfishes scavenge dead or dying animals on the cold, dark seafloor. When threatened, a hagfish exudes an enormous amount of slime from special glands on the sides of its body (Figure 17.30a). Recently, hagfishes have become endangered because their skin is used to make "eel-skin" belts, purses, and boots. Unlike hagfishes, lampreys are vertebrates. Some lampreys are parasites that use their jawless mouths as suckers to attach to the sides of large fish and draw blood (Figure 17.30b).

We know from the fossil record that the first jawed vertebrates were fishes that evolved about 470 million years ago. They had two pairs of fins, making them agile swimmers. Some early fishes were active predators up to 10 m in length that could chase prey and bite off chunks of flesh. Even today, most fishes are carnivores.

Cartilaginous fishes, such as sharks and rays, have a flexible skeleton made of cartilage (Figure 17.30c). Most sharks are adept predators because they are fast swimmers with streamlined bodies, acute senses, and powerful jaws. A shark does not have keen eyesight, but its sense of smell is very sharp. In addition, special electrosensors on the head can detect minute electrical fields produced by muscle contractions in nearby animals. Sharks also have a **lateral line system**, a row of sensory organs running along each side of the body. Sensitive to changes in water pressure, the lateral line system enables a shark to detect minor vibrations caused by animals swimming in its neighborhood. There are about 750 living species of cartilaginous fishes, nearly all of them marine.

Most fishes—about 27,000 species—are **bony fishes** with skeletons reinforced by calcium (Figure 17.30d). They also have a lateral line system, a keen sense of smell, and excellent eyesight. On each side of the head, a protective flap called the **operculum** (plural, *opercula*) covers a chamber housing the gills, feathery external organs that extract oxygen from water. Movement of the operculum allows the fish to breathe without swimming. By contrast, sharks lack opercula and must swim to pass water over their gills. The need to move water over the gills is why a shark must keep moving to stay alive. Also unlike sharks, bony fishes have an organ that helps keep them buoyant—the **swim bladder**, a gas-filled sac. Thus, many bony fishes can conserve energy by remaining almost motionless, in contrast to sharks, which sink if they stop swimming. Some bony fishes have a connection between the swim bladder and the digestive tract that enables them to gulp air and extract oxygen from it when the oxygen level in the water gets too low.

Bony fishes are common in the seas and in freshwater habitats. Most bony fishes, including trout, bass, perch, and tuna, are **ray-finned fishes**. Their fins are supported by thin, flexible skeletal rays. A second evolutionary branch includes the **lobe-finned fishes**. The lobe-fins are named for their muscular fins supported by stout bones. One lineage of lobe-finned fishes, **lungfishes**, lives today in the Southern Hemisphere. Lungfishes inhabit stagnant ponds and swamps, surfacing to gulp air into their lungs. A second lineage of lobe-finned fishes is represented today by the coelacanth, a deep-sea dweller that was believed to have been extinct for millions of years before being "rediscovered" in 1938. Coelacanths may use their fins to waddle along the seafloor. A third lineage of lobe-finned fishes migrated out of fresh water and adapted to life on land, playing a key role in the evolution of amphibians, the first terrestrial vertebrates. ✓

☑CHECKPOINT

A shark has a _____ skeleton, whereas a tuna has a _____ skeleton.

Answer: cartilaginous; bony

▼ Figure 17.30 **Fish diversity.**

(a) **Hagfish** (inset: slime)

(b) **Lamprey** (inset: mouth)

(c) **Shark, a cartilaginous fish**

Lateral line

Operculum

(d) **Bony fish**

Amphibians

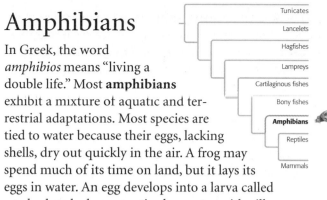

In Greek, the word *amphibios* means "living a double life." Most **amphibians** exhibit a mixture of aquatic and terrestrial adaptations. Most species are tied to water because their eggs, lacking shells, dry out quickly in the air. A frog may spend much of its time on land, but it lays its eggs in water. An egg develops into a larva called a tadpole, a legless, aquatic algae-eater with gills, a lateral line system resembling that of fishes, and a long-finned tail. In changing into a frog, the tadpole undergoes a radical metamorphosis (**Figure 17.31a**). When a young frog crawls onto shore and begins life as a terrestrial insect-eater, it has four legs, air-breathing lungs instead of gills, external eardrums, and no lateral line system. But even as adults, amphibians are most abundant in damp habitats, such as swamps and rain forests. This is partly because amphibians depend on their moist skin to supplement lung function in exchanging gases with the environment. Thus, even those frogs that are adapted to relatively dry habitats spend much of their time in humid burrows or under piles of moist leaves. The amphibians of today, including frogs and salamanders, account for about 12% of all living vertebrates, or about 6,000 species (**Figure 17.31b**).

Amphibians were the first vertebrates to colonize land. They descended from fishes that had lungs and fins with muscles and skeletal supports strong enough to enable some movement, however clumsy, on land (**Figure 17.32**). The fossil record chronicles the evolution of four-limbed amphibians from fishlike ancestors. Terrestrial vertebrates—amphibians, reptiles, and mammals—are collectively called **tetrapods**, which means "four feet." ☑

▼ Figure 17.31 **Amphibian diversity.**

(a) Tadpole and adult golden palm tree frog

Red-eyed tree frog
(b) Frogs and salamanders: the two major groups of amphibians

Texas barred tiger salamander

▼ Figure 17.32 **The origin of tetrapods.**

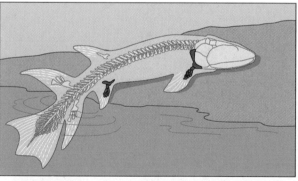

Lobe-finned fish. Fossils of some lobe-finned fishes have skeletal supports extending into their fins.

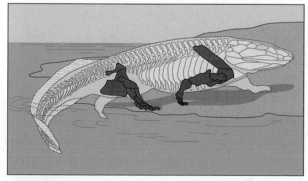

Early amphibian. Fossils of early amphibians have limb skeletons that probably functioned in helping them move on land.

Reptiles

Reptiles (including birds) and mammals are **amniotes**. The evolution of amniotes from an amphibian ancestor included many adaptations for living on land. The adaptation that gives the group its name is the **amniotic egg**, a fluid-filled egg inside of which the embryo develops. The amniotic egg functions as a self-contained "pond" that enables amniotes to complete their life cycle on land.

The **reptiles** include snakes, lizards, turtles, crocodiles, alligators, and birds, along with a number of extinct groups, including most of the dinosaurs (**Figure 17.33**). The bull snake in Figure 17.33 displays two reptilian adaptations to living on land: scaled waterproof skin, preventing dehydration in dry air, and amniotic eggs with shells, providing a watery, nutritious internal environment where the embryo can develop. These adaptations allowed reptiles to break their ancestral ties to aquatic habitats. Reptiles cannot breathe through their dry skin and so obtain most of their oxygen through their lungs.

Tunicates
Lancelets
Hagfishes
Lampreys
Cartilaginous fishes
Bony fishes
Amphibians
Reptiles
Mammals

Nonbird Reptiles

Nonbird reptiles are sometimes referred to as "cold-blooded" animals because they do not use their metabolism extensively to control body temperature. Reptiles do regulate body temperature, but largely through behavioral adaptations. For example, many lizards regulate their internal temperature by basking in the sun when the air is cool and seeking shade when the air is too warm. Because lizards and other nonbird reptiles absorb external heat rather than generating much of their own, they are said to be **ectotherms**, a term more accurate than "cold-blooded." By heating directly with solar energy rather than through the metabolic breakdown of food, a nonbird reptile can survive on less than 10% of the calories required by a mammal of equivalent size.

As successful as reptiles are today, they were far more widespread, numerous, and diverse during the Mesozoic era, which is sometimes known as the "age of reptiles." Reptiles diversified extensively during that era, producing a dynasty that lasted until about 65 million years ago. Dinosaurs, the most diverse reptile group, included the largest animals ever to inhabit land. Some were gentle giants that lumbered about while browsing vegetation. Others were voracious carnivores that chased their larger prey on two legs.

The age of reptiles began to fade about 70 million years ago. Around that time, the global climate became cooler and more variable (see Chapter 14). This was a period of mass extinctions that claimed all the dinosaurs by about 65 million years ago, except for one lineage. That lone surviving lineage is represented today by the reptilian group we know as birds. ☑

☑**CHECKPOINT**

What is an amniotic egg?

Answer: a shelled egg surrounding fluid that contains an embryo

▼ Figure 17.33 **Reptile diversity.**

Snake. Like all snakes, bull snakes lay eggs. Bull snakes are commonly found in North American prairies, where their coloration helps them blend into the background.

Turtle. A fully grown Galápagos tortoise can weigh more than 600 pounds and live for more than 150 years.

Lizard. The leopard gecko, a popular pet, is native to the deserts of the Middle East.

Birds

Although **birds** were previously placed in their own class—class Aves—recent genetic evidence shows that they are reptiles, having evolved from a lineage of small, two-legged dinosaurs called theropods. But today birds look quite different from reptiles because of their feathers and other distinctive flight equipment. Almost all of the 10,000 living bird species are airborne. The few flightless species, including the ostrich and the penguin, evolved from flying ancestors.

Appreciating the avian world is all about understanding flight. Almost every element of bird anatomy is modified in some way that enhances flight. The bones have a honeycombed structure that makes them strong but light (the wings of airplanes have the same basic construction). For example, a huge seagoing species called the frigate bird has a wingspan of more than 2 m, but its whole skeleton weighs only about 113 g (4 oz). Another adaptation that reduces the weight of birds is the absence of some internal organs found in other vertebrates. Female birds, for instance, have only one ovary instead of a pair. Also, today's birds are toothless, an adaptation that trims the weight of the head (preventing uncontrolled nosedives). Birds do not chew food in the mouth but grind it in the gizzard, a chamber of the digestive tract near the stomach.

Flying requires a great expenditure of energy and an active metabolism. Unlike other reptiles, birds are **endotherms**, meaning they use their own metabolic heat to maintain a warm, constant body temperature.

A bird's most obvious flight equipment is its wings. Bird wings are airfoils that illustrate the same principles of aerodynamics as the wings of an airplane **(Figure 17.34)**. A bird's flight motors are its powerful breast muscles, which are anchored to a keel-like breastbone. It is mainly these flight muscles that we call "white meat" on poultry. Some birds, such as eagles and hawks, have wings adapted for soaring on air currents and flap their wings only occasionally. Other birds, including hummingbirds, excel at maneuvering but must flap continuously to stay aloft. Feathers are made of the protein that forms the scales of reptiles. Feathers may have functioned first as insulation, helping birds retain body heat, only later being adapted as flight gear. ☑

☑**CHECKPOINT**

Birds differ from other reptiles in their main source of body heat, with birds being _____ and other reptiles being _____.

Answer: endotherms; ectotherms

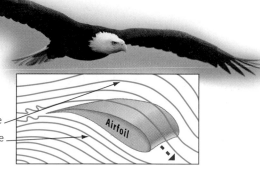

► Figure 17.34 **The aerodynamics of a bald eagle in flight.** Both birds and airplanes owe their "lift" to changes in air pressure caused by the shape of their wings.

Lower air pressure

Higher air pressure

Airfoil

Crocodile. Found throughout central and southern Africa, the Nile crocodile can grow up to 20 feet long and weigh nearly 2,000 pounds.

Birds. These large birds, called red-crowned cranes, are highly endangered, with fewer than 2,000 left in the wild, mostly in China.

Dinosaur. This *Herrarasaurus* skeleton is from a carnivorous biped discovered in Argentina.

Mammals

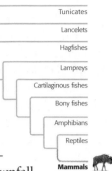

There are two major lineages of amniotes: one that led to the reptiles and one that produced the mammals. The first **mammals** arose about 200 million years ago and were probably small, nocturnal insect-eaters. Mammals became much more diverse after the downfall of the dinosaurs. Most mammals are terrestrial. These include nearly 1,000 species of winged mammals, the bats. And about 80 species of dolphins, porpoises, and whales are totally aquatic. The blue whale, an endangered mammal that grows to lengths of nearly 30 m, is the largest animal that has ever lived. Two features—hair and mammary glands that produce milk that nourishes the young—are mammalian hallmarks. The main function of hair is to insulate the body and help maintain a warm, constant internal temperature; mammals, like birds, are endotherms.

The major groups of mammals are monotremes, marsupials, and eutherians (**Figure 17.35**). The duck-billed platypus and the echidna are the only existing species of **monotremes**, egg-laying mammals. The platypus lives along rivers in eastern Australia and on the nearby island of Tasmania. The female usually lays two eggs and incubates them in a leaf nest. After hatching, the young nurse by licking up milk secreted onto the mother's fur.

Most mammals are born rather than hatched. During pregnancy in marsupials and eutherians, the embryos are nurtured inside the mother by an organ called the **placenta**. Consisting of both embryonic and maternal tissues, the placenta joins the embryo to the mother within the uterus. The embryo is nurtured by maternal blood that flows close to the embryonic blood system in the placenta.

Marsupials, the so-called pouched mammals, include kangaroos, koalas, and opossums. These mammals have a brief pregnancy and give birth to tiny embryonic offspring that complete development while attached to the mother's nipples. The nursing young are usually housed in an external pouch on the mother's abdomen. Nearly all marsupials live in Australia, New Zealand, and North and South America. Australia has been a marsupial sanctuary for much of the past 60 million years. Australian marsupials have diversified extensively, filling terrestrial habitats that on the other continents are occupied by eutherian mammals.

Eutherians are also called **placental mammals** because their placentas provide a more intimate and longer-lasting association between the mother and her developing young than do marsupial placentas. Eutherians make up almost 95% of the 5,300 species of living mammals. Dogs, cats, cows, rodents, rabbits, bats, and whales are all examples of eutherian mammals. One of the eutherian groups is the primates, which include monkeys, apes, and humans. ☑

▼ Figure 17.35 **Mammalian diversity.**

MAJOR GROUPS OF MAMMALS

Monotremes (hatched from eggs)	**Marsupials** (embryonic at birth)	**Eutherians** (fully developed at birth)

Monotremes, such as this echidna, are the only mammals that lay eggs (inset).

The young of marsupials are born very early in their development. The newborn kangaroo (inset) will finish its growth while nursing from a nipple in its mother's pouch.

In eutherians (placental mammals), young develop within the uterus of the mother. There they are nurtured by the flow of blood through the dense network of vessels in the placenta. This newborn wildebeest is coated by remnants of the placenta.

The Human Ancestry

We have now traced animal phylogeny to the **primates**, the mammalian group that includes *Homo sapiens* and its closest kin. To understand what that means, we must follow our ancestry back to the trees, where some of our most treasured traits originated.

The Evolution of Primates

Primate evolution provides a context for understanding human origins. The fossil record supports the hypothesis that primates evolved from insect-eating mammals during the late Cretaceous period, about 65 million years ago. Those early primates were small, arboreal (tree-dwelling) mammals. Thus, primates were first distinguished by characteristics that were shaped, through natural selection, by the demands of living in the trees. For example, primates have limber shoulder joints, which make it possible to swing from branch to branch. The agile hands of primates can hang on to branches and manipulate food. Nails have replaced claws in many primate species, and the fingers are very sensitive. The eyes of primates are close together on the front of the face. The overlapping fields of vision of the two eyes enhance depth perception, an obvious advantage when

swinging in trees. Excellent eye-hand coordination is also important for arboreal maneuvering. Parental care is essential for young animals in the trees. Mammals devote more energy to caring for their young than most other vertebrates, and primates are among the most attentive parents of all mammals. Most primates have single births and nurture their offspring for a long time. Although humans do not live in trees, we retain in modified form many traits that originated there.

Taxonomists divide the primates into three main groups (**Figure 17.36**). The first includes lorises, pottos, and lemurs. These primates live in Madagascar, Africa, and southern Asia. Tarsiers, small nocturnal tree-dwellers found only in Southeast Asia, form the second group of primates. The third group of primates, **anthropoids**, includes monkeys, apes, and humans. All monkeys in the New World (the Americas) are arboreal and are distinguished by prehensile (grasping) tails that function as an extra appendage for swinging. (If you see a monkey in a zoo swinging by its tail, you know it's from the New World.) Although some Old World (African and Asian) monkeys are also arboreal, their tails are not prehensile. And many Old World monkeys, including baboons, macaques, and mandrills,

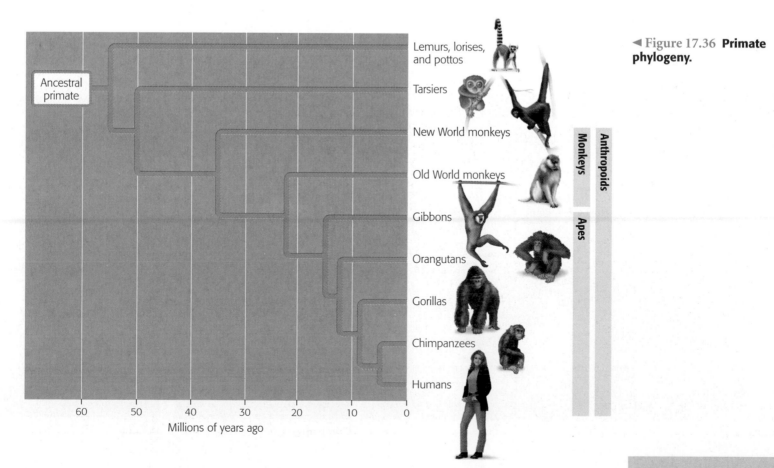

◀ Figure 17.36 **Primate phylogeny.**

Lemurs, lorises, and pottos

Tarsiers

New World monkeys

Old World monkeys

Gibbons

Orangutans

Gorillas

Chimpanzees

Humans

Ancestral primate

Monkeys

Anthropoids

Apes

60   50   40   30   20   10   0
Millions of years ago

are mainly ground-dwellers. Anthropoids also have a fully opposable thumb; that is, they can touch the tips of all four fingers with their thumb.

Our closest anthropoid relatives are the nonhuman apes: gibbons, orangutans, gorillas, and chimpanzees. They live only in tropical regions of the Old World. Except for some gibbons, apes are larger than monkeys, with relatively long arms, short legs, and no tail. Although all apes are capable of living in trees, only gibbons and orangutans are primarily arboreal. Gorillas and chimpanzees are highly social. Apes have larger brains proportionate to body size than monkeys, and their behavior is more adaptable. And, of course, the apes include humans. **Figure 17.37** shows examples of primates.

▼ Figure 17.37 **Primate diversity.**

Ring-tailed lemur   **Tarsier**

Black spider monkey
(New World monkey)

Orangutan (ape)

Patas monkey (Old World monkey)

Gibbon (ape)

Gorilla (ape)   **Chimpanzee** (ape)   **Human**

The Emergence of Humankind

Humanity is one very young twig on the tree of life. In the continuum of life spanning 3.5 billion years, the fossil record and molecular systematics indicate that humans and chimpanzees have shared a common African ancestry for all but the last 5–7 million years (see Figure 17.36). Put another way, if we compressed the history of life to a year, the human branch has existed for only 18 hours.

Some Common Misconceptions

Certain misconceptions about human evolution persist in the minds of many, long after these myths have been debunked by the fossil evidence. Let's first dispose of the myth that our ancestors were chimpanzees or any other modern ape. Chimpanzees and humans represent two divergent branches of the anthropoid tree that evolved from a common, less specialized ancestor. Chimps are not our parent species, but more like our phylogenetic siblings or cousins.

Another misconception envisions human evolution as a ladder with a series of steps leading directly from an ancestral anthropoid to *Homo sapiens*. This is often illustrated as a parade of fossil **hominins** (members of the human family) becoming progressively more modern as they march across the page. If human evolution is a parade, then it is a disorderly one, with many splinter groups having traveled down dead ends. At times in hominin history, several human species coexisted (**Figure 17.38**). Human phylogeny is more like a multi-branched bush than a ladder, with our species being the tip of the only twig that still lives.

One more myth is that various human characteristics, such as upright posture and an enlarged brain, evolved in unison. A popular image is of early humans as half-stooped, half-witted cave-dwellers. In fact, the fossil record reveals that different human features evolved at different rates, with erect posture, or bipedalism, leading the way. Our pedigree includes ancestors who walked upright but had ape-sized brains.

After dismissing some of the folklore on human evolution, however, we must admit that many questions about our ancestry have not yet been resolved.

Human Evolution

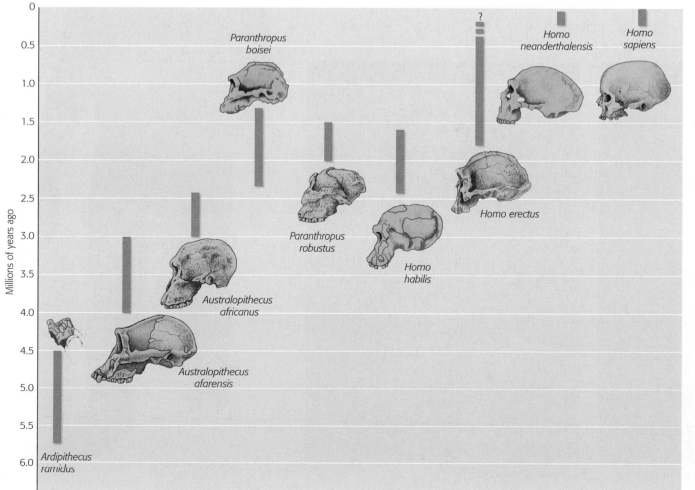

◄ **Figure 17.38 A time line of human evolution.** Notice that there have been times when two or more hominin species coexisted. The skulls are all drawn to the same scale so you can compare the sizes of craniums and hence brains.

Australopithecus and the Antiquity of Bipedalism

Before there was the genus *Homo*, several hominin species of the genus *Australopithecus* walked the African savanna. Fossil evidence indicates that bipedalism first arose in *Australopithecus afarensis* at least 4 million years ago (Figure 17.39).

One of the most complete fossil skeletons of the species *Australopithecus afarensis* dates to about 3.2 million years ago in East Africa. Nicknamed Lucy by her discoverers, the individual was a female, only about 3 feet tall and with a head about the size of a softball. Lucy and her kind lived in savanna areas and may have subsisted on nuts and seeds, bird eggs, and whatever animals they could catch or scavenge from kills made by more efficient predators such as large cats and dogs.

Homo habilis and the Evolution of Inventive Minds

Enlargement of the human brain is first evident in fossils from East Africa dating to about 2.4 million years ago. Thus, the fundamental human trait of an enlarged brain evolved a few million years after the other major human trait—bipedalism. As evolutionary biologist Stephen Jay Gould put it, "Mankind stood up first and got smart later."

Anthropologists have found skulls with brain capacities intermediate in size between those of the latest *Australopithecus* species and those of *Homo sapiens*. Simple handmade stone tools are sometimes found with the larger-brained fossils, which have been dubbed *Homo habilis* ("handy man"). After walking upright for about 2 million years, humans were finally beginning to use their manual dexterity and big brains to invent tools that enhanced their hunting, gathering, and scavenging on the African savanna.

▼ Figure 17.39 **The antiquity of upright posture.**

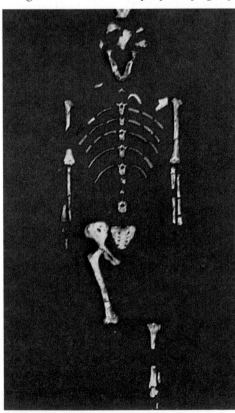

(a) *Australopithecus afarensis* **skeleton**

(b) **Ancient footprints**

(c) **Model of an** *Australopithecus afarensis* **male**

Homo erectus and the Global Dispersal of Humanity

The first species to extend humanity's range from Africa to other continents was *Homo erectus*, perhaps a descendant of *Homo habilis*. Skeletons of *Homo erectus* dating to 1.8 million years ago found in the former Soviet republic of Georgia represent the oldest known fossils of hominins outside Africa. *Homo erectus* migrated to populate many regions of Asia and Europe, eventually moving as far as Indonesia. Remember the "hobbits" (*Homo floresiensis*) from the Biology and Society section? Some anthropologists speculate that they may actually be a long-isolated group of *Homo erectus*.

Homo erectus was taller than *Homo habilis* and had a larger brain capacity. During the 1.5 million years the species existed, the *Homo erectus* brain increased to as large as 1,200 cm^3, comparable to modern humans. Intelligence enabled humans to continue succeeding in Africa and also to survive in the colder climates of the north. *Homo erectus* resided in huts or caves, built fires, made clothes from animal skins, and designed stone tools. In anatomical and physiological adaptations, *Homo erectus* was poorly equipped for life outside the tropics, but made up for the deficiencies with cleverness and social cooperation.

Some African, Asian, European, and Australasian (from Indonesia, New Guinea, and Australia) populations of *Homo erectus* gave rise to regionally diverse descendants that had even larger brains. Among these descendants of *Homo erectus* was *Homo neanderthalensis* (the Neanderthals), who lived in Europe, the Middle East, and Asia from about 200,000 years ago to about 30,000 years ago. (They are named Neanderthals because their fossils were first found in the Neander Valley of Germany.) Compared with us, Neanderthals had slightly heavier browridges and less pronounced chins, but their brains were similar to ours. Neanderthals were skilled toolmakers, and they participated in burials and other rituals that required abstract thought. From ancient bones, scientists can tell the shape of Neanderthal bodies. But given their prehistoric extinction, is it possible to infer their coloration? ☑

THE PROCESS OF SCIENCE: Human Evolution

What Did Neanderthals Look Like?

To try to put a face to Neanderthals, an international team of geneticists began with the **observation** that a gene called *mc1r* has a large effect on the hair and skin pigmentation of modern humans. This led the scientists to **question** what form of the gene would be found in Neanderthals. They formed the **hypothesis** that determining the DNA sequence of this gene would lend insight into the physical appearance of Neanderthals.

Their **experiment** used the polymerase chain reaction (see Chapter 12) to sequence the *mc1r* pigment gene in DNA isolated from Neanderthal bones. Their **results** were surprising: The Neanderthal gene contained a mutation not found in any modern humans. Next, the scientists genetically engineered modern human cells to carry the Neanderthal version of *mc1r*. They found that the mutation impaired the activity of a plasma membrane receptor protein that helps regulate the balance between skin pigments. Anyone with two identical alleles for this mutation (estimated to include at least 1 in 100 Neanderthals and possibly many more) would have red hair and pale skin (**Figure 17.40**).

Interestingly, this coloration is seen in modern Europeans who live in the Neanderthal's old territory. In less sunny regions, pale skin facilitates the production of the essential nutrient vitamin D. Thus, the Neanderthals and *Homo sapiens* who coexisted in Europe may have looked more alike than previously realized.

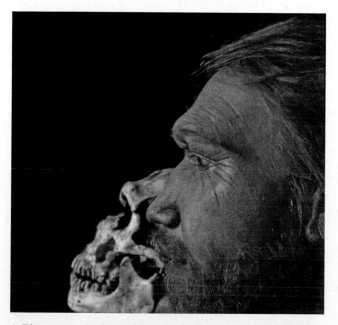

▲ Figure 17.40 **An artist's rendering of a Neanderthal.**

The Origin and Dispersal of *Homo sapiens*

The oldest known fossils of our own species, *Homo sapiens*, were discovered in Ethiopia and date from 160,000 to 195,000 years ago. These early humans lacked the heavy browridges of *Homo erectus* and *Homo neanderthalensis* and were more slender, suggesting that they belong to a distinct lineage. The Ethiopian fossils support molecular evidence about the origin of humans: DNA studies strongly suggest that all living humans can trace their ancestry back to a single African *Homo sapiens* woman who lived 160,000 to 200,000 years ago.

Fossil evidence suggests that our species emerged from Africa in one or more waves, spreading first into Asia and then to Europe and Australia. The oldest fossils of *Homo sapiens* outside Africa date back about 50,000 years. The date of the first arrival of humans in the New World is uncertain, although the generally accepted evidence suggests a minimum of 15,000 years ago. ☑

Cultural Evolution

Certain uniquely human traits have allowed for the development of human societies. The primate brain continues to grow after birth, and the period of growth is longer for a human than for any other primate. The extended period of human development also lengthens the time parents care for their offspring, which contributes to the child's ability to benefit from the experiences of earlier generations. This is the basis of **culture**—social transmission of accumulated knowledge, customs, beliefs, and art over generations. The major means of this transmission is language, spoken and written.

There have been three major stages of cultural evolution. The first stage began with nomads who hunted and gathered food on the African grasslands 2 million years ago. They made tools, organized communal activities, and divided labor. Beautiful ancient art, such as the 30,000-year-old cave paintings shown in **Figure 17.41**, is just one example of our cultural roots in early societies. The second main stage of cultural evolution came with the development of agriculture in Africa, Eurasia, and the Americas about 10,000 to 15,000 years ago. Along with agriculture came permanent settlements and the first cities. The third major stage in our cultural evolution was the Industrial Revolution, which began in the 1700s. Since then, the development of new technology has escalated exponentially; the span of time from the flight of the Wright brothers to Neil Armstrong's walk on the moon was less than an average American's life span. And it took less than a decade for the Internet to transform commerce, communication, and education.

Through all this cultural evolution, from simple hunter-gatherers to high-tech societies, we have not changed biologically in any significant way. We are probably no more intelligent than our cave-dwelling ancestors. The know-how to build skyscrapers, computers, and spacecraft is stored not in our genes but in the cumulative product of hundreds of generations of human experience, passed along by parents, teachers, books, and electronic media.

▼ **Figure 17.41 Art history goes way back—and so does our fascination with and dependence on animal diversity.** Early artists created these remarkable images beginning about 30,000 years ago. Three cave explorers found this prehistoric art gallery in a cavern in southern France.

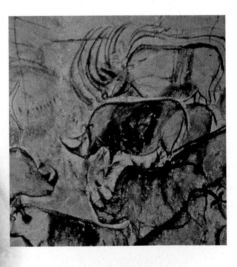

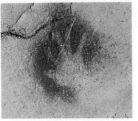

Nothing has had a greater impact on life on Earth than one species: *Homo sapiens*. The global consequences of human evolution have been enormous (Figure 17.42). Cultural evolution made *Homo sapiens* a new force in the history of life—a species that could defy its physical limitations and shortcut biological evolution. We do not have to wait to adapt to an environment through natural selection; we simply change the environment to meet our needs.

We are the most numerous and widespread of all large animals, and wherever we go we bring environmental change faster than many species can adapt; the rate of extinctions during the 1900s was 50 times greater than the average for the past 100,000 years. In the next unit, on ecology, we'll examine the interactions of humans—as well as other species—with the environment.

▶ Figure 17.42 **A clear-cut forest, an example of environmental change caused by *Homo sapiens*.**

EVOLUTION CONNECTION: Human Evolution

Recent Human Evolution

In the Biology and Society section, you learned about the discovery of a possible new human species called *Homo floresiensis*. And in the Process of Science section, you read about a genetic study involving our closest hominin relative, *Homo neanderthalensis*. Such investigations point out how evolution has shaped our own species after our split with the chimpanzees.

Biologists have identified a number of genes that appear to be undergoing rapid evolution in humans. Among them are genes involved in defense against malaria and tuberculosis and at least one gene regulating brain size. Another rapidly evolving human gene is called *FOXP2*. Several lines of evidence suggest that *FOXP2* functions in speech, a characteristic that obviously distinguishes humans from chimpanzees (Figure 17.43). For example, humans with mutations in this gene generally have severe speech impairment. Additionally, there is evidence that the *FOXP2* gene is expressed in the brains of zebra finches and canaries at the time when these songbirds are learning their songs. But the human version of *FOXP2* is unique, perhaps partially explaining our unique linguistic abilities. Further research has determined that the human form of *FOXP2* likely arose within the last 100,000 years, a time period that matches the emergence of *Homo sapiens*. Interestingly, the experiment described in the Process of Science section found that Neanderthals had the same version of *FOXP2* that we do.

▲ Figure 17.43 **Our closest primate relative: the chimpanzee.** The ability to speak is one trait that clearly differentiates humans from chimpanzees.

Chapter Review

The Origins of Animal Diversity

What Is an Animal?

Animals are eukaryotic, multicellular, heterotrophic organisms that obtain nutrients by ingestion. Most animals reproduce sexually and develop from a zygote to a blastula to a gastrula. After the gastrula stage, some develop directly into adults, whereas others pass through a larval stage.

Early Animals and the Cambrian Explosion

Animals probably evolved from a colonial, flagellated protist more than 700 million years ago. At the beginning of the Cambrian period, 542 million years ago, animal diversity increased rapidly.

Animal Phylogeny

Major branches of animal evolution are defined by two key evolutionary differences: the presence or absence of tissues and radial versus bilateral body symmetry. A body cavity (coelom) at least partly lined by mesoderm evolved in a number of later branches.

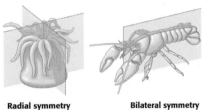

Radial symmetry   **Bilateral symmetry**

Major Invertebrate Phyla

This tree shows the eight major invertebrate phyla, as well as chordates, which include a few invertebrates.

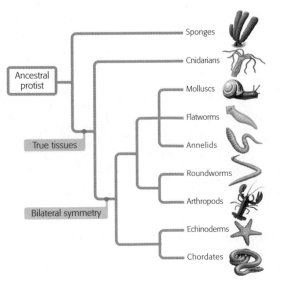

Sponges

Sponges (multiple phyla) are stationary animals with porous bodies but no true tissues. They feed by drawing water through pores in the sides of the body and trapping food particles in mucus.

Cnidarians

Cnidarians (phylum Cnidaria) have radial symmetry, a gastrovascular cavity with a single opening, and tentacles with stinging cnidocytes. The body is either a stationary polyp or a floating medusa.

Molluscs

Molluscs (phylum Mollusca) are soft-bodied animals often protected by a hard shell. The body has three main parts: a muscular foot, a visceral mass, and a fold of tissue called the mantle.

MOLLUSCS		
Gastropods	**Bivalves**	**Cephalopods**

Flatworms

Flatworms (phylum Platyhelminthes) are the simplest bilateral animals. They may be free-living (such as planarians) or parasitic (such as tapeworms).

Annelids

Annelids (phylum Annelida) are segmented worms with complete digestive tracts. They may be free-living or parasitic.

Roundworms

Roundworms, also called nematodes (phylum Nematoda), are unsegmented and cylindrical with tapered ends. They may be free-living or parasitic.

Arthropods

Arthropods (phylum Arthropoda) are segmented animals with an exoskeleton and specialized, jointed appendages.

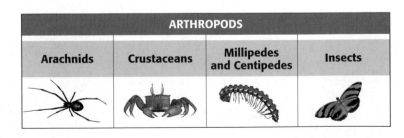

ARTHROPODS			
Arachnids	**Crustaceans**	**Millipedes and Centipedes**	**Insects**

Echinoderms

Echinoderms (phylum Echinodermata) are stationary or slow-moving marine animals that lack body segments and possess a unique water vascular system. Bilaterally symmetrical larvae usually change to radially symmetrical adults. Echinoderms have a bumpy endoskeleton.

Vertebrate Evolution and Diversity

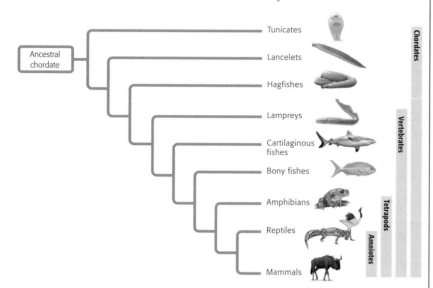

Characteristics of Chordates

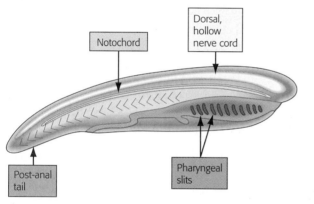

Tunicates and lancelets are invertebrate chordates. The vast majority of chordates are vertebrates, possessing a cranium and backbone.

Fishes

Hagfishes have skulls but lack vertebrae. Lampreys are jawless vertebrates. Cartilaginous fishes, such as sharks, are mostly predators with powerful jaws and a flexible skeleton made of cartilage. Bony fishes have a stiff skeleton reinforced by calcium. Bony fishes are further classified into ray-finned fishes and lobe-finned fishes (including lungfishes).

Amphibians

Amphibians are tetrapod vertebrates that usually deposit their eggs (lacking shells) in water. Aquatic larvae typically undergo a radical metamorphosis into the adult stage. Their moist skin requires that amphibians spend much of their adult life in humid environments.

Reptiles

Reptiles are amniotes, vertebrates that develop in a fluid-filled egg enclosed by a shell. Reptiles include terrestrial ectotherms with lungs and waterproof skin covered by scales. Scales and amniotic eggs enhanced reproduction on land. Birds are endothermic reptiles with wings, feathers, and other adaptations for flight.

Mammals

Mammals are endothermic vertebrates with hair and mammary glands. There are three major groups: Monotremes lay eggs; marsupials use a placenta but give birth to tiny embryonic offspring that usually complete development while attached to nipples inside the mother's pouch; and eutherians, or placental mammals, use their placenta in a longer-lasting association between the mother and her developing young.

MAMMALS		
Monotremes	**Marsupials**	**Eutherians**

The Human Ancestry

The Evolution of Primates

The first primates were small, arboreal mammals that evolved from insect-eating mammals about 65 million years ago. Anthropoids consist of New World monkeys (with prehensile tails), Old World monkeys (without prehensile tails), apes, and humans.

The Emergence of Humankind

Chimpanzees and humans evolved from a common ancestor about 5–7 million years ago. Upright posture evolved in several hominin species of the genus *Australopithecus* at least 4 million years ago. Enlargement of the human brain in *Homo habilis* came later, about 2.4 million years ago. *Homo erectus* was the first species to extend humanity's range from its birthplace in Africa to other continents. *Homo erectus* gave rise to regionally diverse descendants, such as the Neanderthals (*H. neanderthalensis*). Current data indicate a relatively recent dispersal of modern Africans that gave rise to today's human diversity. Human cultural evolution began in Africa with wandering hunter-gatherers, progressed to the development of agriculture and the Industrial Revolution, and continues today with accelerating technological change.

SELF-QUIZ

1. Bilateral symmetry in the animal kingdom is best correlated with
   a. an ability to sense equally in all directions.
   b. the presence of a skeleton.
   c. motility and active predation and escape.
   d. development of a true coelom.
2. The cavity between your outer body wall and your digestive tract that is fully lined by mesoderm is an example of a true _____. Roundworms have a _____, a body cavity not completely lined by mesoderm.

3. Identify which of the following categories includes all others in the list: arthropod, arachnid, insect, butterfly, crustacean, millipede.

4. The oldest group of tetrapods is the _____.

5. Reptiles are much more extensively adapted to life on land than amphibians because reptiles
   a. have a complete digestive tract.
   b. lay eggs that are enclosed in shells.
   c. are endothermic.
   d. go through a larval stage.

6. What is the name of the phylum to which humans belong? For what anatomical structure is the phylum named? Where in your body is a derivative of this anatomical structure found?

7. Fossils suggest that the first major trait distinguishing human primates from other primates was _____.

8. Which of the following types of animals is not included in the human ancestry? (*Hint*: See Figure 17.29.)
   a. a bird
   b. a bony fish
   c. an amphibian
   d. a primate

9. Put the following list of species in order, from the oldest to the most recent: *Homo erectus*, *Australopithecus* species, *Homo habilis*, *Homo sapiens*.

10. Match each of the following animals to its phylum:
    a. human   1. Echinodermata
    b. leech   2. Arthropoda
    c. sea star   3. Cnidaria
    d. lobster   4. Chordata
    e. sea anemone   5. Annelida

Answers to the Self-Quiz questions can be found in Appendix D.

THE PROCESS OF SCIENCE

11. Imagine that you are a marine biologist. As part of your exploration, you dredge up an unknown animal from the seafloor. Describe some of the characteristics you should look at to determine the phylum to which the creature should be assigned.

12. Many people describe themselves as vegetarians. Strictly speaking, vegetarians eat only plant products. Most vegetarians are not, in fact, that strict. Interview acquaintances who describe themselves as vegetarians and determine which taxonomic groups they avoid eating (see Figures 17.5 and 17.29). Try to generalize about their diet. For example, do they avoid eating vertebrates but eat some invertebrates? Do they avoid only birds and mammals? Do they eat dairy products or eggs?

13. Some researchers think that drying and cooling of the climate caused expansion of the African savanna and that this environment favored upright walking in early humans. Why might bipedalism be advantageous in the savanna? How might an erect posture relate to the evolution of a larger brain?

BIOLOGY AND SOCIETY

14. Coral reefs harbor a greater diversity of animals than any other environment in the sea. Australia's Great Barrier Reef has been protected as a marine reserve and is a mecca for scientists and nature enthusiasts. Elsewhere, such as in Indonesia and the Philippines, coral reefs are in danger. Many reefs have been depleted of fish, and runoff from the shore has covered coral with sediment. Nearly all the changes in the reefs can be traced back to human activities. What kinds of activities do you think might be contributing to the decline of the reefs? What are some reasons to be concerned about this decline? Do you think the situation is likely to improve or worsen in the future? Why? What might the local people do to halt the decline? Should the more industrialized countries help? Why or why not?

15. The human body has not changed much in the last 100,000 years, but human culture has changed a great deal. As a result of our culture, we change the environment at a rate far greater than the rate at which many species, including our own, can evolve. What evidence of rapid environmental change do you see regularly? What aspects of human culture are responsible for these changes? Do you see any evidence of a decrease in the rate of human-caused environmental changes?

Unit 4
Ecology

Chapter 18: **An Introduction to Ecology and the Biosphere**

Chapter Thread: **Global Climate Change**

Chapter 19: **Population Ecology**

Chapter Thread: **Invasive Species**

Chapter 20: **Communities and Ecosystems**

Chapter Thread: **The Loss of Biodiversity**

18 An Introduction to Ecology and the Biosphere

Polar bear on arctic ice.
Global climate change is shrinking the permanent sea ice that polar bears use as hunting platforms.

CHAPTER CONTENTS

■■□ Chapter Thread: **GLOBAL CLIMATE CHANGE**

BIOLOGY AND SOCIETY: Global Climate Change

Penguins and Polar Bears in Peril

The scientific debate is over. The great majority of scientists now agree that the global climate is changing. In the past few years, climate change has also become a hot topic for the general public. Articles and documentaries appear regularly, and it seems as if every magazine on the newsstand—including such unlikely publications as *Elle* and *Sports Illustrated*—has featured at least one cover story on climate change. What do we know about climate change now, and what can we expect for the future?

Average global temperatures have risen 0.8°C (about 1.4°F) over the past century, mostly over the last 30 years. The northernmost regions of the Northern Hemisphere and the Antarctic Peninsula have heated up the most. In parts of Alaska, for example, winter temperatures have risen by 5–6°F. The permanent arctic sea ice is shrinking; each summer brings thinner ice and more open water. Polar bears, which stalk their prey on ice and need to store up body fat for the warmer months when there is no ice, are showing signs of starvation as their winter hunting grounds melt away. At the other end of the planet, diminishing sea ice near the Antarctic Peninsula limits the access of Adélie penguins to their food supply, and spring blizzards of unprecedented frequency and severity are taking a heavy toll on their eggs and chicks. But the warming trend isn't the whole story of climate change. Precipitation patterns have also changed, bringing longer and more intense drought to some regions. In other areas, a greater proportion of the total rainfall is falling in torrential downpours that cause flooding.

Any predictions that scientists make now about the future ecological impact of global climate change are based on incomplete information. Much remains to be discovered about species diversity and about the complex interactions of organisms with each other and with their environments. There is overwhelming evidence that human enterprises are responsible for the changes that are occurring. How we respond to this crisis will determine whether circumstances improve or worsen. And the process begins with understanding the basic concepts of ecology, which we start to explore in this chapter.

An Overview of Ecology

In your study of biology so far, you have learned about the diversity of life on Earth and about the molecular and cellular structures and processes that make life tick. **Ecology**, the scientific study of the interactions between organisms and their environments, offers a different perspective on life—biology from the skin out, so to speak.

Humans have always had an interest in other organisms and their environments. As hunters and gatherers, prehistoric people had to learn where and when game and edible plants could be found in greatest abundance. Naturalists, from Aristotle to Darwin and beyond, made the process of observing and describing organisms in their natural habitats an end in itself rather than simply a means of survival. Extraordinary insight can still be gained from this discovery-based approach of watching nature and recording its structure and processes (Figure 18.1). As you might expect, field research, whether discovery-based natural history or hypothesis-driven science, is fundamental to ecology. But ecologists also test hypotheses using laboratory experiments, where conditions can be simplified and controlled. And some ecologists take a theoretical approach, devising mathematical and computer models, which enable them to simulate large-scale experiments that are impossible to conduct in the field.

▼ Figure 18.1 **Discovery science in a rain forest canopy.**

(a) Establishing a canopy research station. This hot-air dirigible is placing a giant rubber raft on the treetops of a tropical rain forest in French Guiana, in northeastern South America.

(b) Studying the canopy. Living and working on this field station 30 m above the forest floor, an international research team studies the canopy environment. French biologist Pierre Grard and two other scientists are cataloging plants and the insects that feed on those plants.

Ecology and Environmentalism

Technological innovations have enabled humans to colonize just about every environment on Earth. Even so, our survival depends on Earth's resources, which have been profoundly altered by human activities (Figure 18.2). Global climate change is just one of the many environmental issues that have stirred public concern in recent decades. Some of our industrial and agricultural practices have contaminated the air, soil, and water. Our relentless quest for land and other resources has endangered a lengthy list of plant and animal species and has even driven some to extinction.

The science of ecology can provide the understanding needed to solve environmental problems. But these problems cannot be solved by ecologists alone, because they require making decisions based on values and ethics. On a personal level, each of us makes daily choices that affect our ecological impact. And legislators and corporations, motivated by environmentally aware voters and consumers, must address questions that have wider implications: How should land use be regulated? Should we try to save all species or just certain ones? What alternatives to environmentally destructive practices can be developed? How can we balance environmental impact with economic needs?

▼ Figure 18.2 **Human impact on the environment.** A man paddles a canoe through a trash-clogged waterway in Manila, capital of the Philippines.

A Hierarchy of Interactions

Many different factors can potentially affect an organism's interaction with the environment. **Biotic factors**—all of the organisms in the area—make up the living component of the environment. Other organisms may compete with an individual for food and other resources, prey upon it, or change its physical and chemical environment. **Abiotic factors** make up the environment's nonliving component and include chemical and physical factors, such as temperature, light, water, minerals, and air. An organism's **habitat**, the specific environment it lives in, includes the biotic and abiotic factors of its surroundings.

When we study the interactions between organisms and their environments, it is convenient to divide ecology into four increasingly comprehensive levels: organismal ecology, population ecology, community ecology, and ecosystem ecology.

Organismal ecology is concerned with the evolutionary adaptations that enable individual organisms to meet the challenges posed by their abiotic environments. The distribution of organisms is limited by the abiotic conditions they can tolerate. For example, amphibians such as the salamander in **Figure 18.3a** are generally restricted to moist environments because their skin does not prevent dehydration (in fact, it plays a significant role in exchanging gases with the environment). You'll encounter other examples of organismal ecology throughout this chapter.

The next level of organization in ecology is the **population**, a group of individuals of the same species living in a particular geographic area. **Population ecology** concentrates mainly on factors that affect population density and growth (**Figure 18.3b**). Chapter 19 focuses on population ecology.

A **community** consists of all the organisms that inhabit a particular area; it is an assemblage of populations of different species. Questions in **community ecology** focus on how interactions between species, such as predation and competition, affect community structure and organization (**Figure 18.3c**).

An **ecosystem** includes all the abiotic factors in addition to the community of species in a certain area. For example, a savanna ecosystem includes not only the organisms, such as diverse plants and animals, but also the soil, water sources, sunlight, and other abiotic factors of the environment. In **ecosystem ecology**, questions concern energy flow and the cycling of chemicals among the various biotic and abiotic factors (**Figure 18.3d**). You'll learn about community and ecosystem ecology in Chapter 20.

The **biosphere** is the global ecosystem—the sum of all the planet's ecosystems, or all of life and where it lives. The most complex level in ecology, the biosphere includes the atmosphere to an altitude of several kilometers, the land down to water-bearing rocks about 1,500 m deep, lakes and streams, caves, and the oceans to a depth of several kilometers. Isolated in space, the biosphere is self-contained, or closed, except that its photosynthetic producers derive energy from sunlight and it loses heat to space. ☑

▼ Figure 18.3 **Examples of questions at different levels of ecology.**

Ecological Hierarchy

(a) Organismal ecology. What range of temperatures can a red salamander tolerate?

(b) Population ecology. What factors limit the number of striped mice that can inhabit a particular area?

(c) Community ecology. How do predators such as this beech marten affect the diversity of rodents in a community?

(d) Ecosystem ecology. What processes recycle vital chemical elements such as nitrogen within a savanna ecosystem in Africa?

☑ **CHECKPOINT**

What does the ecosystem level of classification have in common with the community level of classification? What does the ecosystem level include that the community level does not?

Answer: all the biotic factors of the area; the abiotic factors of the area

Living in Earth's Diverse Environments

Whether you have seen the world by traveling or through television and movies, you have probably noticed that there are striking regional patterns in the distribution of life. For example, some terrestrial areas, such as the tropical forests of South America and Africa, are home to plentiful plant life, whereas other areas, such as deserts, are relatively barren. Coral reefs are alive with vibrantly colored organisms; other parts of the ocean appear empty by comparison.

The distribution of life varies on a local scale, too. In the aerial view of an Alaskan wilderness in **Figure 18.4**, we can see a mixture of forest, small lakes, a meandering river, and open meadows. Within these different environments, variation occurs on a smaller scale. For example, we would find that each lake has several different habitats, and each habitat has a characteristic community of organisms.

▼ Figure 18.4 **Local variation of the environment in an Alaskan wilderness.**

Abiotic Factors of the Biosphere

Patterns in the distribution of life mainly reflect differences in the abiotic factors of the environment. In this section, we look at several major abiotic factors that influence where organisms live.

Energy Source

All organisms require a usable source of energy to live. Solar energy from sunlight, captured by chlorophyll during the process of photosynthesis, powers most ecosystems. In the image shown in **Figure 18.5**, colors are keyed to the relative abundance of chlorophyll. Green areas on land indicate high densities of plant life. Orange areas on land, including the Sahara region of Africa and much of the western United States, are much less productive. Green regions of the ocean contain an abundance of algae and photosynthetic bacteria compared to darker regions.

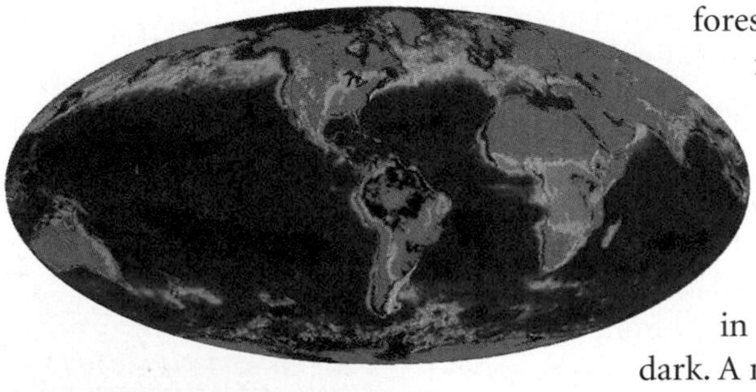

Lack of sunlight is seldom the most important factor limiting plant growth for terrestrial ecosystems (although shading by trees does create intense competition for light among plants growing on forest floors). In many aquatic environments, however, light cannot penetrate beyond certain depths. As a result, most photosynthesis in a body of water occurs near the surface. Surprisingly, life also thrives in environments that are completely dark. A mile or more below the ocean's

surface lies the unique world of hydrothermal vents, sites near the adjoining edges of giant plates of Earth's crust where molten rock and hot gases surge upward from Earth's interior. Towering chimneys, sometimes as high as 30 m, emit scalding water and hot gases. These are ecosystems powered by chemoautotrophic bacteria (see Figure 15.12) that derive energy from the oxidation of inorganic chemicals such as hydrogen sulfide (**Figure 18.6**). Bacteria with similar metabolic talents support communities of cave-dwelling organisms.

▼ Figure 18.5 **Distribution of life in the biosphere.** In this image of Earth, colors are keyed to the relative abundance of chlorophyll, which correlates with the regional densities of photosynthetic organisms.

▼ Figure 18.6 **A deep-sea hydrothermal vent.** "Black smokers" west of Vancouver Island spew plumes of hot gases from Earth's interior. Giant tube worms (inset), annelids that may grow to 2 m long, are members of the vent community.

Temperature

Temperature is an important abiotic factor because of its effect on metabolism. Few organisms can maintain a sufficiently active metabolism at temperatures close to 0°C, and temperatures above 45°C destroy the enzymes of most organisms. Still, extraordinary adaptations enable some species to live outside this temperature range. For example, bacteria and archaea living in hot springs have enzymes that function optimally at extremely high temperatures (Figure 18.7). At the other extreme, communities consisting of bacteria, algae, and small invertebrates inhabit sea ice at both poles.

▲ Figure 18.7 **Home of hot prokaryotes.** "Heat-loving" archaea thrive in this pool at Waiotapu geothermal area in the volcanic zone of New Zealand's North Island, where temperatures may exceed 80°C (176°F).

Water

Water is essential to all life. For terrestrial organisms, the primary threat is drying out in the air. Many land species have watertight coverings that reduce water loss. A waxy coating on the leaves and other aerial parts of most plants helps prevent dehydration. Aquatic organisms are surrounded by water, but they face problems of water balance if their own solute concentration does not match that of their surroundings (see Figure 5.14).

Nutrients

The distribution and abundance of photosynthetic organisms, including plants, algae, and photosynthetic bacteria, depend on the availability of inorganic nutrients such as compounds of nitrogen and phosphorus. Plants obtain these nutrients from the soil. Soil structure, pH, and nutrient content often play major roles in determining the distribution of plants. In many aquatic ecosystems, low levels of nitrogen and phosphorus limit the growth of algae and photosynthetic bacteria.

Other Aquatic Factors

Several abiotic factors are important in aquatic, but not terrestrial, ecosystems. While terrestrial organisms have a plentiful supply of oxygen from the air, aquatic organisms must depend on oxygen dissolved in water. This is a critical factor for many species of fish. Cold, fast-moving water has a higher oxygen content than warm or stagnant water (Figure 18.8). Salinity (saltiness), currents, and tides also play a role in many aquatic ecosystems.

Other Terrestrial Factors

Some abiotic factors affect terrestrial, but not aquatic, ecosystems. For example, wind is often an important abiotic factor on land. Wind increases an organism's rate of water loss by evaporation. The resulting increase in evaporative cooling can be advantageous on a hot summer day, but it can cause dangerous wind chill in the winter. In some ecosystems, frequent occurrences of natural disturbances such as storms or fire play a role in the distribution of organisms. ☑

▶ Figure 18.8 **Trout stream in Aspen, Colorado.** Fish such as trout and salmon need the higher oxygen content found in cold, fast-moving water.

☑**CHECKPOINT**

_____ energy is such an important abiotic factor because _____ provides most of the organic fuel and building material for the organisms of most ecosystems.

Answer: *Solar (or light); photosynthesis*

The Evolutionary Adaptations of Organisms

The ability of organisms to live in Earth's diverse environments demonstrates the close relationship between the fields of ecology and evolutionary biology. Charles Darwin was an ecologist (although he predated the word *ecology*). It was the geographic distribution of organisms and their exquisite adaptations to specific environments that provided Darwin with evidence for evolution. Evolutionary adaptation via natural selection results from the interactions of organisms with their environments, which brings us back to our definition of ecology. Thus, events that occur in the short term, during the course of an individual's lifetime, may translate into effects over the longer scale of evolutionary time. For example, because the availability of water affects a plant's growth and ultimately its reproductive success, precipitation has an impact on the gene pool of a plant population. After a period of below-average rainfall, drought-resistant individuals may be more prevalent in a plant population. We'll look at some evolutionary responses to biotic interactions in Chapter 20. ☑

☑CHECKPOINT

How are the fields of ecology and evolution linked?

Answer: The process of evolutionary adaptation via natural selection results from the interactions of organisms with their environments.

Adjusting to Environmental Variability

The abiotic factors in a habitat may vary from year to year, seasonally, or over the course of a day. An individual's abilities to adjust to environmental changes that occur during its lifetime are themselves adaptations refined by natural selection. For instance, if you see a bird on a cold day, it may look unusually fluffy (Figure 18.9). Small muscles in the skin raise the bird's feathers, a physiological response that traps insulating pockets of air. Some species of birds adjust to seasonal cold by growing heavier feathers. And some bird species respond to the onset of cold weather by migrating to warmer regions—a behavioral response. Note that these responses occur during the lifetime of an individual, so they do not qualify as evolution, which is change in a population over time.

▼ Figure 18.9 **A gray-headed chickadee demonstrating its physiological response to cold weather.**

Physiological Responses

Like birds, mammals can create a temporary layer of insulation on cold days by contracting skin muscles, in this case attached to hairs. (Our own muscles do this, too, but we just get "goose bumps" instead of a furry insulation.) The blood vessels in the skin also constrict, which slows the loss of body heat. In both cases, the adjustment occurs in just seconds.

A gradual, though still reversible, physiological adjustment that occurs in response to an environmental change is called **acclimation**. For example, suppose you moved from Boston, which is essentially at sea level, to the mile-high city of Denver. One physiological response to the lower oxygen supply in your new environment would be a gradual increase in the number of your red blood cells, which transport O_2 from your lungs to other parts of your body. Acclimation can take days or weeks. This is why high-altitude climbers, such as those attempting to scale Mount Everest, need extended stays at a high-elevation base camp before proceeding to the summit.

The ability to acclimate is generally related to the range of environmental conditions a species naturally experiences. Species that live in very warm climates, for example, usually cannot acclimate to extreme cold. Among vertebrates, birds and mammals can generally tolerate the greatest temperature extremes because, as endotherms, they use their metabolism to regulate internal temperature. In contrast, ectothermic reptiles can only tolerate a more limited range of temperatures (Figure 18.10).

▼ Figure 18.10 **The number of lizard species in different regions of the contiguous United States.** Notice that there are fewer and fewer lizard species in more northern regions. This reflects lizards' ectothermic physiology, which depends on environmental heat for keeping the body warm enough for the animal to be active.

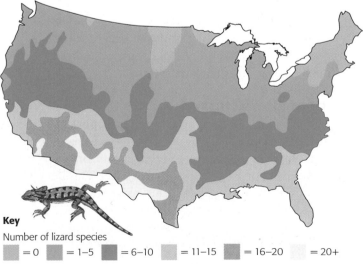

Key
Number of lizard species
■ = 0   ■ = 1–5   ■ = 6–10   ■ = 11–15   ■ = 16–20   ■ = 20+

Anatomical Responses

Many organisms respond to environmental challenge with some type of change in body shape or structure. When the change is reversible, the response is an example of acclimation. Many mammals, for example, grow a heavier coat of fur in response to cold, and shed it when summer comes. Sometimes fur or feather color changes seasonally as well, camouflaging the animal against winter snow and summer vegetation (Figure 18.11).

Other anatomical changes are irreversible over the lifetime of an individual. Environmental variation can affect growth and development so much that there may be remarkable differences in body shape within a population. You can see an example in Figure 18.12, which shows the "flagging" that wind causes in certain trees. In general, plants are more anatomically changeable than animals. Rooted and unable to move to a better location, plants rely entirely on their anatomical and physiological responses to survive environmental fluctuations.

▲ Figure 18.11 **The arctic fox in winter and summer coats.**

▼ Figure 18.12 **Wind as an abiotic factor that shapes trees.** The mechanical disturbance of the prevailing wind hinders limb growth on the windward side of these fir trees on Oregon's Mount Hood, while limbs on the other side grow normally. This anatomical response is an evolutionary adaptation that reduces the number of limbs that are broken during strong winds.

Behavioral Responses

In contrast to plants, most animals can respond to an unfavorable change in the environment by moving to a new location. Such movement may be fairly localized. For example, many desert ectotherms, including reptiles, maintain a reasonably constant body temperature by shuttling between sun and shade. Some animals are capable of migrating great distances in response to such environmental cues as the changing seasons. Many migratory birds overwinter in Central and South America, returning to northern latitudes to breed during summer. And humans, with their large brains and available technology, have an especially rich range of behavioral responses available to them (Figure 18.13). ☑

▼ Figure 18.13 **Behavioral responses have expanded the geographic range of humans.** Dressing for the weather is a thermoregulatory behavior unique to humans.

Biomes

The abiotic factors you learned about in the previous section are largely responsible for the distribution of life on Earth. (You'll learn about the role of biotic factors in species distribution in Chapter 20.) Using various combinations of these factors, ecologists have categorized Earth's environments into biomes. A **biome** is a major terrestrial or aquatic life zone, characterized by vegetation type in terrestrial biomes or the physical environment in aquatic biomes. In this section, we'll briefly survey the aquatic biomes, followed by the terrestrial biomes.

Aquatic biomes, which occupy roughly 75% of Earth's surface, are determined by their salinity and other physical factors. Freshwater biomes (lakes, streams and rivers, and wetlands) typically have a salt concentration of less than 1%. The salt concentrations of marine biomes (oceans, intertidal zones, coral reefs, and estuaries) are generally around 3%.

Freshwater Biomes

Freshwater biomes cover less than 1% of Earth, and they contain a mere 0.01% of its water. But they harbor a disproportionate share of biodiversity—an estimated 6% of all described species. Moreover, we depend on freshwater biomes for drinking water, crop irrigation, sanitation, and industry.

Freshwater biomes fall into two broad groups: standing water, which includes lakes and ponds, and flowing water, such as rivers and streams. The difference in water movement results in profound differences in ecosystem structure.

Lakes and Ponds

Standing bodies of water range from small ponds only a few square meters in area to large lakes, such as North America's Great Lakes, that are thousands of square kilometers (Figure 18.14).

◄ Figure 18.14 **Satellite view of the Great Lakes.**

In lakes and large ponds, the communities of plants, algae, and animals are distributed according to the depth of the water and its distance from shore (Figure 18.15). Shallow water near shore and the upper layer of water away from shore make up the **photic zone**, so named because light is available for photosynthesis. Microscopic algae and cyanobacteria grow in the photic zone, joined by rooted plants and floating plants such as water lilies in the photic area near shore. If a lake or pond is deep enough or murky enough, it has an **aphotic zone**, where light levels are too low to support photosynthesis.

At the bottom of all aquatic biomes is the **benthic realm**. Made up of sand and organic and inorganic sediments, the benthic realm is occupied by communities of organisms that are collectively called benthos. (If you have ever waded barefoot into a pond or lake, you have felt the benthos squish between your toes.) Dead material that "rains" down from the productive surface waters of the photic zone is a major source of food for the benthos.

The mineral nutrients nitrogen and phosphorus typically regulate the amount of phytoplankton growth in a lake or pond. Many lakes and ponds are affected by large inputs of nitrogen and phosphorus from sewage and runoff from fertilized lawns and farms. These nutrients often produce heavy growth of algae, which reduces light penetration. When the algae die and decompose, a pond or lake can suffer serious oxygen depletion, killing fish that are adapted to high-oxygen conditions. ☑

▼ Figure 18.15 **Zones in a lake.**

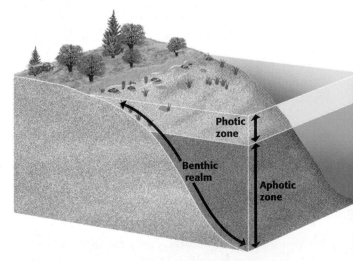

Photic zone

Benthic realm

Aphotic zone

▲ Figure 18.16 **A stream in the Appalachian Mountains.**

Rivers and Streams

Rivers and streams, which are bodies of flowing water, generally support quite different communities of organisms than lakes and ponds (Figure 18.16). A river or stream changes greatly between its source (perhaps a spring or snowmelt in the mountains) and the point at which it empties into a lake or the ocean. Near a source, the water is usually cold, low in nutrients, and clear. The channel is often narrow, with a swift current that does not allow much silt to accumulate on the bottom. The current also inhibits the growth of phytoplankton; most of the organisms found here are supported by the photosynthesis of algae attached to rocks or by organic material (such as leaves) carried into the stream from the surrounding land. The most abundant benthic animals are usually insects that eat algae, leaves, or one another. Trout are often the predominant fishes, locating their food, including insects, mainly by sight in the clear water.

Downstream, a river or stream typically widens and slows. There the water is usually warmer and may be murkier because of sediments and phytoplankton suspended in it. Worms and insects that burrow into mud are often abundant, as are waterfowl, frogs, and catfish and other fishes that find food more by scent and taste than by sight.

Many streams and rivers have been affected by pollution from human activities. We have also altered rivers by constructing dams used for flood control, to provide reservoirs of drinking water, or to generate hydroelectric power. In many cases, dams have completely changed the downstream ecosystems, altering the rate and volume of water flow and affecting fish and invertebrate populations (Figure 18.17).

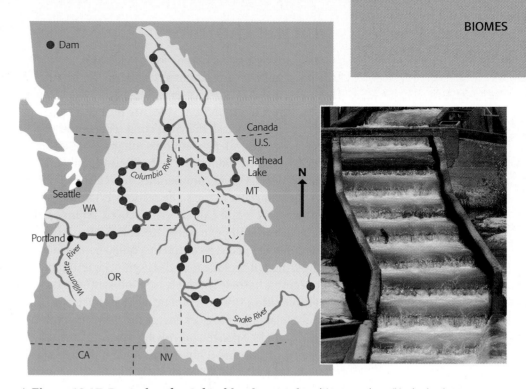

▲ Figure 18.17 **Damming the Columbia River Basin.** This map shows only the largest of the 250 dams that have altered freshwater ecosystems throughout the Pacific Northwest. These great concrete obstacles make it difficult for salmon to swim upriver to their breeding streams, though many dams now have "fish ladders" that provide detours (inset).

Wetlands

A **wetland** is a transitional biome between an aquatic ecosystem and a terrestrial one. Freshwater wetlands include swamps, bogs, and marshes (Figure 18.18). Covered with water either permanently or periodically, wetlands support the growth of aquatic plants and are among the richest of biomes in terms of species diversity. They provide water storage areas that reduce flooding and improve water quality by trapping pollutants such as metals and organic compounds in their sediments. The recognition of their ecological and economic value has led to government and private efforts to protect and restore wetlands.

► Figure 18.18 **A marsh in the Pocono Mountains, Pennsylvania.**

Marine Biomes

Gazing out over a vast ocean, you might think that it is the most uniform environment on Earth. But marine habitats can be as different as night and day. The deepest ocean, where hydrothermal vents are located, is perpetually dark. In contrast, the vivid coral reefs nearer the surface are utterly dependent on sunlight. Habitats near shore are different from those in mid-ocean, and the seafloor hosts different communities than the open waters.

As in freshwater biomes, the seafloor is known as the benthic realm (Figure 18.19). The **pelagic realm** of the oceans includes all open water. In shallow areas, such as the submerged parts of continents, called continental shelves, the photic zone includes both pelagic and benthic regions. In these sunlit areas, photosynthesis by **phytoplankton** (photosynthetic algae and bacteria) and multicellular algae provides energy for a diverse community of animals. Sponges, burrowing worms, clams, sea anemones, crabs, and echinoderms inhabit the benthos. **Zooplankton** (free-floating animals, including many microscopic ones), fishes, marine mammals, and many other types of animals are abundant in the pelagic photic zone.

The **coral reef** biome occurs in the photic zone of warm tropical waters in scattered locations around the globe (Figure 18.20). A coral reef is built up slowly by successive generations of coral animals—a diverse group of cnidarians that secrete a hard external skeleton—and by multicellular algae encrusted with limestone. Unicellular algae live within the coral's cells, providing the coral with food. The physical structure and productivity of coral reefs support a huge variety of invertebrates and fishes.

The photic zone extends down a maximum of 200 m in the ocean. Although there is not enough light for photosynthesis between 200 and 1,000 m, some light does reach these depths of the aphotic zone. This dimly lit world, sometimes called the twilight zone, is dominated by a fascinating variety of small fishes and crustaceans. Food sinking from the photic zone provides some sustenance for these animals. In addition, many of them migrate to the surface at night to feed. Some fishes in the twilight zone have enlarged eyes, enabling them to see in the very dim light, and light-emitting organs that attract mates and prey.

Below 1,000 m, the ocean is completely and permanently dark. Adaptation to this environment has produced many bizarre-looking creatures. Most of the benthic organisms here are deposit feeders, animals that consume dead organic material on the seafloor. Crustaceans, polychaete worms, sea anemones, and echinoderms such as sea cucumbers, sea stars, and sea

▼ Figure 18.19 **Ocean life.** (Zone depths and organisms not drawn to scale.)

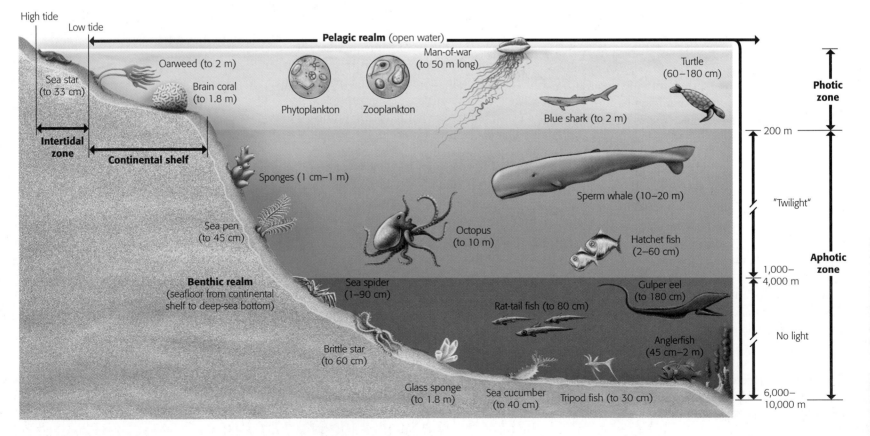

▲ Figure 18.20 **A coral reef in the Red Sea off the coast of Egypt.**

▼ Figure 18.21 **An intertidal zone of Sitka Sound, in southeast Alaska.**

urchins are common. Food is scarce, however. The density of animals is low except at hydrothermal vents, the prokaryote-powered ecosystems mentioned earlier (see Figure 18.6).

The marine environment also includes distinctive biomes where the ocean interfaces with land or with fresh water. In the **intertidal zone**, where the ocean meets land, the shore is pounded by waves during high tide and exposed to the sun and drying winds during low tide (**Figure 18.21**). The rocky intertidal zone is home to many sedentary organisms, such as algae, barnacles, and mussels, which attach to rocks and are thus prevented from being washed away. On sandy beaches, suspension-feeding worms, clams, and predatory crustaceans bury themselves in the ground.

Figure 18.22 shows an **estuary**, a transition area between a river and the ocean. The saltiness of estuaries ranges from nearly that of fresh water to that of the ocean. With their waters enriched by nutrients from the river, estuaries, like freshwater wetlands, are among the most productive areas on Earth. Oysters, crabs, and many fishes live in estuaries or reproduce in them. Estuaries are also crucial nesting and feeding areas for waterfowl. Mudflats and salt marshes are extensive coastal wetlands that often border estuaries.

For centuries, people viewed the ocean as a limitless resource, harvesting its bounty and using it as a dumping ground for wastes. These practices caused little damage when Earth's human population was small, but their negative impact is now great. From worldwide declines in commercial fish species to dying coral reefs, to beaches closed by pollution, danger signs abound. Because of their proximity to land, estuaries are especially vulnerable. Many have been completely replaced by development on landfill. Other threats include nutrient pollution, contamination by pathogens or toxic chemicals, alteration of freshwater inflow, and introduction of non-native species. Coral reefs have suffered from many of the same problems. Excessive fishing has also upset the species balance in some reef communities and greatly reduced diversity. The widespread, massive deaths of reef-building corals in some regions have been attributed to rising sea surface temperatures; ocean acidification is also a serious threat. A 2008 assessment of coral reefs found that 19% had already fallen victim to a combination of factors and warned that most of the remaining reefs could disappear by 2040 if current trends continue. ☑

▼ Figure 18.22 **An estuary on the edge of Chesapeake Bay, Maryland.**

☑CHECKPOINT

_____, the small photosynthetic organisms inhabiting the _____ zone of the pelagic realm, provide most of the food for oceanic life.

Answer: Phytoplankton; photic

How Climate Affects Terrestrial Biome Distribution

Terrestrial biomes are determined primarily by climate, especially temperature and rainfall. Before we survey these biomes, let's look at the broad patterns of global climate that help explain their locations.

Earth's global climate patterns are largely the result of the input of radiant energy from the sun and the planet's movement in space. Because of its curvature, Earth receives an uneven distribution of solar energy (Figure 18.23). The equator receives the greatest intensity of solar radiation. As it is heated by the direct rays of the sun, air at the equator rises. It then cools, forms clouds, and drops rain (Figure 18.24). This largely explains why rain forests are concentrated in the **tropics**—the region from the Tropic of Cancer to the Tropic of Capricorn.

After losing moisture over equatorial zones, dry high-altitude air masses spread away from the equator until they cool and descend again at latitudes of about 30° north and south. Many of the world's great deserts—the Sahara in North Africa and the Arabian on the Arabian Peninsula, for example—are centered at these latitudes because of the dry air they receive.

Latitudes between the tropics and the Arctic Circle in the north and the Antarctic Circle in the south are called **temperate zones**. Generally, these regions have milder climates than the tropics or the polar regions. Notice in Figure 18.24 that some of the descending dry air heads into the latitudes above 30°. At first these air masses pick up moisture, but they tend to drop it as they cool at higher latitudes. This is why the north and south temperate zones tend to be relatively wet. Coniferous forests dominate the landscape at the wet but cool latitudes around 60° north.

Proximity to large bodies of water and the presence of landforms such as mountain ranges also affect climate. Oceans and large lakes moderate climate by absorbing heat when the air is warm and releasing heat to cold air. Mountains affect climate in two major ways. First, air temperature drops as elevation increases. As a result, driving up a tall mountain offers a quick tour of several biomes. Figure 18.25 shows the scenery you might encounter on a journey from the scorching lowlands of the Sonoran desert to a cool coniferous forest at an elevation of 11,000 feet above sea level.

▼ Figure 18.23 **Uneven heating of Earth.**

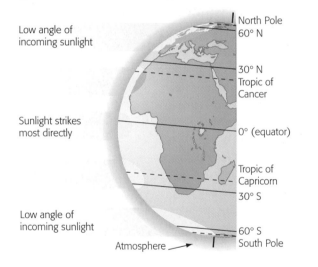

▼ Figure 18.24 **How uneven heating of Earth produces various climates.**

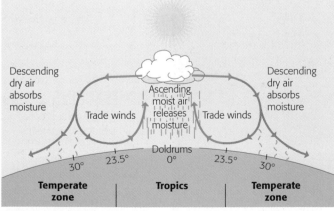

► Figure 18.25 **Effect of altitude on vegetation.** The zones shown are typical of the Sonoran Desert region in southwestern North America.

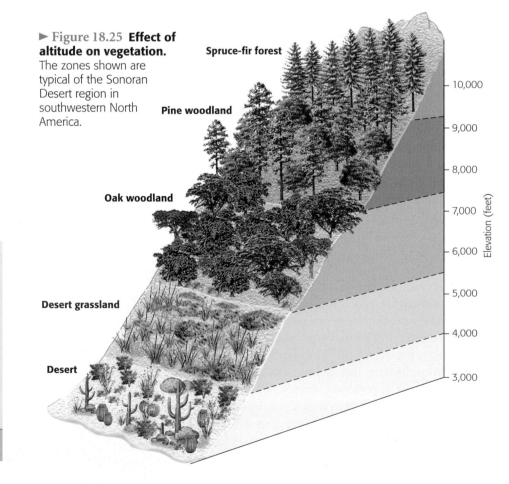

Second, mountains can block the flow of cool, moist air from a coast, causing radically different climates on opposite sides of a mountain range. In the example shown in **Figure 18.26**, moist air moves in off the Pacific Ocean and encounters the Coast Range in the state of Washington. Air flows upward, cools at higher altitudes, and drops a large amount of rainfall. The biological community in this wet region is a temperate rain forest. Precipitation increases again farther inland as the air moves up and over the Cascade Range. On the eastern side of the Cascades, there is little precipitation in the area known as a rain shadow. As a result, eastern Washington is very arid, almost qualifying as a desert. ☑

▼ Figure 18.26 **How mountains affect rainfall.**

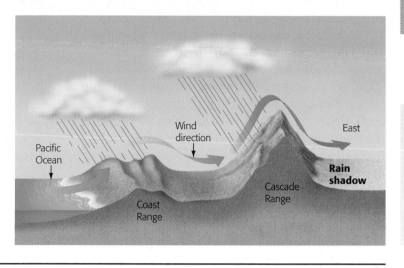

☑**CHECKPOINT**

Why is there so much rainfall in the tropics?

Answer: Air at the equator rises as it is warmed by direct sunlight. As the air rises, it cools. This causes cloud formation and rainfall because cool air holds less moisture than warm air.

Terrestrial Biomes

Terrestrial ecosystems are grouped into biomes primarily on the basis of their vegetation type **(Figure 18.27)**. By providing food, shelter, nesting sites, and much of the organic material for the decomposers that recycle mineral nutrients, plants build the foundation for the communities of animals and other organisms typical of each biome. The geographic distribution of plants, and thus of biomes, largely depends on climate, with temperature and rainfall often the key factors determining the kind of biome that exists in a particular region. If the climate in two geographically separate areas is similar, the same type of biome may occur in both. Coniferous forests,

for instance, extend in a broad band across North America, Europe, and Asia. There is local variation within each biome, giving the vegetation a patchy, rather than a uniform, appearance. For example, in northern coniferous forests, snowfall may break branches and small trees, causing openings where broadleaf trees such as aspen and birch can grow. Local storms and fires also create openings in many biomes.

Each biome is characterized by a type of biological community, rather than an assemblage of particular species. For example, the groups of species living in the deserts of southwestern North America and in the Sahara Desert of Africa are different, but both groups are adapted to desert conditions. Organisms in widely

► Figure 18.27 **Map of the major terrestrial biomes.** Although this map has sharp boundaries, biomes actually grade into one another. We'll use smaller versions of this map, highlighted by color coding, during our closer look at the terrestrial biomes in the next several pages.

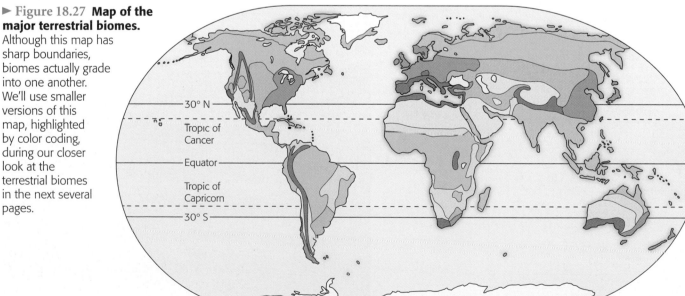

Key

Tropical forest

Savanna

Desert

Chaparral

Temperate grassland

Temperate broadleaf forest

Coniferous forest

Arctic tundra

High mountains (coniferous forest and alpine tundra)

Polar ice

separated biomes may look alike because of convergent evolution, the appearance of similar traits in independently evolved species living in similar environments.

Figure 18.28 is a climograph, a visual representation of the differences in precipitation and temperature ranges that characterize terrestrial biomes. The *x*-axis shows the range of annual average precipitation for the biome, and the *y*-axis displays the biome's range of annual average temperature. By studying the climograph plots, we can see how these abiotic factors compare in different biomes. For example, organisms that live in temperate broadleaf forests must have adaptations that allow them to cope with a wide range of temperatures, while organisms in tropical forests experience very little temperature variation. Although the range of precipitation in temperate broadleaf forests is similar to that of northern coniferous forests, the lower range of temperatures in northern coniferous forests reveals a significant difference in the abiotic environments of these two biomes.

Today, concern about global warming is generating intense interest in the effect of climate on vegetation patterns. Using powerful new tools, such as satellite imagery, scientists are documenting latitudinal shifts in biome borders, decreases in snow and ice coverage, and changes in the length of the growing season. At the same time, many natural biomes have been fragmented and altered by human activity. We'll discuss both of these issues after we survey the major terrestrial biomes, beginning near the equator and generally approaching the poles.

✓**CHECKPOINT**

Why are climbing plants common in tropical rain forests?

Answer: Climbing is a plant adaptation for reaching sunlight in a closed canopy, where little sunlight reaches the forest floor.

Tropical Forest

Tropical forests occur in equatorial areas where the temperature is warm and days are 11–12 hours long year-round. The type of vegetation is determined primarily by rainfall. Tropical rain forests, like the one shown in **Figure 18.29**, receive 200–400 cm (79–157 in.) of rain per year.

The layered structure of tropical rain forests provides many different habitats. Treetops form a closed canopy over one or two layers of smaller trees and a shrub understory. Few plants grow in the deep shade of the forest floor. Many trees are covered by woody vines growing toward the light. Other plants, such as orchids, gain access to sunlight by growing on the branches or trunks of tall trees. Scattered trees reach full sunlight by towering above the canopy. Many of the animals also dwell in trees, where food is abundant. Monkeys, birds, insects, snakes, bats, and frogs find food and shelter many meters above the ground.

Rainfall is less plentiful in other tropical forests. Tropical dry forests predominate in lowland areas that have a prolonged dry season or scarce rainfall at any time. The plants found there are a mixture of thorny shrubs and trees and succulents. In regions with distinct wet and dry seasons, tropical deciduous trees are common. ✓

▼ Figure 18.29 **Tropical rain forest in Borneo.**

▼ Figure 18.28 **A climograph for some major biomes in North America.**

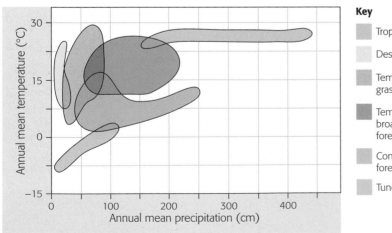

Key
- Tropical forest
- Desert
- Temperate grassland
- Temperate broadleaf forest
- Coniferous forest
- Tundra

Savanna

Savannas, such as the one shown in **Figure 18.30**, are dominated by grasses and scattered trees. The temperature is warm year-round. Rainfall averages 30–50 cm (roughly 12–20 in.) per year, with dramatic seasonal variation.

Fire, caused by lightning or human activity, is an important abiotic factor in the savanna. The grasses survive burning because the growing points of their shoots are below ground. Other plants have seeds that sprout rapidly after a fire. Poor soil and lack of moisture, along with fire and grazing animals, prevent the establishment of most trees. The luxuriant growth of grasses and small broadleaf plants during the rainy season provides a rich food source for plant-eating animals.

Many of the world's large grazing mammals and their predators inhabit savannas. African savannas are home to zebras and many species of antelope, as well as to lions and cheetahs. Several species of kangaroo are the dominant grazers of Australian savannas. Oddly, though, the large grazers are not the dominant plant-eaters in savannas. That distinction belongs to insects, especially ants and termites. Other animals include burrowers such as mice, moles, gophers, and ground squirrels.

Desert

Deserts are the driest of all biomes, characterized by low and unpredictable rainfall—less than 30 cm (about 12 in.) per year. Some deserts are very hot, with daytime soil surface temperatures above 60°C (140°F) and large daily temperature fluctuations. Other deserts, such as those west of the Rocky Mountains and the Gobi Desert, spanning northern China and southern Mongolia, are relatively cold. Air temperatures in cold deserts may fall below −30°C (−22°F).

Desert vegetation typically includes water-storing plants such as cacti and deeply rooted shrubs. Various snakes, lizards, and seed-eating rodents are common inhabitants. Arthropods such as scorpions and insects also thrive in the desert. Evolutionary adaptations of desert plants and animals include a remarkable array of mechanisms that conserve water. For example, the "pleated" stem of saguaro cacti (**Figure 18.31**) enables the plants to expand when they absorb water during wet periods. Some desert mice *never* drink, deriving all their water from the metabolic breakdown of the seeds they eat. Protective adaptations that deter feeding by mammals and insects, such as spines on cacti and poisons in the leaves of shrubs, are common in desert plants. ☑

☑**CHECKPOINT**
1. How does the savanna climate vary seasonally?
2. What abiotic factor characterizes deserts?

Answers: 1. Temperature stays about the same year-round, but rainfall varies dramatically. 2. Rainfall is low and unpredictable.

▼ **Figure 18.30 Savanna in Kenya.**

▼ **Figure 18.31 Sonoran desert.**

Chaparral

The climate that supports **chaparral** vegetation results mainly from cool ocean currents circulating offshore, producing mild, rainy winters and hot, dry summers. As a result, this biome is limited to small coastal areas, some in California (**Figure 18.32**). The largest region of chaparral surrounds the Mediterranean Sea; in fact, Mediterranean is another name for this biome. Dense, spiny, evergreen shrubs dominate chaparral. Annual plants are also common during the wet winter and spring months. Animals characteristic of the chaparral are deer, fruit-eating birds, seed-eating rodents, and lizards and snakes.

Chaparral vegetation is adapted to periodic fires caused by lightning. Many plants contain flammable chemicals and burn fiercely, especially where dead brush has accumulated. After a fire, shrubs use food reserves stored in the surviving roots to support rapid shoot regeneration. Some chaparral plants produce seeds that will germinate only after a hot fire. The ashes of burned vegetation fertilize the soil with mineral nutrients, promoting regrowth of the plant community. Houses do not fare as well. The firestorms that race through the densely populated canyons of Southern California can be devastating to the human inhabitants.

Temperate Grassland

Temperate grasslands have some of the characteristics of tropical savannas, but they are mostly treeless, except along rivers or streams, and are found in regions of relatively cold winter temperatures. Rainfall, averaging between 25 and 75 cm per year (approximately 10–30 in.), with frequent severe droughts, is too low to support forest growth. Periodic fires and grazing by large mammals also prevent invasion by woody plants. These grazers include the bison and pronghorn in North America, the wild horses and sheep of the Asian steppes, and kangaroos in Australia. As in the savanna, however, the dominant plant-eaters are invertebrates, especially grasshoppers and soil-dwelling nematodes.

Without trees, many birds nest on the ground. Many small mammals, such as rabbits, voles, ground squirrels, prairie dogs, and pocket gophers, dig burrows to escape predators. Temperate grasslands like the one shown in **Figure 18.33** once covered much of central North America.

Because grassland soil is both deep and rich in nutrients, these habitats provide fertile land for agriculture. Most grassland in the United States has been converted to cropland or pasture, and very little natural prairie exists today. ☑

▼ Figure 18.32 **Chaparral in California.**

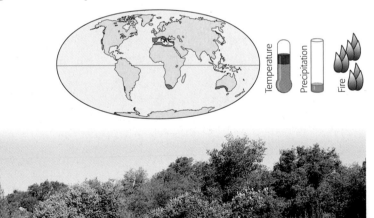

▼ Figure 18.33 **Temperate grassland in South Dakota.**

Temperate Broadleaf Forest

Temperate broadleaf forests occur throughout midlatitudes where there is sufficient moisture to support the growth of large trees. Annual precipitation is relatively high, ranging from 75–150 cm (30–60 in.), and typically distributed evenly around the year. Annual temperature varies over a wide range, with hot summers and cold winters. In the Northern Hemisphere, dense stands of deciduous trees are trademarks of temperate forests, such as the one pictured in **Figure 18.34**. Deciduous trees drop their leaves before winter, when temperatures are too low for effective photosynthesis and water lost by evaporation is not easily replaced from frozen soil.

Numerous invertebrates live in the soil and the thick layer of leaf litter that accumulates on the forest floor. Some vertebrates, such as mice, shrews, and ground squirrels, burrow for shelter and food, while others, including many species of birds, live in the trees. Predators include bobcats, foxes, black bears, and mountain lions. Many mammals that inhabit these forests enter a dormant winter state called hibernation, and some bird species migrate to warmer climates.

Virtually all the original temperate broadleaf forests in North America were destroyed by logging or cleared for agriculture or development. These forests tend to recover after disturbance, however, and today we see deciduous trees growing in undeveloped areas over much of their former range.

Coniferous Forest

Cone-bearing evergreen trees such as pine, spruce, fir, and hemlock dominate **coniferous forests** in the Northern Hemisphere. (Other kinds of conifers grow in parts of South America, Africa, and Australia.) The northern coniferous forest, or **taiga** (**Figure 18.35**), is the largest terrestrial biome on Earth, stretching in a broad band across North America and Asia south of the Arctic Circle. Taiga is also found at cool, high elevations in more temperate latitudes, for example, in much of the mountainous region of western North America. The taiga is characterized by long, snowy winters and short, wet summers that are sometimes warm. The slow decomposition of conifer needles in the thin, acidic soil makes few nutrients available for plant growth. The conical shape of many conifers prevents too much snow from accumulating on their branches and breaking them. Animals of the taiga include moose, elk, hares, bears, wolves, grouse, and migratory birds. The Asian taiga is home to the dwindling number of Siberian tigers that remain in the wild.

The **temperate rain forests** of coastal North America (from Alaska to Oregon) are also coniferous forests. Warm, moist air from the Pacific Ocean supports this unique biome, which, like most coniferous forests, is dominated by a few tree species, typically hemlock, Douglas fir, and redwood. These forests are heavily logged, and the old-growth stands of trees are rapidly disappearing. ☑

☑ CHECKPOINT

1. How does the loss of leaves function as an adaptation of deciduous trees to cold winters?
2. What type of trees are characteristic of the taiga?

Answers: 1. by reducing loss of water from the trees when that water cannot be replaced because of frozen soil. 2. conifers such as pine, spruce, fir, and hemlock

▼ **Figure 18.34**
Temperate broadleaf forest in Great Smokies National Park, North Carolina.

Temperature   Precipitation

▼ **Figure 18.35**
Northern coniferous forest in Finland, with the sky lit by the northern lights.

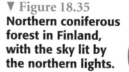

Temperature   Precipitation

Tundra

Tundra covers expansive areas of the Arctic between the taiga and polar ice. **Permafrost** (permanently frozen subsoil), bitterly cold temperatures, and high winds are responsible for the absence of trees and other tall plants in the arctic tundra shown in **Figure 18.36**. The arctic tundra receives very little annual precipitation. However, water cannot penetrate the underlying permafrost, so melted snow and ice accumulate in pools on the shallow topsoil during the short summer.

Tundra vegetation includes small shrubs, grasses, mosses, and lichens. When summer arrives, annual plants grow quickly and flower in a rapid burst. Caribou, musk oxen, wolves, and small rodents called lemmings are among the mammals found in the arctic tundra. Many migratory animals use the tundra as a summer breeding ground. During the brief but productive warm season, the marshy ground supports the aquatic larvae of insects, providing food for migratory waterfowl, and clouds of mosquitoes often fill the tundra air.

High winds and cold temperatures create plant communities called alpine tundra on very high mountaintops at all latitudes, including the tropics. Although these communities are similar to arctic tundra, there is no permafrost beneath alpine tundra.

Polar Ice

Polar ice covers the land at high latitudes north of the arctic tundra in the Northern Hemisphere and in Antarctica in the Southern Hemisphere (**Figure 18.37**). The temperature in these regions is extremely cold year-round, and precipitation is very low. Only a small portion of these landmasses is free of ice or snow, even during the summer. Nevertheless, small plants, such as mosses and lichens, eke out a living, and invertebrates such as nematodes, mites, and wingless insects called springtails inhabit the frigid soil. Nearby sea ice provides feeding platforms for large animals such as polar bears (in the Northern Hemisphere), penguins (in the Southern Hemisphere), and seals. Seals, penguins, and other marine birds visit the land to rest and breed. The polar marine biome provides the food that sustains these birds and mammals. In the Antarctic, penguins feed at sea, eating a variety of fish, squid, and small shrimplike crustaceans known as krill. Antarctic krill, an important food source for many species of fish, seals, squid, seabirds, and filter-feeding whales as well as penguins, depend on sea ice for breeding and as a refuge from predators. As the amount and duration of sea ice decline as a consequence of global climate change, krill habitat is shrinking. ☑

▼ Figure 18.36 **Arctic tundra in central Alaska.**

▼ Figure 18.37 **Polar ice in Antarctica.**

The Water Cycle

Biomes are not self-contained units. Rather, all parts of the biosphere are linked by the global water cycle, illustrated in **Figure 18.38**, and by nutrient cycles, which you will learn about in Chapter 20. Consequently, events in one biome may reverberate throughout the biosphere.

As you learned earlier in this chapter, water and air move in global patterns driven by solar energy. Precipitation and evaporation continuously move water between the land, oceans, and the atmosphere. Water also evaporates from plants, which pull it from the soil in a process called transpiration.

Over the oceans, evaporation exceeds precipitation. The result is a net movement of water vapor to clouds that are carried by winds from the oceans across the land. On land, precipitation exceeds evaporation and transpiration. Excess precipitation forms systems of surface water (such as lakes and rivers) and groundwater, all of which flow back to the sea, completing the water cycle.

Just as the water draining from your shower carries dead skin cells from your body along with the day's grime, the water washing over and through the ground carries traces of the land and its history. For example, water flowing from land to the sea carries with it silt (fine soil particles) and chemicals such as fertilizers and pesticides.

Erosion from coastal development has caused silt to muddy the waters of some coral reefs, dimming the light available to the photosynthetic algae that power the reef community. Chemicals in surface water may travel hundreds of miles by stream and river to the ocean, where currents then carry them even farther from their point of origin. For instance, traces of pesticides and chemicals from industrial wastes have been found in marine mammals in the Arctic and in deep-sea octopuses and squid. Airborne pollutants such as nitrogen oxides and sulfur oxides, which combine with water to form acid precipitation, are also distributed by the water cycle.

Human activity also affects the global water cycle itself in a number of important ways. One of the main sources of atmospheric water is transpiration from the dense vegetation making up tropical rain forests. The destruction of these forests changes the amount of water vapor in the air. This, in turn, will likely alter local, and perhaps global, weather patterns. Pumping large amounts of groundwater to the surface for irrigation affects the water cycle, too. This practice can increase the rate of evaporation over land and may deplete groundwater supplies. We'll consider some of these environmental impacts in the next section. ✔

▼ Figure 18.38 **The global water cycle.**

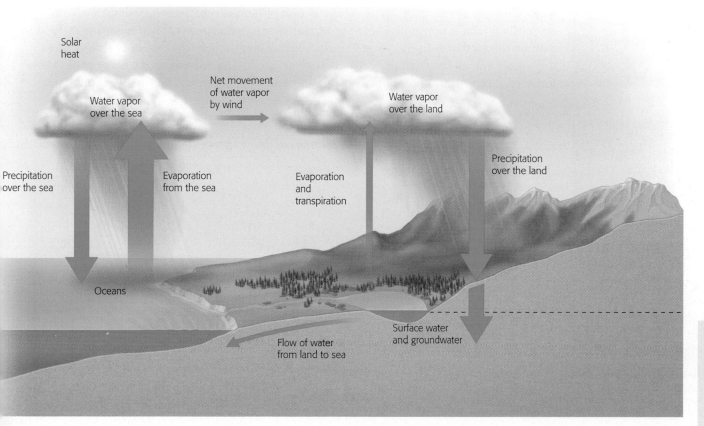

✔CHECKPOINT

What is the main way that living organisms contribute to the water cycle?

Answer: Plants move water from the ground to the air via transpiration.

Human Impact on Biomes

For hundreds of years, people have been using increasingly effective technologies to capture or produce food, to extract resources from the environment, and to build cities. It is now clear that the environmental costs of these enterprises are staggering. In this section, you'll see some examples of how human activities are affecting forest and freshwater resources. Throughout the remainder of this unit, you'll learn about the role of ecological knowledge in achieving **sustainability**, the goal of developing, managing, and conserving Earth's resources in ways that meet the needs of people today without compromising the ability of future generations to meet their needs.

▼ **Figure 18.39 Satellite photos of the Rondonia area of the Brazilian rain forest.**

1975. In 1975, the forest in this remote region was virtually intact.

2001. Same area in 2001, after a paved highway through the region brought loggers and farmers. The "fishbone" pattern marks the new network of roads carved through the forest.

Forests

The map in Figure 18.27 shows the terrestrial biomes that would be expected to flourish under the prevailing climatic conditions. However, about three-quarters of Earth's land surface has been altered by thousands of years of human occupation. Most of the land that we've appropriated is used for agriculture; another hefty chunk is covered by the asphalt and concrete of development. Changes in vegetation are especially dramatic in regions like tropical forests that escaped large-scale human intervention until recently. Satellite photos of a small area in Brazil show how thoroughly a landscape can be altered in a short amount of time (Figure 18.39).

Every year, more and more forested land is cleared for agriculture (Figure 18.40). You might think that this land is needed to feed new mouths as the human population continues to grow, but that's not entirely the case. Unsustainable agricultural practices have degraded much of the world's cropland so severely that it is unusable. Researchers estimate that replacing worn-out farmland accounts for up to 80% of the deforestation occurring today. Forests are also being lost to logging, mining, and air pollution, problems that are hitting coniferous forests especially hard. (As we mentioned in the previous section, most temperate broadleaf forests were replaced by human enterprises long ago.) Land that hasn't been directly converted to food production and living space also bears the imprint of our presence. Roads penetrate regions that are otherwise unaltered, bringing pollution to the wilderness, providing avenues for new diseases to emerge, and slicing vast tracts of biome into segments that are too small to support a full array of species.

Land uses that provide resources such as food, fuel, and shelter are clearly beneficial to us. But natural ecosystems also provide services that support the human population—purification of air and water, nutrient cycling, and recreation, to name just a few. We'll return to the topic of ecosystem services in Chapter 20.

Fresh Water

The impact of human activities on freshwater ecosystems may pose an even greater threat to life on Earth—including ourselves—than the damage to terrestrial ecosystems. Freshwater ecosystems are being polluted by large amounts of nitrogen and phosphorus compounds

▼ **Figure 18.40 Deforestation by cutting and burning trees in the Amazon basin, Brazil.**

that run off from heavily fertilized farms or from live-stock feedlots. A wide variety of other pollutants, such as industrial wastes, also contaminate freshwater habitats, drinking water, and groundwater. Some regions of the world face dire shortages of water as a result of the overuse of groundwater for irrigation, extended droughts (partially caused by global climate change), or poor water management practices.

Las Vegas, the population center of Clark County, Nevada, is one example of a city whose water resources are increasingly stressed by drought and overuse. Figure 18.41a is a satellite photo of Las Vegas in 1973, when the population of Clark County was 319,400. Figure 18.41b shows the same area less than 30 years later, when the population had swelled to 1,620,748. In contrast to the disappearance of greenery in the photos of Brazilian rain forest, the mark of human activities in Figure 18.41b is the notable expansion of greenery—the result of watering lawns and golf courses. Las Vegas is situated in a high valley in the Mojave Desert. Where does it get the water to turn barren desert into green fields?

Las Vegas taps underground aquifers for some water, but its main water supply is Lake Mead. Lake Mead is an enormous reservoir formed by the Hoover Dam on the Colorado River, which in turn receives almost all of its water from snowmelt in the Rocky Mountains. With decreased annual snowfall, attributable largely to global warming, the flow of the Colorado has greatly diminished. The water level in Lake Mead has dropped drastically (Figure 18.42), and parched cities and farms farther downstream are pleading for more water.

It is clear that even with restrictions on water use, Las Vegas must look elsewhere to supply its needs.

Among other options, Las Vegas is eyeing the abundant supply of groundwater in the northern end of the valley where it lies. Although sparsely populated, that area is home to many ranchers whose livelihoods depend on the groundwater. It is also home to numerous endangered species. Not surprisingly, environmentalists and residents of the north valley are resisting efforts to pipe its groundwater to Las Vegas.

Nevada is just one of many places where the hard realities of climate change are beginning to affect daily life. Battles over water resources are shaping up throughout the arid West and Southwest of the United States, where changing precipitation patterns due to global warming are projected to continue the drought for many years to come. Record drought also hit the southeastern United States in 2006, draining reservoirs, threatening water supplies, and providing fuel for massive wildfires. Conditions there, however, have begun to improve.

While policymakers are dealing with current crises and planning how to manage resources in the future, researchers are seeking methods of sustainable agriculture and water use. Basic ecological research is an essential component of ensuring that sufficient food and water will be available for people now—and for the generations to come. You'll learn more about the goal of sustainability in the next two chapters. But first, let's take a closer look at a major threat to sustainability: global climate change. ☑

▼ Figure 18.42 **Low water level in Lake Mead.** The white "bathtub ring" is caused by mineral deposits on rocks that were once submerged.

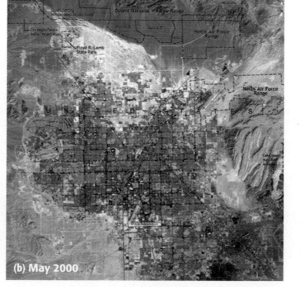

▼ Figure 18.41 **Satellite photos of Las Vegas, Nevada.**

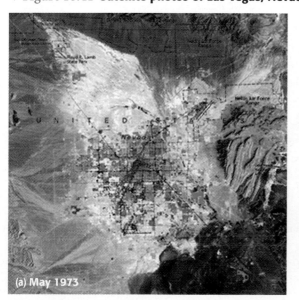

(a) May 1973

(b) May 2000

☑**CHECKPOINT**

1. What human activity has been responsible for the greatest loss of natural land cover?
2. Why is decreased snowfall in the Rocky Mountains a concern for people who live in Las Vegas?

Answers: 1. agriculture 2. Snowmelt from the Rockies flows into the Colorado River, which supplies water for Las Vegas residents.

Global Warming

CHECKPOINT

Why are gases such as CO_2 and methane called greenhouse gases?

Answer: They allow solar radiation to pass through the atmosphere but they prevent the heat from reflecting back out, much like the glass of a greenhouse retains the sun's heat inside the building.

Global Climate Change

Rising concentrations of carbon dioxide (CO_2) and certain other gases in the atmosphere are changing global climate patterns. This was the overarching conclusion of the assessment report released by the Intergovernmental Panel on Climate Change (IPCC) in 2007. Thousands of scientists and policymakers from more than 100 countries participated in producing the report, which is based on data published in hundreds of scientific papers. Thus, there is no debate among scientists regarding whether climate change is occurring. In this section, you'll learn why it is occurring, how it is affecting the biosphere, and what you can do about it.

The Greenhouse Effect and Global Warming

Why is Earth's atmosphere becoming warmer? A useful analogy is a greenhouse, which is used to grow plants when the weather outside is too cold. Its transparent glass or plastic walls allow solar radiation to pass through, but some of the heat that accumulates inside the building is trapped by the glass. Similarly, certain gases in Earth's atmosphere are transparent to solar radiation but absorb or reflect heat. Some of these so-called **greenhouse gases** are natural, including CO_2, water vapor, and methane. Others, such as chlorofluorocarbons (CFCs, found in some aerosol sprays and refrigerants), are synthetic. As **Figure 18.43** shows, greenhouse gases act as a blanket that traps heat in the atmosphere. For this reason, the increase in global temperatures is often called the **greenhouse effect**. This natural heating effect is highly beneficial. Without it, the average air temperature on Earth would be a frigid $-18°C$ ($-2.4°F$), far too cold for most life as we know it. However, increasing the insulation that the blanket provides is making Earth overly warm.

The signature effect of rapidly increasing greenhouse gases is the steady increase in the average global temperature, which has risen $0.8°C$ ($1.4°F$) over the last 100 years, with 75% of that increase occurring over the last three decades. Further increases of $2-4.5°C$ are likely by the end of this century, according to the 2007 IPPC report. Ocean temperatures are also rising, in deeper layers as well as at the surface. But the temperature increases are not distributed evenly around the globe. Warming is greater over land than sea, and the largest increases are in the northernmost regions of the Northern Hemisphere and in the Antarctic Peninsula (**Figure 18.44**). ☑

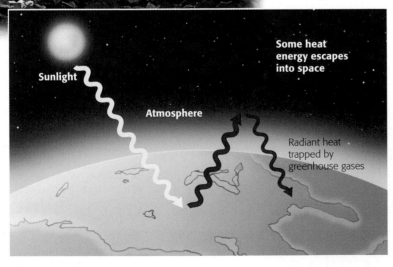

▼ **Figure 18.43 The greenhouse effect.** The atmosphere traps heat in the same way that glass keeps heat inside a greenhouse.

▼ Figure 18.44 **Temperature differences between the periods 2000–2005 and 1951–1981 (in °C).** The largest temperature increases are shown in dark red. Gray indicates regions for which no data are available.

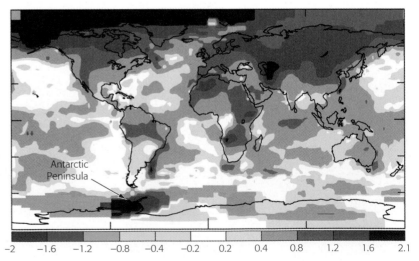

The Accumulation of Greenhouse Gases

After many years of data collection and debate, the vast majority of scientists are confident that human activities have caused the rising concentrations of greenhouse gases. Major sources of emissions include agriculture, landfills, and the burning of fossil fuels (oil, coal, and natural gas).

Let's take a closer look at CO_2, the dominant greenhouse gas. For 650,000 years, the atmospheric concentration of CO_2 did not exceed 300 parts per million (ppm); the concentration before the Industrial Revolution was 280 ppm. Today, atmospheric CO_2 is about 385 ppm and still rising (**Figure 18.45**). The levels of other greenhouse gases have increased dramatically, too. As you saw in Figure 6.2, CO_2 is removed from the atmosphere by the process of photosynthesis and stored in organic molecules such as carbohydrates. These molecules are eventually broken down by cellular respiration, releasing CO_2. Overall, uptake of CO_2 by photosynthesis roughly equals the release of CO_2 by cellular respiration (**Figure 18.46**). However, extensive deforestation has significantly decreased the incorporation of CO_2 into organic material. At the same time, CO_2 is

flooding into the atmosphere from the burning of fossil fuels and wood, a process that releases CO_2 from organic material much more rapidly than cellular respiration.

CO_2 is also exchanged between the atmosphere and the surface waters of the oceans. For decades, the oceans have acted as massive sponges, soaking up considerably more CO_2 than they have released. But now, the excess CO_2 has made the oceans more acidic, a change that could have a profound effect on marine communities. Many marine animals, including corals, many species of plankton, and many molluscs, will be unable to build their shells or exoskeletons as ocean acidification worsens. Their demise would spell trouble for the creatures that feed on them and ultimately for marine ecosystems around the world. ✔

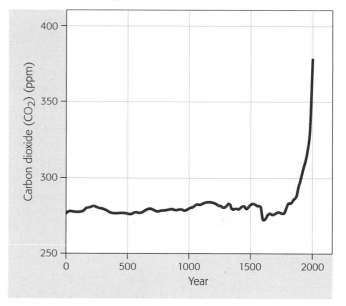

▼ Figure 18.45 **Atmospheric concentration of CO_2.** Notice that the concentration was relatively stable until the Industrial Revolution, which began in the late 1700s.

◄ Figure 18.46
How CO_2 enters and leaves the atmosphere.

Atmosphere

Photosynthesis   Respiration

Combustion of
fossil fuel

Ocean

☑CHECKPOINT

What is the major source of CO_2 released by human activities?

Answer: burning fossil fuels

How Does Climate Change Affect Species Distribution?

As you've learned in this chapter, abiotic factors of the environment are fundamental determinants of where organisms live. It stands to reason that changes in temperature and precipitation patterns will have a significant effect on the distribution of life. With rising temperatures, the ranges of many species have already shifted toward the poles or to higher elevations. For example, shifts in the ranges of many bird species have been reported; the Inuit peoples living north of the Arctic Circle have sighted birds such as robins in the region for the first time.

Let's examine how a team of ecologists investigated the impact of climate change on European butterflies.

Using the **observation** that the average temperature in Europe has risen 0.8°C and that butterflies are sensitive to temperature change, the researchers asked the **question**: Have the ranges of butterflies changed in response to the temperature changes? This question led to the **hypothesis** that butterfly range boundaries are shifting in line with the warming trend. The researchers **predicted** that butterfly species will be establishing new populations to the north of their former ranges, and populations at the southern edges of their ranges will become extinct. The **experiment** involved analyzing historical data on the ranges of 35 species of butterflies in Europe. The **results** showed that more than 60% of the species have pushed their northern range boundaries poleward over the last century, some by as much as 150 miles. The southern boundaries have simultaneously contracted for some species, but not for others.

Figure 18.47 shows the range shift for the species *Argynnis paphia*.

While some organisms have the dispersal ability and the room for northward population shifts, species that live on mountaintops or in polar regions have nowhere to go. For example, researchers in Costa Rica have reported the disappearance of 20 species of frogs and toads as warmer Pacific Ocean temperatures reduce the dry-season mists in their mountain habitats. And as we mentioned in the Biology and Society section, animals at both poles are also at risk.

▶ Figure 18.47 **Northward shift of *Argynnis paphia*.** On the map, orange represents the butterfly's range in 1970; its 1997 range is shown in light green.

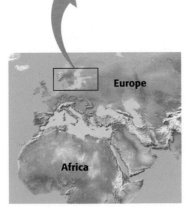

Argynnis paphia (silver-washed fritillary butterfly)

Effects of Climate Change on Ecosystems

The sentiment expressed by the poet John Donne that "no man is an island" is equally true for every other species in nature: Every species needs others to survive. Climate change is knocking some of these interactions out of sync. In many plants and animals, life cycle events are triggered by warming temperatures. Across the Northern Hemisphere, the warm temperatures of spring are arriving earlier. Satellite images show earlier greening of the landscape, and flowering occurs sooner. A variety of species, including birds and frogs, have begun their breeding seasons earlier. But for other species, the environmental cue that spring has arrived is daylength, which is not affected by climate change. Consequently, plants may bloom before pollinators have emerged, or eggs may hatch before a dependable food source is available.

The combined effects of climate change on forest ecosystems in western North America have spawned catastrophic wildfire seasons (**Figure 18.48**). In these regions, spring snowmelt in the mountains releases water into streams that sustain forest moisture levels over the summer dry season. With the earlier arrival of spring, snowmelt begins earlier and dwindles away before the dry season ends. As a result, the fire season is lasting longer. Meanwhile, bark beetles, which bore into conifers to lay their eggs, have benefited from global warming. Healthy trees can fight off the pests, but drought-stressed trees are too weak to resist (**Figure 18.49**). As a bonus for the beetles, the warmer weather allows them to reproduce twice a year rather than once. In turn, vast numbers of dead trees add fuel to a fire. Wildfires burn longer, and the number of acres burned has increased dramatically.

The map of terrestrial biomes (see Figure 18.27), which is primarily determined by temperature and rainfall, is also changing. Melting permafrost is shifting the boundary of the tundra northward as shrubs and conifers are able to stretch their ranges into the previously frozen ground. Prolonged droughts are extending the boundaries of deserts. Scientists also predict that great expanses of the Amazonian rain forest will gradually become savanna as increased temperatures dry out the soil.

Global climate change has significant consequences for humans, too, as changing temperature and precipitation patterns affect food production, the availability of fresh water, and the structural integrity of buildings and roads. All projections point to an even greater impact in the future. Unlike other species, however, humans can take action to reduce greenhouse gas emissions and maybe even reverse the warming trend. ✔

▼ **Figure 18.49 Pines in Southern California infested by bark beetles.** Red or light-colored foliage indicates dead or dying trees. Green trees are still healthy.

▼ **Figure 18.48 A forest fire near South Lake Tahoe, California, June 2007.**

✔CHECKPOINT

Why does the earlier arrival of spring temperatures affect life cycle events for some organisms?

Answer: In some organisms, life cycle events are triggered by warm temperatures.

Looking to Our Future

Emissions of greenhouse gases are accelerating. From 2000 to 2005, global CO_2 emissions increased *four times faster* than in the preceding ten-year span. At this rate, further climate change is inevitable. With effort and ingenuity, however, we may be able to begin reducing emissions. International negotiations are under way on a new climate change treaty to succeed the 1997 Kyoto Protocol, which required industrialized nations to reduce their greenhouse gas emissions.

Given the vast scope and complexity of the problem, you might think that the actions of a single individual would have little impact on greenhouse gas emissions. But it was the collective activities of individuals that caused—and are still causing—emissions to rise. The amount of greenhouse gas emitted as a result of the actions of a single individual is that person's **carbon footprint** (from the fact that the most important greenhouse gas is CO_2). A carbon footprint can be estimated using a set of rough calculations; several different calculators are available online.

Home energy use is one major contributor to the carbon footprint. It's easy to reduce your energy use by turning off the lights, TV, and other electronics when they aren't in use. You can also switch to energy-efficient lightbulbs. In addition, you can unplug items that consume electricity even when they are not in use; examples are cell phone chargers, videogame consoles, and printers.

Transportation is another significant part of the carbon footprint. If you have a car, keep it well maintained, consolidate trips, share rides with friends, and use alternative means of transportation whenever possible.

Manufactured goods are a third category in the carbon footprint. Every item you purchase generated its own carbon footprint in the process of going from raw materials to store shelves. You can reduce your carbon emissions by not buying unnecessary goods and by recycling—or better yet, reusing—items instead of putting them in the trash (**Figure 18.50**). Landfills are the largest human-related source of methane, a greenhouse gas more potent than CO_2. The methane is released by bacteria that decompose (break down) landfill waste.

Changes in your eating habits can also shrink your carbon footprint. Like landfills, the digestive system of cattle depends on methane-producing bacteria. The methane released by cattle and by the bacteria that decompose their manure accounts for about 20% of methane emissions in the United States. Thus, replacing beef and dairy products in your diet with fish, chicken, eggs, and vegetables reduces your carbon footprint. In addition, eating locally grown fresh foods may lower the greenhouse gas emissions that result from food processing and transportation (**Figure 18.51**). Many websites offer additional suggestions for reducing your carbon footprint. ✓

☑CHECKPOINT

What is a personal carbon footprint?

Answer: the amount of carbon that a person is responsible for emitting

▼ Figure 18.50 **Turning trash into treasure.** If you have things that you no longer need, don't trash them when you move out. Environmentally conscious students at Davidson College in North Carolina collect discarded items for local charities.

▼ Figure 18.51 **Eating locally grown foods may reduce your carbon footprint—and they taste good, too!**

EVOLUTION CONNECTION: Global Climate Change

Climate Change as an Agent of Natural Selection

Environmental change has always been a part of life; in fact, it is a key ingredient of evolutionary change. Will evolutionary adaptation counteract the negative effects of climate change on organisms? Researchers have documented microevolutionary shifts in a few populations, including red squirrels, a few bird species, and a tiny mosquito (**Figure 18.52a**). It appears that some populations, especially those with high genetic variability and short life spans, may avoid extinction by means of evolutionary adaptation. However, evolutionary adaptation is unlikely to save long-lived species such as polar bears and penguins (**Figure 18.52b**) that are experiencing rapid habitat loss. The rate of climate change is incredibly fast compared to major climate shifts in evolutionary history. If climate change continues on its present course, thousands of species—the Intergovernmental Panel on Climate Change estimates as many as 30% of plants and animals—will face extinction by midcentury.

▼ Figure 18.52 **Which species will survive climate change?**

(a) Pitcher plant mosquito. The pitcher plant mosquito, pictured inside the carnivorous plant for which it is named, may be able to evolve quickly enough.

(b) Adélie penguin. Adélie penguins, which live on the rapidly warming Antarctic Peninsula, are not likely to make it.

Chapter Review

SUMMARY OF KEY CONCEPTS

(MB) Go to the Study Area at **www.masteringbiology.com** for practice quizzes, my**e**Book, BioFlix™ 3-D animations, MP3 Tutor Sessions, videos, current events, and more.

An Overview of Ecology

Ecology is the scientific study of interactions between organisms and their environments. The environment includes abiotic (nonliving) and biotic (living) components. Ecologists use observation, experiments, and computer models to test hypothetical explanations of these interactions.

Ecology and Environmentalism

Human activities have had an impact on all parts of the biosphere. Ecology provides the basis for understanding and addressing these environmental problems.

A Hierarchy of Interactions

Ecologists study interactions at four increasingly complex levels.

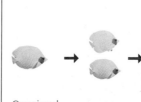

Organismal ecology (individual)

Population ecology (group of individuals)

Community ecology (all organisms in a particular area)

Ecosystem ecology (all organisms and abiotic factors)

Living in Earth's Diverse Environments

The biosphere is an environmental patchwork in which abiotic factors affect the distribution and abundance of organisms.

Abiotic Factors of the Biosphere

These include the availability of sunlight, water, nutrients, and temperature. In aquatic habitats, dissolved oxygen, salinity, current, and tides are also important. Additional factors in terrestrial environments include wind and fire.

The Evolutionary Adaptations of Organisms

Adaptation via natural selection results from the interactions of organisms with their environments.

Adjusting to Environmental Variability

Organisms also have adaptations that enable them to cope with environmental variability, including physiological, behavioral, and anatomical responses to changing conditions.

Biomes

A biome is a major terrestrial or aquatic life zone.

Freshwater Biomes

Freshwater biomes include lakes, ponds, rivers, streams, and wetlands. Lakes vary, depending on depth, with regard to light penetration, temperature, nutrients, oxygen levels, and community structure. Rivers change greatly from their source to the point at which they empty into a lake or an ocean.

Marine Biomes

Marine life is distributed into distinct realms (benthic and pelagic) and zones (photic, aphotic, and intertidal) according to the depth of the water, degree of light penetration, distance from shore, and open water versus bottom. The coral reef biome, which occurs in warm tropical waters above the continental shelf, has an abundance of biological diversity. A biome found near hydrothermal vents in the deep ocean is powered by chemical energy from Earth's interior instead of sunlight. Estuaries, located where a freshwater river or stream merges with the ocean, are one of the most biologically productive environments on Earth.

How Climate Affects Terrestrial Biome Distribution

The geographic distribution of terrestrial biomes is based mainly on regional variations in climate. Climate is largely determined by the uneven distribution of solar energy on Earth. Proximity to large bodies of water and the presence of landforms such as mountains also affect climate.

Terrestrial Biomes

Most biomes are named for major physical or climatic features and for their predominant vegetation. The major terrestrial biomes include tropical forest, savanna, desert, chaparral, temperate grassland, temperate broadleaf forest, coniferous forest, tundra, and polar ice. If the climate in two geographically separate areas is similar, the same type of biome may occur in both.

The Water Cycle

The global water cycle links aquatic and terrestrial biomes. Human activities are disrupting the water cycle.

Human Impact on Biomes

Land use by humans has altered vast tracts of forest and degraded the services provided by natural ecosystems. Unsustainable agricultural practices have depleted cropland fertility. Human activities have polluted freshwater ecosystems, which are vital for life. Agriculture, population growth, drought, and declining snowfall are all factors in the rapid depletion of freshwater resources in some regions.

Global Climate Change

The Greenhouse Effect and Global Warming

So-called greenhouse gases, including CO_2 and methane, increase the amount of heat retained in Earth's atmosphere. The accumulation of these gases has caused increases in the average global temperature.

The Accumulation of Greenhouse Gases

Human activities, especially the burning of fossil fuels, are responsible for the rise in greenhouse gases over the past century. Release of CO_2 has exceeded the amount that can be absorbed by natural processes.

Effects of Climate Change on Ecosystems

Climate change is disrupting interactions between species. Devastating wildfires are among the effects of climate change in certain ecosystems. Climate change is also shifting biome boundaries.

Looking to Our Future

Each person has a carbon footprint—that person's responsibility for a portion of global greenhouse gas emissions. We can take action to reduce our carbon footprints.

SELF-QUIZ

1. Place these levels of ecological study in order, from the least to the most comprehensive: community ecology, ecosystem ecology, organismal ecology, population ecology.

2. Name several abiotic factors that might affect the community of organisms living inside a home fish tank.

3. The formation of goose bumps on your skin in cold weather is an example of a (an) _____ response, while seasonal migration is an example of a (an) _____ response.

4. Which of the following sea creatures might be described as a pelagic animal of the aphotic zone?
   a. a coral reef fish
   b. a giant clam near a deep-sea hydrothermal vent
   c. an intertidal snail
   d. a deep-sea squid
   e. a species of phytoplankton

5. Identify the following biomes on the climograph below: tundra, coniferous forest, desert, grassland, temperate broadleaf forest, and tropical rain forest.

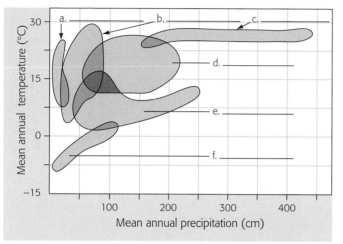

6. We are on a coastal hillside on a hot, dry summer day among evergreen shrubs that are adapted to fire. We are most likely standing in a _____ biome.

7. What three abiotic factors account for the rarity of trees in arctic tundra?

8. What human activity is responsible for the greatest amount of deforestation?

9. What is the greenhouse effect? How is the greenhouse effect related to global warming?

10. The recent increase in atmospheric CO_2 concentration is mainly a result of an increase in
  a. primary productivity.
  b. the absorption of heat radiating from Earth.
  c. the burning of fossil fuels and wood.
  d. cellular respiration by the increasing human population.

11. What populations of organisms are most likely to survive climate change via evolutionary adaptation?

Answers to the Self-Quiz questions can be found in Appendix D.

THE PROCESS OF SCIENCE

12. Design a laboratory experiment to measure the effect of water temperature on the population growth of a certain phytoplankton species from a pond.

13. Some people are not convinced that human-induced global climate change is a real phenomenon. Using your knowledge of the scientific process from Chapter 1 and the information from this chapter, develop arguments you could use to explain the scientific basis for saying that global climate change is truly occurring and that humans are responsible for it.

BIOLOGY AND SOCIETY

14. Near Lawrence, Kansas, there was a rare patch of the original North American temperate grassland that had never been plowed. It was home to numerous native grasses, annual plants, and grassland animals. Among the species present were two endangered plants. Environmental activists thought the area should be set aside as a nature preserve, and they started to raise money to save the patch of land. In 1990, the owner of the land plowed it, stating that there are no federal laws protecting endangered plants on private grasslands and that he did not want to be told what he could do with his property. What issues and values are in conflict in this situation? How could this story have had a more satisfactory ending for all concerned? What would you have done if you were an environmental activist? If you were the farmer?

15. Land clearing for agricultural use is responsible for most of the deforestation that is currently taking place. Agriculture also pumps enormous amounts of groundwater for irrigation and pollutes surface waters with nitrogen via runoff. But agriculture also supplies the world's food. Are the environmental consequences of agriculture simply the price we must pay to eat? What are the long-term costs? Suggest some ways to alleviate the environmental impact of agriculture.

16. Research your country's per capita (per person) carbon emissions. Calculate your own carbon footprint and compare it with the average for your country. Make a list of actions you are willing to take to reduce your personal carbon footprint. What actions could you take to persuade others to reduce their carbon footprints?

17. In the summer of 2007, unprecedented drought conditions brought the city of Atlanta, Georgia, within weeks of running out of water. Most of Atlanta's water comes from Lake Lanier, a reservoir that the Army Corps of Engineers created by damming the Chattahoochee River. With Lake Lanier drying up from lack of rainfall, Georgia petitioned the Corps, which manages the dam, to reduce the amount of water released downstream. The Corps refused, citing its obligation under the Endangered Species Act to protect the habitats of a species of sturgeon (a fish) and two species of mussel (a mollusc). Objections were also raised by Alabama and Florida, where hundreds of towns, recreational facilities, and power plants depend on the water released downstream. Some people thought that Atlanta authorities had brought the water shortage on themselves by allowing developers to build without considering whether adequate water was available. Florida also argued that reduction of freshwater inflows would harm its oyster fisheries. How would you prioritize the competing claims on the water from Lake Lanier? Who should allocate scarce water resources? How can cities and states plan more wisely for future shortages?

Population Ecology

Rabbit plague.
Methods to control the
rabbit population
included driving the
rabbits into pens to be
killed.

BIOLOGY AND SOCIETY: Invasive Species

Multiplying Like Rabbits

In 1859, 12 pairs of European rabbits were released on a ranch in southern Australia by an Englishman who wanted to hunt familiar game. The animals quickly became a nuisance. In 1865, 20,000 rabbits were killed on that ranch alone. By 1900, several hundred *million* rabbits were distributed over most of the continent. The rabbit invasion was a catastrophe in several ways. Their activities destroyed farm and grazing land by eating vegetation down to the roots. Especially in arid regions, the loss of plant cover led to soil erosion. Extensive underground rabbit burrows made grazing treacherous for cattle and sheep. The rabbits also competed directly with native marsupials, such as kangaroos and bilbies, for food and living space.

The rabbits were soon joined by another sportsman's favorite game animal, the European red fox. This adaptable predator also thrived in Australia, thanks to the seemingly endless supply of rabbits. The fox isn't a fussy eater, however. As they followed the spread of rabbits across the continent, red foxes also ate several species of native birds and small mammals to extinction, and they continue to threaten other native species today.

Rabbits, foxes, and a long list of other non-native animal and plant species are still entrenched in Australia, damaging the environment and costing hundreds of millions of dollars in economic losses. Eradication is impossible; the best that can be done is to minimize the impact of these species. Nor is Australia alone with this problem. For as long as humans have traveled the globe, they have carried—intentionally or accidentally—thousands of species to new habitats. Like the European rabbits and foxes, many of these non-native species have established populations that spread far and wide, leaving environmental havoc in their wake. As you learned at the end of Chapter 17, humans have also multiplied and spread far from our point of origin, and we have radically changed our environment. As you explore population ecology in this chapter, you'll learn about trends in human population growth and other applications of this area of ecological research. Along the way, you'll find out why those rabbits multiplied so freely and whether they have yet been brought under control.

An Overview of Population Ecology

In Chapter 18, you learned about the factors that characterize an environment, how organisms respond to those factors, and how environmental factors determine the distribution of organisms in biomes. In this chapter, we turn our attention to another of the four hierarchical levels of ecology: **population ecology**, the study of factors that affect population density and growth.

Ecologists generally define a **population** as a group of individuals of a single species that occupy the same general area. Members of a population are likely to interact and breed with one another. They rely on the same resources and are influenced by the same environmental factors. Population ecology focuses on the factors that influence a population's density (number of individuals per unit area or volume), structure (such as relative numbers of individuals of different ages), size (number of individuals), and growth rate (rate of change in population size) (Figure 19.1).

Population ecology also plays a key role in applied research. For example, it provides critical information for identifying and saving endangered species. Population ecology is being used to develop sustainable fisheries throughout the world and to manage wildlife

populations. Studying the population ecology of pests and pathogens provides insight into controlling how they spread. Population ecologists also study human population growth, one of the most critical environmental issues of our time.

Let's consider what a snapshot of a population might look like. The first question is, Which individuals are included in this population? A population's geographic boundaries may be natural, as with certain species of largemouth bass in an isolated lake. But ecologists often define a population's boundaries in more arbitrary ways that fit their research questions. For example, an ecologist studying the contribution of asexual reproduction to the population growth of sea anemones might define a population as all the anemones of one species in a particular tide pool. Another researcher studying the effects of hunting on deer might define a population as all the deer within a particular state. Yet another researcher, attempting to determine which segment of the human population will be most affected by the AIDS epidemic, might study the HIV infection rate of the human population in one nation or throughout the world.

▼ Figure 19.1 **An ecologist studying a population of black-browed albatross in the Falkland Islands, east of the tip of South America.**

Population Density

Our snapshot of a population would include **population density**, the number of individuals of a species per unit area or volume of the habitat: the number of largemouth bass per cubic kilometer (km³) of a lake, for example, or the number of oak trees per square kilometer (km²) in a forest, or the number of nematodes per cubic meter (m³) in the forest's soil. In rare cases, an ecologist can actually count all the individuals within the boundaries of the population. For example, we could count the total number of oak trees (say, 200) in a forest covering 50 km². The population density would be the total number of trees divided by the area, or 4 trees per square kilometer (4/km²).

In most cases, however, it is impractical or impossible to count all individuals in a population. Instead, ecologists use a variety of sampling techniques to estimate population density. For example, they might estimate the density of alligators in the Florida Everglades based on a count of individuals in a few sample plots of 1 km² each. Generally speaking, the larger the number and size of sample plots, the more accurate the estimates. Population densities may also be estimated not by counts of organisms but by indirect indicators, such as number of bird nests or rodent burrows (**Figure 19.2**).

Keep in mind that population density is not a constant number. It changes when individuals are born or die and when new individuals enter the population (immigration) or leave it (emigration).

Population Age Structure

The **age structure** of a population—the distribution of individuals in different age-groups—reveals information that is not apparent from population density. For instance, age structure can provide insight into the history of a population's survival or reproductive success and how it relates to environmental factors. **Figure 19.3** shows the age structure of males in a population of cactus finches on the Galápagos Islands in 1987. (See Figure 13.11 for other examples of Galápagos finches.) Four-year-old birds, born in 1983, made up almost half the population, while no 2- or 3-year-olds were present at all. Why was there such dramatic variation? For their food, cactus finches depend on vegetation, which in turn depends on rainfall. The 1983 baby boom resulted from unusually wet weather that produced abundant food. Severe droughts in 1984 and 1985 limited the food supply, preventing reproduction and causing many deaths. As we'll see later in this chapter, age structure is also a useful tool for predicting future changes in a population. ☑

▼ Figure 19.2 **An indirect census of a prairie dog population.** We could get a rough estimate of the number of prairie dogs in this colony in South Dakota by counting the number of mounds constructed by the rodents and then multiplying by the number of animals using a typical burrow.

▼ Figure 19.3 **Age structure for the males in a population of large cactus finches (inset) on one of the Galápagos Islands in 1987.**

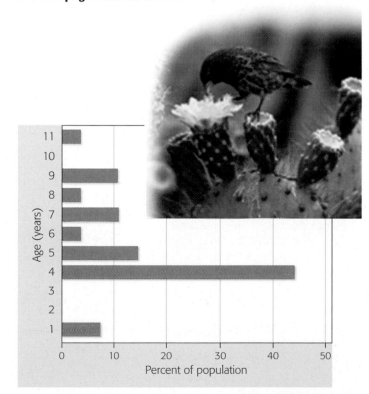

Life Tables and Survivorship Curves

Life tables track survivorship, the chance of an individual in a given population surviving to various ages. The life insurance industry uses life tables to predict how long, on average, a person of a given age will live. Starting with a population of 100,000 people, Table 19.1 shows the number who are expected to be alive at the beginning of each age interval, based on death rates in 2004. For example, 93,735 out of 100,000 people are expected to live to age 50. The chance of surviving to age 60, shown in the last column of the same row, is 0.939; about 94% of 50-year-olds will reach the age of 60. The chance of surviving to age 90, however, is only 0.412. Population ecologists have adopted this technique and constructed life tables to help them understand the structure and dynamics of various plant and animal species. By identifying the most vulnerable stage of the life cycle, life table data may also help conservationists develop effective measures to protect species whose populations are declining.

Ecologists represent life table data graphically in a **survivorship curve**, a plot of the number of individuals still alive at each age in the maximum life span (Figure 19.4). By using a percentage scale instead of actual ages on the *x*-axis, we can compare species with widely varying life spans on the same graph. The curve for the human population (red) shows that most people survive to the older age intervals. Ecologists refer to the shape of this curve as Type I survivorship. Species that

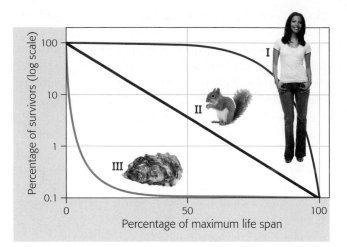

▲ Figure 19.4 **Three idealized types of survivorship curves.**

exhibit a Type I curve—humans and many other large mammals—usually produce few offspring but give them good care, increasing the likelihood that they will survive to maturity.

In contrast, a Type III curve (blue) indicates low survivorship for the very young, followed by a period when survivorship is high for those few individuals who live to a certain age. Species with this type of survivorship curve usually produce very large numbers of offspring but provide little or no care for them. Some species of fish, for example, produce millions of eggs at a time, but most of these offspring die as larvae from predation or other causes. Many invertebrates, such as oysters, also have Type III survivorship curves.

A Type II curve (black) is intermediate, with survivorship constant over the life span. That is, individuals are no more vulnerable at one stage of the life cycle than another. This type of survivorship has been observed in some invertebrates, lizards, and rodents. ☑

Life History Traits as Evolutionary Adaptations

A population's pattern of survivorship is a key feature of its **life history**, the set of traits that affect an organism's schedule of reproduction and survival. Let's take a closer look now at how natural selection affects the reproductive patterns that evolve in populations.

Key life history traits include the age at first reproduction, the frequency of reproduction, the number of offspring, and the amount of parental care given. For a given population in a particular environment, natural selection will favor the combination of life history traits that maximizes an individual's output of viable, fertile

Table 19.1	Life Table for the U.S. Population in 2004		
	Number Living at Start of Age Interval	Number Dying During Interval	Chance of Surviving Interval
Age Interval	(N)	(D)	1 − (D/N)
0−10	100,000	871	0.991
10−20	99,129	419	0.996
20−30	98,709	933	0.991
30−40	97,776	1,259	0.987
40−50	96,517	2,781	0.971
50−60	93,735	5,697	0.939
60−70	88,038	11,847	0.865
70−80	76,191	22,267	0.708
80−90	53,925	31,706	0.412
90+	22,219	22,219	0.000

☑**CHECKPOINT**

Sea turtles lay their eggs in nests on sandy beaches. Although a single nest may contain as many as 200 eggs, only a small proportion of the hatchlings survive the journey from the beach to the open ocean. Once sea turtles have matured, however, the mortality rate is low. What type of survivorship do sea turtles exhibit?

Answer: Type III

offspring. In other words, life history traits, like anatomical features, are shaped by evolutionary adaptation.

Natural selection cannot maximize all life history traits simultaneously because an organism has limited time, energy, and nutrients. For example, an organism that gives birth to a large number of offspring will not be able to provide a great deal of parental care. Consequently, the combination of life history traits represents trade-offs that balance the demands of reproduction and survival. Because selection pressures vary, life histories are very diverse. Nevertheless, ecologists have observed some patterns that are useful for understanding how natural selection influences life history characteristics.

One life history pattern is typified by small-bodied, short-lived animals (for example, insects and small rodents) that develop and reach sexual maturity rapidly, have a large number of offspring, and offer little or no parental care. In plants, "parental care" can be measured by the amount of nutritional material stocked in each seed. Many small, nonwoody plants (dandelions, for example) produce thousands of tiny seeds. Such organisms have an **opportunistic life history**, one that enables the plant or animal to take immediate advantage of favorable conditions. In general, populations with this life history pattern exhibit a Type III survivorship curve.

In contrast, some organisms have an **equilibrial life history**, a pattern of developing and reaching sexual maturity slowly and producing few, well-cared-for offspring. Organisms that have an equilibrial life history are typically larger-bodied, longer-lived species (for example, bears and elephants). Populations with this life history pattern exhibit a Type I survivorship curve. Plants with comparable life history traits include certain trees. For example, coconut palms produce relatively few seeds that are well stocked with nutrient-rich material. Table 19.2 compares key traits of opportunistic and equilibrial life history patterns.

What accounts for the differences in life history patterns?

Some ecologists hypothesize that the potential survival rate of the offspring and the likelihood that the adult will live to reproduce again are the critical factors. In a harsh, unpredictable environment, where the adult may have just one good shot at reproduction, it may be an advantage to invest in quantity rather than quality. On the other hand, in an environment where favorable conditions are more dependable, an adult is more likely to survive to reproduce again. Seeds are more likely to fall on fertile ground and newly emerged animals are more likely to survive to adulthood. In that case, it may be more advantageous for the adult to invest its energy in producing a few well-cared-for offspring at a time.

Of course, there is much more diversity in life history patterns than the two extremes described here. Nevertheless, the contrasting patterns are useful for understanding the interactions between life history traits and our next topic, population growth. ✓

☑**CHECKPOINT**

How does the term *opportunistic* capture the key characteristics of that life history pattern?

Answer: An opportunistic life history is characterized by an ability to produce a large number of offspring very rapidly when the environment affords a temporary opportunity for a burst of reproduction.

Dandelions have an opportunistic life history.

Elephants have an equilibrial life history.

Table 19.2	Some Life History Characteristics of Opportunistic and Equilibrial Populations	
Characteristic	**Opportunistic Populations (such as many wildflowers)**	**Equilibrial Populations (such as many large mammals)**
Climate	Relatively unpredictable	Relatively predictable
Maturation time	Short	Long
Life span	Short	Long
Death rate	Often high	Usually low
Number of offspring per reproductive episode	Many	Few
Number of reproductions per lifetime	Usually one	Often several
Timing of first reproduction	Early in life	Later in life
Size of offspring or eggs	Small	Large
Parental care	Little or none	Often extensive

Population Growth Models

Population size fluctuates as new individuals are born or immigrate into an area and others die or emigrate out of an area. Some populations—for example, trees in a mature forest—are relatively constant over time. Other populations change rapidly, even explosively. Consider a single bacterium that divides every 20 minutes. There would be two bacteria after 20 minutes, four after 40 minutes, eight after 60 minutes, and so on. In just 12 hours, the population would approach 70 billion cells. If reproduction continued for a day and a half—a mere 36 hours—there would be enough bacteria to form a layer a foot deep over the entire Earth. Population ecologists use idealized models to investigate how the size of a particular population may change over time under different conditions. We'll describe two simple mathematical models that illustrate fundamental concepts of population growth.

The Exponential Population Growth Model: The Ideal of an Unlimited Environment

Exponential population growth describes the expansion of a population in an ideal, unlimited environment. In the exponential model, the population size of each new generation is calculated by multiplying the current population size by a constant factor that represents the number of births minus the number of deaths. Let's look at how such a population grows. In **Figure 19.5**, we begin with a population of 20 rabbits, indicated on the *y*-axis. Each month, there are more rabbit births than deaths; as a result, the population size increases each month.

Notice in Figure 19.5 that each increase is larger than the previous one—the *rate* of population growth changes with population size. The increasing rate of population growth produces a J-shaped curve that is typical of exponential growth. The slope of the curve represents the growth rate of the population. At the outset, when the population is small, the curve is almost flat—over the first 4 months, the population increases by a total of only 37 individuals, an average of 9.25 births per month. By the end of 7 months, 105 rabbits have been added to the population and the growth rate has increased to an average of 15 births per month. The largest growth is seen in the period from 10 to 12 months, when an average of 85 rabbits are born each month. Thus, the exponential growth model explains how a few dozen rabbits can multiply into millions and overrun a continent.

Exponential population growth is common in certain situations. For example, a disturbance such as a fire, flood, hurricane, drought, or cold snap may suddenly reduce the size of a population. Organisms that have opportunistic life history patterns can rapidly take advantage of the lack of competition and quickly recolonize the habitat by exponential population growth. Human activity can also be a major cause of disturbance, and plants and animals with opportunistic life history traits commonly occupy road cuts, freshly cleared fields and woodlots, and poorly maintained lawns. However, no natural environment can sustain exponential growth indefinitely. ☑

▼ Figure 19.5 **Exponential growth of a rabbit population.**

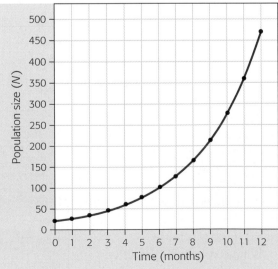

☑**CHECKPOINT**

Why does the exponential model of population growth produce a curve shaped like a J?

Answer: The rate of population growth increases as population size increases.

The Logistic Population Growth Model: The Reality of a Limited Environment

Most natural environments do not have an unlimited supply of the resources needed to sustain population growth. Environmental factors that hold population growth in check are called **limiting factors**. Limiting factors ultimately restrict the number of individuals that can occupy a habitat. Ecologists define **carrying capacity** as the maximum population size that a particular environment can sustain. In **logistic population growth**, the growth rate decreases as the population size approaches carrying capacity. When the population is at carrying capacity, the growth rate is zero.

You can see the effect of limiting factors in the graph in **Figure 19.6**, which shows the growth of a population of fur seals on St. Paul Island, off the coast of Alaska. (For simplicity, only the mated bulls were counted. Each has a harem of females, as shown in the photograph.) Before 1925, the seal population on the island remained low—between 1,000 and 4,500 mated bulls—because of uncontrolled hunting. After hunting was controlled, the population increased rapidly until about 1935, when it began to level off and started fluctuating around a population size of about 10,000 bull seals—the carrying capacity for St. Paul Island. In this instance, the main limiting factor was the amount of space suitable for breeding territories.

The carrying capacity for a population varies, depending on the species and the resources available in the habitat. For example, carrying capacity might be considerably less than 10,000 for a fur seal population on a smaller island with fewer breeding sites. Even in one location, it is not a fixed number. Organisms interact with other organisms in their communities, including predators, pathogens, and food sources, and these interactions may affect carrying capacity. Changes in abiotic factors may also increase or decrease carrying capacity. In any case, the concept of carrying capacity expresses an essential fact of nature: Resources are finite.

Ecologists hypothesize that selection for organisms exhibiting equilibrial life history patterns occurs in environments where the population size is at or near carrying capacity. Because competition for resources is keen under these circumstances, organisms gain an advantage by allocating energy to their own survival and to the survival of their descendants.

Figure 19.7 compares logistic growth (blue) with exponential growth (red). As you can see, the logistic curve is J-shaped at first, but gradually levels off to resemble an S shape as carrying capacity is reached. Both the logistic model and the exponential model are theoretical ideals of population growth. No natural population fits either one perfectly. However, these models are useful starting points for studying population growth. Ecologists use them to predict how populations will grow in certain environments and as a basis for constructing more complex models. ✓

▼ Figure 19.6 **Logistic growth of a seal population.**

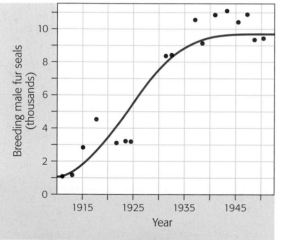

Figure 19.6 graph: Breeding male fur seals (thousands) vs. Year

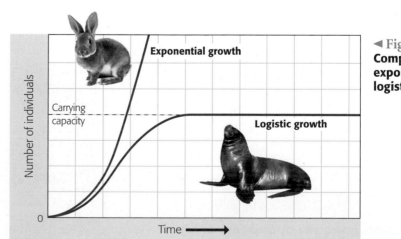

◄ Figure 19.7 **Comparison of exponential and logistic growth.**

Figure 19.7: Number of individuals vs. Time. Exponential growth, Logistic growth, Carrying capacity.

Regulation of Population Growth

Now let's take a closer look at how population growth is regulated in nature. What stops a population from continuing to increase after reaching carrying capacity?

Density-Dependent Factors

The major biological assumption of the logistic model is that increasing population density reduces the resources available for individual organisms, ultimately limiting population growth. The logistic model is a description of **intraspecific competition**—competition between individuals of the same species for the same limited resources. As population size increases, competition becomes more intense, and the growth rate declines in proportion to the intensity of competition. Thus, the growth rate of the population depends on population density. A **density-dependent factor** is a population-limiting factor whose effects intensify as the population increases in density. For example, as a limited food supply is divided among more and more individuals, birth rates may decline. **Figure 19.8a** shows the results of a bird population study where such a decline occurred. Clutch size (the number of eggs a female bird lays at one time) declined as the population density—the number of competitors—increased.

Density-dependent factors often depress a population's growth by increasing the death rate. In a laboratory experiment with flour beetles, for example, survivorship declined with increasing population density (**Figure 19.8b**). In a natural setting, plants that are growing close together may experience increased mortality. In an animal population, the death rate may climb as a result of increased disease transmission under crowded conditions or the accumulation of toxic waste products. Predation may also be an important cause of density-dependent mortality. A predator may concentrate on and capture more of a particular kind of prey as that prey becomes abundant.

A limited resource may be something other than food or nutrients. In many vertebrates that defend a territory, the availability of space may limit reproduction. For instance, the number of nesting sites on rocky islands may limit the population size of oceanic birds such as gannets, which maintain breeding territories (**Figure 19.9**).

▼ Figure 19.8 **Density-dependent regulation of population growth.**

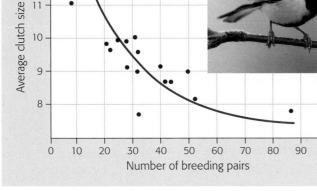

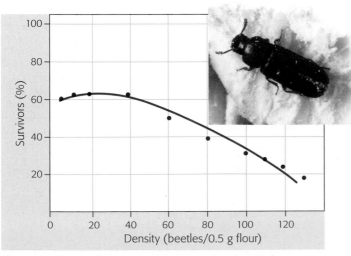

(a) **Decreasing birth rate with increasing density in a population of great tits**

▼ Figure 19.9 **Space as a limiting resource in a population of gannets.**

(b) **Decreasing survival rates with increasing density in a population of flour beetles**

Density-Independent Factors

In many natural populations, abiotic factors such as weather may limit or reduce population size well before other limiting factors become important. A population-limiting factor whose intensity is unrelated to population density is called a **density-independent factor**. If we look at the growth curve of such a population, we see something like exponential growth followed by a rapid decline rather than a leveling off. **Figure 19.10** shows this effect for a population of aphids, insects that feed on the sugary sap of plants. These and many other insects undergo virtually exponential growth in the spring and then rapid die-off when the weather turns hot and dry in the summer. A few individuals may remain, allowing population growth to resume if favorable conditions return. In some populations of insects—many mosquitoes and grasshoppers, for instance—the adults die off entirely, leaving behind eggs that will initiate population growth the following year. In addition to seasonal changes in the weather, environmental disturbances, such as fire, floods, and storms, can affect a population's size regardless of its density.

Over the long term, most populations are probably regulated by a complex interaction of density-dependent and density-independent factors. Although some populations remain fairly stable in size and are presumably close to a carrying capacity that is determined by biotic factors such as competition or predation, most populations for which we have long-term data do fluctuate.

Population Cycles

Some populations of insects, birds, and mammals undergo dramatic fluctuations in density with remarkable regularity. "Booms" characterized by rapid exponential growth are followed by "busts," during which the population falls back to a minimal level. A striking example is the boom-and-bust growth cycles of lemming populations that occur every three to four years. (Lemmings are small rodents that live in the tundra.) Some researchers hypothesize that natural changes in the lemmings' food supply may be the underlying cause. Another hypothesis is that stress from crowding during the "boom" may cause the "bust" by reducing reproduction via physiological mechanisms.

Figure 19.11 illustrates another example—the cycles of snowshoe hare and lynx. The lynx is one of the main predators of the snowshoe hare in the far northern forests of Canada and Alaska. About every ten years, both hare and lynx populations show a rapid increase followed by a sharp decline. What causes these boom-and-bust cycles? Since ups and downs in the two populations seem to almost match each other on the graph, does this mean that changes in one directly affect the other? For the hare cycles, there are three main hypotheses. First, cycles may be caused by winter food shortages that result from overgrazing. Second, cycles may be due to predator-prey interactions. Many predators other than lynx, such as coyotes, foxes, and great-horned owls, eat hares, and together these predators might overexploit their prey. Third, cycles may be affected by a combination of food resource limitation and excessive predation. Recent field studies support the hypothesis that the ten-year cycles of the snowshoe hare are largely driven by excessive predation, but are also influenced by fluctuations in the hare's food supplies. Long-term studies are the key to unraveling the complex causes of such population cycles. ☑

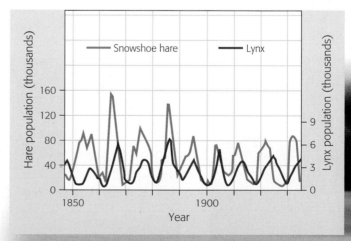

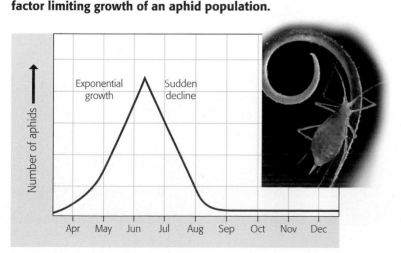

☑**CHECKPOINT**

List some density-dependent factors that limit population growth.

Answer: food and nutrient limitations, insufficient space for territories or nests, increase in disease and predation, accumulation of toxins

▼ **Figure 19.11 Population cycles of the snowshoe hare and the lynx.**

▼ **Figure 19.10 Weather change as a density-independent factor limiting growth of an aphid population.**

Applications of Population Ecology

In Chapter 18, you learned the extent to which people have converted Earth's natural ecosystems to ecosystems that are managed to produce goods and services for our benefit. In some cases, we try to increase populations of organisms that we wish to harvest and decrease populations of organisms that we consider pests. Other efforts are aimed at saving populations that are perilously close to extinction. Principles of population ecology help guide us toward these various resource management goals.

Conservation of Endangered Species

The U.S. Endangered Species Act defines an **endangered species** as one that is "in danger of extinction throughout all or a significant portion of its range." **Threatened species** are defined as those that are likely to become endangered in the foreseeable future. Endangered and threatened species are characterized by highly reduced population sizes. The challenge for conservationists is to determine the circumstances responsible for the population decline and try to remedy the situation.

The case of the endangered red-cockaded woodpecker provides an example of how conservationists rescued a species on the brink of extinction by identifying and correcting the factors responsible for reducing the population (**Figure 19.12**). The red-cockaded woodpecker requires longleaf pine forests, where it drills its nest

holes in mature, living pine trees. Originally found throughout the southeastern United States, the numbers of red-cockaded woodpeckers declined as suitable habitats were lost to logging and agriculture. Moreover, we have altered the composition of many of the remaining forests by suppressing the fires that are a natural occurrence in these ecosystems. Research has revealed that breeding birds tend to abandon nests when vegetation among the pines is thick and higher than about 15 feet. Apparently, the birds require a clear flight path between their home trees and the neighboring feeding grounds.

The recent recovery of the red-cockaded woodpecker from near-extinction to sustainable populations is largely due to recognizing and providing the key factors that regulate the bird's population growth. For example, controlled fires that reduce forest undergrowth help maintain mature pine trees and thus woodpecker populations as well. ✔

Sustainable Resource Management

A goal of wildlife managers, fishery biologists, and foresters is to gather the maximum amount of crop while sustaining the productivity of the resource for future harvests. In terms of population growth, this means maintaining a high population growth rate to replenish the population. According to the logistic growth model, the fastest growth rate occurs when the

✔CHECKPOINT

What was a key factor in the recovery of red-cockaded woodpecker populations?

Answer: removal of understory vegetation by controlled burns

▼ Figure 19.12 **Habitat of the red-cockaded woodpecker.**

A red-cockaded woodpecker perches at the entrance to its nest in a long-leaf pine tree.

High, dense undergrowth impedes the woodpeckers' access to feeding grounds.

Low undergrowth offers birds a clear flight path between nest sites and feeding grounds.

population size is at roughly half the carrying capacity of the habitat. Theoretically, a resource manager should achieve the best results by harvesting the population down to this level. However, the logistic model assumes that growth rate and carrying capacity are stable over time. Calculations based on these assumptions, which are not realistic for some populations, may lead to unsustainably high harvest levels that ultimately deplete the resource. In addition, human economic and political pressures often outweigh ecological concerns, and scientific information is frequently insufficient.

Fish, the only wild animals still hunted on a large scale, are particularly vulnerable to overharvesting. For example, in the northern Atlantic cod fishery, estimates of cod stocks were too high, and the practice of discarding young cod (not of legal size) at sea caused a higher mortality rate than was predicted. The fishery collapsed in 1992 and has not recovered (Figure 19.13).

Until the 1970s, marine fisheries concentrated on species such as cod that inhabit the continental shelves. As these resources dwindled, attention turned to deeper waters, most commonly the continental slopes below 600 m (see Figure 18.19). In many of these new fisheries, however, catches are initially high, but then quickly fall off as stocks are depleted. Deeper waters are colder, and food is relatively scarce. Fish that are adapted to this environment, such as Chilean sea bass and orange roughy, typically grow more slowly, take longer to reach maturity, and have a lower reproductive rate than continental shelf species. Sustainable catch rates can't be estimated without knowing these essential life history traits for the target species. In addition, knowledge of population ecology alone is not sufficient; sustainable fisheries also require knowledge of community and ecosystem characteristics. ✔

☑CHECKPOINT

Explain why managers often try to maintain populations of fish and game species at about half their carrying capacity.

Answer: to protect wildlife from overharvesting yet maintain lower population levels so that growth rate is high

▼ Figure 19.13 **Collapse of northern cod fishery off Newfoundland.** As of 2009, the fishery had not recovered.

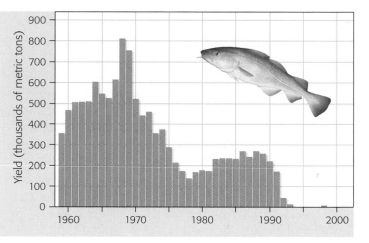

Invasive Species

Organisms that are introduced into non-native habitats, such as the rabbits featured in the Biology and Society section, have many similarities to crop pests. An **invasive species** is a non-native species that has spread far beyond the original point of introduction and causes environmental or economic damage by colonizing and dominating suitable habitats. In the United States alone, there are hundreds of invasive species, including plants, mammals, birds, fishes, arthropods, and molluscs. Worldwide, there are thousands more. Regardless of where you live, an invasive plant or animal is probably living nearby. Invasive species are a leading cause of local extinctions, a topic we'll return to in Chapter 20. And the economic costs of invasive species are enormous— an estimated $137 billion a year in the United States.

Not every organism that is introduced to a new habitat is successful, and not every species that survives in its new habitat becomes invasive. There is no single explanation for why any non-native species turns into a destructive pest, but invasive species typically exhibit an opportunistic life history pattern. A female European rabbit, for example, can begin to breed at the age of 4 months and produce five or more litters of four or five offspring per year.

The life history traits of cheatgrass (*Bromus tectorum*), an invasive plant of the arid western United States, have enabled its spectacular success as well. Its seeds were accidentally carried into the United States with grain from Asia and then spread by livestock. Currently, cheatgrass covers more than 60 million acres of rangeland that was formerly dominated by native grasses and sagebrush, and it claims an estimated 4,000 more acres every day. Cheatgrass seeds sprout during fall rains, and the roots continue to grow underground through the winter. Already established when the warm spring weather arrives, dense clumps of cheatgrass deprive native species or crops of soil moisture and mineral nutrients (Figure 19.14). It also produces seeds earlier and in greater abundance than its competitors. After

▼ Figure 19.14 **A reddish sea of cheatgrass threatens to overwhelm native sagebrush (green).**

cheatgrass seeds mature in early summer, the plants become extremely dry and flammable, creating abundant fuel that is easily ignited by lightning or a stray spark. Cheatgrass fires are more intense and occur much more frequently than the fires that native plants have evolved to tolerate. After a few fire cycles, the native plants are gone, robbing more than 150 species of birds and mammals of the food and shelter they derive from sagebrush. Global climate change is also hastening the transition of rangeland into fields of cheatgrass. Studies have shown that cheatgrass responds to increased CO_2 levels by growing faster and accumulating more tissue, which in turn becomes more fuel for the fires that extend its domain.

Many invasive animals also deprive native species of resources. For example, European starlings (**Figure 19.15**) aggressively displace woodpeckers, bluebirds, and swallows from their nesting sites, even pulling out nestlings to make room for their own offspring. These destructive birds were first released in New York in 1890. From the original introduction of 100 or so birds, it took just 50 years for starlings to spread throughout North America and reach an estimated population of over 200 million. Flocks numbering in the tens or hundreds of thousands devour grain and fruits from fields, feedlots, and orchards. In urban areas, the acidic droppings from dense starling populations are corrosive to buildings, and foul smelling, too. ✔

✔CHECKPOINT

What distinguishes invasive species from organisms that are introduced to non-native habitats but do not become invasive?

Answer: Invasive species spread far from where they were introduced, and they cause environmental or economic damage.

▼ Figure 19.15 **A large flock of starlings (inset) settling into a tree to roost for the night.**

Biological Control of Pests

The absence of biotic factors such as pathogens, predators, or herbivores that suppress population growth rate by increasing mortality may also contribute to the success of invasive species. Accordingly, efforts to eliminate or control these troublesome organisms often focus on **biological control**, the intentional release of a natural enemy to attack a pest population. Agricultural researchers have long been interested in identifying potential biological agents to control insects, weeds, and other organisms that reduce crop yield.

It was biological control that finally brought the Australian rabbit population down to manageable levels. After fruitless attempts to stem the tide by poisoning, destroying rabbit colonies, and even herding rabbits into corrals and shooting hundreds at a time, the Australian government introduced a lethal virus into the rabbits' environment. The campaign, begun in 1950, was initially successful at driving down the population size. However, following the familiar pattern of antibiotic resistance in bacteria and pesticide resistance in insects, natural selection favored rabbits that were resistant to the deadly virus. As the percentage of resistant rabbits in the population increased, the population size quickly rebounded. The virus evolved, too. Genetic strains of the virus that were not fatal or that killed their hosts more slowly left more descendants. Evolutionary changes such as these, in which an adaptation of one species leads to a counteradaptation in a second species, are known as **coevolution**. Over the course of many generations, the rabbit population became more resistant to the disease and the virus became less lethal. The government managed to stave off a complete resurgence of the rabbit population by introducing new viral strains, but in 1995, they had to switch to a different pathogen to maintain control.

Coevolution is just one potential pitfall of biological control. Among other potential problems, an imported control agent may turn out to be as invasive as its target. Thus, rigorous research is needed to assess the safety and efficacy of potential biological control agents.

Can Biological Control Defeat Kudzu?

Let's take a brief look at the search for a biological agent to control kudzu, an invasive vine known as "the plant that ate the South" (Figure 19.16). In the 1930s, the U.S. Department of Agriculture distributed plantings of this Asian import to help control erosion along road cuts and irrigation canals. Today, kudzu covers an estimated 12,000 square miles. With its formidable growth rate of up to a foot per day, kudzu climbs over forest trees and blankets the ground in dense greenery. The shoots die back in winter but quickly regenerate from the roots in the spring. Cold winters have limited its acquisition of new territory—the roots don't survive being frozen. However, as global climate change brings warmer winters, kudzu is advancing farther north.

Unlike many invasive species, kudzu does have natural enemies in the United States, but it easily outgrows the damage they inflict. Researchers are investigating the possibility that one of these native pathogens or insects

could be manipulated to provide effective control. Several possibilities have already been tested and discarded. For example, experiments showed that larvae of a species of moth have a prodigious appetite for kudzu. Further investigation, however, proved that the larvae actually preferred soybeans—an important crop species closely related to kudzu. At present, a fungal pathogen called *Myrothecium verrucaria* appears to be a promising candidate.

Researchers chose to test *M. verrucaria* because of **observations** that it causes severe disease in other weeds belonging to the same family as kudzu. Preliminary tests in a greenhouse, in controlled environment chambers, and in small outdoor plantings established that *M. verrucaria* kills kudzu when a high enough concentration of *M. verrucaria* spores is sprayed on the plants along with a "wetting agent" (a soap-like substance that reduces the surface tension of water). These findings led them to ask the **question**, Will *M. verrucaria* treatment work on an established stand of kudzu in a natural setting? Their **hypothesis** was that the *M. verrucaria* treatment that was most effective in the small outdoor plantings would also be most effective in a natural setting. Their **prediction** was that the greatest kudzu mortality would result from the treatment that sprayed the highest concentration of spores in combination with a wetting agent. The **results** of this field experiment, which are shown in Figure 19.17, support the hypothesis. However, these experiments are only the initial steps toward biological control of kudzu. A great deal of research is needed to ensure that the method is safe, effective, and practical.

▼ Figure 19.16 **Kudzu (*Pueraria lobata*).** Unseen in this photo are the plants that have died for lack of sunlight beneath the large kudzu leaves.

▼ Figure 19.17 **Biological control of a natural infestation of kudzu with the fungus *Myrothecium verrucaria*.**

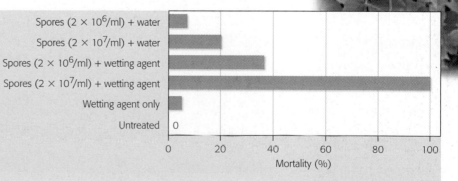

Integrated Pest Management

In contrast to enterprises such as fisheries, which harvest resources from natural ecosystems, agricultural operations create their own highly managed ecosystems. A typical crop population consists of genetically similar individuals (a monoculture) planted in close proximity to each other—a banquet laid out for the many plant-eating animals and pathogenic bacteria, viruses, and fungi in the community. The tilled, fertile ground nurtures weeds as well as crops. Thus, farmers wage an eternal war against pests that compete with their crop for soil minerals, water, and light; that siphon nutrients from the growing plants; or that consume their leaves, roots, fruits, or seeds. At home, you may be engaged in combat against pests on a smaller scale as you attempt to eradicate the weeds, insects, fungi, and bacteria that attack your lawn and garden or the mosquitoes that make your summer evenings miserable.

Like invasive species, most crop pests have an opportunistic life history pattern that enables them to rapidly take advantage of a favorable habitat. The history of agriculture abounds with examples of devastating pest outbreaks. For example, in the 1840s, a protist that causes a disease called potato rot wiped out the entire potato crop of Ireland in just a few years, causing the "Great Famine." Plant-eating insects can cause massive damage as well (**Figure 19.18**). When synthetic herbicides and insecticides such as DDT were developed in the 1940s, they quickly became the method of choice in agriculture. However, chemical solutions to pest problems bring numerous problems of their own. These chemicals are pollutants that can be carried great distances by air or water currents. And as you saw in Figure 13.14, natural selection may result in populations that are not affected by a pesticide. Furthermore, most insecticides kill both the pest and their natural predators. Because prey species often have a higher reproductive rate than predators, pest populations rapidly rebound before their predators can reproduce. There may be other unintended damage as well, such as killing pollinators that are essential for both agricultural and natural ecosystems.

Integrated pest management (IPM) uses a combination of biological, chemical, and cultural methods for sustainable control of agricultural pests. Researchers are also investigating IPM approaches to invasive species. IPM relies on knowledge of the population ecology of the pest and its associated predators and parasites, as well as plant growth dynamics. In contrast to traditional methods of pest control, IPM advocates tolerating a low level of pests rather than attempting total eradication. Thus, many pest control measures are aimed at lowering the habitat's carrying capacity for the pest population by using pest-resistant varieties and mixed-species plantings and rotating crops to deprive the pest of a dependable food source. Biological control is also used when possible. Pesticides are applied when necessary, but adherence to the principles of IPM prevents the overuse of chemicals. ✔

▼ Figure 19.18 **Plants infested with Japanese beetles (left) and grasshoppers (right).**

✔CHECKPOINT

Why is integrated pest management considered a more sustainable method of pest control than the use of chemicals alone?

Answer: Use of pesticides may result in resistant pest populations, which makes the pesticides less effective. Also, pesticide use is unsustainable in the long run because of problems with pollution.

Human Population Growth

Now that we have examined the regulation of population growth in other organisms, what about our own species? Let's begin by looking at the history of the human population and then consider some current and future trends in population growth.

The History of Human Population Growth

In the time it takes to read this sentence, approximately 26 babies will be born somewhere in the world and 10 people will die. An imbalance between births and deaths is the cause of population growth (or decline), and as the line graph in **Figure 19.19** shows, the human population is expected to continue increasing for at least the next several decades. The bar graph in Figure 19.19 tells a different part of the story. The number of people added to the population each year has been declining since the 1980s. How do we explain these patterns of human population growth?

Let's begin with the rise in world population from approximately 480 million people in 1500 to the current population of more than 6.7 billion. In the exponential population growth model introduced earlier in this chapter, we assumed that the *rate* of the population growth was constant—births and deaths were roughly equal. As a result, population growth depended only on the size of the existing population. Throughout most of human history, this assumption held true. Although parents had many children, mortality was also high. As a result, human population growth was initially very slow. (If we extended the *x*-axis of Figure 19.19 back in

time to year 1, when the population was roughly 300 million, the line would be almost flat for 1,500 years.) The 1 billion mark was not reached until the early 1800s. As economic development in Europe and the United States led to advances in nutrition and sanitation, and later, medical care, humans took control of their population's growth rate. At first, the death rate decreased while the birth rate remained the same. The net rate of increase rose, and population growth began to pick up steam by the beginning of the 1900s. By midcentury, improvements in nutrition, sanitation, and health care had spread to the developing world, spurring growth at a breakneck pace as birth rates far outstripped death rates.

As the world population skyrocketed from 2 billion in 1927 to 3 billion just 33 years later, some scientists became alarmed. They feared that Earth's carrying capacity would be reached and that density-dependent factors would maintain that population size through human suffering and death. But the overall growth rate peaked in 1962. In the more developed nations, advanced medical care continued to improve survivorship, but effective contraceptives held down the birth rate. As a result, the overall growth rate of the world's population began a downward trend as the difference between birth rate and death rate decreased.

Because economic development has occurred at different times in different regions, worldwide population growth rates reflect a mosaic of the changes occurring in different countries. In the most developed nations, the overall growth rates are near zero (**Table 19.3**). In the developing world, on the other hand, death rates have dropped, but high birth rates persist. As a result, these populations are growing rapidly. Of the nearly 82 million people added to the world each year, almost 80 million are in developing nations. ☑

☑CHECKPOINT

Why was the growth rate of the world's population so high during most of the 1900s? What accounts for the recent decrease in the growth rate of the world's population?

Answer: In the 1900s, the death rate decreased dramatically due to improved sanitation and health care, but the birth rate remained high. As a result, the overall population growth rate was high. The recent decrease in growth rate is the result of lower birth rates in some regions.

▼ Figure 19.19 **Five centuries of human population growth, projected to 2050.**

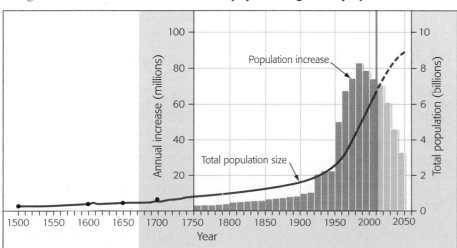

Table 19.3	Population Trends in 2008		
Population	**Birth Rate per 1,000**	**Death Rate per 1,000**	**Growth Rate (%)**
World	21	8	1.2
More developed countries	12	10	0.2
Less developed countries	23	8	1.5

Age Structures

Age structures, which were introduced at the beginning of this chapter, are helpful for predicting a population's future growth. **Figure 19.20** shows the estimated and projected age structures of Mexico's population in 1985, 2010, and 2035. In these diagrams, the area to the left of each vertical line represents the number of males in each age-group; females are represented on the right side of the line. The three different colors represent the portion of the population in their prereproductive years (0–14), prime reproductive years (15–44), and postreproductive years (45 and older). Within each of these broader groups, each horizontal bar represents the population in a 5-year age-group.

In 1985, each age-group was larger than the one above it, indicating a high birth rate. The pyramidal shape of this age structure is typical of a population that is growing rapidly. In 2010, the population's growth rate is lower—notice that the youngest (bottommost) three age-groups are roughly the same size. However, the population continues to be affected by its earlier expansion. This situation, which results from the increased proportion of women of child-bearing age in the population, is known as **population momentum**. Girls 0–14 years old in the 1985 age structure (outlined in pink) are in their reproductive prime in 2010, and girls who are 0–14 years old in 2010 (outlined in blue) will carry the legacy of rapid growth forward to 2035. Putting the brakes on a rapidly expanding population is like stopping a freight train; the actual event takes place

long after the decision to do it was made. Even when fertility is reduced to replacement rate (an average of two children per female), the total population size will continue to increase for several decades. Thus, the percentage of individuals under the age of 15 gives a rough idea of future growth. In the developing countries, about 30% of the population is in this age-group. In contrast, roughly 17% of the population of developed nations is under the age of 15. Population momentum also explains why the total population size in Figure 19.19 continues to increase even though fewer people are added to the population each year.

Age structure diagrams may also indicate social conditions. For instance, an expanding population has an increasing need for schools, employment, and infrastructure. A large elderly population requires that extensive resources be allotted to health care. Let's look at trends in the age structure of the United States from 1983 to 2033 (**Figure 19.21**). The noticeable bulge in the 1983 population (yellow screen) corresponds to the "baby boom" that lasted for about two decades after World War II ended in 1945. The large number of children swelled school enrollments, prompting construction of new schools and creating a demand for teachers. On the other hand, graduates who were born near the end of the boom faced stiff competition for jobs. Because they make up such a large segment of the population, boomers have had an enormous influence on social, economic, and political trends. They also

▼ Figure 19.20 **Population momentum in Mexico.**

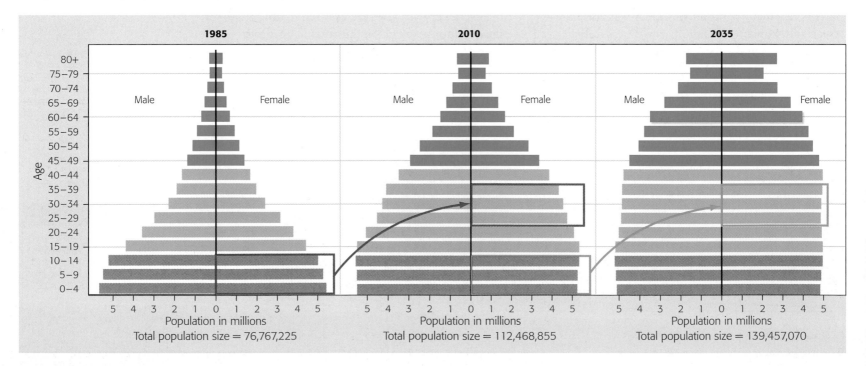

▼ Figure 19.21 **Age structures for the United States in 1983, 2008, and 2033 (projected).**

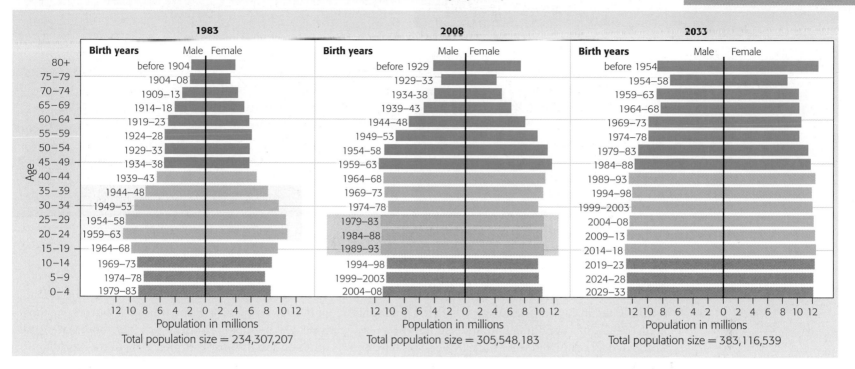

produced a boomlet of their own, seen in the 0–4 age group in 1983 and the bump (pink screen) in the 2008 age structure.

Where are the baby boomers now? The leading edge is reaching retirement age and will soon place pressure on programs such as Medicare and Social Security. In 2008, 60% of the U.S. population was between 20 and 64, the ages most likely to be in the workforce, and 13% was over 65. In 2033, the percentages are projected to be 54 and 19.7, respectively. In part, the increase in the elderly population results from people living longer. The percentage of the population over 80, which was 2.4% in 1983, is projected to rise to 5.6% in 2033 and 8%—close to 34 million people—in 2050. ✔

Our Ecological Footprint

Figure 19.19 shows that the world's population is growing exponentially, but at a slower rate than it did in the last century. The rate of increase, as well as population momentum, predicts that the populations of most developing nations will continue to increase for the foreseeable future. The U.S. Census Bureau projects a global population of 8 billion by 2025 and 9.5 billion by the middle of the century. But these numbers are only part of the story. Trillions of bacteria can live in a petri dish *if* they have sufficient resources. What is Earth's carrying capacity for the human population? Are there sufficient resources to sustain 8 or 9 billion people?

To accommodate all the people expected to live on our planet in the coming decades and improve the diets of those who are currently malnourished or undernourished, world food production must increase dramatically. As you learned in Chapter 18, cropland is being depleted and our current levels of water use are not sustainable. Overgrazing by the world's growing herds of livestock is turning vast areas of grassland into desert. And as additional natural resources are commandeered to support the expanding human population, many thousands of other species are expected to become extinct.

An **ecological footprint** is one approach to understanding resource availability and usage. The footprint is an estimate of the amount of land required to provide the raw materials an individual or a population consumes, including food, fuel, water, housing, and waste disposal. When the total area of ecologically productive land on Earth is divided by the global population, we each have a share of about 2.1 hectares (1 ha = 2.47 acres). In 2005, the average ecological footprint for the world's population was 2.7 ha. We have already overshot the planet's capacity to sustain us.

The ecological footprint of the United States (9.4 ha per person) is almost twice what its own land and

▲ Figure 19.22 **Families in India (left) and the United States (right) display their possessions.**

resources can support (5 ha per person); in other words, it has an enormous ecological deficit. Looking at Figure 19.22, it is not difficult to understand why. Compared to a family in rural India, Americans have an abundance of possessions. We also consume a disproportionate amount of food and fuel. By this measure, the ecological impact of affluent nations such as the United States is potentially as damaging as the unrestrained population growth in the developing world. So the problem is not just overpopulation, but overconsumption.

Figure 19.23 is a graphic representation of the disparity in consumption throughout the world. Countries are drawn in proportion to the amount of resources they consume. Regions depicted in red and pink have large ecological deficits—the average personal footprint is far greater than the global average.

The world's richest countries, with 20% of the global population, use 86% of the world's resources, leaving just 14% of global resources—energy, food, water, and other essentials—for the other 80% of the world's population to share. Indeed, the poorest 20% of the population accounts for just 1.3% of resource consumption. Some researchers estimate that providing everyone with the same standard of living as the United States would require the resources of three more planet Earths.

If you would like to learn about your personal resource consumption, several versions of online quizzes can provide a rough estimate of your ecological footprint. Like the carbon footprint calculators described in Chapter 18, these tools are useful for learning how to reduce your environmental impact. ☑

▼ Figure 19.23 **World map with area corresponding to ecological footprint.**

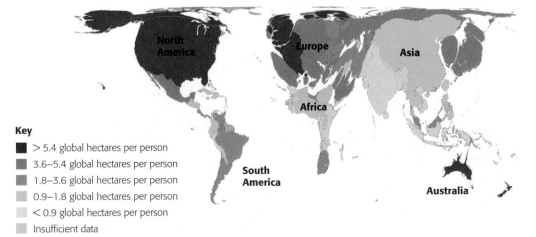

Key
■ > 5.4 global hectares per person
■ 3.6–5.4 global hectares per person
■ 1.8–3.6 global hectares per person
■ 0.9–1.8 global hectares per person
■ < 0.9 global hectares per person
■ Insufficient data

☑CHECKPOINT

How does an individual's large ecological footprint affect Earth's carrying capacity?

Answer: The more resources required to sustain an individual, the lower Earth's carrying capacity will be. (That is, Earth can sustain fewer people if each of those people consumes a large share of available resources.)

Humans as an Invasive Species

The magnificent pronghorn antelope (*Antilocapra americana*) is the descendant of ancestors that roamed the open plains and shrub deserts of North America millions of years ago (Figure 19.24). With strides that cover 6 m (20 ft) or more at its top speed of 97 km/h (60 mph), it is easily the fastest mammal on the continent. The pronghorn's speed is more than a match for its major predator, the wolf, which typically takes adults that have been weakened by age or illness. What selection pressure promoted such extravagant speed? Ecologists hypothesize that the pronghorn's ancestors were running from the now-extinct American cheetah, a fleet-footed predator that bore some similarities to the more familiar African cheetah.

Cheetahs were not the only danger in the pronghorn's environment. During the Pleistocene epoch, which lasted from 1.8 million to 10,000 years ago, North America was also home to an intimidating list of other predators: lions, jaguars, saber-toothed cats with canines up to 7 inches long, and towering short-faced bears, which stood 11 feet tall and weighed $\frac{3}{4}$ of a ton. There were plenty of potential prey for these fearsome predators, including massive ground sloths, bison with horns that spread 10 feet, elephant-like mammoths, a variety of horses and camels, and several species of pronghorns.

Of all these species of large mammals, only *Antilocapra americana* remained at the end of the Pleistocene. The others went extinct during a relatively brief period of time that coincided with the spread of humans throughout North America. Although the cause of the extinctions has been hotly disputed, many scientists think that the human invasion, combined with climate change at the end of the last ice age, were responsible. Taken together, changes in the biotic and abiotic environments happened too rapidly for an evolutionary response by these large mammals.

The role of humans in the Pleistocene extinctions was merely a preview of things to come. The human population continues to increase, colonizing almost every corner of the globe. Like other invasive species, we change the environment of the other organisms that share our habitats. As the scope and speed of human-induced environmental changes increase, extinctions are occurring at an accelerating pace. This rapid loss of biodiversity will be our unifying thread in the next chapter.

▼ Figure 19.24
A pronghorn antelope racing across the North American plains.

Chapter Review

An Overview of Population Ecology

A population consists of members of a species living in the same place at the same time. Population ecology focuses on the factors that influence a population's size, growth rate, density, and structure.

Population Density

Population density, the number of individuals of a species per unit area or volume, can be estimated by a variety of sampling techniques.

Population Age Structure

A graph showing the distribution of individuals in different age-groups often provides useful information about the population.

Life Tables and Survivorship Curves

A population's pattern of mortality is a key feature of life history. A life table tracks survivorship and mortality in a population. Survivorship curves can be classified into three general types, depending on the rate of mortality over the entire life span.

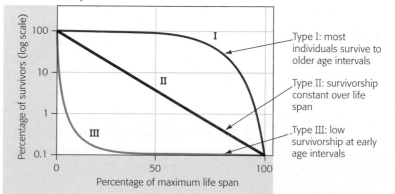

Type I: most individuals survive to older age intervals

Type II: survivorship constant over life span

Type III: low survivorship at early age intervals

Life History Traits as Evolutionary Adaptations

Life history traits are shaped by evolutionary adaptation; they may vary within a species and may change as the environmental context changes. Most populations probably fall between the extreme opportunistic life histories (reach sexual maturity rapidly; produce many offspring; little or no parental care) of many insects and the equilibrial life histories (develop slowly and produce few, well-cared-for offspring) of many larger-bodied species.

Population Growth Models

The Exponential Population Growth Model: The Ideal of an Unlimited Environment

Exponential population growth is the accelerating increase that occurs when growth is unlimited. The exponential model predicts that the larger a population becomes, the faster it grows. Exponential growth in nature is generally a short-lived consequence of organisms being introduced to a new or underexploited environment.

The Logistic Population Growth Model: The Reality of a Limited Environment

Logistic population growth occurs when growth is slowed by limiting factors. The logistic model predicts that a population's growth rate will be small when the population size is either small or large, and highest when the population is at an intermediate level relative to the carrying capacity.

Regulation of Population Growth

Over the long term, most population growth is limited by a mixture of density-independent and density-dependent factors. Density-dependent factors intensify as a population increases in density, increasing the death rate, decreasing the birth rate, or both. Density-independent factors affect the same percentage of individuals regardless of population size. Some populations have regular boom-and-bust cycles.

Applications of Population Ecology

Conservation of Endangered Species

Endangered and threatened species are characterized by very small population sizes. One approach to conservation is identifying and attempting to supply the critical combination of habitat factors needed by the population.

Sustainable Resource Management

According to the logistic growth model, a population's highest growth rate occurs when the population size is half the carrying capacity. For some populations, harvesting at this level produces sustainable yields. However, some populations don't meet the assumptions of the logistic model. Thorough scientific knowledge is needed to estimate yields that are truly sustainable.

Invasive Species

Invasive species are non-native organisms that spread far beyond their original point of introduction and cause environmental and economic damage. Typically, invasive species have an opportunistic life history pattern, and many are superior to native species at obtaining resources.

Biological Control of Pests

Biological control, the intentional release of a natural enemy to attack a pest population, is sometimes effective against invasive species. However, coevolution may eventually lessen the effectiveness of biological control. Researchers are currently investigating whether a fungal pathogen could be used to control kudzu, an invasive plant.

Integrated Pest Management

Crop scientists have developed integrated pest management (IPM) strategies, combinations of biological, chemical, and cultural methods, to deal with agricultural pests. IPM may prove useful at combating invasive species, too.

Human Population Growth

The History of Human Population Growth

The human population grew rapidly during the 1900s and currently stands at more than 6.7 billion. A shift from high birth and death rates to low birth and death rates has lowered the rate of growth in more developed countries. In developing nations, death rates have dropped, but birth rates are still high.

Age Structures

The age structure of a population affects its future growth. The wide base of the age structure of Mexico in 1985—the 0–14 age-group—predicts continued population growth in the next generation. Population momentum is the continued growth that occurs (despite fertility having been reduced to replacement rate) as a result of girls in the 0–14 age-group of a previously expanding population reaching their childbearing years. Age structures may also indicate social and economic trends, as in the age structure on the right below.

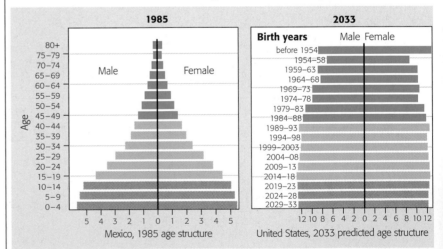

Our Ecological Footprint

The ecological footprint represents the amount of land per person needed to support a nation's resource needs. The carrying capacity of the world may already be smaller than the population's ecological footprint. There is a huge disparity between resource consumption in more developed and less developed nations.

SELF-QUIZ

1. What two values would you need to know to figure out the human population density of your community?

2. If members of a species produce a large number of offspring but provide minimal parental care, then a Type _____ survivorship curve is expected. In contrast, if members of a species produce few offspring and provide them with long-standing care, then a Type _____ survivorship curve is expected.

3. Label the following characteristics as typical of opportunistic (O) or equilibrial (E) life history patterns:
   _____ produce many offspring per reproductive episode
   _____ typical of habitats that have unpredictable climate or frequent disturbances
   _____ life span typically long
   _____ maturation time short
   _____ often provide extensive care to offspring

4. Use this graph of the idealized exponential and logistic growth curves to complete the following.

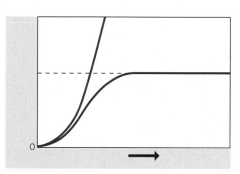

   a. Label the axes and curves on the graph.
   b. What does the dotted line represent?
   c. For each curve, indicate and explain where population growth is the most rapid.
   d. Which of these curves better represents global human population growth?

5. Which of the following shows the effects of a density-dependent limiting factor?
   a. A forest fire kills all the pine trees in a patch of forest.
   b. Early rainfall triggers the explosion of a locust population.
   c. Drought decimates a wheat crop.
   d. Rabbits multiply, and their food supply begins to dwindle.

6. Which life history pattern is typical of invasive species?

7. Skyrocketing growth of the human population since the beginning of the Industrial Revolution appears to be mainly a result of
   a. migration to thinly settled regions of the globe.
   b. better nutrition boosting the birth rate.
   c. a drop in the death rate due to better nutrition and health care.
   d. the concentration of humans in cities.

8. How does the age structure of the human population explain the current surplus in the Social Security fund?

9. According to the ecological footprint study produced in 2005,
   a. the carrying capacity of the world is 10 billion.
   b. the carrying capacity of the world would increase if all people ate more meat.
   c. the current demand on global resources by more developed countries is less than the resources available in those countries.
   d. the United States has a larger ecological footprint than its own resources can provide.
   e. nations with the largest ecological footprints have the fastest population growth rates.

Answers to the Self-Quiz questions can be found in Appendix D.

THE PROCESS OF SCIENCE

10. If researchers establish that *Myrothecium verrucaria* is an effective biological control agent against kudzu, they must then demonstrate that the pathogen will not harm desirable species such as soybeans (a close relative of kudzu). Describe experiments that could fulfill this purpose.

11. Propose a hypothesis to explain why North American mammals such as cheetahs, short-faced bears, and ground sloths did not survive the changes in their environment through evolutionary adaptation.

BIOLOGY AND SOCIETY

12. The mountain gorilla, spotted owl, giant panda, snow leopard, and grizzly bear are all endangered by human encroachment on their environments. Another thing these animals have in common is that they all have equilibrial life history traits. Why might they be more easily endangered than animals with opportunistic life history traits? What general type of survivorship curve would you expect these species to exhibit? Explain your answer.

13. Many people regard the rapid population growth of less developed countries as our most serious environmental problem. Others think that resource consumption in more developed countries is actually a greater threat to the environment. What kinds of problems result from these two issues? Which do you think is the greater environmental threat?

20 Communities and Ecosystems

Enjoying nature.
A kayaker and humpback whale share the waters of the Inside Passage in southeast Alaska.

CHAPTER CONTENTS

BIOLOGY AND SOCIETY: The Loss of Biodiversity

Does Biodiversity Matter?

As the human population has expanded, hundreds of species have become extinct and thousands more are endangered or threatened with extinction. These changes represent a loss in biological diversity, or *biodiversity*. Biodiversity loss goes hand in hand with the disappearance of natural ecosystems. Only about a quarter of Earth's land surfaces remain untouched by human alterations, and our activities have had an enormous impact on aquatic biomes as well. Most of us are only aware of the extent of biodiversity loss when it is brought to our attention by media headlines or biology textbooks. And frankly, it may be hard for us to imagine why it matters. In industrialized countries, clean water flows from our taps; food, clothing, and other necessities are readily available; and wastes vanish into plumbing fixtures or landfills. Life is good.

Most people appreciate the direct benefits provided by certain ecosystems. For example, you are probably aware that we use resources, such as water, wood, and fish and shellfish, that come from natural or near-natural ecosystems. You may know that 25% of all prescription drugs contain substances derived from plants. And many people enjoy activities such as hiking or whitewater rafting in pristine ecosystems, while others enjoy nature in less strenuous ways. But human well-being also depends on less obvious services provided by ecosystems whose biodiversity is intact. Healthy ecosystems purify air and water, decompose wastes, and recycle nutrients. Wetlands buffer coastal populations against hurricanes, reduce the impact of flooding rivers, and filter pollutants. Natural vegetation helps retain fertile soil and prevent landslides. Other ecosystem services include the control of agricultural pests by natural predators and pollination of crops. Some scientists have attempted to assign an economic value to these benefits, arriving at an average annual value of ecosystem services at $33 trillion. In contrast, the global gross national product for the same year was $18 trillion. Although rough, these estimates make the important point that we cannot afford to take biodiversity for granted.

In this chapter, we'll examine the interactions among organisms and how those relationships determine the features of communities. On a larger scale, we'll explore the dynamics of ecosystems. Finally, we'll consider how scientists are working to save biodiversity.

The Loss of Biodiversity

☑**CHECKPOINT**

How does the loss of genetic diversity endanger a population?

Answer: A population with decreased genetic diversity has less ability to evolve in response to environmental change.

As mentioned previously, **biodiversity** is short for biological diversity, the variety of living things that you learned about in Unit 3. It includes genetic diversity, species diversity, and ecosystem diversity. Thus, the loss of biodiversity encompasses more than just the fate of individual species.

Genetic Diversity

When species are lost, so are their unique genes. The genetic diversity within populations of a species is the raw material that makes microevolution and adaptation to the environment possible. If local populations are lost and the number of individuals in a species declines, so do the genetic resources for that species. Severe reduction in genetic variation threatens the survival of a species. The enormous genetic diversity of all the organisms on Earth has great potential benefit for people, too. Many researchers and biotechnology leaders are enthusiastic about the potential that genetic "bioprospecting" holds for future development of new medicines, industrial chemicals, and other products. Bioprospecting may also hold the key to the world's food supply. For example, researchers are currently scrambling to stop the spread of a deadly new strain of wheat stem rust, a fungal pathogen that has devastated harvests in eastern Africa and central Asia. At least 75% of the wheat varieties planted worldwide are susceptible to this pathogen, but researchers hope to find a resistance gene in the wild relatives of wheat (**Figure 20.1**).

Species Diversity

In view of the damage we are doing to the biosphere, ecologists believe that we are pushing species toward extinction at an alarming rate. The present rate of species loss may be as much as 1,000 times higher than at any time in the past 100,000 years. Several researchers estimate that at the current rate of destruction, over half of all currently living plant and animal species will be gone by the end of this century. **Figure 20.2** shows two recent victims. Here are some examples of where things stand:

- Approximately 12% of the 9,934 known bird species and 20% of the 5,416 known mammalian species in the world are threatened with extinction.

- About 20% of the known freshwater fishes in the world have either become extinct during human history or are seriously threatened.

- Roughly 32% of all known amphibian species are either near extinction or endangered.

- Of the approximately 20,000 known plant species in the United States, 200 species have become extinct since dependable records have been kept, and 730 species are endangered or threatened. ☑

▼ Figure 20.2 **Recent additions to the list of human-caused extinctions.**

A Chinese river dolphin. Formerly a resident of the Yangtze River, the Chinese river dolphin was the first large animal to become extinct since the 1950s.

Golden toads. These small frogs, unique to high-altitude Costa Rican cloud forests, have not been seen since 1989. The males shown here are gathered near a puddle to wait for mates.

Ecosystem Diversity

Ecosystem diversity is the third component of biological diversity. Recall from Chapter 18 that an ecosystem includes both the organisms and the abiotic factors in a particular area. Because of the network of interactions among populations of different species within an ecosystem, the local extinction of one species can have a negative effect on the entire ecosystem. In addition, **ecosystem services**, functions performed by an ecosystem that directly or indirectly benefit people, have been lost. These vital services include air and water purification, climate regulation, and erosion control. In Figure 18.39, you saw an all-too-common example of ecosystem damage and loss—the destruction of tropical forests to make room for and support the growing human population. Like tropical forests, coral reefs are rich in species diversity (Figure 20.3). An estimated 20% of the world's coral reefs have been destroyed by human activities, and 24% are in imminent danger of collapse. Scientists predict that another 26% of coral reefs will succumb in the next few decades if they are not protected. ☑

Causes of Declining Biodiversity

Ecologists have identified four main factors responsible for the loss of biodiversity: habitat destruction and fragmentation, invasive species, overexploitation, and pollution. The ever-expanding size and dominance of the human population are at the root of all four factors.

Habitat Destruction

The massive destruction and fragmentation of habitats caused by agriculture, urban development, forestry, and mining pose the single greatest threat to biodiversity (Figure 20.4). According to the International Union for the Conservation of Nature, which compiles information on the conservation status of species worldwide, habitat destruction is implicated in the decline of 73% of the species in modern history that have become extinct, endangered, vulnerable, or rare. We'll take a closer look at the consequences of habitat fragmentation later in this chapter.

Invasive Species

Ranking second behind habitat loss as a cause of biodiversity loss is the introduction of invasive species. In Chapter 19, we saw how uncontrolled population growth of human-introduced species to non-native habitats has caused havoc when the introduced species have competed with, preyed on, or parasitized native species. The lack of interactions with other species that could keep the newcomers in check is often a key factor in a non-native species becoming invasive.

▼ Figure 20.4 **Habitat destruction.** In a controversial method known as mountaintop removal, mining companies blast the tops off of mountains and then scoop out coal. The earth removed from the mountain is dumped into a neighboring valley.

▼ Figure 20.3 **Coral reef, a colorful display of biodiversity.**

☑CHECKPOINT

When ecosystems are destroyed, the services they provide are lost. What are some examples of ecosystem services?

Answer: Possible answers are mentioned in the Biology and Society section as well as on this page. Services include air and water purification; climate regulation; erosion control; recreation; and resources used by people, such as wood, water, and food.

Overexploitation

The marine fisheries we talked about in Chapter 19 demonstrate how people can overexploit wildlife by harvesting at rates that exceed the ability of populations to rebound (**Figure 20.5**). Tigers, the American bison, and Galápagos tortoises are among the terrestrial species whose numbers have been drastically reduced by excessive commercial harvesting, poaching, or sport hunting. Overharvesting also threatens some plants, including rare trees such as mahogany and rosewood that produce valuable wood.

Pollution

Acid precipitation is a threat to forest and aquatic ecosystems. Aquatic ecosystems also may be polluted by toxic chemicals, nutrients, or other contaminants. Air or water pollution is a contributing factor in declining populations of hundreds of species worldwide. ☑

▲ Figure 20.5 **Bluefin tuna ready for sale.** Bluefin is the tastiest tuna for sushi and sashimi, but the surging popularity of those dishes has left the species plummeting toward extinction—and the price skyrocketing. A single fish can fetch more than $100,000.

Community Ecology

On your next walk through a field or woodland, or even across campus or your own backyard, observe the variety of species present. You may see birds in trees, butterflies on flowers, dandelions in the grass of a lawn, or lizards darting for cover as you approach. Each of these organisms interacts with other organisms as it goes about looking for food, nesting sites, living space, or shelter. An organism's biotic environment includes not just individuals from its own population, but also populations of other species living in the same area. Such an assemblage of species living close enough together for potential interaction is called a **community**. In **Figure 20.6**, the lion, the zebra, the hyena, the vultures, the plants, and the unseen microbes are all members of an ecological community in Kenya.

Interspecific Interactions

Our study of communities begins with **interspecific interactions**—that is, interactions between species. Interspecific interactions can be classified according to the effect on the populations concerned, which may be helpful (+) or harmful (−). In some cases, two populations in a community vie for a resource such as food or space. The effect of this interaction is generally negative for both species (−/−)—neither species has access to the full range of resources offered by the habitat. On the other hand, some interspecific interactions benefit both parties (+/+). For example, the interactions between flowers and their pollinators are mutually beneficial. In a third type of interspecific interaction, one species exploits another species as a source of food. The effect of this interaction is clearly beneficial to one population and harmful to the other (+/−). In the next several pages, you will learn more about these interspecific interactions and how they affect communities. We will also look at some examples of interspecific interactions as powerful agents of natural selection.

▼ Figure 20.6 **Diverse species interacting in a Kenyan savanna community.**

Interspecific Competition (−/−)

In the logistic model of population growth (see Figure 19.6), increasing population density reduces the amount of resources available for each individual. Intraspecific (within-species) competition for limited resources ultimately limits population growth. In **interspecific competition** (between-species competition), the population growth of a species may be limited by the population densities of competing species as well as by the density of its own population.

What determines whether populations in a community compete with each other? Each species has an **ecological niche**, defined as the sum of its use of the biotic and abiotic resources in its environment. For example, the ecological niche of a small bird called the Virginia's warbler (Figure 20.7a) includes its nest sites and nest-building materials, the insects it eats, and climatic conditions such as the amount of precipitation and the temperature and humidity that enable it to survive. In other words, the ecological niche encompasses everything the Virginia's warbler needs for its existence. The ecological niche of the orange-crowned warbler (Figure 20.7b) includes some of the same resources used by the Virginia's warbler. Thus, when these two species inhabit the same area, they are competitors.

Ecologists investigated the effects of interspecific competition between populations of these two birds in a community in central Arizona. When they removed either Virginia's warblers or orange-crowned warblers from the study site, members of the remaining species were significantly more successful in raising their offspring. Thus, interspecific competition had a direct, negative effect on reproductive fitness.

If the ecological niches of two species are too similar, they cannot coexist in the same place. Ecologists call this the **competitive exclusion principle**, a concept introduced by Russian ecologist G. F. Gause, who demonstrated this effect with an elegant series of experiments. Gause used two closely related species of protists, *Paramecium caudatum* and *P. aurelia*. First, he established the carrying capacity for each species separately under the conditions used to culture (grow) them in the laboratory (Figure 20.8, top graph). Then he cultured the two species in the same habitat. Within two weeks, the *P. caudatum* population had crashed (bottom graph). Gause concluded that the requirements of these two species were so similar that the superior competitor—in this case, *P. aurelia*—deprived *P. caudatum* of essential resources. ☑

▼ Figure 20.8 **Competitive exclusion in laboratory populations of *Paramecium.***

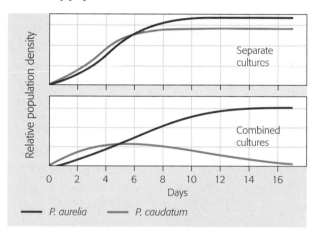

Paramecium aurelia

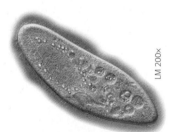

Paramecium caudatum

▼ Figure 20.7 **Species that use similar resources.**

(a) **Virginia's warbler**

(b) **Orange-crowned warbler**

Mutualism (**+/+**)

In **mutualism**, both species benefit from an interaction. For example, symbiotic root-fungus associations called mycorrhizae (see Figure 16.2) are mutualistic. Mutualism can also occur between species that are not symbiotic, such as flowers and their pollinators. Coral reef ecosystems depend on the mutualism between certain species of coral animals and unicellular algae. Reefs are constructed by successive generations of colonial corals that secrete an external calcium carbonate skeleton. The formation of the skeleton must outpace erosion and competition for space from fast-growing seaweeds. Massive coral reefs are made possible by the millions of algae that live in the cells of each coral polyp (Figure 20.9). In turn, the reef provides the food, shelter, and living space that support the splendid diversity of the reef community. The sugars that the algae produce by photosynthesis provide at least half of the energy used by the coral animals. In return, the algae gain a secure shelter that allows access to light. They also use the coral's waste products, including CO_2 and ammonia, a valuable source of nitrogen.

▲ Figure 20.9 **Mutualism.** Coral polyps are inhabited by unicellular algae.

Predation (**+/−**)

Predation refers to an interaction in which one species (the predator) kills and eats another (the prey). Because predation has such a negative impact on the reproductive success of the prey, numerous adaptations for predator avoidance have evolved in prey populations through natural selection. For example, some prey species, like the pronghorn antelope, run fast enough to escape their predators (see the Evolution Connection section in Chapter 19). Others, like rabbits, flee into shelters. Still other prey species rely on mechanical defenses, such as the porcupine's sharp quills or the hard shells of clams and oysters.

Adaptive coloration is a type of defense that has evolved in many species of animals. Camouflage, called **cryptic coloration**, makes potential prey difficult to spot against its background (Figure 20.10). **Warning coloration**, bright patterns of yellow, red, or orange in combination with black, often marks animals with effective chemical defenses. Predators learn to associate these color patterns with undesirable consequences, such as a noxious taste or painful sting, and avoid potential prey with similar markings. The vivid colors of the poison dart frog (Figure 20.11), an inhabitant of Costa Rican rain forests, warn of noxious chemicals in the frog's skin.

A prey species may also gain significant protection through mimicry, a "copycat" adaptation in which one species looks like another. For example, the pattern of alternating red, yellow, and black rings of the harmless scarlet king snake resembles the bold color pattern of the venomous eastern coral snake (Figure 20.12). Some insects have combined protective coloration with adaptations of body structures in elaborate disguises. For instance, there are insects that resemble twigs, leaves, and bird droppings. Some

▼ Figure 20.10
Cryptic coloration.
Camouflage conceals the Argentine horned frog from predators.

► Figure 20.11
Warning coloration of a poison dart frog.

▲ Figure 20.12 **Mimicry in snakes.** The color pattern of the nonvenomous scarlet king snake (left) is similar to that of the venomous eastern coral snake (right).

even do a passable imitation of a vertebrate. For example, the colors on the dorsal side of a hawk moth caterpillar are an effective camouflage, but when disturbed, the caterpillar flips over to reveal the snakelike eyes of its ventral side (**Figure 20.13**). Eyespots that resemble vertebrate eyes are common in several groups of moths and butterflies. A flash of these large "eyes" startles would-be predators. In other species, an eyespot may deflect a predator's attack away from vital body parts.

Herbivory (+/−)

Herbivory is the consumption of plant parts or algae by an animal. Although herbivory is not usually fatal, a plant whose body parts have been partially eaten by an animal must expend energy to replace the loss. Consequently, numerous defenses against herbivores have evolved in plants. Spines and thorns are obvious antiherbivore devices, as anyone who has plucked a rose from a thorny rosebush or brushed against a spiky cactus knows. Chemical toxins are also very common in plants. Like the chemical defenses of animals, toxins in plants

are distasteful, and herbivores learn to avoid them. Among such chemical weapons are the poison strychnine, produced by a tropical vine called *Strychnos toxifera*; morphine, from the opium poppy; nicotine, produced by the tobacco plant; mescaline, from peyote cactus; and tannins, from a variety of plant species. Other defensive compounds that are not toxic to humans but may be distasteful to herbivores are responsible for the familiar flavors of peppermint, cloves, and cinnamon (**Figure 20.14**). Some plants even produce chemicals that cause abnormal development in insects that eat them. Chemical companies have taken advantage of the poisonous properties of certain plants to produce pesticides. For example, nicotine is used as an insecticide. ☑

☑**CHECKPOINT**

People find most bitter-tasting foods objectionable. Why do you suppose we have taste receptors for bitter-tasting chemicals?

Answer: Taste receptors sensitive to bitter chemicals presumably enabled the ancestors of humans to identify potentially toxic plants when they foraged for food and thus to survive longer.

▼ Figure 20.13 **An insect mimicking a snake.** When disturbed, the hawk moth larva flips over (left) and resembles a snake (right).

▼ Figure 20.14 **Flavorful plants.**

Peppermint. Parts of the peppermint plant yield a pungent oil.

Cloves. The cloves used in cooking are the flower buds of this plant.

Cinnamon. Cinnamon comes from the inner bark of this tree.

Parasites and Pathogens (+/−)

Both plants and animals may be victimized by parasites or pathogens. These interactions are beneficial to one species (the parasite or pathogen) and harmful to the other, known as the host. A **parasite** lives on or in a **host** from which it obtains nourishment. Pathogens are disease-causing bacteria, viruses, fungi, or protists that can be thought of as microscopic parasites. You may recall learning about several invertebrate parasites in Chapter 17, including flukes and tapeworms and a variety of roundworms, which live inside a host organism's body. External parasites include arthropods such as ticks, lice, mites, and mosquitoes, which attach to their victims temporarily to feed on blood or other body fluids. Plants are also attacked by parasites, including roundworms and aphids, tiny insects that tap into the phloem to suck plant sap (see Figure 19.10). In any parasite population, reproductive success is greatest for

individuals that are best at locating and feeding on their hosts. For example, some aquatic leeches first locate a host by detecting movement in the water and then confirm its identity based on the host's body temperature and chemical cues on its skin.

Non-native pathogens, whose impact can be rapid and dramatic, have provided some opportunities to investigate the effects of pathogens on communities. In one example, ecologists studied the consequences of the epidemic of chestnut blight. The loss of chestnuts, massive canopy trees that once dominated many forest communities in North America, had a significant impact on community composition and structure. Trees such as oaks and hickories that had formerly competed with chestnuts became more numerous; overall, the diversity of tree species increased. Dead chestnut trees also furnished niches for other organisms, such as insects, cavity-nesting birds, and, eventually, decomposers. ☑

Trophic Structure

Now that we have looked at how populations in a community interact with one another, let's consider the community as a whole. The feeding relationships among the various species in a community are referred to as its **trophic structure**. A community's trophic structure determines the passage of energy and nutrients from plants and other photosynthetic organisms to herbivores and then to predators. The sequence of food transfer between trophic levels is called a **food chain**.

Figure 20.15 compares a terrestrial food chain and an aquatic food chain. Starting at the bottom, the trophic level that supports all others consists of autotrophs, which ecologists call **producers**. Photosynthetic producers use light energy to power the synthesis of organic compounds. Plants are the main producers on land. In water, the producers are mainly photosynthetic protists and cyanobacteria, collectively called phytoplankton. Multicellular algae and aquatic plants are also important producers in shallow waters. In a few communities, such as those surrounding hydrothermal vents, the producers are chemosynthetic prokaryotes.

All organisms in trophic levels above the producers are heterotrophs, or **consumers**, and all consumers depend directly or indirectly on the output of producers. **Herbivores**, which eat plants, algae, or phytoplankton, are the **primary consumers**. Primary consumers on land include grasshoppers and many other insects, snails, and certain vertebrates, such as grazing mammals and birds that eat seeds and fruits. In aquatic environments, primary consumers include a variety of

▼ **Figure 20.15 Examples of food chains.** The arrows trace the transfer of food from producers through various levels of consumers in terrestrial and aquatic ecosystems.

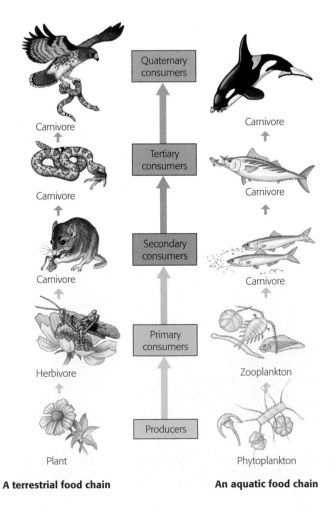

A terrestrial food chain　　　An aquatic food chain

zooplankton (mainly protists and microscopic animals such as small shrimps) that eat phytoplankton.

Above the primary consumers, the trophic levels are made up of **carnivores**, which eat the consumers from the level below. On land, **secondary consumers** include many small mammals, such as the mouse shown in Figure 20.15 eating an herbivorous insect, and a great variety of birds, frogs, and spiders, as well as lions and other large carnivores that eat grazers. In aquatic ecosystems, secondary consumers are mainly small fishes that eat zooplankton. Higher trophic levels include **tertiary consumers**, such as snakes that eat mice and other secondary consumers. Most communities have secondary and tertiary consumers. As the figure indicates, some also have a higher level, **quaternary consumers**. These include hawks in terrestrial ecosystems and killer whales in the marine environment.

Figure 20.15 shows only those consumers that eat living organisms. Some consumers derive their energy from **detritus**, the dead material left by all trophic levels, including animal wastes, plant litter, and dead organisms. **Detritivores**, which are often called scavengers, consume detritus. They may occupy any consumer level, depending on the source of the detritus. A great variety of animals, including earthworms, many rodents and insects, and vultures, are detritivores. Detritivores in aquatic communities include catfish, crabs, and crayfish. **Decomposers**, consisting of prokaryotes, fungi, and protists, secrete enzymes that digest molecules in organic material and convert them to inorganic forms (**Figure 20.16**). Enormous numbers of microscopic decomposers in the soil and in the mud at the bottom of lakes and oceans break down most of the community's organic materials to inorganic compounds that plants or phytoplankton can use. Decomposition of organic materials is essential for all communities and, indeed, for the continuation of life on Earth.

▼ Figure 20.16 **Fungi decomposing a dead log.**

Biological Magnification

As you learned in Chapter 18, toxic human-made chemicals, including pesticides and industrial wastes, frequently find their way into aquatic ecosystems. Organisms don't metabolize these chemicals, when these chemicals are consumed, they remain in the body. **Figure 20.17** shows the fate of chemicals called PCBs (organic compounds used in electrical equipment until 1977) in a Great Lakes food chain. Zooplankton feed on phytoplankton that have been contaminated by PCBs from the water. Because each individual consumes many phytoplankton, the concentration of PCBs is higher in zooplankton than in phytoplankton. For the same reason, the concentration of PCBs increases in the individuals that occupy each successive trophic level. This accumulation of toxins in the tissues of consumers in a food chain is called **biological magnification**. Thus, the top-level predators—herring gulls in Figure 20.17—have the highest concentrations of PCBs in the food chain and are the organisms most severely affected by any toxic compounds in the environment. In the research results shown in Figure 20.17, the concentration of PCBs in herring gulls eggs in the Great Lakes food chain was almost 5,000 times higher than that measured in phytoplankton. Fewer of these contaminated eggs hatched, resulting in a decline in the reproductive success of herring gulls.

▼ Figure 20.17 **Biological magnification of PCBs in a Great Lakes food chain in the early 1960s.**

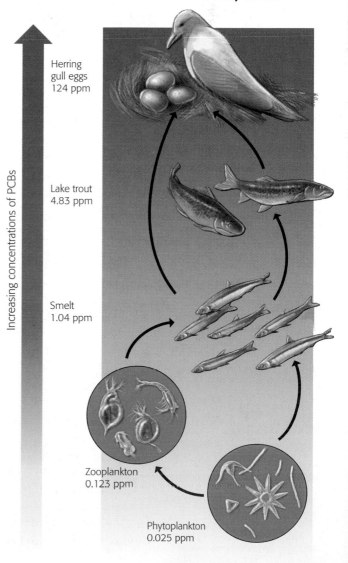

Increasing concentrations of PCBs

Herring gull eggs 124 ppm

Lake trout 4.83 ppm

Smelt 1.04 ppm

Zooplankton 0.123 ppm

Phytoplankton 0.025 ppm

Food Webs

Actually, few, if any, communities are so simple that they are characterized by a single unbranched food chain. Several types of primary consumers usually feed on the same plant species, and one species of primary consumer may eat several different plants. Such branching of food chains occurs at the other trophic levels as well. For example, adult frogs, which are secondary consumers, eat several insect species that may also be eaten by various birds. In addition, some consumers feed at several different trophic levels. An owl, for instance, may eat mice, which are mainly primary consumers that may also eat certain invertebrates. But an owl may also feed on snakes, which are strictly carnivorous. **Omnivores**, including humans, eat producers as well as consumers of different levels. Thus, the feeding relationships in a community are usually woven into elaborate **food webs**.

Figure 20.18 shows a simplified food web. An actual food web would involve many more organisms at each trophic level, and most of the animals would have a more diverse diet than shown in the figure. Note that some species feed at more than one trophic level and that the same species may be eaten by more than one consumer. ☑

▶ Figure 20.18 **A simplified food web for a Sonoran desert community.** As in the food chains of Figure 20.15, the arrows in this web indicate "who eats whom," the direction of nutrient transfers. We also continue the color coding introduced in Figure 20.15 for the trophic levels and food transfers.

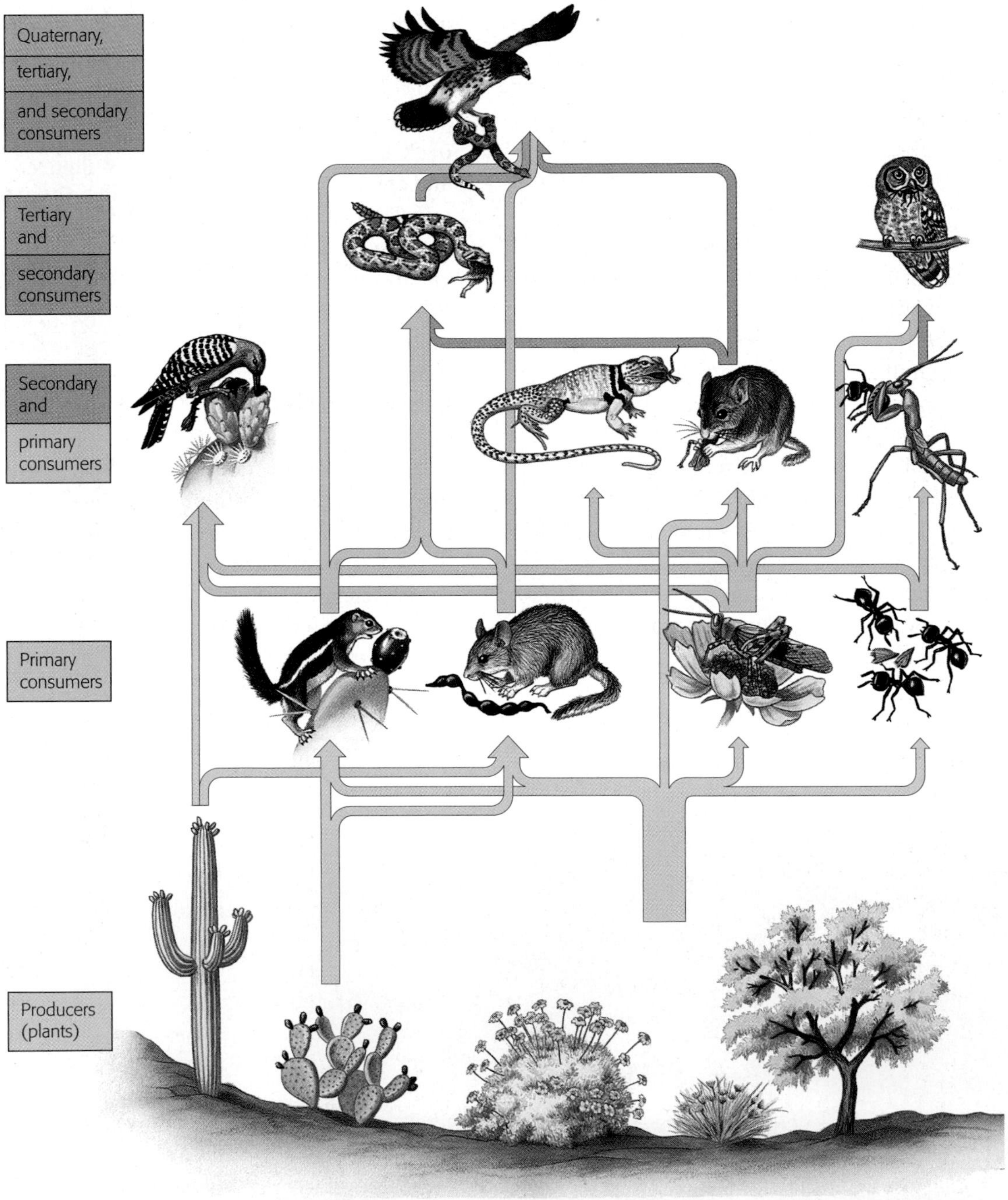

Quaternary, tertiary, and secondary consumers

Tertiary and secondary consumers

Secondary and primary consumers

Primary consumers

Producers (plants)

Species Diversity in Communities

The **species diversity** of a community—the variety of species that make up the community—has two components. The first component is **species richness**, or the number of different species in the community. The other component is the **relative abundance** of the different species, the proportional representation of a species in a community. To understand why both components are important for describing species diversity, imagine walking through the woodlands shown in **Figure 20.19**. On the path through woodland A, you would pass by four different species of trees, but most of the trees you encounter would be the same species. Now imagine walking on a path through woodland B. You would see the same four species of trees that you saw in woodland A—the species richness of the two woodlands is the same. However, woodland B might seem more diverse to you because no single species predominates. As **Figure 20.20** shows, the relative abundance of one species in woodland A is much higher than the relative abundances of the other three species. In woodland B, all four species are equally abundant. As a result, species diversity is greater in woodland B. Because plants provide food and shelter for many animals, a diverse plant community promotes animal diversity.

Although the abundance of a dominant species such as a forest tree can have an impact on the diversity of other species in the community, a nondominant species may also exert control over community composition. A **keystone species** is a species whose impact on its community is much larger than its total mass or abundance indicates. The term *keystone species* was derived from the wedge-shaped stone at the top of an arch that locks the other pieces in place. If the keystone is removed, the arch collapses. A keystone species occupies an ecological niche that holds the rest of its community in place.

To investigate the role of a potential keystone species in a community, ecologists compare diversity when the species is present or absent. Experiments by Robert Paine in the 1960s were among the first to provide evidence of the keystone species effect. Paine manually removed a predator, a sea star of the genus *Pisaster* (**Figure 20.21**), from experimental areas within the intertidal zone of the Washington coast. The result was that *Pisaster*'s main prey, a mussel, outcompeted many of the other shoreline organisms (algae, barnacles, and snails, for instance) for the important resource of space on the rocks. The number of different organisms present in experimental areas dropped from more than 15 species to fewer than 5 species.

Ecologists have identified other species that play a key role in ecosystem structure. For instance, the decline of sea otters off the western coast of Alaska allowed populations of sea urchins, their main prey, to increase. The abundance of urchins, which consume seaweeds such as kelp, has resulted in the loss of many of the kelp "forests" (see Figure 15.25) and the diversity of marine life that they support. In many ecosystems, however, ecologists are just beginning to understand the complex relationships among species; the value of an individual species may not be apparent until it is gone. ✔

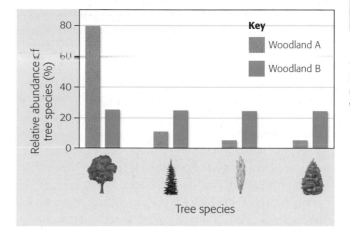

◄ **Figure 20.20**
Relative abundance of tree species in woodlands A and B.

▼ **Figure 20.21** **A *Pisaster* sea star eating its favorite food, a mussel.**

▼ **Figure 20.19** **Which woodland is more diverse?**

Woodland A

Woodland B

✔CHECKPOINT

How could a community appear to have relatively little diversity even though it is rich in species?

Answer: if one or a few of the diverse species accounted for almost all the organisms in the community, with the other species being rare

Disturbances in Communities

Most communities are constantly changing in response to disturbances. **Disturbances** are episodes that damage biological communities, at least temporarily, by destroying organisms and altering the availability of resources such as mineral nutrients and water. Examples of natural disturbances are storms, fire, floods, and droughts, but humans are by far the most significant agents of disturbance today.

Understanding the effects of disturbance is especially important because of the potential impact on the human population. For instance, one consequence of human-caused disturbance is the emergence of previously unknown infectious diseases. Three-quarters of emerging diseases have jumped to humans from another vertebrate species. In many cases, people come into contact with the pathogens through activities such as clearing land for agriculture, road building, or hunting in previously isolated ecosystems. HIV, which may have been transmitted to humans from the blood of primates butchered for food, is probably the best-known example. Others include potentially fatal hemorrhagic fevers, such as Ebola. Habitat destruction may also cause pathogen-carrying animals to venture closer to human dwellings in search of food.

Small-scale natural disturbances often have positive effects on a biological community. For example, when a large tree falls in a windstorm, it creates new habitats (Figure 20.22). More light may now reach the forest floor, giving small seedlings the opportunity to grow; or the depression left by its roots may fill with water and be used as egg-laying sites by frogs, salamanders, and numerous insects. ☑

☑ CHECKPOINT

Why might the effect of a small-scale natural disturbance, such as a forest tree felled by wind, be considered positive?

Answer: A fallen tree alters abiotic factors such as water and sunlight in a small patch of the forest and provides new habitats for other organisms. The loss of one tree is not a major disruption to the forest community as a whole (unless it is a rare tree).

Ecological Succession

Communities change drastically following a severe disturbance that strips away vegetation and even soil. The disturbed area may be colonized by a variety of species, which are gradually replaced by a succession of other species, in a process called **ecological succession**.

When ecological succession begins in a virtually lifeless area with no soil, it is called **primary succession** (Figure 20.23). Examples of such areas are cooled lava flows on volcanic islands and the rubble left by a retreating glacier. Often the only life-forms initially present are autotrophic bacteria. Lichens and mosses, which grow from windblown spores, are commonly the first multicellular photosynthesizers to colonize the area. Soil develops gradually as rocks weather and organic matter accumulates from the decomposed remains of the early colonizers. Lichens and mosses are eventually overgrown by grasses and shrubs that sprout from seeds blown in from nearby areas or carried in by animals. Finally, the area is colonized by plants that become the community's prevalent form of vegetation. Primary succession can take hundreds or thousands of years.

▼ Figure 20.22 **A small-scale disturbance.** When this tree fell during a windstorm, its root system and the surrounding soil uplifted, resulting in a depression that filled with water. The dead tree, the root mound, and the water-filled depression are new habitats.

▼ Figure 20.23 **Primary succession under way on a lava flow in Valley of Fires State Park, New Mexico.**

Secondary succession occurs where a disturbance has destroyed an existing community but left the soil intact. For example, secondary succession occurs as areas recover from floods or fires (**Figure 20.24**). Disturbances that lead to secondary succession are also caused by human activities. Even before colonial times, humans were clearing the temperate deciduous forests of eastern North America for agriculture and settlements. Some of this land was later abandoned as the soil was depleted of its chemical nutrients or the residents moved west to new territories. Whenever human intervention stops, secondary succession begins. ☑

▼ Figure 20.24 **Secondary succession after a fire.**

☑**CHECKPOINT**

What is the main abiotic factor that distinguishes primary from secondary succession?

Answer: absence of soil (primary succession) versus presence of soil (secondary succession) at the onset of succession

Ecosystem Ecology

In addition to the community of species in a given area, an **ecosystem** includes all the abiotic factors, such as energy, soil characteristics, and water, described in Chapter 18. Perhaps you have seen or even made a terrarium like the one in **Figure 20.25**. Such a microcosm exhibits the two major processes that sustain all ecosystems: energy flow and chemical cycling. **Energy flow** is the passage of energy through the components of the ecosystem. **Chemical cycling** is the use and reuse of chemical elements such as carbon and nitrogen within the ecosystem.

Energy enters the terrarium in the form of sunlight (yellow arrows). Plants (producers) convert light energy to chemical energy through the process of photosynthesis. Animals (consumers) take in some of this chemical energy in the form of organic compounds when they eat the plants. Detritivores and decomposers in the soil obtain chemical energy when they feed on the dead remains of plants and animals. Every use of chemical energy by organisms involves a loss of some energy to the surroundings in the form of heat (red arrows). Because so much of the energy captured by photosynthesis is lost as heat, this ecosystem would run out of energy if it were not powered by a continuous inflow of energy from the sun.

In contrast to energy flow, chemical cycling (blue arrows in Figure 20.25) involves the transfer of materials within the ecosystem. While most ecosystems have a constant input of energy from sunlight or another source, the supply of the chemical elements used to construct molecules is limited. Chemical elements such as carbon and nitrogen are cycled between the abiotic component of the ecosystem, including the air, water, and soil, and the biotic component of the ecosystem (the community). Plants acquire these elements in inorganic

Energy Flow in Ecosystems

form from the air and soil and fix them into organic molecules. Animals, such as the snail in Figure 20.25, consume some of these organic molecules. When the plants and animals become detritus, decomposers return most of the elements to the soil and air in inorganic form. Some elements are also returned to the air and soil as the by-products of plant and animal metabolism.

In summary, both energy flow and chemical cycling involve the transfer of substances through the trophic levels of the ecosystem. However, energy flows through, and ultimately out of, ecosystems, whereas chemicals are recycled within and between ecosystems.

▼ Figure 20.25 **A terrarium ecosystem.** Though it is small and artificial, this sealed terrarium illustrates the two major ecosystem processes: energy flow and chemical cycling.

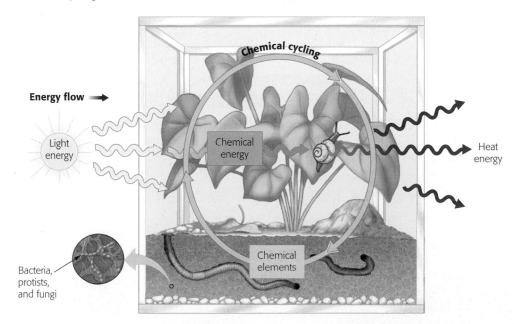

Energy Flow in Ecosystems

All organisms require energy for growth, maintenance, reproduction, and, in many species, locomotion. In this section, we take a closer look at energy flow through ecosystems. Along the way, we'll answer two key questions: What limits the length of food chains? and How do lessons about energy flow apply to people's use of resources?

Primary Production and the Energy Budgets of Ecosystems

Each day, Earth receives about 10^{19} kcal of solar energy, the energy equivalent of about 100 million atomic bombs. Most of this energy is absorbed, scattered, or reflected by the atmosphere or by Earth's surface. Of the visible light that reaches plants, algae, and cyanobacteria, only about 1% is converted to chemical energy by photosynthesis.

Ecologists call the amount, or mass, of living organic material in an ecosystem the **biomass**. The rate at which an ecosystem's producers convert solar energy to the chemical energy stored in biomass is called **primary production**. The primary production of the entire biosphere is roughly 165 billion tons of biomass per year.

Different ecosystems vary considerably in their primary production (**Figure 20.26**) as well as in their contribution to the total production of the biosphere. Tropical rain forests are among the most productive terrestrial ecosystems and contribute a large portion of the planet's overall production of biomass. Coral reefs also

have very high production, but their contribution to global production is small because they cover such a small area. Interestingly, even though the open ocean has very low production, it contributes the most to Earth's total net primary production because of its huge size—it covers 65% of Earth's surface area. Whatever the ecosystem, primary production sets the spending limit for the energy budget of the entire ecosystem because consumers must acquire their organic fuels from producers. Now let's see how this energy budget is divided among the different trophic levels in an ecosystem's food web. ✔

Ecological Pyramids

When energy flows as organic matter through the trophic levels of an ecosystem, much of it is lost at each link in the food chain. Consider the transfer of organic matter from plants (producers) to herbivores (primary consumers). In most ecosystems, herbivores manage to eat only a fraction of the plant material produced, and they can't digest all of what they do consume. For example, a caterpillar feeding on leaves passes about half the energy in the leaves as feces (**Figure 20.27**). Another 35% of the energy is expended in cellular respiration. Only about 15% of the energy in the caterpillar's food is transformed into caterpillar biomass. Only this biomass (and the energy it contains) is available to the consumer that eats the caterpillar.

Figure 20.28, called a **pyramid of production**, illustrates the cumulative loss of energy with each transfer in a food chain. Each tier of the pyramid represents all of the organisms in one trophic level, and the width of each tier indicates how much of the chemical energy of the tier below is actually incorporated into the organic matter of that trophic level. Note that producers convert

▼ Figure 20.26 **Primary production of different ecosystems.** Primary production is the amount of biomass created by the producers of an ecosystem over a unit of time, in this case a year. Aquatic ecosystems are color-coded blue in these histograms; terrestrial ecosystems are green.

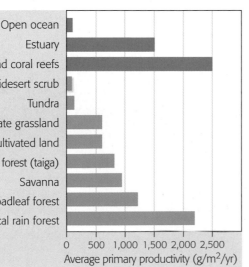

▶ Figure 20.27
What becomes of a caterpillar's food?
Only about 15% of the calories of plant material this herbivore consumes will be stored as biomass available to the next link in the food chain.

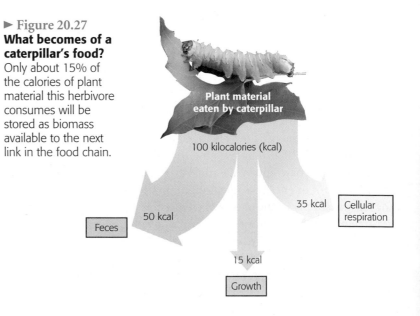

▼ Figure 20.28 **An idealized pyramid of production.**

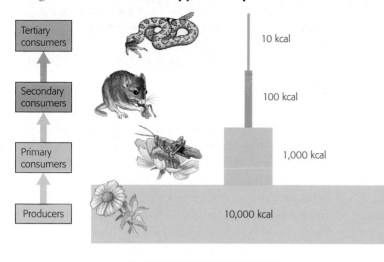

Tertiary consumers	10 kcal
Secondary consumers	100 kcal
Primary consumers	1,000 kcal
Producers	10,000 kcal

1,000,000 kcal of sunlight

there is simply not enough energy at the very top of an ecological pyramid to support another trophic level. There are, for example, no nonhuman predators of lions, eagles, and killer whales; the biomass in populations of these top-level consumers is insufficient to support yet another trophic level with a reliable source of nutrition.

Ecosystem Energetics and Human Resource Use

The dynamics of energy flow apply to the human population as much as to other organisms. As omnivores, we eat both plant material and meat. When we eat grain or fruit and vegetables, we are primary consumers; when we eat beef or other meat from herbivores, we are secondary consumers. When we eat fish such as trout and salmon (which eat insects and other small animals), we are tertiary or quaternary consumers.

only about 1% of the energy in the sunlight available to them to primary production. In this generalized pyramid, 10% of the energy available at each trophic level becomes incorporated into the next higher level. Actual efficiencies of energy transfer are usually in the range of 5–20%. In other words, 80–95% of the energy at one trophic level never reaches the next.

An important implication of this stepwise decline of energy in a trophic structure is that the amount of energy available to top-level consumers is small compared with that available to lower-level consumers. Only a tiny fraction of the energy stored by photosynthesis flows through a food chain to a tertiary consumer, such as a snake feeding on a mouse. This explains why top-level consumers such as lions and hawks require so much geographic territory; it takes a lot of vegetation to support trophic levels so many steps removed from photosynthetic production. You can also understand why most food chains are limited to three to five levels;

The two production pyramids in **Figure 20.29** are based on the same generalized model used to construct Figure 20.28. The pyramid on the left shows energy flow from primary producers (represented by corn) to human vegetarians. In the pyramid on the right, the primary producers are eaten by cattle; people then eat the beef, making them secondary consumers. The human population has about ten times more energy available to it when people eat corn than when they process the same amount of corn through another trophic level and eat corn-fed beef.

Eating meat of any kind is both economically and environmentally expensive. Producing meat for human consumption usually requires that more land be used for growing grain, which also means greater use of fertilizers and pesticides and more water for irrigation. ☑

▼ Figure 20.29 **Food energy available to the human population at different trophic levels.**

Trophic level

Secondary consumers	
Primary consumers	
Producers	

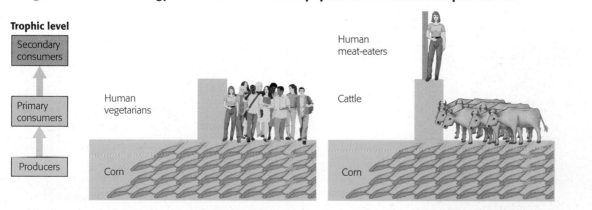

Human vegetarians

Corn

Human meat-eaters

Cattle

Corn

Chemical Cycling in Ecosystems

The sun (or in some cases Earth's interior) supplies ecosystems with a continual input of energy, but aside from an occasional meteorite, there are no extraterrestrial sources of chemical elements. Life, therefore, depends on the recycling of chemicals. While an organism is alive, much of its chemical stock changes continuously, as nutrients are acquired and waste products are released. Atoms present in the complex molecules of an organism at the time of its death are returned to the environment by the action of decomposers, replenishing the pool of inorganic nutrients that plants and other producers use to build new organic matter (Figure 20.30). In a sense, each living thing only borrows an ecosystem's chemical elements, returning what is left in its body after it dies. Let's take a closer look at how chemicals cycle between organisms and the abiotic components of ecosystems.

BioFlix™
The Carbon Cycle

The General Scheme of Chemical Cycling

Because chemical cycles in an ecosystem involve both biotic and abiotic components, they are called **biogeochemical cycles**. Figure 20.31 is a general scheme for the cycling of a nutrient within an ecosystem. Note that the cycle has an **abiotic reservoir** (white box) where a chemical accumulates or is stockpiled outside of living organisms. The atmosphere, for example, is an abiotic reservoir for carbon. Phosphorus, on the other hand, is available only from the soil. The water of aquatic ecosystems contains dissolved carbon, nitrogen, and phosphorus compounds.

Let's trace our way around our general biogeochemical cycle. ❶ Producers incorporate chemicals from the abiotic reservoir into organic compounds. ❷ Consumers feed on the producers, incorporating some of the chemicals into their own bodies. ❸ Both producers and consumers release some chemicals back to the environment in waste products. ❹ Decomposers play a central role by breaking down the complex organic molecules in detritus such as plant litter, animal wastes, and dead organisms. The products of this metabolism are inorganic molecules that replenish the abiotic reservoirs. Geologic processes such as erosion and the weathering of rock also contribute to the abiotic reservoirs. Producers use the inorganic molecules from abiotic reservoirs as raw materials for synthesizing new organic molecules (carbohydrates and proteins, for example), and the cycle continues.

Biogeochemical cycles can be local or global. The chemical phosphorus, for example, cycles almost entirely within local areas, at least over the short term, as you'll see on the next page. Soil is the main reservoir for nutrients in a local cycle. In contrast, for those chemicals that spend part of their time in gaseous form—carbon and nitrogen are examples—the cycling is essentially global. For instance, some of the carbon a plant acquires from the air may have been released into the atmosphere by the respiration of a plant or animal on another continent.

Now let's examine three important biogeochemical cycles more closely: the cycles for carbon, phosphorus, and nitrogen. As you study the cycles, look for the four basic steps we described, as well as the geologic processes that may move chemicals around and between ecosystems. In all the diagrams, the main abiotic reservoirs appear in white boxes. ☑

☑CHECKPOINT

What is the role of abiotic reservoirs in biogeochemical cycles? What are the three major abiotic reservoirs?

Answer: Abiotic reservoirs are sources of the nutrients used by producers for synthesizing organic molecules. The atmosphere, soil, and water are the three major abiotic reservoirs.

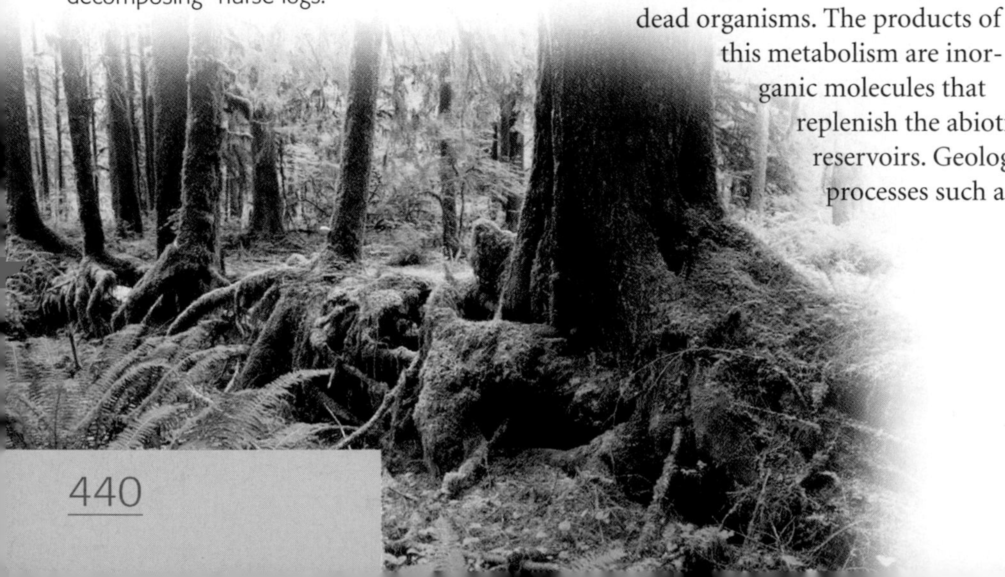

▼ Figure 20.30 **Plant growth on fallen tree.** In the temperate rain forest of Olympic National Park, in Washington State, plants—including other trees—quickly take advantage of the mineral nutrients supplied by decomposing "nurse logs."

▼ Figure 20.31 **General scheme for biogeochemical cycles.**

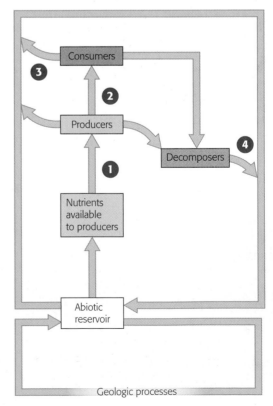

The Carbon Cycle

You saw a simple version of the carbon cycle in Figure 18.46, where you learned that the abiotic reservoirs of carbon include the atmosphere, fossil fuels, and dissolved carbon compounds in the oceans. The reciprocal metabolic processes of photosynthesis and cellular respiration, also familiar from previous chapters, are mainly responsible for the cycling of carbon between the biotic and abiotic worlds (**Figure 20.32**). **1** Photosynthesis removes CO_2 from the atmosphere and incorporates it into organic molecules, which are **2** passed along the food chain by consumers. **3** Cellular respiration returns CO_2 to the atmosphere. **4** Decomposers break down the carbon compounds in detritus; that carbon, too, is eventually released as CO_2. On a global scale, the return of CO_2 to the atmosphere by respiration closely balances its removal by photosynthesis. As you saw in Chapter 18, however, increasing levels of CO_2 caused by **5** the burning of wood and fossil fuels (coal and petroleum) is contributing to global climate change.

The Phosphorus Cycle

Organisms require phosphorus as an ingredient of nucleic acids, phospholipids, and ATP and (in vertebrates) as a mineral component of bones and teeth. In contrast to the carbon cycle and the other major biogeochemical cycles, the phosphorus cycle does not have an atmospheric component. Rocks are the only source of phosphorus for terrestrial ecosystems; in fact, rocks that have high phosphorus content are mined for fertilizer.

At the center of **Figure 20.33**, **1** the weathering (breakdown) of rock gradually adds inorganic phosphate (PO_4^{3-}) to the soil. **2** Plants absorb dissolved phosphate from the soil and assimilate it by building the phosphorus atoms into organic compounds. **3** Consumers obtain phosphorus in organic form from plants. **4** Phosphates are returned to the soil by the action of decomposers on animal waste and the remains of dead plants and animals. **5** Some of the phosphates drain from terrestrial ecosystems into the sea, where they may settle and eventually become part of new rocks. Phosphorus removed from the cycle in this way will not be available to living organisms until **6** geologic processes uplift the rocks and expose them to weathering.

Because weathering is generally a slow process, the amount of phosphates available to plants in natural ecosystems is often quite low and may be a limiting factor. Farmers and gardeners often use crushed phosphate rock, bone meal (finely ground bones from slaughtered livestock or fish), or guano to add phosphorus to the soil. Guano, the droppings of sea birds and bats, is mined from densely populated colonies or caves, where meters-deep deposits have accumulated.

▼ Figure 20.32 **The carbon cycle.**

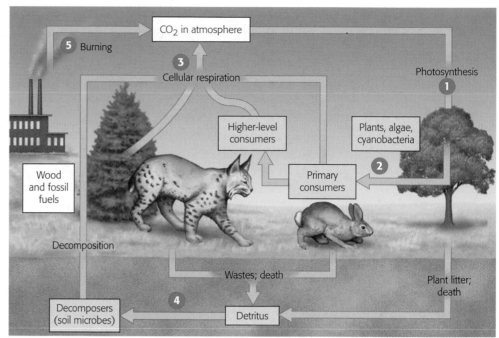

▼ Figure 20.33 **The phosphorus cycle.**

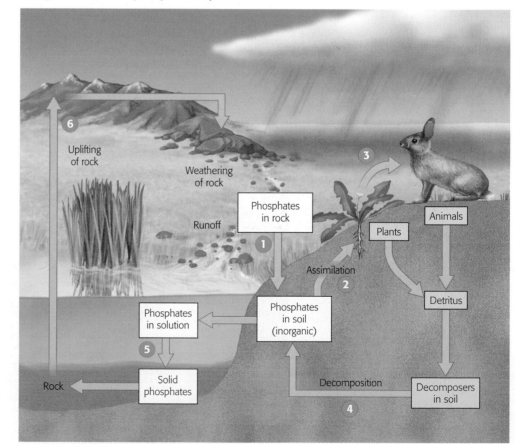

The Nitrogen Cycle

As an ingredient of proteins and nucleic acids, nitrogen is essential to the structure and functioning of all organisms. In particular, it is a crucial and often limiting plant nutrient. Nitrogen has two abiotic reservoirs, the atmosphere and the soil. The atmospheric reservoir is huge; almost 80% of the atmosphere is nitrogen gas (N_2). However, plants cannot assimilate nitrogen in the form of N_2. The process of **nitrogen fixation** converts gaseous N_2 to ammonia and nitrates, which can be used by plants. A small amount of nitrogen is fixed by high-energy processes such as lightning strikes. However, most of the nitrogen available in natural ecosystems comes from biological fixation performed by certain bacteria. Without these organisms, the reservoir of usable soil nitrogen would be extremely limited.

Figure 20.34 illustrates the actions of two types of nitrogen-fixing bacteria. ❶ Some bacteria live symbiotically in the roots of certain species of plants, supplying their hosts with a direct source of usable nitrogen. The largest group of plants with this mutualistic relationship is the legumes, a family that includes peanuts, soybeans, and kudzu. ❷ Free-living nitrogen-fixing bacteria in soil or water convert N_2 to ammonium (NH_4^+). The N_2 is available in air pockets in the soil and dissolved in water.

❸ After nitrogen is "fixed," some of the NH_4^+ is taken up and used by plants. ❹ Nitrifying bacteria in the soil also convert some of the NH_4^+ to nitrate (NO_3^-), ❺ which is more readily assimilated by plants. Plants use the nitrogen they assimilate to synthesize molecules such as amino acids, which are then incorporated into proteins.

❻ When an herbivore (represented here by a rabbit) eats a plant, it digests the proteins into amino acids and then uses the amino acids to build the proteins it needs. Higher-order consumers get nitrogen from the nitrogen-containing organic molecules of their prey. Also, nitrogen-containing waste products are formed during protein metabolism; consumers excrete some nitrogen as well as incorporating it into their body tissues. The urine that rabbits and other mammals excrete contains urea, a nitrogen-containing substance that is widely used as fertilizer.

Organisms that are not consumed eventually die and become detritus, which is decomposed by bacteria and fungi. ❼ Decomposition releases NH_4^+ from organic compounds back into the soil, replenishing the soil reservoir. Under low-oxygen conditions, however, ❽ soil bacteria known as denitrifying bacteria strip the oxygen atoms from NO_3^-, releasing N_2 back into the atmosphere and depleting the soil reservoir of usable nitrogen.

Although not shown in the figure, some NH_4^+ and NO_3^- are made in the atmosphere by chemical reactions involving N_2 and ammonia gas (NH_3). The ions produced by these chemical reactions reach the soil in precipitation and dust, which are a crucial source of nitrogen for plants in some ecosystems. ☑

☑CHECKPOINT

What are the abiotic reservoirs of nitrogen? In what form does nitrogen occur in each reservoir?

Answer: atmosphere: N_2; soil: NH_4^+ and NO_3^-

▼ Figure 20.34 **The nitrogen cycle.**

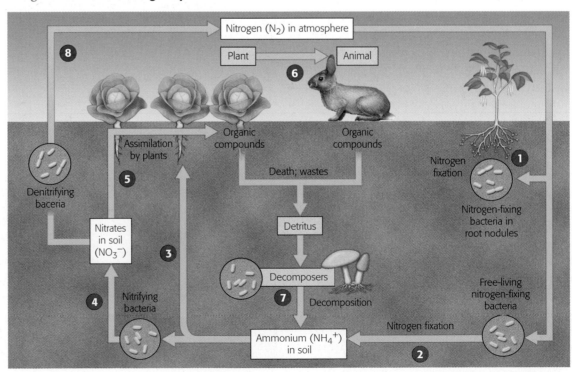

Nutrient Pollution

Low nutrient levels, especially of phosphorus and nitrogen, often limit the growth of algae and cyanobacteria in aquatic ecosystems. Nutrient pollution occurs when human activities add excess amounts of these chemicals to aquatic ecosystems.

In many areas, phosphate pollution comes from agricultural fertilizers. Phosphates are also a common ingredient in pesticides. Other major sources of phosphates include outflow from sewage treatment facilities and runoff of animal waste from livestock feedlots (where hundreds of animals are penned together). Phosphate pollution of lakes and rivers results in heavy growth of algae and cyanobacteria (**Figure 20.35**). Microbes consume a great deal of oxygen as they decompose the extra biomass, a process that depletes the water of oxygen. These changes lead to reduced diversity of aquatic species and a much less appealing body of water.

Sewage treatment facilities may discharge large amounts of dissolved inorganic nitrogen compounds into rivers or streams when extreme conditions (such as unusual storms) or malfunctioning equipment prevent them from meeting water quality standards. Agricultural sources of nitrogen include feedlots and the large amounts of inorganic nitrogen fertilizers that are routinely applied to crops. Lawns and golf courses also receive sizable applications of nitrogen fertilizers. Crop and lawn plants take up some of the nitrogen compounds,

and denitrifiers convert some to atmospheric N_2, but nitrate is not bound tightly by soil particles and is easily washed out of the soil by rain or irrigation. As a result, chemical fertilizers often exceed the soil's natural recycling capacity.

In an example of how far-reaching this problem can be, nitrogen runoff from Midwestern farm fields has been linked to an annual summer "dead zone" in the Gulf of Mexico (**Figure 20.36**). Vast algal blooms extend outward from where the Mississippi River deposits its nutrient-laden waters. As the algae die, decomposition of the huge quantities of biomass diminishes the supply of dissolved oxygen over an area that ranges from 13,000 km^2 (roughly the area of Connecticut) to 22,000 km^2 (about the area of Michigan). Oxygen depletion disrupts benthic communities, displacing fish and invertebrates that can move and killing organisms that are attached to the substrate. More than 400 recurring and permanent coastal dead zones totaling approximately 245,000 km^2 have been documented worldwide. ☑

☑CHECKPOINT

How does the excessive addition of mineral nutrients to a pond eventually result in the loss of most fish in the pond?

Answer: Overfertilization by nutrient pollution initially causes population explosions of algae and the organisms that feed on them. The respiration of so much life, especially of the microbes decomposing all the organic refuse, consumes most of the lake's oxygen, which the fish require.

▼ Figure 20.35 **Algal growth resulting from nutrient pollution.** The flat green area in this photo is not a lawn, but rather the surface of a polluted pond.

▼ Figure 20.36 **The Gulf of Mexico dead zone.**

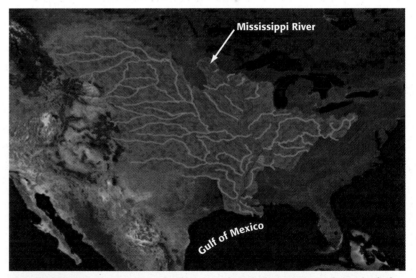

Mississippi River

Gulf of Mexico

Light blue lines represent rivers draining into the Mississippi River (shown in dark blue). Nitrogen runoff carried by these rivers ends up in the Gulf of Mexico. In the images below, red and orange indicate high concentrations of phytoplankton. Bacteria feeding on dead phytoplankton deplete the water of oxygen, creating a "dead zone."

Gulf of Mexico

Gulf of Mexico

Summer

Winter

Conservation and Restoration Biology

As we have seen in this unit, many of the environmental problems facing us today have been caused by human enterprises. But the science of ecology is not just useful for telling us how things have gone wrong. Ecological research is also the foundation for finding solutions to these problems and for reversing the negative consequences of ecosystem alteration. Thus, we end the ecology unit by highlighting the beneficial applications of ecological research.

Conservation biology is a goal-oriented science that seeks to understand and counter the loss of biodiversity. Conservation biologists recognize that biodiversity can be sustained only if the evolutionary mechanisms that have given rise to species and communities of organisms continue to operate. Thus, the goal is not simply to preserve individual species but to sustain ecosystems, where natural selection can continue to function, and to maintain the genetic variability on which natural selection acts. The new and expanding field of **restoration ecology** uses ecological principles to develop methods of returning degraded areas to their natural state.

Biodiversity "Hot Spots"

Conservation biologists are applying their understanding of population, community, and ecosystem dynamics in establishing parks, wilderness areas, and other legally protected nature reserves. Choosing locations for protection often focuses on **biodiversity hot spots**. These relatively small areas have a large number of endangered and threatened species and an exceptional concentration of **endemic species**, species that are found nowhere else. Together, the "hottest" of Earth's biodiversity hot spots, shown in Figure 20.37, total less than 1.5% of Earth's land surface but are home to a third of all species of plants and vertebrates. For example, all lemurs are endemic to Madagascar, a large island off the eastern coast of Africa that is home to more than 50 species of lemurs. In fact, almost all of the mammals, reptiles, amphibians, and plants that inhabit Madagascar are endemic. There are also hot spots in aquatic ecosystems, such as certain river systems and coral reefs. Because biodiversity hot spots can also be hot spots of extinction, they rank high on the list of areas demanding strong global conservation efforts.

Concentrations of species provide an opportunity to protect many species in very limited areas. However, the "hot spot" designation tends to favor the most noticeable organisms, especially vertebrates and plants. Invertebrates and microorganisms are often overlooked. Furthermore, species endangerment is a global problem, and focusing on hot spots should not detract from efforts to conserve habitats and species diversity in other areas. Finally, even the protection of a nature reserve does not shield organisms from the effects of climate change or other threats, such as invasive species or infectious disease. The golden toad you saw in Figure 20.2 went extinct despite living in a protected reserve—a victim of changing weather patterns, airborne pollution, and disease. To stem the tide of biodiversity loss, we will have to address environmental problems globally as well as locally. ☑

▼ Figure 20.37 **Earth's terrestrial biodiversity hot spots (purple).**

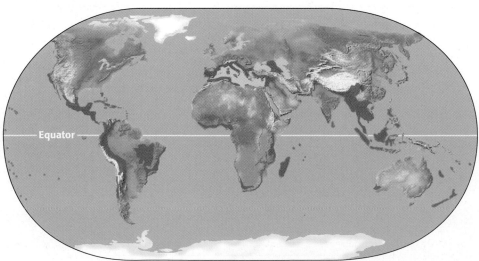

Conservation at the Ecosystem Level

Most conservation efforts in the past have focused on saving individual species, and this work continues. (You have already learned about one example, the red-cockaded woodpecker; see Figure 19.12.) More and more, however, conservation biology aims at sustaining the biodiversity of entire communities and ecosystems. On an even broader scale, conservation biology considers the biodiversity of whole landscapes. Ecologically, a **landscape** is a regional assemblage of interacting ecosystems, such as an area with forest, adjacent fields, wetlands, streams, and streamside habitats. **Landscape ecology** is the application of ecological principles to the study of land-use patterns. Its goal is to make ecosystem conservation a functional part of the planning for land use.

Edges between ecosystems are prominent features of landscapes, whether natural or altered by people (**Figure 20.38**). Such edges have their own sets of physical conditions—such as soil type and surface features—that differ from the ecosystems on either side of them. Edges also may have their own type and amount of disturbance. For instance, the edge of a forest often has more blown-down trees than a forest interior because the edge is less protected from strong winds. Because of their specific physical features, edges also have their own communities of organisms. Some organisms thrive in edges because they require resources found only there. For instance, whitetail deer browse on woody shrubs found in edge areas between woods and fields, and their populations often expand when forests are logged or interrupted by development.

Edges can have both positive and negative effects on biodiversity. A recent study in a tropical rain forest in western Africa indicated that natural edge communities are important sites of speciation. On the other hand, landscapes where human activities have produced edges often have fewer species.

Another important landscape feature, especially where habitats have been severely fragmented, is the **movement corridor**, a narrow strip or series of small clumps of suitable habitat connecting otherwise isolated patches. In places where there is extremely heavy human impact, artificial corridors are sometime constructed (**Figure 20.39**). Corridors can promote dispersal and help sustain populations, and they are especially important to species that migrate between different habitats seasonally. But a corridor can also be harmful—as, for example, in the spread of disease, especially among small subpopulations in closely situated habitat patches. ✔

▼ Figure 20.38 **Edges between ecosystems within a landscape.**

Natural edges. Grasslands border forest ecosystems in Yellowstone National Park.

Edges created by human activity. Pronounced edges (roads) surround clear-cuts in this photograph of a heavily logged rain forest in Malaysia.

▶ Figure 20.39 **An artificial corridor.** This bridge over a road provides an artificial corridor for animals in Banff National Park, Canada.

How Does Tropical Forest Fragmentation Affect Biodiversity?

Long-term studies are essential for learning how we might best conserve biodiversity and other natural resources. One site for such studies is the Biological Dynamics of Forest Fragmentation Project (BDFFP), a 1,000-km^2 ecological "laboratory" located deep in the Amazonian forest of Brazil. When the project was begun in 1979, laws required landowners who cleared forest for ranching or other agricultural operations to leave scattered tracts of untouched forest. Biologists recruited some of these land-owners to create reserves in isolated fragments of 1 ha (roughly 2.5 acres), 10 ha, and 100 ha. Figure 20.40 shows some of these forest "islands." Before each area was isolated from the main forest, a small army of specialists inventoried the organisms present and measured the trees.

Hundreds of researchers have used the BDFFP sites to investigate the effects of forest fragmentation on all levels of ecological study. The initial **observations** for these investigations were gleaned from the results of other ecological studies—for example, the effects of fragmentation in temperate forests or differences in the biodiversity of small islands compared to mainland ecosystems. These observations led many researchers to ask the **question**, How does fragmentation of tropical forests affect species diversity within the fragments? Based on previous research on numerous species, a reasonable **hypothesis** might be: Species diversity declines with the size of the forest fragment. An ecologist studying large predators such as jaguars and pumas, which require large hunting territories, might therefore make the **prediction** that predators will only be found in the largest areas. The BDFFP is unique because it allows researchers to test their predictions with new observations that compare species diversity in forest fragments with a comparable control: species diversity in the same area when it was intact. An undisturbed area of 25,400 acres is also available for comparison with fragmented areas. In addition, data can be collected over a period of years or even decades.

Since the BDFFP was established, scientists have studied many different groups of plants and animals. In general, the **results** have shown that fragmentation of forest into smaller pieces leads to a decline in species diversity. Species richness decreases as a result of local extinctions of many species of large mammals, insects, and insectivorous birds. The population density of remaining species often declines. Researchers also documented edge effects, such as those described in the previous section. Changes in abiotic factors along the fragment edges, including increased wind disturbance, higher temperature, and decreased soil moisture, played a role in community alterations. For example, ecologists found that tree mortality was higher than normal at fragment edges. They also observed changes in the composition of communities of invertebrates that inhabit the soil and leaf litter.

▼ Figure 20.40 **Fragments of forest in the Amazon that were created as part of the Biological Dynamics of Forest Fragmentation Project.** The patch of forest on the right is one hectare.

Restoring Ecosystems

One of the major strategies in restoration ecology is **bioremediation**, the use of living organisms to detoxify polluted ecosystems. For example, bacteria have been used to clean up old mining sites and oil spills (see Figure 15.18). Researchers are also investigating the potential of using plants to remove toxic substances such as heavy metals and organic pollutants (for example, PCBs) from contaminated soil (Figure 20.41).

Some restoration projects have the broader goal of returning ecosystems to their natural state, which may involve replanting vegetation, fencing out non-native animals, or removing dams that restrict water flow. Hundreds of restoration projects are currently under way in the United States. One of the most ambitious endeavors is the Kissimmee River Restoration Project in south central Florida.

The Kissimmee River was once a meandering shallow river that wound its way from Lake Kissimmee southward into Lake Okeechobee. Periodic flooding of the river covered a wide floodplain during about half of the year, creating wetlands that provided habitat for large numbers of birds, fishes, and invertebrates. And as the floods deposited the river's load of nutrient-rich silt on the floodplain, they boosted soil fertility and maintained the water quality of the river.

Between 1962 and 1971, the U.S. Army Corps of Engineers converted the 166-km wandering river to a straight canal 9 m deep, 100 m wide, and 90 km long. This project, designed to allow development on the floodplain, drained approximately 31,000 acres of wetlands, with significant negative impacts on fish and wetland bird populations. Without the marshes to help filter and reduce agricultural runoff, the river transported phosphates and other excess nutrients from Lake Okeechobee to the Everglades ecosystem to the south.

The restoration project involves removing water control structures such as dams, reservoirs, and channel modifications and filling in about 35 km of the canal (Figure 20.42). The first phase of the project was completed in 2004. The photo shows a section of the Kissimmee canal that has been plugged, diverting flow into the remnant river channels. Birds and other wildlife have returned in unexpected numbers to the 11,000 acres of wetlands that have been restored. The marshes are filled with native vegetation, and game fishes again swim in the river channels. ☑

☑CHECKPOINT

The water in the Kissimmee River eventually flows into the Everglades. How will the Kissimmee River Restoration Project affect water quality in the Everglades ecosystem?

Answer: Wetlands filter agricultural runoff, which prevents nutrient pollution from flowing downstream. By restoring this ecosystem service, the project will improve water quality in the Everglades.

▼ Figure 20.41 **Bioremediation using plants.** A researcher from the U.S. Department of Agriculture investigates the use of canola plants to reduce toxic levels of selenium in contaminated soil.

▼ Figure 20.42 **The Kissimmee River Restoration Project.**

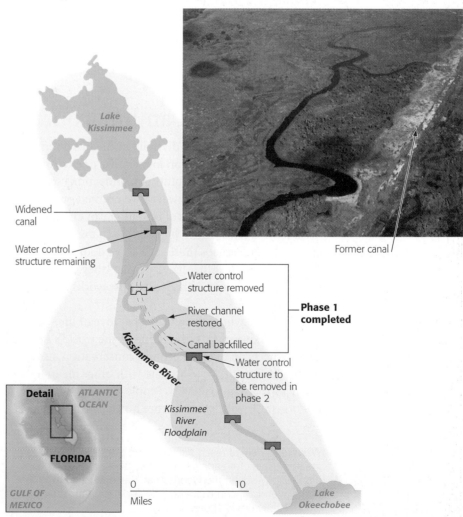

The Goal of Sustainable Development

The demand for the "provisioning" services of ecosystems, such as food, wood, and water, is increasing as the world population grows and becomes more affluent. Although these demands are currently being met, they are satisfied at the expense of other critical ecosystem services, such as climate regulation and protection against natural disasters. Clearly, we have set ourselves and the rest of the biosphere on a precarious path into the future. How can we best manage Earth's resources to ensure that all generations inherit an adequate supply of natural and economic resources and a relatively stable environment?

Many nations, scientific societies, and private foundations have embraced the concept of **sustainable development**. The Ecological Society of America, the world's largest organization of ecologists, endorses a research agenda called the Sustainable Biosphere Initiative. The goal of this initiative is to acquire the ecological information necessary for the responsible development, management, and conservation of Earth's resources. The research agenda includes the search for ways to sustain the productivity of natural and artificial ecosystems and studies of the relationship between biological diversity, global climate change, and ecological processes.

Sustainable development depends on more than continued research and application of ecological knowledge. It also requires that we connect the life sciences with the social sciences, economics, and humanities. Conservation of biodiversity is only one side of sustainable development; the other key factor is improving the human condition. Public education and the political commitment and cooperation of nations are essential to the success of this endeavor.

An awareness of our unique ability to alter the biosphere and jeopardize the existence of other species, as well as our own, may help us choose a path toward a sustainable future. The risk of a world without adequate natural resources for all its people is not a vision of the distant future. It is a prospect for your children's lifetime, or perhaps even your own. But although the current state of the biosphere is grim, the situation is far from hopeless. Now is the time to aggressively pursue greater knowledge about the diversity of life on our planet and to work toward long-term sustainability. ✓

✓ CHECKPOINT

What is meant by sustainable development?

Answer: development that ensures an adequate supply of natural and economic resources for future generations

EVOLUTION CONNECTION: The Loss of Biodiversity

Biophilia and an Environmental Ethic

For millions of years, the diversity of life has flourished via evolutionary adaptation in response to environmental change. In the Evolution Connections sections in Chapters 18 and 19, however, we pointed out that for many species, the pace of evolution can't match the breakneck speed at which humans are changing the environment. Perhaps those species are doomed to extinction. On the other hand, perhaps they can be saved by one human characteristic that works in their favor: biophilia.

Biophilia, which literally means "love of life," is a term that Edward O. Wilson, one of the world's foremost experts on biodiversity and conservation, uses for the human desire to affiliate with other life in its many forms (**Figure 20.43**). People develop close relationships with pets, nurture houseplants, invite

▶ Figure 20.43 **Doing what comes naturally.** Edward O. Wilson has helped teach scientists and the general public a greater respect for Earth's biodiversity. This photograph finds biophiliac Wilson in the woods near Massachusetts's Walden Pond, a landscape immortalized by another great naturalist and writer, Henry David Thoreau.

avian visitors with backyard feeders, and flock to zoos, gardens, and nature parks (Figure 20.44). Our attraction to pristine landscapes with clean water and lush vegetation is also testimony to our biophilia. Wilson proposes that our biophilia is innate, an evolutionary product of natural selection acting on a brainy species whose survival depended on a close connection to the environment and a practical appreciation of plants and animals. We evolved in natural environments rich in biodiversity, and we still have an affinity for such settings. Our behavior reflects remnants of our ancestral attachment to nature and the diversity of life.

It will come as no surprise that many biologists have embraced the concept of biophilia. After all, these are people who have turned their passion for nature into careers. But biophilia strikes a chord with biologists for another reason. If biophilia is evolutionarily embedded in our genomes, then there is hope that we can become better custodians of the biosphere. If we all pay more attention to our biophilia, a new environmental ethic could catch on among individuals and societies. And that ethic is a resolve never to knowingly allow a single species to become extinct as a result of our actions or any ecosystem to be destroyed as long as there are reasonable ways to prevent it. Yes, we should be motivated to preserve biodiversity because we depend on it for food, medicine, building materials, fertile soil, flood control, habitable climate, drinkable water, and breathable air. But maybe we can also work harder to prevent the extinction of other forms of life just because it is the ethical thing for us to do. Again, Wilson sounds the call: "Right now, we're pushing the species of the world through a bottleneck. We've got to make it a major moral principle to get as many of them through this as possible. It's the challenge now and for the rest of the century. And there's one good thing about our species: We like a challenge!"

Biophilia is a fitting capstone for this unit. Modern biology is the scientific extension of our human tendency to feel connected to and curious about all forms of life. We are most likely to save what we appreciate, and we are most likely to appreciate what we understand. We hope that our discussion of biodiversity has deepened your biophilia and broadened your education.

▼ Figure 20.44 **Biophilia.** Whether we seek other organisms in their own habitats or invite them into ours, we clearly find pleasure in the diversity of life.

Chapter Review

SUMMARY OF KEY CONCEPTS

The Loss of Biodiversity

THE COMPONENTS OF BIODIVERSITY		
Genetic Diversity	**Species Diversity**	**Ecosystem Diversity**
Loss of genetic diversity threatens the survival of a species and eliminates potential benefits to people.	The current rate of species extinctions is extremely high compared to the rate of natural extinctions over the past 100,000 years.	Destruction of ecosystems results in the loss of essential ecosystem services.

Causes of Declining Biodiversity

Habitat destruction is the leading cause of extinctions. Invasive species, overexploitation, and pollution are also significant factors.

Community Ecology

Interspecific Interactions

Populations in a community interact in a variety of ways that can be generally categorized as being beneficial (**+**) or harmful (**−**) to the populations. Because **+/−** interactions (exploitation of one species by another species) may have such a negative impact on the individual that is harmed, defensive evolutionary adaptations are common.

Trophic Structure

The trophic structure of a community defines the feeding relationships among organisms. These relationships are sometimes organized into food chains or food webs. Toxins may accumulate by the process of biological magnification as they are passed up a food chain to the top predators.

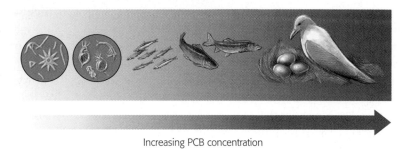

Increasing PCB concentration

Species Diversity in Communities

Diversity within a community includes species richness and relative abundance of different species. A keystone species is a species that has a great impact on the composition of the community despite relatively low abundance or biomass.

Disturbances in Communities

Disturbances are episodes that damage communities, at least temporarily, by destroying organisms or altering the availability of resources such as mineral nutrients and water. Humans are the most significant cause of disturbances today.

Ecological Succession

The sequence of changes in a community after a disturbance is called ecological succession. Primary succession occurs where a community arises in a virtually lifeless area with no soil. Secondary succession occurs where a disturbance has destroyed an existing community but left the soil intact.

INTERACTIONS BETWEEN SPECIES IN A COMMUNITY					
Interspecific Interaction	**Effect on Species 1**	**Effect on Species 2**	**Interspecific Interaction**	**Effect on Species 1**	**Effect on Species 2**
Competition	**−**	**−**	Exploitation	**+**	**−**
			Predation	**+**	**−**
Mutualism	**+**	**+**	Herbivory	**+**	**−**
			Parasites and Pathogens		

Ecosystem Ecology

Energy Flow in Ecosystems

An ecosystem is a biological community and the abiotic factors with which the community interacts. Energy must flow continuously through an ecosystem, from producers to consumers and decomposers. Chemical elements can be recycled between an ecosystem's living community and the abiotic environment. Trophic relationships determine an ecosystem's routes of energy flow and chemical cycling.

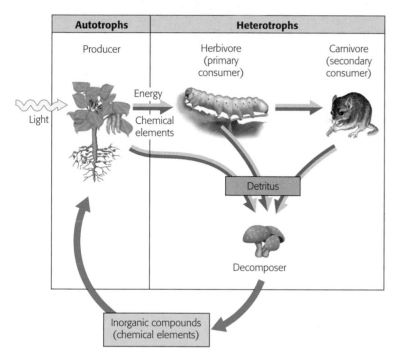

Primary production is the rate at which plants and other producers build biomass. Ecosystems vary considerably in their productivity. Primary production sets the spending limit for the energy budget of the entire ecosystem because consumers must acquire their organic fuels from producers.

In a food chain, only about 10% of the biomass at one trophic level is available to the next, resulting in a pyramid of production:

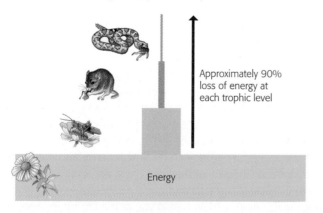

Approximately 90% loss of energy at each trophic level

When people eat producers instead of consumers, less photosynthetic production is required, which reduces the impact to the environment.

Chemical Cycling in Ecosystems

Biogeochemical cycles involve biotic and abiotic components. Each circuit has an abiotic reservoir through which the chemical cycles. Some chemical elements require "processing" by certain microorganisms before they are available to plants as inorganic nutrients. A chemical's specific route through an ecosystem varies with the element and the trophic structure of the ecosystem. Phosphorus is not very mobile and is cycled locally. Carbon and nitrogen spend part of their time in gaseous form and are cycled globally. Runoff of nitrogen and phosphorus, especially from agricultural land, causes algal blooms in aquatic ecosystems, lowering water quality and sometimes depleting the water of oxygen.

Conservation and Restoration Biology

Biodiversity "Hot Spots"

Conservation biology is a goal-oriented science that seeks to counter the loss of biodiversity. The front lines for conservation biology are biodiversity "hot spots," relatively small geographic areas that are especially rich in endangered species.

Conservation at the Ecosystem Level

Increasingly, conservation biology aims at sustaining the biodiversity of entire communities, ecosystems, and landscapes. Edges between ecosystems are prominent features of landscapes, with positive and negative effects on biodiversity. Corridors can promote dispersal and help sustain populations.

Restoring Ecosystems

In some cases, toxic substances such as heavy metals can be removed from an ecosystem by microbes or plants. Ecologists are working to revitalize some ecosystems by planting native vegetation, removing barriers to wildlife, and other means. The Kissimmee River Restoration Project is an attempt to undo the ecological damage done when the river was engineered into straight channels.

The Goal of Sustainable Development

Balancing human needs with the health of the biosphere, sustainable development has the goal of long-term prosperity of human societies and the ecosystems that support them.

Evolution Connection: Biophilia and an Environmental Ethic

Biophilia is a term used by biologist E. O. Wilson to describe the innate affinity humans have for nature and for other organisms. Wilson—and many other biologists—hope that our biophilia will help us develop an environmental ethic of preserving biodiversity.

SELF-QUIZ

1. Currently, the number one cause of biodiversity loss is _____.

2. According to the concept of competitive exclusion,
   a. two species cannot coexist in the same habitat.
   b. extinction or emigration is the only possible result of competitive interactions.
   c. intraspecific competition results in the success of the best-adapted individuals.
   d. two species cannot share the same niche in a community.

3. The concept of trophic structure of a community emphasizes the
   a. prevalent form of vegetation.
   b. keystone species concept.
   c. feeding relationships within a community.
   d. species richness of the community.

4. Match each organism with its trophic level (you may choose a level more than once).
   a. alga   1. decomposer
   b. grasshopper   2. producer
   c. zooplankton   3. tertiary consumer
   d. eagle   4. secondary consumer
   e. fungus   5. primary consumer

5. Why are the top predators in food chains most severely affected by pesticides such as DDT?

6. Over a period of many years, grass grows on a sand dune, then shrubs grow, and then eventually trees grow. This is an example of ecological _____.

7. According to the pyramid of production, why is eating grain-fed beef a relatively inefficient means of obtaining the energy trapped by photosynthesis?

8. Local conditions, such as heavy rainfall or the removal of plants, may limit the amount of nitrogen, phosphorus, or calcium available to a particular terrestrial ecosystem, but the amount of carbon available to the ecosystem is seldom a problem. Why?

9. A _____ is a local grouping of interacting ecosystems with several adjacent habitats.

10. Movement corridors are
    a. the routes taken by migratory animals.
    b. strips or clumps of habitat that connect isolated fragments.
    c. landscapes that include several different ecosystems.
    d. edges or boundaries between ecosystems.
    e. buffer zones that protect the long-term viability of protected areas.

Answers to the Self-Quiz questions can be found in Appendix D.

THE PROCESS OF SCIENCE

11. An ecologist studying desert plants performed the following experiment. She staked out two identical plots that included a few sagebrush plants and numerous small annual wildflowers. She found the same five wildflower species in similar numbers in both plots. Then she enclosed one of the plots with a fence to keep out kangaroo rats, the most common herbivores in the area. After two years, four species of wildflowers were no longer present in the fenced plot, but one wildflower species had increased dramatically. The unfenced control plot had not changed significantly in species composition. Using the concepts discussed in the chapter, what do you think happened?

12. Imagine that you have been chosen as the biologist for the design team implementing a self-contained space station to be assembled in orbit. It will be stocked with the organisms you choose, creating an ecosystem that will support you and five other people for two years. Describe the main functions you expect the organisms to perform. List the types of organisms you would select, and explain why you chose them.

13. Biologists in the United States are concerned that populations of many migratory songbirds, such as warblers, are declining. Evidence suggests that some of these birds might be victims of pesticides. Most of the pesticides implicated in songbird mortality have not been used in the United States since the 1970s. Suggest a hypothesis to explain the current decline in songbird numbers. Design an experiment that could test your hypothesis.

BIOLOGY AND SOCIETY

14. Some organizations are starting to envision a sustainable society—one in which each generation inherits sufficient natural and economic resources and a relatively stable environment. The Worldwatch Institute, an environmental policy organization, estimates that we must reach sustainability by the year 2030 to avoid economic and environmental collapse. In what ways is our current system not sustainable? What might we do to work toward sustainability, and what are the major roadblocks to achieving it? How would your life be different in a sustainable society?

15. The Biological Dynamics of Forest Fragmentation Project, which has contributed so much to our understanding of threats to biodiversity, is itself endangered. Encouraged by Brazilian government agencies, urban sprawl and intensive forest settlement are closing in on the study site. Activities such as clear-cutting, burning, hunting, and logging threaten the integrity of the surrounding forest. Researchers at the BDFFP, which is jointly operated by a Brazilian research agency and the Smithsonian Tropical Institute, hope that they can bring attention to the problem through the Brazilian media and pressure the government into protecting the project. How would you argue the importance of protecting the BDFFP against ecologically destructive activities in a letter to a newspaper editor or to a government official?

Appendix A Metric Conversion Table

MEASUREMENT	UNIT AND ABBREVIATION	METRIC EQUIVALENT	APPROXIMATE METRIC-TO-ENGLISH CONVERSION FACTOR	APPROXIMATE METRIC-TO-ENGLISH CONVERSION FACTOR
Length	1 kilometer (km)	= 1,000 (10^3) meters	1 km = 0.6 mile	1 mile = 1.6 km
	1 meter (m)	= 100 (10^2) centimeters	1 m = 1.1 yards	1 yard = 0.9 m
		= 1,000 millimeters	1 m = 3.3 feet	1 foot = 0.3 m
			1 m = 39.4 inches	
	1 centimeter (cm)	= 0.01(10^{-2}) meter	1 cm = 0.4 inch	1 foot = 30.5 cm
				1 inch = 2.5 cm
	1 millimeter (mm)	= 0.001 (10^{-3}) meter	1 mm = 0.04 inch	
	1 micrometer (µm)	= 10^{-6} meter (10^{-3} mm)		
	1 nanometer (nm)	= 10^{-9} meter (10^{-3} µm)		
	1 angstrom (Å)	= 10^{-10} meter (10^{-4} µm)		
Area	1 hectare (ha)	= 10,000 square meters	1 ha = 2.5 acres	1 acre = 0.4 ha
	1 square meter (m²)	= 10,000 square centimeters	1 m² = 1.2 square yards	1 square yard = 0.8 m²
			1 m² = 10.8 square feet	1 square foot = 0.09 m²
	1 square centimeter (cm²)	= 100 square millimeters	1 cm² = 0.16 square inch	1 square inch = 6.5 cm²
Mass	1 metric ton (t)	= 1,000 kilograms	1 t = 1.1 tons	1 ton = 0.91 t
	1 kilogram (kg)	= 1,000 grams	1 kg = 2.2 pounds	1 pound = 0.45 kg
	1 gram (g)	= 1,000 milligrams	1 g = 0.04 ounce	1 ounce = 28.35 g
			1 g = 15.4 grains	
	1 milligram (mg)	= 10^{-3} gram	1 mg = 0.02 grain	
	1 microgram (µg)	= 10^{-6} gram		
Volume *(solids)*	1 cubic meter (m³)	= 1,000,000 cubic centimeters	1 m³ = 1.3 cubic yards	1 cubic yard = 0.8 m³
	1 cubic centimeter (cm³ or cc)	= 10^{-6} cubic meter	1 m³ = 35.3 cubic feet	1 cubic foot = 0.03 m³
	1 cubic millimeter (mm³)	= 10^{-9} cubic meter (10^{-3} cubic centimeter)	1 cm³ = 0.06 cubic inch	1 cubic inch = 16.4 cm³
Volume *(liquids and gases)*	1 kiloliter (kL or kl)	= 1,000 liters	1 kL = 264.2 gallons	1 gallon = 3.79 L
	1 liter (L)	= 1,000 millimeters	1 L = 0.26 gallon	1 quart = 0.95 L
			1 L = 1.06 quarts	
	1 milliliter (mL or ml)	= 10^{-3} liter	1 mL = 0.03 fluid ounce	1 quart = 946 mL
		= 1 cubic centimeter	1 mL = approx. $\frac{1}{4}$ teaspoon	1 pint = 473 mL
			1 mL = approx. 15–16 drops	1 fluid ounce = 29.6 mL
				1 teaspoon = approx. 5 mL
Volume *(liquids and gases)*	1 microliter (µl or µL)	= 10^{-6} liter (10^{-3} milliliters)		
Time	1 second (s)	= $\frac{1}{60}$ minute		
	1 millisecond (ms)	= 10^{-3} second		
Temperature	Degrees Celsius (°C)		°F = $\frac{9}{5}$°C + 32	°C = $\frac{5}{9}$(°F − 32)

Appendix B Periodic Table of Elements

Periodic Table of Elements

Representative elements

Alkali metals

Alkaline earth metals

Halogens Noble gases

Transition elements

Period number	Group 1A	Group 2A	3B	4B	5B	6B	7B	8B	8B	8B	1B	2B	Group 3A	Group 4A	Group 5A	Group 6A	Group 7A	18 Group 8A
1	H 1.008							10	11	12			13	14	15	16	17	He 4.003
2	Li 6.941	Be 9.012											B 10.81	C 12.01	N 14.01	O 16.00	F 19.00	10 Ne 20.18
3	11 Na 22.99	12 Mg 24.31											13 Al 26.98	14 Si 28.09	15 P 30.97	16 S 32.07	17 Cl 35.45	18 Ar 39.95
4	19 K 39.10	20 Ca 40.08	21 Sc 44.96	22 Ti 47.87	23 V 50.94	24 Cr 52.00	25 Mn 54.94	26 Fe 55.85	27 Co 58.93	28 Ni 58.69	29 Cu 63.55	30 Zn 65.41	31 Ga 69.72	32 Ge 72.64	33 As 74.92	34 Se 78.96	35 Br 79.90	36 Kr 83.80
5	37 Rb 85.47	38 Sr 87.62	39 Y 88.91	40 Zr 91.22	41 Nb 92.91	42 Mo 95.94	43 Tc (98)	44 Ru 101.1	45 Rh 102.9	46 Pd 106.4	47 Ag 107.9	48 Cd 112.4	49 In 114.8	50 Sn 118.7	51 Sb 121.8	52 Te 127.6	53 I 126.9	54 Xe 131.3
6	55 Cs 132.9	56 Ba 137.3	57* La 138.9	72 Hf 178.5	73 Ta 180.9	74 W 183.8	75 Re 186.2	76 Os 190.2	77 Ir 192.2	78 Pt 195.1	79 Au 197.0	80 Hg 200.6	81 Tl 204.4	82 Pb 207.2	83 Bi 209.0	84 Po (209)	85 At (210)	86 Rn (222)
7	87 Fr (223)	88 Ra (226)	89† Ac (227)	10 Rf (261)	10 Db (262)	10 Sg (266)	10 Bh (264)	10 Hs (269)	10 Mt (268)	11 Ds (271)	11 — (272)	11 — (285)	11 — (284)	11 — (289)	11 — (288)			

*Lanthanides	58 Ce 140.1	59 Pr 140.9	60 Nd 144.2	61 Pm (145)	62 Sm 150.4	63 Eu 152.0	64 Gd 157.3	65 Tb 158.9	66 Dy 162.5	67 Ho 164.9	68 Er 167.3	69 Tm 168.9	70 Yb 173.0	71 Lu 175.0
†Actinides	90 Th 232.0	91 Pa 231.0	92 U 238.0	93 Np (237)	94 Pu (244)	95 Am (243)	96 Cm (247)	97 Bk (247)	98 Cf (251)	99 Es 252	10 Fm 257	10 Md 258	10 No 259	10 Lr 260

Metals   Metalloids   Nonmetals

NAME	SYMBOL	NAME	SYMBOL	NAME	SYMBOL	NAME	SYMBOL	NAME	SYMBOL
Actinium	Ac	Cobalt	Co	Iodine	I	Osmium	Os	Silicon	Si
Aluminum	Al	Copper	Cu	Iridium	Ir	Oxygen	O	Silver	Ag
Americium	Am	Curium	Cm	Iron	Fe	Palladium	Pd	Sodium	Na
Antimony	Sb	Darmstadtium	Ds	Krypton	Kr	Phosphorus	P	Strontium	Sr
Argon	Ar	Dubnium	Db	Lanthanum	La	Platinum	Pt	Sulfur	S
Arsenic	As	Dysprosium	Dy	Lawrencium	Lr	Plutonium	Pu	Tantalum	Ta
Astatine	At	Einsteinium	Es	Lead	Pb	Polonium	Po	Technetium	Tc
Barium	Ba	Erbium	Er	Lithium	Li	Potassium	K	Tellurium	Te
Berkelium	Bk	Europium	Eu	Lutetium	Lu	Praseodymium	Pr	Terbium	Tb
Beryllium	Be	Fermium	Fm	Magnesium	Mg	Promethium	Pm	Thallium	Tl
Bismuth	Bi	Fluorine	F	Manganese	Mn	Protactinium	Pa	Thorium	Th
Bohrium	Bh	Francium	Fr	Meitnerium	Mt	Radium	Ra	Thulium	Tm
Boron	B	Gadolinium	Gd	Mendelevium	Md	Radon	Rn	Tin	Sn
Bromine	Br	Gallium	Ga	Mercury	Hg	Rhenium	Re	Titanium	Ti
Cadmium	Cd	Germanium	Ge	Molybdenum	Mo	Rhodium	Rh	Tungsten	W
Calcium	Ca-	Gold	Au	Neodymium	Nd	Rubidium	Rb	Uranium	U
Californium	Cf	Hafnium	Hf	Neon	Ne	Ruthenium	Ru	Vanadium	V
Carbon	C	Hassium	Hs	Neptunium	Np	Rutherfordium	Rf	Xenon	Xe
Cerium	Ce	Helium	He	Nickel	Ni	Samarium	Sm	Ytterbium	Yb
Cesium	Cs	Holmium	Ho	Niobium	Nb	Scandium	Sc	Yttrium	Y
Chlorine	Cl	Hydrogen	H	Nitrogen	N	Seaborgium	Sg	Zinc	Zn
Chromium	Cr	Indium	In	Nobelium	No	Selenium	Se	Zirconium	Zr

Appendix C Credits

PHOTO CREDITS

UNIT OPENERS: Unit I Dr. Dennis Kunkel/Phototake NYC; **Unit II** William James Warren/Science Faction/Corbis; **Unit III** Gavin Kincome/SPL/Getty Images; **Unit IV** Michael Quinton/Minden Pictures.

CHAPTER 1: Chapter opening photo Michael Boyny/ PhotoLibrary; **1.1a** Nikreates/Alamy; **1.1b** P. Narayan/The Image Bank/AGE Fotostock; **1.1c** Gallo Images-Anthony Bannister/Getty Images; **1.1d** Michael Lander/NordicPhotos/ AGE Fotostock; **1.1e** Kim Taylor and Jane Burton/Dorling Kindersley; **1.1f** Steve Bloom Images/Alamy; **1.1g** Gerry Ellis/Minden Pictures; **1.2.1** NASA/Goddard Space Flight Center; **1.2.2** Michio Hoshino/Minden Pictures; **1.2.4** Kevin Schafer/Corbis; **1.2.7** Manfred Kage/Peter Arnold; **1.3** Christopher Elwell/shutterstock; **1.4 left** S. C. Holt/Biological Photo Service; **right** Dr. Gopal Murti/Visuals Unlimited; **1.6** Maximilian Stock Ltd/Photo Researchers; **1.7** Martin Strmiska/ AGE Fotostock; **1.8 top to bottom** Steve Gschmeissner/SPL/ Photo Researchers; Ralph Robinson/Visuals Unlimited; Neil Fletcher/Dorling Kindersley; Malcom Coulson/Dorling Kindersley; Stockbyte/Getty Images; D. P. Wilson/Photo Researchers; **1.9** James L. Amos/Corbis; **1.11 top left** Mike Hayward/Alamy; **top right** Neg./Transparency no. 330300. Courtesy Dept. of Library Services, American Museum of Natural History; **bottom** Tui De Roy/Minden Pictures; **1.13a left to right** iStockphoto; Paul Rapson/Alamy; Foodcollection.com/Alamy; Suzannah Skelton/iStockphoto; iStockphoto; iStockphoto; **1.13b left** Chris Collins/Corbis; **right** Eric Baccega/Nature Picture Library; **1.14 left** Michael Nichols/National Geographic Image Collection; **right** Tim Ridley/Dorling Kindersley, Courtesy of the Jane Goodall Institute, Clarendon Park, Hampshire; **1.17** Jim Newcomb; **1.18** AP Photo; **1.19 left** Simon Fraser/SPL/Photo Researchers; **right** S. Lowry/University of Ulster/Getty Images.

CHAPTER 2: Chapter opening photo moodboard/Corbis; **2.1 top** Clive Streeter/Dorling Kindersley; **bottom two** Dorling Kindersley; **2.3 left to right** Ivan Polunin/Bruce Coleman; Lukasz Panek/iStockphoto; Enrico Fianchini/ iStockphoto; Howard Shooter/Dorling Kindersley; **2.9** Doug Allan/Minden Pictures; **2.10 left** Stephen Alvarez/NGS Image Collection; **right** R. Kessel-Shih/Visuals Unlimited; **2.11** Piet Munsterman/Foto Natura/Minden Pictures; **2.12** NBAE/ Getty Images; **2.13** Claudio Baldini/iStockphoto; **2.14** Kristin Piljay; **p. 31 bottom** John Kelly/Getty Images; **2.16 top to bottom** Beth Van Trees/shutterstock; Monika Wisniewska/ iStockphoto; Feng Yu/iStockphoto; Jakub Semeniuk/ iStockphoto; **2.17** J M Hall-Spencer; **2.18** NASA.

CHAPTER 3: Chapter opening photos, main Ian O'Leary/Dorling Kindersley; **inset** Dr. Tim Evans/SPL/Photo Researchers; **3.3 left** James M Phelps, Jr./shutterstock; **right** Dorling Kindersley; **3.5** Gerry Ellis/PhotoLibrary; **3.7** Andy Crawford/Dorling Kindersley; **3.8** Sandra Caldwell/ iStockphoto; **3.9 top to bottom** Biophoto Associates/Photo Researchers; Courtesy of Dr. L. M. Beidler; Biophoto Associates/Photo Researchers; **3.10** Martyn F. Chillmaid/ SPL/Photo Researchers; **3.12 left box** Hannamariah/ shutterstock; iStockphoto; Valentin Mosichev/shutterstock; Thomas M. Perkins/shutterstock; **right box** Dorling

Kindersley; Oleksandr Staroseltsev/iStockphoto; Robyn Mackenzie/shutterstock; vladm/shutterstock; Multiart/ shutterstock; iStockphoto; **3.13 left** EcoPrint/shutterstock; **right** Stockbyte/Getty Images; **3.14 left** Pool Getty Images/AP Wide World Photos; **right** Peter Foley/epa/ Corbis; **3.15 left to right** Glow Images/Alamy; Dorling Kindersley; Dave King/Dorling Kindersley; Harry Taylor/ Dorling Kindersley; Dorling Kindersley; Steve Gschmeissner/ SPL/Photo Researchers; Kristin Piljay; **3.19** Stanley Flegler/ Visuals Unlimited; **3.28** Brochard/Jupiter Images.

CHAPTER 4: Chapter opening photo SPL/Photo Researchers; **4.1 left to right** Dennis Kunkel/Phototake USA; Dennis Kunkel/Phototake USA; Microworks Color/Phototake USA; **4.2** Sinclair Stammers/SPL/Photo Researchers; **4.4** NIBSC/ SPL/Photo Researchers; **4.7** CDC/SPL/Photo Researchers; **4.8 both** Courtesy of Richard Rodewald/Biological Photo Service; **4.10** Hybrid Medical Animation/Photo Researchers; **4.11** D. W. Fawcett/Photo Researchers; **4.13** Barry King/BPS; **4.15** SPL/Photo Researchers; **4.16** Daniel S. Friend; **4.17a** Cabisco/Visuals Unlimited; **4.17b** Dr. Henry C. Aldrich/ Visuals Unlimited; **4.18** Garry Cole/BPS; **4.19:** Courtesy of W.P. Wergin and E.H. Newcomb, University of Wisconsin/ BPS; **4.20** Courtesy of Daniel S. Friend, Harvard Medical School; **4.21a** M. Schliwa/Visuals Unlimited; **4.21b** Dr. D. J. Patterson/SPL/Photo Researchers; **4.22 left to right** Dr. Dennis Kunkel/Phototake NYC; Steve Gschmeissner/ SPL/Photo Researchers; Charles Daghlian/Photo Researchers; **4.23 top** The Advertising Archives; **bottom** Centers for Disease Control and Prevention.

CHAPTER 5: Chapter opening photo J. W. Shuler/SPL/ Photo Researchers; **5.1** Alain Lecocq/Sygma/Corbis; **5.3a** Dave King/Dorling Kindersley; **5.3b** Russell Sadur/Dorling Kindersley; **5.8** PDB ID: 1BGM, Jacobson, R.H., Zhang, X.J., DuBose, R.F., Matthews, B.W.: Three-dimensional structure of beta-galactosidase from E. coli. Nature 369 pp761 (1994); **5.15** David Cook/blueshiftstudios/Alamy; **5.19** Jim Cummins/ Corbis; **5.20** Peter B. Armstrong.

CHAPTER 6: Chapter opening photo David Stoecklein/ Corbis; **6.1** Pete Oxford/Minden Pictures; **6.2 left to right** Elena Elisseeva/iStockphoto; Eric Isselée/shutterstock; Ultrashock/shutterstock; **6.3** Jim Cummins/Getty Images; **6.14** Chris Mole/shutterstock; **6.16 left** stanislaff/ iStockphoto; **right** stocksnapp/shutterstock; **6.17** José Carlos Pires Pereira/iStockphoto.

CHAPTER 7: Chapter opening photo Jamie Grill/Getty Images; **7.1 left to right** Dorling Kindersley; Jeff Rotman/ Nature Picture Library; Susan M. Barns, Ph.D.; **7.2 top to bottom** John Fielding/Dorling Kindersley; M. Eichelberger/ Visuals Unlimited; Courtesy of W.P. Wergin and E.H. Newcomb, University of Wisconsin/BPS; **7.6** Jamie Grill/Getty Images; **7.7** Pete Turner/The Image Bank/Getty Images; **7.8** Photos by LQ/Alamy; **7.14 left** Dinodia/ Pixtal/AGE Fotostock; **right** ImageDJ/Jupiter Images.

CHAPTER 8: Chapter opening photo, main Alan Watson/ Dorling Kindersley; **inset** National Tropical Botanical Garden; **8.1 left to right** Dr Gopal Murti/Photo Researchers; Claude Cortier/Photo Researchers; Biophoto Associates/ Photo Researchers; Roger Steene/Image Quest; Eric J. Simon;

8.2 top to bottom Eric Isselée/shutterstock; Christian Musat/shutterstock; Tony Wear/shutterstock; Edyta Pawlowska/iStockphoto; Eric Isselée/shutterstock; Milton H. Gallardo; **8.3** Ed Reschke/Peter Arnold; **8.4 top** A.L. Olins, University of Tennessee/BPS; **bottom** Biophoto Associates/ Photo Researchers; **8.7 all** Conly L. Rieder, Ph.D.; **8.8a** David M. Phillips/Visuals Unlimited; **8.8b** Carolina Biological Supply/Phototake NYC; **8.10** Chris Carroll/Uppercut/Getty Images; **8.11** CNRI/SPL/Photo Researchers; **8.12** Eric McNatt/ Getty Images; **8.14** Ed Reschke/Peter Arnold; **8.17** David M. Phillips/Photo Researchers; **8.19** Photograph courtesy Dr. David Mark Welch; **8.22 top** Lauren Shear/Photo Researchers; **bottom** CNRI/SPL/Photo Researchers; **8.24** Grant Heilman Photography; **p. 143** Carolina Biological Supply/Phototake.

CHAPTER 9: Chapter opening photo Magdalena Rehova/ Alamy; **9.1** Hulton Archive/Getty Images; **9.4** Patrick Lynch/ Alamy; **9.5** Bettmann/Corbis; **9.8** Martin Shields/Photo Researchers; **9.9a left to right** Tracy Morgan/Dorling Kindersley; Tracy Morgan/Dorling Kindersley; Eric Isselée/ shutterstock; Andrew Johnson/iStockphoto; **9.10 left** Tracy Morgan/Dorling Kindersley; **right** Eric Isselée/shutterstock; **9.12 top left to right** James Woodson/Getty Images; Adrianna Williams/zefa/Corbis; Jose Luis Pelaez, Inc./Blend Images/ Getty Images; **9.12 bottom left to right** John Evans/ iStockphoto; Rubberball/Getty Images; Bruce Talbot/DK Stock/Getty Images; **9.13 left to right** Comstock Images/ Jupiter Images; Bruce Talbot/DK Stock/Getty Images; Jose Luis Pelaez/Getty Images; **9.15** Michael Ciesielski Photography; **9.16** Rex Rystedt/Time & Life Pictures/Getty Images; **9.17** Eric Isselée/shutterstock; **9.20** Mauro Fermariello/Photo Researchers; **9.21 left** Oliver Meckes/Nicole Ottawa/Photo Researchers; **right** Eye of Science/Photo Researchers; **9.23:** Eric J. Simon; **9.30 top left** 4x6/iStockphoto; **top center** Yuri Arcurs/iStockphoto; **bottom left** Dave King/Dorling Kindersley; **bottom center** Jo Foord/Dorling Kindersley; **right** Andrew Syred/Photo Researchers; **9.32:** FPG International/Taxi/Getty Images; **9.33 top left to bottom right** Jerry Young/Dorling Kindersley; Nikolay Titov/ iStockphoto; Dave King/Dorling Kindersley; Dave King/ Dorling Kindersley; Tracy Morgan/Dorling Kindersley; Dave King/Dorling Kindersley; Jerry Young/Dorling Kindersley; Dave King/Dorling Kindersley; Dave King/Dorling Kindersley; Tracy Morgan/Dorling Kindersley; Le Loft 1911/shutterstock; **p. 171** USDA/APHIS Animal And Plant Health Inspection Service.

CHAPTER 10: Chapter opening main photo Somos/Veer/Getty Images; **inset** James Cavallini/SPL/Photo Researchers; **10.3 left** Barrington Brown/Photo Researchers; **right top and bottom** Courtesy of the Library of Congress; **10.9** Dr. Arthur Siegelman/Visuals Unlimited; **10.12** Dr. Masaru Okabe, Research Institute for Microbial Diseases; **10.23** Ron Dahlquist/Pacific Stock; **10.24** Stephen Welstead/Corbis; **10.25** Oliver Meckes/Eye of Science/Photo Researchers; **10.26** Russell Kightley/SPL/Photo Researchers; **10.27** N. Thomas/Photo Researchers; **10.29** Hazel Appleton, Centre for Infections/Health Protection Agency/SPL/Photo Researchers; **10.30** Jeff Zelevansky/Reuters/Corbis; **10.32** National Institute for Biological Standards and Control

17.4 left Publiphoto/Photo Researchers; **right** Chip Clark; **17.8** James D. Watt/Image Quest Marine; **17.9 top left** Sue Daly/Nature Photo Library; **top middle** Jan Van Arkel/Foto Natura/Minden Pictures; **top right** Kim Taylor/ Nature Photo Library; **bottom:** Jordi Chias/Image Quest Marine; **17.12 top left** Georgette Douwma/Nature Photo Library; **bottom left** Jez Tryner/Image Quest Marine; **center left** Harold W. Pratt/BPS; **center right** Marevision/AGE Fotostock; **right** Chris Newbert/Minden Pictures; **17.13 top** E. R. Degginger/Photo Researchers; **bottom left** Geoff Brightling /Dorling Kindersley; **bottom right** Dennis Kunkel/Phototake NYC; **17.14 left to right** A.N.T./Photoshot/ NHPA Limited; Carol Buchanan/Photolibrary; WILDLIFE/ Peter Arnold; **p. 346** Astrid & Hanns-Frieder Michler/Photo Researchers; **17.16 left to right** Steve Gschmeissner/Photo Researchers; Eye of Science/Photo Researchers; Reproduced by permission from Howard Shiang, D.V.M., Journal of the American Veterinary Medical Association 163:981, Oct. 1973; **17.17 top to bottom** Callan Morgan/Peter Arnold; Piotr Naskrecki/Minden Pictures; George Grall/National Geographic Stock; ZSSD/Minden Pictures; **17.18** Dorling Kindersley; **17.19 clockwise from top center** Dorling Kindersley; Andrew Syred Photo Researchers; Larry West/ Photo Researchers; Callan Morgan/Peter Arnold; Herbert Hopfensperger/agcfotostock; **17.20 top** Dave King/Dorling Kindersley; **center left to right** Piotr Naskrecki/Minden Pictures; Tom McHugh/Photo Researchers; Nature's Images/ Photo Researchers; **bottom left to right** Barry Mansell/ Minden Pictures; Nancy Sefton/SPL/Photo Researchers; **17.21 left** George Grall/National Geographic Stock; **right** Tom McHugh/Photo Researchers; **17.22** Radius Images/ Alamy; **17.23 top center** Piotr Naskrecki/Minden Pictures; **top left** ZSSD/Minden Pictures; **middle left** Piotr Naskrecki/ Minden Pictures; **bottom left** Michael Durham/Minden Pictures; **bottom center** Mark Moffett/Minden Pictures; **bottom right** Mitsuhiko Imamori/Minden Pictures; **middle right** Cordier - Huguet/AGE Fotostock; **top right** Piotr Naskrecki/Minden Pictures; **17.24 top series** John Shaw/Tom Stack and Associates; **bottom** Chris Sharp/Photolibrary; **17.25 top left** Roger Steene/Image Quest Marine; **inset** Gary Milburn/Tom Stack and Associates; **top right** Jose B. Ruiz/ Nature Photo Library; **bottom left** tbkmedia.de/Alamy; **bottom middle** Fred Bavendam/Minden Pictures; **bottom right** Roger Steene/Image Quest Marine; **17.26** Colin Keates/Dorling Kindersley; **17.28 left** Runk/Schoenberger/ Grant Heilman Photography; **right** Roger Steene/Image Quest Marine; **17.30a** Tom McHugh/Photo Researchers; **inset** Brandon D. Cole/Corbis; **17.30b** Breck P. Kent/Animals Animals - Earth Scenes; **inset** Breck P. Kent/Animals Animals - Earth Scenes; **17.30c** Carol Buchanan/F1 ONL/agefotostock; **17.30c** Peter Scoones/Nature Photo Library; **17.31a left** Michael & Patricia Fogden/Minden Pictures; **right** Michael & Patricia Fogden/Minden Pictures; **17.31b left** Michael Durham/Minden Pictures; **right** Jack Goldfarb/agefotostock; **17.33 p. 358 left** Robert and Linda Mitchell; **middle** Michael Durham/Minden Pictures; **right** Tui De Roy/Minden Pictures; **17.33 p. 359 left** Jerry Young/Dorling Kindersley; **middle** DLILLC/Corbis; **right** Miguel Periera/Dorling Kindersley; **17.34** Adam Jones/Digital Vision/Getty Images; **17.35 left** Reg Morrison/Minden Pictures; **inset** D. Parer and

E. Parer Cook/Auscape; **17.35 middle** blickwinkel/Alamy; **inset** Mitsuaki Iwago/Minden Pictures; **17.35 right** Mitsuaki Iwago/Minden Pictures; **17.37 top row left to right** Pete Oxford/Minden Pictures; Siegfried Grassegger/Photolibrary; Arco Images GmbH/Alamy; Juan Carlos Muñoz/agefotostock; Anup Shah/Nature Photo Library; **17.37 left center** P. Wegner/ Arco Images/Peter Arnold; **17.37 bottom left to right** Ingo Arndt/Nature Photo Library; Anup Shah/Nature Photo Library; John Kelly/Getty Images; **17.39a** The Cleveland Museum of Natural History; **17.39b** John Reader/SPL/Photo Researchers; **17.39c** Javier Trueba/Madrid Scientific Films/ Photo Researchers; **17.40** Philippe Plailly & Atelier Daynes/ Photo Researchers; **17.41 left & top right** Jean Clottes/Sygma/ Corbis; **bottom right** Jean-Marie Chauvet/Sygma/Corbis; **17.42** Steven Poe/UNEP/Still Pictures/Peter Arnold; **17.43** Susan Kuklin/Photo Researchers.

CHAPTER 18: Chapter opening photo Donovan Reese/Getty Images; **18.1a-b** Raphael Gaillarde/Liaison/Getty Images; **18.2** JAY DIRECTO/AFP/Getty Images; **18.3a** Barry Mansell/ Nature Picture Library; **18.3b** Ingrid Van Den Berg/Animals Animals; **18.3c** JUNIORS BILDARCHIV/AGE Fotostock; **18.3d** Jeremy Woodhouse/Getty Images; **18.4** Stephen Krasemann/Photo Researchers; **18.5** NASA; **18.6** Verena Tunnicliffe, University of Victoria; **inset** Woods Hole Oceanographic Institution; **18.7** Wilfried Krecichwost/Getty Images; **18.8** Gregg Adams/Getty Images; **18.9** Antti Leinonen/ Jupiter Images; **18.11 left** Daniel Cox/Photolibrary; **right** J & C Sohns/AGE Fotostock; **18.12** Brian Parker/Tom Stack & Associates; **18.13** Robert Stainforth/Alamy; **18.14** Worldsat International Inc./SPL/Photo Researchers; **18.16** Ishbukar Yalilfatar/shutterstock; **18.17** Kevin Schafer/Alamy; **18.18** GADOMSKI, MICHAEL/Animals Animals/Earth Scenes; **18.20** Digital Vision/Getty Images; **18.21** Carr Clifton/ Minden Pictures; **18.22** James Randklev/Image Bank/Getty Images; **18.29** Frans Lanting/Minden Pictures; **18.30** Yva Momatiuk & John Eastcott/Minden Pictures; **18.31** Juan Carlos Muñoz/AGE Fotostock; **18.32** John D. Cunningham/ Visuals Unlimited; **18.33** Jake Rajs/Getty Images; **18.34** Kennan Ward/Corbis; **18.35** Jorma Luhta/Nature Picture Library; **18.36** Darrell Gulin/Corbis; **18.37** Gordon Wiltsie/ NGS Image Collection; **18.39 both** UNEP/GRID-Sioux Falls; **18.40** Paul Edmondson/Getty Images; **18.41 both** UNEP/ GRID-Sioux Falls; **18.42** Jim West/Alamy; **18.43** Mark Turner/Jupiter Images; **18.47** Chris Martin Bahr/SPL/Photo Researchers; **18.48** Bryan Patrick/Sacramento Bee/MCT; **18.49** David McNew/Getty Images; **18.50** AP Photo; **18.51** Chris Cheadle/Photolibrary **18.52a** University of Oregon **18.52b** Volvox/Jupiter Images.

CHAPTER 19: Chapter opening photo John Carnemolla/ Corbis; **19.1** Luciano Candisani/Minden Pictures; **19.2** Jim Brandenburg/Minden Pictures; **19.3** Kevin Schafer/Peter Arnold; **19.4 left to right** Roger Phillips/Dorling Kindersley; Jane Burton/Dorling Kindersley; Yuri Arcurs/shutterstock; **Table 19.2 left** Achim Prill/iStockphoto; **right** Anke van Wyk/ shutterstock; **19.5** Edwin Giesbers/Foto Natura/Minden Pictures; **19.6** Roy Corral/Corbis; **19.7 left** Joshua Lewis/ shutterstock; **right** WizData, inc./shutterstock; **19.8a** Marcin Perkowski/shutterstock; **19.8b** Nigel Cattlin/Alamy; **19.9** Wolfgang Kaehler/Corbis; **19.10** Meul/ARCO/Nature Picture

Library; **19.11** Alan Carey/Photo Researchers; **19.12 left to right** Rob Curtis/The Early Birder; David Sieren/Visuals Unlimited; USDA Forest Service; **19.13:** Dan Burton/Nature Picture Library; **19.14** Ric Ergenbright/Corbis; **19.15** Jan Tove Johansson/Pictor International, Ltd./PictureQuest; **inset** David Tipling/Nature Picture Library; **19.16** Dennis, David M./Animals Animals/Earth Scenes; **19.17** Matt Meadows/Peter Arnold; **19.18 left** Scott Camazine/Alamy; **right** B. Runk/S. Schoenberger/Grant Heilman Photography; **19.22 left** Peter Ginter; **right** Peter Menzel/Peter Menzel Photography; **19.24** franzfoto.com/Alamy.

CHAPTER 20: Chapter opening photo Alaskastock; **20.1** D. Cavagnaro/Visuals Unlimited; **20.2 left** Mark Carwardine/ Still Pictures/Peter Arnold; **right** Michael & Patricia Fogden/ Minden Pictures; **20.3** Fred Bavendam/Minden Pictures; **20.4** University of Maine; **20.5** AFP/Getty Images; **20.6** Richard D. Estes/Photo Researchers; **20.7a** Doug Backlund; **20.7b** Terry Sohl; **20.8 both** M. I. Walker/SPL/Photo Researchers; **20.9** Jurgen Freund/naturepl.com; **20.10** Joe McDonald/Corbis; **20.11** Mark Moffett/Minden Pictures; **20.12 left** E. R. Degginger/Photo Researchers; **right** Breck P. Kent; **20.13 left** Stephen J. Krasemann/Photo Researchers; **right** Peter J. Mayne; **20.14 top left** BILDAGENTUR-ONLINE/ TH FOTO/SPL/Photo Researchers; **top right** Luca Invernizzi Tettoni/Photolibrary; **bottom** ImageState/Alamy Images; **20.16** Michael Sedam/Photolibrary; **20.21** William E. Townsend/Photo Researchers; **20.22** Todd Sieling/Corvus Consulting; **20.23** David Muench/Corbis; **20.24** Scott T. Smith/Corbis; **20.30** Gerry Ellis/Minden Pictures; **20.35** Michael Marten/SPL/Photo Researchers; **20.36 all** NASA; **20.38 top** Yann Arthus-Bertrand/Corbis; **bottom** James P. Blair/National Geographic Society; **20.39** Joel Sartore/NGS Image Collection; **20.40** R. O. Bierregaard, Jr., Biology Department, University of North Carolina, Charlotte; **20.41** USDA; **20.42** Photo provided by Kissimmee Division staff, South Florida Water Management District (WPB); **20.43** Frans Lanting/Minden Pictures; **20.44 left** Comstock Images/ Getty Images; **top right** Nordicphotos/Alamy; **bottom center** Frans Lanting/Minden Pictures; **bottom right** Image Source/ Corbis; **p. 450 Biodiversity table left to right** D. Cavagnaro/ Visuals Unlimited; Michael & Patricia Fogden/Minden Pictures; Fred Bavendam/Minden Pictures; **p. 450 Interactions table left top to bottom** Terry Sohl; Doug Backlund; Jurgen Freund/Nature Picture Library; **right top to bottom** James Balog/Getty Images; Scott Camazine/ Alamy; Renaud Visage/Getty Images.

ILLUSTRATION AND TEXT CREDITS

The following figures are adapted from C. K. Mathews and K. E. van Holde, *Biochemistry*, 2nd ed. Copyright © 1996 Pearson Education, Inc., publishing as Pearson Benjamin Cummings: **4.9, 6.11,** and **8.4.**

CHAPTER 4: 4.20: Adapted from W. M. Becker, L. J. Kleinsmith, and J. Hardin, *The World of the Cell*. Copyright ©

2003 Pearson Education, Inc., publishing as Pearson Benjamin Cummings.

CHAPTER 5: 5.3: Data from S. E. Gebhardt and R. G. Thomas, *Nutritive Values of Foods* (USDA, 2002); S. A. Plowman and D. L. Smith, *Exercise Physiology for Health, Fitness, and Performance,* 2nd ed. Copyright © 2003 Pearson Education, Inc., publishing as Pearson Benjamin Cummings.

CHAPTER 6: 6.5: Copyright © 2002 from *Molecular Biology of the Cell,* 4th ed. by Bruce Alberts et al., fig. 2.69, p. 92. Garland Science/Taylor & Francis Books, Inc. Reproduced by permission of Garland Science/Taylor & Francis LLC.

CHAPTER 7: 7.12: Adapted from Richard and David Walker, *Energy, Plants and Man,* fig. 4.1, p. 69. Oxygraphics. Copyright © Richard Walker. Used courtesy of Richard Walker, http://www.oxygraphics.co.uk.

CHAPTER 10: Page 194: Text quotation by Joshua Lederberg, from Barbara J. Culliton, "Emerging Viruses, Emerging Threat," *Science, 247,* p. 279, 1/19/1990. Copyright © 1990 American Association for the Advancement of Science.

CHAPTER 11: Table 11.2: Data from the American Cancer Society website, "Cancer Facts & Figures 2008" at www.cancer.org/docroot/STT/stt_0.asp; **11.10:** Adapted with permission from an illustration by William McGinnis, UCSD.

CHAPTER 13: 13.15: Data for graph from K. V. Young et al., "How the horned lizards got its horns," in *Science, 304,*

4/2/2004, p. 65. American Association for the Advancement of Science; **13.30:** Adapted from A. C. Allison, "Abnormal Hemoglobins and Erythrovute Enzyme-Deficiency Traits," in *Genetic Variation in Human Populations,* by G. A. Harrison, ed. (Oxford: Elsevier Science, 1961).

CHAPTER 14: 14.26: Adapted from Hickman, Roberts, and Larson. 1997, *Zoology,* 10/e, Wm. C. Brown, fig. 31.1.

CHAPTER 15: 15.17: Adapted from G. J. Tortora, B. R. Funke, and C. L. Case, *Microbiology: An Introduction,* 9th ed. Copyright © 2007 Pearson Education, Inc., publishing as Pearson Benjamin Cummings.

CHAPTER 16: Chapter opening poem: "Evil Tendencies Cancel" from *The Poetry of Robert Frost* edited by Edward Connery Lathem. Copyright 1969 by Henry Holt and Company. Copyright 1938 by Robert Frost, copyright 1964 by Lesley Frost Ballantine; **Table 16.1:** Adapted from Randy Moore et al., *Botany,* 2nd ed. Dubuque, IA: Brown, 1998, Table 2.2, p. 37.

CHAPTER 17: 17.38: *H. neanderthalensis* adapted from *The Human Evolution Coloring Book. P. boisei* drawn from a photo by David Bill.

CHAPTER 18: 18.25: From J. H. Withgott and S. R. Brennan, *Environment: The Science Behind the Stories,* 3rd ed., fig, 6.30, p 169. Copyright © 2008 Pearson Education, Inc., Prentice Hall; **18.27:** Adapted from Heinrich Walter and Siegmar-Walter Breckle 2003. *Walter's Vegetation of the Earth,* fig. 16,

p. 36. Springer-Verlag, © 2003; **18.44:** From "Global Temperature Change," J. Hansen et al., *PNAS,* Sept. 26, 2006, Vol. 103, No. 39, Fig. 1B. Copyright 2006 National Academy of Sciences, U.S.A. Reprinted with permission; **18.45:** Adapted from *Climate Change 2007: The Physical Science Basis,* Intergovernmental Panel on Climate Change, FAQ 2.1, Fig. 1. Reprinted by permission of IPCC, c/o World Meteorological Organization.

CHAPTER 19: 19.13: Data from Fisheries and Oceans, Canada, 1999; **19.19:** Data from United Nations, the World at Six Billion, 2007; **19.20 and 19.21:** Data from U. S. Census Bureau International Data Base; **19.23:** Adapted from information at phtbb.org/natural/footprint; **Table 19.1:** Data from Centers for Disease Control and Prevention website, "United States Life Tables, 2003," *National Vital Statistics Report,* vol. 50, no. 14, April 19, 2006. www.cdc.gov; **Table 19.2:** Adapted from E. R. Pianka, *Evolutionary Ecology,* 6th ed., p. 186. Copyright © 2000 Pearson Education, Inc., publishing as Pearson Benjamin Cummings; **Table 19.3:** Data from Population Reference Bureau, 2008, www.prb.org.

CHAPTER 20: 20.18: From Robert Leo Smith, *Ecology and Field Biology,* 4th ed. Copyright © 1990 by Robert Leo Smith. Reprinted by permission of Addison Wesley Longman, Inc.; **20.37:** From N. Myers et al., "Biodiversity hotspots for conservation priorities," *Nature,* Vol. 403, p. 853, 2/24/2000. Reprinted by permission from Macmillan Publishers Ltd. Copyright © 2000 Nature Publishing, Inc.

Appendix D Self-quiz Answers

CHAPTER 1

1. b (some living organisms are single-celled)
2. atom, molecule, cell, tissue, organ, organism, population, ecosystem, biosphere; the cell
3. Photosynthesis cycles nutrients by converting the carbon in carbon dioxide into sugar, which is then consumed by other organisms. Additionally, the oxygen in water is released as oxygen gas. Photosynthesis contributes to energy flow by converting sunlight into chemical energy, which is then also consumed by other organisms, and by producing heat.
4. a4, b1, c3, d2.
5. On average, those individuals with heritable traits best suited to the local environment produce the greatest number of offspring that survive and reproduce. This increases the frequency of those traits in the population over time. The result is the accumulation of evolutionary adaptations.
6. Without a control group, you don't know if the experimental outcome is due to the variable you are trying to test or to some other variable.
7. d
8. c
9. evolution
10. a3, b2, c1, d4

CHAPTER 2

1. electrons; neutrons; protons
2. Nitrogen-14 has an atomic number of 7 and a mass number of 14. The radioactive isotope, nitrogen-16, has an atomic number of 7 and a mass number of 16.
3. Organisms incorporate radioactive isotopes of an element into their molecules just as they do the nonradioactive isotopes, and researchers can detect the presence of the radioactive isotopes.
4. 2
5. Each carbon atom has only three covalent bonds instead of the required four.
6. The positively charged hydrogen regions would repel each other.
7. d
8. a
9. The positive and negative poles cause adjacent water molecules to become attracted to each other, forming hydrogen bonds. The properties of water such as cohesion, temperature regulation, and water's ability to act as a solvent all arise from this atomic stickiness.
10. The cola is an aqueous solution, with water as the solvent, sugar as the main solute, and the CO_2 making the solution acidic.

CHAPTER 3

1.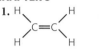

2. dehydration; hydrolysis
3. b
4. fatty acid; glycerol
5. b
6. c
7. d
8. Hydrophobic amino acids are most likely to be found within the interior of a protein, far from the watery environment.
9. a
10. starch (or glycogen or cellulose); nucleotide
11. Both DNA and RNA are polynucleotides; both have the same phosphate group along the backbone; and both use A, C, and G bases. But DNA uses T while RNA uses U as a base; the sugar differs between them; and DNA is usually double-stranded while RNA is usually single-stranded.

CHAPTER 4

1. d
2. about 0.075 mm, which equals 75 μm
3. b
4. A membrane is fluid because its components are not locked into place. A membrane is mosaic because it contains a variety of embedded proteins.
5. endomembrane system
6. smooth ER; rough ER
7. rough ER; Golgi apparatus; plasma membrane
8. Both organelles use membranes to organize enzymes, and both provide energy to the cell. But chloroplasts use pigments to capture energy from sunlight in photosynthesis, whereas mitochondria release energy from glucose using oxygen in cellular respiration. Chloroplasts are only in photosynthetic plants and protists, whereas mitochondria are in almost all eukaryotic cells.
9. a3, b1, c5, d2, e4
10. nucleus, nuclear pores, ribosomes, rough ER, Golgi apparatus

CHAPTER 5

1. You convert the chemical energy from food to the kinetic energy of your upward climb. At the top of the stairs, some of the energy has been stored as potential energy because of your higher elevation. The rest has been converted to heat.
2. Energy; entropy
3. 10,000 g (or 10 kg); remember that 1 Calorie on a food label equals 1,000 calories of heat energy.
4. The three phosphate groups store chemical energy, a form of potential energy. The release of a phosphate group makes some of this potential energy available to cells to perform work.
5. Hydrolases are enzymes that participate in hydrolysis reactions, breaking down large molecules into the smaller molecules that make them up.
6. An inhibitor's binding to another site on the enzyme can cause the enzyme's active site to change shape.
7. b
8. a
9. *Hypertonic* and *hypotonic* are relative terms. A solution that is hypertonic to tap water could be hypotonic to seawater. In using these terms, you must provide a comparison, as in "The solution is hypertonic to the cell's cytoplasm."
10. Passive transport moves atoms or molecules along their concentration gradient (from higher to lower concentration), while active transport moves them against their concentration gradient.
11. b
12. signal transduction pathway

CHAPTER 6

1. d
2. Plants produce organic molecules by photosynthesis. Consumers must acquire organic material by consuming it rather than making it.
3. In breathing, your lungs exchange CO_2 and O_2 between your body and the atmosphere. In cellular respiration, your cells consume the O_2 in extracting energy from food and release CO_2 as a waste product.
4. the electron transport chain
5. glucose; NAD^+
6. O_2
7. The majority of the energy provided by cellular respiration is generated during the electron transport chain. Shutting down that pathway will deprive cells of energy very quickly.
8. b
9. glycolysis
10. b
11. Because fermentation supplies only 2 ATP per glucose molecule compared with 38 from cellular respiration, the yeast will have to consume 19 times as much glucose to produce the same amount of ATP.

CHAPTER 7

1. thylakoids; stroma
2. The Calvin cycle, which consumes the NADPH and ATP, occurs in the stroma.
3. inputs: a, d, e; outputs: b, c
4. Green light is reflected by chlorophyll, not absorbed, and therefore cannot drive photosynthesis.
5. H_2O
6. c
7. The reactions of the Calvin cycle require the outputs of the light reactions (ATP and NADPH).
8. In hot, dry environments, most plants close their stomata, which saves water but decreases the amount of CO_2 available inside the leaves.
9. C_4 and CAM plants can close their stomata and save water without shutting down photosynthesis.
10. c

CHAPTER 8

1. c
2. They have identical genes (DNA).
3. They are in the form of very long, thin strands.
4. b
5. prophase and telophase
6. a) 1; 1, b) 1; 2, c) 2; 4, d) 2n; n, e) individually; by homologous pair, f) identical; unique, g) repair, growth, asexual reproduction; gamete formation
7. b

8. 39

9. Prophase II or metaphase II; it cannot be during meiosis I because then you would see an even number of chromosomes; it cannot be during a later stage in meiosis II because then you would see the sister chromatids separated.

10. benign; malignant

11. 16 ($2n = 8$, so $n = 4$, and $2^n = 2^4 = 16$)

12. Nondisjunction would create just as many gametes with an extra copy of chromosome 3 or 16, but extra copies of chromosome 3 or 16 are probably fatal.

CHAPTER 9

1. genotype; phenotype

2. Statement **a.** is the law of independent assortment; statement **b.** is the law of segregation.

3. c

4. c

5. d

6. d

7. d

8. $\frac{1}{4}$ ($\frac{1}{2}$ chance the child will be male times $\frac{1}{2}$ chance that he will inherit the X carrying the disease allele)

9. The parental-type gametes are *WS* and *ws*. Recombinant gametes are *Ws* and *wS*, produced by crossing over.

10. Height appears to result from polygenic inheritance, like human skin color. See Figure 9.22.

11. The brown allele appears to be dominant, the white allele recessive. The brown parent appears to be homozygous dominant, *BB*, and the white mouse is homozygous recessive, *bb*. The F_1 mice are all heterozygous, *Bb*. If two of the F_1 mice are mated, $\frac{3}{4}$ of the F_2 mice will be brown.

12. The best way to find out whether a brown F_2 mouse is homozygous dominant or heterozygous is to do a testcross: Mate the brown mouse with a white mouse. If the brown mouse is homozygous, all the offspring will be brown. If the brown mouse is heterozygous, you would expect half the offspring to be brown and half to be white.

13. Freckles is dominant, so Tim and Jan must both be heterozygous. There is a $\frac{3}{4}$ chance that they will produce a child with freckles, a $\frac{1}{4}$

chance that they will produce a child without freckles. The probability that the next two children will have freckles is $\frac{3}{4} \times \frac{3}{4} = \frac{9}{16}$.

14. Half their children will be heterozygous and have elevated cholesterol levels. There is a $\frac{1}{4}$ chance that their next child will be homozygous, *hh,* and have an extremely high cholesterol level, like Katerina.

15. The bristle-shape alleles are sex-linked, carried on the X chromosome. Normal bristles is dominant (*F*) and forked is recessive (*f*). The genotype of the female parent is $X^F X^f$. The genotype of the male parent is $X^F Y$. Their female offspring are $X^F X^f$; their male offspring are $X^f Y$.

16. The mother is a heterozygous carrier, and the father is normal. See Figure 9.32 for a pedigree. $\frac{1}{4}$ of their children will be boys suffering from hemophilia; $\frac{1}{4}$ will be female carriers.

17. For a woman to be colorblind, she must inherit X chromosomes bearing the colorblindness allele from both parents. Her father has only one X chromosome, which he passes on to all his daughters, so he must be colorblind. A male only needs to inherit the colorblindness allele from a carrier mother; both his parents are usually phenotypically normal.

CHAPTER 10

1. polynucleotides; nucleotides; sugar (deoxyribose); phosphate; nitrogenous base

2. b

3. Each daughter DNA molecule will have half the radioactivity of the parent molecule, since one polynucleotide from the original parental DNA molecule winds up in each daughter DNA molecule.

4. CAU; GUA; histidine (His)

5. A gene is the polynucleotide sequence with information for making one polypeptide. Each codon—a triplet of bases in DNA or RNA—codes for one amino acid. Transcription occurs when RNA polymerase produces mRNA using one strand of DNA as a template. A ribosome is the site of translation, or polypeptide synthesis, and tRNA

molecules serve as interpreters of the genetic code. Each tRNA molecule has an amino acid attached at one end and a three-base anticodon at the other end. Beginning at the start codon, mRNA moves relative to the ribosome a codon at a time. A tRNA with a complementary anticodon pairs with each codon, adding its amino acid to the polypeptide chain. The amino acids are linked by peptide bonds. Translation stops at a stop codon, and the finished polypeptide is released. The polypeptide folds to form a functional protein, sometimes in combination with other polypeptides.

6. a3, b3, c1, d2, e2 and 3

7. d

8. d

9. The genetic material of these viruses is RNA, which is replicated inside the infected cell by special enzymes encoded by the virus. The viral genome (or its complement) serves as mRNA for the synthesis of viral proteins.

10. reverse transcriptase

11. The process of reverse transcription occurs only in infections by RNA-containing retroviruses like HIV. Cells do not require reverse transcriptase (their RNA molecules do not undergo reverse transcription), so reverse transcriptase can be knocked out without harming the human host.

CHAPTER 11

1. c

2. operon

3. b

4. a

5. DNA polymerase and other proteins required for transcription do not have access to tightly packed DNA.

6. the ability of these cells to produce entire organisms through cloning

7. nuclear transplantation

8. which genes are active in a particular sample of cells

9. b

10. embryonic tissue (ES cells), umbilical cord blood, and bone marrow (adult stem cells)

11. Proto-oncogenes are normal genes involved in the control of the cell

cycle. Mutation or viruses can cause them to be converted to oncogenes, or cancer-causing genes. Proto-oncogenes are necessary for normal control of cell division.

12. Master control genes, called homeotic genes, regulate many other genes during development.

CHAPTER 12

1. c, d, b, a

2. because it does not contain introns

3. vector

4. Such an enzyme creates DNA fragments with "sticky ends," single-stranded regions whose unpaired bases can hydrogen-bond to the complementary sticky ends of other fragments created by the same enzyme.

5. PCR

6. Different people tend to have different numbers of repeats at each STR site. DNA fragments prepared from the STR sites of different people will thus have different lengths, causing them to migrate to different locations on a gel.

7. a

8. b

9. Chop the genome into fragments using restriction enzymes, clone and sequence each fragment, and reassemble the short sequences into a continuous sequence for every chromosome.

10. c, b, a, d

CHAPTER 13

1. b

2. c

3. Gradualism is the idea that large changes on Earth can result from the accumulation of small changes over a very long time. Darwin applied this idea to suggest that species evolve through the slow accumulation of small changes over time.

4. *Bb*: 0.42; *BB*: 0.49; *bb*: 0.09

5. The fitness of an individual (or of a particular genotype) is measured by the relative number of alleles that it contributes to the gene pool of the next generation compared with the contribution of others. Thus the number of fertile offspring produced determines an individual's fitness.

6. e

7. c

8. Both types of events result in populations small enough for significant sampling error in the gene pool for the first few generations. A bottleneck reduces the size of an existing population in a given location. The founder effect occurs when a new, small population colonizes a new territory.
9. stabilizing selection
10. b, c, d
11. the prevalence of malaria

CHAPTER 14
1. Microevolution is a change in the gene pool of a population, often associated with adaptation. Macroevolution is marked by major changes in the history of life, including the origin of new species, and these changes are often noticeable enough to be evident in the fossil record.
2. b
3. prezygotic: a, b, c, e; postzygotic: d
4. because a small gene pool is more likely to be changed substantially by genetic drift and natural selection
5. exaptations
6. d
7. d
8. species, genus, family, order, class, phylum, kingdom, domain
9. 2.6
10. Archaea and Bacteria

CHAPTER 15
1. g, d, f, c, b, a, e
2. d, c, a, b, e
3. DNA polymerase is a protein, which must be transcribed from a gene. But a DNA gene requires DNA polymerase to be replicated. This creates a paradox about which came first—DNA or protein. But RNA can act as both an information storage molecule and an enzyme, suggesting that dual-role RNA may have preceded both DNA and proteins.
4. Exotoxins are poisons secreted by pathogenic bacteria; endotoxins are components of the outer membrane of pathogenic bacteria.
5. Autotrophs make their own organic compounds from CO_2, while heterotrophs must obtain at least one type of organic compound from another organism.
6. They can form endospores.
7. Both you and a mushroom are chemoheterotrophs.
8. because the first eukaryotes were protists, and these ancient protists were ancestral to all other eukaryotes, including plants, fungi, animals, and modern protists
9. c
10. b

CHAPTER 16
1. cuticle
2. flowers
3. a. sporophyte b. cones; angiosperms c. fruit
4. b
5. b
6. a fern
7. Because the plants do not lose their leaves during autumn and winter, the leaves are already fully developed for photosynthesis when the short growing season begins in spring.
8. vascular plant
9. a
10. a ripened ovary of a flower that protects and aids in the dispersal of seeds contained in the fruit
11. algae; fungi
12. A fungus digests its food externally by secreting digestive juices into the food and then absorbing the small nutrients that result from digestion.

In contrast, humans and most other animals ingest relatively large pieces of food and digest the food within their bodies.

CHAPTER 17
1. c
2. coelom; pseudocoelom
3. arthropod
4. amphibians
5. b
6. chordata; notochord; cartilage disks between your vertebrae
7. bipedalism
8. a
9. *Australopithecus* species, *Homo habilis*, *Homo erectus*, *Homo sapiens*
10. a4, b5, c1, d2, e3

CHAPTER 18
1. organismal ecology, population ecology, community ecology, ecosystem ecology
2. light, water temperature, chemicals added
3. physiological; behavioral
4. d
5. a. desert; b. grassland; c. tropical rain forest; d. temperate broadleaf forest; e. coniferous forest; f. tundra
6. chaparral
7. permafrost, very cold winters, and high winds
8. agriculture
9. Carbon dioxide and other gases in the atmosphere absorb heat energy radiating from Earth and reflect it back toward Earth. This is called the greenhouse effect. As the carbon dioxide concentration in the atmosphere increases, more heat is retained, causing global warming.
10. c
11. populations of organisms that have high genetic variability and short life spans

CHAPTER 19
1. the number of people and the land area in which they live
2. III; I
3. O; O; E; O; E
4. a. The x-axis is time; the y-axis is the number of individuals; the red curve represents exponential growth; the blue curve represents logistic growth. b. carrying capacity c. In exponential growth, the population growth rate continues to increase as long as the population size increases. In logistic growth, the population grows fastest when it is about $\frac{1}{2}$ the carrying capacity. d. exponential growth curve, though the worldwide growth rate is slowing
5. d
6. opportunistic
7. c
8. The large population segment, the baby boomers, are currently in the workforce and are in their peak earning years.
9. d

CHAPTER 20
1. habitat destruction
2. d
3. c
4. a2, b5, c5, d3 or d4, e1
5. because the pesticides become concentrated in their prey
6. succession
7. Only 10% of the energy trapped by photosynthesis is turned into biomass by the plant, and only 10% of that energy is turned into the meat of a grazing animal. Therefore, eating grain-fed beef obtains only about 1% of the energy captured by photosynthesis.
8. Many nutrients come from the soil, but carbon comes from the air.
9. landscape
10. b

Appendix E Student Media

Mastering BIOLOGY™

The following student media are found in the Study Area of www.masteringbiology.com.

	Resources for Visual Learners		Resources for Auditory Learners	Relevant Biology Topics in the News and Media	Resources to Study and Practice Before a Test	Resources to Build Graphing Skills
	BIOFLIX	VIDEO TUTOR SESSIONS	MP3 TUTOR SESSIONS	DISCOVERY CHANNEL VIDEOS	ACTIVITIES	YOU DECIDE & GRAPHIT!
Available for all chapters				Current Events with *NY Times* Articles, RSS Feeds, and *Scientific American* Podcasts	Chapter Quizzes including Pre-Tests, Post-Tests, and Cumulative Quiz Flashcards	
Chapter 1		Survey of Biodiversity Phylogenetic Trees	The Process of Science	Antibiotics	The Levels of Life Card Game Energy Flow and Chemical Cycling Classification Schemes Darwin and the Galápagos Islands Science, Technology, and Society: DDT	You Decide: What Can We Do About Antibiotic-Resistant Bacteria? GraphIt!: An Introduction to Graphing
Chapter 2			The Properties of Water		The Structure of Atomic Nucleus Electron Arrangement Build an Atom Ionic Bonds Covalent Bonds Hydrogen Bonds Polarity of Water Cohesion of Water Acids, Bases, and pH	
Chapter 3		DNA Structure	Protein Structure and Function DNA Structure		Diversity of Carbon-Based Molecules Functional Groups Making and Breaking Polymers Models of Glucose Carbohydrates Lipids Protein Functions Protein Structure Nucleic Acid Functions Nucleic Acid Structure	You Decide: Low-Fat or Low-Carb Diets—Which Is Healthier?

	Resources for Visual Learners		Resources for Auditory Learners	Relevant Biology Topics in the News and Media	Resources to Study and Practice Before a Test	Resources to Build Graphing Skills
	BioFlix	**VIDEO tutor sessions**	**MP3 tutor sessions**			
	BIOFLIX	**VIDEO TUTOR SESSIONS**	**MP3 TUTOR SESSIONS**	**DISCOVERY CHANNEL VIDEOS**	**ACTIVITIES**	**YOU DECIDE & GRAPHIT!**
Chapter 4	Tour of an Animal Cell Tour of a Plant Cell		Cell Organelles	Cells	Metric System Review Prokaryotic Cell Structure and Function Comparing Cells Build an Animal Cell and a Plant Cell Membrane Structure Role of the Nucleus and Ribosomes in Protein Synthesis The Endomembrane System Build a Chloroplast and a Mitochondrion Cilia and Flagella Review: Animal Cell Structure and Function Review: Plant Cell Structure and Function	
Chapter 5	Membrane Transport		Basic Energy Concepts	Cells	Energy Transformations The Structure of ATP How Enzymes Work Membrane Structure Diffusion Facilitated Diffusion Osmosis and Water Balance in Cells Active Transport Exocytosis and Endocytosis Cell Signaling	
Chapter 6	Cellular Respiration		Cellular Respiration Part 1: Glycolysis Cellular Respiration Part 2: Citric Acid Cycle and Electron Transport	Space Plants Tasty Bacteria	Build a Chemical Cycling System Overview of Cellular Respiration Glycolysis The Citric Acid Cycle Electron Transport Fermentation	
Chapter 7	Photosynthesis		Photosynthesis	Space Plants Trees	The Plants in Our Lives The Sites of Photosynthesis Overview of Photosynthesis Light Energy and Pigments The Light Reactions The Calvin Cycle Photosynthesis in Dry Climates	

	Resources for Visual Learners		Resources for Auditory Learners	Relevant Biology Topics in the News and Media	Resources to Study and Practice Before a Test	Resources to Build Graphing Skills
	BIOFLIX	**VIDEO TUTOR SESSIONS**	**MP3 TUTOR SESSIONS**	**DISCOVERY CHANNEL VIDEOS**	**ACTIVITIES**	**YOU DECIDE & GRAPHIT!**
Chapter 8	Mitosis Meiosis	Mitosis and Meiosis	Mitosis Meiosis Comparison of Mitosis and Meiosis	Cells Fighting Cancer	Asexual and Sexual Life Cycles The Cell Cycle Mitosis and Cytokinesis Animation Mitosis and Cytokinesis Video Causes of Cancer Human Life Cycle Meiosis Animation Origins of Genetic Variation	
Chapter 9		Sex-Linked Pedigrees	Chromosomal Basis of Inheritance	Colored Cotton Novelty Gene	Monohybrid Cross Dihybrid Cross Gregor's Garden Incomplete Dominance Linked Genes and Crossing Over Sex-Linked Genes	
Chapter 10	DNA Replication Protein Synthesis	DNA Structure	DNA to RNA to Protein	Emerging Diseases Vaccines	The Hershey-Chase Experiment DNA and RNA Structure DNA Double Helix DNA Replication: An Overview Overview of Protein Synthesis Transcription RNA Processing Translation Simplified Viral Reproductive Cycle Phage Lytic Cycle Phage Lysogenic and Lytic Cycles Retrovirus (HIV) Reproductive Cycle	
Chapter 11			Control of Gene Expression	Cloning Fighting Cancer	The *lac* Operon in *E. coli* Overview: Control of Gene Regulation Control of Transcription Post-Transcriptional Control Mechanisms Review: Control of Gene Regulation Signal-Transduction Pathway Causes of Cancer	You Decide: Do Cell Phones Cause Brain Cancer? You Decide: Is Second-Hand Smoke Dangerous?
Chapter 12		DNA Profiling Techniques	DNA Technology	DNA Forensics Transgenics	Applications of DNA Technology Restriction Enzymes Cloning a Gene in Bacteria DNA Fingerprinting Gel Electrophoresis of DNA Analyzing DNA Fragments Using Gel Electrophoresis The Human Genome Project: Genes on Human Chromosome 17 Making Decisions About DNA Technology: Golden Rice	

	Resources for Visual Learners		Resources for Auditory Learners	Relevant Biology Topics in the News and Media	Resources to Study and Practice Before a Test	Resources to Build Graphing Skills
	BioFlix BIOFLIX	**VIDEO tutor sessions** VIDEO TUTOR SESSIONS	**MP3 tutor sessions** MP3 TUTOR SESSIONS	DISCOVERY CHANNEL VIDEOS	ACTIVITIES	YOU DECIDE & GRAPHIT!
Chapter 13	Mechanisms of Evolution		Natural Selection	Antibiotics Charles Darwin	The Voyage of the *Beagle*: Darwin's Trip Around the World Darwin and the Galápagos Islands Reconstructing Forelimbs Genetic Variation from Sexual Recombination Causes of Microevolution	
Chapter 14		Phylogenetic Trees	Speciation	Mass Extinctions Charles Darwin	Overview of Macroevolution Polyploid Plants Allometric Growth A Scrolling Geologic Record Classification Schemes	You Decide: Can We Prevent Species Extinction?
Chapter 15		Survey of Biodiversity	Microbial Life	Antibiotics Bacteria Early Life Tasty Bacteria	The History of Life Prokaryotic Cell Structure and Function Classification of Prokaryotes	
Chapter 16		Survey of Biodiversity Phylogenetic Trees	Evolution of Plants Alternation of Generations	Colored Cotton Fungi Leafcutter Ants Plant Pollination Trees	Terrestrial Adaptations of Plants Highlights of Plant Phylogeny Moss Life Cycle Fern Life Cycle Pine Life Cycle Angiosperm Life Cycle Madagascar and the Biodiversity Crisis Fungal Reproduction and Nutrition	
Chapter 17		Survey of Biodiversity Phylogenetic Trees	Human Evolution	Invertebrates	Animal Phylogenetic Tree Characteristics of Invertebrates Characteristics of Chordates Primate Diversity Human Evolution	
Chapter 18			Ecological Hierarchy Global Warming	Rain Forests	Science, Technology, and Society: DDT Adaptations to Biotic and Abiotic Factors Aquatic Biomes Terrestrial Biomes Water Pollution from Nitrates The Greenhouse Effect	You Decide: Does Human Activity Cause Global Warming? GraphIt!: Forestation Change GraphIt!: Global Fresh Water Resources GraphIt!: Municipal Solid Waste Trends in the U.S. GraphIt!: Atmospheric CO_2 and Temperature Changes GraphIt!: Prospects for Renewable Energy

	Resources for Visual Learners		Resources for Auditory Learners	Relevant Biology Topics in the News and Media	Resources to Study and Practice Before a Test	Resources to Build Graphing Skills
	BIOFLIX	VIDEO TUTOR SESSIONS	MP3 TUTOR SESSIONS	DISCOVERY CHANNEL VIDEOS	ACTIVITIES	YOU DECIDE & GRAPHIT!
Chapter 19	Population Ecology			Introduced Species Emerging Disease	Techniques for Estimating Population Density and Size Investigating Survivorship Curves Madagascar and the Biodiversity Crisis Introduced Species: Fire Ants Science, Technology, and Society: DDT Human Population Growth Analyzing Age-Structure Pyramids	You Decide: Can We Prevent Species Extinction? GraphIt!: Global Fisheries and Overfishing GraphIt!: Age Pyramids and Population Growth GraphIt!: Municipal Solid Waste Trends in the U.S.
Chapter 20	The Carbon Cycle		Energy Flow in Ecosystems	Introduced Species Leafcutter Ants	Introduced Species: Fire Ants Interspecific Interactions Food Webs Exploring Island Biogeography Primary Succession Energy Flow and Chemical Cycling Pyramids of Production The Carbon Cycle The Nitrogen Cycle Water Pollution from Nitrates Madagascar and the Biodiversity Crisis Conservation Biology Review	You Decide: Can We Prevent Species Extinction? GraphIt!: Global Fisheries and Overfishing GraphIt!: Animal Food Production Efficiency and Food Policy GraphIt!: Forestation Change GraphIt!: Global Fresh Water Resources GraphIt!: Species Area Effect and Island Biogeography

Glossary

A

abiotic factor (ā'-bī-ot'-ik)
A nonliving component of an ecosystem, such as air, water, light, minerals, or temperature.

abiotic reservoir
The part of an ecosystem where a chemical, such as carbon or nitrogen, accumulates or is stockpiled outside of living organisms.

ABO blood groups
Genetically determined classes of human blood that are based on the presence or absence of carbohydrates A and B on the surface of red blood cells. The ABO blood group phenotypes, also called blood types, are A, B, AB, and O.

absorption
The uptake of small nutrient molecules by an organism's own body. In animals, absorption is the third main stage of food processing, following digestion; in fungi, it is acquisition of nutrients from the surrounding medium.

acclimation (ak-li-mā-shun)
Physiological adjustment that occurs gradually, though still reversibly, in response to an environmental change.

achondroplasia (uh-kon'-druh-plā'-zhuh)
A form of human dwarfism caused by a single dominant allele. The homozygous condition is lethal.

acid
A substance that increases the hydrogen ion (H^+) concentration in a solution.

activation energy
The amount of energy that reactants must absorb before a chemical reaction will start.

activator
A protein that switches on a gene or group of genes by binding to DNA.

active site
The part of an enzyme molecule where a substrate molecule attaches (by means of weak chemical bonds); typically, a pocket or groove on the enzyme's surface.

active transport
The movement of a substance across a biological membrane against its concentration gradient, aided by specific transport proteins and

requiring input of energy (often as ATP).

adenine (A) (ad'-uh-nēn)
A double-ring nitrogenous base found in DNA and RNA.

ADP
Adenosine diphosphate (a-den'-ō-sēn dī-fos'-fāt). A molecule composed of adenosine and two phosphate groups. The molecule ATP is made by combining a molecule of ADP with a third phosphate in an energy-consuming reaction.

adult stem cell
A cell present in adult tissues that generates replacements for nondividing differentiated cells.

aerobic (ār-ō'-bik)
Containing or requiring molecular oxygen (O_2).

age structure
The relative number of individuals of each age in a population.

AIDS
Acquired immunodeficiency syndrome; the late stages of HIV infection, characterized by a reduced number of T cells; usually results in death caused by opportunistic infections.

alga (al'-guh)
(plural, **algae**) An informal term that describes a great variety of protists, most of which are unicellular or colonial photosynthetic autotrophs with chloroplasts. Heterotrophic and multicellular protists closely related to unicellular autotrophs are also regarded as algae.

allele (uh-lē'-ul)
An alternative version of a gene.

allopatric speciation
The formation of a new species as a result of an ancestral population becoming isolated by a geographic barrier. sympatric speciation.

alternation of generations
A life cycle in which there is both a multicellular diploid form, the sporophyte, and a multicellular haploid form, the gametophyte; a characteristic of plants and multicellular green algae.

alternative RNA splicing
A type of regulation at the RNA-processing level in which different mRNA molecules are produced from the same primary transcript,

depending on which RNA segments are treated as exons and which as introns.

amino acid (uh-mēn'-ō)
An organic molecule containing a carboxyl group, an amino group, a hydrogen atom, and a variable side chain; serves as the monomer of proteins.

amniote
Member of a clade of tetrapods that has an amniotic egg containing specialized membranes that protect the embryo. Amniotes include mammals and reptiles (including birds).

amniotic egg (am'-nē-ot'-ik)
A shelled egg in which an embryo develops within a fluid-filled amniotic sac and is nourished by yolk. Produced by reptiles (including birds) and egg-laying mammals, it enables them to complete their life cycles on dry land.

amoeba (uh-mē'-buh)
A type of protist characterized by great structural flexibility and the presence of pseudopodia.

amphibian
Member of a class of vertebrate animals that includes frogs and salamanders.

anaerobic (an'-ār-ō'-bik)
Lacking or not requiring molecular oxygen (O_2).

analogy
The similarity of structure between two species that are not closely related, attributable to convergent evolution.

anaphase
The third stage of mitosis, beginning when sister chromatids separate from each other and ending when a complete set of daughter chromosomes has arrived at each of the two poles of the cell.

angiosperm (an'-jē-ō-sperm)
A flowering plant, which forms seeds inside a protective chamber called an ovary.

animal
A eukaryotic, multicellular, heterotrophic organism that obtains nutrients by ingestion.

annelid (an'-uh-lid)
A segmented worm. Annelids include earthworms, polychaetes, and leeches.

anther
A sac in which pollen grains develop, located at the tip of a flower's stamen.

anthropoid (an'-thruh-poyd)
A member of a primate group made up of the apes (gibbons, orangutans, gorillas, chimpanzees, and bonobos), monkeys, and humans.

anticodon (an'-tī-kō'-don)
On a tRNA molecule, a specific sequence of three nucleotides that is complementary to a codon triplet on mRNA.

aphotic zone (ā-fō'-tik)
The region of an aquatic ecosystem beneath the photic zone, where light levels are too low for photosynthesis to take place.

apicomplexan (ap'-ē-kom-pleks'-un)
A type of parasitic protozoan. Some apicomplexans cause serious human disease.

aqueous solution (ā'-kwē-us)
A solution in which water is the solvent.

arachnid
A member of a major arthropod group that includes spiders, scorpions, ticks, and mites.

Archaea (ar-kē'-uh)
One of two prokaryotic domains of life, the other being Bacteria.

archaean
(plural, **archaea**) An organism that is a member of the domain Archaea.

arthropod (ar'-thruh-pod)
A member of the most diverse phylum in the animal kingdom; includes the horseshoe crab, arachnids (for example, spiders, ticks, scorpions, and mites), crustaceans (for example, crayfish, lobsters, crabs, and barnacles), millipedes, centipedes, and insects. Arthropods are characterized by a chitinous exoskeleton, molting, jointed appendages, and a body formed of distinct groups of segments.

asexual reproduction
The creation of genetically identical offspring by a single parent, without the participation of gametes (sperm and egg).

atherosclerosis (ath'-uh-rō'-skluh-rō'-sis)
A cardiovascular disease in which growths called plaques develop on the inner walls of the arteries, narrowing the passageways through which blood can flow.

atom
The smallest unit of matter that retains the properties of an element.

atomic number
The number of protons in each atom of a particular element.

ATP
Adenosine triphosphate (a-den′-ō-sē n trī-fos′-fāt). A molecule composed of adenosine and three phosphate groups; the main energy source for cells.

ATP synthase
A protein cluster, found in a cellular membrane (including the inner membrane of mitochondria, the thylakoid membrane of chloroplasts, and the plasma membrane of prokaryotes), that uses the energy of a hydrogen ion concentration gradient to make ATP from ADP. An ATP synthase provides a port through which hydrogen ions (H^+) diffuse.

autosome
A chromosome not directly involved in determining the sex of an organism; in mammals, for example, any chromosome other than X or Y.

autotroph (ot′-ō-trōf)
An organism that makes its own food from inorganic ingredients, thereby sustaining itself without eating other organisms or their molecules. Plants, algae, and photosynthetic bacteria are autotrophs.

B

bacillus (buh-sil′-us)
(plural, **bacilli**) A rod-shaped prokaryotic cell.

Bacteria
One of two prokaryotic domains of life, the other being Archaea.

bacteriophage (bak-tē r′-ē-ō-fāj)
A virus that infects bacteria; also called a phage.

bacterium
(plural, **bacteria**) An organism that is a member of the domain Bacteria.

base
A substance that decreases the hydrogen ion (H^+) concentration in a solution.

benign tumor
An abnormal mass of cells that remains at its original site in the body.

benthic realm
A seafloor or the bottom of a freshwater lake, pond, river, or stream. The benthic realm is occupied by communities of organisms known as benthos.

bilateral symmetry
An arrangement of body parts such that an organism can be divided equally by a single cut passing longitudinally through it. A bilaterally symmetrical organism has mirror-image right and left sides.

binary fission
A means of asexual reproduction in which a parent organism, often a single cell, divides into two individuals of about equal size.

binomial
A two-part latinized name of a species; for example, *Homo sapiens*.

biodiversity hot spot
A small geographic area that contains a large number of threatened or endangered species and an exceptional concentration of endemic species (those found nowhere else).

biodiversity
The variety of living things; includes genetic diversity, species diversity, and ecosystem diversity.

biogenesis
The principle that all life arises by the reproduction of preexisting life.

biogeochemical cycle
Any of the various chemical circuits occurring in an ecosystem, involving both biotic and abiotic components of the ecosystem.

biogeography
The study of the geographic distribution of species.

biological community
See community.

biological control
The intentional release of a natural enemy to attack a pest population.

biological magnification
The accumulation of persistent chemicals in the living tissues of consumers in food chains.

biological species concept
The definition of a species as a population or group of populations whose members have the potential in nature to interbreed and produce fertile offspring.

biology
The scientific study of life.

biomass
The amount, or mass, of living organic material in an ecosystem.

biome (bī′-ōm)
A major terrestrial or aquatic life zone, characterized by vegetation type in terrestrial biomes or the physical environment in aquatic biomes.

biophilia
The human desire to affiliate with other life in its many forms.

bioremediation
The use of living organisms to detoxify and restore polluted and degraded ecosystems.

biosphere
The global ecosystem; the entire portion of Earth inhabited by life; all of life and where it lives.

biotechnology
The manipulation of living organisms to perform useful tasks. Today, biotechnology often involves DNA technology.

biotic factor (bī-ot′-ik)
A living component of a biological community; any organism that is part of an individual's environment.

bird
Member of a group of reptiles with feathers and adaptations for flight.

bivalve
A member of a group of molluscs that includes clams, mussels, scallops, and oysters.

blastula (blas′-tuh-luh)
An embryonic stage that marks the end of cleavage during animal development; a hollow ball of cells in many species.

body cavity
A fluid-filled space separating the digestive tract from the outer body wall.

body segmentation
Subdivision of an animal's body into a series of repeated parts called segments.

bony fish
A fish that has a stiff skeleton reinforced by calcium salts.

bottleneck effect
Genetic drift resulting from a drastic reduction in population size.

biology
The scientific study of life.

bryophyte (brī′-uh-fīt)
A type of plant that lacks xylem and phloem; a nonvascular plant. Bryophytes include mosses and their close relatives.

buffer
A chemical substance that resists changes in pH by accepting hydrogen ions from or donating hydrogen ions to solutions.

C

C_3 plant
A plant that uses the Calvin cycle for the initial steps that incorporate CO_2 into organic material, first forming a three-carbon compound.

C_4 plant
A plant that prefaces the Calvin cycle with reactions that incorporate CO_2 into four-carbon compounds, the end product of which supplies CO_2 for the Calvin cycle.

calorie
The amount of energy that raises the temperature of 1 g of water by 1°C.

Calvin cycle
The second of two stages of photosynthesis; a cyclic series of chemical reactions that occur in the stroma of a chloroplast, using the carbon in CO_2 and the ATP and NADPH produced by the light reactions to make the energy-rich sugar molecule G3P, which is later used to produce glucose.

CAM plant
A plant that uses the following adaptation for photosynthesis in arid conditions: Carbon dioxide entering open stomata during the night is converted to organic compounds, which release CO_2 for the Calvin cycle during the day, when stomata are closed.

cancer
A malignant growth or tumor caused by abnormal and uncontrolled cell division.

cap
Extra nucleotides added to the beginning of an RNA transcript in the nucleus of a eukaryotic cell.

carbohydrate (kar′-bō-hī′-drūl)
A biological molecule consisting of simple single-monomer sugars

(monosaccharides), two-monomer sugars (disaccharides), and other multi-unit sugars (polysaccharides).

carbon footprint
The amount of greenhouse gas emitted as a result of the actions of a person, nation, or other entity.

carcinogen (kar-sin'-uh-jin)
A cancer-causing agent, either high-energy radiation (such as X-rays or UV light) or a chemical.

carnivore
An animal that mainly eats other animals. herbivore; omnivore.

carpel (kar'-pul)
The egg-producing part of a flower, consisting of a stalk with an ovary at the base and a stigma, which traps pollen, at the tip.

carrier
An individual who is heterozygous for a recessively inherited disorder and who therefore does not show symptoms of that disorder.

carrying capacity
The maximum population size that a particular environment can sustain.

cartilaginous fish (kar-ti-laj'-uh-nus)
A fish that has a flexible skeleton made of cartilage.

case study
An in-depth examination of an actual investigation.

cell cycle
An ordered sequence of events (including interphase and the mitotic phase) that extends from the time a eukaryotic cell is first formed from a dividing parent cell until its own division into two cells.

cell cycle control system
A cyclically operating set of proteins that triggers and coordinates events in the eukaryotic cell cycle.

cell division
The reproduction of a cell.

cell junction
A structure that connects animal cells to one another in a tissue.

cell plate
A membranous disk that forms across the midline of a dividing plant cell. During cytokinesis, the cell plate grows outward, accumulating more cell wall material and eventually fusing into a new cell wall.

cell theory
The theory that all living things are composed of cells and that all cells come from other cells.

cellular differentiation
Specialization in the structure and function of cells that occurs during the development of an organism; results from selective activation and deactivation of the cells' genes.

cellular respiration
The aerobic harvesting of energy from food molecules; the energy-releasing chemical breakdown of food molecules, such as glucose, and the storage of potential energy in a form that cells can use to perform work; involves glycolysis, the citric acid cycle, the electron transport chain, and chemiosmosis.

cellular slime mold
A type of protist that has unicellular amoeboid cells and a multicellular reproductive body in its life cycle.

cellulose (sel'-yu-lōs)
A large polysaccharide composed of many glucose monomers linked into cable-like fibrils that provide structural support in plant cell walls. Because cellulose cannot be digested by animals, it acts as roughage, or fiber, in the diet.

centipede
A carnivorous terrestrial arthropod that has one pair of long legs for each of its numerous body segments, with the front pair modified as poison claws.

central vacuole (vak'-yu-ōl)
A membrane-enclosed sac occupying most of the interior of a mature plant cell, having diverse roles in reproduction, growth, and development.

centromere (sen'-trō-mēr)
The region of a chromosome where two sister chromatids are joined and where spindle microtubules attach during mitosis and meiosis. The centromere divides at the onset of anaphase during mitosis and anaphase II of meiosis.

centrosome (sen'-trō-sōm)
Material in the cytoplasm of a eukaryotic cell that gives rise to microtubules; important in mitosis and meiosis; functions as a microtubule-organizing center.

cephalopod
A member of a group of molluscs that includes squids and octopuses.

chaparral (shap-uh-ral')
A terrestrial biome limited to coastal regions where cold ocean currents circulate offshore, creating mild, rainy winters and long, hot, dry summers; also known as the Mediterranean biome. Chaparral vegetation is adapted to fire.

character
A heritable feature that varies among individuals within a population, such as flower color in pea plants.

charophyte (kār'-uh-fīt')
A member of the green algal group that shares features with land plants. Charophytes are considered the closest relatives of land plants; modern charophytes and modern plants likely evolved from a common ancestor.

chemical bond
An attraction between two atoms resulting from a sharing of outer-shell electrons or the presence of opposite charges on the atoms. The bonded atoms gain complete outer electron shells.

chemical cycling
The use and reuse of chemical elements such as carbon within an ecosystem.

chemical energy
Energy stored in the chemical bonds of molecules; a form of potential energy.

chemical reaction
A process leading to chemical changes in matter, involving the making and/or breaking of chemical bonds.

chemotherapy (kē'-mo-thār'-uh-pē)
Treatment for cancer in which drugs are administered to disrupt cell division of the cancer cells.

chiasma (kī-az'-muh)
(plural, **chiasmata**) The microscopically visible site where crossing over has occurred between chromatids of homologous chromosomes during prophase I of meiosis.

chlorophyll (klor'-ō-fil)
A light-absorbing pigment in chloroplasts that plays a central role in converting solar energy to chemical energy.

chlorophyll a (klor'-ō-fil ā)
A green pigment in chloroplasts that participates directly in the light reactions.

chloroplast (klō'-rō-plast)
An organelle found in plants and photosynthetic protists. Enclosed by two concentric membranes, a chloroplast absorbs sunlight and uses it to power the synthesis of organic food molecules (sugars).

chordate (kōr'-dāt)
An animal that at some point during its development has a dorsal, hollow nerve cord, a notochord, pharyngeal slits, and a post-anal tail. Chordates include lancelets, tunicates, and vertebrates.

chromatin (krō'-muh-tin)
The combination of DNA and proteins that constitutes chromosomes; often used to refer to the diffuse, very extended form taken by the chromosomes when a eukaryotic cell is not dividing.

chromosome (krō'-muh-sōm)
A gene-carrying structure found in the nucleus of a eukaryotic cell and most visible during mitosis and meiosis; also, the main gene-carrying structure of a prokaryotic cell. Each chromosome consists of one very long threadlike DNA molecule and associated proteins. *See also* chromatin.

chromosome theory of inheritance
A basic principle in biology stating that genes are located on chromosomes and that the behavior of chromosomes during meiosis accounts for inheritance patterns.

ciliate (sil'-ē-it)
A type of protozoan that moves and feeds by means of cilia.

cilium (sil'-ē-um)
(plural, **cilia**) A short appendage that propels some protists through the water and moves fluids across the surface of many tissue cells in animals.

citric acid cycle
The metabolic cycle that is fueled by acetyl CoA formed after glycolysis in cellular respiration. Chemical reactions in the cycle complete the metabolic breakdown of glucose molecules to carbon dioxide. The cycle occurs in the matrix of mitochondria and supplies most of the NADH molecules that carry energy to the electron transport chains. Also referred to as the Krebs cycle.

clade

An ancestral species and all its descendants—a distinctive branch in the tree of life.

cladistics *(kluh-dis´-tiks)*

The study of evolutionary history; specifically, the scientific search for clades, taxonomic groups composed of an ancestral species and all its descendants.

class

In classification, the taxonomic category above order.

cleavage furrow

The first sign of cytokinesis during cell division in an animal cell; a shallow groove in the cell surface near the old metaphase plate.

clone

As a verb, to produce genetically identical copies of a cell, organism, or DNA molecule. As a noun, the collection of cells, organisms, or molecules resulting from cloning; also (colloquially), a single organism that is genetically identical to another because it arose from the cloning of a somatic cell.

cnidarian *(nī-dār´-ē-an)*

An animal characterized by cnidocytes, radial symmetry, a gastrovascular cavity, and a polyp or medusa body form. Cnidarians includes hydras, jellies, sea anemones, and corals.

coccus *(kok´-us)*

(plural, **cocci**) A spherical prokaryotic cell.

codominance

The expression of two different alleles of a gene in a heterozygote.

codon *(kō´-don)*

A three-nucleotide sequence in mRNA that specifies a particular amino acid or polypeptide termination signal; the basic unit of the genetic code.

coelom *(sē´-lōm)*

A body cavity completely lined by tissue derived from mesoderm.

coevolution

Evolutionary change in which an adaptation in one species leads to a counteradaptation in a second species.

cohesion *(kō-hē´-zhun)*

The attraction between molecules of the same kind.

community

All the organisms inhabiting and potentially interacting in a particular area; an assemblage of populations of different species.

community ecology

The study of how interactions between species affect community structure and organization.

comparative anatomy

The comparison of body structures in different species.

competitive exclusion principle

The concept that populations of two species cannot coexist in a community if their niches are nearly identical. Using resources more efficiently and having a reproductive advantage, one of the populations will eventually outcompete and eliminate the other.

complementary DNA (cDNA)

A DNA molecule made in vitro using mRNA as a template and the enzyme reverse transcriptase. A cDNA molecule therefore corresponds to a gene but lacks the introns present in the DNA of the genome.

complete digestive tract

A digestive tube with two openings, a mouth and an anus.

compound

A substance containing two or more elements in a fixed ratio; for example, table salt (NaCl) consists of one atom of the element sodium (Na) for every atom of chlorine (Cl).

concentration gradient

An increase or decrease in the density of a chemical substance within a given region. Cells often maintain concentration gradients of hydrogen ions across their membranes. When a gradient exists, the ions or other chemical substances involved tend to move from where they are more concentrated to where they are less concentrated.

conifer *(kon´-nuh-fer)*

A gymnosperm, or naked-seed plant, most of which produce cones.

coniferous forest *(kō-nif´-rus)*

A terrestrial biome characterized by conifers, cone-bearing evergreen trees.

conservation biology

A goal-oriented science that seeks to understand and counter the loss of biodiversity.

conservation of energy

The principle that energy can neither be created nor destroyed.

consumer

An organism that obtains its food by eating plants or by eating animals that have eaten plants.

consumer

An organism that obtains its food by eating plants or by eating animals that have eaten plants.

controlled experiment

A component of the process of science whereby a scientist carries out two parallel tests, an experimental test and a control test. The experimental test differs from the control by one factor, the variable.

convergent evolution

Adaptive change resulting in nonhomologous (analogous) similarities among organisms. Species from different evolutionary lineages come to resemble each other (evolve analogous structures) as a result of living in very similar environments.

coral reef

Tropical marine biome characterized by the hard skeletal structures secreted primarily by the resident cnidarians.

covalent bond *(kō-vā´-lent)*

An attraction between atoms that share one or more pairs of outer-shell electrons.

crista *(kris´-tuh)*

(plural, **cristae**) A fold of the inner membrane of a mitochondrion. Enzyme molecules embedded in cristae make ATP.

cross

The cross-fertilization of two different varieties of an organism or of two different species; also called hybridization.

crossing over

The exchange of segments between chromatids of homologous chromosomes during prophase I of meiosis.

crustacean

A member of a major arthropod group that includes lobsters, crayfish, crabs, shrimps, and barnacles.

cryptic coloration

Adaptive coloration that makes an organism difficult to spot against its background.

culture

The accumulated knowledge, customs, beliefs, arts, and other human products that are socially transmitted over the generations.

cuticle *(kyū´-tuh-kul)*

(1) In animals, a tough, nonliving outer layer of the skin. (2) In plants, a waxy coating on the surface of stems and leaves that helps retain water.

cystic fibrosis *(sis′-tik fī-brō´-sis)*

A genetic disease that occurs in people with two copies of a certain recessive allele; characterized by an excessive secretion of mucus and consequent vulnerability to infection; fatal if untreated.

cytokinesis *(sī-tō-kuh-nē´-sis)*

The division of the cytoplasm to form two separate daughter cells. Cytokinesis usually occurs during telophase of mitosis, and the two processes (mitosis and cytokinesis) make up the mitotic (M) phase of the cell cycle.

cytoplasm *(sī´-tō-plaz´-um)*

Everything inside a eukaryotic cell between the plasma membrane and the nucleus; consists of a semifluid medium and organelles; can also refer to the interior of a prokaryotic cell.

cytosine (C) *(sī´-tuh-sēn)*

A single-ring nitrogenous base found in DNA and RNA.

cytoskeleton

A meshwork of fine fibers in the cytoplasm of a eukaryotic cell; includes microfilaments, intermediate filaments, and microtubules.

D

decomposer

An organism that secretes enzymes that digest molecules in organic material and convert them to inorganic form.

dehydration reaction *(dē-hī-drā´-shun)*

A chemical process in which a polymer forms when monomers are linked by the removal of water molecules. One molecule of water is removed for each pair of monomers linked. A dehydration reaction is the opposite of a hydrolysis reaction.

denaturation *(dē-nā´-chuh-rā´-shun)*

A process in which a protein unravels, losing its specific conformation and hence function; can be caused by changes in pH or salt concentration or by high temperature; also refers to the separation of the two strands of the DNA double helix, caused by similar factors.

density-dependent factor
A limiting factor whose effects intensify with increasing population density.

density-independent factor
A limiting factor whose occurrence and effects are not related to population density.

desert
A terrestrial biome characterized by low and unpredictable rainfall (less than 30 cm per year).

detritivore (di-trī'-tuh-vor)
An organism that consumes dead organic matter (detritus).

detritus (di-trī'-tus)
Dead organic matter.

diatom (dī'-uh-tom)
A unicellular photosynthetic alga with a unique glassy cell wall containing silica.

diffusion
The spontaneous movement of particles of any kind down a concentration gradient; that is, movement of particles from where they are more concentrated to where they are less concentrated.

dihybrid cross (dī'-hī'-brid)
An mating of individuals differing at two genetic loci.

dinoflagellate (dī'-nō-flaj'-uh-let)
A unicellular photosynthetic alga with two flagella situated in perpendicular grooves in cellulose plates covering the cell.

diploid (dip'-loid)
Containing two sets of chromosomes (homologous pairs) in each cell, one set inherited from each parent; referring to a 2n cell.

directional selection
Natural selection that acts in favor of the individuals at one end of a phenotypic range.

disaccharide (dī-sak'-uh-rīd)
A sugar molecule consisting of two monosaccharides linked by a dehydration reaction.

discovery science
The process of scientific inquiry that focuses on describing nature. *See also* hypothesis-driven science.

disruptive selection
Natural selection that favors extreme over intermediate phenotypes.

disturbance
In an ecological sense, a force that damages a biological community, at least temporarily, by destroying organisms and altering the availability of resources needed by organisms in the community. Disturbances, such as fires and storms, play a pivotal role in structuring many biological communities.

DNA Deoxyribonucleic acid (dē-ok'-sēn-rī'-bō-nū-klā'-ik).
The genetic material that organisms inherit from their parents; a double-stranded helical macromolecule consisting of nucleotide monomers with a deoxyribose sugar, a phosphate group, and the nitrogenous bases adenine (A), cytosine (C), guanine (G), and thymine (T). gene.

DNA ligase (lī'-gās)
An enzyme, essential for DNA replication, that catalyzes the covalent bonding of adjacent DNA nucleotides; used in genetic engineering to paste a specific piece of DNA containing a gene of interest into a bacterial plasmid or other vector.

DNA microarray
A glass slide containing thousands of different kinds of single-stranded DNA fragments arranged in an array (grid). Tiny amounts of DNA fragments, representing different genes, are attached to the glass slide. These fragments are tested for hybridization with various samples of cDNA molecules, thereby measuring the expression of thousands of genes at one time.

DNA polymerase (puh-lim'-er-ās)
An enzyme that assembles DNA nucleotides into polynucleotides using a preexisting strand of DNA as a template.

DNA profiling
A procedure that analyzes an individual's unique collection of genetic markers using PCR and gel electrophoresis. DNA profiling can be used to determine whether two samples of genetic material were derived from the same individual.

DNA technology
Methods used to study or manipulate genetic material.

domain
A taxonomic category above the kingdom level. The three domains of life are Archaea, Bacteria, and Eukarya.

dominant allele
In a heterozygote, the allele that determines the phenotype with respect to a particular gene.

dorsal, hollow nerve cord
One of the four hallmarks of chordates; the chordate brain and spinal cord.

double fertilization
In flowering plants, the formation of both a zygote and a cell with a triploid nucleus, which develops into the endosperm.

double helix
The form assumed by DNA in living cells, referring to its two adjacent polynucleotide strands wound into a spiral shape.

double helix
The form of native DNA, referring to its two adjacent polynucleotide strands wound into a spiral shape.

Down syndrome
A human genetic disorder resulting from the presence of an extra chromosome 21; characterized by heart and respiratory defects and varying degrees of mental retardation.

E

earthworm
A type of annelid, or segmented worm, which extracts nutrients from soil.

echinoderm (ih-kī'-nuh-derm)
Member of a group of slow-moving or sessile marine animals characterized by a rough or spiny skin, a water vascular system, typically an endoskeleton, and radial symmetry in adults. Echinoderms include sea stars, sea urchins, and sand dollars.

ecological footprint
An estimate of the amount of land required to provide the raw materials an individual or a

population consumes, including food, fuel, water, housing, and waste disposal.

ecological niche
The sum total of a species' use of the biotic and abiotic resources of its habitat.

ecological succession
The process of biological community change resulting from disturbance; transition in the species composition of a biological community, often following a flood, fire, or volcanic eruption. *See also* primary succession; secondary succession.

ecology
The scientific study of the interactions between organisms and their environments.

ecosystem (ē'-kō-sis-tem)
All the organisms in a given area, along with the nonliving (abiotic) factors with which they interact; a biological community and its physical environment.

ecosystem ecology
The study of energy flow and the cycling of chemicals among the various biotic and abiotic factors in an ecosystem.

ecosystem services
Functions performed by an ecosystem that directly or indirectly benefit people.

ectotherm (ek'-tō-therm)
An animal that warms itself mainly by absorbing heat from its surroundings.

electromagnetic spectrum
The full range of radiation, from the very short wavelengths of gamma rays to the very long wavelengths of radio signals.

electron
A subatomic particle with a single unit of negative electrical charge. One or more electrons move around the nucleus of an atom

electron microscope (EM)
An instrument that focuses an electron beam through or onto the surface of a specimen. An electron microscope achieves a thousandfold greater resolving power than a light microscope; the most powerful EM can distinguish objects as small as 0.2 nm (2×10^{-10} m).

electron transport
A redox (oxidation-reduction) reaction in which one or more electrons are transferred to carrier molecules. A series of such reactions, called an electron transport chain, can release the energy stored in high-energy molecules such as glucose. *See also* electron transport chain.

electron transport chain
A series of electron carrier molecules that shuttle electrons during the redox reactions that release energy used to make ATP; located in the inner membrane of mitochondria, the thylakoid membrane of chloroplasts, and the plasma membrane of prokaryotes.

element
A substance that cannot be broken down into other substances by chemical means. Scientists recognize 92 chemical elements occurring in nature.

embryonic stem cell (ES cell)
Any of the cells in the early animal embryo that differentiate during development to give rise to all the kinds of specialized cells in the body.

emerging virus
A virus that has appeared suddenly or has recently come to the attention of medical scientists.

endangered species
As defined in the U.S. Endangered Species Act, a species that is in danger of extinction throughout all or a significant portion of its range.

endemic species
A species whose distribution is limited to a specific geographic area.

endocytosis (en′-dō-sī-tō′-sis)
The movement of materials into the cytoplasm of a cell via vesicles or vacuoles.

endomembrane system
A network of organelles that partitions the cytoplasm of eukaryotic cells into functional compartments. Some of the organelles are structurally connected to each other, whereas others are structurally separate but functionally connected by the traffic of vesicles between them.

endoplasmic reticulum (ER) (reh-tik′-yuh-lum)
An extensive membranous network in a eukaryotic cell, continuous with the outer nuclear membrane and composed of ribosome-studded (rough) and ribosome-free (smooth) regions. *See also* rough ER; smooth ER.

endoskeleton
A hard interior skeleton located within the soft tissues of an animal; found in all vertebrates and a few invertebrates (such as echinoderms).

endosperm
In flowering plants, a nutrient-rich mass formed by the union of a sperm cell with the diploid central cell of the embryo sac during double fertilization; provides nourishment to the developing embryo in the seed.

endospore
A thick-coated, protective cell produced within a prokaryotic cell exposed to harsh conditions.

endosymbiosis (en′-dō-sim-bē-ō′-sis)
Symbiotic relationship in which one species resides within another species. The mitochondria and chloroplasts of eukaryotic cells probably evolved from symbiotic associations between small prokaryotic cells living inside larger ones.

endotherm
An animal that derives most of its body heat from its own metabolism.

endotoxin
A poisonous component of the outer membrane of certain bacteria.

energy
The capacity to perform work, or to move matter in a direction it would not move if left alone.

energy flow
The passage of energy through the components of an ecosystem.

enhancer
A eukaryotic DNA sequence that helps stimulate the transcription of a gene at some distance from it. An enhancer functions by means of a transcription factor called an activator, which binds to it and then to the rest of the transcription apparatus. *See* silencer.

entropy (en′-truh-pē)
A measure of disorder, or randomness. One form of disorder is heat, which is random molecular motion.

enzyme (en′-zīm)
A protein that serves as a biological catalyst, changing the rate of a chemical reaction without itself being changed in the process.

enzyme inhibitor
A chemical that interferes with an enzyme's activity by changing the enzyme's shape, either by plugging up the active site or binding to another site on the enzyme.

equilibrial life history (ē-kwi-lib′-rē-ul)
The pattern of reaching sexual maturity slowly and producing few offspring but caring for the young; often seen in long-lived, large-bodied species.

estuary (es′-chuh-wār-ē)
The area where a freshwater stream or river merges with seawater.

Eukarya (yū-kār′-yuh)
The domain of eukaryotes, organisms made up of eukaryotic cells; includes all of the protists, plants, fungi, and animals.

eukaryote (yū-kār′-ē-ōt)
An organism characterized by eukaryotic cells. *See also* eukaryotic cell.

eukaryotic cell (yū-kār-ē-ot′-ik)
A type of cell that has a membrane-enclosed nucleus and other membrane-enclosed organelles. All organisms except bacteria and archaea are composed of eukaryotic cells.

eutherian (yū-thē′-ē-un)
See placental mammal.

evaporative cooling
A property of water whereby a body becomes cooler as water evaporates from it.

evo-devo
Evolutionary developmental biology, which studies the evolution of developmental processes in multicellular organisms.

evolution
Descent with modification; genetic change in a population or species over generations; the heritable changes that have produced Earth's diversity of organisms.

evolutionary adaptation
A population's increase in the frequency of traits suited to the environment.

evolutionary tree
A branching diagram that reflects a hypothesis about evolutionary relationships between groups of organisms.

exocytosis (ek′-sō-sī-tō′-sis)
The movement of materials out of the cytoplasm of a cell via membranous vesicles or vacuoles.

exon (ek′-son)
In eukaryotes, a coding portion of a gene. *See also* intron.

exoskeleton
A hard, external skeleton that protects an animal and provides points of attachment for muscles.

exotoxin
A poisonous protein secreted by certain bacteria.

exponential population growth
A model that describes the expansion of a population in an ideal, unlimited environment.

extracellular matrix
A substance in which the cells of an animal tissue are embedded; consists of protein and polysaccharides.

F

F_1 generation
The offspring of two parental (P generation) individuals. F_1 stands for first filial.

F_2 generation
The offspring of the F_1 generation. F_2 stands for second filial.

facilitated diffusion
The passage of a substance across a biological membrane down its concentration gradient, aided by specific transport proteins.

family
In classification, the taxonomic category above genus.

fat
A large lipid molecule made from an alcohol called glycerol and three fatty acids; a triglyceride. Most fats function as energy-storage molecules.

feedback regulation
A method of metabolic control in which the end product of a metabolic pathway acts as an inhibitor of an enzyme within that pathway.

fermentation
The anaerobic harvest of food by some cells.

fern
Any of a group of seedless vascular plants.

fertilization
The union of a haploid sperm cell with a haploid egg cell, producing a zygote.

filament
In a flowering plant, the stalk of a stamen.

fitness
The contribution an individual makes to the gene pool of the next generation relative to the contribution of other individuals in the population.

flagellate *(flaj´-uh-lit)*
A protist (protozoan) that moves by means of one or more flagella.

flagellum *(fluh-jel´-um)*
(plural, **flagella**) A long appendage that propels protists through the water and moves fluids across the surface of many tissue cells in animals. A cell may have one or more flagella.

flatworm
A bilateral animal with a thin, flat body form, a gastrovascular cavity with a single opening, and no body cavity. Flatworms include planarians, flukes, and tapeworms.

flower
In an angiosperm, a short stem with four sets of modified leaves, bearing structures that function in sexual reproduction.

fluid mosaic
A description of membrane structure, depicting a cellular membrane as a mosaic of diverse protein molecules embedded in a fluid bilayer made of phospholipid molecules.

food chain
The sequence of food transfers between the trophic levels of a community, beginning with the producers.

food vacuole *(vak´-ū-ōl)*
(1) A tiny sac in a eukaryotic cell's cytoplasm that engulfs nutrients. (2) The simplest type of digestive compartment.

food web
A network of interconnecting food chains.

foram
A marine protozoan that secretes a shell and extends pseudopodia through pores in its shell; short for foraminifer.

forensics
The scientific analysis of evidence for crime scene investigations and other legal proceedings.

fossil
A preserved imprint or remains of an organism that lived in the past.

fossil fuel
An energy deposit formed from the fossilized remains of long-dead plants and animals.

fossil record
The ordered sequence of fossils as they appear in the rock layers, marking the passing of geologic time.

founder effect
The genetic drift resulting from the establishment of a small, new population whose gene pool differs from that of the parent population.

fruit
A ripened, thickened ovary of a flower, which protects dormant seeds and aids in their dispersal.

functional group
The atoms that form the chemically reactive part of an organic molecule.

fungus
(plural, **fungi**) A chemoheterotrophic eukaryote that digests its food externally and absorbs the resulting small nutrient molecules. Most fungi consist of a netlike mass of filaments called hyphae. Molds, mushrooms, and yeasts are examples of fungi.

G

gametangium *(gam´-uh-tan´-jē-um)*
(plural, **gametangia**) A reproductive organ that houses and protects the gametes of a plant.

gamete *(gam´-ēt)*
A sex cell; a haploid egg or sperm. The union of two gametes of opposite sex (fertilization) produces a zygote.

gametophyte *(guh-mē´-tō-fīt)*
The multicellular haploid form in the life cycle of organisms undergoing alternation of generations; results from a union of spores andmitotically produces haploid gametes that unite and grow into the sporophyte generation.

gastropod
A member of the largest group of molluscs, including snails and slugs.

gastrovascular cavity
A digestive compartment with a single opening that serves as both the entrance for food and the exit for undigested wastes; may also function in circulation, body support, and gas exchange. Jellies and hydras are examples of animals with a gastrovascular cavity.

gastrula *(gas´-trū-luh)*
The embryonic stage resulting from gastrulation in animal development. Most animals have a gastrula made up of three layers of cells: ectoderm, endoderm, and mesoderm.

gel electrophoresis *(jel´ ē-lek´-trō-fōr-ē´-sis)*
A technique for sorting macromolecules. A mixture of molecules is placed on a gel between a positively charged electrode and a negatively charged one; negative charges on the molecules are attracted to the positive electrode, and the molecules migrate toward that electrode. The molecules separate in the gel according to their rates of migration.

gene
A unit of inheritance in DNA (or RNA, in some viruses) consisting of a specific nucleotide sequence that programs the amino acid sequence of a polypeptide. Most of the genes of a eukaryote are located in its chromosomal DNA; a few are carried by the DNA of mitochondria and chloroplasts.

gene cloning
The production of multiple copies of a gene.

gene expression
The process whereby genetic information flows from genes to proteins; the flow of genetic information from the genotype to the phenotype: DNA → RNA → protein.

gene flow
The gain or loss of alleles from a population by the movement of individuals or gametes into or out of the population.

gene pool
All the genes in a population at any one time.

gene regulation
The turning on and off of specific genes within a living organism.

genetic code
The set of rules giving the correspondence between nucleotide triplets (codons) in mRNA and amino acids in protein.

genetic drift
A change in the gene pool of a population due to chance.

genetic engineering
The direct manipulation of genes for practical purposes.

genetic marker
(1) An allele tracked in a genetic study. (2) A specific section of DNA that contains a particular allele; may contain specific restriction sites (points where restriction enzymes cut the DNA) that occur only in DNA that contains the allele.

genetic recombination
The production of offspring with gene combinations that differ from that found in either parent.

genetically modified (GM) organism
An organism that has acquired one or more genes by artificial means. If the gene is from another organism, typically of another species, the recombinant organism is also known as a transgenic organism.

genetics
The scientific study of heredity (inheritance).

genomic library *(juh-nō´-mik)*
The entire collection of DNA segments from an organism's genome. Each segment is usually carried by a plasmid or phage.

genomics
The study of whole sets of genes and their interactions.

genotype *(jē´-nō-tīp)*
The genetic makeup of an organism.

genus *(jē´-nus)*
(plural, **genera**) In classification, the taxonomic category above species; the first part of a species' binomial; for example, *Homo.*

geologic time scale
A time scale established by geologists that reflects a consistent sequence of geologic periods, grouped into four divisions: Precambrian, Paleozoic, Mesozoic, and Cenozoic.

germinate
To initiate growth, as in a plant seed.

glycogen (glī'-kō-jen)
A complex, extensively branched polysaccharide made up of many glucose monomers; serves as an energy-storage molecule in liver and muscle cells.

glycolysis (glī-kol'-uh-sis)
The multistep chemical breakdown of a molecule of glucose into two molecules of pyruvic acid; the first stage of cellular respiration in all organisms; occurs in the cytoplasmic fluid.

Golgi apparatus (gol'-jē)
An organelle in eukaryotic cells consisting of stacks of membranous sacs that modify, store, and ship products of the endoplasmic reticulum.

granum (gran'-um)
(plural, **grana**) A stack of hollow disks formed of thylakoid membrane in a chloroplast. Grana are the sites where light energy is trapped by chlorophyll and converted to chemical energy during the light reactions of photosynthesis.

granum (gran'-um)
(plural, **grana**) A stack of hollow disks formed of thylakoid membrane in a chloroplast. Grana are the sites where light energy is trapped by chlorophyll and converted to chemical energy during the light reactions of photosynthesis.

green alga
One of a group of photosynthetic protists that includes unicellular, colonial, and multicellular species. Green algae are the photosynthetic protists most closely related to plants.

greenhouse effect
The warming of the atmosphere caused by CO_2, CH_4, and other gases that absorb heat radiation and slow its escape from Earth's surface.

greenhouse gas
Any of the gases in the atmosphere that absorb heat radiation, including CO_2, methane, water vapor, and synthetic chlorofluorocarbons.

growth factor
A protein secreted by certain body cells that stimulates other cells to divide.

guanine (G) (gwa'-nēn)
A double-ring nitrogenous base found in DNA and RNA.

gymnosperm (jim'-nō-sperm)
A naked-seed plant. Its seed is said to be naked because it is not enclosed in a fruit.

H

habitat
A place where an organism lives; a specific environment in which an organism lives.

haploid
Containing a single set of chromosomes; referring to an *n* cell.

Hardy-Weinberg equilibrium
The condition describing a nonevolving population (one that is in genetic equilibrium).

heat
The amount of kinetic energy contained in the movement of the atoms and molecules in a body of matter. Heat is energy in its most random form.

hemophilia (hē'-muh-fil'-ē-uh)
A human genetic disease caused by a sex-linked recessive allele and characterized by excessive bleeding following injury.

herbivore
An animal that eats mainly plants, algae, or phytoplankton. carnivore; omnivore.

herbivory
The consumption of plant parts or algae by an animal.

heredity
The transmission of traits from one generation to the next.

heterotroph (het'-er-ō-trōf)
An organism that cannot make its own organic food molecules from inorganic ingredients and must obtain them by consuming other organisms or their organic products; a consumer or a decomposer in a food chain.

heterozygous (het'-er-ō-zī'-gus)
Having two different alleles for a given gene.

histone (his'-tōn)
A small protein molecule associated with DNA and important in DNA packing in the eukaryotic chromosome.

HIV
Human immunodeficiency virus; the retrovirus that attacks the human immune system and causes AIDS.

homeotic gene (hō'-mē-ot'-ik)
A master control gene that determines the identity of a body structure of a developing organism, presumably by controlling the developmental fate of groups of cells. (In plants, such genes are called organ identity genes.)

hominin (hah'-mi-nin)
A species on the human branch of the evolutionary tree; a member of the family Hominidae, including *Homo sapiens* and our ancestors.

homologous chromosomes (hō-mol'-uh-gus)
The two chromosomes that make up a matched pair in a diploid cell. Homologous chromosomes are of the same length, centromere position, and staining pattern and possess genes for the same characteristics at corresponding loci. One homologous chromosome is inherited from the organism's father, the other from the mother.

homology (hō-mol'-uh-jē)
Anatomical similarity due to common ancestry.

homozygous (hō'-mō-zī'-gus)
Having two identical alleles for a given gene.

host
An organism that is exploited by a parasite or pathogen.

human gene therapy
A recombinant DNA procedure intended to treat disease by altering an afflicted person's genes.

Human Genome Project
An international collaborative effort that sequenced the DNA of the entire human genome.

Huntington's disease
A human genetic disease caused by a dominant allele; characterized by uncontrollable body movements and degeneration of the nervous system; usually fatal 10 to 20 years after the onset of symptoms.

hybrid
The offspring of parents of two different species or of two different varieties of one species; the offspring of two parents that differ in one or more inherited traits; an individual that is heterozygous for one or more pairs of genes.

hydrocarbon
A chemical compound composed only of the elements carbon and hydrogen.

hydrogen bond
A type of weak chemical bond formed when a partially positive hydrogen atom from one polar molecule is attracted to the partially negative atom in another molecule (or in another part of the same molecule).

hydrogenation
The process of converting unsaturated fats to saturated fats by the addition of hydrogen.

hydrolysis (hī-drol'-uh-sis)
A chemical process in which macromolecules are broken down by the chemical addition of water molecules to the bonds linking their monomers; an essential part of digestion. A hydrolysis reaction is the opposite of a dehydration reaction.

hydrophilic (hī'-drō-fil'-ik)
"Water-loving"; pertaining to polar, or charged, molecules (or parts of molecules), which are soluble in water.

hydrophobic (hī'-drō-fō'-bik)
"Water-fearing"; pertaining to nonpolar molecules (or parts of molecules), which do not dissolve in water.

hypercholesterolemia (hī'-per-kō-les'-tur-ah-lēm'-ē-uh)
An inherited human disease characterized by an excessively high level of cholesterol in the blood.

hypertonic
In comparing two solutions, referring to the one with the greater concentration of solutes.

hypha (hī'-fuh)
(plural, **hyphae**) One of many filaments making up the body of a fungus.

hypothesis (hī-poth'-uh-sis)
(plural, **hypotheses**) A tentative explanation that a scientist proposes for a specific phenomenon that has been observed.

hypothesis-driven science
The process of scientific inquiry that uses the steps of the scientific method to answer questions about nature. *See also* discovery science; scientific method.

hypotonic
In comparing two solutions, referring to the one with the lower concentration of solutes.

I

inbreeding
The mating of close relatives.

incomplete dominance
A type of inheritance in which the phenotype of a heterozygote (*Aa*) is intermediate between the phenotypes of the two types of homozygotes (*AA* and *aa*).

induced fit
The interaction between a substrate molecule and the active site of an enzyme, which changes shape slightly to embrace the substrate and catalyze the reaction.

insect
An arthropod that usually has three body segments (head, thorax, and abdomen), three pairs of legs, and one or two pairs of wings.

interphase
The phase in the eukaryotic cell cycle when the cell is not actually dividing. During interphase, cellular metabolic activity is high, chromosomes and organelles are duplicated, and cell size may increase. Interphase accounts for 90% of the cell cycle. *See also* mitosis.

interspecific competition
Competition between populations of two or more species that require similar limited resources.

interspecific interaction
Any interaction between members of different species.

intertidal zone (*in′-ter-tīd′-ul*)
A shallow zone where the waters of an estuary or ocean meet land.

intraspecific competition
Competition between individuals of the same species for the same limited resources.

intron (*in′-tron*)
In eukaryotes, a nonexpressed (noncoding) portion of a gene that is excised from the RNA transcript. *See also* exon.

invasive species
A non-native species that has spread far beyond the original point of

introduction and causes environmental or economic damage by colonizing and dominating suitable habitats.

invertebrate
An animal that does not have a backbone.

ion (*ī′-on*)
An atom or molecule that has gained or lost one or more electrons, thus acquiring an electrical charge.

ionic bond (*ī-on′-ik*)
An attraction between two ions with opposite electrical charges. The electrical attraction of the opposite charges holds the ions together.

isomer (*ī′-sō-mer*)
One of two or more molecules with the same molecular formula but different structures and thus different properties.

isotonic (*ī-sō-ton′-ik*)
Having the same solute concentration as another solution.

isotope (*ī′-sō-tōp*)
A variant form of an atom. Isotopes of an element have the same number of protons and electrons but different numbers of neutrons.

K

karyotype (*kār′-ē-ō-tīp*)
A display of micrographs of the metaphase chromosomes of a cell, arranged by size and centromere position.

keystone species
A species whose impact on its community is much larger than its biomass or abundance indicates.

kinetic energy (*kuh-net′-ik*)
Energy of motion. Moving matter performs work by transferring its motion to other matter, such as leg muscles pushing bicycle pedals.

kingdom
In classification, the broad taxonomic category above phylum.

Krebs cycle
See citric acid cycle.

L

lancelet
One of a group of bladelike invertebrate chordates.

landscape
A regional assemblage of interacting ecosystems.

landscape ecology
The application of ecological principles to the study of land-use patterns; the scientific study of the biodiversity of interacting ecosystems.

larva (*lar′-vuh*)
(plural, **larvae**) A free-living, sexually immature form in some animal life cycles that may differ from the adult in morphology, nutrition, and habitat.

lateral line system
A row of sensory organs along each side of a fish's body. Sensitive to changes in water pressure, it enables a fish to detect minor vibrations in the water.

law of independent assortment
A general rule of inheritance, first proposed by Gregor Mendel, that states that when gametes form during meiosis, each pair of alleles for a particular character segregate (separate) independently of each other pair.

law of segregation
A general rule of inheritance, first proposed by Gregor Mendel, that states that the two alleles in a pair segregate (separate) into different gametes during meiosis.

leech
A type of annelid, or segmented worm, that typically lives in fresh water.

lichen (*lī′-ken*)
A mutually beneficial symbiotic association between a fungus and an alga or between a fungus and a cyanobacterium.

life
The set of common characteristics that distinguish living organisms, including such properties and processes as order, regulation, growth and development, energy utilization, response to the environment, reproduction, and the capacity to evolve over time.

life cycle
The entire sequence of stages in the life of an organism, from the adults of one generation to the adults of the next.

life history
The traits that affect an organism's schedule of reproduction and survival.

life table
A listing of survivals and deaths in a population in a particular time period and predictions of how long, on average, an individual of a given age will live.

light microscope (LM)
An optical instrument with lenses that refract (bend) visible light to magnify images and project them into a viewer's eye or onto photographic film.

light reactions
The first of two stages in photosynthesis, the steps in which solar energy is absorbed and converted to chemical energy in the form of ATP and NADPH. The light reactions power the sugar-producing Calvin cycle but produce no sugar themselves.

lignin (*lig′-nin*)
A chemical that hardens the cell walls of plants. Lignin makes up most of what we call wood.

limiting factor
An environmental factor that restricts the number of individuals that can occupy a particular habitat, thus holding population growth in check.

linkage map
A map of a chromosome showing the relative positions of genes.

linked genes
Genes located close enough together on a chromosome that they are usually inherited together.

lipid
An organic compound consisting mainly of carbon and hydrogen atoms linked by nonpolar convalent bonds and therefore mostly hydrophobic and insoluble in water. Lipids include fats, waxes, phospholipids, and steroids.

lobe-finned fish
A bony fish with strong, muscular fins supported by bones.

locus
(plural, **loci**) The particular site where a gene is found on a chromosome. Homologous chromosomes have corresponding gene loci.

logistic population growth
A model that describes population growth that decreases as population size approaches carrying capacity.

lungfish
A bony fish that generally inhabits stagnant waters and gulps air into lungs connected to a pharynx.

lysogenic cycle (lī-sō-jen'-ik)
A bacteriophage reproductive cycle in which the viral genome is incorporated into the bacterial host chromosome as a prophage. New phages are not produced, and the host cell is not killed or lysed unless the viral genome leaves the host chromosome.

lysosome (lī'-sō-sōm)
A digestive organelle in eukaryotic cells; contains enzymes that digest the cell's food and wastes.

lytic cycle (lit'-ik)
A viral reproductive cycle resulting in the release of new viruses by lysis (breaking open) of the host cell.

M

macroevolution
Evolutionary change on a grand scale, encompassing the origin of new species, the origin of evolutionary novelty, diversification, and mass extinction.

macromolecule
A giant molecule in a living organism. Examples include proteins, polysaccharides, and nucleic acids.

magnification
An increase in the apparent size of an object.

malignant tumor
An abnormal tissue mass that spreads into neighboring tissue and to other parts of the body; a cancerous tumor.

mammal
Member of a class of endothermic amniotes that possesses mammary glands and hair.

mantle
In molluscs, the outgrowth of the body surface that drapes over the animal. The mantle produces the shell and forms the mantle cavity.

marsupial (mar-sū'-pē-ul)
A pouched mammal, such as a kangaroo, opossum, or koala.

Marsupials give birth to embryonic offspring that complete development while housed in a pouch and attach to nipples on the mother's abdomen.

mass
A measure of the amount of material in an object.

mass number
The sum of the number of protons and neutrons in an atom's nucleus.

matrix
The thick fluid contained within the inner membrane of the mitochondrion.

matter
Anything that occupies space and has mass.

medusa (med-ū'-suh)
(plural, **medusae**) One of two types of cnidarian body forms; a floating, umbrella-like body form; also called a jelly.

meiosis (mī-ō'-sis)
In a sexually reproducing organism, the division of a single diploid cell into four haploid daughter cells. Meiosis and cytokinesis produce haploid gametes from diploid cells in the reproductive organs of the parents.

messenger RNA (mRNA)
The type of ribonucleic acid that encodes genetic information from DNA and conveys it to ribosomes, where the information is translated into amino acid sequences.

metabolism (muh-tab'-uh-liz-um)
The total of all the chemical reactions in an organism.

metamorphosis (met'-uh-mōr'-fuh-sis)
The transformation of a larva into an adult.

metaphase (met'-eh-fāz)
The second stage of mitosis. During metaphase, the centromeres of all the cell's duplicated chromosomes are lined up on an imaginary plate equidistant between the poles of the mitotic spindle.

metastasis (muh-tas'-tuh-sis)
The spread of cancer cells beyond their original site.

microevolution
A change in a population's gene pool over a succession of generations; evolutionary changes in species over relatively brief periods of geologic time.

microtubule
The thickest of the three main kinds of fibers making up the cytoskeleton of a eukaryotic cell; a straight, hollow tube made of globular proteins called tubulins. Microtubules form the basis of the structure and movement of cilia and flagella.

millipede
A terrestrial arthropod that has two pairs of short legs for each of its numerous body segments and that eats decaying plant matter.

mitochondrion (mī'-tō-kon'-drē-on)
(plural, **mitochondria**) An organelle in eukaryotic cells where cellular respiration occurs. Enclosed by two concentric membranes, it is where most of the cell's ATP is made.

mitosis (mī'-tō-sis)
The division of a single nucleus into two genetically identical daughter nuclei. Mitosis and cytokinesis make up the mitotic (M) phase of the cell cycle.

mitotic (M) phase
The phase of the cell cycle when mitosis divides the nucleus and distributes its chromosomes to the daughter nuclei and cytokinesis divides the cytoplasm, producing two daughter cells.

mitotic spindle
A spindle-shaped structure formed of microtubules and associated proteins that is involved in the movement of chromosomes during mitosis and meiosis. (A spindle is shaped roughly like a football.)

modern synthesis
A comprehensive theory of evolution that incorporates genetics and includes most of Darwin's ideas, focusing on populations as the fundamental units of evolution.

molecular biology
The study of the molecular basis of heredity; molecular genetics.

molecule
A group of two or more atoms held together by covalent bonds.

mollusc (mol'-lusk)
A soft-bodied animal characterized by a muscular foot, mantle, mantle cavity, and radula. Molluscs include gastropods (snails and slugs), bivalves (clams, oysters, and

scallops), and cephalopods (squids and octopuses).

monohybrid cross
A mating of individuals differing at one genetic locus.

monomer (mon'-uh-mer)
A chemical subunit that serves as a building block of a polymer.

monosaccharide (mon'-uh-sak'-uh-rīd)
The smallest kind of sugar molecule; a single-unit sugar; also known as a simple sugar. Monosaccharides are the building blocks of more complex sugars and polysaccharides.

monotreme (mon'-uh-trēm)
An egg-laying mammal, such as the duck-billed platypus.

moss
Any of a group of seedless nonvascular plants.

movement corridor
A series of small clumps or a narrow strip of quality habitat (usable by organisms) that connects otherwise isolated patches of quality habitat.

mutagen (myū'-tuh-jen)
A chemical or physical agent that interacts with DNA and causes a mutation.

mutation
A change in the nucleotide sequence of DNA; a major source of genetic diversity.

mutualism
An interspecific interaction in which both partners benefit.

mycelium (mī-sē'-lē-um)
(plural, **mycelia**) The densely branched network of hyphae in a fungus.

mycorrhiza (mī'-kō-rī'-zuh)
(plural, **mycorrhizae**) A mutually beneficial symbiotic association of a plant root and fungus.

N

NADH
An electron carrier (a molecule that carries electrons) involved in cellular respiration and photosynthesis. NADH carries electrons from glucose and other fuel molecules and deposits them at the top of an electron transport chain. NADH is generated during glycolysis and the citric acid cycle.

NADPH

An electron carrier (a molecule that carries electrons) involved in photosynthesis. Light drives electrons from chlorophyll to NADP$^+$, forming NADPH, which provides the high-energy electrons for the reduction of carbon dioxide to sugar in the Calvin cycle.

natural selection

A process in which organisms with certain inherited characteristics are more likely to survive and reproduce than are organisms with other characteristics; differential reproductive success.

nematode (nem'-uh-tōd)

An animal characterized by a pseudocoelom, a cylindrical, wormlike body form, and a complete digestive tract; also called a roundworm.

neutron

An electrically neutral particle (a particle having no electrical charge), found in the nucleus of an atom.

nitrogen fixation

The conversion of atmospheric nitrogen (N_2) to nitrogen compounds (NH_4, NO_3) that plants can absorb and use.

nondisjunction

An accident of meiosis or mitosis in which a pair of homologous chromosomes or a pair of sister chromatids fail to separate at anaphase.

notochord (nō'-tuh-kord)

A flexible, cartilage-like, longitudinal rod located between the digestive tract and nerve cord in chordate animals, present only in embryos in many species.

nuclear envelope

A double membrane, perforated with pores, that encloses the nucleus and separates it from the rest of the eukaryotic cell.

nuclear transplantation

A technique in which the nucleus of one cell is placed into another cell that already has a nucleus or in which the nucleus has been previously destroyed. The cell is them stimulated to grow, producing an embryo that is a genetic copy of the nucleus donor.

nucleic acid (nū-klā'-ik)

A polymer consisting of many nucleotide monomers; serves as a blueprint for proteins and, through the actions of proteins, for all cellular structures and activities. The two types of nucleic acids are DNA and RNA.

nucleic acid probe (nū-klā'-ik)

In DNA technology, a labeled single-stranded nucleic acid molecule used to find a specific gene or other nucleotide sequence within a mass of DNA. The probe hydrogen-bonds to the complementary sequence in the targeted DNA.

nucleolus (nū-klē'-ō-lus)

A structure within the nucleus of a eukaryotic cell where ribosomal RNA is made and assembled with proteins to make ribosomal subunits; consists of parts of the chromatin DNA, RNA transcribed from the DNA, and proteins imported from the cytoplasm.

nucleosome (nū'-klē-ō-sōm)

The bead-like unit of DNA packing in a eukaryotic cell; consists of DNA wound around a protein core made up of eight histone molecules.

nucleotide (nū'-klē-ō-tīd)

An organic monomer consisting of a five-carbon sugar covalently bonded to a nitrogenous base and a phosphate group. Nucleotides are the building blocks of nucleic acids.

nucleus

(plural, **nuclei**) (1) An atom's central core, containing protons and neutrons. (2) The genetic control center of a eukaryotic cell.

O

omnivore

An animal that eats both plants and animals. *See also* carnivore; herbivore.

oncogene (on'-kō-jēn)

A cancer-causing gene; usually contributes to malignancy by abnormally enhancing the amount or activity of a growth factor made by the cell.

operator

In prokaryotic DNA, a sequence of nucleotides near the start of an operon to which an active repressor can attach. The binding of repressor prevents RNA polymerase from attaching to the promoter and transcribing the genes of the operon.

operculum (ō-per'-kyū-lum)

(plural, **opercula**) A protective flap on each side of a bony fish's head that covers a chamber housing the gills.

operon (op'-er-on)

A unit of genetic regulation common in prokaryotes; a cluster of genes with related functions, along with the promoter and operator that control their transcription.

opportunistic life history

The pattern of reproducing when young and producing many offspring that receive little or no parental care; often seen in short-lived, small-bodied species.

order

In classification, the taxonomic category above family.

organelle (ōr-guh-nel')

A membrane-enclosed structure with a specialized function within a eukaryotic cell.

organic compound

A chemical compound containing the element carbon and usually synthesized by cells.

organismal ecology

The study of the evolutionary adaptations that enable individual organisms to meet the challenges posed by their abiotic environments.

osmoregulation

The control of the gain or loss of water and dissolved solutes in an organism.

osmosis (oz-mō'-sis)

The diffusion of water across a selectively permeable membrane.

ovary

(1) In animals, the female gonad, which produces egg cells and reproductive hormones. (2) In flowering plants, the base of a carpel in which the egg-containing ovules develop.

ovule (ō'-vyūl)

A reproductive structure in a seed plant, containing the female gametophyte and the developing egg. An ovule develops into a seed.

oxidation

The loss of electrons from a substance involved in a redox reaction; always accompanies reduction.

P

P generation

The parent individuals from which offspring are derived in studies of inheritance. P stands for parental.

paedomorphosis (pē'-duh-mōr'-fuh-sis)

The retention in the adult of features that were juvenile in ancestral species.

parasite

An organism that lives in or on another organism (host) from which it obtains nourishment; an organism that benefits at the expense of another organism, which is harmed in the process.

passive transport

The diffusion of a substance across a biological membrane without any input of energy.

pathogen

A disease-causing virus or organism.

pedigree

A family tree representing the occurrence of heritable traits in parents and offspring across a number of generations.

pelagic realm (puh-laj'-ik)

The region of an ocean occupied by water; the open ocean.

peptide bond

The covalent linkage between two amino acid units in a polypeptide, formed by a dehydration reaction between two amino acids.

permafrost

Continuously frozen subsoil found in the arctic tundra.

petal

A modified leaf of a flowering plant. Petals are the often colorful parts of a flower that advertise it to insects and other pollinators.

pH scale

A measure of the relative acidity of a solution, ranging in value from 0 (most acidic) to 14 (most basic). pH stands for potential hydrogen and refers to the concentration of hydrogen ions (H$^+$).

phage (fāj)

See bacteriophage.

phagocytosis (fag′-ō-sī-tō′-sis)
Cellular "eating"; a type of endocytosis whereby a cell engulfs large molecules, other cells, or particles into its cytoplasm.

pharyngeal slit (fuh-rin′-jē-ul)
A gill structure in the pharynx, found in chordate embryos and some adult chordates.

phenotype (fē′-nō-tīp)
The expressed traits of an organism.

phloem (flō′-um)
The portion of a plant's vascular system that conveys sugars, nutrients, and hormones throughout a plant. Phloem is made up of live food-conducting cells.

phospholipid (fos′-fō-lip′-id)
A molecule that is a constituent of the inner bilayer of biological membranes, having a hydrophilic head and a hydrophobic tail.

phospholipid bilayer
A double layer of phospholipid molecules (each molecule consisting of a phosphate group bonded to two fatty acids) that is the primary component of all cellular membranes.

photic zone (fō′-tik)
Shallow water near shore or the upper layer of water away from the shore; region of an aquatic ecosystem where sufficient light is available for photosynthesis.

photon (fō′-ton)
A fixed quantity of light energy. The shorter the wavelength of light, the greater the energy of a photon.

photosynthesis (fō′-tō-sin′-thuh-sis)
The process by which plants, algae, and some bacteria transform light energy to chemical energy stored in the bonds of sugars made from carbon dioxide and water.

photosystem
A light-harvesting unit of a chloroplast's thylakoid membrane; consists of several hundred antenna molecules, a reaction-center chlorophyll, and a primary electron acceptor.

phylogenetic tree (fī′-lō-juh-net′-ik)
A branching diagram that represents a hypothesis about evolutionary relationships between organisms.

phylum (fī′-lum)
(plural, **phyla**) In classification, the taxonomic category above class and below kingdom. Members of a phylum all have a similar general body plan.

phytoplankton (fī′-tō-plank′-ton)
Algae and photosynthetic bacteria that drift passively in aquatic environments.

pinocytosis (pī′-nō-sī-tō′-sis)
Cellular "drinking"; a type of endocytosis in which the cell takes fluid and dissolved solutes into small membranous vesicles.

placenta
(pluh-sen′-tuh) In most mammals, the organ that provides nutrients and oxygen to the embryo and helps dispose of its metabolic wastes; formed of the embryo's chorion and the mother's endometrial blood vessels.

placental mammal
(pluh-sen′-tul) Mammal whose young complete their embryonic development in the uterus, nourished via the mother's blood vessels in the placenta; also called a eutherian.

plankton
Communities oforganisms, mostly microscopic, that drift passively in ponds, lakes, and oceans.

plant
A multicellular eukaryote that carries out photosynthesis.

plasma membrane
The thin layer of lipids and proteins that sets a cell off from its surroundings and acts as a selective barrier to the passage of ions and molecules into and out of the cell; consists of a phospholipid bilayer in which proteins are embedded.

plasmid
A small ring of self-replicating DNA separate from the chromosome(s). Plasmids are found in prokaryotes and yeasts.

plasmodial slime mold (plaz-mō′-dē-ul)
A type of protist named for an amoeboid plasmodial feeding stage in its life cycle.

plasmolysis (plaz-mol′-uh-sis)
A phenomenon that occurs in plant cells in a hypertonic environment. The cell loses water and shrivels, and its plasma membrane pulls away from the cell wall, usually killing the cell.

pleiotropy (plī′-uh-trō-pē)
The control of more than one phenotypic character by a single gene.

polar ice
A terrestrial biome that includes regions of extremely cold temperature and low precipitation located at high latitudes north of the arctic tundra and in Antarctica.

polar molecule
A molecule containing polar covalent bonds (having opposite charges on opposite ends).

pollen grain
In a seed plant, the male gametophyte that develops within the anther of a stamen. It houses cells that will develop into sperm.

polychaete
A type of annelid, or segmented worm, that typically lives on the seafloor.

polygenic inheritance (pol′-ē-jen′-ik)
The additive effect of two or more genes on a single phenotypic characteristic.

polymer (pol′-uh-mer)
A large molecule consisting of many identical or similar molecular units, called monomers, covalently joined together in a chain.

polymerase chain reaction (PCR) (puh-lim′-uh-rās)
A technique used to obtain many copies of a DNA molecule or many copies of part of a DNA molecule. A small amount of DNA mixed with the enzyme DNA polymerase, DNA nucleotides, and a few other ingredients replicates repeatedly in a test tube.

polynucleotide (pol′-ē-nū′-klē-ō-tīd)
A polymer made up of many nucleotides covalently bonded together.

polyp (pol′-ip)
One of two types of cnidarian body forms; a sessile, columnar, hydra-like body.

polypeptide
A chain of amino acids linked by peptide bonds.

polysaccharide (pol′-ē-sak′-uh-rīd)
A carbohydrate polymer consisting of many monosaccharides (sugars) linked by covalent bonds.

population
A group of interacting individuals belonging to one species and living in the same geographic area at the same time.

population density
The number of individuals of a species per unit area or volume of the habitat.

population ecology
The study of how members of a population interact with their environment, focusing on factors that influence population density and growth.

population ecology
The study of how members of a population interact with their environment, focusing on factors that influence population density and growth.

population momentum
In a population in which the fertility rate averages two children per female (replacement rate), the continuation of population growth as girls reach their reproductive years.

post-anal tail
A tail posterior to the anus, found in chordate embryos and most adult chordates.

postzygotic barrier (pōst′-zī-got′-ik)
A reproductive barrier that operates if interspecies mating occurs and forms hybrid zygotes.

potential energy
Stored energy; the energy that an object has due to its location and/or arrangement. Water behind a dam and chemical bonds both possess potential energy.

predation
An interaction between species in which one species, the predator, kills and eats the other, the prey.

prezygotic barrier (prē′-zī-got′-ik)
A reproductive barrier that impedes mating between species or hinders fertilization of eggs if members of different species attempt to mate.

primary consumer
An organism that eats only autotrophs; an herbivore.

primary electron acceptor
A molecule in the reaction center of a photosystem that traps the light-excited electron from the reaction center chlorophyll.

primary production
The amount of solar energy converted to chemical energy (organic compounds) by autotrophs in an ecosystem during a given time period.

primary structure

The first level of protein structure; the specific sequence of amino acids making up a polypeptide chain.

primary succession

A type of ecological succession in which a biological community begins in an area without soil. *See also* secondary succession.

primate

Member of the mammalian group that includes lorises, pottos, lemurs, tarsiers, monkeys, apes, and humans.

prion

An infectious form of protein that may multiply by converting related proteins to more prions. Prions cause several related diseases in different animals, including scrapie in sheep, mad cow disease, and Creutzfeldt-Jakob disease in humans.

producer

An organism that makes organic food molecules from CO_2, H_2O, and other inorganic raw materials: a plant, alga, or autotrophic bacterium; the trophic level that supports all others in a food chain or food web.

product

An ending material in a chemical reaction.

prokaryote *(prō-kār′-ē-ōt)*

An organism characterized by prokaryotic cells. *See also* prokaryotic cell.

prokaryotic cell *(prō-kār′-ē-ot′-ik)*

A type of cell lacking a nucleus and other organelles; found only in the domains Bacteria and Archaea.

promoter

A specific nucleotide sequence in DNA, located at the start of a gene, that is the binding site for RNA polymerase and the place where transcription begins.

prophage *(prō′-fāj)*

Phage DNA that has inserted into the DNA of a prokaryotic chromosome.

prophase

The first stage of mitosis. During prophase, duplicated chromosomes condense to form structures visible with a light microscope, and the mitotic spindle forms and begins moving the chromosomes toward the center of the cell.

protein

A biological polymer constructed from amino acid monomers.

proteomics

The systematic study of the full protein sets (proteomes) encoded by genomes.

protist *(prō′-tist)*

Any eukaryote that is not a plant, animal, or fungus.

proton

A subatomic particle with a single unit of positive electrical charge, found in the nucleus of an atom.

proto-oncogene *(prō′-tō-on′-kō-jēn)*

A normal gene that can be converted to a cancer-causing gene.

protozoan *(prō′-tō-zō′-un)*

A protist that lives primarily by ingesting food; a heterotrophic, animal-like protist.

provirus

Viral DNA that inserts into a host genome.

pseudocoelom *(sū′-dō-sē′-lōm)*

A body cavity that is not completely lined by tissue derived from mesoderm.

pseudopodium *(sū′-dō-pō′-dē-um)*

(plural, **pseudopodia**) A temporary extension of an amoeboid cell. Pseudopodia function in moving cells and engulfing food.

punctuated equilibria

The term describing long periods of little change, or equilibrium, punctuated by abrupt periods of speciation.

Punnett square

A diagram used in the study of inheritance to show the results of random fertilization.

pyramid of production

A diagram depicting the cumulative loss of energy with each transfer in a food chain.

Q

quaternary consumer *(kwot′-er-nār-ē)*

An organism that eats tertiary consumers.

R

radial symmetry

An arrangement of the body parts of an organism like pieces of a pie around an imaginary central axis. Any slice passing longitudinally through a radially symmetrical organism's central axis divides it into mirror-image halves.

radiation therapy

Treatment for cancer in which parts of the body that have cancerous tumors are exposed to high-energy radiation to disrupt cell division of the cancer cells.

radioactive isotope

An isotope whose nucleus decays spontaneously, giving off particles and energy.

radiometric dating

A method for determining the age of fossils and rocks from the ratio of a radioactive isotope to the nonradioactive istope(s) of the same element in the sample.

radula *(rad′-yū-luh)*

A file-like organ found in many molluscs, used to scrape up or shred food.

ray-finned fish

A bony fish having fins supported by thin, flexible skeletal rays. All but one living species of bony fishes are ray-fins. *See* lobe-finned fish.

reactant

A starting material in a chemical reaction.

reaction center

In a photosystem in a chloroplast, the chlorophyll *a* molecule and primary electron acceptor that trigger the light reactions of photosynthesis. The chlorophyll donates an electron excited by light energy to the primary electron acceptor, which passes an electron to an electron transport chain.

receptor-mediated endocytosis *(en′-dō-sī-tō′-sis)*

The movement of specific molecules into a cell by the inward budding of vesicles. The vesicles contain proteins with receptor sites specific to the molecules being taken in.

recessive allele

In heterozygotes, the allele that has no noticeable effect on the phenotype.

recombinant DNA

A DNA molecule carrying genes derived from two or more sources, often from different species.

recombination frequency

With respect to two given genes, the number of recombinant progeny from a mating divided by the total number of progeny. Recombinant progeny carry combinations of alleles different from that seen in either of the parents as a result of independent assortment of chromosomes and crossing over.

red-green colorblindness

A category of common sex-linked human disorders involving several genes on the X chromosome and characterized by a malfunction of light-sensitive cells in the eyes; affects mostly males but also homozygous females.

redox reaction

Short for oxidation-reduction reaction; a chemical reaction in which electrons are lost from one substance (oxidation) and added to another (reduction). Oxidation and reduction always occur together.

reduction

The gain of electrons by a substance involved in a redox reaction; always accompanies oxidation.

regeneration

The regrowth of body parts from pieces of an organism.

relative abundance

The proportional representation of a species in a biological community; one component of species diversity.

repetitive DNA

Nucleotide sequences that are present in many copies in the DNA of a genome. The repeated sequences may be long or short and may be located next to each other or dispersed in the DNA.

repressor

A protein that blocks the transcription of a gene or operon.

reproductive barrier

Anything that prevents individuals of closely related species from interbreeding, even when populations of the two species live together.

reproductive cloning

Using a somatic cell from a multicellular organism to make one or more genetically identical individuals.

reptile

Member of the clade of amniotes that includes snakes, lizards, turtles, crocodiles, alligators, birds, and a number of extinct groups (most of the dinosaurs).

resolving power
A measure of the clarity of an image; the ability of an optical instrument to show two objects as separate.

restoration ecology
A field of ecology that develops methods of returning degraded ecosystems to their natural state.

restriction enzyme
A bacterial enzyme that cuts up foreign DNA at one very specific nucleotide sequence, thus protecting bacteria against intruding DNA from phages and other organisms. Restriction enzymes are used in DNA technology to cut DNA molecules in reproducible ways.

restriction fragment
A molecule of DNA produced from a longer DNA molecule cut up by a restriction enzyme; used in genome mapping and other applications.

retrovirus
An RNA virus that reproduces by means of a DNA molecule. It reverse-transcribes its RNA into DNA, inserts the DNA into a cellular chromosome, and then transcribes more copies of the RNA from the viral DNA. HIV and a number of cancer-causing viruses are retroviruses.

reverse transcriptase *(tran-skrip'-tās)*
An enzyme that catalyzes the synthesis of DNA on an RNA template.

ribosomal RNA (rRNA) *(rī'-buh-sōm'-ul)*
The type of ribonucleic acid that, together with proteins, makes up ribosomes; the most abundant type of RNA.

ribosome *(rī'-buh-sōm)*
A cellular structure consisting of RNA and protein organized into two subunits and functioning as the site of protein synthesis in the cytoplasm. The ribosomal subunits are constructed in the nucleolus.

ribozyme *(rī'-bō-zīm)*
An enzymatic RNA molecule that catalyzes chemical reactions.

RNA Ribonucleic acid *(rī'-bō-nū-klā'-ik)*.
A type of nucleic acid consisting of nucleotide monomers with a ribose sugar, a phosphate group, and the nitrogenous bases adenine (A), cytosine (C), guanine (G), and uracil (U); usually single-stranded; functions in protein synthesis and as the genome of some viruses.

RNA polymerase *(puh-lim'-uh-rās)*
An enzyme that links together the growing chain of RNA nucleotides during transcription, using a DNA strand as a template.

RNA splicing
The removal of introns and joining of exons in eukaryotic RNA, forming an mRNA molecule with a continuous coding sequence; occurs before mRNA leaves the nucleus.

root
The underground organ of a plant. Roots anchor the plant in the soil, absorb and transport minerals and water, and store food.

rough ER (rough endoplasmic reticulum) *(reh-tik'-yuh-lum)*
A network of interconnected membranous sacs in a eukaryotic cell's cytoplasm. Rough ER membranes are studded with ribosomes that make membrane proteins and secretory proteins. The rough ER constructs membrane from phospholipids and proteins.

roundworm
A nematode. *See* nematode.

rule of multiplication
A rule stating that the probability of a compound event is the product of the separate probabilities of the independent events.

S

saturated
Pertaining to fats and fatty acids whose hydrocarbon chains contain the maximum number of hydrogens and therefore have no double covalent bonds. Saturated fats and fatty acids solidify at room temperature.

savanna
A terrestrial biome dominated by grasses and scattered trees. Frequent fires and seasonal drought are significant abiotic factors.

scanning electron microscope (SEM)
A microscope that uses an electron beam to study the surface architecture of a cell or other specimen.

science
Any method of learning about the natural world that follows the scientific method. *See* discovery science; hypothesis-driven science.

scientific method
Scientific investigation involving the observation of phenomena, the formulation of a hypothesis concerning the phenomena, experimentation to demonstrate the truth or falseness of the hypothesis, and results that validate or modify the hypothesis.

seaweed
A large, multicellular marine alga.

secondary consumer
An organism that eats primary consumers.

secondary succession
A type of ecological succession that occurs where a disturbance has destroyed an existing biological community but left the soil intact. *See also* primary succession.

seed
A plant embryo packaged with a food supply within a protective covering.

sepal *(sē'-pul)*
A modified leaf of a flowering plant. A whorl of sepals encloses and protects the flower bud before it opens.

sex chromosome
A chromosome that determines whether an individual is male or female; in mammals, for example, the X or Y chromosome.

sex-linked gene
A gene located on a sex chromosome.

sexual dimorphism
Distinction in appearance based on secondary sexual characteristics, noticeable differences not directly associated with reproduction or survival.

sexual reproduction
The creation of offspring by the fusion of two haploid sex cells (sperm and egg), forming a diploid zygote.

sexual selection
A form of natural selection in which individuals with certain characteristics are more likely than other individuals to obtain mates.

shoot
The aerial organs of a plant, consisting of stem and leaves. Leaves are the main photosynthetic organ of most plants.

short tandem repeat (STR)
DNA consisting of tandem (in a row) repeats of a short sequence of nucleotides.

sickle-cell disease
A genetic disorder in which the red blood cells have abnormal hemoglobin molecules and take on an abnormal shape.

signal transduction pathway
A series of molecular changes that converts a signal on a target cell's surface to a specific response inside the cell.

silencer
A eukaryotic DNA sequence that inhibits the start of gene transcription; may act analogously to an enhancer, binding a repressor.

sister chromatid *(krō'-muh-tid)*
One of the two identical parts of a duplicated chromosome. While joined, two sister chromatids make up one chromosome; chromatids are eventually separated during mitosis or meiosis II.

smooth ER (smooth endoplasmic reticulum) *(reh-tik'-yuh-lum)*
A network of interconnected membranous tubules in a eukaryotic cell's cytoplasm. Smooth ER lacks ribosomes. Enzymes embedded in the smooth ER membrane function in the synthesis of certain kinds of molecules, such as lipids.

solute *(sol'-yūt)*
A substance that is dissolved in a solution.

solution
A liquid consisting of a homogeneous mixture of two or more substances: a dissolving agent, the solvent, and a substance that is dissolved, the solute.

solvent
The dissolving agent in a solution. Water is the most versatile known solvent.

somatic cell *(sō-mat'-ik)*
Any cell in a multicellular organism except a sperm or egg cell or a cell that develops into a sperm or egg; a body cell.

speciation *(spē'-sē-ā'-shun)*
The formation of new species.

species
A group of populations whose members possess similar anatomical characteristics and have the ability to interbreed. *See also* biological species concept.

species diversity

The variety of species that make up a biological community; the number and relative abundance of species in a biological community.

species richness

The total number of different species in a community; one component of species diversity.

sponge

An aquatic stationary animal characterized by a highly porous body, choanocytes, and no true tissues.

spontaneous generation

The incorrect notion that life can emerge from nonliving matter.

spore

(1) In plants and algae, a haploid cell that can develop into a multicellular individual without fusing with another cell. (2) In prokaryotes, protists, and fungi, any of a variety of thick-walled life cycle stages capable of surviving unfavorable environmental conditions.

sporophyte *(spor'-uh-fīt)*

The multicellular diploid form in the life cycle of organisms undergoing alternation of generations; results from a union of gametes and meiotically produces haploid spores that grow into the gametophyte generation.

stabilizing selection

Natural selection that favors intermediate variants by acting against extreme phenotypes.

stamen *(stā'-men)*

A pollen-producing part of a flower, consisting of a stalk (filament) and an anther.

starch

A storage polysaccharide found in the roots of plants and certain other cells; a polymer of glucose.

start codon *(kō'-don)*

On mRNA, the specific three-nucleotide sequence (AUG) to which an initiator tRNA molecule binds, starting translation of genetic information.

steroid *(stir'-oyd)*

A type of lipid whose carbon skeleton is in the form of four fused rings: three 6-sided rings and one 5-sided ring, Examples are cholesterol, testosterone, and estrogen.

stigma *(stig'-muh)*

(plural, **stigmata**) The sticky tip of a flower's carpel that traps pollen grains.

stoma *(stō'-muh)*

(plural, **stomata**) A pore surrounded by guard cells in the epidermis of a leaf. When stomata are open, CO_2 enters the leaf, and water and O_2 exit. A plant conserves water when its stomata are closed.

stop codon

In mRNA, one of three triplets (UAG, UAA, UGA) that signal gene translation to stop.

STR analysis

A method of DNA profiling that compares the lengths of STR sequences at specific sites in the genome.

stroma *(strō'-muh)*

A thick fluid enclosed by the inner membrane of a chloroplast. Sugars are made in the stroma by the enzymes of the Calvin cycle.

style

The stalk of a flower's carpel, with the ovary at the base and the stigma at the top.

substrate

(1) A specific substance (reactant) on which an enzyme acts. Each enzyme recognizes only the specific substrate of the reaction it catalyzes. (2) A surface in or on which an organism lives.

succession

See ecological succession; primary succession; secondary succession.

sugar-phosphate backbone

The alternating chain of sugar and phosphate to which DNA and RNA nitrogenous bases are attached.

survivorship curve

A plot of the number of individuals that are still alive at each age in the maximum life span; one way to represent age-specific mortality.

sustainability

The goal of developing, managing, and conserving Earth's resources in ways that meet the needs of people today without compromising the ability of future generations to meet their needs.

sustainable development

The long-term prosperity of human societies and the ecosystems that support them.

swim bladder

A gas-filled internal sac that helps bony fishes maintain buoyancy.

symbiosis *(sim'-bē-ō'-sis)*

An interaction between organisms of different species in which one species, the symbiont, lives in or on another species, the host.

sympatric speciation

The formation of a new species as a result of a change that produces a reproductive barrier between the changed population (mutants) and the parent population. Sympatric speciation occurs without a geographic barrier. *See also* allopatric speciation.

systematics

A discipline of biology that focuses on classifying organisms and determining their evolutionary relationships.

T

taiga *(tī'-guh)*

The northern coniferous forest, characterized by long, snowy winters and short, wet summers. Taiga extends across North America and Eurasia, to the southern border of the arctic tundra; it is also found just below alpine tundra on mountainsides in temperate zones.

tail

Extra nucleotides added at the end of an RNA transcript in the nucleus of a eukaryotic cell.

taxonomy

The branch of biology concerned with identifying, naming, and classifying species.

telophase

The fourth and final stage of mitosis, during which daughter nuclei form at the two poles of a cell. Telophase usually occurs together with cytokinesis.

temperate broadleaf forest

A terrestrial biome located throughout midlatitude regions where there is sufficient moisture to support the growth of large, broadleaf deciduous trees.

temperate grassland

Terrestrial biome located in the temperate zone and characterized by low rainfall and nonwoody vegetation. Tree growth is hindered by occasional fires and periodic severe drought.

temperate rain forest

Coniferous forests of coastal North America (from Alaska to Oregon) supported by warm, moist air from the Pacific Ocean.

temperate zones

Latitudes between the tropics and the Arctic Circle in the north and the Antarctic Circle in the south; regions with milder climates than the tropics or polar regions.

temperature

A measure of the intensity of heat, reflecting the average kinetic energy or speed of molecules.

terminator

A special sequence of nucleotides in DNA that marks the end of a gene. It signals RNA polymerase to release the newly made RNA molecule, which then departs from the gene.

tertiary consumer *(ter'-shē-ār-ē)*

An organism that eats secondary consumers.

testcross

The mating between an individual of unknown genotype for a particular character and an individual that is homozygous recessive for that same character.

tetrapod

A vertebrate with four limbs. Tetrapods include mammals, amphibians, and reptiles (including birds).

theory

A widely accepted explanatory idea that is broad in scope and supported by a large body of evidence.

therapeutic cloning

The cloning of human cells by nuclear transplantation for therapeutic purposes, such as the replacement of body cells that have been irreversibly damaged by disease or injury. nuclear transplantation; reproductive cloning.

threatened species

As defined in the U.S. Endangered Species Act, a species that is likely to become endangered in the foreseeable future throughout all or a significant portion of its range.

three-domain system

A system of taxonomic classification based on three basic groups: Bacteria, Archaea, and Eukarya.

thylakoid (thī'-luh-koyd)
One of a number of disk-shaped membranous sacs inside a chloroplast. Thylakoid membranes contain chlorophyll and the enzymes of the light reactions of photosynthesis. A stack of thylakoids is called a granum.

thymine (T) (thī'-mēn)
A single-ring nitrogenous base found in DNA.

trace element
An element that is essential for the survival of an organism but is needed in only minute quantities.

trait
A variant of a character found within a population, such as purple flowers in pea plants.

trans fat
An unsaturated fatty acid produced by the partial hydrogenation of vegetable oils and present in hardened vegetable oils, most margarines, commercial baked foods, and many fried foods.

transcription
The synthesis of RNA on a DNA template.

transcription factor
In the eukaryotic cell, a protein that functions in initiating or regulating transcription. Transcription factors bind to DNA or to other proteins that bind to DNA.

transfer RNA (tRNA)
A type of ribonucleic acid that functions as an interpreter in translation. Each tRNA molecule has a specific anticodon, picks up a specific amino acid, and conveys the amino acid to the appropriate codon on mRNA.

transgenic organism
An organism that contains genes from another organism, typically of another species.

translation
The synthesis of a polypeptide using the genetic information encoded in an mRNA molecule. There is a change of "language" from nucleotides to amino acids.

transmission electron microscope (TEM)
A microscope that uses an electron beam to study the internal structure of thinly sectioned specimens.

transport protein
A membrane protein that helps move substances across a cell membrane.

transport vesicle
A tiny membranous sphere in a cell's cytoplasm carrying molecules produced by the cell. The vesicle buds from the endoplasmic reticulum or Golgi apparatus and eventually fuses with another organelle or the plasma membrane, releasing its contents.

triglyceride (trī-glis'-uh-rīd)
A dietary fat, which consists of a molecule of glycerol linked to three molecules of fatty acid.

trisomy 21
See Down syndrome.

trophic structure (trō'-fik)
The feeding relationships among the various species in a community.

tropical forest
A terrestrial biome characterized by warm temperatures year-round.

tropics
Region between the Tropic of Cancer and the Tropic of Capricorn; latitudes between 23.5° north and south.

true-breeding
Referring to organisms for which sexual reproduction produces offspring with inherited traits identical to those of the parents. The organisms are homozygous for the characteristics under consideration.

tumor
An abnormal mass of cells that forms within otherwise normal tissue.

tumor-suppressor gene
A gene whose product inhibits cell division, thereby preventing uncontrolled cell growth.

tundra
A terrestrial biome characterized by bitterly cold temperatures. Plant life is limited to dwarf woody shrubs, grasses, mosses, and lichens. Arctic tundra has permanently frozen subsoil (permafrost); alpine tundra, found at high elevations, lacks permafrost.

tunicate
One of a group of sessile invertebrate chordates.

U

unsaturated
Pertaining to fats and fatty acids whose hydrocarbon chains lack the maximum number of hydrogen atoms and therefore have one or more double covalent bonds. Unsaturated fats and fatty acids do not solidify at room temperature.

uracil (U) (yū'-ruh-sil)
A single-ring nitrogenous base found in RNA.

V

vaccine
A harmless variant or derivative of a pathogen (a disease-causing virus or organism) used to stimulate a host organism's immune system to mount a long-term defense against the pathogen.

vacuole (vak'-ū-ōl)
A membrane-enclosed sac, part of the endomembrane system of a eukaryotic cell, having diverse functions.

vascular tissue
Plant tissue consisting of cells joined into tubes that transport water and nutrients throughout the plant body. Xylem and phloem make up vascular tissue.

vector
A piece of DNA, usually a plasmid or a viral genome, that is used to move genes from one cell to another.

vertebrate (ver'-tuh-brāt)
A chordate animal with a backbone. Vertebrates include lampreys, cartilaginous fishes, bony fishes, amphibians, reptiles (including birds), and mammals.

vestigial structure
A structure of marginal, if any, importance to an organism. Vestigial structures are historical remnants of structures that had important functions in ancestors.

virus
A microscopic particle capable of infecting cells of living organisms and inserting its genetic material. Viruses have a very simple structure and are generally not considered to be alive because they do not display all of the characteristics associated with life.

W

warning coloration
The bright color pattern, often yellow, red, or orange in combination with black, of animals that have effective chemical defenses.

water vascular system
In echinoderms, a radially arranged system of water-filled canals that branch into extensions called tube feet. The system provides movement and circulates water, facilitating gas exchange and waste disposal.

wavelength
The distance between crests of adjacent waves, such as those of the electromagnetic spectrum.

wetland
An ecosystem intermediate between an aquatic ecosystem and a terrestrial ecosystem. Wetland soil is saturated with water permanently or periodically.

wild-type trait
The trait most commonly found in nature.

X

X chromosome inactivation
In female mammals, the inactivation of one X chromosome in each somatic cell. Once X inactivation occurs in a given cell (during embryonic development), all descendants of that cell will have the same copy of the X chromosome inactivated.

xylem (zī'-lum)
The portion of a plant's vascular system that provides support and conveys water and inorganic nutrients from the roots to the rest of the plant. Xylem consists mainly of vessel elements and/or tracheids, water-conducting cells.

Z

zooplankton
In aquatic environments, free-floating animals, including many microscopic ones.

zygote (zī'-gōt)
The fertilized egg, which is diploid, that results from the union of haploid gametes (sperm and egg) during fertilization.

Index

Page numbers with *f* indicate figure, *t* indicate table, and those in bold indicate page where listed as key term.

H